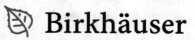

Frontiers in Mathematics

This series is designed to be a repository for up-to-date research results which have been prepared for a wider audience. Graduates and postgraduates as well as scientists will benefit from the latest developments at the research frontiers in mathematics and at the "frontiers" between mathematics and other fields like computer science, physics, biology, economics, finance, etc. All volumes are online available at SpringerLink.

Cristodor Ionescu

Classes of Good Noetherian Rings

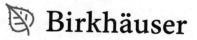 Birkhäuser

Cristodor Ionescu
Simion Stoilow Institute of Mathematics of the
Romanian Academy
Bucharest, Romania

ISSN 1660-8046   ISSN 1660-8054   (electronic)
Frontiers in Mathematics
ISBN 978-3-031-22291-7   ISBN 978-3-031-22292-4   (eBook)
https://doi.org/10.1007/978-3-031-22292-4

Mathematics Subject Classification: 13-02, 13HXX, 13FXX

This book is published under the imprint Birkhäuser, www.birkhauser-science.com by the registered company
Springer Nature Switzerland AG
The registered company address is: Gewerbestrasse 11, 6330 Cham, Switzerland

To the memory of my teachers,
Alexandru Brezuleanu and Nicolae Radu

Preface

Commutative algebra is one of the most beautiful and interesting domains of algebra, and Noetherian local rings are an important object of study in commutative algebra. They constitute a delightful chapter by themselves as well, but moreover they are a main tool in algebraic geometry, number theory and complex analytic geometry.

The roots of this theory can be traced back to the work of Emmy Noether (1882–1935), in the 1920s. Since then, Noetherian rings became a central object in algebra. Another significant moment is the work of Wolfgang Krull (1899–1971), who introduced the notion of dimension of a ring, nowadays called Krull dimension and developed the methods of localization and completion. One of the most important moments for the evolution of the theory is the contribution of Yasuo Akizuki (1902–1984), in the 1930s. In particular, he obtained an example of a Noetherian integral domain whose integral closure was not a finite module. In the 1940s, Claude Chevalley (1909–1984) and Oscar Zariski (1899–1986) applied the theory of Noetherian local rings in algebraic geometry. In particular, Zariski in the 1950s worked on the problem whether the completion of a normal Noetherian local ring remains normal. In the same period, huge developments of this theory were being done by Masayoshi Nagata (1927–2008). He created the theory of Henselian rings, constructed several counterexamples, most of them being a permanent source of inspiration for similar constructions. In particular, he produced a normal local domain whose completion is not normal, thus solving Zariski's problem. He also initiated the theory of Japanese and pseudo-geometric (or Nagata) rings.

Probably the most influential person in this field was the German-born French, but for a long period stateless person using a Nansen passport, Fields Medallist Alexander Grothendieck (1928–2014). In the 1960s, he completely reconsidered algebraic geometry developing the theory of schemes, in which Noetherian local rings play an important role. Essentially, in Grothendieck's work, algebraic geometry and commutative algebra were fully twinned. Grothendieck created the theory of smoothness, the theory of formal fibres and the theory of excellent rings, among others. In this way, he brought to light classes of Noetherian rings behaving well from geometric points of view.

Another extremely important moment in the evolution of commutative algebra was the use of homological tools. This was due essentially to the work of Maurice Auslander (1926–1994), David Buchsbaum (1929–2021) and Jean-Pierre Serre (b. 1926) and it emerged at the end of the 1950s.

This monograph intends to present some classes of Noetherian rings and morphisms of Noetherian rings that showed up in this branch of mathematics. It is intended to continue the themes of the excellent books of H. Matsumura [248] and [249] and A. Brezuleanu-N. Radu [68]. The aim is to provide the reader with a treatment, as exhaustive as possible, of several classes of Noetherian rings and morphisms of Noetherian local rings: (quasi)-excellent rings, Nagata rings, rings with good formal fibres, rings with Artin's approximation property, regular morphisms, reduced morphisms, normal morphisms and complete intersection morphisms. The material of this book will be useful to PhD students and to researchers in commutative algebra, algebraic and arithmetic geometry and number theory.

There are several features that represent the author's aim and choice in preparing this book:

(1) To present all important existing characterizations and criteria for the considered classes of Noetherian local rings and morphisms between them. For example, the characterization of regular morphisms in characteristic $p > 0$ due to André and Radu, the connection between excellent rings and rings with Artin's approximation property, and so on.

(2) To present, with complete proof and in full generality, Popescu's Theorem on Neron Desingularization and the structure of regular morphisms. Since this is a very difficult task, we chose to follow closely the proof clarified by Swan [361], as it is arranged by Majadas and Rodicio [231].

(3) To present some of the most recent developments in the theory of fibres of morphisms.

The first chapter is a preparatory one. It contains generalities about the properties **P** that will be considered, about fibres of morphisms and the **P**-locus of a Noetherian ring. In Sect. 1.3, there are some preparations about inductive limits and in Sect. 1.4, there are the properties of algebras over a field. It contains also some preparations needed for the presentation of several counter examples, following J. Nishimura [271]. In the final section, the definitions and main properties of **P**-rings and **P**-morphisms are described.

The second chapter is mainly dedicated to the class of Nagata rings. It contains generalities about Japanese and Nagata rings, as well as their main properties and important examples. The third section emphasizes, via Zariski-Nagata theorem (2.3.6) and Theorem 2.3.12, the connection between Nagata rings, reduced morphisms and the openness of the normal locus. Some remarks about local rings with reduced or integral domain completion are pointed out.

The objects of study in Chap. 3 are the classes of excellent rings and of regular morphisms. A consistent section is devoted to the chain conditions in Noetherian rings, paying special attention to (universal) catenarity, formal equidimensionality and connections between them, including Grothendieck-Ratliff characterization of universally catenary rings. A famous result, André's Theorem on the localization of formal smoothness (3.2.9) is also presented in this chapter. This theorem will be a crucial ingredient for many further considerations. Characterizations of regular morphisms in positive characteristic given by André and Radu and the fact that F-finite rings are excellent (Kunz and Seydi) are included here. An entire section of the third chapter is devoted to the presentation of the theory of the universally finite module of differentials invented by G. Scheja (1932–2014) and U. Storch (1940–2017). Their result about excellent rings in characteristic zero as well as some extensions in positive characteristic are the main goal of this section. There is a consistent section on F-finite rings containing some recent results, such as Gabber's Theorem 3.5.30, showing that any F-finite ring is a factor ring of a regular ring. The final section is dedicated to the study of inductive limits of some special classes of rings, such as for example Nagata or excellent rings.

The subject of Chap. 4 is the localization and lifting theorems. The story begins effectively with André's theorem on the localization of formal smoothness mentioned above. This was the starting point for several important results about the localization and lifting theorem for **P**-rings in the semilocal case. The recent uniform proof of the localization theorem given by Murayama, Gabber's recent result on the completion of quasi-excellent rings and the counterexample of Nishimura of a ring whose good properties of the formal fibres are lost by completion are contained in this chapter. The chapter contains also a section dedicated to geometric tools used, as well as Flenner's proof of the second theorem of Bertini for local rings.

The next chapter is dedicated to the structure of regular morphisms. The main result is Popescu's theorem on Néron desingularization (5.2.56). This theorem is proved in full generality, and the proof is spread on several pages, following closely the presentation of Swan [361], as detailed in the book of Majadas and Rodicio [231]. As a consequence, together with Rotthaus' result (5.3.13), it allows to identify the class of local rings with Artin's approximation property with the class of Henselian excellent local rings.

Chapter 6 presents several results about other classes of rings. The first section is a rather quick presentation of a tool that can be of big help, the Cohen factorization of local morphisms. The second section is dedicated to morphisms with complete intersection fibres and the third one to the connection between Hochschild homology and regular morphisms. The final section deals with some interesting and not so well-known results about morphisms and rings whose fibres are normal.

Due to the fact that the book is a sequel of quite advanced monographs, the background for reading this book is very consistent. Not only the knowledge of a basic text on commutative algebra [18, 248] and homological algebra [79, 324, 393] is needed but also the basic results and techniques about formal smoothness, module of differentials [131–134, 249] and even André (co)-homology [4, 6, 295] are needed. Thus, it is assumed that the reader is acquainted with these topics.

The book ends with a consistent bibliography rather than only references. It contains monographs, textbooks and research papers which are somehow connected with the topics of the present monograph. The reader will notice that there are not only general textbooks in commutative algebra but also in homological algebra or even some textbooks dedicated mainly to algebraic geometry, number theory or singularity theory. As regarding research papers, the bibliography includes not only papers containing results directly used or presented in the book but also articles that appeared during the past years and having as main subject topics considered in the book. Some of them are maybe overtaken by newer results. We have mentioned them, in order to give an idea of the historical development of the theory. We also included in the bibliography some papers, connected with the considered topics, whose content, from the technical point of view, could not be presented in the book. They are possible suggestions of further study for the interested readers.

Finally, I would like to thank my colleagues and friends Florin Ambro, Mihai Epure and Marius Vlădoiu for the help they provided me in clarifying some issues and correcting some errors. I thank also some other people who made critical remarks about this book and did not want that I mention their names. I thank the editors for their patience in waiting that I finish this project and the Birkhäuser and Straive teams for their very professional support in the editing process.

Bucharest, Romania   Cristodor Ionescu

Conventions

The sets of nonnegative integers, integers, rationals and real and complex numbers are denoted, respectively, by $\mathbb{N}, \mathbb{Z}, \mathbb{Q}, \mathbb{R}$ and $\mathbb{C}$.

Throughout this book, all the rings considered will be commutative, with unit and Noetherian. In some special situations, when the property of being Noetherian or commutative can be forgotten, we shall explicitly mention this possibility.

If A is a local ring with maximal ideal $\mathfrak{m}$ and residue field $k = A/\mathfrak{m}$, we shall denote this by $(A, \mathfrak{m}, k)$. If $(A, \mathfrak{m}, k)$ and $(B, \mathfrak{n}, K)$ are local rings, a morphism u from A to B will always mean a local morphism, that is $u(\mathfrak{m}) \subseteq \mathfrak{n}$.

If A is a commutative ring, by $Q(A)$ we will mean the total ring of fractions of A. In particular, if A is an integral domain, $Q(A)$ will be the field of fractions of A.

If A is an integral domain with field of fractions K, then the integral closure of A will be usually denoted by A' or A'_K. If L is a field extension of K, the integral closure of A in L will usually be denoted by A'_L.

If A is a ring, we denote by $J(A)$ and $N(A)$ the Jacobson radical, resp. the nilradical of A and by $A_{red} := A/N(A)$, the quotient of A by its nilradical.

If A is a semilocal ring, by the radical topology on A we will mean the adic topology given by the powers of $J(A)$.

If A is a ring, on $\mathrm{Spec}(A)$, which is the set of all prime ideals of A, we will consider the Zariski topology and on subsets of $\mathrm{Spec}(A)$ the induced topology. We will denote by $\mathrm{Max}(A)$ the set of the maximal ideals of A and by $\mathrm{Min}(A)$ the set of minimal prime ideals of A. If $\mathfrak{a}$ is an ideal of A, then we will denote $V(\mathfrak{a}) := \{\mathfrak{p} \in \mathrm{Spec}(A) \mid \mathfrak{p} \supseteq \mathfrak{a}\}$ and $D(\mathfrak{a}) := \mathrm{Spec}(A) \setminus V(\mathfrak{a}) = \{\mathfrak{p} \in \mathrm{Spec}(A) \mid \mathfrak{p} \not\supseteq \mathfrak{a}\}$.

If $u : A \to B$ is ring morphism, we denote by $\Omega_{B/A}$ the module of differentials of B over A [131, §20]. If A is the prime ring of B, we simply denote $\Omega_{B/A} := \Omega_B$.

Contents

Fibres of Noetherian Rings

This chapter is mainly containing preparatory notions and results. The properties **P** of Noetherian local rings that will be considered in the next chapters, together with their main properties, are listed. In Sect. 1.3 we list several properties of inductive (direct) limits that will be useful later. The bases of several counterexamples that will be mentioned later are built in Sect. 1.5. In the final section, the **P**-rings and **P**-morphisms, notions that will play a central role, are described.

1.1 Properties of Noetherian Local Rings

We will consider properties **P** that make sense for Noetherian local rings (e.g. **P**= regular, normal, Cohen-Macaulay, etc.). We shall give examples of such properties later. Let **P** be such a property.

Notation 1.1.1 For a ring A we put $\mathbf{P}(A):=\{\mathfrak{p} \in \operatorname{Spec}(A) \mid A_{\mathfrak{p}} \text{ has the property } \mathbf{P}\}$ and $\mathbf{NP}(A):= \operatorname{Spec}(A)\setminus \mathbf{P}(A)$. The sets $\mathbf{P}(A)$ and $\mathbf{NP}(A)$ are called the **P-locus** and, respectively, the **P-singular locus** of A.

We consider a set of conditions about the property **P**:

$(\mathbf{A_1})$ Any Noetherian regular local ring has the property **P**.

$(\mathbf{A_2})$ **P** is a local property, in the sense that if a Noetherian local ring A has the property **P**, then $A_{\mathfrak{p}}$ has **P** for all $\mathfrak{p} \in \operatorname{Spec}(A)$.

$(\mathbf{A_3})$ If $u : A \to B$ is a flat local morphism of Noetherian local rings and if B has the property **P**, then A has **P**.

$(\mathbf{A_4})$ Let $u : (A, \mathfrak{m}, k) \to B$ be a flat local morphism of Noetherian local rings. If A and the closed fibre of u, $B \otimes_A k$ have **P**, then B has **P**.

C. Ionescu, *Classes of Good Noetherian Rings*, Frontiers in Mathematics, https://doi.org/10.1007/978-3-031-22292-4_1

(A_5) Let $u : A \to B$ be a flat local morphism of Noetherian local rings. If A and all the
fibres of u have **P**, then B has **P**.

(A_6) If A is a complete Noetherian local ring, then $\mathbf{P}(A)$ is open in $\mathrm{Spec}(A)$.

(A_7) Let $(A, \mathfrak{m}, k)$ be a Noetherian local ring and let $x \in \mathfrak{m}$ be a non-zero-divisor. If
A/xA has **P**, then A has **P**.

Remark 1.1.2 It is clear that if a property **P** satisfies (A_4), then **P** satisfies also (A_5). The
converse is not true as we will see in Remark 1.1.27.

Which are the properties **P** that we shall consider? We list the main properties of
Noetherian local rings that can be considered, and we shall briefly discuss to what extent
they satisfy the conditions we have introduced. Condition (A_6) will be studied separately
in Sect. 1.2 (see Corollary 1.2.48). We start with the property which is by far the most
interesting and important one.

(1) **P** = *regular*

Definition 1.1.3 A Noetherian local ring $(A, \mathfrak{m}, k)$ is called a *regular local ring*, if
$\dim(A) = \dim_k(\mathfrak{m}/\mathfrak{m}^2)$.

Remark 1.1.4 For a Noetherian local ring $(A, \mathfrak{m}, k)$, the integer $\dim_k(\mathfrak{m}/\mathfrak{m}^2)$ is called the
embedding dimension of A and is denoted by $\mathrm{edim}(A)$. In general, for any Noetherian
local ring $(A, \mathfrak{m}, k)$, it is true that $\dim(A) \leq \mathrm{edim}(A)$.

The property of being a regular local ring corresponds to the geometric notion of a
simple point of an algebraic variety. The properties of regular local rings are carefully
studied in many textbooks on commutative algebra, as, for example, [18, 57, 248, 249], etc.
The following assertion can be traced back in any of these references.

Theorem 1.1.5 *The property* **P** = *regular satisfies the conditions* (A_1) − (A_5) *and* (A_7).

Proof (A_1) is obvious. (A_2) − (A_4) and (A_7) are proved in [249, Th. 19.3, Th. 23.7 and
Th. 14.2].   □

Example 1.1.6 In particular, we denote

$$\mathrm{Reg}(A) := \{\mathfrak{p} \in \mathrm{Spec}(A) \mid A_\mathfrak{p} \text{ is a regular local ring }\}$$

$$\mathrm{Sing}(A) = \mathrm{Spec}(A) \setminus \mathrm{Reg}(A)$$

Remark 1.1.7 In [6, Prop. VI.26], one can find the characterization of regular local rings in terms of the André-Quillen (co)-homology, characterization that will be useful later. Actually, it is shown that for a Noetherian local ring $(A, \mathfrak{m}, k)$, the following conditions are equivalent:

(i) A is a regular local ring.
(ii) $H_2(A, k, k) = 0$.
(iii) $H^2(A, k, k) = 0$.
(iv) For any $t \neq 1$ and for any k-module E, we have $H_t(A, k, E) = 0$.
(v) For any $t \neq 1$ and for any k-module E, we have $H^t(A, k, E) = 0$.

(2) **P** = *Complete intersection*

Definition 1.1.8 A Noetherian local ring $(A, \mathfrak{m}, k)$ is called a **complete intersection**, if $\widehat{A}$, the completion of A with respect to the topology of the maximal ideal, can be expressed as $\widehat{A} \cong B/\mathfrak{a}$, where B is a regular local ring and $\mathfrak{a}$ is an ideal generated by a regular sequence.

Remark 1.1.9 In [6, Prop. VI.27, Cor. X.20 and Th. XVII.13] and [22], it is proved that complete intersection local rings can also be characterized in terms of the André-Quillen (co)-homology. Let $(A, \mathfrak{m}, k)$ be a Noetherian local ring. The following conditions are equivalent:

(i) A is a complete intersection.
(ii) $H_3(A, k, k) = 0$.
(iii) $H^3(A, k, k) = 0$.
(iv) $H_4(A, k, k) = 0$.
(v) $H^4(A, k, k) = 0$.
(vi) For any $t \geq 3$ and for any k-module E, we have $H_t(A, k, E) = 0$.
(vii) For any $t \geq 3$ and for any k-module E we have $H^t(A, k, E) = 0$.
(viii) There exists $n \in \mathbb{N}$ such that $H_t(A, k, k) = 0$, $\forall t \geq n$.

Notation 1.1.10 Let $(A, \mathfrak{m}, k)$ be a Noetherian local ring and $x = (x_1, \ldots, x_n)$ be a minimal system of generators of $\mathfrak{m}$. We denote by $K_\bullet(x, A)$ the Koszul complex on x and by $H_i(K_\bullet(x, A))$, $i \geq 0$, the Koszul homology, that is, the homology groups of the Koszul complex. It is well-known that the Koszul homology does not depend on the system of generators, so that we can denote $H_i(K_\bullet(A)) := H_i(K_\bullet(x, A))$.

Definition 1.1.11 If $(A, \mathfrak{m}, k)$ is a Noetherian local ring, we define

$$\mathrm{cid}(A) := \mathrm{rk}_k H_1(K_\bullet(A)) - \mathrm{edim}(A) + \dim(A).$$

The integer $\mathrm{cid}(A)$ is called the **complete intersection defect** of A.

Remark 1.1.12 It is well-known (cf. [77, Th. 2.3.3]) that a local ring A is a complete intersection if and only if $\mathrm{cid}(A) = 0$.

We will use the following facts:

Lemma 1.1.13

(i) *Let $u : (A, \mathfrak{m}, k) \to (B, \mathfrak{n}, K)$ be a flat morphism of Noetherian local rings. Then*

$$\mathrm{cid}(B) = \mathrm{cid}(A) + \mathrm{cid}(B/\mathfrak{m}B).$$

(ii) *Let A be a Noetherian local ring and $\mathfrak{p} \in \mathrm{Spec}(A)$. Then*

$$\mathrm{cid}(A_{\mathfrak{p}}) \leq \mathrm{cid}(A) \ and \ \mathrm{cid}(\widehat{A}) = \mathrm{cid}(A).$$

Proof (i) is proved in [20, Prop. 3. 6]. The first relation in (ii) is proved in [20, Prop. 3.8], and the second one follows from (i). □

Theorem 1.1.14 *The property P = complete intersection satisfies the conditions $(\mathbf{A_1})$ – $(\mathbf{A_5})$ and $(\mathbf{A_7})$.*

Proof $(\mathbf{A_1})$ is obviously satisfied. $(\mathbf{A_2})$ – $(\mathbf{A_5})$ follow from 1.1.13. $(\mathbf{A_7})$ follows from [20, Prop. 3.10]. □

(3) $\mathbf{P}$ = *Gorenstein*

Definition 1.1.15 A Noetherian local ring $(A, \mathfrak{m}, k)$ is called a **Gorenstein ring** if $\mathrm{id}_A(A) < \infty$.

The main properties of Gorenstein rings can be found, for example, in [249] or [77].

Theorem 1.1.16 *The property P = Gorenstein satisfies the conditions $(\mathbf{A_1})$ – $(\mathbf{A_5})$ and $(\mathbf{A_7})$.*

Proof $(\mathbf{A_1})$ can be found in [249, Prop. 18.1]. $(\mathbf{A_2})$ – $(\mathbf{A_4})$ are proved in [249, Prop. 18.2, 23.4] and $(\mathbf{A_7})$ in [77, Prop. 3.1.19]. □

(4) $\mathbf{P} = Cohen\text{-}Macaulay$

Definition 1.1.17 A Noetherian local ring $(A, \mathfrak{m}, k)$ is called a **Cohen-Macaulay ring** if $\mathrm{depth}_A(A) = \dim(A)$.

Remark 1.1.18 In general, in any Noetherian local ring, it is true that $\mathrm{depth}_A(A) \leq \dim(A)$.

Definition 1.1.19 If $(A, \mathfrak{m}, k)$ is a Noetherian local ring, we define

$$\mathrm{cmd}(A) := \dim(A) - \mathrm{depth}_A(A).$$

The integer $\mathrm{cmd}(A)$ is called the **Cohen-Macaulay defect** of A.

Definition 1.1.20 If $(A, \mathfrak{m}, k)$ is a Noetherian local ring such that $\mathrm{depth}(A) = t$, we define

$$\mathrm{type}(A) := \dim_k \mathrm{Ext}_A^t(k, A).$$

The integer $\mathrm{type}(A)$ is called the **Cohen-Macaulay type** of A.

Remark 1.1.21

(i) It is clear that a Noetherian local ring A is Cohen-Macaulay if and only if $\mathrm{cmd}(A) = 0$.
(ii) (cf. P. Roberts [316]) A Noetherian local ring A is a Gorenstein ring if and only if $\mathrm{type}(A) = 1$.

Later we will need the following facts:

Lemma 1.1.22

(i) Let $u : (A, \mathfrak{m}, k) \to (B, \mathfrak{n}, K)$ be a flat morphism of Noetherian local rings. Then

$$\mathrm{cmd}(B) = \mathrm{cmd}(A) + \mathrm{cmd}(B/\mathfrak{m}B).$$

If moreover B is a Cohen-Macaulay ring, then

$$\mathrm{type}(B) = \mathrm{type}(A) \cdot \mathrm{type}(B/\mathfrak{m}B).$$

(ii) Let A be a Noetherian local ring and $\mathfrak{p} \in \mathrm{Spec}(A)$. Then

$$\mathrm{cmd}(A_\mathfrak{p}) \leq \mathrm{cmd}(A) \text{ and } \mathrm{cmd}(\widehat{A}) = \mathrm{cmd}(A).$$

If moreover A is a Cohen-Macaulay ring, then

$$\text{type}(A_{\mathfrak{p}}) \leq \text{type}(A) \ and \ \text{type}(\widehat{A}) = \text{type}(A).$$

Proof (i) and the relations in (ii) related to $\mathfrak{p}$ are proved in [249, Th. 15.1 and Th. 23.3] and [158, Prop. 6.16]. The inequalities in (ii) related to $\widehat{A}$ follow from (i).   □

A careful study of Cohen-Macaulay rings is done in [77]. One can find the main properties of Cohen-Macaulay rings in many textbooks in commutative algebra, for example, [100, 249], etc.

Theorem 1.1.23 *The property P = Cohen-Macaulay satisfies the conditions* $(\mathbf{A_1})$-$(\mathbf{A_5})$ *and* $(\mathbf{A_7})$.

Proof $(\mathbf{A_1})$ is proved in [77, Prop. 2.2.6]. $(\mathbf{A_2}) - (\mathbf{A_4})$ and $(\mathbf{A_7})$ are proved in [77, Th. 2.1.3 and Th. 2.1.7].   □

(5) The properties of Serre $P = (R_k)$ and $P = (S_k), k \in \mathbb{N}$.

Definition 1.1.24 Let k be a natural number. We say that a Noetherian local ring $(A, \mathfrak{m}, K)$ **has the property** $(\mathbf{R_k})$, if for any prime ideal $\mathfrak{p} \in \text{Spec}(A)$ such that $\text{ht}(\mathfrak{p}) \leq k$, $A_{\mathfrak{p}}$ is a regular local ring.
We say that $(A, \mathfrak{m}, K)$ **has the property** $(\mathbf{S_k})$, if for any prime ideal $\mathfrak{p} \in \text{Spec}(A)$, we have the relation $\text{depth}(A_{\mathfrak{p}}) \geq \min(k, \text{ht}(\mathfrak{p}))$.

Remark 1.1.25

(i) A ring with the property (R_k) is also called, for obvious geometric reasons, **regular in codimension k**.
(ii) Clearly if a ring A has the property (R_k), resp. (S_k), then it has the property (R_l), resp. (S_l), for any $l \leq k$.
(iii) It is obvious that a ring A is a regular local ring if and only if A has the property (R_k), for all $k \in \mathbb{N}$. Also, it is clear from the definitions that a Noetherian local ring A is Cohen-Macaulay if and only if A has the property (S_k), for all $k \in \mathbb{N}$.

Theorem 1.1.26 *The property* $P = (R_k)$ *satisfies the conditions* $(\mathbf{A_1})$, $(\mathbf{A_2})$, $(\mathbf{A_3})$ *and* $(\mathbf{A_5})$.

Proof $(\mathbf{A_1}) - (\mathbf{A_3})$ are easy to prove. $(\mathbf{A_5})$ is proved in [248, (21.E), Corollary]. As for $(\mathbf{A_9})$, it is proved in [132, Prop. 5.12.5].   □

Remark 1.1.27

(i) The property $\mathbf{P} = (R_k)$ does not satisfy $(\mathbf{A_4})$. In fact, the ring A constructed in Example 1.6.2 has the property (R_1), and $\widehat{A}$ has no property (R_0).
(ii) It is also known (cf. [132, Rem. 5.12.6]) that $\mathbf{P} = (R_k)$ doesn't satisfy $(\mathbf{A_7})$, even for the class of catenary rings.

Theorem 1.1.28 *The property $\mathbf{P} = (S_k)$ satisfies the conditions $(\mathbf{A_1})$, $(\mathbf{A_2})$, $(\mathbf{A_3})$, $(\mathbf{A_5})$ and $(\mathbf{A_7})$.*

Proof $(\mathbf{A_1}) - (\mathbf{A_3})$ are easy to prove. $(\mathbf{A_5})$ is proved in [248, (21.E), Corollary]. Let us prove $(\mathbf{A_7})$. Let $\mathfrak{q} \in \mathrm{Spec}(A)$ be such that $\mathrm{depth}(A_\mathfrak{q}) = n \leq k$. If $x \in \mathfrak{q}$, then $\mathrm{depth}(A/xA)_\mathfrak{q} = n - 1 < k$, so by assumption we have $\mathrm{ht}(\mathfrak{q}/xA) = n - 1$ and it follows that $\mathrm{ht}(\mathfrak{q}) = n$. If $x \notin \mathfrak{q}$, let us take $\mathfrak{p} \in \mathrm{Min}(\mathfrak{q} + xA)$. Then the ideal $(\mathfrak{p} + xA)A_\mathfrak{q}$ is $\mathfrak{q}A_\mathfrak{q}$-primary and $\mathrm{depth}(A_\mathfrak{p}) \leq \mathrm{depth}(A_\mathfrak{q}) + 1 = n + 1$. Since $\mathrm{depth}((A/xA)_\mathfrak{q}) \leq n$, it follows that $\mathrm{ht}(\mathfrak{p}/xA) \leq n$; hence $\mathrm{ht}(\mathfrak{p}) \leq n + 1$. As x is a regular element, we get $\mathrm{ht}(\mathfrak{q}) \leq n$.   □

Remark 1.1.29 It is not known, at least to the author, whether the property (S_k) satisfies the axiom $(\mathbf{A_4})$.

The following result shows that the property (S_k) has something more in common with (R_k).

Proposition 1.1.30 *Let A be a Noetherian local ring and k a natural number. The following are equivalent:*

(i) A has the property (S_k).
(ii) $A_\mathfrak{p}$ is Cohen-Macaulay for any $\mathfrak{p} \in \mathrm{Spec}(A)$ such that $\mathrm{depth}(A_\mathfrak{p}) \leq k - 1$.

Proof $(i) \Rightarrow (ii)$: Let $\mathfrak{p} \in \mathrm{Spec}(A)$ such that $\mathrm{depth}(A_\mathfrak{p}) \leq k - 1$. Then obviously $\mathrm{depth}(A_\mathfrak{p}) \geq \mathrm{ht}(\mathfrak{p})$, that is, $A_\mathfrak{p}$ is Cohen-Macaulay.
$(ii) \Rightarrow (i)$: If $\mathrm{depth}(A_\mathfrak{p}) \leq k-1$, by assumption $A_\mathfrak{p}$ is Cohen-Macaulay and consequently

$$\mathrm{depth}(A_\mathfrak{p}) = \mathrm{ht}(\mathfrak{p}) = \min\{k, \mathrm{ht}(\mathfrak{p})\}.$$

If $\mathrm{depth}(A_\mathfrak{p}) \geq k$, then clearly $\mathrm{depth}(A_\mathfrak{p}) \geq \min\{k, \mathrm{ht}(\mathfrak{p})\}$. Finally it follows that A has the property (S_k).   □

Remark 1.1.31 The properties (R_k) and (S_k) were introduced by J.P. Serre; therefore they are usually called Serre's properties.

Combining some of the properties above, one can define new ones, some of them already used and important in the study of Noetherian rings. Let us first note the following:

Lemma 1.1.32 *Let A be a Noetherian ring and k a natural number. The following conditions are equivalent:*

(i) *A satisfies the properties (R_k) and (S_{k+1}).*

(ii) *For any prime ideal $\mathfrak{p} \in \mathrm{Spec}(A)$, such that $\mathrm{depth}(A_\mathfrak{p}) \leq k$, $A_\mathfrak{p}$ is a regular local ring.*

Proof $(i) \Rightarrow (ii)$: Let $\mathfrak{p} \in \mathrm{Spec}(A)$. If $\mathrm{depth}(A_\mathfrak{p}) \leq k$, from the property (S_{k+1}), it follows that $\mathrm{depth}(A_\mathfrak{p}) = \mathrm{ht}(\mathfrak{p})$. Hence, from the property (R_k), we get that $A_\mathfrak{p}$ is regular. $(ii) \Rightarrow (i)$: Let again $\mathfrak{p} \in \mathrm{Spec}(A)$. If $\mathrm{ht}(\mathfrak{p}) \leq k$, then $\mathrm{depth}(A_\mathfrak{p}) \leq \mathrm{ht}(\mathfrak{p}) \leq k$, and it follows that $A_\mathfrak{p}$ is regular, that is, A satisfies (R_k). Moreover, this implies that $\mathrm{ht}(\mathfrak{p}) = \mathrm{depth}(A_\mathfrak{p})$, so that A satisfies also (S_{k+1}). $\qquad\square$

Lemma 1.1.33 *Let A be a Noetherian ring and x be a regular element in the Jacobson radical of A. If A/xA has the properties (R_k) and (S_{k+1}), then A has the properties (R_k) and (S_{k+1}).*

Proof Let $\mathfrak{p} \in \mathrm{Spec}(A)$ such that $\mathrm{ht}(\mathfrak{p}) = n \leq k$. If $x \in \mathfrak{p}$, since $\mathrm{ht}(\mathfrak{p}/xA) = n - 1$, it follows that $(A/xA)_\mathfrak{p}$ is a regular ring, and from Theorem 1.1.5 it follows that $A_\mathfrak{p}$ is regular. If $x \notin \mathfrak{p}$, take $\mathfrak{q} \in \mathrm{Min}(\mathfrak{p} + xA)$. It follows that $\mathrm{depth}((A/xA)_\mathfrak{q}) \leq n$. Then $\mathrm{ht}(\mathfrak{q}/xA) \leq n$, because A/xA has the property (S_{k+1}). We have that $(A/xA)_\mathfrak{q}$ is regular, and from the fact that A/xA satisfies (R_k), it follows that $A_\mathfrak{q}$ is regular. Consequently also $A_\mathfrak{p}$ is regular. $\qquad\square$

Remark 1.1.34 The rings constructed in Examples 1.6.1 and 1.6.2 show that the properties (R_k) and (S_{k+1}) do not satisfy $(\mathbf{A_4})$.

For the following well-known definition, let us note that it works also for non-Noetherian rings.

Definition 1.1.35 A ring A is called **reduced** if A does not have non-zero nilpotent elements.

Theorem 1.1.36 *Let A be a Noetherian ring. The following are equivalent:*

(i) *A is a reduced ring.*

(ii) *A satisfies the properties (R_0) and (S_1).*

(iii) *$A_\mathfrak{p}$ is a regular local ring for any prime ideal $\mathfrak{p}$ of A such that $\mathrm{depth}(A_\mathfrak{p}) = 0$.*

Proof $(i) \Leftrightarrow (ii)$: Obvious.

$(ii) \Leftrightarrow (iii)$: Follows from 1.1.32. $\qquad\qquad\qquad\qquad\qquad\qquad\qquad\qquad\square$

Remark 1.1.37 The property **P** = reduced does not satisfy $(\mathbf{A_4})$. In fact in Example 1.6.1, we shall see an example of a reduced Noetherian local ring whose completion is not reduced.

Remark 1.1.38 Recall that an integral domain (not necessarily Noetherian) A is called integrally closed [249, pag. 64], if any element from its field of fractions integral over A is actually from A.

Definition 1.1.39 A ring A, not necessarily Noetherian, is called **normal** if $A_{\mathfrak{p}}$ is an integrally closed domain, for any $\mathfrak{p} \in \operatorname{Spec}(A)$.

Remark 1.1.40 ([58, ch. 10, §1, no. 1, Prop. 14]) Let A be a Noetherian ring. The following are equivalent:

   (i) A is normal.
  (ii) A is reduced and is integrally closed in its total quotient ring.
 (iii) A is isomorphic with a finite direct product of integrally closed rings.

Remark 1.1.41

   (i) A Noetherian ring A is integrally closed if and only if A is a normal domain.
  (ii) A local normal domain is integrally closed.

Example 1.1.42 In particular, we denote

$$\operatorname{Nor}(A) := \{\mathfrak{p} \in \operatorname{Spec}(A) \mid A_{\mathfrak{p}} \text{ is a normal ring } \}$$

$$\operatorname{NNor}(A) = \operatorname{Spec}(A) \setminus \operatorname{Nor}(A).$$

Theorem 1.1.43 *Let A be a Noetherian ring. The following are equivalent:*

   (i) A is a normal ring.
  (ii) A satisfies the properties (R_1) and (S_2).
 (iii) $A_{\mathfrak{p}}$ is a regular local ring for any prime ideal $\mathfrak{p}$ of A such that $\operatorname{depth}(A_{\mathfrak{p}}) \leq 1$.

Proof $(i) \Leftrightarrow (ii)$: [248, Th. 39].

$(ii) \Leftrightarrow (iii)$: Follows from Lemma 1.1.32. $\qquad\qquad\qquad\qquad\qquad\qquad\square$

Remark 1.1.44 From Example 1.6.2 it will follow that **P** = normal does not satisfy $(\mathbf{A_4})$.

Given a property **P** satisfying the conditions (A_1), (A_2), (A_3) and (A_4) and a natural number k, in analogy with the properties (R_k) and (S_k), one can construct from **P** two properties that we will denote by $(\mathbf{P_k})$ and $(\mathbf{P'_k})$. These properties are constructed from the given property **P** in a similar way the properties (R_k) are constructed from *regular* and, respectively, the properties *normal* or *reduced* are constructed from *regular* (see 1.1.36, 1.1.43), or in the way the property (S_k) is connected with the *Cohen-Macaulay* property (see Proposition 1.1.30).

Definition 1.1.45 Let **P** be a property satisfying the properties (A_1), (A_2), (A_3) and (A_4) and let k a natural number. We say that a Noetherian ring **A has the property** $(\mathbf{P_k})$ if for any prime ideal $\mathfrak{p} \in \mathrm{Spec}(A)$ such that $\mathrm{ht}(\mathfrak{p}) \leq k$, $A_\mathfrak{p}$ has the property **P**. We say that a Noetherian ring **A has the property** $(\mathbf{P'_k})$ if for any prime ideal $\mathfrak{p} \in \mathrm{Spec}(A)$ such that $\mathrm{depth}(A_\mathfrak{p}) \leq k$, $A_\mathfrak{p}$ has the property **P**.

Lemma 1.1.46 *Let **P** be a property satisfying properties (A_1), (A_2), (A_3) and (A_4) and let k a natural number. Then $(\mathbf{P_k})$ and $(\mathbf{P'_k})$ satisfy the properties (A_1), (A_2), (A_3) and (A_5).*

Proof Since **P** satisfies (A_1) and (A_2), it follows at once that $(\mathbf{P_k})$ and $(\mathbf{P'_k})$ satisfy (A_1) and (A_2). To prove (A_3), let $u : A \to B$ be a flat morphism of local rings such that B has $(\mathbf{P_k})$ and let $\mathfrak{p}$ be a prime ideal of A such that $\mathrm{ht}(\mathfrak{p}) \leq k$. Let $\mathfrak{q}$ be a minimal prime ideal of $\mathfrak{p}B$. Then $\mathrm{ht}(\mathfrak{q}) = \mathrm{ht}(\mathfrak{p}) \leq k$; hence $B_\mathfrak{q}$ has the property **P**, and since **P** satisfies (A_3), it follows that $A_\mathfrak{p}$ has **P**. For (A_5) let $\mathfrak{q}$ be a prime ideal of B such that $\mathrm{ht}(\mathfrak{q}) \leq k$ and let $\mathfrak{p} = \mathfrak{q} \cap A$. Then $\mathrm{ht}(\mathfrak{p}) + \mathrm{ht}(\mathfrak{q}/\mathfrak{p}B) = \mathrm{ht}(\mathfrak{q}) \leq k$; hence $\mathrm{ht}(\mathfrak{p}) \leq k$ and $\mathrm{ht}(\mathfrak{q}/\mathfrak{p}B) \leq k$. Thus $A_\mathfrak{p}$ and $(B/\mathfrak{p}B)_\mathfrak{q}$ have the property **P**; hence by (A_4) for **P**, we get $B_\mathfrak{q}$ has **P**. The proof for the property $(\mathbf{P'_k})$ is similar.   □

If a property **P** satisfies the condition (A_7), what happens to the properties $(\mathbf{P'_k})$ and $(\mathbf{P_k})$? Do they continue to satisfy the same condition? For $(\mathbf{P'_k})$ this is true.

Lemma 1.1.47 ([172, Lemma 3.1]) *Let **P** be a property satisfying (A_7) and k a natural number. Then the property $(\mathbf{P'_k})$ satisfies (A_7).*

Proof Let $\mathfrak{p} \in \mathrm{Spec}(A)$ be such that $\mathrm{depth}(A_\mathfrak{p}) = r \leq k$. There are two cases. If $x \in \mathfrak{p}$, then $\mathrm{depth}(A_\mathfrak{p}/xA_\mathfrak{p}) = r - 1 \leq k$; hence $A_\mathfrak{p}/xA_\mathfrak{p}$ has the property **P**. By assumption it follows that $A_\mathfrak{p}$ has the property **P**. If $x \notin \mathfrak{p}$, let $\mathfrak{q}$ be a minimal prime over-ideal of $\mathfrak{p}+xA$. Then $(\mathfrak{p} + xA)A_\mathfrak{q}$ is a $\mathfrak{q}A_\mathfrak{q}$-primary ideal. Thus

$$\text{depth}(A_q) \leq \text{depth}(A_p) + 1 = r + 1 \leq k + 1,$$

hence

$$\text{depth}(A_q/xA_q) \leq \text{depth}(A_p/xA_p) + 1 = r \leq k.$$

It follows that A_q/xA_q has the property $\mathbf{P}$; hence A_q has the property $\mathbf{P}$. Consequently A_p has the property $\mathbf{P}$. □

For the property $(\mathbf{P_k})$, this is no longer true, as one can see from Remark 1.1.27.

Example 1.1.48

(i) If $\mathbf{P}$ = regular, it follows from Theorems 1.1.36 and 1.1.43 that $(\mathbf{P'_0})$ = reduced, $(\mathbf{P'_1})$ = normal and $(\mathbf{P_k}) = (R_k)$. Also, if $\mathbf{P}$ = CM, then from Proposition 1.1.30 we have that $(\mathbf{P'_{k-1}}) = (S_k)$. For example, if $\mathbf{P}$=Gorenstein, then $(\mathbf{P'_k})$ is called k-Gorenstein [112, 281, 307, 391] or (G_k) [283], and $(\mathbf{P_k})$ is called (T_k) [125, 281].

(ii) There are some other properties of Noetherian local rings that can be considered. For example, the properties denoted cid $\leq k$ or cmd $\leq k$ (see, e.g. [27]) that means that for a local ring A, one has cid$(A) \leq k$, resp. cmd$(A) \leq k$. We let to the reader the pleasure to check that the entries of the Table 1.1 are correct.

In Table 1.1 we present several properties and how they satisfy the conditions considered at the beginning of the section. It contains the results from the next section about the condition $\mathbf{A_6}$.

Table 1.1 List of properties and conditions they satisfy

	A1	A2	A3	A4	A5	A6	A7
Regular	Yes	Yes	Yes	Yes	Yes	Yes	Yes
Complete intersection	Yes	Yes	Yes	Yes	Yes	Yes	Yes
Gorenstein	Yes	Yes	Yes	Yes	Yes	Yes	Yes
Cohen-Macaulay	Yes	Yes	Yes	Yes	Yes	Yes	Yes
(R_k)	Yes	Yes	Yes	No	Yes	Yes	No
(S_k)	Yes	Yes	Yes	No	Yes	Yes	Yes
$(R_k) + (S_{k+1})$	Yes	Yes	Yes	No	Yes	Yes	Yes
Reduced	Yes	Yes	Yes	No	Yes	Yes	Yes
Normal	Yes	Yes	Yes	No	Yes	Yes	Yes
Normal and (R_k)	Yes	Yes	Yes	No	Yes	Yes	No
(CI_k)	Yes	Yes	Yes	No	Yes	Yes	Yes
(T_k)	Yes	Yes	Yes	No	Yes	Yes	Yes
k-Gorenstein	Yes	Yes	Yes	No	Yes	Yes	Yes
cid $\leq n$	Yes	Yes	Yes	No	No	Yes	Yes
cmd $\leq n$	Yes	Yes	Yes	No	No	Yes	Yes

1.2   The P-Locus of a Noetherian Ring

We have defined in Notation 1.1.1 the **P**-locus of a Noetherian ring, where **P** is one of the properties considered before. As this set is a subset of the topological space $\mathrm{Spec}(A)$, we are interested in its topological properties. We start with an example, showing that this set doesn't always have good properties.

Proposition 1.2.1 (Hochster [163, Prop. 1]) *Let k be a field and suppose that we have a family of k-algebras $(A_i)_{i \in \mathbb{N}}$ and for each $i \in \mathbb{N}$ a non-zero prime ideal $\mathfrak{p}_i \in \mathrm{Spec}(A_i)$. Let $A' := \bigotimes_{i \in \mathbb{N}} A_i$ and $S := A' - (\bigcup_{i \in \mathbb{N}} \mathfrak{p}_i A')$, which is obviously a multiplicative system in A' so that we can consider $A := S^{-1} A'$. Assume that for every $i \in \mathbb{N}$ and for any field L containing k, the ring $L \otimes_k A_i$ is a Noetherian domain, and $L \otimes_k \mathfrak{p}_i$ is a prime ideal. Then:*

  (i) *A' is an integral domain and $\mathfrak{p}_i A'$ is a prime ideal, for any $i \in \mathbb{N}$.*
 (ii) *$\mathrm{Max}(A) = \{\mathfrak{p}_i A \mid i \in \mathbb{N}\}$ and $\mathfrak{p}_i \neq \mathfrak{p}_j$ for $i \neq j$.*
(iii) *Any non-zero element of A belongs to only finitely many maximal ideals of A.*
 (iv) *For every $i \in \mathbb{N}$, if $L_i := Q(\bigotimes_{j \neq i} A_j)$ and $\mathfrak{q}_i := \mathfrak{p}_i(L_i \otimes_k A_i)$, then*

$$A_{\mathfrak{p}_i A} = (L_i \otimes_k A_i)_{\mathfrak{q}_i}.$$

 (v) *A is a Noetherian ring.*

Proof (i) For any natural number $n \in \mathbb{N}$, let $B_n := \bigotimes_{1 \leq j \leq n} A_j$. We have a canonical morphism $B_n \to B_{n+1}$, which is injective by flatness. Then $A' = \varinjlim_{n \in \mathbb{N}} B_n$. It follows that also the canonical morphisms $B_n \to A'$ are injective, that is, $A' = \bigcup_{n \in \mathbb{N}} B_n$. Hence, to show that A' is a domain, it suffices to show that all the rings B_n are integral domains. Let $L := Q(B_{n-1})$. Then $B_n = B_{n-1} \otimes_k A_i$ is a subring of $L \otimes_k A_i$ which is an integral domain by assumption. By induction on n, we get that B_n is a domain. Replacing A_i with $A_i/\mathfrak{p}_i$, we get in the same way that $\mathfrak{p}_i A'$ is a prime ideal.
(ii) Let $i \in \mathbb{N}$ and let T_i be a basis of $\mathfrak{p}_i$ as a k-vector space. There exists a basis of A_i as a k-vector space containing T_i, say T_i'. Let $T \subseteq A'$ be the set of finite tensor products of elements of T_i. Then T is a basis of A' as a k-vector space. If we consider the set of products of elements from T' containing an element of T_i', this set is clearly a basis of $\mathfrak{p}_i A'$ as a k-vector space. Let $r \in A'$, $r \neq 0$ and $H(r) := \{i \in \mathbb{N} \mid r \in \mathfrak{p}_i A'\}$. If $r \in B_n$, then $H(r) \subseteq \{1, \ldots, n\}$. Hence $H(r)$ is a finite set. Moreover, if $r_1, r_2 \in A'$, there exists $m \in \mathbb{N}$ such that $r_1, r_2 \in B_m$. Let $b_1, \ldots, b_t \in T$ be the non-zero elements occurring in r_1 with non-zero coefficients and $c_1, \ldots, c_s \in T$ be the non-zero elements occurring in r_2 with non-zero coefficients. Let $i \in \mathbb{N}$, $i > m$, and fix an element $x \in T_i'$.

Then the elements of T occurring with non-zero coefficients in $r_1 + xr_2$ are $b_1, \ldots, b_t$ and $xc_1, \ldots, xc_s$. Consequently, $H(r_1 + xr_2) = H(r_1) \cap H(r_2)$. In this way we obtain that for $r_1, \ldots, r_m \in A'$ there is $r \in A'r_1 + \cdots + A'r_m$ such that $H(r) = H(r_1) \cap \cdots \cap H(r_s)$. On the other hand, it is clear that $r \in S$ if and only if $H(r) = \emptyset$. Let $\mathfrak{a}$ be a proper ideal of A and let $i \in \bigcap_{r \in \mathfrak{a} \cap A'} H(r) \neq \emptyset$. Then $\mathfrak{a} \subseteq \mathfrak{p}_i A$. Assume that $\mathfrak{p}_i A \notin \mathrm{Max}(A)$. Then there is $j \neq i$ such that $\mathfrak{p}_i A' \subseteq \mathfrak{p}_j A'$. But any element $x \in T_i$, viewed as an element of A', is not in $\mathfrak{p}_j A'$. Contradiction! Hence $\mathfrak{p}_i A \in \mathrm{Max}(A)$. The assertion is proved.

(iii) Follows from the above considerations.

(iv) Follows from the fact that $(A_i' - (0)) \cap (\mathfrak{p}_i A') = \emptyset$.

(v) From (iv) we get that $A_{\mathfrak{p}_i A}$ is a Noetherian ring for any i. From (iii) and [266, E1.1], it follows that A is a Noetherian ring. $\qquad \square$

Remark 1.2.2

(i) Any subset of $\mathrm{Spec}(A)$ containing infinitely many maximal ideals is dense in $\mathrm{Spec}(A)$.

(ii) If for any $i \in \mathbb{N}$, A_i is a subring of a polynomial ring over k generated by a finite number of forms of positive degree and $\mathfrak{p}_i$ is the ideal generated by these forms, the assumptions of Proposition 1.2.1 are fulfilled. Moreover, in this case, the localizations of A are affine algebras over k.

Proof (i) If $D(f)$ is an open principal set, only finitely many maximal ideals are not in $D(f)$.

(ii) If $A_i \subseteq k[X_1, \ldots, X_n]$, then for any L, field extension of k, by flatness we have $B_i := L \otimes_k A_i \subseteq L[X_1, \ldots, X_n]$. Clearly B_i is generated by the same forms as A_i. $\qquad \square$

Let us first note a useful result.

Lemma 1.2.3 *Let A be a Noetherian ring. The following conditions are equivalent:*

(i) $P(A)$ is not open.

(ii) The set $\mathrm{Min}(\mathbf{NP}(A))$ is infinite.

Proof $(i) \Rightarrow (ii)$: Suppose that $\mathrm{Min}(\mathbf{NP}(A)) = \{\mathfrak{p}_1, \ldots, \mathfrak{p}_s\}$. Let $\mathfrak{a} := \bigcap_{i=1}^{s} \mathfrak{p}_i$. Then $V(\mathfrak{a}) = \bigcup_{i=1}^{s} V(\mathfrak{p}_i)$. We will show that $V(\mathfrak{a}) = \mathbf{NP}(A)$, and this will complete the proof of this implication. First, if $\mathfrak{q} \in V(\mathfrak{a})$, this means that $\mathfrak{q} \supseteq \mathfrak{a} = \bigcap_{i=1}^{s} \mathfrak{p}_i$, and it results that there exists some index $i \in \{1, \ldots, s\}$ such that $\mathfrak{q} \in V(\mathfrak{p}_i)$. From $(\mathbf{A_2})$ it follows that $\mathfrak{p}_i \in \mathbf{P}(A)$. Conversely, if $\mathfrak{q} \in \mathbf{NP}(A)$, there exists $i \in \{1, \ldots, s\}$ such that $\mathfrak{q} \supseteq \mathfrak{p}_i$. This means that $\mathfrak{q} \in V(\mathfrak{p}_i) \subseteq V(\mathfrak{a})$.

$(ii) \Rightarrow (i)$: Suppose that $\mathbf{NP}(A)$ is closed. Then there is some ideal $\mathfrak{a}$ such that $\mathbf{NP}(A) = V(\mathfrak{a})$. But then $\text{Min}(\mathbf{NP}(A)) = \text{Min}(\mathfrak{a})$ is a finite set. $\qquad\square$

Example 1.2.4 (Hochster [163]) There exists a Noetherian ring A, such that $\text{Reg}(A)$ and $\text{Nor}(A)$ are not open in $\text{Spec}(A)$.

Proof Let $\{X_i, Y_i, \ i \geq 1\}$ be variables and $A_i := k[X_i^2, X_i^3]$, $\mathfrak{p}_i = (X_i^2, X_i^3)A_i$. Let A be the ring constructed from A_i and $\mathfrak{p}_i$ as in Proposition 1.2.1. Thus

$$A_{\mathfrak{p}_i A} = L_i[X_i^2, X_i^3]_{\mathfrak{q}_i},$$

where $\mathfrak{q}_i = (X_i^2, X_i^3)L_i[X_i^2, X_i^3]$. Then $A_{\mathfrak{p}_i A}$ is a local ring of dimension 1 which is not regular and consequently not normal. If $\mathfrak{q} := (0)$, then $\mathfrak{q} \in \text{Reg}(A)$ but $\mathfrak{q} \in \overline{\text{Sing}(A)}$. It follows that $\text{Reg}(A)$ is not an open set of $\text{Spec}(A)$. The same proof works for $\text{Nor}(A)$. $\quad\square$

Motivated at least by the example above, in this section we study the topological properties of the **P**-locus of a Noetherian ring, introduced in 1.1.1, where **P** is a property of local rings as the properties studied in Sect. 1.1. Let us consider such a property **P**. We assume from the beginning that **P** verifies the axioms $(\mathbf{A_1})$, $(\mathbf{A_2})$, $(\mathbf{A_3})$ and $(\mathbf{A_5})$. One of the main problems is to establish if the given property **P** satisfies also $(\mathbf{A_6})$.

Notation 1.2.5 If **P** is a property as above and A is a Noetherian ring, we shall denote

$$i(\mathbf{P}(A)) = \begin{cases} A, & \text{if } \mathbf{P}(A) = \text{Spec}(A) \\ \bigcap_{\mathfrak{p} \in \mathbf{NP}(A)} \mathfrak{p}, & \text{if } \mathbf{P}(A) \neq \text{Spec}(A) \end{cases}$$

Remark 1.2.6

(i) $i(\mathbf{P}(A))$ is a radical ideal.
(ii) $\overline{\mathbf{NP}(A)} = V(i(\mathbf{P}(A)))$.

One of our main problems will be to decide whether the **P**-locus of a ring is open. To help us we have the following result that characterizes open subsets of $\text{Spec}(A)$.

Lemma 1.2.7 *Let A be a Noetherian ring and let U be a subset of $\text{Spec}(A)$. The following conditions are equivalent:*

(i) *U is open.*
(ii) (a) *If $\mathfrak{p} \in U$ and $\mathfrak{q} \in \text{Spec}(A)$ is such that $\mathfrak{q} \subseteq \mathfrak{p}$, then $\mathfrak{q} \in U$ (we say that U is stable at generalizations).*
   (b) *If $\mathfrak{p} \in U$, then U contains a non-empty open set of $V(\mathfrak{p})$.*

Proof $(i) \Rightarrow (ii)$: Obvious.

$(ii) \Rightarrow (i)$: Let $F := \mathrm{Spec}(A) \setminus U$ and let $\mathfrak{p}_1, \ldots, \mathfrak{p}_s$ be the generic points of $\overline{F}$. This means that there is an ideal $\mathfrak{a}$ in A such that $\overline{F} = V(\mathfrak{a})$ and $\mathrm{Min}(A/\mathfrak{a}) = \{\mathfrak{p}_1, \ldots, \mathfrak{p}_s\}$. From (ii) we get that $\mathfrak{p}_i \notin U$, $\forall i = 1, \ldots, s$. Then $\mathfrak{p}_i \in F$, $\forall i = 1, \ldots, s$, and this implies that $F = \overline{F}$.   □

Definition 1.2.8   A ring A is called a **P-0 ring** if $\mathbf{P}(A)$ contains a non-empty open set. A is called a **P-1 ring** if $\mathbf{P}(A)$ is an open subset of $\mathrm{Spec}(A)$ and a **P-2 ring** if every A-algebra of finite type is **P**-1.

The proof of the following lemma is obvious.

Lemma 1.2.9   *Let A be a ring which is **P**-2. Then any A-algebra of finite type and any ring of fractions of A are **P**-2.*

Lemma 1.2.10 (Greco [122, Cor. 2.2])   *Let $u : A \to B$ be an injective morphism of finite type of integral domains and let $\mathbf{P}$ be a property satisfying $(\mathbf{A_3})$. If B is **P**-0, then A is **P**-0.*

Proof   By generic freeness [132, Lemme 6.9.2] or [100, Th. 14.4], there exists $\alpha \in A$ such that B_α is a free A_α-module. Thus the induced map $u_\alpha : A_\alpha \to B_\alpha$ is flat and of finite type; hence by [248, (6.I), Th. 8], it follows that $\mathrm{Spec}(u_\alpha)$ is open.   □

Proposition 1.2.11 (Greco [122, Cor. 2.4])   *Let $u : A \to B$ be a finite-type ring morphism such that $\mathrm{Spec}(u)$ is surjective and let $\mathbf{P}$ be a property satisfying $(\mathbf{A_3})$. Then A is **P**-2 if B is **P**-2.*

Proof   Obviously if A is **P**-2, then B is **P**-2. Assume that B is **P**-2 and let C be a finitely generated A-algebra which is a domain. By assumption $D = C \otimes_A B \neq 0$. Let $\mathfrak{p} \in \mathrm{Spec}(D)$ and $E = D/\mathfrak{p}$. Then E is **P**-0 and we apply Lemma 1.2.10.   □

We shall consider the following condition on a Noetherian ring A, studied initially by Nagata, condition proved to be useful in many situations.

Definition 1.2.12   We say that the property **P** satisfies **Nagata's criterion**, shortly (**NC**), if **P** satisfies the following condition: *If A is a Noetherian ring, such that for every $\mathfrak{p} \in \mathbf{P}(A)$, the set $\mathbf{P}(A/\mathfrak{p})$ contains a non-empty open set of $\mathrm{Spec}(A/\mathfrak{p})$, then $\mathbf{P}(A)$ is open in $\mathrm{Spec}(A)$.*

Our next task is to see whether the properties considered in Sect. 1.1 satisfy (**NC**) or not. We begin with a very useful lemma.

Lemma 1.2.13 (Massaza [245, Lemma 1]) *Let A be a Noetherian ring. The following conditions are equivalent:*

(i) $P(A)$ *is not open.*

(ii) *There exists a prime ideal* $\mathfrak{q} \in P(A)$ *such that* $\mathfrak{q}$ *can be written* $\mathfrak{q} = \bigcap_{i \in \mathbb{N}} \mathfrak{p}_i$, *where*

$\mathfrak{p}_i \in Min(NP(A))$ *for any* $i \in \mathbb{N}$ *and such that for every infinite subset* $J \subseteq \mathbb{N}$ *we have* $\bigcap_{j \in J} \mathfrak{p}_j = \mathfrak{q}$.

Proof $(i) \Rightarrow (ii)$: Suppose that $NP(A)$ is closed and that $Min(NP(A)) = \{\mathfrak{p}_l\}_{l \in L}$, where $|L| \geq \aleph_0$. Consider a countable subfamily $\{\mathfrak{p}_i\}_{i \in \mathbb{N}}$, and put $\mathfrak{a} := \bigcap_{i \in \mathbb{N}} \mathfrak{p}_i$. We have the following facts:

(a) Since $\mathfrak{a}$ is a radical ideal in a Noetherian ring, it can be written as a finite intersection of prime ideals $\mathfrak{a} = \mathfrak{q}_1 \cap \cdots \cap \mathfrak{q}_s$, such that $\mathfrak{q}_i$ does not contain $\mathfrak{q}_j$ for any indices $i, j \in \{1, \ldots, s\}$ such that $i \neq j$.

(b) For all indices $i \in \{1, \ldots, s\}$, $\mathfrak{q}_i$ is the intersection of those $\mathfrak{p}'_j$s containing it. Indeed, let $\{\mathfrak{p}_{i_h}\}_{h \in H_i}$ be the family of the ideals from $\{\mathfrak{p}_j\}_{j \in \mathbb{N}}$ containing $\mathfrak{q}_i$. Then we have

$$\mathfrak{q}_i \subseteq \bigcap_{h \in H_i} \mathfrak{p}_{i_h} := \mathfrak{a}_i, \quad \mathfrak{a} = \bigcap_{i=1}^{s} \mathfrak{q}_i = \bigcap_{i=1}^{s} \mathfrak{a}_i.$$

Since $\bigcap_{i=1}^{s} \mathfrak{a}_i \subseteq \mathfrak{q}_j$, there is an index i_0 such that $\mathfrak{a}_{i_0} \subseteq \mathfrak{q}_j$ and then

$$\mathfrak{q}_{i_0} \subseteq \mathfrak{a}_{i_0} \subseteq \mathfrak{q}_i,$$

and from this it clearly follows

$$\mathfrak{q}_{i_0} = \mathfrak{a}_{i_0} = \mathfrak{q}_j,$$

proving the assertion.

In order to finish the proof of this implication, observe that since the family of the primes $\mathfrak{p}_i$ is an infinite one, there exists a prime ideal $\mathfrak{q}_j$ which is the intersection of infinitely many $\mathfrak{p}'_i$s. This means that we can suppose that $\bigcap_{i \in \mathbb{N}} \mathfrak{p}_i := \mathfrak{q} \in Spec(A)$. Moreover, if there is an infinite subfamily $\mathcal{F}$ of $\{\mathfrak{p}_i\}_{i \in \mathbb{N}}$ whose intersection is strictly bigger than $\mathfrak{q}$, we consider the family $\mathcal{F}$ instead of the family $\{\mathfrak{p}_i\}_{i \in \mathbb{N}}$, obtaining as above a new prime $\mathfrak{q}_1 \supset \mathfrak{q}$. Going on with this procedure, we obtain a strictly ascending chain of prime ideals, which must stop since the ring A is Noetherian. The last prime in this chain is the prime ideal $\mathfrak{q}$ that we are looking for.

$(ii) \Rightarrow (i)$: Clear. $\square$

The following theorem enables us to prove that certain properties satisfy **(NC)**.

Theorem 1.2.14 (Massaza [245, Teorema 1]) *Let A be a Noetherian ring and let* ***P*** *be a property satisfying the following conditions:*

(i) For any $\mathfrak{q} \in \mathbf{P}(A)$, *there exists an element* $f \in A \setminus \mathfrak{q}$ *such that for any prime ideal* $\mathfrak{p} \in D(f) \cap V(\mathfrak{q}) \cap \mathrm{Min}(\mathbf{NP}(A))$, *we have* $\mathfrak{p}A_{\mathfrak{p}}/\mathfrak{q}A_{\mathfrak{p}} \in \mathbf{NP}(A)$.
(ii) For any prime $\mathfrak{p} \in \mathbf{P}(A)$, *the set* $\mathbf{P}(A/\mathfrak{p})$ *contains a non-empty open set of* $\mathrm{Spec}(A/\mathfrak{p})$. *Then* $\mathbf{P}(A)$ *is open.*

Proof Suppose that $\mathbf{P}(A)$ is not open. Then, by Lemma 1.2.13, there exists a prime $\mathfrak{q} \in \mathbf{P}(A)$ such that

$$\mathfrak{q} = \bigcap_{i \in \mathbb{N}} \mathfrak{p}_i, \ \ \mathfrak{p}_i \in \mathrm{Min}(\mathbf{NP}(A)).$$

Let $f \in A \setminus \mathfrak{q}$ be such that

$$\forall \, \mathfrak{p} \in D(f) \cap V(\mathfrak{q}) \cap \mathrm{Min}(\mathbf{NP}(A)), \ \mathfrak{p}A_{\mathfrak{p}}/\mathfrak{q}A_{\mathfrak{p}} \in \mathbf{NP}(A).$$

We observe that f is not contained in infinitely many of the ideals $\mathfrak{p}_i$. Indeed, otherwise, since $\mathfrak{q}$ is the intersection of any infinite subfamily of the family $\{\mathfrak{p}_i\}_{i \in \mathbb{N}}$, from (i) it would follow that $f \in \mathfrak{q}$. It results that, restricting if necessary the family $\{\mathfrak{p}_i\}_{i \in \mathbb{N}}$, we can suppose that $f \notin \mathfrak{p}_i, \forall i \in \mathbb{N}$. But then it follows easily that $\mathbf{P}(A)$ doesn't contain any open set. This is a contradiction, because, if we put $\overline{\mathfrak{p}}_i := \mathfrak{p}_i/\mathfrak{q}$, the set $\{\overline{\mathfrak{p}}_i\}$ is not contained in any proper closed set. Indeed, if

$$V(\overline{\mathfrak{p}}_i) \subset V(\overline{\mathfrak{P}}_1 \cap \cdots \cap \overline{\mathfrak{P}}_m), \ i \in \mathbb{N}$$

it follows that in A we have

$$V(\mathfrak{p}_i) \subset V(\mathfrak{P}_1 \cap \cdots \cap \mathfrak{P}_m), \ i \in \mathbb{N},$$

that is,

$$\mathfrak{P}_1 \cap \cdots \cap \mathfrak{P}_m \subseteq \mathfrak{p}_i.$$

Then

$$\mathfrak{P}_1 \cap \cdots \cap \mathfrak{P}_m \subseteq \bigcap_{i \in \mathbb{N}} \mathfrak{p}_i = \mathfrak{q},$$

and furthermore

$$\overline{\mathfrak{P}}_1 \cap \cdots \cap \overline{\mathfrak{P}}_m = \overline{(0)}.$$

This implies that

$$V(\overline{\mathfrak{P}}_1 \cap \cdots \cap \overline{\mathfrak{P}}_m) = \mathrm{Spec}(A/\mathfrak{q}).$$

$\square$

Lemma 1.2.15 *Let A be a Noetherian ring and let $\mathfrak{p} \in \mathrm{Spec}(A)$. If $x_1, \ldots, x_r \in A$ is a regular sequence on $A_\mathfrak{p}$, then there exists $f \notin \mathfrak{p}$ such that $x_1, \ldots, x_r$ is a regular sequence on A_f.*

Proof We argue by induction on r. If $r = 1$, then

$$\mathrm{Ass}_{A_\mathfrak{p}}(A_\mathfrak{p}) = \{\mathfrak{q} \in \mathrm{Ass}_A(A) \mid \mathfrak{q} \subseteq \mathfrak{p}\}.$$

But x_1 is regular in $A_\mathfrak{p}$ if and only if

$$x_1 \notin \bigcup_{\mathfrak{q} \in \mathrm{Ass}(A_\mathfrak{p})} \mathfrak{q} = \bigcup_{\mathfrak{q} \in \mathrm{Ass}(A) \cap V(\mathfrak{p})} \mathfrak{q}.$$

Suppose that $\{\mathfrak{p}_1, \ldots, \mathfrak{p}_n\}$ are those associated primes of (0) such that $\mathfrak{p}_i$ is not contained in $\mathfrak{p}$, $i = 1, \ldots, n$, and let $\mathfrak{a} := \mathfrak{p}_1 \cap \cdots \cap \mathfrak{p}_n$. In the case that all associated primes are contained in $\mathfrak{p}$, we take $\mathfrak{a} = A$. It is clear that $\mathfrak{a}$ is not contained in $\mathfrak{p}$, since otherwise one of the primes $\mathfrak{p}_i$ would be contained in $\mathfrak{p}$. Take then $f \in \mathfrak{a} \setminus \mathfrak{p}$. Clearly x_1 is regular in A_f. Suppose now that the assertion is true for $r - 1$. Let $h \notin \mathfrak{p}$ be such that $x_1, \ldots, x_{r-1}$ is a regular sequence on A_h and let $B := A_h/(x_1, \ldots, x_{r-1})A_h$ and $\overline{\mathfrak{p}} := \mathfrak{p}B$. Since x_r is regular on $B_{\overline{\mathfrak{p}}}$, there exists $\overline{g} \notin \overline{\mathfrak{p}}$ such that $\overline{x}_r$ is regular on $B_{\overline{g}}$, that is, on $A_{hg}/(x_1, \ldots, x_{r-1})A_{hg}$. Consequently $x_1, \ldots, x_r$ is a regular sequence on A_{hg}. $\square$

The main result concerning Nagata's criterion belongs of course to Nagata.

Theorem 1.2.16 (Nagata [265, Lemma 4]) *The property $P = $ regular satisfies (NC).*

Proof Let $\mathfrak{q} \in \mathrm{Reg}(A)$ and let $a_1, \ldots, a_s$ be a regular system of parameters in $A_\mathfrak{q}$, in particular $\mathfrak{q}A_\mathfrak{q} = (a_1, \ldots, a_s)A_\mathfrak{q}$. Then by Lemma 1.2.15, there exists $g \notin \mathfrak{q}$ such that $a_1, \ldots, a_s$ is a regular sequence in A_g, and moreover we may assume that $\mathfrak{q}A_g = a_1A_g + \cdots + a_sA_g$. Hence, changing A with A_g, we may assume that $(a_1, \ldots, a_s) = \mathfrak{q}$ and is generated by a regular sequence. By assumption $\mathrm{Reg}(A/\mathfrak{q})$ contains a non-empty open subset Y of $V(\mathfrak{q}) = \mathrm{Spec}(A/\mathfrak{q})$. Let $\mathfrak{p} \in \mathrm{Spec}(A)$ be such that $\mathfrak{p}/\mathfrak{q} \in Y$. This means that $A_\mathfrak{p}/\mathfrak{q}A_\mathfrak{p}$ is a regular local ring. But $a_1, \ldots, a_s$ is a A-regular sequence; hence $\mathfrak{q}A_\mathfrak{p}$

is generated by an $A_{\mathfrak{p}}$-regular sequence and $A_{\mathfrak{p}}$ is a regular local ring. This shows that $Y \subseteq V(\mathfrak{q}) \cap \mathrm{Reg}(A)$. $\square$

Theorem 1.2.17 *The property* $P = (S_k)$ *satisfies* (**NC**).

Proof Take $\mathfrak{q} \in S_k(A)$, that is, $\mathrm{depth}(A_{\mathfrak{q}}) \geq \min\{k, \mathrm{ht}(\mathfrak{q})\}$. There are two cases to consider:

Case 1. $\mathrm{ht}(\mathfrak{q}) \leq k$; hence $\mathrm{depth}(A_{\mathfrak{q}}) = \mathrm{ht}(\mathfrak{q})$. Then by [132, Prop. 6.10.6], there exists $f \notin \mathfrak{q}$ such that for any prime ideal $\mathfrak{p} \in D(f) \cap V(\mathfrak{q})$ we have the relations

$$\dim(A_{\mathfrak{p}}) = \dim(A_{\mathfrak{q}}) + \dim(A_{\mathfrak{p}}/\mathfrak{q}A_{\mathfrak{p}})$$

and

$$\mathrm{depth}(A_{\mathfrak{p}}) = \mathrm{depth}(A_{\mathfrak{q}}) + \mathrm{depth}(A_{\mathfrak{p}}/\mathfrak{q}A_{\mathfrak{p}}).$$

Let $\mathfrak{p} \in D(f) \cap V(\mathfrak{q}) \cap NS_k(A)$, that is, $\mathrm{depth}(A_{\mathfrak{p}}) < \min\{k, \mathrm{ht}(\mathfrak{p})\}$. If $\mathrm{ht}(\mathfrak{p}) \leq k$, then $\mathrm{ht}(\mathfrak{q}) = \mathrm{depth}(A_{\mathfrak{q}})$ and $\mathrm{ht}(\mathfrak{p}) > \mathrm{depth}(A_{\mathfrak{p}})$. It follows that

$$\mathrm{depth}(A_{\mathfrak{p}}/\mathfrak{q}A_{\mathfrak{p}}) = \mathrm{depth}(A_{\mathfrak{p}}) - \mathrm{depth}(A_{\mathfrak{q}}) < \dim(A_{\mathfrak{p}}) - \dim(A_{\mathfrak{q}}) = \dim(A_{\mathfrak{p}}/\mathfrak{q}A_{\mathfrak{p}})$$

and then $A_{\mathfrak{p}}/\mathfrak{q}A_{\mathfrak{p}}$ is not (S_k).
If $\mathrm{ht}(\mathfrak{p}) > k$, it follows that $\mathrm{depth}(A_{\mathfrak{p}}) < k$. Then we have

$$\mathrm{depth}(A_{\mathfrak{p}}/\mathfrak{q}A_{\mathfrak{p}}) = \mathrm{depth}(A_{\mathfrak{p}}) - \mathrm{depth}(A_{\mathfrak{q}}) < \min\{k, \dim(A_{\mathfrak{p}}/\mathfrak{q}A_{\mathfrak{p}})\}$$

and it follows again that $A_{\mathfrak{p}}/\mathfrak{q}A_{\mathfrak{p}}$ has not the property (S_k).

Case 2. $\mathrm{ht}(\mathfrak{q}) > k$. Then we have also $\mathrm{ht}(\mathfrak{p}) > k$. If we take a sequence $\{x_1, \ldots, x_r\}$ which is regular on $A_{\mathfrak{q}}$, from Lemma 1.2.15, it follows that there exists $f \notin \mathfrak{q}$, such that $\{x_1, \ldots, x_r\}$ is also a regular sequence on A_f. Now if $\mathfrak{p} \in D(f) \cap V(\mathfrak{q})$, we obtain that $A_{\mathfrak{p}}$ has the property (S_k). $\square$

Next we investigate if the property $P =$ complete intersection satisfies **NC**. For this we will use the characterization of complete intersections in terms of André-Quillen homology (1.1.9).

Lemma 1.2.18 (Greco and Marinari [125, Lemma 3.2]) *Let* $(A, \mathfrak{m}, k)$ *be a Noetherian local ring and let* $\mathfrak{p} \in \mathrm{Spec}(A)$ *be such that:*

(i) $H_3(A, A/\mathfrak{p}, A/\mathfrak{p}) = H_4(A, A/\mathfrak{p}, A/\mathfrak{p}) = 0$.
(ii) $H_i(A, A/\mathfrak{p}, A/\mathfrak{p})$ *is a free* $A/\mathfrak{p}$-*module for* $i = 0, 1, 2$.

Then A is a complete intersection if and only if $A/\mathfrak{p}$ is a complete intersection.

Proof From [4, Prop. 16.1] it follows that there is a spectral sequence

$$\mathrm{Tor}_i^{A/\mathfrak{p}}(H_j(A, A/\mathfrak{p}, A/\mathfrak{p}), k) \underset{i}{\Rightarrow} H_{i+j}(A, A/\mathfrak{p}, k).$$

For $i + j = 3$, there are only two possibilities:
If $i > 0$, then $j < 3$, and by the assumption (i) $H_j(A, A/\mathfrak{p}, A/\mathfrak{p})$ is free, hence $\mathrm{Tor}_i^{A/\mathfrak{p}}(H_j(A, A/\mathfrak{p}, A/\mathfrak{p}), k) = 0$.
If $i = 0$, then $j = 3$, and consequently in this case too, we have the relation $\mathrm{Tor}_i^{A/\mathfrak{p}}(H_j(A, A/\mathfrak{p}, A/\mathfrak{p}), k) = 0$. All this gives that $H_3(A, A/\mathfrak{p}, k) = 0$. In the same way, we prove that $H_4(A, A/\mathfrak{p}, k) = 0$. Now consider the morphisms

$$A \to A/\mathfrak{p} \to k$$

and the associated Jacobi-Zariski exact sequence [6, Th. V.1]

$$\cdots \to 0 = H_4(A, A/\mathfrak{p}, k) \to H_4(A, k, k) \to H_4(A/\mathfrak{p}, k, k) \to H_3(A, A/\mathfrak{p}, k) = 0 \to \cdots$$

It follows that

$$H_4(A, k, k) \cong H_4(A/\mathfrak{p}, k, k)$$

and the assertion of the lemma follows from Remark 1.1.9.   □

Theorem 1.2.19 (Greco and Marinari [125, Th. 3.1]) *The property $P = $ complete intersection satisfies (NC).*

Proof Let $\mathfrak{p} \in \mathrm{CI}(A) = \{\mathfrak{p} \in \mathrm{Spec}(A) \mid A_\mathfrak{p}$ is a complete intersection$\}$. Then $H_i(A_\mathfrak{p}, k(\mathfrak{p}), k(\mathfrak{p})) = 0$, for $i = 3, 4$, and then by [4, Prop. 20.2, 20.3 and 20.4], we have

$$0 = H_i(A_\mathfrak{p}, k(\mathfrak{p}), k(\mathfrak{p})) \cong H_i(A, k(\mathfrak{p}), k(\mathfrak{p})) \cong H_i(A, A/\mathfrak{p}, k(\mathfrak{p})) \cong$$

$$\cong H_i(A, A/\mathfrak{p}, A/\mathfrak{p}) \otimes_A A_\mathfrak{p}, \ i = 3, 4.$$

But by [6, Prop. IV.55], we know that $H_i(A, A/\mathfrak{p}, A/\mathfrak{p})$ are finite $A/\mathfrak{p}$-modules for $i \geq 0$, so that there is an element $h \notin \mathfrak{p}$ such that

$$H_3(A_h, A_h/\mathfrak{p}A_h, A_h/\mathfrak{p}A_h) = H_4(A_h, A_h/\mathfrak{p}A_h, A_h/\mathfrak{p}A_h) = (0),$$

and by the theorem on generic flatness [248, Th. 53], we get that there exists an element $g \notin \mathfrak{p}$ such that $H_i(A_g, A_g/\mathfrak{p}A_g, A_g/\mathfrak{p}A_g)$ is a free $A/\mathfrak{p}$-module, for $0 \leq i < 3$. Let $f = h \cdot g$. It follows that for any $\mathfrak{q} \in D(f) \cap V(\mathfrak{p})$ we have

$$H_3(A_\mathfrak{q}, A_\mathfrak{q}/\mathfrak{p}A_\mathfrak{q}, A_\mathfrak{q}/\mathfrak{p}A_\mathfrak{q}) = H_4(A_\mathfrak{q}, A_\mathfrak{q}/\mathfrak{p}A_\mathfrak{q}, A_\mathfrak{q}/\mathfrak{p}A_\mathfrak{q}) = (0)$$

and that

$$H_i(A_\mathfrak{q}, A_\mathfrak{q}/\mathfrak{p}A_\mathfrak{q}, A_\mathfrak{q}/\mathfrak{p}A_\mathfrak{q}) \text{ is free for } 0 \leq i \leq 2.$$

From Lemma 1.2.18, it follows that $A_\mathfrak{q}/\mathfrak{p}A_\mathfrak{q} \cong (A/\mathfrak{p})_\mathfrak{q}$ is a complete intersection. $\square$

In order to prove the same result for **P**=Gorenstein, we need a slightly different approach, also due to Greco and Marinari [125].

Lemma 1.2.20 *Let A be a Noetherian ring, M a finitely generated A-module and $\mathfrak{p}$ a minimal element of $\mathrm{Ass}_A(M)$. Then there is an element $g \notin \mathfrak{p}$ and a normal series $(\widetilde{M}_i)_{0 \leq i \leq m}$ of the A_g-module M_g such that*

$$\widetilde{M}_i/\widetilde{M}_{i+1} \cong A_g/\mathfrak{p}A_g, \text{ for any } i = 0, \ldots, m-1.$$

Proof From [56, Ch. IV, §1, no. 4, Th. 1 and Th. 2], it follows that there exists a normal series $(M_i)_{0 \leq i \leq n}$ of M such that

$$M_i/M_{i+1} \cong A/\mathfrak{p}_i, \text{ for any } i = 0, \ldots, n-1,$$

where $\mathfrak{p}_i \in \mathrm{Supp}(M)$, for $i = 0, \ldots, n-1$, and that there is $j \in \{0, \ldots, n-1\}$ such that $\mathfrak{p}_j = \mathfrak{p}$. Let $\mathfrak{a} := \bigcap_{\mathfrak{p}_i \in \{\mathfrak{p}_0, \ldots, \mathfrak{p}_{n-1}\} \setminus \mathfrak{p}_j} \mathfrak{p}_i$ and take $g \in \mathfrak{a} \setminus \mathfrak{p}$. Then $((M_i)_g)_{0 \leq i \leq n}$ is a normal series for the A_g-module M_g such that

$$(M_i)_g/(M_{i+1})_g \cong (A/\mathfrak{p}_i)_g \cong \begin{cases} (0), & \text{if } \mathfrak{p}_i \neq \mathfrak{p} \\ (A/\mathfrak{p})_g, & \text{if } \mathfrak{p}_i = \mathfrak{p}. \end{cases}$$

Now it is clear that one can extract a normal series $(\widetilde{M}_i)_{0 \leq i \leq m}$, $m \leq n$ with the desired property. $\square$

Lemma 1.2.21 *Let A be a Noetherian ring such that $\mathrm{Ass}_A(A) = \{\mathfrak{p}\}$ and suppose that $\mathrm{Ext}_A^1(A/\mathfrak{p}, A) = 0$. Then there exists an element $g \notin \mathfrak{p}$, such that*

$$\mathrm{Ext}_{A_g}^i(A_g/\mathfrak{p}A_g, A_g) = (0), \text{ for any } i > 0.$$

Proof If $\mathfrak{p} = (0)$ there is nothing to prove. If $\mathfrak{p} \neq (0)$, then $\mathrm{Ass}_A(\mathfrak{p}) \neq \emptyset$, and since $\mathrm{Ass}_A(\mathfrak{p}) \subseteq \mathrm{Ass}_A(A)$, it follows that $\mathrm{Ass}_A(\mathfrak{p}) = \{\mathfrak{p}\}$. By Lemma 1.2.20 there exists $g \notin \mathfrak{p}$ such that the A_g-module $\mathfrak{p}A_g$ has a normal series without repetitions, $(\widetilde{\mathfrak{p}}_i)_{0 \leq i \leq m}$ and with

$$\widetilde{\mathfrak{p}}_i / \widetilde{\mathfrak{p}}_{i+1} \cong A_g / \mathfrak{p}A_g, \text{ for any } i = 0, \ldots, m-1.$$

Consider the m exact sequences of A-modules

$$0 \to \widetilde{\mathfrak{p}}_m = (0) \to \widetilde{\mathfrak{p}}_{m-1} \to A_g / \mathfrak{p}A_g \to 0$$

$$\cdots$$

$$0 \to \widetilde{\mathfrak{p}}_1 \to \widetilde{\mathfrak{p}}_0 = \mathfrak{p}A_g \to A_g / \mathfrak{p}A_g \to 0.$$

Since $\mathrm{Ext}_A^1(A/\mathfrak{p}, A) = (0)$, we obtain successively

$$\mathrm{Ext}_{A_g}^1(\widetilde{\mathfrak{p}}_{m-1}, A_g) = \cdots = \mathrm{Ext}_{A_g}^1(\mathfrak{p}A_g, A_g) = (0).$$

Now from the exact sequence

$$0 \to \mathfrak{p}A_g \to A_g \to A_g / \mathfrak{p}A_g \to 0$$

we get $\mathrm{Ext}_{A_g}^2(A_g / \mathfrak{p}A_g, A_g) = (0)$. To conclude we use induction on i. Suppose that $\mathrm{Ext}_{A_g}^i(A_g / \mathfrak{p}A_g, A_g) = 0$. Using the same exact sequences as above we obtain successively

$$\mathrm{Ext}_{A_g}^i(\widetilde{\mathfrak{p}}_{m-1}, A_g) = \cdots = \mathrm{Ext}_{A_g}^i(\mathfrak{p}A_g, A_g) = (0)$$

and using the exact sequence

$$0 \to \mathfrak{p}A_g \to A_g \to A_g / \mathfrak{p}A_g \to 0$$

we get $\mathrm{Ext}_{A_g}^{i+1}(A_g / \mathfrak{p}A_g, A_g) = (0)$. □

The following lemma is similar to Lemma 1.2.18.

Lemma 1.2.22 *Let* $(A, \mathfrak{m}, k)$ *be a Noetherian local ring and let* $\mathfrak{p} \in \mathrm{Min}(A)$ *be such that* $\dim(A/\mathfrak{p}) = \dim(A)$. *Suppose that:*

(i) $\mathrm{Ext}_A^i(A/\mathfrak{p}, A) = 0$, *for any* $i \geq 1$.
(ii) $\mathrm{Hom}_A(A/\mathfrak{p}, A) \cong A/\mathfrak{p}$.

Then A is a Gorenstein ring if and only if A/$\mathfrak{p}$ is a Gorenstein ring.

Proof Let $0 \to A \to I^\bullet$ be a minimal injective resolution of A. From (i) and (ii), it follows that $\mathrm{Hom}(A/\mathfrak{p}, I^\bullet)$ is an injective resolution of $A/\mathfrak{p}$. We have:

$$\mathrm{Ext}^i_A(k, A) \cong H_i(\mathrm{Hom}_A(k, I^\bullet)) \cong H^i(\mathrm{Hom}_{A/\mathfrak{p}}(k, \mathrm{Hom}_A(A/\mathfrak{p}, I^\bullet))) \cong$$

$$\cong \mathrm{Ext}^i_{A/\mathfrak{p}}(k, A/\mathfrak{p}).$$

To obtain the assertion, we apply the fact [249, Th. 18.8] that a ring A is Gorenstein if and only if for all $\mathfrak{p} \in \mathrm{Spec}(A)$, $\dim_{k(\mathfrak{p})}(\mathrm{Ext}^i_{A_\mathfrak{p}}(k(\mathfrak{p}), A_\mathfrak{p})) = \delta_{i,\mathrm{ht}(\mathfrak{p})}$. $\qquad\square$

Theorem 1.2.23 (Greco and Marinari [125, Th. 1.4]) *The property **P** = Gorenstein satisfies (**NC**).*

Proof If $\mathrm{Gor}(A) := \{\mathfrak{p} \in \mathrm{Spec}(A) \mid A_\mathfrak{p} \text{ is Gorenstein}\} = \emptyset$, there is nothing to prove. Suppose that $\mathrm{Gor}(A) \neq \emptyset$ and let $\mathfrak{p} \in \mathrm{Gor}(A)$. By Lemma 1.2.7, since **P** = Gorenstein satisfies (**A₂**), it is enough to show that $\mathrm{Gor}(A) \cap V(\mathfrak{p})$ contains a non-empty open set. Because $A_\mathfrak{p}$ is Gorenstein, there are elements $x_1, \ldots, x_r \in \mathfrak{p}$ such that the sequence $\{\frac{x_1}{1}, \ldots, \frac{x_r}{1}\}$ is a regular sequence of length $r = \mathrm{ht}(\mathfrak{p})$ in $\mathfrak{p}A_\mathfrak{p}$. From Lemma 1.2.15 we obtain that there exists an element $f \notin \mathfrak{p}$ such that $\mathbf{x}_f = \{\frac{x_1}{1}, \ldots, \frac{x_r}{1}\}$ is a regular sequence in $\mathfrak{p}A_f$ and $\mathfrak{p}A_f/\mathbf{x}A_f \in \mathrm{Min}(A_f/\mathbf{x}A_f)$. Pick $h_f \in (\bigcup_{q \in \mathrm{Min}(A_f/\mathbf{x}A_f)} q) \setminus (\mathfrak{p}A_f/\mathbf{x}A_f)$. It follows that $C := (A_f/\mathbf{x}A_f)_{h_f}$ has a unique associated prime. Changing A with C, we may assume that $\mathfrak{p}$ is the unique associated prime ideal of A. But then $A_\mathfrak{p}$ is a Gorenstein ring of dimension 0 so that

$$(0) = \mathrm{Ext}^1_{A_\mathfrak{p}}(k(\mathfrak{p}), A_\mathfrak{p}) \cong \mathrm{Ext}^1_A(A/\mathfrak{p}, A) \otimes_A A_\mathfrak{p}.$$

Since all of the above modules are finitely generated, there is an element $h \notin \mathfrak{p}$ such that $\mathrm{Ext}^1_{A_h}(A_h/\mathfrak{p}A_h, A_h) = (0)$. By Lemma 1.2.21 there exists an element $g \notin \mathfrak{p}$ such that $\mathrm{Ext}^i_{A_{gh}}(A_{gh}/\mathfrak{p}A_{gh}, A_{gh}) = (0)$, for any $i > 0$. Replacing A with A_{gh}, we can furthermore assume that $\mathrm{Ext}^i_A(A/\mathfrak{p}, A) = (0)$, for any $i > 0$.

We have $\mathrm{Hom}_{A_\mathfrak{p}}(k(\mathfrak{p}), A_\mathfrak{p}) \cong k(\mathfrak{p})$, so that arguing as above we may also assume that $\mathrm{Hom}_A(A/\mathfrak{p}, A) \cong A/\mathfrak{p}$. Now $\mathrm{Gor}(A/\mathfrak{p})$ contains a non-empty open set. This means that there exists an element $s \notin \mathfrak{p}$ such that $A_s/\mathfrak{p}A_s$ is Gorenstein. Replacing A with A_s, we may also assume that $A/\mathfrak{p}$ is Gorenstein, and clearly we may also assume that A is local. Using Lemma 1.2.22 it follows that A is Gorenstein. $\qquad\square$

Theorem 1.2.24 *The property P = Cohen-Macaulay satisfies (NC).*

Proof In order to prove the assertion, we apply Theorem 1.2.14. For any prime ideal $\mathfrak{q}$ such that $A_\mathfrak{q}$ is Cohen-Macaulay, that is, $\dim(A_\mathfrak{q}) = \mathrm{depth}(A_\mathfrak{q})$, by assumption $CM(A/\mathfrak{q})$ contains a non-empty open set, that is, there exists an element $f \notin \mathfrak{q}$ such that for any prime ideal $\mathfrak{p} \in D(f) \cap V(\mathfrak{q})$, we have

$$\dim(A_\mathfrak{q}/\mathfrak{p}A_\mathfrak{q}) = \mathrm{depth}(A_\mathfrak{q}/\mathfrak{p}A_\mathfrak{q}).$$

By [132, Prop. 6.10.6] there exists an element $g \notin \mathfrak{q}$ such that for any prime ideal $\mathfrak{p} \in D(g) \cap V(\mathfrak{q})$ we have the relations

$$\dim(A_\mathfrak{p}) = \dim(A_\mathfrak{q}) + \dim(A_\mathfrak{p}/\mathfrak{q}A_\mathfrak{p})$$

and

$$\mathrm{depth}(A_\mathfrak{p}) = \mathrm{depth}(A_\mathfrak{q}) + \mathrm{depth}(A_\mathfrak{p}/\mathfrak{q}A_\mathfrak{p}).$$

Then, if $\mathfrak{p} \in \mathrm{Spec}(A_{fg})$, from the above relations, it follows that $A_\mathfrak{p}$ is Cohen-Macaulay. $\square$

We want to study to what extent the condition (NC) is preserved from a given property P to the property $\mathbf{P_k}$.

Notation 1.2.25 If k is a natural number, then we denote by

$$\mathbf{P}_k(A) := \{\mathfrak{p} \in \mathrm{Spec}(A) \mid A_\mathfrak{p} \text{ has the property } \mathbf{P_k}\}.$$

Lemma 1.2.26 *Let A be a Noetherian ring, k be a natural number and P be a property of Noetherian local rings. Then*

$$\mathrm{Min}(\mathbf{NP_k}(A)) = \{\mathfrak{q} \in \mathrm{Min}(\mathbf{NP}(A)) \mid \mathrm{ht}(\mathfrak{q}) \le k\}.$$

Proof Clearly $\mathfrak{q} \notin \mathbf{P_k}(A)$ if and only if there exists $\mathfrak{q}_1 \subseteq \mathfrak{q}$ such that $\mathfrak{q}_1 \notin \mathbf{P}(A)$ and $\mathrm{ht}(\mathfrak{q}_1) \le k$. But then $\mathfrak{q}_1 \notin \mathbf{P}_k(A)$. It follows that $\mathfrak{q} \in \mathrm{Min}(\mathbf{NP_k}(A))$ if and only if $\mathfrak{q} = \mathfrak{q}_1 \in \mathrm{Min}(\mathbf{NP}(A))$. $\square$

Lemma 1.2.27 *Let A be a Noetherian ring, k be a natural number and P be a property of Noetherian local rings. If $\mathbf{P}(A)$ is open, then also $\mathbf{P}_k(A)$ is open.*

Proof If $\mathbf{P}_k(A)$ is not open, then by Lemmata 1.2.13 and 1.2.26, we obtain that $\bigcap\limits_{i\in\mathbb{N}} \mathfrak{p}_i = \mathfrak{p} \in \mathbf{P}_k(A)$, where $\mathfrak{p}_i \in \mathrm{Min}(\mathbf{NP}(A)) \subseteq \mathrm{Min}(\mathbf{NP}_k(A))$. It follows that $\mathfrak{p} \in \mathbf{P}(A)$ and then $\mathbf{P}(A)$ is not open by Lemma 1.2.13. Contradiction!   □

Corollary 1.2.28 *If A is a ring such that $\mathrm{Reg}(A)$ is open, then, for any $k \in \mathbb{N}$, the set $R_k(A)$ is open.*

Notation 1.2.29 Let $\mathbf{P}^{(1)}, \ldots, \mathbf{P}^{(n)}$ be properties of Noetherian local rings. We consider the property $\mathbf{P} = \mathbf{P}^{(1)} + \cdots + \mathbf{P}^{(n)}$. This means that a ring A has $\mathbf{P}$ if and only if A has all the properties $\mathbf{P}^{(1)}, \ldots, \mathbf{P}^{(n)}$. It follows that

$$\mathbf{P}(A) = \{\mathfrak{p} \in \mathrm{Spec}(A) \mid A_\mathfrak{p} \text{ has the properties } \mathbf{P}^{(1)}, \ldots, \mathbf{P}^{(n)}\} =$$

$$= \mathbf{P}^{(1)}(A) \cap \cdots \cap \mathbf{P}^{(n)}(A).$$

Example 1.2.30 For example, if $\mathbf{P}^{(1)} = (R_1)$ and $\mathbf{P}^{(2)} = (S_2)$, then $\mathbf{P} = \mathbf{P}^1 + \mathbf{P}^2 =$ *normal.*

Proposition 1.2.31 (Massaza [245, Lemma 4]) *Let $\mathbf{P}^{(1)}, \ldots, \mathbf{P}^{(n)}$ be properties of Noetherian local rings and consider the property $\mathbf{P} := \mathbf{P}^{(1)} + \cdots + \mathbf{P}^{(n)}$. Suppose that each of the properties $\mathbf{P}^{(i)}$ satisfies the following condition: For any $\mathfrak{q} \in \mathbf{P}^{(i)}(A)$, there exists $f \in A \setminus \mathfrak{q}$ such that for any $\mathfrak{p} \in D(f) \cap V(\mathfrak{q}) \cap \mathrm{Min}(\mathbf{NP}^{(i)}(A))$, $\mathfrak{p}/\mathfrak{q} \in \mathbf{NP}^{(i)}(A/\mathfrak{q})$. Then the same condition is satisfied by the property $\mathbf{P}$, too.*

Proof Since $\mathbf{P}(A) = \mathbf{P}^{(1)}(A) \cap \cdots \cap \mathbf{P}^{(n)}(A)$, it follows that for any prime ideal $\mathfrak{q} \in \mathbf{P}(A)$, there exists elements $f_1, \ldots, f_n \in A \setminus \mathfrak{q}$, such that

$$\forall\, \mathfrak{p} \in D(f_i) \cap V(\mathfrak{q}) \cap \mathrm{Min}(\mathbf{NP}^{(i)}(A)), \ \mathfrak{p}/\mathfrak{q} \in \mathbf{NP}^{(i)}(A/\mathfrak{q}).$$

Let $f := f_1 \cdots \cdots f_n$ and let $\mathfrak{p} \in V(\mathfrak{q}) \cap D(f) \cap \mathrm{Min}(\mathbf{NP}(A))$. Then there exists $1 \leq i \leq n$ such that $\mathfrak{p} \notin \mathbf{P}^{(i)}(A/\mathfrak{q})$. But then $\mathfrak{p}/\mathfrak{q} \in \mathbf{NP}(A/\mathfrak{q}) \subseteq \mathbf{NP}^{(i)}(A/\mathfrak{q})$.   □

Proposition 1.2.32 (Massaza [245, Lemma 5]) *Let k be a natural number and let $\mathbf{P}$ be a property of Noetherian local rings that satisfies the following condition: For any $\mathfrak{q} \in \mathbf{P}(A)$ there exists $f \in A \setminus \mathfrak{q}$ such that for any $\mathfrak{p} \in D(f) \cap V(\mathfrak{q}) \cap \mathrm{Min}(\mathbf{NP}(A))$, $\mathfrak{p}/\mathfrak{q} \in \mathbf{NP}(A/\mathfrak{q})$. Then the same condition is satisfied by the property $\mathbf{P}_k$, too.*

Proof Let $\mathfrak{q} \in \mathbf{P}_k(A)$, that is, contained in some $\mathfrak{p} \in \mathrm{Min}(\mathbf{NP}_k(A))$. Then from Lemma 1.2.26, it follows that $\mathfrak{p} \in \mathrm{Min}(\mathbf{NP}(A))$ and consequently $\mathfrak{q} \in \mathbf{P}(A)$ and $\mathrm{ht}(\mathfrak{q}) \leq k$.

By assumption, there exists $f \notin \mathfrak{q}$ such that

$$\forall\, \mathfrak{p} \in D(f) \cap V(\mathfrak{q}) \cap \mathrm{Min}(\mathbf{NP}(A)),\ \ \mathfrak{p}/\mathfrak{q} \notin \mathbf{P}(A/\mathfrak{q}).$$

Applying again Lemma 1.2.26, it follows that, if $\mathfrak{p} \in D(f) \cap V(\mathfrak{q}) \cap \mathrm{Min}(\mathbf{NP}_k(A))$, then $\mathfrak{p}/\mathfrak{q} \notin \mathbf{P}(A/\mathfrak{q})$ and $\mathrm{ht}(\mathfrak{p}/\mathfrak{q}) \leq k$. Hence $\mathfrak{p}/\mathfrak{q} \notin \mathbf{P}_k(A/\mathfrak{q})$.   □

Corollary 1.2.33 (Massaza [245, Prop. 4]) *Let k be a natural number, and suppose that P is one of the following properties:*

  (i) P = *regular.*
 (ii) P = *complete intersection.*
(iii) P = *Cohen-Macaulay.*
      Then $\mathbf{P}_k$ satisfies the condition (NC).

Proof Follows from Theorem 1.2.14 and Proposition 1.2.32.   □

Proposition 1.2.34 *Let k be a natural number and let $\mathbf{P}$ be a property of Noetherian local rings that satisfies the following condition: For any $\mathfrak{q} \in \mathbf{P}(A)$, there exists $f \in A \setminus \mathfrak{q}$ such that for any $\mathfrak{p} \in D(f) \cap V(\mathfrak{q}) \cap \mathrm{Min}(\mathbf{NP}(A))$, $\mathfrak{p}/\mathfrak{q} \in \mathbf{NP}(A/\mathfrak{q})$.*
Then the same condition is satisfied by the property $\mathbf{P}'_k$.

Proof Follows from the fact that $\mathbf{P}'_k = \mathbf{P}_k \cup (S_{k+1})$, the proof of Theorem 1.2.17 and Proposition 1.2.32.   □

Corollary 1.2.35 *Let k be a natural number, and suppose that P is one of the following properties:*

  (i) P = *regular.*
 (ii) P = *complete intersection.*
(iii) P = *Cohen-Macaulay.*
      Then $\mathbf{P}'_k$ satisfies the condition (NC).

Proof Follows from Theorem 1.2.14 and Proposition 1.2.34.   □

Corollary 1.2.36 *The properties normal, (R_k), (S_k) satisfy (NC).*

Our next task is to investigate the property **P-2**. The next theorem is essentially due to Nagata [265, Th. 1] who proved it in the case **P**=regular. The present form is an adaptation of Valabrega [387].

Theorem 1.2.37 (Valabrega [387, Prop. 1]) *Let A be a Noetherian ring and let P be a property of Noetherian local rings satisfying the axioms $(\mathbf{A_1}) - (\mathbf{A_4})$ and the condition (NC). The following are equivalent:*

(i) *A is a **P-2** ring.*

(ii) *Any finite A-algebra is a **P-1** ring.*

(iii) *For every $\mathfrak{q} \in \mathrm{Spec}(A)$ and for every field L, finite extension of $k(\mathfrak{q})$, there is a finite A-algebra B, containing $A/\mathfrak{q}$ and having L as field of fractions, such that B is a **P-0** ring.*

Proof $(i) \Rightarrow (ii) \Rightarrow (iii)$: Obvious.

$(iii) \Rightarrow (i)$. Let $\mathfrak{q} \in \mathrm{Spec}(A)$ and consider a field L, finite extension of $k(\mathfrak{q})$ and B a finite A-algebra which is a **P-0** ring and has L as fraction field. Then there exist elements $b_1, \ldots, b_r \in B$ such that $\{b_1, \ldots, b_r\}$ is a basis of L as a vector space over $k(\mathfrak{q})$. Moreover there exists a non-zero element $f \in A/\mathfrak{q}$, such that B_f is a finite free $(A/\mathfrak{q})_f$-module having $\{b_1, \ldots, b_r\}$ as a basis. Then, since B_f has the property **P**, by the axiom $(\mathbf{A_4})$ we get that $(A/\mathfrak{q})_f$ has the property **P**; hence $A/\mathfrak{q}$ is **P-0**. By the condition (NC), it follows that A is **P-1** and also $A/\mathfrak{p}$ is **P-1**, for every $\mathfrak{p} \in \mathrm{Spec}(A)$. By using (NC) it is enough to show that, if C is a finitely generated A-algebra which is a domain, then C is **P-0**. Let $\mathfrak{q}$ be the kernel of the morphism $A \to C$. We can replace A by $A/\mathfrak{q}$, thus assuming that A is contained in C. Considering a suitable open set of $\mathrm{Spec}(A)$, we can moreover assume that A has the property **P**. Denote by K and L the fraction fields of A, respectively C. There are two cases:

Case 1: The extension $K \subseteq L$ is separable. Choose a separating transcendence basis of L over K, say $\{t_1, \ldots, t_n\}$, whose elements are in C. Let $E := A[t_1, \ldots, t_n]$, $M := K(t_1, \ldots, t_n)$. Then E has the property **P**, because it is a polynomial ring over a ring having the property **P**. Replacing A by E and K by M, we can assume that the extension $K \subseteq L$ is separable algebraic. Moreover we can choose a basis $\{e_1, \ldots, e_m\}$ of L over K, whose elements belong to C and an element $f \in A$ such that C_f is a finite free A_f-module having $\{e_1, \ldots, e_m\}$ as a basis. Let us replace A and C by A_f resp. C_f. From the separability assumption, we have that $d := \det(\mathrm{tr}_{L/K}(e_i e_j)) \neq 0$. Suppose that $\mathfrak{P} \in \mathrm{Spec}(C) \cap V(d)$ and let $\mathfrak{p} := \mathfrak{P} \cap A$. Then the canonical image of d in $C \otimes k(\mathfrak{p})$ is not zero in $k(\mathfrak{p})$. This implies that $C \otimes k(\mathfrak{p})$ is a product of fields; hence $C_{\mathfrak{P}}/\mathfrak{p}C_{\mathfrak{P}}$ is a field. But the morphism $A_{\mathfrak{p}} \to C_{\mathfrak{P}}$ is faithfully flat, hence by $(\mathbf{A_4})$ we get that $C_{\mathfrak{P}}$ has the property **P**, hence C_d has the property **P**.

Case 2: $\mathrm{ch}(L) = p > 0$. There exists a finite purely inseparable field extension N of K, such that the field extension $N \subseteq L(N)$ is a separable one. Then by assumption there is $F \subseteq N$ such that F is **P-0**, and by Case 1 it follows that also $F[C]$ is **P-0**. Since $F[C]$ is a finite C-algebra, from Lemma 1.2.10 it follows that also C is **P-0**.   □

Corollary 1.2.38 *The Theorem 1.2.37 is valid for **P**=regular, complete intersection, Gorenstein, Cohen-Macaulay.*

Similar results can be proved for some other properties that do not satisfy ($\mathbf{A_4}$). We start with the property (R_k). The proof is almost similar to the previous one.

Theorem 1.2.39 (Massaza and Valabrega [246, Teorema 2]) *Let A be a Noetherian ring and k a natural number. The following conditions are equivalent:*

(i) *A is a $(R_k) - 2$ ring.*
(ii) *Any finite A-algebra is a $(R_k) - 1$ ring.*
(iii) *For every $\mathfrak{q} \in \mathrm{Spec}(A)$ and for every field L, finite extension of $k(\mathfrak{q})$, there is a finite A-algebra B, containing $A/\mathfrak{q}$ and having L as field of fractions, such that B is a $(R_k) - 0$ ring.*

Proof We have only to prove that $(iii) \Rightarrow (i)$. Let $\mathfrak{q} \in \mathrm{Spec}(A)$ and consider a field L, finite extension of $k(\mathfrak{q})$ and B a finite A-algebra which is a $(R_k) - 0$ ring and has L as fraction field. Then there exist elements $b_1, \ldots, b_r \in B$ such that $\{b_1, \ldots, b_r\}$ is a basis of L as a vector space over $k(\mathfrak{q})$. Moreover there exists a non-zero element $f \in A/\mathfrak{q}$, such that B_f is a finite free $(A/\mathfrak{q})_f$-module having $\{b_1, \ldots, b_r\}$ as a basis. Then, since B_f has the property (R_k), by the axiom ($\mathbf{A_4}$), we get that $(A/\mathfrak{q})_f$ has the property (R_k); hence $A/\mathfrak{q}$ is $(R_k) - 0$. By the condition (**NC**), it follows that A is $(R_k) - 1$ and also $A/\mathfrak{p}$ is $(R_k) - 1$, for every $\mathfrak{p} \in \mathrm{Spec}(A)$. By using (**NC**) it is enough to show that, if C is a finitely generated A-algebra which is a domain, then C is $(R_k - 0)$. Let $\mathfrak{q}$ be the kernel of the morphism $A \to C$. We can replace A by $A/\mathfrak{q}$, thus assuming that A is contained in C. Considering a suitable open set of $\mathrm{Spec}(A)$, we can moreover assume that A has the property (R_k). Denote by K and L the fraction fields of A resp. C. There are two cases:
Case 1: The extension $K \subseteq L$ is separable. Choose a separating transcendence basis of L over K, say $\{t_1, \ldots, t_n\}$, whose elements are in C. Let $E := A[t_1, \ldots, t_n]$, $M := K(t_1, \ldots, t_n)$. Then E has the property (R_k), because it is a polynomial ring over a ring having the property (R_k). Replacing A by E and K by M, we can assume that the extension $K \subseteq L$ is separable algebraic. Moreover we can choose a basis $\{e_1, \ldots, e_m\}$ of L over K, whose elements belong to C and an element $f \in A$ such that C_f is a finite free A_f-module having $\{e_1, \ldots, e_m\}$ as a basis. Let us replace A and C by A_f resp. C_f. From the separability assumption, we have that $d := \det(\mathrm{tr}_{L/K}(e_i e_j)) \neq 0$. Suppose that $\mathfrak{P} \in \mathrm{Spec}(C) \cap V(d)$, such that $\mathrm{ht}(\mathfrak{P}) \leq k$, and let $\mathfrak{p} := \mathfrak{P} \cap A$. Then the canonical image of d in $C \otimes k(\mathfrak{p})$ is not zero in $k(\mathfrak{p})$. This implies that $C \otimes k(\mathfrak{p})$ is a product of fields; hence $C_\mathfrak{P}/\mathfrak{p}C_\mathfrak{P}$ is a field. But the morphism $A_\mathfrak{p} \to C_\mathfrak{P}$ is faithfully flat and $A_\mathfrak{p}$ is regular; hence by ($\mathbf{A_4}$) we get that $C_\mathfrak{P}$ is regular, hence C_d has the property (R_k).
Case 2: $\mathrm{ch}(L) = p > 0$. There exists a finite purely inseparable field extension M of K, such that $M \subseteq L(M)$ is a separable extension. Then by assumption there is $E \subseteq M$ so that E is (R_k)-0, and by Case 1 it follows that also $E[C]$ is (R_k)-0. Since $E[C]$ is a finite C-algebra, from Lemma 1.2.10, it follows that also C is (R_k)-0. $\qquad\square$

Remark 1.2.40 Let us remark that what is proved in Theorem 1.2.39 is actually a characterization of a $\mathbf{P}_k - 2$ ring, where **P**=regular. One sees immediately that analogously we can obtain similar characterizations for the properties **P**=Cohen-Macaulay, Gorenstein, complete intersection.

The property **P**=normal has some special features from this point of view.

Theorem 1.2.41 (Grothendieck [132, Prop. 6.13.7]) *Let A be a Noetherian ring. The following conditions are equivalent:*

  (i) *A is a* Nor -2 *ring.*
  (ii) *Any finite A-algebra is a* Nor -1 *ring.*
  (iii) *For every* $\mathfrak{q} \in \mathrm{Spec}(A)$ *and for every field L, finite extension of $k(\mathfrak{q})$, there is a finite A-algebra B, containing $A/\mathfrak{q}$ and having L as field of fractions, such that B is a* Nor -0 *ring.*

Proof We have only to prove that $(iii) \Rightarrow (i)$. Let $\mathfrak{q} \in \mathrm{Spec}(A)$ and consider a field L, finite extension of $k(\mathfrak{q})$ and B a finite A-algebra which is a Nor-0 ring and has L as fraction field. Then there exist elements $b_1, \ldots, b_r \in B$ such that $\{b_1, \ldots, b_r\}$ is a basis of L as a vector space over $k(\mathfrak{q})$. Moreover there exists a non-zero element $f \in A/\mathfrak{q}$, such that B_f is a finite free $(A/\mathfrak{q})_f$-module having $\{b_1, \ldots, b_r\}$ as a basis. Then, since B_f is normal, by the axiom (**A₄**) we get that $(A/\mathfrak{q})_f$ is normal; hence $A/\mathfrak{q}$ is Nor-0. By the condition (**NC**), it follows that A is Nor-1 and also $A/\mathfrak{p}$ is Nor-1, for every $\mathfrak{p} \in \mathrm{Spec}(A)$. By using (**NC**) it is enough to show that, if C is a finitely generated A-algebra which is a domain, then C is Nor-0. Let $\mathfrak{q}$ be the kernel of the morphism $A \to C$. We can replace A by $A/\mathfrak{q}$, thus assuming that A is contained in C. Considering a suitable open set of $\mathrm{Spec}(A)$, we can moreover assume that A is normal. Denote by K and L the fraction fields of A, respectively, C. There are two cases:

Case 1: The extension $K \subseteq L$ is separable. Choose a separating transcendence basis of L over K, say $\{t_1, \ldots, t_n\}$, whose elements are in C. Let $E := A[t_1, \ldots, t_n]$, $M := K(t_1, \ldots, t_n)$. Then E is normal, being a polynomial ring over a normal ring. Replacing A by E and K by M, we can assume that the extension $K \subseteq L$ is separable algebraic. Moreover we can choose a basis $\{e_1, \ldots, e_m\}$ of L over K, whose elements belong to C and an element $f \in A$ such that C_f is a finite free A_f-module having $\{e_1, \ldots, e_m\}$ as a basis. Let us replace A and C by A_f resp. C_f. From the separability assumption, we have that $d := \det(\mathrm{tr}_{L/K}(e_i e_j)) \neq 0$. Suppose that $\mathfrak{P} \in \mathrm{Spec}(C) \cap V(d)$, such that $\mathrm{depth}(C_\mathfrak{P}) \leq 1$, and let $\mathfrak{p} := \mathfrak{P} \cap A$. Then the canonical image of d in $C \otimes k(\mathfrak{p})$ is not zero in $k(\mathfrak{p})$. This implies that $C \otimes k(\mathfrak{p})$ is a product of fields; hence $C_\mathfrak{P}/\mathfrak{p}C_\mathfrak{P}$ is a field. But the morphism $A_\mathfrak{p} \to C_\mathfrak{P}$ is faithfully flat; hence $\mathrm{depth}(A_\mathfrak{p}) \leq 1$ so that $A_\mathfrak{p}$ is regular, hence by (**A₄**) we get that $C_\mathfrak{P}$ is regular. Therefore C_d is normal.

Case 2: $\mathrm{ch}(L) = p > 0$. There exists a finite purely inseparable field extension M of K, such that $M \subseteq L(M)$ is a separable extension. Then by assumption there is $E \subseteq M$ so that E is (Nor)-0, and by case 1 it follows that also $E[C]$ is Nor-0. Since $E[C]$ is a finite C-algebra, from 1.2.10 it follows that also C is Nor-0.   □

Remark 1.2.42 Looking closer at Theorem 1.2.41, one sees that actually the proof works as a characterization of $\mathbf{P}'_k - 2$ rings, where $\mathbf{P}$ is one of the properties from 1.2.37: regular, Cohen-Macaulay, Gorenstein, complete intersection. The reader is encouraged to carry out the details. In particular, it gives a characterization of $\mathbf{P}$-2 rings similar to Theorem 1.2.37 for the property $\mathbf{P} = (S_k)$.

Example 1.2.43

  (i) Any algebra of finite type over a field k is a Reg -2 ring. This follows from the Jacobian criterion of regularity [202, Prop.7.7].
 (ii) Any Dedekind domain with perfect fraction field is Reg -2.
(iii) It must be noted that the property of being a Reg -2 ring is not local, that is, there exists a ring A such that $A_\mathfrak{p}$ is Reg -2 for any prime ideal $\mathfrak{p}$, while A itself is not. Such a ring is the one constructed in Example 1.2.4, taking account of Remark 1.2.2, ii).

The next result gives another important class of examples.

Theorem 1.2.44 (Nagata [265, Th. 3]) *Let A be a Noetherian complete semilocal ring. Then A is Reg-2.*

Proof Any finite A-algebra is a finite direct product of local complete Noetherian ring; hence from Theorem 1.2.37, it is enough to show that any Noetherian complete local domain $(A, \mathfrak{m}, k)$ is Reg-0. Let $p := \text{char}(A)$.
If $p = 0$, from Cohen's structure theorem [249, Th. 29.4], there exists an injective morphism $C \hookrightarrow A$, where C is a Noetherian complete regular local ring. From 1.2.10 it follows that A is Reg-0.
Assume that $p > 0$. Then again from Cohen's structure theorem, it follows that A is isomorphic to a factor ring of $k[[X_1, \ldots, X_n]]$, say $A \cong k[[X_1, \ldots, X_n]]/\mathfrak{a}$, where k is a coefficient field. Therefore A is Reg-1 by the Jacobian criterion of Nagata [248, (29.P)] or [131, Th. 22.7.3].   □

Notation 1.2.45 If we have two different properties $\mathbf{P}$ and $\mathbf{Q}$ of local rings, we will say that $\mathbf{P}$ *implies* $\mathbf{Q}$ if any ring having the property $\mathbf{P}$ has also the property $\mathbf{Q}$. For example, $\mathbf{P}$=Gorenstein implies $\mathbf{Q}$=Cohen-Macaulay.

Proposition 1.2.46 *Let* **P** *and* **Q** *be properties of Noetherian local rings and let* A *be a* **P**-*2 ring. Assume that:*

(i) **Q** *satisfies* (**NC**).
(ii) **P** *implies* **Q**.
    Then **Q**(A) *is open and* A *is a* **Q**-*2 ring.*

Proof Let $\mathfrak{p} \in \mathbf{Q}(A)$. Then $\mathbf{P}(A/\mathfrak{p}) \subseteq \mathbf{Q}(A/\mathfrak{p})$ and $\mathbf{P}(A/\mathfrak{p})$ is a non-empty open subset of $\mathrm{Spec}(A/\mathfrak{p})$. It follows that $\mathbf{Q}(A/\mathfrak{p})$ contains a non-empty open set. From (**NC**) we get that $\mathbf{Q}(A)$ is open.
Let $u : A \to B$ be an A-algebra of finite type. Then B is a **P**-**2** ring. By Proposition 1.2.46 it follows that $\mathbf{Q}(A)$ is open. Therefore A is a **Q**-**2** ring. $\qquad\square$

Corollary 1.2.47 *Let* A *be a Noetherian complete semilocal ring. Then* A *is* $(R_k) - 2$, *CM-2,* (S_k)-*2, Gor-2, CI-2 and Nor-2.*

Proof Follows from Theorem 1.2.44 and Proposition 1.2.46. $\qquad\square$

Corollary 1.2.48 *Let* **P** *be one of the properties considered in Sect. 1.1 (i.e.* **P**=*regular, normal, reduced, Gorenstein, complete intersection,* (R_k), (S_k), *etc.). Then* **P** *satisfies axiom* (**A₆**).

1.3 Some Useful Results on Inductive Limits

This section gathers some facts about inductive limits of rings and modules that will be used later in the book. For the main results and notions about inductive limits (or direct limits or colimits), one can see [324, Section 5.2].

Notation 1.3.1 In order to fix the notations, if $I = (I, \leq)$ is a partially ordered set, we recall that an *inductive (or direct) system* means a collection $(E_i, \varphi_{ij})_{i,j \in I, i \leq j}$ where for any indices $i \leq j$, the map $\varphi_{ij} : E_i \to E_j$ is a ring (or module) homomorphisms, such that $\varphi_{ii} = 1_{E_i}$ for any $i \in I$ and for any $i \leq j \leq k$ we have the relation $\varphi_{jk}\varphi_{ij} = \varphi_{ik}$ [324, Definition, page 237]. The *inductive limit* (or *direct limit* or *colimit*) is a ring (or module) L, together with a family of morphisms $\phi_j : E_j \to L$ such that $\phi_j \circ \phi_{ij} = \phi_i$ whenever $i \leq j$ and satisfying a universal condition [324, Definition, page 238]. The inductive system will be shortly denoted by $(E_i, \varphi_{ij})_{i \in I}$, the inductive limit of the system by $\varinjlim_i E_i$ or by $\varinjlim E_i$ when no confusion can arise and $\varphi_j : E_j \to \varinjlim_i E_i$ will denote the canonical morphisms. The inductive system will be called *filtered* if the partially ordered set I is filtered, that is, for any $i, j \in I$, there is some $k \in I$ such that $i \leq k$ and $j \leq k$.

The following basic fact is actually well-known [135, Ch. 0, Prop. 6.1.2].

Proposition 1.3.2 *Let* $(A_i, \varphi_{ij})_{i \in I}$ *be a filtered inductive system of ring homomorphisms, let* $\mathfrak{a}_i \subseteq A_i$ *be an ideal of* A_i *for any index* $i \in I$ *such that* $\varphi_{ij}(\mathfrak{a}_i) \subseteq \mathfrak{a}_j, \forall i \leq j$ *and let* $A = \varinjlim_i A_i$. *Then:*

(i) $(\mathfrak{a}_i, \varphi_{ij}|_{\mathfrak{a}_i})_{i \in I}$ *is an inductive system of sets,* $\mathfrak{a} = \varinjlim_i \mathfrak{a}_i$ *is an ideal of* A *and* $A/\mathfrak{a} \cong$
$\varinjlim_i A_i/\mathfrak{a}_i$.

(ii) *If moreover* $\mathfrak{a}_i = \varphi_{ij}^{-1}(\mathfrak{a}_j), \forall i \leq j$, *then* $\mathfrak{a}_i = \varphi_i^{-1}(\mathfrak{a}), \forall i \in I$, *where* $\varphi_i : A_i \to A$ *is the canonical morphism.*

(iii) *Conversely, for any ideal* $\mathfrak{a}$ *of* A, *let* $\mathfrak{a}_i = \varphi_i^{-1}(\mathfrak{a})$. *Then* $(\mathfrak{a}_i, \varphi_{ij}|_{\mathfrak{a}_i})$ *is an inductive system and* $\mathfrak{a} = \varinjlim_i \mathfrak{a}_i$.

Proof (i) and (ii): The direct limit of A_i-modules is an A-module; hence $\mathfrak{a}$ is an ideal of A. Since $\varinjlim$ is an exact functor in the category **Ab** of abelian groups [54, Ch. II, §6, no. 2, Prop. 3], it follows that $A/\mathfrak{a} \cong \varinjlim_i A_i/\mathfrak{a}_i$. Moreover $\mathfrak{a} = \bigcup_{i \in I} \varphi_i(\mathfrak{a}_i)$. Let $x_i \in A_i$ such that $\varphi_i(x_i) \in \mathfrak{a}$. There exists $j \geq i$ such that $\varphi_i(x_i) = \varphi_j(y_j)$ for some $y_j \in \mathfrak{a}_j$. Then $\varphi_j(\varphi_{ij}(x_i)) = \varphi_j(y_j)$; hence there exists $l \geq j$ such that $\varphi_{jl}(\varphi_{ij}(x_i)) = \varphi_{jl}(y_j)$. It follows that $\varphi_{il}(x_i) \in \mathfrak{a}_l$. Thus $x_i \in \mathfrak{a}_i$ and $\mathfrak{a}_i = \varphi_{il}^{-1}(\mathfrak{a}_l)$.
(iii) Follows from (i). $\square$

Corollary 1.3.3 *Let* $(A_i, \varphi_{ij})_{i \in I}$ *be a filtered inductive system of ring homomorphisms, let* $A = \varinjlim_i A_i$ *and for any* $i \in I$ *let* $\mathfrak{n}_i = N(A_i)$ *be the nilradical of* A_i. *Then the nilradical of* A *is* $N(A) = \varinjlim_i \mathfrak{n}_i$ *and* $A_{red} = \varinjlim_i (A_i)_{red}$. *In particular, if* A_i *is a reduced ring for any* $i \in I$, *then* A *is reduced.*

Proof Clearly $\varphi_{ij}(\mathfrak{n}_i) \subseteq \mathfrak{n}_j$ and $\varinjlim_i \mathfrak{n}_i$ is a nilpotent ideal of A, whence $\varinjlim_i \mathfrak{n}_i \subseteq N(A)$.

Conversely, let $x \in A$ such that $x^k = 0$ and assume that $x = \varphi_i(x_i), x_i \in A_i$. Then $\varphi_i(x_i^k) = 0$ and there exists $j \geq i$ such that $\varphi_{ij}(x_i^k) = 0$. If $x_j = \varphi_{ij}(x_i)$, we get that $x_j^k = 0$, that is, $x_j \in \mathfrak{n}_j$. As $x = \varphi_j(x_j)$ we have that $\varinjlim_i \mathfrak{n}_i = N(A)$. The other assertions follow at once. $\square$

Proposition 1.3.4 *Let* $(A_i, \varphi_{ij})_{i \in I}$ *be a filtered inductive system of ring homomorphisms and let* $A = \varinjlim_i A_i$. *If* A_i *is an integral domain for any* $i \in I$, *then* A *is an integral domain.*

Proof Let $x, y \in A$ be such that $xy = 0$. There exists $i \in I$ and $x_i, y_i \in A_i$ such that $x = \varphi_i(x_i)$, $y = \varphi_i(y_i)$. Hence $\varphi_i(x_i y_i) = 0$. It follows that there exists $j \geq i$ such that, if $x_j = \varphi_{ij}(x_i)$ and $y_j = \varphi_{ij}(y_i)$, we have that $x_j y_j = 0$; hence $x_j = 0$ or $y_j = 0$. But $x = \varphi_j(x_j)$ and $y = \varphi_j(y_j)$; hence $x = 0$ or $y = 0$. □

Proposition 1.3.5 *Let $(A_i, \varphi_{ij})_{i \in I}$ be a filtered inductive system of ring homomorphisms and let $A = \varinjlim_i A_i$. For any index $i \in I$, let $\mathfrak{p}_i$ be a prime ideal of A_i such that $\varphi_{ij}(\mathfrak{p}_i) \subseteq \mathfrak{p}_j, \forall i \leq j$, and let $\mathfrak{p} = \varinjlim_i \mathfrak{p}_i$. Then:*

(i) $\mathfrak{p} \in \mathrm{Spec}(A)$.

(ii) if moreover $\mathfrak{p}_i = \varphi_{ij}^{-1}(\mathfrak{p}_j), \forall j \leq i$, then $A_{\mathfrak{p}} \cong \varinjlim_i (A_i)_{\mathfrak{p}_i}$;

(iii) Assume that $(A, \mathfrak{p})$ is a local ring and that φ_i is injective for any $i \in I$ and let $\mathfrak{p}_i = \varphi_i^{-1}(\mathfrak{p})$. Then $A \cong \varinjlim_i (A_i)_{\mathfrak{p}_i}$, and the canonical morphisms $(A_i)_{\mathfrak{p}_i} \to A$ are injective.

Proof (i) Follows from Propositions 1.3.2 and 1.3.4.

(ii) Set $S_i = A \setminus \mathfrak{p}_i$ and $S = A \setminus \mathfrak{p}$. Then $(S_i, \varphi_{ij}|_{S_i})_{i \in I}$ is an inductive system of sets, and $S = \varinjlim S_i$ is a multiplicative system. It follows that $((A_i)_{\mathfrak{p}_i})_{i \in i}$ is an inductive system of local rings and local morphisms, and the canonical morphisms $(A_i)_{\mathfrak{p}_i} \to A_{\mathfrak{p}}$ form an inductive system of morphisms having as limit the surjective morphism ψ : $\varinjlim_i ((A_i)_{\mathfrak{p}_i}) \to A_{\mathfrak{p}}$. We have to show that ψ is injective. Let $x_i \in A_i$ and $s_i \in S_i$ be such that $\frac{\varphi_i(x_i)}{\varphi_i(s_i)} = 0$. There exists $t \in S$ such that $t\varphi_i(x_i) = 0$. Changing if necessary the index i, we may assume that $t = \varphi_i(t_i)$ for some $t_i \in S_i$. We obtain that $\varphi_i(t_i x_i) = 0$; hence there exists $j \geq i$ such that $\varphi_{ij}(t_i x_i) = 0 \in A_j$. It follows that $\frac{\varphi_{ij}(x_i)}{\varphi_{ij}(s_i)} = 0 \in (A_j)_{\mathfrak{p}_j}$, whence $\frac{x_i}{s_i} = 0$ in $\varinjlim_i (A_i)_{\mathfrak{p}_i}$.

(iii) Since $A_{\mathfrak{p}} = A$, from ii), we obtain that $A = \varinjlim_i (A_i)_{\mathfrak{p}_i}$. Moreover the elements of S_i are invertible in A; hence $A_{\mathfrak{p}_i} = A$. As φ_i is injective, by flatness it follows that the morphism $(A_i)_{\mathfrak{p}_i} \to A$ is injective. □

Corollary 1.3.6 *Let $(A_i, \varphi_{ij})_{i \in I}$ be a filtered inductive system of ring homomorphisms and let $A = \varinjlim_i A_i$. Assume that for any $i \in I$, A_i is an integral domain and that φ_{ij} is injective for any $i, j \in I, i \leq j$. Then:*

(i) If A_i' is the integral closure of A_i and A' is the integral closure of A, then $A' = \varinjlim_i A_i'$.

(ii) If A_i is integrally closed for any $i \in I$, then A is integrally closed.

Proof (i) Let $K_i := Q(A_i)$ be the field of fractions of A_i. Then Proposition 1.3.5 tells us that A is an integral domain with field of fractions $Q(A) := K = \varinjlim_i K_i$ and that

$A' = \varinjlim_i A_i'$ is a subring of K, integral over A. Let $z \in K$ be an element integral over A

and let $z^n + \sum_{i=1}^n c_i z^{n-i} = 0$ with $c_1, \ldots, c_n \in A$ being an equation of integral dependence of z over A. Then there exist an index $j \in I$ and elements $z_j \in K_j$ and $c_{ij} \in A_j$, such that z, respectively c_j are the images of z_j and respectively c_{ij}. It follows that $z_j \in A_j'$; hence $z \in A'$.

(ii) Follows from (i). $\square$

It is well-known and very easy to see that an inductive limit of Noetherian rings is not necessarily a Noetherian ring. Just look at the non-Noetherian ring $\mathbb{Z}[X_1, \ldots, X_n, \ldots] = \bigcup_{n \geq 0} \mathbb{Z}[X_1, \ldots, X_n]$. Actually any ring can be easily described as an inductive limit of Noetherian rings. On the other hand, sometimes we need to know that an inductive limit of Noetherian rings is Noetherian. The following result will be useful. It is a strengthening, given by Ogoma [278, Theorem], of a well-known fact [130, Lemma 10.3.1.3], where the morphisms φ_{ij} were supposed to be flat.

Proposition 1.3.7 *Let* $((A_i, \mathfrak{m}_i, k_i), \varphi_{ij})_{i \in I}$ *be a filtered inductive system of local rings and morphisms and let* $A := \varinjlim_i A_i$, $\mathfrak{m} := \varinjlim_i \mathfrak{m}_i$ *and* $k := \varinjlim_i k_i$. *Then:*

  (i) $(A, \mathfrak{m}, k)$ *is a local ring.*
  (ii) If moreover $\mathfrak{m}_j = \mathfrak{m}_i A_j$ *for all* $i < j$, *then* $\mathfrak{m} = \mathfrak{m}_i A$, *for any* $i \in I$.
  (iii) If moreover A_i *is Noetherian for any* $i \in I$, *then* A *is Noetherian and the morphism*
      $\varphi_i : A_i \to A$ *is flat, for any* $i \in I$.

Proof (i) Since $\varphi_{ij}(\mathfrak{m}_i) \subseteq \mathfrak{m}_j$, it follows that $(\mathfrak{m}_i, \varphi_{ij}|_{\mathfrak{m}_i})_{i \in I}$ is an inductive system. By Proposition 1.3.2 we get that $\mathfrak{m}$ is an ideal of A and that $k = \varinjlim_i k_i$. Let $x \notin \mathfrak{m}$. There exist an index $i \in I$ and an element $x_i \in A_i$ such that $x = \varphi_i(x_i)$, where $\varphi_i : A_i \to A$ is the canonical morphism. Then $x_i \notin \mathfrak{m}_i$; hence there exists $y_i \in A_i$ such that $x_i y_i = 1$. It follows that $y = \varphi_i(y_i)$ is such that $xy = 1$. Thus A is a local ring with maximal ideal $\mathfrak{m}$. Since the inductive limit is an exact functor, it follows that the residue field of A is $\varinjlim_i k_i = k$.

(ii) By the hypothesis we get that $\mathfrak{m}_i \otimes_{A_i} A_j \to \mathfrak{m}_j$ is surjective. Since the inductive limit is an exact functor commuting with the tensor product, we obtain that $\mathfrak{m} = \mathfrak{m}_i A$.

(iii) The assumptions imply that the morphism $\mathrm{gr}_{\mathfrak{m}_i}(A_i) \otimes_{A_i} A_j \to \mathrm{gr}_{\mathfrak{m}_j}(A_j)$ is surjective; therefore the morphisms $\mathrm{gr}_{\mathfrak{m}_i}(A_i) \to \mathrm{gr}_{\mathfrak{m}_j}(A_j), i \leq j$ induce the surjective morphisms $\mathrm{gr}_{\mathfrak{m}_i}(A_i) \otimes_{A_i} k \to \mathrm{gr}_{\mathfrak{m}_j}(A_j) \otimes_{A_j} k$. This means that the system $\{\mathrm{gr}_{\mathfrak{m}_i}(A_i) \otimes_{A_i} k\}_{i \in I}$ is a filtered inductive system of Noetherian rings whose morphisms are surjective. Hence there exists $i_0 \in I$ such that for any $i, j \geq i_0$ the morphism $\mathrm{gr}_{\mathfrak{m}_i}(A_i) \otimes_{A_i} k \to \mathrm{gr}_{\mathfrak{m}_j}(A_j)$ is an isomorphism. By [248, (20.C), Th. 49] the morphisms φ_{ij} are flat for any indices $i_0 \leq i \leq j$. Because φ_{ij} is flat, it follows that φ_i is flat and local, hence faithfully flat for any index $i \in I$. Since A_i is Noetherian, the $\mathfrak{m}_i$-adic topology on A_i is separated. Let $x \in \bigcap_{n=0}^{\infty} \mathfrak{m}^n$. There exists an index $l \in I$ and an element $x_l \in A_l$, such that $\varphi_l(x_l) = x$. Since

$$\varphi_l^{-1}(\mathfrak{m}^n) = \varphi_l^{-1}(\mathfrak{m}_l^n A) = \mathfrak{m}_l^n \text{ for any natural number } n, \text{ it follows that } x_l \in \bigcap_{n=0}^{\infty} \mathfrak{m}_l^n = (0).$$

Therefore $x = 0$, that is, the $\mathfrak{m}$-adic topology on A is separated, so that if $\widehat{A}$ is the $\mathfrak{m}$-adic completion of A, we have that $A \subseteq \widehat{A} = \varprojlim_n A/\mathfrak{m}^n$. Because φ_i is flat, we get

$$\mathfrak{m}^n/\mathfrak{m}^{n+1} = (\mathfrak{m}_i^n/\mathfrak{m}_i^{n+1}) \otimes_{A_i} A = \mathfrak{m}_i^n/\mathfrak{m}_i^{n+1} \otimes_{k_i} (k_i \otimes_{A_i} A) = \mathfrak{m}_i^n/\mathfrak{m}_i^{n+1} \otimes_{k_i} k.$$

But $\dim_{k_i} \mathfrak{m}_i^n/\mathfrak{m}_i^{n+1} < \infty$; hence $\dim_k \mathfrak{m}^n/\mathfrak{m}^{n+1} < \infty$, so that $\widehat{A}$ is Noetherian. On the other hand, $\widehat{A}/\mathfrak{m}^n \widehat{A} \cong A/\mathfrak{m}^n \cong A_i/\mathfrak{m}_i^n \otimes_{A_i} A$, whence $A/\mathfrak{m}^n$ is flat over $A_i/\mathfrak{m}_i^n$. This implies that $A_i \to \widehat{A}$ is flat. Since $\mathfrak{m}\widehat{A} \neq \widehat{A}$, we get that $A \to \widehat{A}$ is faithfully flat; hence A is Noetherian. $\qquad\square$

Proposition 1.3.8 *Let $(A_i, \varphi_{ij})_{i \in I}$ be a filtered inductive system of ring morphisms and let $A = \varinjlim_i A_i$. Assume that:*

(i) A_i is a normal ring for any $y \in I$.
(ii) For any $i, j \in I$ such that $i \leq j$ and for any minimal prime $\mathfrak{p}_j \in \mathrm{Min}(A_j)$, the ideal $\varphi_{ij}^{-1}(\mathfrak{p}_j)$ is a minimal prime ideal of A_i.
Then A is a normal ring.

Proof Let $\mathfrak{p} \in \mathrm{Spec}(A)$ and for any index $j \in I$ let $\mathfrak{p}_j := \phi_j^{-1}(\mathfrak{p})$. Since A_i is normal, there exists a unique $\mathfrak{q}_i \in \mathrm{Min}(A_i)$, such that $\mathfrak{q}_i \subseteq \mathfrak{p}_i$. Then, for any i and j such that $i \leq j$, we have $\mathfrak{q}_j \cap A_i \in \mathrm{Min}(A_i)$ and $\mathfrak{q}_j \cap A_i \subseteq \mathfrak{p}_i$; hence $\mathfrak{q}_j \cap A_i = \mathfrak{q}_i$. Let $B_i := A_i/\mathfrak{q}_i$ and let $\psi_{ij} : B_i \to B_j$ be the morphism induced by φ_{ij}. Then $(B_i, \psi_{ij})_{i \in I}$ is a filtered inductive system of integrally closed domains. It follows by Corollary 1.3.6 that $B := \varinjlim_i B_i$ is an integrally closed domain. Let $\mathfrak{r}_i = \mathfrak{p}_i/\mathfrak{q}_i$. Then $(B_i)_{\mathfrak{r}_i} = (A_i)_{\mathfrak{p}_i}$. As $A_\mathfrak{p} = \varinjlim_i (A_i)_{\mathfrak{p}_i} = \varinjlim_i (B_i)_{\mathfrak{r}_i}$ and again by Corollary 1.3.6, it follows that $A_\mathfrak{p}$ is an integrally closed domain. $\qquad\square$

Lemma 1.3.9 *Let $(A_i, \varphi_{ij})_{i \in I}$ be a filtered inductive system of ring morphisms, let $A := \varinjlim_i A_i$ and let M a finitely presented A-module. Then there exists an index $i \in I$ and an A_i-module of finite presentation M_i, such that $M \cong M_i \otimes_{A_i} A$.*

Proof Let $M = A^r/N$, where $N \subseteq A^r$ is a finitely generated submodule. Since $A^r = \varinjlim_i A_i^r$ and N is finitely generated, there exist an index $i \in I$ and a finitely generated A_i-submodule $N_i \subseteq A_i^r$ such that N is the canonical image of $N_i \otimes_{A_i} A$. Since the tensor product is right-exact, we get that $M \cong (A_i^r/N_i) \otimes_{A_i} A$. □

Proposition 1.3.10 *Let* $(A_i, \varphi_{ij})_{i \in I}$ *be a filtered inductive system of ring morphisms and let* $A = \varinjlim_i A_i$. *Assume that:*

(i) A_i *is a Noetherian regular ring for any* $i \in I$.
(ii) φ_{ij} *is a flat morphism for any indices* $i, j \in I$ *such that* $i \leq j$.
(iii) A *is Noetherian.*
   Then A *is a regular ring.*

Proof Let $\mathfrak{p} \in \operatorname{Spec}(A)$ and put $\mathfrak{p}_i = \varphi_i^{-1}(\mathfrak{p})$ for any $i \in I$. Then $\mathfrak{p} = \varinjlim_i \mathfrak{p}_i$ and $A_{\mathfrak{p}} = \varinjlim_i (A_i)_{\mathfrak{p}_i}$; hence we may assume that A_i and consequently A are local rings. It suffices to show that if M and N are finitely generated A-modules, there exists $l_0 \in \mathbb{N}$ such that $\operatorname{Tor}_l^A(M, N) = 0$, for any $l \geq l_0$. By Lemma 1.3.9 there exists an index $j \in I$ and A_j-modules of finite presentation M_j and N_j such that $M = M_j \otimes_{A_j} A$ and $N = N_j \otimes_{A_j} A$. But φ_j is flat; hence $\operatorname{Tor}_l^A(M, N) = \operatorname{Tor}_l^{A_j}(M_j, N_j) \otimes_{A_j} A$. This concludes the proof. □

Corollary 1.3.11 *Let* $(A_i, \varphi_{ij})_{i \in I}$ *be a filtered inductive system of ring morphisms, let* $A := \varinjlim_{i \in I} A_i$ *and let* k *a natural number. Let* $\varphi_i : A_i \to A$ *be the canonical morphisms. Assume that:*

(i) A_i *is Noetherian and has the property* (R_k) *for any* $i \in I$.
(ii) φ_{ij} *is flat for any* $i \leq j$.
(iii) $\operatorname{Spec}(\varphi_i)$ *is a surjective homeomorphism for any* $i \in I$.
   Then A *has the property* (R_k).

Proof Let $\mathfrak{p} \in \operatorname{Spec}(A)$ and $\mathfrak{p}_i = \varphi_i^{-1}(\mathfrak{p})$. Then $\mathfrak{p} = \varinjlim_i \mathfrak{p}_i$ and $A_{\mathfrak{p}} = \varinjlim_i (A_i)_{\mathfrak{p}_i}$, whence $\operatorname{Spec}(A_{\mathfrak{p}}) \to \operatorname{Spec}(A_i)_{\mathfrak{p}_i}$ is a surjective homeomorphism, for any $i \in I$. Thus $\operatorname{ht}(\mathfrak{p}) = \operatorname{ht}(\mathfrak{p}_i)$, for any $i \in I$. If $\operatorname{ht}(\mathfrak{p}) \leq k$, then $(A_i)_{\mathfrak{p}_i}$ is regular. By Proposition 1.3.10 we obtain that $A_{\mathfrak{p}}$ is regular. □

Proposition 1.3.12 *Let* $(A_i, \varphi_{ij})_{i \in I}$ *be a filtered inductive system of ring homomorphisms, let* $A = \varinjlim_i A_i$ *and let* k *a natural number. Assume that:*

(i) *A is a Noetherian ring.*

(ii) φ_{ij} *is flat for any* $i \leq j$.

(iii) A_i *is Cohen-Macaulay (resp.* (S_k), *Gorenstein, complete intersection) for any* $i \in I$. *Then* A *is Cohen-Macaulay (resp.* (S_k), *Gorenstein, complete intersection).*

Proof Let $\mathfrak{p} \in \mathrm{Spec}(A)$. For any index $i \in I$, put $\mathfrak{p}_i := \varphi_i^{-1}(\mathfrak{p})$, $k_i := k(\mathfrak{p}_i)$ and $C_i := (A_i)_{\mathfrak{p}_i}$. Let also $k := k(\mathfrak{p})$ and $C := A_{\mathfrak{p}}$. Then $C = \varinjlim_i C_i$. As A is Noetherian, there exists $i_0 \in I$ such that $\mathfrak{p} = \mathfrak{p}_i C$ for any $i \geq i_0$. It follows that $C \otimes_{C_i} k_i = k$ for $i \geq i_0$. Since the canonical morphism $\psi_i : C_i \to C$ is flat, for any $i \geq i_0$, we have $\dim(C) = \dim(C_i)$ and $\mathrm{depth}(C_i) = \mathrm{depth}(C)$. This implies the assertion for Cohen-Macaulay and (S_k). We have $\mathfrak{p} = \varinjlim_i \mathfrak{p}_i$. So we may assume that $(A_i, \mathfrak{m}_i)$ and $(A, \mathfrak{m})$ are local rings. If $\mathfrak{m}_i := \varphi_i^{-1}(\mathfrak{m})$, by faithful flatness it results that $\mathfrak{m}_i$ is the maximal ideal of A_i. Moreover there exists $i_0 \in I$ such that $\mathfrak{m}_i A = \mathfrak{m}$, for any $i \geq i_0$. Let now $i \geq i_0$. Then the canonical morphism of the inductive limit $\varphi_i : A_i \to A$ is faithfully flat and $A/\mathfrak{m}_i A = A/\mathfrak{m}$; hence A_i and $A/\mathfrak{m}_i A$ are Gorenstein (resp. complete intersection). Then A is Gorenstein (resp. complete intersection). $\qquad\square$

Lemma 1.3.13 *Let* A_0 *be a ring,* $(A_i, \varphi_{ij})_{i \in I}$ *be an inductive system of* A_0-*algebras,* $A = \varinjlim A_i$ *and* B *be an* A-*algebra of finite presentation. Then there exist an index* $i \in I$ *and an* A_i-*algebra of finite presentation* B_i *such that* $B \cong B_i \otimes_{A_i} A$.

Proof We have $B \cong A[T_1, \ldots, T_n]/\mathfrak{a}$, where $\mathfrak{a} = (f_1, \ldots, f_m)$ is a finitely generated ideal. There exists an index $i \in I$ such that all the coefficients of the polynomials $f_1, \ldots, f_m$ are images of elements of A_i by the canonical morphism φ_i. Hence there exist $f_{1i}, \ldots, f_{mi} \in A_i[T_1, \ldots, T_n]$, such that $f_s = \varphi_i(f_{si})$, for $s = 1, \ldots, m$. Let $\mathfrak{a}_i := (f_{1i}, \ldots, f_{mi})$. Then

$$\mathfrak{a} = \varphi_i(\mathfrak{a}_i \otimes_{A_i} A) \subseteq A[T_1, \ldots, T_n] \cong A_i[T_1, \ldots, T_n] \otimes_{A_i} A.$$

Thus we can take $B_i := A_i[T_1, \ldots, T_n]/\mathfrak{a}_i$. $\qquad\square$

Lemma 1.3.14 *Let* $(A_i, \varphi_{ij})_{i \in I}$ *be an inductive system of ring morphisms and let* $A = \varinjlim_i A_i$. *Assume that* A *is a Noetherian ring, that* A_i *is an Artinian ring for any* $i \in I$ *and* *that $A_i^1, \ldots, A_i^{t_i}$ are its Artinian local components. Then:*

(i) *A is an Artinian ring.*

(ii) *If moreover for any $i, j \in I$ such that $i \leq j$, the morphism φ_{ij} is injective, there exists an index $j_0 \in I$ such that for any $j \geq j_0$, we have $t_j = t_{j_0}$, $\varphi_{j_0 j}(A_{j_0}^s) \subseteq (A_j^s)$, for any $s = 1, \ldots, t_j$ and $\varinjlim_{i \geq j_0} A_i^s$, for $s = 1, \ldots, t_j$ are the Artinian local components of A.*

Proof

(i) The hypothesis says that $A_i = \prod_{i=1}^{t_i} A_i^s$, where $A_i^1, \ldots, A_i^{t_i}$ are Artinian local rings [18, Th. 8.7]. Since A is Noetherian, we have only to prove that $\dim(A) = 0$. Assume there exists $\mathfrak{p} \in \mathrm{Spec}(A) \setminus \mathrm{Max}(A)$ and let $\mathfrak{m} \in \mathrm{Max}(A)$ be such that $\mathfrak{p} \subsetneq \mathfrak{m}$. Consider an element $x \in \mathfrak{m} \setminus \mathfrak{p}$, and for any index $i \in I$, let $\mathfrak{m}_i = \varphi^{-1}(\mathfrak{m})$ and $\mathfrak{p}_i = \varphi^{-1}(\mathfrak{p})$. Hence for any $i \in I$, we have $\mathfrak{p}_i \subseteq \mathfrak{m}_i$ and since A_i is Artinian, we get $\mathfrak{p}_i = \mathfrak{m}_i$. Let $i_0 \in I$ be such that $x \in A_{i_0}$. Then obviously $x \in \mathfrak{m}_{i_0} \setminus \mathfrak{p}_{i_0}$. Contradiction!

(ii) Let $i, j \in I$ be such that $i \leq j$ and let $\mathfrak{p} \in \mathrm{Spec}(A_i)$. Let also $S = A_i \setminus \mathfrak{p}$. Consider an ideal $\mathfrak{q} \subset A_j$ such that $\mathfrak{q}$ is maximal between the ideals of A_j disjoint from S. Then $\mathfrak{q} \in \mathrm{Spec}(A_j)$. Clearly $\varphi^{-1}(\mathfrak{q}) \supseteq \mathfrak{p}$ and since A_i is Artinian, it follows that $\varphi^{-1}(\mathfrak{q}) = \mathfrak{p}$. Thus the morphisms $\psi_{ij} := \mathrm{Spec}(\varphi_{ij})$ are all surjective. Let t_i be the number of Artinian local components of A_i and t be the number of Artinian local components of A. Then clearly $t_i \leq t$ for any $i \in I$, and as ψ_{ij} are surjective, it is also clear that $t_i \leq t_j$ for any $i \leq j$. Hence there is $j_0 \in I$ such that for any $j \geq i \geq j_0$, the morphism ψ_{ij} is bijective. Now the assertion follows. □

Proposition 1.3.15 (Marot [240, Prop. 2.2]) *Let $(A_i, \varphi_{ij})_{i \in I}$ be an inductive system of Noetherian rings and $A = \varinjlim_i A_i$. For any index $i \in I$, consider an ideal $\mathfrak{a}_i \subsetneq A_i$. Let $B_i = \widehat{(A_i, \mathfrak{a}_i)}$ be the completion of A_i in the $\mathfrak{a}_i$-adic topology and let $\alpha_i : A_i \to B_i$ be the canonical morphisms. Assume that $\varphi_{ij}(\mathfrak{a}_i) \subseteq \mathfrak{a}_j$, for any indices $i \leq j$. Then:*

(i) *For any indices $i, j \in I$ such that $i \leq j$, there exist uniquely determined continuous morphisms $\psi_{ij} : B_i \to B_j$, such that $\psi_{ij} \circ \alpha_i = \alpha_j \circ \varphi_{ij}$.*

(ii) *$(B_i, \psi_{ij})_{i \in I}$ is an inductive system of Noetherian rings.*

(iii) *There exists a unique morphism $\alpha : A \to B := \varinjlim_i B_i$, such that $\alpha \circ \varphi_i = \psi_i \circ \alpha_i$, $\forall i \in I$.*

(iv) *If $\mathfrak{a} = \varinjlim \mathfrak{a}_i$ and $C = \widehat{(A, \mathfrak{a})}$ is the completion of A in the $\mathfrak{a}$-adic topology, there exists a unique canonical morphism*

$$\psi : B = \varinjlim_i (\varprojlim_n A_i / \mathfrak{a}_i^n) \to C := \varprojlim_n (\varinjlim_i A_i / \mathfrak{a}_i^n).$$

Proof (i) The existence of ψ_{ij} follows by the universal properties of completion:

$$
\begin{array}{ccc}
A_i & \xrightarrow{\phi_{ij}} & A_j \\
\alpha_i \downarrow & & \downarrow \alpha_j \\
B_i & \xdashrightarrow{\psi_{ij}} & B_j
\end{array}
$$

By assumption, φ_{ij} is continuous if for any $i \in I$ we consider the $\mathfrak{a}_i$-adic topologies on A_i. Hence ψ_{ij} is continuous.

(ii) Obvious.

(iii) and (iv)

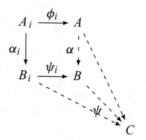

The existence of α and ψ follows from the universal property of inductive limits. □

Remark 1.3.16 B is not necessarily Noetherian. It is so in a particular case, shown in the next proposition.

Proposition 1.3.17 (Marot [240, Prop. 2.2]) *With the notations from 1.3.15, assume moreover that:*

(i) A is a Noetherian ring.

(ii) φ_{ij} is flat for any $i \leq j$.

Then B is a Noetherian Zariski ring for the $\mathfrak{a}B$-adic topology and $(B, \mathfrak{a}\widehat{B}) \cong C$.

Proof Since B_i is $\mathfrak{a}_i B_i$-adically complete, we have $\mathfrak{a}_i B_i \subseteq J(B_i)$. Also since $\mathfrak{a}B = \varinjlim \mathfrak{a}_i B_i$, we see that $\mathfrak{a}B \subseteq J(B)$. Since $A_i/\mathfrak{a}_i \cong B_i/\mathfrak{a}_i B_i$ for any index $i \in I$, by passing to inductive limit, we get that $A/\mathfrak{a} \cong B/\mathfrak{a}B \cong C/\mathfrak{a}C$. Let $\mathfrak{m} \in \mathrm{Max}(B)$. Then $\mathfrak{a}B \subseteq \mathfrak{m}$ and from the above isomorphism, we obtain that there exists a unique $\mathfrak{n} \in \mathrm{Max}(A)$, $\mathfrak{n} \supseteq \mathfrak{a}$, such that $\mathfrak{m} = \mathfrak{n}B$. Clearly $\mathfrak{n}C \in \mathrm{Max}(C)$; hence $\mathfrak{m}C = \mathfrak{n}C \neq C$. On the other hand, the assumption ii) implies that the canonical morphisms $\varphi_i : A_i \to A$ are flat; hence the canonical morphisms $\rho_i : B_i \to C$ are flat. Passing to inductive limit, we get that ψ is flat.

Since for any $\mathfrak{m} \in \mathrm{Max}(B)$ we have $\mathfrak{m}C \neq C$, it follows that ψ is faithfully flat. Since C is Noetherian, this implies that B is Noetherian and concludes the proof.   □

Corollary 1.3.18 *With the notations from 1.3.15, assume moreover that:*

(i) A is a Noetherian ring.
(ii) φ_{ij} is flat $\forall i \leq j$.

Let $\mathfrak{m} \in \mathrm{Max}(A)$ and let $\mathfrak{m}_i = \varphi_i^{-1}(\mathfrak{m})$. Then $\varinjlim_i \widehat{(A_i)_{\mathfrak{m}_i}}$ is a Noetherian local ring and its completion in the topology of the maximal ideal is canonically isomorphic to $\widehat{A_{\mathfrak{m}}}$.

1.4 Algebras over a Field and Base Change

Let us first consider the following old and well-known example.

Example 1.4.1 (Zariski [397, Ex. 3]) Let k be a field of characteristic $p > 2$ and $K :=$ $k(T^p)$, where T is an indeterminate. Consider the polynomial $f(X, Y) := X^2 - Y^p - T^p \in$ $K[X, Y]$. Then $A := K[X, Y]/(f)$ is a Noetherian normal ring of dimension 1, hence a regular ring. Let us consider the finite field extension $L := K(T)$. Then we have

$$A \otimes_k L \cong K[X, Y]/(X^2 - Y^p - T^p) \otimes_K K(T) \cong$$

$$\cong K(T)[X, Y]/(X^2 - Y^p - T^p) \cong$$

$$\cong K(T)[X, Y]/(X^2 - (Y - T)^p) \cong$$

$$\cong K(T)[x, y], \ x^2 = (y - T)^p,$$

which is clearly not a regular ring. Moreover $A \otimes_k L$ is not even a normal ring, because the element $v := \frac{x}{y-T}$ is integral over $A \otimes_K L$ and is not in $A \otimes_K L$. This shows that the property of an algebra over a field is not always preserved by extending the base field.

In view of the above example, it is natural to consider the following definition. *It originated when Claude Chevalley and André Weil pointed out to Oscar Zariski (1947) that, over non-perfect fields, the Jacobian criterion for a simple point of an algebraic variety is not equivalent to the condition that the local ring is regular* (Wikipedia).
Let **P** be a property which makes sense for Noetherian local rings, as in Sect. 1.1.

Definition 1.4.2 Let k be a field and A a Noetherian k-algebra. We say that A is a **geometrically P k-algebra** if for any field L, finite extension of k, the ring $B \otimes_A L$ has the property **P**.

Remark 1.4.3 It is obvious that the ring $B \otimes_A L$ in the above definition is a Noetherian ring, so it makes sense to speak about the property **P** for this ring.

Let us first of all investigate geometrically reduced algebras over a field. In this case, since reducedness is not necessarily a property of Noetherian rings, we can consider more general situations.

Notation 1.4.4 If k is a field of characteristic $p > 0$, we call the **perfect closure** of k, the field $k^{perf} := \bigcup_{n=0}^{\infty} k^{p^{-n}}$.

Proposition 1.4.5 *Let k be a field and A a k-algebra. The following are equivalent:*

(i) $A \otimes_k B$ is a reduced ring, for any reduced k-algebra B.
(ii) $A \otimes_k K$ is reduced, for any field extension K of k.
(iii) $A \otimes_k \overline{k}$ is reduced, where $\overline{k}$ is the algebraic closure of k.
(iv) $A \otimes_k L$ is reduced, for any field L, finite purely inseparable extension of k.
(v) $A \otimes_k k^{p^{-1}}$ is reduced.
(vi) $A \otimes_k k^{perf}$ is reduced.
(vii) A is reduced and for any minimal prime ideal $\mathfrak{p}$ of A, $k(\mathfrak{p})$ is a separable extension of k.

Proof (i)$\Rightarrow$(ii)$\Rightarrow$(iii): Clear.
(iii)$\Rightarrow$iv): Since $A \otimes_k L \subseteq A \otimes_k \overline{k}$ and the latter ring is reduced, it follows that the first one is reduced too.
(iv)$\Rightarrow$(vii): Since $A \cong A \otimes_k k$, by assumption, it is reduced. By [132, Prop. 4.3.5], in order to show that $k(\mathfrak{p})$ is a separable extension of k, we have to show that for any finite purely inseparable extension L of k, the ring $k(\mathfrak{p}) \otimes_k L$ is reduced. By assumption the ring $A \otimes_k L$ is reduced and $k(\mathfrak{p}) \otimes_k L$ is a ring of fractions of it, hence reduced.
(vii)$\Rightarrow$i): Since A is reduced, the intersection of all the minimal prime ideals of A is 0. Then

$$A \subseteq \prod_{\mathfrak{p} \in \mathrm{Min}(A)} A/\mathfrak{p} \subseteq \prod_{\mathfrak{p} \in \mathrm{Min}(A)} k(\mathfrak{p}).$$

In the same way, we can consider B as a subalgebra of a product of field extensions of k, say $\prod_j L_j$. Then $A \otimes_k B$ is a subalgebra of $\prod_{\mathfrak{p}} k(\mathfrak{p}) \otimes_k \prod_j L_j$. But [132, Prop. 4.3.5] tells that this is a reduced ring.

(iii)$\Rightarrow$(vi)$\Rightarrow$(v): k^{perf} is contained in $\overline{k}$ and $k^{p^{-1}}$ is contained in k^{perf}.

(v)$\Rightarrow$(vii): Since $A \otimes_k k^{p^{-1}}$ is reduced, it follows that A is reduced. But then, for any minimal prime ideal $\mathfrak{p}$ of A, the ring $A_{\mathfrak{p}}$ is reduced, hence a field. We conclude as in the proof of (vii)$\Rightarrow$(i). $\square$

Now let us see that in order to check that an algebra over a field is geometrically **P**, it is enough to tensorize with purely inseparable extensions.

Lemma 1.4.6 *Let k be a field and K and L be two field extensions of k, such that one of them is a separable extension of k. If K is an extension of finite type of k, then $K \otimes_k L$ is a Noetherian regular ring.*

Proof (Majadas [228]) Let $A := K \otimes_k L$. The ring A is Noetherian, because K is an extension of finite type of k. Let $\mathfrak{p}$ be a prime ideal of A. By [6, Prop. V.21] we have an exact sequence

$$0 = H_2(K, k(\mathfrak{p}), k(\mathfrak{p})) \oplus H_2(L, k(\mathfrak{p}), k(\mathfrak{p})) \to H_2(A_{\mathfrak{p}}, k(\mathfrak{p}), k(\mathfrak{p})) \to$$

$$\to H_1(k, k(\mathfrak{p}), k(\mathfrak{p})) \xrightarrow{\delta} H_1(K, k(\mathfrak{p}), k(\mathfrak{p})) \oplus H_1(L, k(\mathfrak{p}), k(\mathfrak{p})).$$

On the other hand, we have two Jacobi-Zariski exact sequences

$$H_1(k, L, k(\mathfrak{p})) \to H_1(k, k(\mathfrak{p}), k(\mathfrak{p})) \xrightarrow{\alpha} H_1(L, k(\mathfrak{p}), k(\mathfrak{p}))$$

and

$$H_1(k, K, k(\mathfrak{p})) \to H_1(k, k(\mathfrak{p}), k(\mathfrak{p})) \xrightarrow{\beta} H_1(K, k(\mathfrak{p}), k(\mathfrak{p})).$$

The separability of K or L over k implies by [6, Prop. VII.11] that at least one of the morphisms α or β is injective; hence δ is injective. This yields that $H_2(A_{\mathfrak{p}}, k(\mathfrak{p}), k(\mathfrak{p})) = 0$, whence $A_{\mathfrak{p}}$ is regular. $\square$

Proposition 1.4.7 *Let k be a field, A a Noetherian k-algebra and L a field, separable extension of finite type of K.*

 (i) *If A is regular, then $A \otimes_k L$ is regular.*
 (ii) *If A is normal, then $A \otimes_k L$ is normal.*

Proof Let $\mathfrak{p}$ be a prime ideal of A. Then $k(\mathfrak{p}) \otimes_k L$ is a regular ring, as follows from Lemma 1.4.6. Now the assertions follow from the fact that the property regular satisfies the axiom ($\mathbf{A_4}$) (cf. Theorem 1.1.5) and the property normal satisfies the axiom ($\mathbf{A_5}$) (cf. 1.1.26 and 1.1.28). $\square$

Proposition 1.4.8 *Let k be a field let A be a Noetherian k-algebra and let n be a natural number. The following are equivalent:*

(i) *A is a geometrically reduced (resp. normal, regular, (R_n)) k-algebra.*
(ii) *For any field K, finite purely inseparable extension of k, the ring $A \otimes_k K$ is a reduced (resp. normal, regular, (R_n)) ring.*

Proof It is enough to show that ii)$\Rightarrow$i). Let K be a finite extension of k. Then there exist a finite extension $K \subseteq L$ and a subextension $k \subseteq k' \subseteq L$ such that k' is finite and purely inseparable over k and such that L is a separable extension of k'. Then by assumption $A \otimes_k k'$ is regular and from Proposition 1.4.7 we get that $A \otimes_k K \cong A \otimes_k k' \otimes_{k'} K$ is regular. For the other properties the proof is the same. $\square$

Proposition 1.4.9 *Let k be a field, L a field separable over k and A an integrally closed k-algebra. If $L \otimes_k A$ is an integral domain, then it is integrally closed.*

Proof Let $K = Q(A)$. Then $L \otimes_k A$ is isomorphic to a k-subalgebra of $L \otimes_A K$ and the rings A and L to subrings of $L \otimes_k A$. Also, if $S = A \setminus \{0\}$, then $L \otimes_k K \cong S^{-1}(L \otimes_k A)$. As $L \otimes_k A$ is a domain, $L \otimes_k K$ is a subring of $Q(L \otimes_k A)$. If $[L : k] < \infty$, we apply Proposition 1.4.7, and we can assume that L is a separable finitely generated extension of k. Let $\{x_1, \ldots, x_d\}$ be a separable transcendence basis of L over k. Since L and K are algebraically disjoint over k in $Q(L \otimes_k A)$, the elements $x_1, \ldots, x_d$ are algebraically independent over k. Hence $A[x_1, \ldots, x_d]$ is integrally closed. Let $B = k[x_1, \ldots, x_d]$ and $T = B \setminus \{0\}$. Then $k_1 := T^{-1}(k[x_1, \ldots, x_d]) = k(x_1, \ldots, x_d) \subseteq L$. We have

$$k_1 \otimes_k A = (T^{-1}B) \otimes_k A = (T^{-1}B) \otimes_B (B \otimes_k A) =$$

$$= T^{-1}(B \otimes_k A) = T^{-1}A[x_1, \ldots, x_d].$$

Thus $k_1 \otimes_k A$ is integrally closed. But L is a separable finite extension of k_1; hence by the previous case, it follows that $L \otimes_k A$ is integrally closed. In the general case, let $z \in Q(L \otimes_k A)$ be integral over $L \otimes_k A$. We have a relation $z^m + b_1 z^{m-1} + \ldots + b_m = 0$, where $b_i \in L \otimes_k A$. Then there exists a subextension $L' \subseteq L$, finitely generated over k, such that $b_1, \ldots, b_m \in L' \otimes_k A$ and $z \in L' \otimes_k K$. By the finitely generated case, it follows that $z \in L' \otimes_k A$; hence $L \otimes_k A$ is integrally closed. $\square$

Corollary 1.4.10 *Let k be a field, A and B be integrally closed k-algebras and $K = Q(A), L = Q(B)$. Assume that $A \otimes_k B$ is an integral domain and that K and L are separable extensions of k. Then $A \otimes_k B$ is integrally closed.*

Proof Clearly K and L are isomorphic to subfields of $Q(A \otimes_k B)$, linearly disjoint over k. From Proposition 1.4.9 it results that $A \otimes_k L$ and $K \otimes_k B$ are integrally closed. But $(A \otimes_k L) \cap (K \otimes_k B) = A \otimes_k B$, whence $A \otimes_k B$ is integrally closed.   □

For some properties **P** of Noetherian local rings, we have stronger results.

Proposition 1.4.11 *Let k be a field and K and L field extensions of k such that $K \otimes_k L$ is a Noetherian ring. Then $K \otimes_k L$ is a complete intersection.*

Proof (Majadas [228]) Let $\mathfrak{q}$ be a prime ideal of the ring $B := K \otimes_k L$ and let $k(\mathfrak{q})$ be the residue field of B in $\mathfrak{q}$. By [6, Prop. V.21 and Prop. VII.4], there exists an exact sequence

$$\ldots \to (0) = H_3(K, k(\mathfrak{q}), k(\mathfrak{q})) \oplus H_3(L, k(\mathfrak{q}), k(\mathfrak{q})) \to H_3(B_\mathfrak{q}, k(\mathfrak{q}), k(\mathfrak{q})) \to$$

$$\to H_2(K, k(\mathfrak{q}), k(\mathfrak{q})) = (0) \to \ldots .$$

Hence $H_3(B_\mathfrak{q}, k(\mathfrak{q}), k(\mathfrak{q})) = 0$, that is, $B_\mathfrak{q}$ is a complete intersection.   □

Theorem 1.4.12 (Toussi and Yassemi [375, Th. 6]) *Let k be a field and A and B be k-algebras such that $A \otimes_k B$ is a Noetherian ring and n a natural number. Suppose that A and B are complete intersection (resp. Gorenstein, Cohen-Macaulay, (S_n)). Then $A \otimes_k B$ is a complete intersection (resp. Gorenstein, Cohen-Macaulay, (S_n)).*

Proof We have the commutative diagram:

$$\begin{array}{ccc} k & \longrightarrow & A \\ \downarrow & & \downarrow \phi \\ B & \longrightarrow & A \otimes_k B \end{array}$$

Obviously ϕ is flat. Assume that we have proved that for all $\mathfrak{p} \in \mathrm{Spec}(A)$, the ring $(A \otimes_k B) \otimes_A k(\mathfrak{p})$ is a complete intersection (resp. Gorenstein, Cohen-Macaulay, (S_n)). Then, if we take $\mathfrak{q} \in \mathrm{Spec}(A \otimes_k B)$, setting $\mathfrak{p} = \mathfrak{q} \cap A$, it follows that the ring $(A \otimes_k B)_\mathfrak{q}/\mathfrak{p}(A \otimes_k B)_\mathfrak{q}$ is a complete intersection (resp. Gorenstein, Cohen-Macaulay, (S_n)), being a localization of $(A \otimes_k B) \otimes_A k(\mathfrak{p})$. In order to prove that $(A \otimes_k B) \otimes_A k(\mathfrak{p})$ is a complete intersection, let us notice that

$$(A \otimes_k B) \otimes_A k(\mathfrak{p}) \cong B \otimes_k k(\mathfrak{p}).$$

We have the flat morphism $\gamma : B \to B \otimes_k k(\mathfrak{p})$. As above, it is enough to show that for any prime ideal $\mathfrak{q}$ of B, the ring $B \otimes_k k(\mathfrak{p}) \otimes_B k(\mathfrak{q})$ is a complete intersection (resp. Gorenstein,

Cohen-Macaulay, (S_k)). But

$$T := A/\mathfrak{p} \otimes_k B/\mathfrak{q} \cong A \otimes_k B/(\mathfrak{p} \otimes B + A \otimes_k \mathfrak{q})$$

is a Noetherian ring, and $k(\mathfrak{p}) \otimes_b k(\mathfrak{q})$ is a fraction ring of T and consequently Noetherian. Now Proposition 1.4.11 gives the assertion.   □

Corollary 1.4.13 *Let k be a field, n a natural number and A a Noetherian k-algebra. Suppose that A is Cohen-Macaulay (resp. Gorenstein, complete intersection, (S_n)). Then A is geometrically Cohen-Macaulay (resp. Gorenstein, complete intersection, (S_n)).*

Proof Follows immediately from 1.4.12.   □

Now let us see some general properties of geometrically **P** algebras over a field.

Proposition 1.4.14 *Let k be a field, A a Noetherian geometrically **P** k-algebra and S a multiplicatively closed system in A. Then $S^{-1}A$ is a geometrically **P** k-algebra.*

Proof It follows from the fact that for any field K, finite extension of k, the ring $(S^{-1}A) \otimes_k K$ is a ring of fractions of $A \otimes_k K$.   □

Proposition 1.4.15 *Let k be a field and A a Noetherian k-algebra. If $A_{\mathfrak{m}}$ is a geometrically **P** k-algebra, for every maximal ideal $\mathfrak{m}$ of A, then A is a geometrically **P** k-algebra.*

Proof Let K be a field, finite purely inseparable extension of k and let $\mathfrak{n} \in \mathrm{Max}(A \otimes_k K)$. Since $A \otimes_k K$ is a finite A-algebra, we get that $\mathfrak{m} := \mathfrak{n} \cap A$ is a maximal ideal of A. Moreover, $(A \otimes_k K)_{\mathfrak{n}}$ is a ring of fractions of $(A_{\mathfrak{m}}) \otimes_k K$, which has the property **P**.   □

Proposition 1.4.16 *Let k be a field, A a geometrically **P** k-algebra and K a field, extension of finite type of k. Then $A \otimes_k K$ is a geometrically **P** K-algebra.*

Proof There exists a finite field extension L of K and a finite purely inseparable extension k' of K contained in L, such that L is a separable extension of k'. Then $A \otimes_k k'$ has by assumption the property **P**, and $A \otimes_k L = A \otimes_k k' \otimes_{k'} L$ has the property **P** by 1.4.5, 1.4.7 and 1.4.12 and by the condition $(\mathbf{A_3})$.   □

We end this section with some characterizations of geometrically regular algebras over a field. These are well-known results, but we mention them here for their beauty and importance. We first note that we use the definition of smoothness from [249].

Definition 1.4.17 Let A be a ring, B be an A-algebra and $\mathfrak{a}$ an ideal of B. We consider on B the $\mathfrak{a}$-adic topology. We say that B is $\mathfrak{a}$-**smooth** over A if given an A-algebra C, an ideal

$\mathfrak{n}$ of C such that $\mathfrak{n}^2 = 0$ and an A-algebra morphism $u : B \to C/\mathfrak{n}$ which is continuous for the discrete topology of $C/\mathfrak{n}$, there exists an A-algebra morphism $v : B \to C$, that is, a lifting of u:

$$
\begin{array}{ccc}
A & \longrightarrow & B \\
\downarrow & \swarrow{v} & \downarrow{u} \\
C & \longrightarrow & C/\mathfrak{n}
\end{array}
$$

If in the above situations there is at most one lifting v, then we call B an $\mathfrak{a}$-**unramified** algebra and if there exists exactly one lifting, we say that the A-algebra B is $\mathfrak{a}$-**étale**. If $\mathfrak{a} = 0$, that is, if on B we consider the discrete topology, then we will call B a **smooth**, resp. **unramified**, resp. **étale** A-algebra. Moreover, if $u : (A, \mathfrak{m}) \to (B, \mathfrak{n})$ is a morphism of Noetherian local rings, when we say that u is **formally smooth**, we understand that we consider the topologies given by the maximal ideals of A and B or what corresponds to $\mathfrak{n}$-smooth in the above definition. The same convention applies to **formally unramified** and **formally étale** morphisms.

The reader can find the main properties and results about formally smooth, formally unramified and formally étale morphisms in [132, 249] and [359]. The proof of the next result can be found in many textbooks on commutative algebra (see, e.g. [249, Th. 28.7] or [231, Cor. 2.6.5]).

Theorem 1.4.18 *Let k be a field and $(A, \mathfrak{m}, K)$ be a Noetherian local k-algebra. The following are equivalent:*

(i) A is geometrically regular over k.
(ii) A is $\mathfrak{m}$-smooth over k.

Proof $(i) \Rightarrow (ii)$ (Faltings [106]): It is clear that A is a regular local ring. If the extension $k \subseteq K$ is separable, $\widehat{A}$ has a coefficient field K containing k [249, Th. 28.3]. Let $x_1, \ldots, x_n$ be a regular system of parameters of $\widehat{A}$ and $\psi : K[[X_1, \ldots, X_n]] \to \widehat{A}$ be the morphism given by $\psi(X_i) = x_i$. Then ψ is surjective, and comparing dimensions we obtain that $K[[X_1, \ldots, X_n]] \cong \widehat{A}$. Then $\widehat{A}$ is $\widehat{\mathfrak{m}}$-smooth over K. Since $k \subseteq K$ is smooth, it follows that $\widehat{A}$ is $\widehat{\mathfrak{m}}$-smooth over k; hence A is $\mathfrak{m}$-smooth over k. If $\operatorname{char}(k) = p > 0$, applying [249, Th. 28.6], we have to prove that the canonical map

$$
\Omega_k \otimes_k K \to \Omega_A \otimes_A K
$$

is injective. Let $x_1, \ldots, x_n \in k$ be p-independent elements. Let $\overline{k}$ be the algebraic closure of k and let $\alpha_1, \ldots, \alpha_n \in \overline{k}$ be such that $\alpha_i^p = x_i$, for $i = 1, \ldots, n$. Put

$k' := k(\alpha_1, \ldots, \alpha_n)$. Then obviously

$$B := A \otimes_k k' = A[Y_1, \ldots, Y_n]/(Y_1^p - x_1, \ldots, Y_n^p - x_n)$$

is a Noetherian local ring. Let $\mathfrak{n}$ be the maximal ideal of B and $L := B/\mathfrak{n}$. The sequences

$$0 \to \mathfrak{n}/\mathfrak{n}^2 \to \Omega_B \otimes_B L \to \Omega_L \to 0$$

and

$$0 \to \mathfrak{m}/\mathfrak{m}^2 \to \Omega_A \otimes_A K \to \Omega_K \to 0$$

are exact due to the separability of K and L over $\mathbb{F}_p$. We consider the commutative diagram with canonical maps

$$
\begin{array}{ccccccccc}
0 & \longrightarrow & \mathfrak{m}/\mathfrak{m}^2 \otimes_k L & \longrightarrow & \Omega_A \otimes_A L & \longrightarrow & \Omega_K \otimes_K L & \longrightarrow & 0 \\
 & & \downarrow{\scriptstyle u} & & \downarrow{\scriptstyle v} & & \downarrow{\scriptstyle w} & & \\
0 & \longrightarrow & \mathfrak{n}/\mathfrak{n}^2 & \longrightarrow & \Omega_B \otimes_B L & \longrightarrow & \Omega_L & \longrightarrow & 0.
\end{array}
$$

By the snake lemma, we obtain a long exact sequence of L-vector spaces

$$0 \to \ker(u) \to \ker(v) \to \ker(w) \to \mathrm{Coker}(u) \to \mathrm{Coker}(v) \to \mathrm{Coker}(w) \to 0.$$

Since A and B are regular local rings and $\dim(A) = \dim(B)$, we have

$$\mathrm{rk}_K(\mathfrak{m}/\mathfrak{m}^2) = \dim(A) = \mathrm{rk}_L(\mathfrak{n}/\mathfrak{n}^2) = \dim(B).$$

It follows that

$$\mathrm{rk}(\ker(u)) < \infty, \ \ \mathrm{rk}(\mathrm{Coker}(u)) < \infty$$

and moreover

$$\mathrm{rk}(\ker(u)) = \mathrm{rk}(\mathrm{Coker}(u)).$$

Since $[L : K] < \infty$, from Cartier's equality [249, Th. 26.10], we get that

$$\mathrm{rk}(\ker(w)) < \infty, \ \ \mathrm{rk}(\mathrm{Coker}(w)) < \infty \ \text{ and } \ \mathrm{rk}(\ker(w)) = \mathrm{rk}(\mathrm{Coker}(w)).$$

Hence from the long exact sequence, we obtain

$$\mathrm{rk}(\ker(v)) = \mathrm{rk}(\mathrm{Coker}(v)).$$

But

$$\mathrm{Coker}(v) \cong \Omega_{B/A} \otimes_B L \cong (BdY_1 + \cdots + BdY_n) \otimes_B L \cong B^n \otimes_B L,$$

so that

$$\mathrm{rk}(\ker(v)) = \mathrm{rk}(\mathrm{Coker}(v)) = n.$$

Let $\mathfrak{a} := (Y_1^p - x_1, \ldots, Y_n^p - x_n)$. We have an exact sequence

$$\mathfrak{a}/\mathfrak{a}^2 \xrightarrow{\delta} \Omega_{A[Y_1,\ldots,Y_n]} \otimes_A B \cong \Omega_A \otimes_A B \oplus \left(\sum_{i=1}^{n} BdY_i \right) \to \Omega_B \to 0.$$

This sequence remains exact after tensorizing with L over B. It follows that

$$\ker(u) = (dx_1 \Omega_A \otimes L + \cdots + dx_n \Omega_A \otimes L).$$

Hence $dx_1, \ldots, dx_n$ are linearly independent over K, when seen as elements of $\Omega_A \otimes_A K$.
$(ii) \Rightarrow (i)$: Let L be a field, finite extension of k. Consider the commutative diagram:

$$
\begin{array}{ccc}
k & \longrightarrow & A \\
\downarrow & & \downarrow \\
L & \longrightarrow & B := A \otimes_k L.
\end{array}
$$

Then B is $\mathfrak{m}B$-smooth over L. For any $\mathfrak{n} \in \mathrm{Max}(B)$, we have $\mathfrak{m}B \subseteq \mathfrak{n}$, so if we put $C := B_\mathfrak{n}$ and $\mathfrak{r} := \mathfrak{n}C$, it follows that $B \to C$ is continuous in the $\mathfrak{m}B$-adic topology on B and the $\mathfrak{r}$-adic topology on C and C is étale over B. From [249, Th. 28.1], it follows that C is $\mathfrak{r}$-smooth over L; hence C is a regular local ring. $\square$

In the positive characteristic case, we have the following very nice characterization of geometrically regular algebras that will be useful later. One can find a different proof of the next result in [231, Th. 2.5.9], using André-Quillen homology.

Theorem 1.4.19 (Grothendieck [131, Th. 22.5.8] and Radu [298, Théorème]) *Let k be a field of characteristic $p > 0$ and $(A, \mathfrak{m}, K)$ be a Noetherian local k-algebra. The* *following are equivalent:*

  (i) A is geometrically regular over k.

  (ii) For any field L, finite extension of k and such that $L^p \subseteq k$, $A \otimes_k L$ is a regular ring.

  (iii) $A \otimes_k k^{p^{-1}}$ is a Noetherian regular local ring.

Proof

$(i) \Rightarrow (ii)$: Follows from 1.4.18.

$(ii) \Rightarrow (iii)$: From the assumption we know that A is a regular ring. Let $B := A \otimes_k k^{p^{-1}}$. We have $B = \varinjlim_i (A \otimes_k k_i)$, where k_i runs over all subfields $k_i \subseteq k^{p^{-1}}$, such that $[k_i : k] < \infty$. Since the rings $A \otimes_k k_i$ are local domains, it follows that B is a local domain. Let $L := Q(A)$ be the field of fractions of A. We consider the subring $C := A^{p^{-1}} \subseteq L^{p^{-1}}$. Since C is isomorphic with A, it follows that C is a regular local ring. For any field k' as above, there exists a canonical morphism $\tau : A \otimes_k k' \to C$. Applying [249, Th. 23.1], since the morphism τ is integral, it follows that τ is flat. Moreover the canonical morphism $g : B \to C$ is the direct limit of these morphisms, hence it is flat. As B and C are local rings, g is even faithfully flat. Since C is a Noetherian ring, it follows that also B is a Noetherian ring. C is a regular local ring; hence also B is a regular local ring.

$(iii) \Rightarrow (i)$: Let $L := k^{p^{-1}}$, $B := A \otimes_k L$ and $\mathfrak{n}$ the maximal ideal of B. Then B is L-smooth relative to k (cf. [248, (30.A), Def.]); therefore the canonical morphism

$$v : \Omega_{L/K} \otimes_L B \to \Omega_{B/k}$$

is left invertible (see [249, Th. 28.4]). Since $L^p = k$ and $k \subseteq B^p$, we get that $\Omega_{L/k} = \Omega_L$ and $\Omega_{B/k} = \Omega_B$; hence the morphism

$$w : \Omega_L \otimes_L B \to \Omega_B$$

is left invertible. Let $\mathbb{F}_p$ be the prime field of k. Then B is L-smooth relative to $\mathbb{F}_p$. Since B is a regular local ring, it follows that B is a $\mathfrak{n}$-smooth $\mathbb{F}_p$-algebra; hence B is a $\mathfrak{n}$-smooth L-algebra. Because $B^p \subseteq A$, the $\mathfrak{n}$-adic and $\mathfrak{m}B$-adic topologies coincide. Then A is $\mathfrak{m}$-smooth over k. $\qquad\square$

Let us finally mention the following important characterization of smoothness in the case of morphisms between Noetherian local rings.

Theorem 1.4.20 *Let $u : (A, \mathfrak{m}, k) \to (B, \mathfrak{n}, K)$ be a morphism of Noetherian local rings. The following are equivalent:*

  (i) u is $\mathfrak{n}$-smooth.

  (ii) u is flat and $B/\mathfrak{m}B$ is geometrically regular over k.

  (iii) u is flat and $B/\mathfrak{m}B$ is $\mathfrak{n}/\mathfrak{m}B$-smooth over k.

Proof See [131, Th. 19.7.1] or [231, Cor. 2.6.5] for a different proof, using André-Quillen homology.   □

1.5   Preparing for Bad Examples

Over the years, a lot of counterexamples were constructed in commutative algebra. Probably the most famous place to find such examples of pathologies is the Appendix of the celebrated book of Nagata [266]. Many other examples were constructed after the publication of Nagata's book by many people (Ferrand-Raynaud [109], Brodmann-Rotthaus [74], Ogoma [276], Brezuleanu-Rotthaus [70], Nishimura [269], Heitmann [155], etc.). We will present a general unified construction, belonging to Nishimura [271], that will allow us to point out many pathological rings, including the classical examples by Akizuki, Nagata, etc. We must mention that there exists another source for a huge quantity of counterexamples, namely, the recent book by W. Heinzer, Ch. Rotthaus and Sylvia Wiegand [154].

Let K_0 be a countable field and K be a purely transcendental field extension of K_0, of countable degree over K_0. Thus, we have

$$K = K_0(a_{ij}, i = 1, \ldots, n, j = 1, 2, \ldots) =$$

$$= K(a_{11}, a_{21}, \ldots, a_{n1}, a_{12}, a_{22}, \ldots, a_{n2}, \ldots, a_{1l}, a_{2l}, \ldots, a_{nl}, \ldots),$$

where $(a_{ij})_{1 \leq i \leq n, j \geq 1}$ is a countable transcendence basis. We can write $K = \bigcup_{l=0}^{\infty} K_l$, where $K_l = K_{l-1}(a_{1l}, \ldots, a_{nl})$, $l \in \mathbb{N}$. Consider $Z_1, \ldots, Z_n$ indeterminates over K and let

$$S_0 := K_0[Z_1, \ldots, Z_n], \quad \mathfrak{N}_0 := (Z_1, \ldots, Z_n)S_0,$$

$$S_l := S_{l-1}[a_{1l}, \ldots, a_{nl}], \quad \mathfrak{N}_l := (Z_1, \ldots, Z_n)S_l, \ l \in \mathbb{N}.$$

Furthermore we denote

$$S := \bigcup_{l=0}^{\infty} S_l = K_0[a_{1l}, \ldots, a_{nl}, l \in \mathbb{N}][Z_1, \ldots, Z_n], \quad \mathfrak{N} := (Z_1, \ldots, Z_n)S.$$

Localizing in the above prime ideals, we set:

$$R_0 := (S_0)_{\mathfrak{N}_0} = K_0[Z_1, \ldots, Z_n]_{(Z_1, \ldots, Z_n)}, \quad \mathfrak{n}_0 := (Z_1, \ldots, Z_n)R_0,$$

$$R_l := (S_l)_{\mathfrak{N}_l} = K_l[Z_1, \ldots, Z_n]_{(Z_1, \ldots, Z_n)}, \quad \mathfrak{n}_l := (Z_1, \ldots, Z_n)R_0,$$

$$R := S_{\mathfrak{N}} = K[Z_1, \ldots, Z_n]_{(Z_1, \ldots, Z_n)}, \quad \mathfrak{n} := (Z_1, \ldots, Z_n)R.$$

Remark 1.5.1 It is clear that $R_l = R_{l-1}(a_{1l}, \ldots, a_{nl})$ and that $(R, \mathfrak{n})$ is a countable regular local ring such that $R = \bigcup_{l=1}^{\infty} R_l = K[Z_1, \ldots, Z_n]_{(Z_1,\ldots,Z_n)}$.

Let now $\mathcal{P}$ be a subset of $\mathfrak{N} \setminus \{0\}$ such that for any non-zero prime ideal $\mathfrak{p}$ of R, there exists at least one $p \in \mathcal{P}$ such that $p \in \mathfrak{p}$. Thus $\mathcal{P}$ is a countable set. We may assume that $Z_1 + \ldots + Z_n \in \mathcal{P}$ and that $\mathcal{P}$ contains infinitely many elements from S_0. Since $\mathcal{P}$ is countable, we can choose a numbering of $\mathcal{P}$, that is, a surjective map $\rho : \mathbb{N} \to \mathcal{P}$, and we denote $p_i := \rho(i)$ for any $i \in \mathbb{N}$. We may assume that $p_1 = Z_1 + \ldots + Z_n$ and that $p_l \in S_{l-2}$ for every $l \geq 2$. We take a strictly increasing sequence of natural numbers $\{\epsilon_i\}_{i \geq 1}$. We can take, for example, $\epsilon_i = i$.
Further, let us define:

$$p_1 = Z_1 + \ldots + Z_n \tag{1.1}$$

$$Z_{i,0} := Z_i \tag{1.2}$$

$$q_l := p_1 \ldots p_l \tag{1.3}$$

$$Z_{i,l} := Z_i + a_{i1}q_1^{\epsilon_1} + a_{i2}q_2^{\epsilon_2} + \ldots + a_{il}q_l^{\epsilon_l}, \, l \geq 1. \tag{1.4}$$

Remark 1.5.2 With the above notations, the ideal $P_l := (Z_{1l}, \ldots, Z_{ml})R$ is a prime ideal of height m, since it is generated by a part of a regular system of parameters.

Lemma 1.5.3 (Heitmann [155, Prop. 1]) *With the above notations, if $m < n$ and the numbering ρ is such that $p_l = \rho(l) \in S_{l-2}$ for every $l \geq 2$, then:*

(i) *$(Z_{1,l}, \ldots, Z_{j,l})S_l$ is a prime ideal of S_l generated by a regular sequence $Z_{1,l}, \ldots, Z_{j,l}$ for every $j = 1, \ldots, m$.*
(ii) *$p_h \notin P_l$, if $h \leq l + 1$.*

Proof We make induction on l. If $l = 0$, we have $(Z_{1,0}, \ldots, Z_{j,0})S_0 = (Z_1, \ldots, Z_j)S_0$, which is a S_0-regular sequence, and since $m < n$, we have $p_1 = Z_1 + \ldots + Z_n \notin (Z_1, \ldots, Z_m)S_0 = P_0 \cap S_0$. Now suppose that $l > 0$ and that the assertions i) and ii) are true for $l - 1$. This means that $(Z_{1,l-1}, \ldots, Z_{j,l-1})S_{l-1}$ is a prime ideal generated by the S_{l-1}-regular sequence $Z_{1,l-1}, \ldots, Z_{j,l-1}$ for every $1 \leq j \leq m$ and that $q_l \notin P_{l-1} \cap S_{l-1} = (Z_{1,l-1}, \ldots, Z_{m,l-1})S_{l-1}$. Thus $Z_{1,l-1}, \ldots, Z_{m,l-1}, q_l$ is a S_{l-1}-regular sequence. Now in this S_{l-1}- regular sequence, since q_l is not zero in the integral domain $S_{l-1}/(Z_{1,l-1}, \ldots, Z_{j-1,l-1})$ for every $j = 1, \ldots, m$, by [190, Th. 118], we can permute q_l and $Z_{j,l-1}$ and still get a regular sequence. Repeating the procedure and applying at the end [249, Th. 16.1], we have that $q_l^{\epsilon_l}, Z_{1,l-1}, \ldots, Z_{m,l-1}$ is a S_{l-1}-regular sequence too. Since $q_l \in S_{l-1}$, we see that

$$Z_{i,l} = \left(Z_i + \sum_{j=1}^{l-1} a_{ij} q_j^{\epsilon_j}\right) + a_{il} q_l^{\epsilon_l} = Z_{i,l-1} + a_{il} q_l^{\epsilon_l}$$

is a linear polynomial in a_{il} with coefficients in S_{l-1}. Hence $(Z_{1,l} \ldots, Z_{j,l}) S_l$ is the kernel of the S_{l-1}-algebra morphism

$$\phi_j : S_{l-1}[a_{nl}, \ldots, a_{1l}] \to S_{l-1}[a_{nl}, \ldots, a_{(j+1)l}]\left[\frac{Z_{1,l-1}}{q_l^{\epsilon_l}}, \ldots, \frac{Z_{j,l-1}}{q_l^{\epsilon_l}}\right]$$

$$\phi(a_{il}) = -\frac{Z_{i,l-1}}{q_l^{\epsilon_l}}, 1 \le i \le l.$$

Therefore using [190, Ch. 3, Ex. 3], we obtain that for $j = 1, \ldots, m$, the ideal $(Z_{1l}, \ldots, Z_{jl}) S_l$ is a prime ideal generated by the S_l-regular sequence $Z_{1,l}, \ldots, Z_{j,l}$ and that $(Z_{1,l}, \ldots, Z_{m,l}) S_l \cap S_{l-1} = (0)$.   □

Notation 1.5.4 Keeping the above notations, let $n, r, m \in \mathbb{N}$ and assume that $m < n$. For each $j \in \{1, \ldots, r\}$, consider a polynomial $F_j := F_j(Z_1, \ldots, Z_m) \in (Z_1, \ldots, Z_m) K_0[Z_1, \ldots, Z_m]$. We also consider a strictly increasing sequence of natural numbers $\{v_i\}_{i \ge 1}$ such that $v_i \le \epsilon_i$, for any $i \in \mathbb{N}$. If we chose $\epsilon_i = i$, then we can also take $v_i = i$. For $j = 1, \ldots, r$ we define

$$\alpha_{j,l} := \frac{1}{q_l^{v_l}} F_j(Z_{1,l}, \ldots, Z_{m,l}) \in Q(R) = K(Z_1, \ldots, Z_n).$$

Lemma 1.5.5 *With the above notations, we have the relations*

$$\alpha_{j,l} = \frac{q_{l+1}^{v_{l+1}}}{q_l^{v_l}} \alpha_{j,l+1} + \frac{q_{l+1}^{v_{l+1}}}{q_l^{v_l}} s_{j,l}, \ s_{j,l} \in S_{l+1}.$$

Proof We have

$$\alpha_{j,l+1} = \frac{1}{q_{l+1}^{v_{l+1}}} F_j(Z_{1,l+1}, \ldots, Z_{m,l+1}) =$$

$$= \frac{1}{q_{l+1}^{v_{l+1}}} F_j(Z_{1,l} + a_{1,l+1} q_{l+1}^{v_{l+1}}, \ldots, Z_{m,l} + a_{m,l+1} q_{l+1}^{v_{l+1}}).$$

□

Notation 1.5.6 Let $B := \bigcup_{l=0}^{\infty} R[\alpha_{1,l}, \ldots, \alpha_{r,l}] \subset Q(R) = K(Z_1, \ldots, Z_n)$.

Lemma 1.5.7 *With the above notations, the ideal* $\mathfrak{M} := (Z_1, \ldots, Z_n)B$ *is a maximal ideal of B.*

Proof We have a canonical injection $\theta : R \rightarrow B = \bigcup_{l=0}^{\infty} R[\alpha_{1,l}, \ldots, \alpha_{r,l}]$. Since by Lemma 1.5.5 we have that $\alpha_{j,l} \in \mathfrak{M}$ for every j and l, we also have a canonical surjective morphism $\overline{\theta} : R/\mathfrak{n} \rightarrow B/\mathfrak{M}$, and it is enough to show that $\mathfrak{M} \neq B$. If not, there exist elements $\beta_1, \ldots, \beta_n \in B$ such that $\beta_1 Z_1 + \ldots + \beta_n Z_n = 1$. Of course we can assume that $\beta_1, \ldots, \beta_n \in R[\alpha_{1,l}, \ldots, \alpha_{r,l}]$ for sufficiently large l. Then there exist $r_1, \ldots, r_n \in R$ and $s \in \mathbb{N}$ such that $q_l^s(\beta_1 - r_1), \ldots, q_l^s(\beta_n - r_n) \in P_l$. Then $q_l^s(r_1 Z_1 + \ldots + r_n Z_n - 1) \in P_l$. But $r_1 Z_1 + \ldots + r_n Z_n - 1$ is invertible in R, hence $q_l \in P_l$. Contradiction! $\square$

Now let us define $A := B_{\mathfrak{M}} \subset Q(R) = K(Z_1, \ldots, Z_n)$ and $\mathfrak{m} = \mathfrak{M}A$. Let us also define

$$\zeta_i := Z_i + a_{i1}q_1^{\epsilon_1} + \ldots + a_{il}q_l^{\epsilon_l} + \ldots = Z_i + \sum_{l=1}^{\infty} a_{il}q_l^{\epsilon_l}, \quad 1 \leq i \leq n$$

and

$$f_j = F_j(\zeta_1, \ldots, \zeta_m) \in K_0[[\zeta_1, \ldots, \zeta_m]] \subset K[[\zeta_1, \ldots, \zeta_n]] = \widehat{R}, \quad 1 \leq j \leq r.$$

Theorem 1.5.8 (Nishimura [271, Th. 1.4]) *Let K_0 be a countable field and K be a purely transcendental extension field of K_0 of countable infinite degree. Let $F_1, \ldots, F_r \in K_0[Z_1, \ldots, Z_m]$ be polynomials without constant term. For any $n > m$, let $(A, \mathfrak{m})$ be the ring defined as above. Then:*

(i) $(A, \mathfrak{m})$ is a Noetherian local domain.

(ii) For any prime ideal $\mathfrak{p} \neq (0)$ of A, the ring $A/\mathfrak{p}$ is a K-algebra essentially of finite type.

(iii) The canonical map

$$\widetilde{\pi} : K[[\zeta_1, \ldots, \zeta_n]]/(F_1(\zeta_1, \ldots, \zeta_m), \ldots, F_r(\zeta_1, \ldots, \zeta_m)) = \widehat{R}/(f_1, \ldots, f_r) \rightarrow \widehat{A}$$

is an isomorphism.

(iv) $\widehat{\mathfrak{p}} := (\widetilde{\pi}(\zeta_1), \ldots, \widetilde{\pi}(\zeta_m))\widehat{A}$ is a prime ideal of $\widehat{A}$ and $\widehat{\mathfrak{p}} \cap A = (0)$.

Proof (i) and (ii): Clearly $(A, \mathfrak{m})$ is a local domain. To show that A is Noetherian, we will show that every prime ideal is finitely generated. Let $\mathfrak{p} \neq (0)$ be a prime ideal of A.

Because $Q(A) = Q(R)$, it follows that $\mathfrak{p} \cap R \neq (0)$; hence there exists $h \in \mathbb{N}$ such that $p_h \in \mathfrak{p} \cap \mathcal{P}$. Lemma 1.5.5 shows that $\alpha_{j,l} \in R + p_l A$ for every $j = 1, \ldots, r$ and every $l \geq 1$. But then we have a canonical surjective morphism $\pi_h : R \to A/p_h A$, and $A/p_h A$ is a K-algebra essentially of finite type. Therefore $\mathfrak{p}/p_h A$ and consequently $\mathfrak{p}$ are finitely generated, and $A/\mathfrak{p}$ is essentially of finite type over K.

(iii) and (iv): The map $\tilde{\pi}$ is induced by the canonical injection $R \hookrightarrow A$ which induces also the canonical surjection $\widehat{\pi} : \widehat{R} \to \widehat{A}$. Let us compute $\ker(\widehat{\pi})$. Let us first observe that $\zeta_i - Z_{i,l} \in q_{l+1}^{\epsilon_{l+1}} \widehat{R}$, $i = 1, \ldots, n$. Therefore,

$$f_j - q_l^{\nu_l} \alpha_{j,l} = F_j(\zeta_1, \ldots, \zeta_m) - F_j(Z_{1,l}, \ldots, Z_{m,l}) =$$

$$= q_{l+1}^{\epsilon_{l+1}} \eta_{j,l}, \ \eta_{j,l} \in \widehat{R}, \ j = 1, \ldots, r.$$

Then $\widehat{\pi}(f_j) \in q_l^{\nu_l} \widehat{A}$ for every $l \in \mathbb{N}$; hence $f_1, \ldots, f_r \in \ker(\widehat{\pi})$. For the converse, first let $\mathfrak{q} = (\zeta_1, \ldots, \zeta_m)\widehat{R}$, which is a prime ideal of height m in $\widehat{R}$. We want to show that $\mathfrak{q} \cap R = (0)$. Suppose $\mathfrak{q} \cap R \neq (0)$. There exists $h \in \mathbb{N}$ such that $p_h \in \mathfrak{q} \cap R$. Then $(p_h, Z_{1,h-1}, \ldots, Z_{m,h-1})\widehat{R} \subseteq \mathfrak{q}$, and $\mathrm{ht}((p_h, Z_{1,h-1}, \ldots, Z_{m,h-1})\widehat{R}) = m + 1$. Contradiction! Hence $\mathfrak{q} \cap R = (0)$. Let $\overline{\mathfrak{q}} \in \mathrm{Ass}(\widehat{R}/(f_1, \ldots, f_r)\widehat{R})$. Then $\overline{\mathfrak{q}} \subseteq \mathfrak{q}$, hence $\overline{\mathfrak{q}} \cap R = (0)$, therefore $\widehat{R}/(f_1, \ldots, f_r)\widehat{R}$ is R-torsion free. The canonical morphism $\sigma : R \to \widehat{R}/(f_1, \ldots, f_r)\widehat{R}$ induces an R-algebra morphism

$$\lambda : A \to Q(R) \otimes_R \widehat{R}/(f_1, \ldots, f_r)\widehat{R},$$

$$\lambda(\alpha_{j,l}) = \alpha_{j,l} \otimes 1 = 1 \otimes (-q_l^{\epsilon_{l+1}-\nu_l} p_{l+1}^{\epsilon_{l+1}} \eta_{j,l}).$$

Then λ factors through $\widehat{R}/(f_1, \ldots, f_r)\widehat{R}$ which is R-torsion-free, and we have the commutative diagram

$$
\begin{array}{ccc}
\widehat{R} & \xrightarrow{=} & \widehat{R} \\
{\scriptstyle \pi}\downarrow & {\scriptstyle \lambda} & \downarrow{\scriptstyle \tilde{\pi}} \\
\widehat{A} & \xrightarrow{\lambda} & \widehat{R}/(f_1, \ldots, f_r)\widehat{R}
\end{array}
$$

Thus $\ker(\widehat{\pi}) \subseteq (f_1, \ldots, f_r)\widehat{R}$; hence $\ker(\widehat{\pi}) = (f_1, \ldots, f_r)\widehat{R}$. This means that we have $\widehat{\mathfrak{p}} = \mathfrak{q}/(f_1, \ldots, f_r)\widehat{R}$. Therefore $\widehat{\mathfrak{p}}$ is a prime ideal of $\widehat{A}$ and $\widehat{\mathfrak{p}} \cap A = (0)$. $\qquad \square$

Corollary 1.5.9 *Let K_0 be a countable field and K be a purely transcendental extension field of K_0 of countable infinite degree. Let $F_1, \ldots, F_r \in K_0[Z_1, \ldots, Z_n]$ be polynomials without constant term. Let $f_{j,l} := F_j(Z_{1,l}, \ldots, Z_{n,l})$ and let $\mathfrak{a}_l := (f_{1,l}, \ldots, f_{r,l})R$. With the notations from above, assume that $p_h \notin \sqrt{\mathfrak{a}_l}$ for every $h \leq l$, for l large enough and* that $\widehat{R}/(F_1(\zeta_1, \ldots, \zeta_n), \ldots, F_r(\zeta_1, \ldots, \zeta_n))\widehat{R}$ is R-torsion-free. Define $(A, \mathfrak{m})$ as above. Then:

(i) $(A, \mathfrak{m})$ is a Noetherian local domain.
(ii) The map

$$\tilde{\theta} : K[[\zeta_1, \ldots, \zeta_n]]/(F_1(\zeta_1, \ldots, \zeta_n) \ldots, F_r(\zeta_1, \ldots, \zeta_n)) = \widehat{R}/(f_1, \ldots, f_r)\widehat{R} \to \widehat{A}$$

induced by the map $\theta : R \to B$ from Lemma 1.5.7, is an isomorphism.
(iii) For any prime ideal $\mathfrak{p} \neq (0)$ of A, the ring $A/\mathfrak{p}$ is a K-algebra essentially of finite type.

Proof One must notice that the proof of 1.5.7 works well to show $\mathfrak{M} \neq B$, using the fact that $p_h \notin \sqrt{\mathfrak{a}_l}$. Moreover the R-torsion-freeness of $\widehat{R}/(f_1, \ldots, f_r)\widehat{R}$ allows the proof of (ii). $\square$

We describe now a variation of the above construction. The big difference is that what we obtain are bad Noetherian local domains with a *distinguished prime element*. Let K_0, K and $K_l, l \geq 1$ be as above. Set

$$S_0 := K_0[X, Z_1, \ldots, Z_n], \quad \mathfrak{N}_0 := (X, Z_1, \ldots, Z_n)S_0,$$

$$S_l := S_{l-1}[a_{1l}, \ldots, a_{nl}], \quad \mathfrak{N}_l := (X, Z_1, \ldots, Z_n)S_l, \ l \in \mathbb{N}$$

$$S := \bigcup_{l=0}^{\infty} S_l = K_0[a_{1l}, \ldots, a_{nl}, l \in \mathbb{N}][X, Z_1, \ldots, Z_n], \quad \mathfrak{N} := (X, Z_1, \ldots, Z_n)S.$$

Localizing in the prime ideals above, we get:

$$R_0 := (S_0)_{\mathfrak{N}_0} = K[X, Z_1, \ldots, Z_n]_{(X, Z_1, \ldots, Z_n)}, \quad \mathfrak{n}_0 := (X, Z_1, \ldots, Z_n)R_0,$$

$$R_l := (S_l)_{\mathfrak{N}_l} = K_l[X, Z_1, \ldots, Z_n]_{(X, Z_1, \ldots, Z_n)}, \quad \mathfrak{n}_l := (X, Z_1, \ldots, Z_n)R_0,$$

$$R := S_{\mathfrak{N}} = K[X, Z_1, \ldots, Z_n]_{(X, Z_1, \ldots, Z_n)}, \quad \mathfrak{n} := (X, Z_1, \ldots, Z_n)R.$$

Remark 1.5.10 It is clear that $R_l = R_{l-1}^*(a_{1l}, \ldots, a_{nl})$ and that $(R, \mathfrak{n})$ is a countable regular local ring such that $R = \bigcup_{l=1}^{\infty} R_l = K[X, Z_1, \ldots, Z_n]_{(X, Z_1, \ldots, Z_n)}$.

Let now $\mathcal{P}^*$ be a subset of $\mathfrak{N} \setminus XS$ such that for any prime ideal $\mathfrak{p}$ of R, such that $\mathfrak{p} \notin \{(0), XR\}$, there exists at least one $p \in \mathcal{P}^*$ with $p \in \mathfrak{p}$. Thus $\mathcal{P}^*$ is a countable set.

We may assume that $Z_1 + \ldots + Z_n \in \mathcal{P}^*$ and that $\mathcal{P}^*$ contains infinitely many elements from S_0.

Let us denote in the following by $\bar{}$ classes modulo X, that is, classes in $\overline{S} := S/XS$, $\overline{R} := R/XR$, etc. Let $\overline{\mathcal{P}} := \{\overline{p} \in \overline{S} \mid p \in \mathcal{P}^*\}$. Then $\overline{\mathcal{P}} \subset \overline{\mathfrak{N}} \setminus \overline{(0)}$ is such that, for any non-zero prime ideal $\overline{\mathfrak{p}} \in \mathrm{Spec}(\overline{R})$, there exists at least one $\overline{p} \in \overline{\mathcal{P}}$ such that $\overline{p} \in \overline{\mathfrak{p}}$. Since $\mathcal{P}^*$ is countable, we can choose a numbering of $\mathcal{P}^*$, that is, a surjective map $\rho^* : \mathbb{N} \to \mathcal{P}^*$ and we denote $p_i := \rho^*(i)$. We may assume that $p_1 = Z_1 + \ldots + Z_n$ and that $p_l \in S_{l-2}$ for every $l \geq 2$.

Remark 1.5.11 Let us observe that, if ρ^* is as above, then the induced map $\overline{\rho} : \mathbb{N} \to \overline{\mathcal{P}}$, $\overline{\rho}(i) = \overline{p}_i$ is a numbering of $\overline{\mathcal{P}}$ such that $\overline{p}_1 = \overline{Z}_1 + \ldots + \overline{Z}_n$ and that $\overline{p}_l \in \overline{S}_{l-2} = S_{l-2}/XS_{l-2}$, $\forall l \geq 2$.

Let us take a strictly increasing sequence of natural numbers $\{\epsilon_i\}_{i \geq 1}$, and as before, let us define:

$$Z_{i,0} := Z_i \tag{1.5}$$

$$q_l := p_1 \ldots p_l \tag{1.6}$$

$$Z_{i,l} := Z_i + a_{i1}q_1^{\epsilon_1} + a_{i2}q_2^{\epsilon_2} + \ldots + a_{il}q_l^{\epsilon_l}, \; l \geq 1. \tag{1.7}$$

Moreover we define:

$$\overline{Z}_{i,0} := \overline{Z}_i \tag{1.8}$$

$$\overline{q}_l := \overline{p}_1 \ldots \overline{p}_l \tag{1.9}$$

$$\overline{Z}_{i,l} := \overline{Z}_i + \overline{a}_{i1}\overline{q}_1^{\epsilon_1} + \overline{a}_{i2}\overline{q}_2^{\epsilon_2} + \ldots + \overline{a}_{il}\overline{q}_l^{\epsilon_l}, \; l \geq 1. \tag{1.10}$$

Remark 1.5.12 With the above notations, the ideal $Q_l := (X, Z_{1l}, \ldots, Z_{ml})R$ is a prime ideal of height $m + 1$ for $l \geq 0$, since it is generated by a part of a regular system of parameters.

Lemma 1.5.13 *With the above notations, if $m < n$ and the numbering ρ^* is such that $p_l = \rho(l) \in S_{l-2}$ for every $l \geq 2$, then:*

(i) $(X, Z_{1,l}, \ldots, Z_{j,l})S_l$ *is a prime ideal of S_l generated by a regular sequence $X, Z_{1,l}, \ldots, Z_{j,l}$ for every $j = 1, \ldots, m$.*

(ii) $p_h \notin Q_l$, *if $h \leq l + 1$.*

Proof Is the same as the proof of Lemma 1.5.3. □

Notation 1.5.14 Keeping the above notations, let $n, r, m \in \mathbb{N}$ and assume that $m < n$. For each $j \in \{1, \ldots, r\}$ consider a polynomial

$$G_j := G_j(X, Z_1, \ldots, Z_m) \in (X.Z_1, \ldots, Z_m)K_0[X, Z_1, \ldots, Z_m].$$

As above, consider a strictly increasing sequence of natural numbers $\{v_i\}_{i \geq 1}$ such that $v_i \leq \epsilon_i$ for all i. Let

$$F_j(Z_1, \ldots, Z_m) := G_j(0, Z_1, \ldots, Z_m) \in K_0[Z_1, \ldots, Z_m] =$$

$$= K_0[X, Z_1, \ldots, Z_m]/XK_0[X, Z_1, \ldots, Z_m]$$

and for $j = 1, \ldots, r$ define

$$\alpha_{j,l} := \frac{1}{q_l^{v_l}} G_j(X, Z_{1,l}, \ldots, Z_{m,l}) \in Q(R) = K(X, Z_1, \ldots, Z_n)$$

and

$$\overline{\alpha}_{j,l} := \frac{1}{\overline{q}_l^{v_l}} G_j(0, \overline{Z}_{1,l}, \ldots, \overline{Z}_{m,l}) = \frac{1}{\overline{q}_l^{v_l}} F_j(\overline{Z}_{1,l} \ldots, \overline{Z}_{m,l}) \in Q(\overline{R}) = K(\overline{Z}_1, \ldots, \overline{Z}_n)$$

Lemma 1.5.15 *With the above notations, we have the relations*

$$\alpha_{j,l} = \frac{q_{l+1}^{v_{l+1}}}{q_l^{v_l}} \alpha_{j,l+1} + \frac{q_{l+1}^{v_{l+1}}}{q_l^{v_l}} s_{j,l}, \ s_{j,l} \in S_{l+1}.$$

Proof We have

$$\alpha_{j,l+1} = \frac{1}{q_{l+1}^{v_{l+1}}} G_j(X, Z_{1,l+1}, \ldots, Z_{m,l+1}) =$$

$$= \frac{1}{q_{l+1}^{v_{l+1}}} G_j(X, Z_{1,l} + a_{1,l+1}q_{l+1}^{v_{l+1}}, \ldots, Z_{m,l} + a_{m,l+1}q_{l+1}^{v_{l+1}}).$$

□

Notation 1.5.16 Let $B := \bigcup_{l=0}^{\infty} R[\alpha_{1,l}, \ldots, \alpha_{r,l}] \subset Q(R) = K(X, Z_1, \ldots, Z_n)$.

Lemma 1.5.17 *With the above notations, the ideal* $\mathfrak{M} := (X, Z_1, \ldots, Z_n)B$ *is a maximal ideal of B.*

Proof Is the same as the proof of Lemma 1.5.7.   □

Now we define, in the same way as before, $A := B_{\mathfrak{M}} \subset Q(R) = K(X, Z_1, \ldots, Z_n)$ and $\mathfrak{m} = \mathfrak{M}A$. Let us define also

$$\zeta_i := Z_i + a_{i1}q_1^{\epsilon_1} + \ldots + a_{il}q_l^{\epsilon_l} + \ldots = Z_i + \sum_{l=1}^{\infty} a_{il}q_l^{\epsilon_l},$$

$$g_j = G_j(x, \zeta_1, \ldots, \zeta_m) \in K_o[[x, \zeta_1, \ldots, \zeta_n]] \subset K[[x, \zeta_1, \ldots, \zeta_n]] = \widehat{R},$$

$$1 \le i \le n, \ 1 \le j \le r,$$

$$\overline{\zeta}_i := \overline{Z}_i + \overline{a}_{i,1}\overline{q}_1^{\epsilon_1} + \ldots + \overline{a}_{i,l}\overline{q}_l^{\epsilon_l} = \overline{Z}_i + \sum_{l=1}^{\infty} \overline{a}_{il}\overline{q}_l^{\epsilon_l},$$

$$f_j = F_j(\overline{\zeta}_1, \ldots, \overline{\zeta}_n) \in K_o[[\overline{\zeta}_1, \ldots, \overline{\zeta}_m]] \subset K[[\overline{\zeta}_1, \ldots, \overline{\zeta}_n]] = \widehat{\overline{R}},$$

$$1 \le i \le n, \ 1 \le j \le r.$$

Theorem 1.5.18 (Nishimura [271, Th. 3.4]) *Let K_0 be a countable field and K be a purely transcendental extension field of K_0 of countable degree. Let $G_1, \ldots, G_r \in K_0[X, Z_1, \ldots, Z_m]$ be polynomials without constant term. Let $F_j(Z_1, \ldots, Z_m) := G_j(0, Z_1, \ldots, Z_m) \in K_0[Z_1, \ldots, Z_m]$, $j = 1, \ldots, r$, and let $U, T_1, \ldots, T_r$ be indeterminates. Define the ring morphisms:*

$$\tilde{\varphi} : K_0[X, Z_1, \ldots, Z_m, U][T_1, \ldots, T_r] \to K_0[X, Z_1, \ldots, Z_m, U][\frac{G_1}{U}, \ldots, \frac{G_r}{U}],$$

$$\tilde{\varphi}(T_j) = \frac{G_j}{U},$$

and

$$\varphi : K_0[Z_1, \ldots, Z_m, U][T_1, \ldots, T_r] \to K_0[Z_1, \ldots, Z_m, U][\frac{G_1}{U}, \ldots, \frac{G_r}{U}],$$

$$\varphi(T_j) = \frac{G_j}{U}.$$

Suppose that $\ker(\varphi) = K_0[Z_1, \ldots, Z_m, U] \otimes_{K_0[X, Z_1, \ldots, Z_m, U]} \ker(\tilde{\varphi})$. *For any* $n > m$, *let* $(A, \mathfrak{m})$ *be the ring defined as above. Then:*

(i) $(A, \mathfrak{m})$ *is a Noetherian local domain with a prime element* $x \in \mathfrak{m}$.

(ii) *For any prime ideal* $\mathfrak{p} \in \mathrm{Spec}(A) \setminus \{xA, (0)\}$, *the ring* $A/\mathfrak{p}$ *is a* K-*algebra essentially of finite type.*

(iii) *The canonical map*

$$\tilde{\pi} : K[[x, \zeta_1, \ldots, \zeta_n]]/(G_1(x, \zeta_1, \ldots, \zeta_m), \ldots, G_r(x, \zeta_1, \ldots, \zeta_m)) = \widehat{R}/(g_1, \ldots, g_r) \to \widehat{A}$$

is an isomorphism.

(iv) $\widehat{\mathfrak{q}} := (\pi(x), \pi(\zeta_1), \ldots, \pi(\zeta_m))\widehat{A}$ *is a prime ideal of* $\widehat{A}$ *and* $\widehat{\mathfrak{p}} \cap A = xA$.

(v) *The canonical map*

$$\tilde{\tilde{\pi}} : K[[\overline{\zeta}_1, \ldots, \overline{\zeta}_n]]/(F_1(\overline{\zeta}_1, \ldots, \overline{\zeta}_m), \ldots, F_r(\overline{\zeta}_1, \ldots, \overline{\zeta}_m)) = \widehat{R/xR}/(f_1, \ldots, f_r) \to \widehat{A}/x\widehat{A}$$

is an isomorphism.

Proof The proof is similar to the proof of Theorem 1.5.8.

(i) and (ii): Let $\mathfrak{p} \neq (0)$ be a prime ideal of A. Then $\mathfrak{p} \cap R \neq (0)$. We have two cases.

Case 1. There exists $s \in \mathbb{N}$ such that $p_s \in \mathfrak{p} \cap \mathcal{P}^*$. Then $\alpha_{j,l} \in R + p_s A$ for every $j = 1, \ldots, r$ and every $l \geq 1$ by Lemma 1.5.15. Thus we have a canonical surjective morphism $\pi_r : R \to A/p_r A$ and then $A/p_r A$ is essentially of finite type over K, whence $\mathfrak{p}$ is finitely generated and $A/\mathfrak{p}$ is essentially of finite type over K.

Case 2. Suppose that $xR = \mathfrak{p} \cap R$. Then from the assumption

$$\ker(\varphi) = K_0[Z_1, \ldots, Z_m, U] \otimes_{K_0[X, Z_1, \ldots, Z_m, U]} \ker(\tilde{\varphi})$$

we have that

$$B/xB \cong \bigcup_{l \in \mathbb{N}} \overline{R}[\overline{\alpha}_{1,l}, \ldots, \overline{\alpha}_{r,l}] \subseteq Q(\overline{R}).$$

Then $\mathfrak{p} = xA$; hence A is Noetherian.

(iii), (iv) and (v): We have canonical surjective morphisms

$$\widehat{\pi} : \widehat{R} \to \widehat{A} \quad \text{and} \quad \widehat{\chi} : \widehat{R}/x\widehat{R} \to \widehat{A}/x\widehat{A}$$

The same argument as in Theorem 1.5.8 shows that

$$g_j \in \ker(\widehat{\pi}), \quad \ker(\widehat{\chi}) = (f_1, \ldots, f_r), \quad (x, \zeta_1, \ldots, \zeta_m)\widehat{R} \cap R = xR.$$

Then $(g_1, \ldots, g_r) \subseteq \ker(\widehat{\pi}) \subseteq (x, g_1, \ldots, g_r)$, and since x is a non-zero-divisor, it follows that $\ker(\widehat{\pi}) = (g_1, \ldots, g_r)$. Therefore $\widehat{\mathfrak{q}} = (x, \zeta_1, \ldots, \zeta_m)\widehat{R}/(g_1, \ldots, g_r)$ is a prime ideal of $\widehat{A}$ and $\widehat{\mathfrak{q}} \cap A = xA$, since $Q(A/xA) = Q(R/xR)$. □

In this case too, we have a result in the case $m = n$, as follows.

Corollary 1.5.19 *Let K_0 be a countable field and K be a purely transcendental extension field of K_0 of countable degree. Let $G_1, \ldots, G_r \in K_0[X, Z_1, \ldots, Z_n]$ be polynomials without constant term. Let $F_j(Z, \ldots, Z_n) = G_j(0, Z_1, \ldots, Z_n) \in K_0[Z_1, \ldots, Z_n]$. Let $g_{j,l} = G_j(x, z_{1,l}, \ldots, z_{n,l})$, $\mathfrak{b}_l = (g_{1,l}, \ldots, g_{r,l})R$ and $f_{j,l} = F_j(\overline{z}_{1,l}, \ldots, \overline{z}_{n,l})$. Suppose that:*

(i) *$p_h \notin \sqrt{\mathfrak{b}_l}$ whenever $h \geq l$, for every sufficiently large l.*
(ii) *$\widehat{R}/(x, G_1(x, \zeta_1, \ldots, \zeta_n), \ldots, G_r(x, \zeta_1, \ldots, \zeta_r))\widehat{R}$ is R/xR-torsion-free.*
(iii) *$R[\frac{g_{1,l}}{q_l^{v_l}}, \ldots, \frac{g_{r,l}}{q_l^{v_l}}]/xR[\frac{g_{1,l}}{q_l^{v_l}}, \ldots, \frac{g_{r,l}}{q_l^{v_l}}] \cong \overline{R}[\frac{f_{1,l}}{\overline{q}_l^{v_l}}, \ldots, \frac{f_{r,l}}{\overline{q}_l^{v_l}}]$ for every $l \geq 0$. Then:*

(i) *$(A, \mathfrak{m})$ is a Noetherian local domain with a prime element x.*
(ii) *The map*

$$\tilde{\theta} : K[[x, \zeta_1, \ldots, \zeta_n]]/(G_1(x, \zeta_1, \ldots, \zeta_n) \ldots, G_r(x, \zeta_1, \ldots, \zeta_n)) \to \widehat{A}$$

is an isomorphism.
(iii) *The map*

$$\tilde{\omega} : K[[\overline{\zeta}_1, \ldots, \overline{\zeta}_n]]/(F_1(\overline{\zeta}_1, \ldots, \overline{\zeta}_n)) \cong \widehat{A/xA} = \widehat{A}/x\widehat{A}$$

is an isomorphism.
(iv) *For any prime ideal $\mathfrak{p} \in \mathrm{Spec}(A) \setminus \{(0), xA\}$, the ring $A/\mathfrak{p}$ is a K-algebra essentially of finite type.*

Let us shortly illustrate the method of this section, by constructing a detailed example. The reader should compare this example with Example 1.2.4.

Example 1.5.20 (Brodmann and Rotthaus [74]) An example of a two-dimensional Noetherian local domain A, whose regular locus is not open.

We shall use the first construction. With the notations as above, let K_0 be a countable field of any characteristic and K be a purely transcendental extension of countable degree, $n = 3$ and $m = 1$. Then, with $R = K[Z_1, Z_2, Z_3]_{(Z_1, Z_2, Z_3)}$ and $F_1(Z_1) = Z_1^c$ with

$c \geq 2$, it follows from Theorem 1.5.8 that the ring constructed is a Noetherian local domain $(A, \mathfrak{m})$ such that:

(i) $\tilde{\pi} : K[[\zeta_1, \zeta_2, \zeta_3]]/(\zeta_1^c) = K[[Z_1, Z_2, Z_3]]/(Z_1^c) \cong \widehat{A}$ is an isomorphism.
(ii) $\widehat{\mathfrak{p}} = (\zeta_1)\widehat{A}$ is a prime ideal of $\widehat{A}$ and $\widehat{\mathfrak{p}} \cap A = (0)$.
(iii) For any non-zero prime ideal $\mathfrak{p}$ of A, the ring $A/\mathfrak{p}$ is a K-algebra essentially of finite type.

We want to prove that $\mathrm{Reg}(A)$ does not contain an open set. Assume that there exists $s \in A$ such that $D(s) := \{\mathfrak{p} \in \mathrm{Spec}(A) \mid s \notin \mathfrak{p}\} \subseteq \mathrm{Reg}(A)$. Clearly $\dim(\widehat{A}/\widehat{\mathfrak{p}}) = \dim(K[[\zeta_1, \zeta_2, \zeta_3]]/(\zeta_1)) = 2$. Since $s \notin \widehat{\mathfrak{p}}$, there exists $t \in A$ such that the images of s and t are a system of parameters in $\widehat{A}/\widehat{\mathfrak{p}}$; hence there exists a prime ideal $\widehat{\mathfrak{q}}$ in $\widehat{A}$, containing $\widehat{\mathfrak{p}}$ and t such that $s \notin \widehat{\mathfrak{q}}$, because s is not contained in all minimal prime ideals of the image of t in $\widehat{A}/\widehat{\mathfrak{p}}$. Let $\mathfrak{q} := \widehat{\mathfrak{q}} \cap A$. Then $\mathfrak{q} \in D(s)$; hence $A_\mathfrak{q}$ is regular and moreover $\mathfrak{q} \neq (0)$. Thus $A/\mathfrak{q}$ is a K-algebra essentially of finite type. It will be seen later that this implies that $\widehat{A}_{\widehat{\mathfrak{q}}}/\mathfrak{q}\widehat{A}_{\widehat{\mathfrak{q}}}$ is regular. But then $\widehat{A}_{\widehat{\mathfrak{q}}}$ is regular and consequently $\widehat{A}_{\widehat{\mathfrak{p}}}$ is regular. Contradiction, because $\widehat{A}_{\widehat{\mathfrak{p}}}$ is not even a domain.

1.6 P-Morphisms and P-Rings

We begin this section by describing two famous counterexamples.

Example 1.6.1 (Nagata [261, Proposition]) There exists a Noetherian local domain whose completion is not reduced.

Proof Let K_0 be a countable field and $n = 2$. Take $b_1, \ldots, b_d \in K_0$ and let $F(Z_1, Z_2) = (Z_1 - b_1 Z_2)^2 \cdot \ldots \cdot (Z_1 - b_d Z_2)^2$. One can check that the assumptions of Corollary 1.5.9 are fulfilled, so that we get a one-dimensional local domain $(A, \mathfrak{m})$ such that $\widehat{A} \cong K[[\zeta_1, \zeta_2]]/((\zeta_1 - b_1 \zeta_2)^2 \cdot \ldots \cdot (\zeta_1 - b_d \zeta_2)^2))$. $\square$

Example 1.6.2 (Nagata [264]) There exists a Noetherian local normal domain whose completion is not normal.

Proof Let $K_0 = \mathbb{Q}, n = 3$ and $m = 1$. Take $F(Z_1, Z_2) = Z_1^2 - Z_2^3$ and apply Theorem 1.5.8. We obtain a two-dimensional local domain $(A, \mathfrak{m})$ such that $\widehat{A} \cong K[[\zeta_1, \zeta_2, \zeta_3]]/(\zeta_1^2 - \zeta_2^3)$. Then $\widehat{A}$ is not normal as it is well-known and $\mathrm{Sing}(\widehat{A}) = V(\zeta_1, \zeta_2)\widehat{A}$. Therefore, since $(\zeta_1, \zeta_2)\widehat{A} \cap A = (0)$, we have that $\mathrm{Reg}(A) = \mathrm{Spec}(A) \setminus \{\mathfrak{m}\}$, and A is the desired example. $\square$

It follows from the above examples that if we consider a given property **P** and a morphism of Noetherian local rings $u : A \to B$, the property **P** is not necessarily preserved from A to B, even if the morphism u is a well-known morphism, as, for example, the completion morphism and has quite good properties, such as flatness. A. Grothendieck [132] was the first who introduced and studied the classes of morphisms behaving well from this point of view. He called them **P**-morphisms and they will be one of the main topics of the rest of this section and also of the rest of this book.

Let **P** be a property which makes sense for Noetherian local rings, as the properties considered in Sect. 1.1 (e.g. **P**= regular, normal, Cohen-Macaulay, etc.).

Definition 1.6.3 A morphism of Noetherian rings $u : A \to B$ is called a **P-morphism**, if u is flat, and if for any prime ideal $\mathfrak{p}$ of A, the $k(\mathfrak{p})$-algebra $B \otimes_A k(\mathfrak{p})$ is geometrically **P**.

Remark 1.6.4 Probably the most important case of Definition 1.6.3 is for **P**=regular. This gives what is usually known as a **regular morphism**: a flat morphism $u : A \to B$ such that for any prime ideal $\mathfrak{p}$ of A, the $k(\mathfrak{p})$-algebra $B \otimes_A k(\mathfrak{p})$ is geometrically regular.

Definition 1.6.5 A Noetherian ring A is called a **P-ring**, or we say that **the formal fibres of A are geometrically P**, if for any prime ideal $\mathfrak{p}$ of A, the canonical completion morphism $A_{\mathfrak{p}} \to \widehat{A_{\mathfrak{p}}}$ is a **P**-morphism.

We must notice that in Definition 1.6.3 of **P**-morphism we assume that the rings A and B are Noetherian and that the morphism u is flat. Of course, when considering a property **P** that makes sense only for Noetherian rings (like e.g. Cohen-Macaulay or regular), we need to consider Noetherian rings. But when working with other properties, for example the property of being reduced, we may consider rings which are not necessarily Noetherian. That is, the reason why we shall sometimes use the following variations:

Definition 1.6.6 Let **P** be a property of commutative rings. A morphism $u : A \to B$ is said to have **geometrically P fibres**, if:

(i) u is flat.
(ii) For any field base change $A \to K$, the ring $K \otimes_A B$ has the property **P**.

Definition 1.6.7 A morphism $u : A \to B$ is said to be an **absolutely P morphism**, if for any base change $A \to C$ such that C verifies **P**, the ring $C \otimes_A B$ verifies **P**.

Remark 1.6.8

(i) It is clear that a flat absolutely **P** morphism has geometrically **P** fibres.
(ii) If $u : A \to B$ has geometrically integral domain fibres, then u is absolutely integral domain. Indeed, if $A \to C$ is an A-algebra, that is, an integral domain, we consider $K = Q(C)$ as the field of fractions of C. Then the flatness shows that $C \otimes_A B$ is an integral domain.

Before proceeding to the study of **P**-morphisms, we show an interesting result concerning the property **P**=reduced. We need a lemma.

Lemma 1.6.9 (Lazarus [217, Th. 3.1]) *Let A be a reduced ring and E be an A-module. If for any finitely generated reduced A-algebra B, the B-module $B \otimes_A E$ is torsion-free, then E is a flat A-module.*

Proof Let $\mathfrak{a} = (x_1, \ldots, x_n)$ be a finitely generated ideal of A. We need to show that the canonical morphism $\lambda : \mathfrak{a} \otimes_A E \to E$ is injective. Let

$$B := \{f \in A[T, U] \mid f = \sum_{i,j} a_{ij} T^i U^j, \ a_{ij} \in \mathfrak{a}^{i-j}, \text{ if } i \geq j\}.$$

Let

$$f = a_{00} + a_{10}T + a_{01}U + a_{11}TU + \ldots + a_{nm}T^n U^m \in B.$$

Then it is easy to see that one must have $f \in A[\mathfrak{a}T, U, TU]$, and since the other inclusion is obvious, we get that $B = A[\mathfrak{a}T, U, TU]$. This shows that B is a finitely generated reduced A-subalgebra of $A[T, U]$, such that $T \notin B$ but $U \in B$. Consider the following A-linear maps:

$$\tau : \mathfrak{a} \to B, \quad \tau(x) = xT \tag{1.11}$$

$$\phi : B \otimes_A E \to B \otimes_A E, \quad \phi(b \otimes e) = U \cdot (b \otimes e) \tag{1.12}$$

$$\psi : E \to B \otimes_A E, \quad \psi(e) = TU \otimes e. \tag{1.13}$$

In the above definition of ϕ, the multiplication $U \cdot (b \otimes e)$ means actually the multiplication by U in the B-module $B \otimes_A E$. We have

$$(\phi \circ (\tau \otimes 1_E))(x \otimes e) = \phi(\tau(x) \otimes e) = \phi(xT \otimes e) = U \cdot (xT \otimes e)$$

and

$$(\psi \circ \lambda)(x \otimes e) = \psi(xe) = TU \otimes xe = U \cdot (xT \otimes e)$$

so that $\phi \circ (\tau \otimes 1_E) = \psi \circ \lambda$. By definition the map τ identifies the ideal $\mathfrak{a}$ with the A-submodule $\mathfrak{a}T$ of B that is a direct factor of B; hence τ is a split injection. Therefore $\tau \otimes 1_E$ is a split injection. On the other hand, U is a regular element of $A[T, U]$, so a fortiori it is a regular element of B. Thus ϕ is injective. This means that $\psi \circ \lambda = \phi \circ (\tau \otimes 1_E)$ is injective and consequently λ is injective. The conclusion now follows. $\square$

Proposition 1.6.10 (Lazarus [217, Prop. 3.2]) *Let* $f : A \to B$ *be an absolutely reduced morphism. If A is reduced, then f is a geometrically reduced morphism.*

Proof Actually we have to prove that f is flat, and for this we apply Lemma 1.6.9. Thus it will be enough to show that for any finitely generated reduced A-algebra C, the tensor product ring $C \otimes_A B$ is a torsion-free C-module. Thus we may assume that $C = A$, and we have to show that B is torsion-free as an A-module. Let S be the set of non-zero-divisors of A, and consider elements $a \in S$ and $b \in B$ such that $ab = 0$. Let $D := A[X, Y]/(Y^2 - aX)$. Then D is a free $A[X]$-module since $Y^2 - aX$ is a monic polynomial in Y; hence D is a free A-module and consequently torsion-free, therefore the morphism $D \to S^{-1}D$ is an injection. But a becomes invertible in $S^{-1}D$ and this implies that $S^{-1}D \cong S^{-1}A[Y]$. Therefore $S^{-1}D$ is reduced and hence its subring D is reduced too. By assumption $E := B[X, Y]/(Y^2 - aX)$ is reduced. But in E we have $y^2 = ax$, where x and y are the classes of X and Y respectively. Then $(yb)^2 = 0$; hence $yb = 0$. But this means that $bY \in (Y^2 - aX)B[X, Y]$; hence $b = 0$. $\qquad\square$

We come back to the notion of **P**-morphism that will be one of our main concern, and we start by presenting several general properties of **P**-morphisms.

Lemma 1.6.11 *If* $u : A \to B$ *is a* **P***-morphism and S is a multiplicatively closed system of A, then the canonical morphism* $S^{-1}u : S^{-1}A \to S^{-1}B$ *is a* **P***-morphism.*

Proof Obvious. $\qquad\square$

Lemma 1.6.12 *Let* **P** *be a property verifying* (A_2), (A_3) *and* (A_5) *and let* $u : A \to B$ *be a morphism of Noetherian local rings. The following are equivalent:*

(i) *u is a* **P***-morphism.*
(ii) *For any* $\mathfrak{q} \in \mathrm{Spec}(B)$, *if* $\mathfrak{p} = \mathfrak{q} \cap A$, *the morphism* $A_{\mathfrak{p}} \to B_{\mathfrak{q}}$ *is flat and the* $k(\mathfrak{p})$-*algebra* $k(\mathfrak{p}) \otimes_{A_{\mathfrak{p}}} B_{\mathfrak{q}}$ *is geometrically* **P***.*
(iii) *For any* $\mathfrak{q} \in \mathrm{Spec}(B)$, *if* $\mathfrak{p} = \mathfrak{q} \cap A$, *the morphism* $A_{\mathfrak{p}} \to B_{\mathfrak{q}}$ *is a* **P***-morphism.*
(iv) *For any* $\mathfrak{n} \in \mathrm{Max}(B)$, *if* $\mathfrak{m} = \mathfrak{n} \cap A$, *the morphism* $A_{\mathfrak{m}} \to B_{\mathfrak{n}}$ *is a* **P***-morphism.*

Proof $(i) \Leftrightarrow (ii) \Leftrightarrow (iv)$ follow from the definitions.
$(iii) \Leftrightarrow (iv)$ Follows from the fact that for any $\mathfrak{q} \in \mathrm{Spec}(B)$, there exists $\mathfrak{n} \in \mathrm{Max}(B)$ such that $\mathfrak{q} \subseteq \mathfrak{n}$. $\qquad\square$

Lemma 1.6.13 *Let* **P** *be a property satisfying* (A_1), (A_2), (A_3) *and* (A_5) *and let* $u : A \to B$ *be a* **P***-morphism. Let* $\mathfrak{q} \in \mathrm{Spec}(B)$ *and* $\mathfrak{p} := \mathfrak{q} \cap A$. *Then* $A_{\mathfrak{p}}$ *has the property* **P** *if and only if* $B_{\mathfrak{q}}$ *has the property* **P***.*

Proof Clearly the canonical induced morphism $A_\mathfrak{p} \to B_\mathfrak{q}$ is a **P**-morphism. Now the assertion follows immediately by the conditions $(\mathbf{A_3})$ and $(\mathbf{A_5})$.   □

Corollary 1.6.14 *Let* **P** *be a property satisfying* $(\mathbf{A_1})$, $(\mathbf{A_2})$, $(\mathbf{A_3})$ *and* $(\mathbf{A_5})$ *and let* $u :$ $A \to B$ *be a* **P***-morphism. Then* A *has the property* **P** *if and only if* B *has the property* **P**.

Proof It follows immediately from Lemma 1.6.13.   □

Corollary 1.6.15 *Let* **P** *be a property satisfying* $(\mathbf{A_2})$, $(\mathbf{A_3})$ *and* $(\mathbf{A_5})$. *Let* $u : A \to B$ *be a* **P***-morphism and* $g = \mathrm{Spec}(u) : \mathrm{Spec}(B) \to \mathrm{Spec}(A)$ *the morphism induced by* u *on the Zariski spectra. Then* $g^{-1}(\mathbf{P}(A)) = \mathbf{P}(B)$.

Proof Follows from Lemma 1.6.13.   □

Lemma 1.6.16 *Let* **P** *be a property satisfying* $(\mathbf{A_1})$, $(\mathbf{A_2})$, $(\mathbf{A_3})$ *and* $(\mathbf{A_5})$. *If* $u : A \to B$ *is a* **P***-morphism and* C *is an* A*-algebra of finite type, then* $u \otimes 1_C : C \to B \otimes_A C$ *is a* **P***-morphism.*

Proof Follows from Proposition 1.4.16.   □

Proposition 1.6.17 *Let* **P** *be a property verifying* $(\mathbf{A_1}) - (\mathbf{A_3})$ *and* $(\mathbf{A_5})$. *Assume that* $u : A \to B$ *and* $v : B \to C$ *are* **P***-morphisms. Then* $v \circ u$ *is a* **P***-morphism.*

Proof Obviously $v \circ u$ is flat. Let $\mathfrak{p} \in \mathrm{Spec}(A)$, $k := k(\mathfrak{p})$ and L a field, finite extension of k. We have the following commutative diagram:

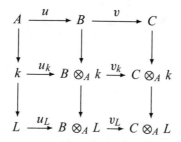

By 1.6.16 we know that v_L is a **P**-morphism. Let $\mathfrak{q} \in \mathrm{Spec}(C \otimes_A L)$ and let $\mathfrak{q}' := \mathfrak{q} \cap (B \otimes_A L)$. It follows that $(B \otimes_A L)_{\mathfrak{q}'}$ has the property **P** and by $(\mathbf{A_5})$ we obtain that $(C \otimes_A L)_\mathfrak{q}$ has **P**.   □

Proposition 1.6.18 *Let P be a property verifying* (**A₃**) *and* (**A₅**). *Suppose that* $u : A \to B$ *and* $v : B \to C$ *are such that* $v \circ u$ *is a P-morphism and* v *is faithfully flat. Then* u *is a P-morphism.* '

Proof It is well-known that u is flat [249, p. 46]. Let $\mathfrak{p} \in \mathrm{Spec}(A)$, $k := k(\mathfrak{p})$ and L a field, finite extension of k. If $\mathfrak{q} \in \mathrm{Spec}(B \otimes L)$, by faithful flatness there exists $\mathfrak{q}' \in \mathrm{Spec}(C \otimes L)$ such that $\mathfrak{q}' \cap (B \otimes L) = \mathfrak{q}$. Moreover $(C \otimes L)_{\mathfrak{q}'}$ has the property **P**. It follows that $(B \otimes L)_{\mathfrak{q}}$ has **P**.   □

Proposition 1.6.19 *Let P be a property of local rings, let A be a semilocal Noetherian ring and suppose that* $\mathrm{Max}(A) = \{\mathfrak{m}_1, \ldots, \mathfrak{m}_r\}$. *Then A is a P-ring if and only if* $A_{\mathfrak{m}_i}$ *is a P-ring, for all* $i = 1, \ldots, r$.

Proof We have $\widehat{A} \cong \bigoplus_{i=1}^{r} \widehat{A}_{\mathfrak{m}_i}$. It follows that if $\mathfrak{p} \in \mathrm{Spec}(A)$, the formal fibre of A in $\mathfrak{p}$ is the direct sum of the formal fibres of $A_{\mathfrak{m}_i}$ in $\mathfrak{p}$, for those indices i for which $\mathfrak{p} \subseteq \mathfrak{m}_i$. The assertion follows.   □

Proposition 1.6.20 *Let P be a property of local rings satisfying* (**A₁**) − (**A₃**) *and* (**A₅**). *Let A be a Noetherian P-ring and* $\mathfrak{a}$ *an ideal of A. Then* $A/\mathfrak{a}$ *is a P-ring.*

Proof Let $\mathfrak{p} \in \mathrm{Spec}(A)$ be such that $\mathfrak{p} \supseteq \mathfrak{a}$. Then the formal fibres of $A_{\mathfrak{p}}$ are geometrically **P** and the formal fibres of $(A/\mathfrak{a})_{\mathfrak{p}} \cong A_{\mathfrak{p}}/\mathfrak{a}A_{\mathfrak{p}}$ are a subset of the formal fibres of $A_{\mathfrak{p}}$.   □

Proposition 1.6.21 *Let P be a property satisfying* (**A₁**), (**A₂**), (**A₃**) *and* (**A₅**). *If A is a semilocal P-ring and B is a finite A-algebra, then B is a P-ring.*

Proof B is a semilocal Noetherian ring and clearly $\widehat{B} \cong B \otimes_A \widehat{A}$. Now we apply Lemma 1.6.16.   □

Lemma 1.6.22 *Let P be a property of local rings satisfying* (**A₂**). *Let* $\phi : A \to B$ *be a morphism of Noetherian rings. The following are equivalent:*

(i) *For any* $\mathfrak{p} \in \mathrm{Spec}(A)$, *for any field k, finite extension of* $k(\mathfrak{p})$ *and for any* $\mathfrak{q} \in \mathrm{Spec}(B \otimes_A k)$ *such that* $\mathfrak{q} \cap A = \mathfrak{p}$, *the ring* $(B \otimes_A k)_{\mathfrak{q}}$ *has the property P.*

(ii) *For any finite A-algebra C which is a domain and for any* $\mathfrak{q} \in \mathrm{Spec}(C \otimes_A B)$ *such that* $\mathfrak{q} \cap B = 0$, *the ring* $(C \otimes_A B)_{\mathfrak{q}}$ *has the property P.*

Proof $(i) \Rightarrow (ii)$: Let us note that $k(\mathfrak{q})$ is a finite extension of $k(\mathfrak{p})$.
$(ii) \Rightarrow (i)$: Let $\mathfrak{p} \in \mathrm{Spec}(A)$ and k a field, finite extension of $k(\mathfrak{p})$. Let $x_1, \ldots, x_n$ be the basis of k over $k(\mathfrak{p})$ such that the elements x_i are integral over $A/\mathfrak{p}$. If $B = A[x_1, \ldots, x_n]$, then B is a finite A-algebra which is a domain and the field of fractions of B is k.   □

Theorem 1.6.23 *Let P be a property of local rings satisfying $(A_1) - (A_3)$ and (A_5). For a Noetherian local ring A, the following are equivalent:*

(i) *A is a P-ring.*

(ii) *For any finite A-algebra B which is a domain and for any $q \in \mathrm{Spec}(\widehat{B})$ such that $q \cap B = 0$, the ring $\widehat{B}_q$ has the property P.*

(iii) *For any finite A-algebra B, if $g : \mathrm{Spec}(\widehat{B}) \to \mathrm{Spec}(B)$ is the morphism induced on the spectra by the canonical completion morphism $B \to \widehat{B}$, we have $g^{-1}(\mathbf{P}(B)) = \mathbf{P}(\widehat{B})$.*

Proof $(i) \Rightarrow (iii)$: Follows immediately from Proposition 1.6.20 and Corollary 1.6.15.
$(i) \Leftrightarrow (ii)$: Follows from Lemma 1.6.22.
$(iii) \Rightarrow (ii)$: Let B be a finite A-algebra which is a domain and let $p = (0) \in \mathrm{Spec}(B)$. From (A_1) we get that $p \in \mathbf{P}(B)$. Now we apply iii). $\qquad\square$

Proposition 1.6.24 *Let P be a property verifying (A_2), (A_3) and (A_5) and let $u : A \to B$ be a P-morphism of Noetherian local rings. If B is a P-ring, then also A is a P-ring.*

Proof We have the following commutative diagram:

$$
\begin{array}{ccc}
A & \xrightarrow{\ u\ } & B \\
{\scriptstyle g}\downarrow & & \downarrow{\scriptstyle h} \\
\widehat{A} & \xrightarrow{\ \widehat{u}\ } & \widehat{B}
\end{array}
$$

From the assumptions and Proposition 1.6.17, it follows that $\widehat{u} \circ g = h \circ u$ is a P-morphism. Since u is flat, also $\widehat{u}$ is flat, hence faithfully flat. From Proposition 1.6.18 we get that g is a P-morphism. $\qquad\square$

Corollary 1.6.25 *Let P be a property verifying (A_2), (A_3) and (A_5). Let A be a Noetherian local P-ring. Assume that $\widehat{A}_q$ is a P-ring for any $q \in \mathrm{Spec}(\widehat{A})$. Then A_p is a P-ring, for any $p \in \mathrm{Spec} A$.*

Proof The canonical morphism $A \to \widehat{A}$ is a P-morphism. Let $p \in \mathrm{Spec}(A)$. Then there exists $q \in \mathrm{Spec}(\widehat{A})$ such that $q \cap A = p$. It follows that $A_p \to \widehat{A}_q$ is a P-morphism. Now we apply Proposition 1.6.24. $\qquad\square$

Lemma 1.6.26 *Let P be a property verifying (A_2). Let $u : A \to B$ be an injective local morphism of Noetherian local domains, such that B is the localization of a finite A-algebra in a maximal ideal. If the formal fibre of A in (0) is geometrically P, then the formal fibre of B in (0) is geometrically P.*

Proof Since B is the localization of a finite A-algebra in a maximal ideal, $\widehat{B} \cong \widehat{A} \otimes_A B$. Also $L := Q(B)$ is a finite extension of $K := Q(A)$. From the assumption $K \otimes_A \widehat{A}$ is a geometrically **P** K-algebra. Hence $L \otimes_B \widehat{B} = L \otimes_A B \otimes_B \widehat{A} = L \otimes_A \widehat{A}$ is a geometrically **P** L-algebra. But this is the formal fibre of B in (0).   □

Corollary 1.6.27 *Let* **P** *be a property verifying* (**A₂**). *Let* A *be a Noetherian local ring. Suppose that for any prime ideal* $\mathfrak{q} \in \operatorname{Spec}(A)$, *there exists a Noetherian local subring* C *of* $A/\mathfrak{q}$ *which is a localization of a finite* A-*algebra and such that the formal fibre of* C *in* (0) *is geometrically* **P**. *Then the formal fibres of* A *are geometrically* **P**.

Our aim now is to prove Theorem 1.6.37 that says that any complete local ring has geometrically regular formal fibres. This result belongs to Grothendieck, and we use the presentation from [359]. In order to simplify notations, we will use the following two definitions [249, p. 241].

Definition 1.6.28 Let K be a field and $L \subseteq K$ be a subfield ok K. We say that L is **cofinite** in K, if K is a finite extension of L.

Definition 1.6.29 If k is a field, a family $(k_\alpha)_{\alpha \in \Lambda}$ of subfields of k is called **directed** if for any $\alpha, \beta \in \Lambda$ there exists $\gamma \in \Lambda$ such that $k_\gamma \subseteq k_\alpha \cap k_\beta$.

The following lemma is well-known [249].

Lemma 1.6.30 *Let* K *be a field, let* $(k_\alpha)_{\alpha \in \Lambda}$ *be a directed family of subfields of* K *and* $k := \bigcap_{\alpha \in \Lambda} k_\alpha$. *Let* V *be a* K-*vector space and* $v_1, \ldots, v_n \in V$ *vectors linearly independent over* k. *Then there exists* $\alpha \in \Lambda$ *such that* $v_1, \ldots, v_n$ *are linearly independent over* k_α.

Proof See [249, Lemma 3, p. 241].   □

We will need several times the notion of p-basis. The reader can find all the necessary results and definitions about this notion in [131, §21] or [248, §38].

Lemma 1.6.31 *Let* k *be a field of characteristic* $p > 0$ *and let* n *and* m *be strictly positive natural numbers. Let* $A := k[[X]][Y] := k[[X_1, \ldots, X_n]][Y_1, \ldots, Y_m]$ *and let* K *be the field of fractions of* A. *Consider the family* $(k_\alpha)_{\alpha \in \Lambda}$ *of all cofinite subfields of* k *containing* k^p *and for any* $\alpha \in \Lambda$, *let* K_α *be the field of fractions of* $A_\alpha := k_\alpha[[X^p]][Y^p]$. *Then:*

(i) *The ring extension* $A_\alpha \subseteq A$ *is finite.*
(ii) *The family* $(K_\alpha)_{\alpha \in \Lambda}$ *is directed.*
(iii) $K^p = \bigcap_{\alpha \in \Lambda} K_\alpha$.

(iv) If $\theta_\alpha : \Omega_K \to \Omega_{K/K_\alpha}$ is the canonical morphism, then $\bigcap\limits_{\alpha \in \Lambda} \ker(\theta_\alpha) = (0)$.

(v) For any finite field extension $K \subseteq L$, we have $L^p = \bigcap\limits_{\alpha \in \Lambda} L^p K_\alpha$.

Proof (i) Since k_α is cofinite in k, it follows that the ring extension $k_\alpha[X^p] \subseteq k[X]$ is finite. It follows easily that $k_\alpha[[X^p]] \subseteq k[[X]]$ is finite; hence A is finite over A_α.

(ii) Follows easily from the fact that the family $(k_\alpha)_\alpha$ is directed.

(iii) Since $A^p \subseteq A_\alpha$, we have $K^p \subseteq \bigcap\limits_{\alpha \in \Lambda} K_\alpha$. Any element $u \in K_\alpha$ for some $\alpha \in \Lambda$ can be written $u = \frac{a}{b^p}$, with $a \in A_\alpha$ and $b \in A, b \neq 0$. Suppose that an element $\frac{f}{g^p} \in K$, $f, g \in A, g \neq 0$ is contained in $\bigcap\limits_{\alpha \in \Lambda} K_\alpha$. Then, for a fixed $\alpha \in \Lambda$, we have $\frac{f}{g^p} = \frac{a}{b^p}$, for some $a \in A_\alpha$ and some $b \in A, b \neq 0$, hence $b^p f \in A_\alpha$. For any $D \in \mathrm{Der}_{A_\alpha}(A)$, we have $D(b^p f) = b^p D(f) = 0$; therefore $D(f) = 0$, because A is an integral domain. If we apply this argument for $\frac{\partial}{\partial x_i}$ and $\frac{\partial}{\partial y_j}$, it follows that $f \in k[[X^p]][Y^p]$. Applying the same to a k_α-derivation $\eta \in \mathrm{Der}_{k_\alpha}(k)$, we get that all the coefficients of f are in k_α, that is $f \in A_\alpha$. But $A^p = \bigcap\limits_{\alpha \in \Lambda} A_\alpha$, whence we are done.

(iv) Let $B = (b_i)_{i \in I}$ be a p-basis of K. Then $(d_K(b_i))_{i \in I}$ is a K-linear basis of Ω_K; hence any non-zero element $\omega \in \Omega_K$ can be uniquely written $\omega = c_{i_1} d_K(b_{i_1}) + \ldots + c_{i_s} d_K(b_{i_s})$, where $c_{i_1}, \ldots, c_{i_s} \in K$. By Lemma 1.6.30 we can take $\alpha \in \Lambda$ such that $\{b_{i_1} \ldots, b_{i_s}\}$ are p-independent over K_α; hence $d_{K/K_\alpha}(b_{i_1}), \ldots, d_{K/K_\alpha}(b_{i_s})$ are linearly independent elements of Ω_{K/K_α}. Then the image of ω in Ω_{K/K_α} is not zero.

(v) Let us first note that if $K \subset M \subset L$ are finite field extensions and we set $M_\alpha = M^p K_\alpha$ and $L_\alpha = L^p K_\alpha = L^p M_\alpha$, we can first prove that $M^p = \bigcap\limits_{\alpha \in \Lambda} M_\alpha$ and afterward that $L^p = \bigcap\limits_{\alpha \in \Lambda} L_\alpha$. This shows that we can assume that $K \subset L$ is a primitive field extension, without proper intermediate subfields. We have to consider two cases:

Case 1. $L = k(\theta)$ is a separable extension of K. Then L is generated by θ^p; hence we may assume that $\theta \in L^p$. Thus, if $d = [L : K]$, we see that

$$L^p = K^p \oplus K^p\theta \oplus \ldots \oplus K^p\theta^{d-1}$$

and

$$L^p K_\alpha = K_\alpha^p \oplus K_\alpha^p\theta \oplus \ldots \oplus K_\alpha^p\theta^{d-1}.$$

Case 2. $L = K(\theta)$ with $\theta^p = t \in K \setminus K^p$. In this case we have

$$L^p = K^p \oplus K^p t \oplus \ldots \oplus K^p t^{p-1}$$

and

$$L^p K_\alpha = K_\alpha^p \oplus K_\alpha^p t \oplus \ldots \oplus K_\alpha^p t^{p-1}.$$

In both cases the conclusion follows.   □

Lemma 1.6.32 *Let A be a regular ring and $f \in A$. Suppose that there exists $D \in \mathrm{Der}(A)$ such that $D(f)$ is a unit in $A/(f)$. Then $A/(f)$ is a regular ring.*

Proof We may assume that $(A, \mathfrak{m}, k)$ is a local ring and it is enough to show that $f \notin \mathfrak{m}^2$. If $f \in \mathfrak{m}^2$, then f can be written $f = \sum_{i=1}^{r} g_i h_i$, where $g_i, h_i \in \mathfrak{m}$. Then by Leibniz rule, $D(f) \in \mathfrak{m}$, contradicting the hypothesis. Thus $f \notin \mathfrak{m}^2$.   □

Lemma 1.6.33 *Let A be a regular ring and $f \in A$. Suppose that there exists $D \in \mathrm{Der}(A)$ such that $D(f)$ is a unit in A. Then for any polynomial $g \in \mathbb{Z}[T]$, the ring $A[T]/(g - f)$ is a regular ring.*

Proof We extend the derivation D on $A[T]$, by $D(T) = 0$. Then $D(g) = 0$, and $D(f)$ is a unit; hence $D(g - f)$ is a unit and we can apply Lemma 1.6.32.   □

Lemma 1.6.34 *[359, More on algebra - Lemma 48.1] Let A be a ring and $B \supseteq A$ be a finite type extension of A such that there exists an element $g \in A$ that is a non-zero-divisor on B with $A_g \cong B_g$. Then, for any derivation $D \in \mathrm{Der}(A)$, there is some $N \geq 0$ such that $g^N D$ extends to B.*

Proof Suppose that $B = A[x_1, \ldots, x_n]$ and let $N \in \mathbb{N}$ be such that $g^{N-1} x_i \in A$, for all $i = 1, \ldots, n$. We will show that $g^N D$ extends to B. However, we know that $g^N D$ extends to A_g. Moreover we have

$$g^N D(x_i) = g^N D(g^{-N+1} g^{N-1} x_i) = -(N-1) g^{N-1} x_i D(g) + g D(g^{N-1} x_i)$$

and both terms of the right side are in A. Therefore

$$g^N D(x_1^{\alpha_1} \ldots x_n^{\alpha_n}) = \sum \alpha_i x_1^{\alpha_1} \ldots x_i^{\alpha_i - 1} \ldots x_n^{\alpha_n} g^N D(x_i) \in B,$$

hence $g^N D$ maps any element of B to an element of B.   □

Proposition 1.6.35 ([359, More on algebra - Lemma 48.5]) *Let p be a prime number, let A be a Noetherian complete local ring and let B be a domain, which is an A-algebra of finite type and such that $p = 0$ in B. Let $f \in B$ be an element which is not in $Q(B)^p$. Then there exists $D \in \mathrm{Der}(B)$ such that $D(f) \neq 0$.*

Proof Replacing if necessary A by its image in B, we may assume that A is a Noetherian complete local domain of characteristic p. By Cohen's structure theorem, there exist a field k and a natural number n such that A is a finite extension of $k[[x_1, \ldots, x_n]]$. We replace A by $k[[x_1, \ldots, x_n]]$. Assume that $B = A[z_1, \ldots, z_s]$. Let K be the fraction field of A and $B_K := B \otimes_A K$. By Noether normalization lemma, there exist elements $y_1, \ldots, y_d \in B_K$ such that $K[y_1, \ldots, y_d] \to B_K$ is a finite injective morphism. Note that we can take $y_1, \ldots, y_d \in B$ because they can be chosen from the ring $\mathbb{Z}[z_1, \ldots, z_s]$. There exist non-zero polynomials $P_1, \ldots, P_s \in K[y_1, \ldots, y_d][T]$ such that $P_1(z_1) = \ldots = P_s(z_s) = 0$ in B_K. Pick an element $g \in A, g \neq 0$, such that $g P_i \in A[y_1, \ldots, y_d][T], i = 1, \ldots, s$, hence $g P_i(z_i) = 0$ in B_K, for all $i = 1, \ldots, s$. After replacing g if necessary, we may assume that $g P_i(z_i) = 0$ in B. Let $v_i := g z_i, i = 1, \ldots, s$, and let $C := A[y_1, \ldots, y_d, v_1, \ldots, v_s] \subseteq B$. Let $Q_i := g^{\deg_T(P_i)} P_i(\frac{T}{g}) \in A[y_1, \ldots, y_d][T]$. Then $Q_i(u_i) = 0, i = 1, \ldots, s$; hence the extension $A[y_1, \ldots, y_d] \subset C$ is finite. Clearly $C_g \subseteq B_g$. But $z_i = \frac{u_i}{g} \in C_g, i = 1, \ldots, s$ generate B_g over A_g; hence the map $C_g \to B_g$ is surjective, that is, $C_g \cong B_g$. Summarizing we found

$$A = k[[x_1, \ldots, x_n]] \subseteq A[y_1, \ldots, y_d] \subseteq C \subseteq B$$

where C is finite over $A[y_1, \ldots, y_d]$ and there exists $g \in A$ such that $C_g \cong B_g$.

We see that $h = g^{p^N} f \in C$, for some large enough N. Obviously h is not a pth power in $Q(C)$. If we can find a derivation $D' \in \operatorname{Der}(C)$ such that $D'(h) \neq 0$, then by Lemma 1.6.34 we know that $D = g^M D'$ extends to B for some $M > 0$. Then

$$D(f) = g^N D'(f) = g^M D'(g^{-p^N} h) = g^{M-p^N} D'(h) \neq 0.$$

This shows that it is enough to prove the assertion in the case B is a finite extension of $A = k[[x_1, \ldots, x_n]][y_1, \ldots, y_m]$. Let $L := Q(B)$ and $d : L \to \Omega_L = \Omega_{L/\mathbb{F}_p}$ be the universal derivation. Suppose first that $df = 0$ in Ω_L. Since $L = \bigcup_{i \in I} K_i$, where K_i are finitely generated field extensions, we may assume that L is a finitely generated field extension of $\mathbb{F}_p$. Let $\alpha_1, \ldots, \alpha_r \in L$ be a transcendence basis such that $\mathbb{F}_p(\alpha_1, \ldots, \alpha_r) \subseteq L$ is a finite separable extension. Then $\Omega_{\mathbb{F}_p(\alpha_1, \ldots, \alpha_r)/\mathbb{F}_p} = \bigoplus_{i=1}^r \mathbb{F}_p(\alpha_1, \ldots, \alpha_r) d\alpha_i$. Note that any rational function whose partial derivatives are zero is a p-th power. Moreover, $\Omega_{L/\mathbb{F}_p} = \bigoplus_{i=1}^r L d\alpha_i$. Now, the minimal polynomial $P(T) \in \mathbb{F}_p(\alpha_1, \ldots, \alpha_r)[T]$ of f over $\mathbb{F}_p(\alpha_1, \ldots, \alpha_r)$ is separable. If $P(T) = T^d + a_1 T^{d-1} + \ldots + a_0$, then $0 = dP(f) = \bigoplus_{i=1}^d f^{d-i} da_i$, this relation being in $\Omega_{L/\mathbb{F}_p}$. This implies that $da_i = 0, i = 1, \ldots, r$ and then $a_i = b_i^p, i = 1, \ldots, r$. Therefore f is a pth power, contradicting the hypothesis. Thus $df \neq 0$ in $\Omega_{L/\mathbb{F}_p}$. By Lemma 1.6.31 there exists a cofinite subfield $\tilde{k} \subseteq k$ containing k^p such that, denoting by $\tilde{A} = \tilde{k}[[x_1^p, \ldots, x_n^p]][y_1^p, \ldots, y_m^p]$ and $\tilde{K} := Q(\tilde{A})$ we have that $df \neq 0$ in $\Omega_{L/\tilde{K}}$. Thus we can find a derivation $\tilde{D} \in \mathrm{Der}_{\tilde{K}}(L)$ with $\tilde{D}(f) \neq 0$. Clearly B is a finite $\tilde{A}$-module generated by say, $b_1, \ldots, b_t \in B$. Then $\tilde{D}(b_i) = \frac{f_i}{g_i}$ with $f_i, g_i \in B$, $g_i \neq 0$ and $D := g_1 \cdot \ldots \cdot g_t \cdot \tilde{D}$ is the desired derivation. $\square$

Proposition 1.6.36 *Let k be a field of characteristic $p > 0$, and let $A := k[[x]][y] = k[[x_1, \ldots, x_n]][y_1, \ldots, y_m]$. Let K be the fraction field of A and $\mathfrak{p} \in \mathrm{Spec}(A)$. Then $K \otimes_A \widehat{A_\mathfrak{p}}$ is a geometrically regular K-algebra.*

Proof Let L be a field, finite purely inseparable extension of K. We have to show that $L \otimes_A \widehat{A_\mathfrak{p}}$ is regular. We proceed by induction on $d := [L : K]$. If $d = 1$, that is $L = K$, since A is regular, $K \otimes_A \widehat{A_\mathfrak{p}}$ is regular as a localization of $\widehat{A_\mathfrak{p}}$. If $d > 1$, let $K \subset M \subset L$ be a subfield such that L is a degree p extension of M obtained by adjoining a p-th root of an element $f \in M$. Let B be a finite A-subalgebra of M having M as field of fractions. Clearing denominators we may assume that $f \in B$. Let $C := B[T]/(T^p - f)$. Then C is finite over B and has L as field of fractions. Since $A \subset B \subset C$ are finite extensions and $K \subset M \subset L$ are purely inseparable, for any element $b \in B$ or $c \in C$, there exists some power $s \in \mathbb{N}$ such that $b^s \in A$ and $c^s \in A$. Thus, there are uniquely determined prime ideals $\mathfrak{r} \in \mathrm{Spec}(B)$ and $\mathfrak{q} \in \mathrm{Spec}(C)$ such that $\mathfrak{r} \cap A = \mathfrak{p}$ and $\mathfrak{q} \cap A = \mathfrak{p}$. By induction the ring $\widehat{A_\mathfrak{p}} \otimes_A M = \widehat{B_\mathfrak{r}} \otimes_B M$ is regular. We have also

$$\widehat{A_\mathfrak{p}} \otimes_A L = \widehat{C_\mathfrak{q}} \otimes_C L = \widehat{B_\mathfrak{r}} \otimes_B M[T]/(T^p - f).$$

By Proposition 1.6.35 there exists $D \in \mathrm{Der}(B)$ such that $D(f) \neq 0$, that is, $g := D(f)$ is invertible in M. Extending successively D through localization, completion and again localization, we finally get a derivation $\tilde{D} \in \mathrm{Der}(\widehat{B_\mathfrak{r}} \otimes_B M)$, such that $\tilde{D}(f) = g$ and g is invertible in $\widehat{B_\mathfrak{r}} \otimes_B M$. By Lemma 1.6.33 the ring $\widehat{A_\mathfrak{p}} \otimes_A L$ is regular and we are done. $\square$

Theorem 1.6.37 (Grothendieck [131, Th. 22.3.3]) *Let $(A, \mathfrak{m}, k)$ be a complete Noetherian local ring. Then the formal fibres of A are geometrically regular.*

Proof We need to show that for any prime ideals $\mathfrak{p} \supset \mathfrak{q}$ of A, the ring

$$\widehat{A_\mathfrak{p}} \otimes_A k(\mathfrak{q}) = \widehat{(A/\mathfrak{q})_\mathfrak{p}} \otimes_{A/\mathfrak{q}} k(\mathfrak{q})$$

is geometrically regular over $k(\mathfrak{q})$. Replacing A with $A/\mathfrak{p}$, we see that it is enough to show that the formal fibre in (0) of the ring $B = A/\mathfrak{q}$ is geometrically regular, that is, we may assume that A is a complete local domain and we have to check the formal fibres in (0). Let K be the fraction field of A. There exists a regular complete local ring R contained in A such that A is a finite R-algebra. It is well-known that we can suppose that R is a formal power series ring over a field or over a Cohen ring. By Lemma 1.6.26 it is enough to prove the conclusion for R, that is, we may assume that A is a formal power series over a field or over a Cohen ring. Since A is regular, it follows that $\widehat{A_{\mathfrak{p}}} \otimes_A K$ is regular. Therefore we are done if $\mathrm{char}(K) = 0$. If $\mathrm{char}(K) = p > 0$, we are in the case $m = 0$ of the Proposition 1.6.36.   □

Theorem 1.6.38 *Let P be a property verifying* $(\mathbf{A_1}) - (\mathbf{A_3})$ *and* $(\mathbf{A_5})$. *Let A be a P-ring and let $\mathfrak{p} \in \mathrm{Spec}(A)$. Then $A_{\mathfrak{p}}$ is a P-ring.*

Proof It follows from Theorem 1.6.37 and Corollary 1.6.25.   □

Corollary 1.6.39 *Let A be a Noetherian ring and P a property of Noetherian local rings verifying* $(\mathbf{A_1}) - (\mathbf{A_3})$ *and* $(\mathbf{A_5})$. *Then A is a P-ring if and only if $A_{\mathfrak{m}} \to \widehat{A_{\mathfrak{m}}}$ is a P-morphism for any $\mathfrak{m} \in \mathrm{Max}(A)$.*

Proof Assume that $A_{\mathfrak{m}} \to \widehat{A_{\mathfrak{m}}}$ is a **P**-morphism for any $\mathfrak{m} \in \mathrm{Max}(A)$. Let $\mathfrak{p} \in \mathrm{Spec}(A)$ and let $\mathfrak{m} \in \mathrm{Max}(A)$ be such that $\mathfrak{p} \subseteq \mathfrak{m}$. Then, by Theorem 1.6.37 and $(\mathbf{A_1})$, we obtain that $\widehat{A_{\mathfrak{m}}}$ is a **P**-ring. From the fact that $A_{\mathfrak{m}} \to \widehat{A_{\mathfrak{m}}}$ is a **P**-morphism and Proposition 1.6.24, it follows that $A_{\mathfrak{m}}$ is a **P**-ring. But then, Theorem 1.6.38 implies that $A_{\mathfrak{p}} \cong (A_{\mathfrak{m}})_{\mathfrak{p}A_{\mathfrak{m}}}$ is a **P**-ring; hence $A_{\mathfrak{p}} \to \widehat{A_{\mathfrak{p}}}$ is a **P**-morphism. It results that A is a **P**-ring. The converse is obvious.   □

Proposition 1.6.40 *Let P be a property verifying* $(\mathbf{A_1}) - (\mathbf{A_3})$ *and* $(\mathbf{A_5})$. *Let A be a Noetherian P-ring, $\mathfrak{a}$ an ideal of A and $\widehat{A} = \widehat{(A, \mathfrak{a})}$ the completion of A in the $\mathfrak{a}$-adic topology. Then $A \to \widehat{A}$ is a P-morphism.*

Proof From Proposition 1.6.12, it follows that it is enough to show that for any $\mathfrak{n} \in \mathrm{Max}(A)$, if $\mathfrak{m} := \mathfrak{n} \cap A$, the morphism $A_{\mathfrak{m}} \to \widehat{A_{\mathfrak{n}}}$ is a **P**-morphism A. But the morphism $A_{\mathfrak{m}} \to \widehat{A_{\mathfrak{n}}}$ is injective, the $\mathfrak{m}A_{\mathfrak{m}}$-adic topology on $A_{\mathfrak{m}}$ is induced by the $\mathfrak{n}\widehat{A_{\mathfrak{n}}}$-adic topology on $\widehat{A_{\mathfrak{n}}}$ and $A_{\mathfrak{m}}$ is dense in $\widehat{A_{\mathfrak{n}}}$ in this topology. It follows from [56, ch. III, §3, n. 4, prop. 8] that

$$\widehat{(A_{\mathfrak{m}}, \mathfrak{m}A_{\mathfrak{m}})} \cong \widehat{(\widehat{A_{\mathfrak{n}}}, \mathfrak{n}\widehat{A_{\mathfrak{n}}})}.$$

So we have the morphisms

$$A_{\mathfrak{m}} \xrightarrow{g} \widehat{A_{\mathfrak{n}}} \xrightarrow{f} \widehat{(A_{\mathfrak{m}})}.$$

Here f is faithfully flat and $f \circ g$ is a **P**-morphism. From Proposition 1.6.18, it follows that g is a **P**-morphism.   □

Lemma 1.6.41 *Let P be a property satisfying $(A_1) - (A_3)$ and (A_5). Let $u : A \to B$ be a flat morphism and $v : A \to C$ a finite A-algebra such that $\ker(v)$ is a nilpotent ideal of A. If the induced morphism $u \otimes_A 1_C : C \to C \otimes_A B$ is a P-morphism, then u is a P-morphism.*

Proof Let $\mathfrak{p} \in \operatorname{Spec}(A)$. By assumption $\mathfrak{p} \supseteq \ker(v)$, hence there exists $\mathfrak{p}' \in \operatorname{Spec}(C)$ such that $v^{-1}(\mathfrak{p}') = \mathfrak{p}$. Replacing A by $A/\mathfrak{p}$, B by $B/\mathfrak{p}B$ and C by $C/\mathfrak{p}'$, we may assume that A and C are integral domains and that v is injective. Then $Q(C)$ is a finite field extension of $Q(A)$, and the assertion follows. □

Proposition 1.6.42 (Greco [122, Prop. 1.3]) *Let $u : A \to B$ be a flat finite morphism such that $\ker(u)$ is a nilpotent ideal of A. The following are equivalent:*

(i) *A is a P-ring.*
(ii) *B is a P-ring.*

Proof Clearly $(i) \Rightarrow (ii)$. To show the converse, let $\mathfrak{p} \in \operatorname{Spec}(A)$ and $\mathfrak{q} \in \operatorname{Spec}(B)$ such that $u^{-1}(\mathfrak{q}) = \mathfrak{p}$. Changing A by $A_\mathfrak{p}$ and B by $B_\mathfrak{p}$, we may assume that A is a local ring, and it is enough to show that $A \to \widehat{A}$ is a P-morphism. By assumption B is a semilocal P-ring; hence $B \to \widehat{B} \cong \widehat{A} \otimes_A B$ is a P-morphism. Now we apply Lemma 1.6.41. □

Corollary 1.6.43 (Greco [122, Cor. 1.4]) *Let $u : A \to B$ be an injective morphism of finite type of integral domains. If B is a P-ring, there exists a non-zero element $f \in A$ such that A_f is a P-ring.*

Proof By Noether normalization lemma, there exist a non-zero element $f \in A$, variables $X = (X_1, \ldots, X_n)$ and a finite injective morphism $A_f[X] \to B$. By Proposition 1.6.42, we get that $A_f[X]$ is a P-ring; hence also $A_f = A_f[X]/(X)$ is a P-ring. □

We want to study the behaviour of P-morphisms with respect to Henselization. For the main properties of Henselian rings and Henselization, one can see [134, 187, 206, 266, 306]. We shall briefly remind the definition and some facts about this topic.

Notation 1.6.44 For a local ring $(A, \mathfrak{m}, k)$ and an element $a \in A$, we denote by $\bar{a}$ its image in k. Also, for a polynomial $f \in A[T]$, we denote by $\bar{f}$ its image in $k[T]$ and by f' the derivative of f with respect to T.

Definition 1.6.45 A local (not necessarily Noetherian) ring $(A, \mathfrak{m}, k)$ is called a **Henselian local ring** if for every monic polynomial $f \in A[T]$ and every root $a \in k$ of $\bar{f}$ such that $\overline{f'}(a) \neq 0$, there exists an element $b \in A$ such that $f(b) = 0$ and $\bar{b} = a$. A Henselian local ring is called **strictly Henselian** if k is separably algebraically closed, that is, k is separably closed in its algebraic closure.

Henselian rings were introduced by the Japanese mathematician Goro Azumaya (1920–2010) who named them after the German mathematician Kurt Hensel (1861–1941).

Example 1.6.46

(i) It is well-known that any field and any complete local ring are Henselian [306, Ch. I, §2, Ex. 1, 3].

(ii) On the other hand, if p is a prime number, the ring $A = \mathbb{Z}_{p\mathbb{Z}}$, the localization of the ring of integers $\mathbb{Z}$ in the prime ideal $p\mathbb{Z}$, is not a Henselian ring. Actually one can consider the polynomial $f(T) = T(T - 1) + p \in A[T]$, that is irreducible over A but is reducible modulo p.

(iii) If k is a field, the local ring $A := k[X_1, \ldots, X_n]_{(X_1, \ldots, X_n)}$ is not a Henselian ring. This can be seen immediately, for example, in the case $n = 1$, by considering the polynomial $f(T) = T^2 + T + X_1$, which is irreducible over A. But $\overline{f} = T(T + 1)$ is obviously not irreducible over k.

For any local ring $(A, \mathfrak{m}, k)$, there is a *smallest* Henselian local ring $(A^h, \mathfrak{m}^h)$ and a morphism $A \to A^h$ with some universal properties. This ring, first considered by M. Nagata (1927–2008), is called the **Henselization** of A. For this notion and its main important properties, one can look at [134, Déf. 18.6.5-Cor. 18.6.15]. There is also a notion of strict Henselization, but we shall not approach it. A generalization of Henselian local rings is the notion of *Henselian couple*, where the maximal ideal of a local ring is replaced by any ideal in a commutative ring.

Definition 1.6.47 A **Henselian couple** is a Noetherian ring A, together with an ideal $\mathfrak{a}$ contained in the Jacobson radical of A, such that for any monic polynomial $f \in A[T]$ and any factorization $\overline{f} = g_0 h_0$, with monic polynomials in $A/\mathfrak{a}[T]$, generating the unit ideal in $A/\mathfrak{a}[T]$, there exists a factorization $f = gh$ in $A[T]$, with monic polynomials, such that $g = g_0 \mod \mathfrak{a}T$ and $h = h_0 \mod \mathfrak{a}T$.

There is also a notion of Henselization of a couple. One can find more details about Henselian couples and about Henselization of a couple in [359, Ch. 15, Section 11] or [306, Ch. IX]. We shall say some words about another notion that appears to be connected with Henselian local rings.

Definition 1.6.48 Let k be a field and $X := (X_1, \ldots, X_n)$ be variables over k. A formal power series $h \in k[[X]]$ is called an **algebraic power series** if there exists a polynomial $P \in k[X][T]$, $P \neq 0$, such that $P(X, h(X)) = 0$.

Remark 1.6.49 We do not ask the polynomial P to be monic in the above definition.

Example 1.6.50

(i) Any polynomial $h \in K[X]$ is obviously an algebraic power series, since we may take
$P(X, T) = h(X) - T$.

(ii) Let $n = 1$ and assume that $\mathrm{char}(k) = 0$. Consider the series $h(X) = (1 + X)^r = \sum_{j=0}^{\infty} \binom{r}{j} X^j$, where $r = \frac{p}{q}$ is a rational number. Let

$$P(X, T) = \begin{cases} T^q - (1 + X)^p, & \text{if } p, q > 0 \\ T^q (1 + X)^{-p} - 1 & \text{if } p < 0 \end{cases}$$

Then one can easily see that h is algebraic.

(iii) Let $n = 1$ and $k = \mathbb{Q}$. Then the formal power series $h = exp(X) := \sum_{j=0}^{\infty} \frac{X^j}{j!}$, as it is
well-known, is not algebraic. Otherwise it would exist $P(X, T) \in \mathbb{Q}[X, T]$, $P \neq 0$,
such that $P(X, exp(X)) = 0$. We may assume that P is irreducible and we can write

$$P(X, T) = P_0(X) + P_1(X)T + \ldots + P_m(X)T^m.$$

It cannot happen that $P_i(1) = 0$ for all $i = 0, \ldots, m$ because then $(X - 1) | P(X, T)$.
Then for $X = 1$ we obtain a relation

$$P_0(1) + P_1(1) \cdot e + \ldots + P_m(1) \cdot e^m = 0, \text{ with } P_0(1), \ldots, P_m(1) \in \mathbb{Q}.$$

But such a relation is impossible.

Notation 1.6.51 For a field k, we will denote by $k\langle X_1, \ldots, X_n \rangle$ the set of algebraic power
series in the variables $X_1, \ldots, X_n$ over k.

Proposition 1.6.52 *Let k be a field. Then $k\langle X_1, \ldots, X_n \rangle$ is a subring of the ring of formal
power series $k[[X_1, \ldots, X_n]]$.*

Proof Let us denote for simplicity $X := (X_1, \ldots, X_n)$. Let $k(X)$ and $k((X))$ be the
fields of fractions of the integral domains $k[X]$ and $k[[X]]$, respectively, and let L be
the algebraic closure of $k(X)$ in $k((X))$. Let $h \in k\langle X \rangle$. Then h is algebraic over $k[X]$;
hence it is algebraic over $k(X)$, that is $h \in L \cap k[[X]]$. Conversely, let $h \in k[[X]] \cap L$
be algebraic over $k(X)$. Therefore there exist $n > 0$ and elements $a_0, \ldots, a_n \in k[X]$,
not all of them 0 and $b_0, \ldots, b_n \in k[X] \setminus \{0\}$, such that $\sum_{j=0}^{n} \frac{a_j}{b_j} h^j = 0$. Multiplying by
$b := b_0 \cdot b_1 \cdot \ldots \cdot b_n \neq 0$ and setting $c_j := \frac{b}{b_j}$, $j = 0, \ldots, n$, we get $\sum_{j=0}^{n} (a_j c_j) h^j = 0$,

that is $h \in k\langle X \rangle$. Finally we have proved that $k\langle X \rangle = k[[X]] \cap L$, which is a subring of $k[[X]]$. □

Remark 1.6.53 The ring $k\langle X \rangle$ is the algebraic closure of $k[X]$ in $k[[X]]$ and also the Henselization of $k[X]_{(X)}$ (cf. [266, ch. VII]).

Theorem 1.6.54 (Grothendieck [134, Th. 18.6.9]) *Let A be a local Noetherian ring and A^h its Henselization. Then, for any $\mathfrak{p} \in \mathrm{Spec}(A)$, we have $A_\mathfrak{p}^h/\mathfrak{p}A_\mathfrak{p}^h \cong \prod_{i=1}^{n} K_i$, where K_i is an algebraic separable extension of $k(\mathfrak{p})$. In particular, the canonical morphism $A \to A^h$ has geometrically regular fibres.*

Proof We know that $A^h = \varinjlim_{i \in I} A_i$, where A_i is a local, essentially étale A-algebra, with the same residue field as A [134, Rém. 18.6.4]. Let $u : A \to A^h$ and $u_i : A_i \to A^h$, $i \in I$, be the canonical morphism. Then u_i is injective, $\forall i \in I$. Moreover,

$$A_\mathfrak{p}^h/\mathfrak{p}A_\mathfrak{p}^h \cong \varinjlim_i (A_i)_\mathfrak{p}/\mathfrak{p}(A_i)_\mathfrak{p}.$$

But A_i is essentially étale over A; hence $(A_i)_\mathfrak{p}/\mathfrak{p}(A_i)_\mathfrak{p}$ is a finite product of fields, finite separable extensions of $k(\mathfrak{p})$. Now we apply Lemma 1.3.14. □

Theorem 1.6.55 (Grothendieck [134, Prop. 18.7.2]) *Let P be a property of Noetherian local rings verifying $(\mathbf{A_1}) - (\mathbf{A_3})$ and $(\mathbf{A_5})$. Let A be a Noetherian local ring and A^h its henselization. Then A is a P-ring if and only if A^h is a P-ring.*

Proof We have

$$A \subseteq A^h \subseteq \widehat{A} \cong \widehat{A^h}.$$

Let $\mathfrak{p} \in \mathrm{Spec}(A)$. Then from Theorem 1.6.54, we get that

$$A_\mathfrak{p}^h/\mathfrak{p}A_\mathfrak{p}^h \cong \prod_{i=1}^{n} K_i,$$

where K_i is an algebraic separable extension of $k(\mathfrak{p})$. Actually, $K_i = k(\mathfrak{q}_i)$, where $\mathfrak{q}_1, \ldots, \mathfrak{q}_n$ are the prime ideals of A^h over $\mathfrak{p}$. It results that the formal fibre of A in $\mathfrak{p}$ is the direct sum of the formal fibres of A^h in $\mathfrak{q}_1, \ldots, \mathfrak{q}_n$. The conclusion follows. □

We close by showing that for some properties **P**, the class of **P**-rings has some special features.

Proposition 1.6.56 *Let P be a property of Noetherian local rings satisfying* $(\mathbf{A_1}) - (\mathbf{A_3})$
and consider the following condition:

$(\mathbf{A_9})$ *For any flat morphism of Noetherian local rings* $u : (R, \mathfrak{m}, k) \to S$, *if S has the
property P, then S/$\mathfrak{m}$S has the property P.*

Let B be a Noetherian local ring with the property P, let $\mathfrak{b}$ *be an ideal of B and let
A := B/$\mathfrak{b}$. Then A is a P-ring.*

Proof We have $\widehat{A} \simeq \widehat{B} \otimes_B A \simeq \widehat{B}/\mathfrak{b}\widehat{B}$; hence the formal fibres of A are exactly the formal
fibres of B in the prime ideals of B containing $\mathfrak{b}$. Let $\mathfrak{p}$ be such a prime ideal, let $S = B \setminus \mathfrak{p}$
and let $C := S^{-1}(\widehat{B}/\mathfrak{b}\widehat{B})$. Let also $\mathfrak{q} \in \mathrm{Spec}(C)$. There exists $\mathfrak{q}' \in \mathrm{Spec}(\widehat{B})$ such that
$\mathfrak{q} = \mathfrak{q}'C$ and $\mathfrak{q}' \cap B = \mathfrak{p}$. Thus we have a local flat morphism $B_{\mathfrak{p}} \to \widehat{B}_{\mathfrak{q}'}$. But $\widehat{B}_{\mathfrak{q}'}$ has the
property $\mathbf{P}$; hence $C_{\mathfrak{q}} \simeq \widehat{B}_{\mathfrak{q}'}/\mathfrak{p}\widehat{B}_{\mathfrak{q}'}$ has also the property $\mathbf{P}$.   □

Corollary 1.6.57 *Let P be one of the properties: complete intersection, Gorenstein and
Cohen-Macaulay. Let A be a Noetherian local ring which is a quotient of a Noetherian
local ring having the property P. Then A is P-ring.*

Proof Indeed the properties $\mathbf{P}$ = complete intersection, $\mathbf{P}$ = Gorenstein and $\mathbf{P}$ = Cohen-
Macaulay satisfy the condition from Proposition 1.6.56 as it follows from [20, Cor. 1.9.2]
and [249, Th. 23.3 and 23.4].   □

Nagata Rings and Reduced Morphisms

2

The origin of the theory of Japanese and Nagata rings is the classical example of Akizuki of a Noetherian integral domain A, whose integral closure is not finite as a module over A. This chapter contains almost all important things about this topic, culminating with the Nagata-Zariski theorem, that makes the connection with the **P**-rings.

2.1   Japanese Rings

As we did several times, we start with an example.

Example 2.1.1 A Noetherian domain A whose integral closure is not A-finite.

Proof Let $(A, \mathfrak{m})$ be the example constructed in Example 1.6.1. Recall that A is a one-dimensional Noetherian local domain whose completion has nilpotent elements. We shall prove that A', the integral closure of A, is not a finite A-module. Suppose that A' is a finite A-module. Then A' is a semilocal Noetherian ring of dimension 1. Let us denote by $\mathfrak{r}$ the Jacobson radical of A'. For any $\mathfrak{n} \in \mathrm{Max}(A')$, the localization $A'_{\mathfrak{n}}$ is a DVR. On the other hand, by [249, Th. 8.15],

$$\widehat{A'} \cong \prod_{\mathfrak{n} \in \mathrm{Max}(A')} \widehat{A'_{\mathfrak{n}}}.$$

This is a finite product of complete DVRs; hence is a reduced ring. On the other hand, by [249, Th. 8.7], we have $\widehat{A'} \cong \widehat{A} \otimes_A A'$. By flatness, $\widehat{A}$ is contained in $\widehat{A} \otimes_A A'$ and therefore is reduced. We have reached a contradiction; hence A' is not a finite A-module.

□

© The Author(s), under exclusive license to Springer Nature Switzerland AG 2023   79
C. Ionescu, *Classes of Good Noetherian Rings*, Frontiers in Mathematics,
https://doi.org/10.1007/978-3-031-22292-4_2

Why is it so important to know if a Noetherian domain has finitely generated integral closure? For many reasons, but just think that normalization is an important tool in Algebraic Geometry and in Algebraic Number Theory. We will study now classes of rings having nice properties from this point of view.

Remark 2.1.2 Remember that for a ring A we denote by A' the integral closure of A in $Q(A)$, the total quotient ring of A. Also, if L is an extension of $Q(A)$, we denote by A'_L the integral closure of A in L.

We start with the following well-known result.

Proposition 2.1.3 *Let A be a Noetherian normal domain with quotient field K and let L be a field, finite separable extension of K. Then A'_L is a finite A-algebra.*

Proof There exists a field M containing L, such that M is a finite Galois extension of K. Changing L with M, we may assume that L is a Galois extension of K. Let $t \in L$ be such that $L = K(t)$ and let G be the Galois group of L over K. Then $\text{ord}(G) = n = [L : K]$. By multiplying with an element from A, we may also assume that $t \in A'_L$. Let $G = \{u_0, \ldots, u_{n-1}\}$ and take an element $x \in A'_L$. Then

$$x = c_0 + c_1 t + \cdots + c_{n-1} t^{n-1}, \ c_j \in K, \ j = 0, \ldots, n-1,$$

hence

$$u_j(x) = c_0 + c_1 u_j(t) + \cdots + c_{n-1}(u_j(t))^{n-1}.$$

Let $d := \det((u_j(t)^i)_{i,j})$. Then $u_j(d^2) = d^2, \ \forall j = 0, \ldots, n-1$. Moreover, $u_j(t)$ are integral over A. It follows that $d^2 \in A'_L \cap K = A$. Since L is separable over K, we have that $d \neq 0$. Consider the matrix $M = (u_j(t)^i)_{i,j}$ and let M_j be the matrix obtained from M by replacing the elements of the i-th column with $u_0(x), \ldots, u_{n-1}(x)$. Let $d_j := \det(M_j)$. Then $d \cdot c_j = d_j$; hence $d \cdot d_j \in B \cap K = A$. It follows that $A'_L \subseteq \frac{1}{d^2} \sum_{i=0}^{n-1} A t^i$, which is a finitely generated A-module. But A is Noetherian; hence also A'_L is a finitely generated A-module. □

Definition 2.1.4 A domain A is called $N - 1$ if the integral closure of A is a finite A-module. The domain A is called **Japanese** if for any field L, finite extension of $Q(A)$, the integral closure of A in L is a finite A-algebra. A ring A is called **universally Japanese** if any finitely generated A-algebra B which is a domain is a Japanese ring.

Note that Definition 2.1.4 is given in a general, not necessarily Noetherian situation, while Proposition 2.1.3 needs Noetherian assumptions. The following result shows that, in the

Noetherian case, that will be our main concern, we can check the defining property of a Japanese ring only for finite purely inseparable field extensions, as Proposition 2.1.3 suggests.

Proposition 2.1.5 *Let A be a Noetherian domain. The following are equivalent:*

(i) *A is Japanese.*

(ii) *For any finite purely inseparable extension L of $K = Q(A)$, A'_L is a finite A-algebra.*

Proof $(i) \Rightarrow (ii)$: Clear.

$(ii) \Rightarrow (i)$: Let L a be a field, finite extension of K. There exists a finite normal extension $K \subset M$ containing L. Then $A'_L \subseteq A'_M$, and since A is Noetherian, it is enough to show that A'_M is a finite A-algebra. Now there exists a subextension $K \subseteq T \subseteq M$ such that T is a finite purely inseparable extension of K and $T \subseteq M$ is separable. By the assumption A'_T is a finite A-algebra and by Proposition 2.1.3, we get that $(A'_T)'_M$ is a finite A'_T-algebra; hence $(A'_T)'_M$ is a finite A-algebra. But $(A'_T)'_M = A'_M$. $\qquad\square$

Corollary 2.1.6 *Let A be a Noetherian domain with perfect fraction field. Then A is Japanese if and only if A' is a finite A-algebra, that is, A is N−1.*

Example 2.1.7

(i) Fields are Japanese rings.

(ii) By Corollary 2.1.6 we see that $\mathbb{Z}$ and $\mathbb{Z}[i]$ are Japanese rings. More generally, any normal domain with perfect fraction field is Japanese.

(iii) The ring A from Example 2.1.1 is not Japanese.

Lemma 2.1.8 *If A is a N−1 ring or a Japanese ring and S is a multiplicatively closed system of A, then $S^{-1}A$ is a N−1 ring, resp. a Japanese ring.*

Proof This follows immediately from the fact that the integral closure commutes with fractions. $\qquad\square$

Proposition 2.1.9 *Let $u : A \to B$ be a finite injective morphism of Noetherian domains. Then A is Japanese if and only if B is Japanese.*

Proof Let $K := Q(A)$, $L := Q(B)$ and let T be a field, finite extension of L. If $C := A'_T$, then $C = B'_T$. If A is Japanese, C is a finite A-algebra and it follows that C is a finite B-algebra. Conversely, suppose that B is Japanese, and let now T be a field, finite extension of K. Let $\mathfrak{p} \in \text{Spec}(T \otimes_K L)$. Then $M := (T \otimes_K L)/\mathfrak{p}$ is a field, finite extension of L. If $C := B'_M$, since B is Japanese, C is a finite extension of B, so that C is finite over A. Let $D := A'_T$. Then $A \subseteq D \subseteq C$ and therefore D is finite over A, because A is Noetherian. $\qquad\square$

Proposition 2.1.10 *Let A be a Noetherian Japanese domain. Then the polynomial ring* $B := A[X_1, \ldots, X_n]$ *is Japanese.*

Proof It suffices to consider the case $n = 1$ and $p = \mathrm{ch}(K) > 0$, where $K := Q(A)$. Let L be a finite purely inseparable extension of $K(X)$. Then there is a natural number s and there are elements $a_1, \ldots, a_m \in A$ such that, if $q := p^s$, then

$$L \subseteq L' := K(a_1^{1/q}, \ldots, a_m^{1/q}, X^{1/q}).$$

Let $C := A[a_1^{\frac{1}{q}}, \ldots, a_m^{\frac{1}{q}}]$. Since A is Japanese and C is a finite A-algebra, by Proposition 2.1.9, it follows that C' is a finite A-algebra. Then $B'' := A'[X^{1/q}]$ is integrally closed and is a finite B-algebra and $B'' = B'_{L'}$. But $B'_L \subseteq B''$ and B is Noetherian, so that B'_L is a finite B-algebra. $\qquad\qquad\square$

Corollary 2.1.11 *If k is a field and A is a domain which is finitely generated over k, then A is Japanese.*

Proof By Noether's normalization lemma, there is a natural number n and a finite injective morphism $k[X_1, \ldots, X_n] \to A$. Now we apply Propositions 2.1.9 and 2.1.10. $\qquad\square$

Example 2.1.12 ([359, Commutative Algebra, Ex. 161.2]) A non-Noetherian Japanese domain.

Proof Let k be a field and $A := k[X_1, X_2, \ldots]$ be the non-Noetherian ring of countably many indeterminates over k. We want to prove that A is a Japanese domain. Let $K := Q(R) = k(X_1, X_2, \ldots)$ be the fraction field of A and let L be a finite field extension of K. Then there is $n \in \mathbb{N}$ and a field M, finite extension of $k(X_1, \ldots, X_n)$ such that $L = M(X_{n+1}, X_{n+2}, \ldots)$. Then $S := k[X_1, \ldots, X_n]'_M$ is a finite $k[X_1, \ldots, X_n]$-algebra by Prop. 2.1.10. The integral closure of A in L is $S[X_{n+1}, X_{n+2}, \ldots] = S \otimes_{k[X_1, \ldots, X_n]} A$; hence is finite over A. $\qquad\qquad\square$

The previous example shows that there are Japanese non-Noetherian domains, so that it makes sense to study also the non-Noetherian situation. However, our main interest being in Noetherian rings we will focus exclusively on the Noetherian situation.

Lemma 2.1.13 *Let A be a ring and $\mathfrak{a}$ be a finitely generated ideal of A. If $A/\mathfrak{a}$ is Noetherian, then $A/\mathfrak{a}^k$ is Noetherian for any $k \in \mathbb{N}$.*

Proof Consider the exact sequence

$$0 \to \mathfrak{a}^k/\mathfrak{a}^{k+1} \to A/\mathfrak{a}^{k+1} \to A/\mathfrak{a}^k \to 0$$

The left-hand term is a finitely generated $A/\mathfrak{a}$-module, so it is a Noetherian A-module. Now we proceed by induction k. □

Lemma 2.1.14 (Nishimura [268, Theorem]) *Let A be a Krull domain and let $y \in A$, $y \neq 0$. Suppose that $A/\mathfrak{p}$ is a Noetherian ring for any prime ideal $\mathfrak{p}$ of A containing y and such that $\mathrm{ht}(\mathfrak{p}) = 1$. Then A is a Noetherian ring.*

Proof We have the decomposition $yA = \mathfrak{p}_1^{(n_1)} \cap \cdots \cap \mathfrak{p}_r^{(n_r)}$, where $\mathfrak{p}_1, \ldots, \mathfrak{p}_r$ are prime ideals of height 1. Then A/yA can be embedded in $A/\mathfrak{p}_1^{(n_1)} \oplus \cdots \oplus A/\mathfrak{p}_r^{(n_r)}$, and by the Eakin-Nagata theorem [249, Th. 3.7], it is enough to prove that $A/\mathfrak{p}^{(n)}$ is Noetherian for any prime ideal $\mathfrak{p}$ of height 1 and for any natural number n. By the approximation theorem for Krull rings [56, ch. 6, §7, n. 2, Th. 1], we can find $x \in K := Q(A)$ such that $\mathfrak{p}^{(n)} = x^n A \cap A, \forall n \in \mathbb{N}$. Let

$$a \in x^n A[x] \cap A, \ a = a_n x^n + \cdots a_{n+t} x^{n+t}, \ a_i \in A.$$

Suppose that we take t minimal with this property. If $t = 0$, then $a \in x^n A$. If $t \neq 0$, multiplying by $b := (a_{n+t})^{n+t-1}$, we obtain that $a_{n+t} x$ is integral over A and therefore $b \in A$. Then we have

$$a = a_n x^n + \cdots + (a_{n+t-1} + b) x^{n+t-1}$$

but this contradicts the minimality of t. Finally we obtain that

$$x^n A[x] \cap A = x^n A \cap A.$$

We have

$$A/\mathfrak{p}^{(n)} = A/(x^n A \cap A) = A/(x^n A[x] \cap A) \cong (A + x^n A[x])/x^n A[x].$$

It follows that the morphism $A/\mathfrak{p}^{(n)} \to A[x]/x^n A[x]$ is finite and injective and $A/\mathfrak{p} \cong A[x]/x A[x]$. Then $A[x]/x A[x]$ is Noetherian, and by Lemma 2.1.13 we have that $A[x]/x^n A[x]$ is Noetherian. Applying again the Eakin-Nagata theorem, it follows that $A/\mathfrak{p}^{(n)}$ is Noetherian. □

Definition 2.1.15 If A is a Noetherian ring and $\mathfrak{a}$ is an ideal of A, an element $x \in A$ is called **integral** over $\mathfrak{a}$ if there exist $n \in \mathbb{N}$ and $a_0, \ldots, a_n \in A$ such that $a_i \in \mathfrak{a}^i$, $i = 0, \ldots, n$ and $x_n + a_1 x^{n-1} + \cdots + a_n = 0$. The set of elements which are integral over $\mathfrak{a}$ is denoted by $\bar{\mathfrak{a}}$ and is called the **integral closure of the ideal** $\mathfrak{a}$.

Remark 2.1.16 For more properties about the integral closure of an ideal, one can see the beautiful monograph [362] by I. Swanson and C. Huneke. In particular, it is shown in [362, Rem. 1.1.3] that $\bar{\mathfrak{a}}$ is an ideal containing $\mathfrak{a}$ and contained in the radical of $\mathfrak{a}$.

Remark 2.1.17 It is easy to see [362, Prop. 1.1.5] that an element $x \in A$ is integral over an ideal $\mathfrak{a}$ if and only if for any $\mathfrak{p} \in \mathrm{Min}(A)$, the image of x in $A/\mathfrak{p}$ is integral over $\mathfrak{a} + \mathfrak{p}/\mathfrak{p}$.

Remark 2.1.18 It is well-known that if A is a Noetherian ring and $\mathfrak{a}$ is an ideal of A, the sets $\{\mathrm{Ass}(A/\mathfrak{a}^n) \mid n \in \mathbb{N}\}$ and $\{\mathrm{Ass}(A/\overline{\mathfrak{a}^n}) \mid n \in \mathbb{N}\}$ are eventually stabilizing to (finite) sets denoted by $A(\mathfrak{a})$ and resp. $A_a(\mathfrak{a})$ (see [257] and [258]). Moreover if A is any Noetherian ring, we have that:

(i) For any ideal $\mathfrak{a}$ of A, $A_a(\mathfrak{a}) \subseteq A(\mathfrak{a})$ [257, Prop. 3.17].
(ii) If $x \in A$ is a non-zero-divisor, $A_a(xA) \subseteq A(xA) = \mathrm{Ass}(A/xA)$ [257, Prop. 3.17].

Let A be a Noetherian ring and $\mathfrak{a}$ an ideal of A. We shall use other two sets of prime ideals connected with $\mathfrak{a}$. For properties and interesting results about these sets of prime ideals, the reader can see the nice textbooks [257] and [258] by S. McAdam.

Definition 2.1.19 Let A be a Noetherian ring and $\mathfrak{a}$ an ideal of A. Then $Q(\mathfrak{a}) = \{\mathfrak{p} \in \mathrm{Spec}(A) \mid \mathfrak{a} \subseteq \mathfrak{p}$ and there exists $\mathfrak{q} \in \mathrm{Spec}(\widehat{A_\mathfrak{p}})$ such that $\mathfrak{p}\widehat{A_\mathfrak{p}} \in \mathrm{Min}(\mathfrak{a}\widehat{A_\mathfrak{p}} + \mathfrak{q})\}$ are called the **quintasymptotic primes** of $\mathfrak{a}$, while $E(\mathfrak{a}) = \{\mathfrak{p} \cap A \mid \mathfrak{p} \in Q(X^{-1}A[\mathfrak{a}X, X^{-1}])\}$ are called the **essential primes** of $\mathfrak{a}$.

Remark 2.1.20 From [258, Lemma 1.1], it follows that if A is any Noetherian ring and $\mathfrak{a}$ is any ideal of A, we have:

(i)

$$\mathrm{Min}(\mathfrak{a}) \subseteq E(\mathfrak{a}) \subseteq A(\mathfrak{a}) \text{ and } A_a(\mathfrak{a}) \cup Q(\mathfrak{a}) \subseteq E(\mathfrak{a}).$$

(ii) $\mathfrak{p} \in E(\mathfrak{a})$ if and only if $\exists\, \mathfrak{q} \in \mathrm{Ass}(A)$ such that $\mathfrak{p}/\mathfrak{q} \in E((\mathfrak{a} + \mathfrak{q})/\mathfrak{q})$.

We shall also use the following fact:

Proposition 2.1.21 ([257, Prop. 1.9]) *Let A be a Noetherian domain, K its field of quotients, L a field, finite extension of K and $\mathfrak{a}$ a finitely generated ideal of A'_L. Then the set $\mathrm{Min}(\mathfrak{a})$ is finite.*

Proof Let us first assume that $L = K$, that is, $B = A'$. Let $\mathfrak{a} = (b_1, \ldots, b_m)$, $b_i \in A'$, $S := A[b_1, \ldots, b_m]$, $\mathfrak{b} = \mathfrak{a}S$. Then S is a Noetherian ring. Let $\mathfrak{q} \in \mathrm{Min}(\mathfrak{a})$. By [257, Cor. 1.8], $\mathfrak{q} \cap A \in A^*(\mathfrak{b})$. But $A^*(\mathfrak{b})$ is a finite set, and only finitely many prime ideals of B lie over a given prime ideal of S; hence $\mathrm{Min}(\mathfrak{a})$ is finite. In the general situation, let B be a finite A-algebra with field of quotients L. Then B is a Noetherian integral domain, and its integral closure is A'_L. Now we apply the first case. $\qquad\square$

Lemma 2.1.22 *Let $f : A \to B$ be a faithfully flat morphism of commutative rings. Assume that A is a domain with field of fractions K. Let $C := B_{\mathrm{red}}$ and let $R = Q(C)$ be the total ring of fractions of C. Then:*

 (i) *The canonical morphism $u : A \to C$ is injective.*
 (ii) *u can be extended to an injective morphism $K \to R$.*
 (iii) *If C' is the integral closure of C in R, then $C' \cap K$ is the integral closure of A.*

Proof

 (i) Let $\mathfrak{n}$ be the nilradical of B so that obviously $\mathfrak{n} \cap A = (0)$. Since $C = B/\mathfrak{n}$, it follows that u is injective.
 (ii) Let $a \in A$, $a \neq 0$. By flatness a is not a zero-divisor in B. Assume that $ax \in \mathfrak{n}$, $x \notin \mathfrak{n}$. Then there is $n \in \mathbb{N}$ such that $a^n x^n = 0$. Since a is not a zero-divisor, we get that $x^n = 0$. Contradiction! It follows that we can extend u to a morphism $K \to R$, which is clearly injective.
 (iii) If A' is the integral closure of A, it is clear that $A' \subseteq C' \cap K$. Conversely, let $x \in C' \cap K$. Then x is integral over C and consequently also integral over B; hence $B[x]$ is a finite B-algebra. Since K is a subring of $B \otimes_A K$, by flatness we see that $B[x]$ is a subring of $B \otimes_A A[x]$. Applying [56, Ch. I, §3, no. 6, Prop. 11], it results that $A[x]$ is an A-module of finite type, so that $x \in A'$. $\qquad\square$

Definition 2.1.23 A morphism of Noetherian rings $u : A \to B$ is called **radicial** if it satisfies the following two conditions:

 (i) The induced morphism $\mathrm{Spec}(u) : \mathrm{Spec}(B) \to \mathrm{Spec}(A)$ is injective.
 (ii) For any $\mathfrak{q} \in \mathrm{Spec}(B)$, the residue field extension $k(\mathfrak{q} \cap A) \subseteq k(\mathfrak{q})$ is purely inseparable.

Remark 2.1.24

 (i) Let $u : A \to B$ be a morphism of Noetherian rings. Then u is a radicial morphism if and only if for any $\mathfrak{p} \in \mathrm{Spec}(A)$, we have that $\mathrm{Spec}(k(\mathfrak{p}) \otimes_A B)$ has at most one element and if $\mathrm{Spec}(k(\mathfrak{p}) \otimes_A B) = \{\mathfrak{q}\}$, then the field extension $k(\mathfrak{p}) \subseteq k(\mathfrak{q})$ is purely inseparable.
 (ii) Essentially the notion of radicial morphism is geometric (see [127, Déf. 3.5.4]), but for our purposes, the above one will be sufficient.

Definition 2.1.25 ([131, 23.2.1]) A local ring A is called **unibranched** if A_{red} is a domain and the integral closure of A_{red} is a local ring. The local ring A is called **geometrically unibranched** if A is unibranched and the field extension $k \subseteq K$ is purely inseparable, where K is the residue field of the integral closure of A_{red}.

Example 2.1.26

  (i) Clearly any integrally closed local domain is unibranched.
 (ii) Let A be a local domain, $K = Q(A)$. Then A is unibranched if and only if any subring C of K which is a finite A-algebra is local. In fact if A is unibranched, for every maximal ideal $\mathfrak{m}$ of C, there exists a maximal ideal $\mathfrak{n}$ of A' such that $\mathfrak{n} \cap C = \mathfrak{m}$. Thus C is local, since A' is local. Conversely, A' is the direct limit of finite A-subalgberas of A'. By assumption these rings are local. It follows easily that also A' is local.
(iii) If A is a local Noetherian domain such that $\widehat{A}$ is a domain, then A is unibranched. To see this, let $K = Q(A)$ and $L = Q(\widehat{A})$ and let $\mathfrak{m}$ be the maximal ideal of A. We have $L = K \otimes_A \widehat{A}$. Let C be a finite A-subalgebra of K. Then $C \otimes_A \widehat{A}$ is a finite $\widehat{A}$-algebra and is the completion of C in the $\mathfrak{m}$-adic topology. But C is a semilocal ring and $\widehat{C}$ is a domain; hence C is local and the maximal ideal of C lies over $\mathfrak{m}$.

Proposition 2.1.27 ([131, Lemma 23.2.2]) *Let A be an integral domain, A' its integral closure and $f : A \to A'$ the canonical morphism. Then f is a radicial morphism if and only if for any $\mathfrak{p} \in \operatorname{Spec}(A)$ the local ring $A_{\mathfrak{p}}$ is geometrically unibranched. If this is the case, then $\operatorname{Spec}(f)$ is a homeomorphism.*

Proof Clearly $(A')_{\mathfrak{p}} = (A_{\mathfrak{p}})'$ for any $\mathfrak{p} \in \operatorname{Spec}(A)$, and any prime ideal of $(A')_{\mathfrak{p}}$ lying over $\mathfrak{p}A_{\mathfrak{p}}$ is maximal. Obviously $\operatorname{Spec}(f)$ is injective if $(A')_{\mathfrak{p}}$ is a local ring and if $A_{\mathfrak{p}}$ is unibranched. Furthermore, f is radicial in $\mathfrak{p}$ if $A_{\mathfrak{p}}$ is geometrically unibranched. Clearly $\operatorname{Spec}(f)$ is injective and closed, hence a homeomorphism.   □

Lemma 2.1.28 (Marot [237, Lemma 1.3]) *Let A be a Noetherian domain with field of fractions K, L a field, finite extension of K and $\mathfrak{a}$ an ideal of A. Then the A-module A'_L is separated in the $\mathfrak{a}$-adic topology.*

Proof We may assume that A is a local ring and $\mathfrak{a} := \mathfrak{m}$ is the maximal ideal. Then from [131, Prop. 23.2.7 and Rem. 23.2.8,ii)], it follows that A'_L is a semilocal ring. If $\mathfrak{m}'$ is the Jacobson radical of A'_L, since $\mathfrak{m}A'_L \subseteq \mathfrak{m}'$, it is enough to show that A'_L is $\mathfrak{m}'$-adically separated. From [131, Prop. 23.2.4], we get that there exists B, a finite sub-A-algebra of A'_L with field of fractions L, such that $(A'_L \otimes_A \widehat{B})_{red} \simeq (\widehat{B})_{red}$ and $B \to A'_L$ is a radicial morphism. Let $\mathfrak{n}$ be the Jacobson radical of the semilocal ring B and let C be the integral closure of $(\widehat{B})_{red}$ in its total quotient ring. From Lemma 2.1.22, it follows that the morphism $f : B \to (\widehat{B})_{red}$ can be extended to an injective morphism $g : A'_L \to C$.

Consider the commutative diagram

$$
\begin{array}{ccc}
B & \xrightarrow{\ f\ } & (\widehat{B})_{\mathrm{red}} \\
{\scriptstyle i}\downarrow & & \downarrow{\scriptstyle j} \\
A'_L & \xrightarrow{\ g\ } & C.
\end{array}
$$

Then g factors as follows:

$$
A'_L \xrightarrow{\ h\ } (A'_L \otimes_A \widehat{B})_{\mathrm{red}} \simeq (\widehat{B})_{\mathrm{red}} \xrightarrow{\ j\ } C.
$$

Since g is injective, it follows that h is injective. Let $X := \mathrm{Max}(B)$, $X' := \mathrm{Max}(A'_L)$ and $Y := \mathrm{Max}((\widehat{B})_{\mathrm{red}})$. It is known that f induces a bijection between Y and X [56, ch. 3, §3, n. 4, Prop. 8]. Also, by Proposition 2.1.27 the radicial morphism i induces a bijection between X' and X. It follows that h induces a bijection between Y and X' and so $\mathfrak{m}' = h^{-1}(\mathfrak{n})$. But h is an injection, and $(\widehat{B})_{\mathrm{red}}$ is a Noetherian ring, so that we get the conclusion. $\qquad\square$

Theorem 2.1.29 ([175]) *Let A be a Noetherian integral domain, $x \in A$, $x \neq 0$. Suppose that:*

(i) A is xA-adically complete.

(ii) $A/\mathfrak{p}$ is Japanese, for any prime ideal $\mathfrak{p} \in A_a(xA)$.

Then A is Japanese.

Proof Let $K := Q(A)$ and L a field, finite extension of K. Let $\mathfrak{q}_1, \ldots, \mathfrak{q}_n$ be the minimal prime over-ideals of xA'_L (cf. Proposition 2.1.21) and let $\mathfrak{p}_i := \mathfrak{q}_i \cap A$ for $i = 1, \ldots, n$. Then by [257, Prop. 3.5], it follows that $\mathfrak{p}_i \in A_a(xA)$ and moreover $[k(\mathfrak{q}_i) : k(\mathfrak{p}_i)] < \infty$. It follows from (ii) that $A'_L/\mathfrak{q}_i$ is Noetherian for all $i = 1, \ldots, n$. By Lemma 2.1.14 also A'_L/xA'_L is Noetherian. Let

$$
\mathfrak{a} := \mathfrak{q}_1 \cap \cdots \cap \mathfrak{q}_n = \sqrt{xA'}.
$$

Since A'_L/xA'_L is Noetherian, there exists $r \in \mathbb{N}$ such that $\mathfrak{a}^r \subseteq xA'_L$. From the exact sequence

$$
0 \to \mathfrak{a}^n/\mathfrak{a}^{n+1} \to A'_L/\mathfrak{a}^{n+1} \to A'_L/\mathfrak{a}^n \to 0
$$

we get by induction that $A'_L/\mathfrak{a}^n$ is a finite A-algebra, $\forall n \in \mathbb{N}$. Then A'_L/xA'_L is a finite A-algebra. But A'_L is separated in the xA'-adic topology by Lemma 2.1.28, and from [249, Th. 8.4], it follows that A'_L is a finite A-algebra. □

Corollary 2.1.30 (Chiriacescu [82]) *Let A be a Noetherian integral domain, $x \in A$, $x \neq 0$. Suppose that:*

(i) A is xA-adically complete.
(ii) $A/\mathfrak{p}$ is Japanese, for any prime ideal $\mathfrak{p} \in \text{Ass}(A/xA)$.

Then A is Japanese.

Proof Follows from Theorem 2.1.29 and Remark 2.1.18, (ii). □

Corollary 2.1.31 *Let A be a Noetherian ring, $x \in A$, $x \neq 0$. Suppose that:*

(i) A is xA-adically complete.
(ii) $A/\mathfrak{p}$ is Japanese, for any prime ideal $\mathfrak{p} \in \text{Ass}(A/xA)$.

Then $A/\mathfrak{q}$ is Japanese for any $\mathfrak{q} \in \text{Min}(A)$.

Proof Let $\mathfrak{q} \in \text{Min}(A)$, $B := A/\mathfrak{q}$, and take $\mathfrak{p}' \in A_a(xA + \mathfrak{q}/\mathfrak{q})$. Then by [257, Prop. 3.18], it follows that $\mathfrak{p}' = \mathfrak{p}/\mathfrak{q}$, where $\mathfrak{p} \in A_a(xA)$. From ii) we get that $B/\mathfrak{p}'$ is Japanese, and we apply 2.1.29. □

Corollary 2.1.32 *Let A be a Noetherian integral domain, $x \in A$, $x \neq 0$. Suppose that:*

(i) A is xA-adically complete.
(ii) $A/\mathfrak{p}$ is Japanese, for any prime ideal $\mathfrak{p} \in E(xA)$.

Then A is Japanese.

Proof Clear from 2.1.29 and 2.1.20. □

The condition (ii) in Theorem 2.1.29 is really weaker than the same one in Corollary 2.1.30. In order to show this, we need some preparation.

Lemma 2.1.33 *Let B be a Noetherian ring and $\mathfrak{m}_1, \mathfrak{m}_2 \in \text{Max}(B)$. Let $p_i : B \to B/\mathfrak{m}_i, i = 1, 2$ be the canonical surjections. Assume that there exists an isomorphism $f : B/\mathfrak{m}_1 \to B/\mathfrak{m}_2$ and let $A := \{b \in B \mid f(p_1(b)) = p_2(b)\}$. Then:*

(i) $\text{Ann}_A(B/A) = \mathfrak{m}_1 \cap \mathfrak{m}_2$.
(ii) B is a finite A-algebra.

Proof

(i) If $\alpha \in \mathfrak{m}_1 \cap A$, then $p_1(\alpha) = 0$; hence $p_2(\alpha) = f(p_1(\alpha)) = 0$, that is, $\alpha \in \mathfrak{m}_2 \cap A$. If $\alpha \in \mathfrak{m}_2 \cap A$, then $p_2(\alpha) = 0$ and $f(p_1(\alpha)) = p_2(\alpha) = 0$, that is, $\alpha \in \mathfrak{m}_1 \cap A$. Moreover, if $\alpha \in \mathfrak{m}_1 \cap \mathfrak{m}_2$, then $p_1(\alpha) = p_2(\alpha) = 0$; hence $\alpha \in A$. Therefore we have $\mathfrak{m}_1 \cap A = \mathfrak{m}_2 \cap A = \mathfrak{m}_1 \cap \mathfrak{m}_2 := \mathfrak{m}$. Let $b \in \mathfrak{m}_1 \setminus \mathfrak{m}_2$ and $x \in \mathfrak{b} := \mathrm{Ann}_A(B/A)$. Then $f(p_1(xb)) = p_2(xb)$. But $b \in \mathfrak{m}_1$; hence $xb \in \mathfrak{m}_1 \cap A \subseteq \mathfrak{m}_2$ and $x \in \mathfrak{m}$. Thus $\mathfrak{b} \subseteq \mathfrak{m}$. The other inclusion is obvious.

(ii) Let $g_i : A/\mathfrak{m} \to B/\mathfrak{m}_i$ and $h_i : B/\mathfrak{m} \to B/\mathfrak{m}_i, i = 1, 2$ be the canonical morphisms. Let $x \in B/\mathfrak{m}$ and $y = f(x)$. Since $B/\mathfrak{m} \cong B/\mathfrak{m}_1 \times B/\mathfrak{m}_2$, there exists an element $z \in B/\mathfrak{m}$ such that $h_1(z) = x$ and $h_2(z) = y$. By the definition of A, we get $z \in A/\mathfrak{m}$ and $g_1(z) = x$. Hence g_1 is surjective. In the same way, it follows that g_2 is surjective. Therefore B is a finite A-algebra. $\qquad\square$

Example 2.1.34 There exists a Noetherian domain A containing an element $x \in A$ such that $A_a(xA) \subsetneq \mathrm{Ass}(A/xA)$ [258, Ex. 2.1].

Proof Indeed, let K be a field and let $B := S^{-1}K[X, Y]$, where S is the multiplicative system $K[X, Y] \setminus \{(X, Y) \cup (X, Y + 1)\}$. Set $\mathfrak{n}_1 := (X, Y)B$ and $\mathfrak{n}_2 := (X, Y + 1)B$. Thus B is a semilocal Noetherian domain having exactly two maximal ideals, namely, $\mathfrak{n}_1$ and $\mathfrak{n}_2$. Let $A := K + \mathfrak{n}_1 \cap \mathfrak{n}_2$ be the ring constructed according to Lemma 2.1.33. Then A is a Noetherian local domain of dimension 2 with maximal ideal $\mathfrak{m} = \mathfrak{n}_1 \cap \mathfrak{n}_2$. Let $\mathfrak{a} := XA$ and $\mathfrak{p} := XB \cap A$. As $\mathfrak{p}$ is a minimal prime over-ideal of $\mathfrak{a}$, obviously $\mathfrak{p} \in A(\mathfrak{a})$. Since $YX \in \mathfrak{m}$ but $Y \notin A$, we have that $YX \in A \setminus XA$. Because $\mathfrak{m}Y \subseteq \mathfrak{m}$, we have that $\mathfrak{m}YX \subseteq XA$. Hence we see that $\mathfrak{m} = (XA : XYA) \in \mathrm{Ass}_A(A/\mathfrak{a}) = A(XA)$; therefore $\mathrm{Ass}_A(A/XA) = \{\mathfrak{p}, \mathfrak{m}\}$. If $\mathfrak{m} \in A_a(\mathfrak{a})$, then by [258, Prop. 1.13,c)] one of the ideals $\mathfrak{n}_1$ or $\mathfrak{n}_2$ would be in $A_a(XB)$. But by [257, Lemma 3.14], this means that these ideals are of height 1 and this is not the case. $\qquad\square$

If A is not complete in the xA-adic topology, we can obtain some results similar to Theorem 2.1.29 and its corrolaries, but we have to put conditions on the fibres of the morphism $A \to \widehat{A}$, where $\widehat{A}$ is the completion of A in the xA-adic topology. The following Lemma is well-known.

Lemma 2.1.35 *Let A be a (not necessarily Noetherian) reduced ring with finitely many minimal prime ideals $\mathfrak{p}_1, \ldots, \mathfrak{p}_n$. Let $R := Q(A)$ be the total quotient ring and for any $i = 1, \ldots, n$ let $A_i := A/\mathfrak{p}_i$ and $L_i := Q(A_i)$. Then:*

(i) $R = L_1 \times \cdots \times L_n$.
(ii) $A' = A'_1 \times \cdots \times A'_n$.

(iii) If moreover $A_\mathfrak{p}$ is a domain for every $\mathfrak{p} \in \mathrm{Spec}(A)$, then $A = A_1 \times \cdots \times A_n$.
(iv) A is normal if and only if $A = A'$.

Proof

(i) Let $S := A \setminus \bigcup\limits_{i=1}^{n} \mathfrak{p}_i$ and $\mathfrak{m}_i := \mathfrak{p}_i R$, $i = 1, \ldots, n$. Then $R = S^{-1}A$ and $\mathrm{Spec}(R) = \{\mathfrak{m}_1, \ldots, \mathfrak{m}_n\}$. Also $\mathfrak{m}_i \neq \mathfrak{m}_j$ for $i \neq j$ and $\bigcap\limits_{i=1}^{n} \mathfrak{m}_i = 0$. Moreover $R/\mathfrak{m}_i = R_{\mathfrak{m}_i} = L_i$, for any $i = 1, \ldots, n$, and the canonical morphism $f : R \to \prod\limits_{i=1}^{n} R/\mathfrak{m}_i$ is injective. Let $\mathfrak{n}_i := \prod\limits_{j \neq i} \mathfrak{m}_j$. If $\mathfrak{n} := \mathfrak{n}_1 + \cdots + \mathfrak{n}_n$, then $\mathfrak{n}$ is not contained in $\mathfrak{n}_j$, for any $j = 1, \ldots, n$. It follows that $\mathfrak{n} = R$ and that there are elements $x_i \in \mathfrak{n}_i$ for each $i = 1, \ldots, n$ such that $x_1 + \cdots + x_n = 1$. This means that $x_1 - 1 \in \mathfrak{m}_i$ and $x_j \in \mathfrak{m}_j$ for $i \neq j$, and consequently f is surjective.

(ii) Follows from (i).

(iii) Let $g : A \to A_1 \times \ldots \times A_n$ be the canonical morphism and let $\mathfrak{n} \in \mathrm{Max}(A)$. Then $\mathfrak{n}$ contains exactly one minimal prime $\mathfrak{p}_j$ and the induced morphism

$$g_\mathfrak{n} : A_\mathfrak{n} \to (A_1 \times \ldots \times A_n)_\mathfrak{n} \cong (A_j)_\mathfrak{n}$$

is an isomorphism. It results that g is an isomorphism.

(iv) If $A = A'$, from (ii) it follows that A is normal. The converse follows immediately from (iii). $\square$

Lemma 2.1.36 *Let $u : A \to B$ be a faithfully flat morphism of Noetherian reduced rings such that B' is finite over B. Then A' is finite over A.*

Proof Since u is flat we clearly have

$$Q(A) \subseteq Q(B) \text{ and } B \subseteq B \otimes_A A' \subseteq B' \subseteq Q(B).$$

Since B is Noetherian, $B \otimes_A A'$ is a finitely generated B-module. It follows that there are elements $a_1, \ldots, a_m \in A'$ such that $B \otimes_A A' = \sum\limits_{i=1}^{m} a_i B$, that is, $(A'/\sum\limits_{i=1}^{m} a_i A') \otimes_A B = 0$. By faithful flatness it results that $A'/\sum\limits_{i=1}^{m} a_i A' = 0$. $\square$

Theorem 2.1.37 *Let A be a Noetherian reduced ring, and consider an element $x \in J(A)$, $x \neq 0$. Let $\widehat{A}$ be the completion of A in the xA-adic topology. Suppose that:*

(i) *The fibres of the morphism $A \to \widehat{A}$ in the minimal prime ideals of A are geometrically reduced.*

(ii) *$A/\mathfrak{p}$ is Japanese for any $\mathfrak{p} \in A_a(xA)$.*

Then $A/\mathfrak{q}$ is Japanese for any $\mathfrak{q} \in \mathrm{Min}(A)$.

Proof Let $\mathfrak{q} \in \mathrm{Min}(A)$, let $B := A/\mathfrak{q}$ and $\mathfrak{p}' \in A_a(xB)$. Then $\mathfrak{p}' = \mathfrak{p}/\mathfrak{q}$ for some prime ideal $\mathfrak{p} \in A_a(xA)$. Assumption (ii) says that $B/\mathfrak{p}'$ is a Japanese ring. Let L be a field, finite extension of $Q(B)$. Suppose that we have shown that $\mathfrak{a} := \sqrt{xB_L'}$ is finitely generated. Then there is $r \in \mathbb{N}$ such that $\mathfrak{a}^r \subseteq xB_L'$. From the exact sequence

$$0 \to \mathfrak{a}^n/\mathfrak{a}^{n+1} \to B_L'/\mathfrak{a}^{n+1} \to B_L'/\mathfrak{a}^n \to 0$$

we get by induction that $B_L'/\mathfrak{a}^r$ is a finite B/xB-algebra. It follows that B_L'/xB_L' is finite over B/xB, and it results that $\widehat{B} \otimes_A B_L'/x(\widehat{B} \otimes_A B_L')$ is finite over $\widehat{B}$, where $\widehat{B}$ is the completion of B in the xB-adic topology. Let B'' be the integral closure of $\widehat{B}$ in $Q(\widehat{B} \otimes_A L)$. Let $\mathfrak{q}_1, \ldots, \mathfrak{q}_n$ be the minimal prime ideals of the reduced ring $\widehat{B} \otimes_A L$. Then by Lemma 2.1.35, we have $B'' = B_1'' \times \cdots \times B_n''$, where B_i'' is the integral closure of $\widehat{B}$ in $k(\mathfrak{q}_i)$, $i = 1, \ldots, n$. Each B_i'' is xA-adically separated, so that B'' is xA-adically separated. On the other hand, we have inclusions

$$\widehat{B} \hookrightarrow \widehat{B} \otimes_A B_L' \hookrightarrow B''$$

so that $\widehat{B} \otimes_A B_L'$ is xA-adically separated. It follows that $\widehat{B} \otimes_A B_L'$ is finite over $\widehat{B}$ and by Lemma 2.1.36 it results that B_L' is finite over B. So it remains to show that $\sqrt{xB_L'}$ is finitely generated. Let $\mathfrak{q} \in \mathrm{Min}(xB_L')$. It follows that $\mathfrak{q} \cap B \in A_a(xB)$ and by (ii) we get that $B/\mathfrak{q} \cap B$ is Japanese. Then $B_L'/\mathfrak{q}$ is finite over $B/\mathfrak{q} \cap B$, so it is a Noetherian ring. Now, by Lemma 2.1.14, we obtain that B_L'/xB_L' is Noetherian and consequently $\sqrt{xB_L'}$ is finitely generated. $\qquad\square$

The above results lead to the following question asked by Chiriacescu [82]:

Question 2.1.38 Let A be a Noetherian domain and $\mathfrak{a}$ an ideal of A. Suppose that:

(i) *A is $\mathfrak{a}$-adically complete.*

(ii) *$A/\mathfrak{p}$ is Japanese for any $\mathfrak{p} \in \mathrm{Ass}(A/\mathfrak{a})$.*

Does it follow that A is Japanese?

We give here a partial answer to this question, obtained in [175] and [176], relying heavily on Theorem 2.1.29.

Lemma 2.1.39 *Let A be a Noetherian ring and $x \in A$. Suppose that:*

(i) A is xA-adically complete.
(ii) $A/\mathfrak{p}$ is Japanese, for any prime ideal $\mathfrak{p} \in E(xA)$.

Then $A/\mathfrak{q}$ is Japanese, for all $\mathfrak{q} \in \mathrm{Ass}(A)$.

Proof Let $\mathfrak{q} \in \mathrm{Ass}(A)$ and let $\mathfrak{p}' \in E((xA + \mathfrak{q})/\mathfrak{q})$. Then $\mathfrak{p}' = \mathfrak{p}/\mathfrak{q}$, where $\mathfrak{p}$ is a prime ideal containing $\mathfrak{q}$. From Remark 2.1.20 it follows that $\mathfrak{p} \in E(xA)$. Let $B := A/\mathfrak{q}$. Then $B/\mathfrak{p}'$ is Japanese by assumption, and by Corollary 2.1.32 it follows that B is Japanese. □

Corollary 2.1.40 *Let A be a Noetherian ring and $x \in A$ be a non-zero-divisor. Suppose that:*

(i) A is xA-adically complete.
(ii) $A/\mathfrak{p}$ is Japanese, for any prime ideal $\mathfrak{p} \in \mathrm{Ass}_A(A/xA)$.

Then $A/\mathfrak{q}$ is Japanese, for all $\mathfrak{q} \in \mathrm{Ass}(A)$.

Proof As x is a non-zero-divisor, it follows that

$$\mathrm{Ass}_A(A/xA) = A_a(xA) \supseteq E(xA).$$

Now we apply Lemma 2.1.39.   □

Theorem 2.1.41 ([175]) *Let A be a Noetherian domain and $\mathfrak{a}$ an ideal of A. Suppose that:*

(i) A is $\mathfrak{a}$-adically complete.
(ii) $A/\mathfrak{a}$ is Japanese for any $\mathfrak{p} \in \mathrm{Ass}(A/\mathfrak{a})$.
(iii) $\mathfrak{a}$ is generated by a regular sequence.

Then A is Japanese.

Proof Let $\mathfrak{a} = (x_1, \ldots, x_n)$. We argue by induction on n. For $n = 1$ we apply Corollary 2.1.30. Let now $\mathfrak{b} := (x_1, \ldots, x_{n-1})$, let $B := A/\mathfrak{b}$ and let $\mathfrak{q} \in \mathrm{Ass}(B/x_n B)$. Then by assumption $B/\mathfrak{q}$ is Japanese. As $\mathfrak{a}$ is generated by a regular sequence, x_n is not a zero-divisor in B, so that by Cor. 2.1.40 we obtain that $B/\mathfrak{r}$ is Japanese, for any $\mathfrak{r} \in \mathrm{Ass}(B)$. By the induction hypothesis, it follows that A is Japanese. □

For an application of Theorem 2.1.41, we use the notion of restricted power series over a ring A, for which we refer to [56, ch. III, §4, no. 2].

Definition 2.1.42 Let A be a commutative topological ring with a linear topology. We say that a formal power series $f = \sum\limits_{i_1,\ldots,i_p} c_{i_1\ldots i_p} T_1^{i_1} \ldots T_p^{i_p}$ is **restricted** if for any V, neighbourhood of 0 in A, there are only finitely many coefficients $c_{i_1,\ldots,i_p}$ of f outside V.

In other words, a restricted power series is a formal series in which the coefficients tend to zero in the topology of A. It is shown in [56, ch. III, §4] that the restricted power series form a ring. In the case that the topology of A is $\mathfrak{a}$-adic, where $\mathfrak{a}$ is an ideal of A, this ring is denoted by $A_{\mathfrak{a}}\{T_1 \ldots, T_p\}$.

Example 2.1.43

(i) If the topology of A is the discrete topology, the restricted power series are exactly the polynomials.

(ii) [56, ch. II, §4, no 2, Prop. 3] If A is separated and complete in the linear topology and $\{\mathfrak{a}_i\}_i$ is a basis of neighbourhoods of 0, then $A\{T_1, \ldots, T_p\}$ is the completion of $A[T_1, \ldots, T_p]$ in the topology given by $\{\mathfrak{a}_i A[T_1, \ldots, T_p]\}_i$.

(iii) If $(K, |\cdot|)$ is a field with a complete non-Archimedean absolute value that is non-trivial and A is the valuation ring, one can consider the topology on A given by the valuation and consequently one can define the restricted power series ring $A\{T_1, \ldots, T_p\}$. Then the ring

$$T_p(K) := K\{T_1, \ldots, T_p\} = A\{T_1, \ldots, T_p\} \otimes_A K$$

is the ring of all power series $\sum\limits_{i_1\ldots i_p} c_{i1\ldots i_p} T_1^{i_1} \ldots T_p^{i_p} \in K[[T_1, \ldots, T_p]]$ such that $|c_{i_1\ldots i_p}|$ tends to 0, when $i_1 + \ldots + i_p$ tend to ∞. This ring is called a **Tate algebra**. This is an important object in the theory of rigid analytic geometry and is also called the **ring of restricted or strictly convergent power series over K** (see [52]).

Corollary 2.1.44 *Let A be a Noetherian domain and $\mathfrak{a}$ an ideal of A. Suppose that:*

(i) A is $\mathfrak{a}$-adically complete.

(ii) $A/\mathfrak{p}$ is Japanese for any $\mathfrak{p} \in \mathrm{Ass}(A/\mathfrak{a})$.

(iii) $\mathfrak{a}$ is generated by a regular sequence.

Then the ring of restricted power series $A_{\mathfrak{a}}\{X\}$ is Japanese.

Proof Let $B := A_{\mathfrak{a}}\{X\}$. Then B is $\mathfrak{a}B$-adically complete and $B/\mathfrak{a}B \cong (A/\mathfrak{a})[X]$. Let $\mathfrak{q} \in \mathrm{Ass}(B/\mathfrak{a}B)$. It follows that $\mathfrak{q} \cap A/\mathfrak{a} = \mathfrak{p}$, where $\mathfrak{p} \in \mathrm{Ass}(A/\mathfrak{a})$. So $A/\mathfrak{p}$ is Japanese and $\mathfrak{q} = \mathfrak{p}(A/\mathfrak{a})[X]$, so that $((A/\mathfrak{a})[X])/\mathfrak{q} = (A/\mathfrak{p})[X]$ is a Japanese ring. Now we apply Theorem 2.1.41. $\square$

2.2   Nagata Rings

Our next aim is to study the class of Nagata rings, also called pseudo-geometric [266]. This class of rings is built from the class of Japanese rings, in such a way that it is stable to the main algebraic operations, especially to extensions of finite type. It also turns out that the class of Nagata rings can be characterized via the fibres of morphisms, thus connecting with the class of **P**-rings, already introduced.

Definition 2.2.1 A Noetherian ring A is called a **Nagata** ring if for any prime ideal $\mathfrak{p}$ of A, the ring $A/\mathfrak{p}$ is Japanese.

Remark 2.2.2 Nagata rings are exactly the Noetherian universally Japanese rings, as it was proved by Nagata [134, Th. 7.7.2]. A nice comparison of the two classes of rings can be found in [359, Algebra, §162].

Example 2.2.3

  (i) Let A be Dedekind domain whose quotient field is perfect. Then by Proposition 2.1.5, it follows that A is a Nagata ring. In particular, if K is a finite extension of $\mathbb{Q}$, then the integral closure of $\mathbb{Z}$ in K is a Nagata ring.
 (ii) Any ring of fractions of a Nagata ring is a Nagata ring. In particular any subring of $\mathbb{Q}$ is a Nagata ring.
(iii) Any algebra of finite type over a field is a Nagata ring. On the other hand, we will see later (Example 2.3.13) that the property of being a Nagata ring is not a local one, that is, there exist Noetherian rings that are not Nagata rings, but are locally Nagata rings.
 (iv) Using 2.3.6 we will see that there are examples of Japanese domains that are not Nagata rings.

Proposition 2.2.4 *Let $u : A \to B$ be a finite morphism of Noetherian rings.*

 (i) If A is a Nagata ring, then B is a Nagata ring.
(ii) If B is a Nagata ring and $\ker(u)$ is a nilpotent ideal of A, then A is a Nagata ring.

Proof It follows at once from 2.1.9.   □

Theorem 2.2.5 (Marot [236]) *Let A be a Noetherian ring, $\mathfrak{a}$ an ideal of A. Suppose that:*

 (i) $A/\mathfrak{a}$ is a Nagata ring.
(ii) A is $\mathfrak{a}$-adically complete.

Then A is a Nagata ring.

Proof (Chiriacescu [82]) Using induction on the number of generators of $\mathfrak{a}$, it is enough to consider the case of a principal ideal $\mathfrak{a} = xA$. Let $\mathfrak{q} \in \mathrm{Spec}(A)$. If $xA \subseteq \mathfrak{q}$, then from (i) it follows that $A/\mathfrak{q}$ is Japanese. If $x \notin \mathfrak{q}$, let $B := A/\mathfrak{q}$ and take $\mathfrak{p} \in \mathrm{Ass}(B/xB)$. There exists $\mathfrak{r} \in \mathrm{Spec}(A)$ such that $x \in \mathfrak{r}$ and $A/\mathfrak{r} = B/\mathfrak{p}$. From (i) it follows that $B/\mathfrak{p}$ is Japanese. Since B is xB-adically complete, Theorem 2.1.29 implies that B is Japanese. $\square$

Corollary 2.2.6 *Let A be a Noetherian ring, $\mathfrak{a}$ a proper ideal of A and $\widehat{A}$ the completion of A in the $\mathfrak{a}$-adic topology. If A is a Nagata ring, then also $\widehat{A}$ is a Nagata ring.*

Proof By Proposition 2.2.4 we have that $A/\mathfrak{a} \cong \widehat{A}/\mathfrak{a}\widehat{A}$ is a Nagata ring. Now we apply Theorem 2.2.5. $\square$

Corollary 2.2.7 *Let A be a Noetherian Nagata ring. Then $A[[X_1, \ldots, X_n]]$ is a Nagata ring.*

Proof It follows immediately by applying Theorem 2.2.5 to the ideal $(X_1, \ldots, X_n)$ of $A[[X_1, \ldots, X_n]]$. $\square$

Corollary 2.2.8 (Nagata) *Let A be a Noetherian complete semilocal ring. Then A is Nagata ring.*

Proof Let $\mathfrak{p} \in \mathrm{Spec}(A)$ and $B := A/\mathfrak{p}$. Then B is a complete local domain, so that B is a finite extension of a ring C which is a formal power series ring over a field or over a Cohen ring. But by Corollary 2.2.7 it follows that C is a Nagata ring. Now apply Proposition 2.2.4. $\square$

Let us finally mention the following useful fact.

Proposition 2.2.9 (Marot [240, Lemma 4.9]) *Let $u : A \to B$ be a faithfully flat reduced morphism. If B is a Nagata ring, then A is a Nagata ring.*

Proof The faithful flatness of u implies however that A is Noetherian. We must prove that for any prime ideal $\mathfrak{p}$ of A, the ring $A/\mathfrak{p}$ is Japanese. Replacing A by $A/\mathfrak{p}$ and B by $B/\mathfrak{p}B$, we may assume that A is a domain and that $\mathfrak{p} = (0)$. Let L be a field, finite extension of $Q(A)$. Let $(C_i)_{i \in I}$ be the filtered family of subrings of L that are finite A-algebras and have L as field of fractions. Since u is flat, the morphism $C_i \otimes_A B \to L \otimes_A B$ is injective for all $i \in I$. But $L \otimes_A B$ is a reduced ring, because u is reduced. Hence also $C_i \otimes_A B$ is reduced. From [131, Prop. 23.2.4], it follows that there exists $i_0 \in I$ such that for any $j \geq i_0$ the morphism $C_{i_0} \otimes_A B \to C_j \otimes_A B$ is bijective. Since u is faithfully flat, it follows that $C_j = C_{i_0}$ for any $j \geq i_0$. Thus A'_L is a finite A-module; hence A is Japanese. $\square$

2.3   The Zariski-Nagata Theorem and Applications

Our next task is to characterize the Nagata local rings with the help of their formal fibres.

Lemma 2.3.1 (Zariski) *Let $u : A \to B$ be a flat morphism of Noetherian rings.*

(i) *Let $\mathfrak{p} \in \mathrm{Reg}(A)$ be such that $\mathfrak{p}B \neq B$ and $B/\mathfrak{p}B$ is reduced. Then $B_{\mathfrak{q}}$ is regular, for any prime ideal $\mathfrak{q} \in \mathrm{Ass}_B(B/\mathfrak{p}B)$.*
(ii) *Let x be an element of A which is A -regular and such that $x \in J(B)$. Suppose that for any prime ideal $\mathfrak{p} \in \mathrm{Ass}_A(A/xA)$ the ring $A_{\mathfrak{p}}$ is regular and $B/\mathfrak{p}B$ is a non-zero reduced ring. Then B is reduced.*

Proof

(i) Let $\mathfrak{q} \in \mathrm{Ass}_B(B/\mathfrak{p}B) = \mathrm{Min}(B/\mathfrak{p}B)$. Then $\mathrm{depth}(B_{\mathfrak{q}}/\mathfrak{p}B_{\mathfrak{q}}) = 0$; hence $B_{\mathfrak{q}}/\mathfrak{p}B_{\mathfrak{q}}$ is regular. Since $A_{\mathfrak{p}}$ is regular, it results that also $B_{\mathfrak{q}}$ is regular.
(ii) First note that x is a B -regular element by assumption. Let $\mathfrak{q} \in \mathrm{Spec}(B)$ be such that $\mathrm{depth}(B_{\mathfrak{q}}) = 0$. Then $x \notin \mathfrak{q}$. Let also $\mathfrak{r}$ be a minimal prime ideal of $\mathfrak{q} + xB$, so that $\mathrm{depth}(B_{\mathfrak{r}}) = 1$. Moreover, we have

$$\mathrm{depth}(B_{\mathfrak{r}}/\mathfrak{p}B_{\mathfrak{r}}) = \mathrm{depth}(B_{\mathfrak{r}}) - \mathrm{depth}(A_{\mathfrak{p}}) = 0$$

and since $B_{\mathfrak{r}}/\mathfrak{p}B_{\mathfrak{r}}$ is reduced, it follows that it is regular. Therefore $B_{\mathfrak{r}}$ is regular and consequently $B_{\mathfrak{q}}$ is regular.   □

Remark 2.3.2 The above lemma can be stated in a more general frame as in [186], but we will not need the generalization.

Definition 2.3.3 A Noetherian semilocal ring will be called **analytically unramified** if $\widehat{A}$, the completion of A in the topology of the Jacobson radical, is a reduced ring.

The interested reader can find a nice section dedicated to analytically unramified rings in the monograph [362, ch. 9].

Lemma 2.3.4 *Let A be a Noetherian semilocal ring and $\widehat{A}$ the completion of A in the radical topology.*

(i) *If A is analytically unramified, then A' is a finite A -algebra.*
(ii) *If A is an analytically unramified domain and $Q(\widehat{A}) \cong L_1 \times \cdots \times L_n$ where L_i are fields, separable extensions of $Q(A)$, then A is Japanese.*

Proof

(i) Let $\operatorname{Min}(\widehat{A}) = \{\mathfrak{p}_1, \ldots, \mathfrak{p}_n\}$. Then $B_i := \widehat{A}/\mathfrak{p}_i$ are complete local domains, so by Corollary 2.2.8 we get that B_i' is a finite B_i-algebra, for all $i = 1, \ldots, n$. Moreover, by Lemma 2.1.35, we have $\widehat{A}' = B_1' \times \cdots \times B_n'$, and this implies that B' is a finite B-algebra. Finally we apply Lemma 2.1.36 to get the conclusion.

(ii) Let L be a field, finite extension of $Q(A)$ and C a subring of L, such that C is a finite A-algebra and $Q(C) = L$. Then $A_L' = C_L'$. But

$$\widehat{C} = C \otimes \widehat{A} \subseteq L \otimes Q(\widehat{A})$$

and this is a reduced ring. Now apply (i). $\qquad\square$

Lemma 2.3.5 *Let A be a semilocal normal Nagata ring. Then A is analytically unramified.*

Proof Let $n = \dim(A)$. If $n = 0$ the assertion is obvious. Let $n > 0$ and suppose that we have proved the assertion for rings of dimension at most $n - 1$. Let $x \in J(A)$ be a non-zero-divisor and let $\mathfrak{p} \in \operatorname{Ass}(A/xA)$. Then $\mathfrak{p}A_\mathfrak{p} \in \operatorname{Ass}(A_\mathfrak{p}/xA_\mathfrak{p})$ and since $A_\mathfrak{p}$ is normal, we have $\operatorname{depth}(A_\mathfrak{p}) = \dim(A_\mathfrak{p}) = 1$ and $A_\mathfrak{p}$ is a regular local ring. Moreover $\dim(A/\mathfrak{p}) \leq n - 1$ and by induction it follows that $\widehat{A/\mathfrak{p}}$ is reduced. From Lemma 2.3.1 we deduce that $\widehat{A}$ is reduced. $\qquad\square$

The following result gives an important characterization of Nagata rings in the (semi)local case. It shows that the class of Nagata rings can be described, in the semilocal case, by means of the formal fibres, notion studied in Sect. 1.6.

Theorem 2.3.6 (Zariski-Nagata) *Let A be a semilocal Noetherian ring. The following conditions are equivalent:*

(i) *A is a Nagata ring.*
(ii) *For any reduced finite A-algebra C, the completion $\widehat{C}$ of C in the radical topology is reduced.*
(iii) *A has geometrically reduced formal fibres.*

Proof $(i) \Rightarrow (ii)$: Let C be a finite reduced A-algebra. By Proposition 2.2.4 it follows that C is a Nagata ring. Let $\operatorname{Min}(C) = \{\mathfrak{p}_1, \ldots, \mathfrak{p}_n\}$ and let $C_i := C/\mathfrak{p}_i$, for each $i = 1, \ldots, n$. Since C is reduced, from Lemma 2.1.35 it follows that $C' = C_1' \times \cdots \times C_n'$ and that C' is a finite C-algebra. Now we have $\widehat{C} \subseteq \widehat{C}'$ and C' is a Nagata ring. To conclude we apply Lemma 2.3.5.

$(ii) \Rightarrow (iii)$: Let $\mathfrak{p} \in \mathrm{Spec}(A)$ and let L be a field, finite extension of $Q(A/\mathfrak{p})$. There exists a finite A-algebra $C \subseteq L$, such that $Q(C) = L$. The ring C is a semilocal domain and $\widehat{C} \cong C \otimes_A \widehat{A}$. It follows that $\widehat{A} \otimes_A L = \widehat{C} \otimes_C L$ is a fraction ring of C. The hypothesis implies that $\widehat{C}$ is a reduced ring, so that $\widehat{C} \otimes_C L$ is reduced and consequently $\widehat{A} \otimes_A L$ is reduced.

$(iii) \Rightarrow (i)$: Follows from Corollary 2.2.8 and Proposition 2.2.9.   □

Corollary 2.3.7 *Let $(A, \mathfrak{m})$ be a Noetherian local ring. Then A is a Nagata ring if and only if the canonical morphism $A \to \widehat{A}$ is reduced.*

Corollary 2.3.8 *Let $(A, \mathfrak{m})$ be a Noetherian local ring. Then A is a Nagata ring if and only if A^h is a Nagata ring.*

Proof It follows from Theorem 1.6.55 and Corollary 2.3.7.   □

Example 2.3.9 (Nagata [264])   A Noetherian Japanese domain that is not a Nagata ring.
    Let K_0 be a countable field of characteristic zero and $n = 2$. Let $c \in \mathbb{N}$ be such that $c \geq 2$ and let $G(X, Z_1) = X + Z_1^c$. Then, by Theorem 1.5.18 we obtain a regular local ring $(A, \mathfrak{m})$ of characteristic zero with a prime element x such that $\widehat{A} \cong K[[x, \zeta_1, \zeta_2]]/(x + \zeta_1^c)$ and $\widehat{A}/x\widehat{A} \cong K[[\zeta_1, \zeta_2]]/(\zeta_1^c)$. Therefore, by Example 2.2.3, (i) and Corollary 2.3.7, A is a Noetherian Japanese domain which is not Nagata.

We want to prove a generalization of Theorem 2.3.6 to the non-semilocal case. For this we need to involve the normal locus of a Noetherian ring.

Lemma 2.3.10 *Let A be a Noetherian domain. If A is Nor-0, then $Nor(A)$ is open in $\mathrm{Spec}(A)$.*

Proof By hypothesis there is a non-zero element $f \in A$, such that $D(f) \subseteq Nor(A)$. Let $E := \{\mathfrak{p} \in \mathrm{Ass}_A(A/fA) \mid \mathfrak{p} \notin Nor(A)\}$ and let $Y := \bigcup_{\mathfrak{p} \in E} V(\mathfrak{p})$. We show that $Nor(A) = \mathrm{Spec}(A) \setminus Y$ and this will conclude the proof.
Let $\mathfrak{q} \in Nor(A)$ and let $\mathfrak{p} \in \mathrm{Ass}_A(A/fA)$, be such that $\mathfrak{q} \in V(\mathfrak{p})$. Then $A_\mathfrak{p}$ is a normal ring, and since $\mathfrak{p}A_\mathfrak{p} \in \mathrm{Ass}_{A_\mathfrak{p}}(A_\mathfrak{p}/fA_\mathfrak{p})$, it follows that $A_\mathfrak{p}$ is a discrete valuation ring. Then $\mathfrak{p} \notin E$, that is, $\mathfrak{q} \in \mathrm{Spec}(A) \setminus Y$.
Conversely suppose that $\mathfrak{q} \in \mathrm{Spec}(A) \setminus Y$. If $f \notin \mathfrak{q}$, then $\mathfrak{q} \in D(f) \subseteq Nor(A)$. Suppose now that $f \in \mathfrak{q}$. Let $\mathfrak{p} \subseteq \mathfrak{q}$ be such that $\mathrm{depth}(A_\mathfrak{p}) \leq 1$. If $f \notin \mathfrak{p}$, then $A_\mathfrak{p}$ is normal and consequently regular. If $f \in \mathfrak{p}$, then $\mathfrak{p} \in \mathrm{Ass}_A(A/fA)$. Since $\mathfrak{q} \notin Y$, it follows that $\mathfrak{p} \notin E$, so that $A_\mathfrak{p}$ is regular.   □

Lemma 2.3.11 *Let A be a Noetherian domain. Suppose that:*

(i) $\mathrm{Nor}(A)$ *is open in* $\mathrm{Spec}(A)$.
(ii) A_{m} *is $N-1$, for any* $\mathrm{m} \in \mathrm{Max}(A)$.

Then A is $N-1$.

Proof Let $A \subseteq D \subseteq K$ be a finite extension of A contained in $K := Q(A)$. Because A_f is normal and D_f is finite over it, it follows that $A_f \cong D_f$ and D_f is normal. By Lemma 2.3.10 we get that $\mathrm{Nor}(D)$ is open in $\mathrm{Spec}(D)$. The canonical morphism $\pi_D : \mathrm{Spec}(D) \to \mathrm{Spec}(A)$ is closed by the going-up property; hence the set $Z_D := \pi_D(\mathrm{Spec}(D) \setminus \mathrm{Nor}(D))$ is closed in $\mathrm{Spec}(A)$.

Let m be a maximal ideal of A and A'_{m} be its integral closure. By assumption ii) we know that A'_{m} is finite over A_{m}. Therefore there exist elements $g_1, \ldots, g_n \in A'$ such that $A'_{\mathrm{m}} = A_{\mathrm{m}}[g_1, \ldots, g_n]$. Let $T_{\mathrm{m}} := A[g_1, \ldots, g_n] \subseteq K$. Then obviously $\mathrm{m} \notin Z_{T_{\mathrm{m}}}$. Let $V := \bigcap_{\mathrm{m} \in \mathrm{Max}(A)} Z_{T_{\mathrm{m}}}$. Since V is closed and does not contain any maximal ideal, it results that $V = \emptyset$. But $\mathrm{Spec}(A)$ is a quasi-compact topological space; hence there are finitely many maximal ideals $\mathrm{m}_1, \ldots, \mathrm{m}_r$ of A such that $\bigcap_{i=1}^{r} Z_{T_{\mathrm{m}_i}} = \emptyset$. Now let $T := A[\mathrm{m}_1, \ldots, \mathrm{m}_r]$, so that T is a finite A-algebra. Let $\theta : \mathrm{Spec}(T) \to \mathrm{Spec}(A)$ be the canonical map. Then $\theta(\mathrm{NNor}(T)) \subseteq Z_{T_{\mathrm{m}_i}}, i = 1, \ldots, r$. It follows that $\mathrm{NNor}(T) = \phi$, that is T is normal and therefore $A' = T$ and is a finite A-algebra.   □

Now we can prove the desired characterization of Nagata rings in the non-semilocal case.

Theorem 2.3.12 *Let A be a Noetherian ring. The following conditions are equivalent:*

(i) A is a Nagata ring.
(ii) A has geometrically reduced formal fibres and A is Nor-2.

Proof $(i) \Rightarrow (ii)$: If $\mathrm{m} \in \mathrm{Max}(A)$ the localization A_{m} is a Nagata ring, so that by Theorem 2.3.6 A_{m} has geometrically reduced formal fibres. Let now B be a finitely generated A-algebra which is a domain and let $\mathfrak{p} \in \mathrm{Spec}(A)$ be such that $C := A/\mathfrak{p} \subseteq B$. There are elements $y_1, \ldots, y_n \in B$, algebraically independent over C, and there is a non-zero element $f \in C$ such that B_f is a $D := C_f[y_1, \ldots, y_n]$-finite algebra [248, §14.G]. By assumption C is Japanese and by Proposition 2.1.10 also D is Japanese. Now from Proposition 2.1.9 we obtain that B_f is Japanese and from Lemma 1.2.10 we get that $\mathrm{Nor}(B_f)$ is open in $\mathrm{Spec}(B)$. From Lemma 2.3.10 it follows that $\mathrm{Nor}(B)$ is open. Now we apply Lemma 2.3.11 to obtain the conclusion.

$(ii) \Rightarrow (i)$: Let $\mathfrak{p} \in \mathrm{Spec}(A)$, L a field, finite extension of $Q(A/\mathfrak{p})$ and let $C :=$ $(A/\mathfrak{p})'_L$. Then there exists a subring B of L such that $Q(B) = L$ and such that B is a finite A-algebra. It follows that $\mathrm{Nor}(B)$ is open and that $B_\mathfrak{n}$ is Japanese for any $\mathfrak{n} \in \mathrm{Max}(B)$. Consequently B'_L is a finite B-algebra. Since $C = B'_L$, it follows that $A/\mathfrak{p}$ is Japanese.

□

We prove now the stability of the property of being a **P**-ring to algebras of finite type. This difficult theorem is due to Grothendieck [132]. Other proofs were given in [64] using separation properties of the module of differentials and in [67] using the Jacobian criterion of formal smoothness. We follow here the presentation from The Stacks Project [359].

Example 2.3.13 As mentioned in Example 2.2.3, there exist locally Nagata rings that are not Nagata. Indeed, the ring constructed in Example 1.2.4 has these properties as follows by Proposition 1.2.1 and Theorem 2.3.12. Recently Heitmann [157] produced such an example which moreover is a PID.

Lemma 2.3.14 *Let $(A, \mathfrak{m})$ be a complete Noetherian local domain with fraction field K of characteristic $p > 0$. Let $\mathfrak{q}$ be a maximal ideal in $A[X]$ such that $\mathfrak{q} \cap A = \mathfrak{m}$ and let $\mathfrak{r} \neq (0)$ be a prime ideal of $A[X]$ contained in $\mathfrak{q}$ such that $\mathfrak{r} \cap A = (0)$. Then $\widehat{A[X]_\mathfrak{q}} \otimes_{A[X]} k(\mathfrak{r})$ is geometrically regular over $k(\mathfrak{r})$.*

Proof Let us first observe that $k(\mathfrak{r})$ is a finite extension of K. Let L be a finite purely inseparable extension of $k(\mathfrak{r})$. Then L is a finite extension of K; hence there exists a finite A-algebra B, with field of fractions L. Denote by $\mathfrak{p}$ the kernel of the morphism

$$B[X] = A[X] \otimes_A B \to k(\mathfrak{r}) \otimes_A B \to L.$$

Thus $k(\mathfrak{p}) = L$. If $\mathfrak{q}_1, \ldots, \mathfrak{q}_s$ are the prime ideals of $B[X]$ lying over $\mathfrak{r}$, then

$$\widehat{A[X]_\mathfrak{q}} \otimes_{A[X]} L = \widehat{A[X]_\mathfrak{q}} \otimes_{A[X]} B[X] \otimes_{B[X]} k(\mathfrak{p}) = \prod_{i=1}^{s} \widehat{B[X]_{\mathfrak{q}_i}} \otimes_{B[X]} k(\mathfrak{p}).$$

Thus it is enough to prove that the components $\widehat{B[X]_{\mathfrak{q}_i}} \otimes_{B[X]} k(\mathfrak{p})$ are regular. This implies that it is enough to show that $\left(\widehat{A[X]_\mathfrak{q}}\right) \otimes_{A[X]} k(\mathfrak{r})$ is regular in the special case $K = k(\mathfrak{r})$. In this case $\mathfrak{r}K[X]$ is generated by $X - f$ for some element $f \in K$ and

$$\left(\widehat{A[X]_\mathfrak{q}}\right) \otimes_{A[X]} k(\mathfrak{r}) = (\widehat{A[X]_\mathfrak{q}} \otimes_A K)/(X - f).$$

Consider $D := \frac{d}{dX} \in \mathrm{Der}(A[X])$. Then D extends to a derivation of $K[X]$ and $D(X - f)$ is invertible in $K[X]$. The morphism $A \to \widehat{A[X]_\mathfrak{q}}$ is formally smooth; hence $\widehat{A[X]_\mathfrak{q}} \otimes_A K$ is a regular ring. Now Lemma 1.6.32 ends the proof.

□

Theorem 2.3.15 (Grothendieck [132, Th. 7.4.4]) *Let A be a Noetherian ring whose formal fibres are geometrically regular and B an A-algebra of finite type. Then B has geometrically regular formal fibres.*

Proof It is enough to consider the case $B = A[X_1, \ldots, X_n]$, and by induction on n, we may assume $n = 1$, that is, $B = A[X]$. Let $\mathfrak{q} \in \mathrm{Max}(B)$. We have to show that the completion morphism $B_\mathfrak{q} \to \widehat{B_\mathfrak{q}}$ is regular. Let $\mathfrak{p} := \mathfrak{q} \cap A$. Replacing A with $A_\mathfrak{p}$, we may assume that $(A, \mathfrak{m})$ is a local ring and that $\mathfrak{q} \cap A = \mathfrak{m}$. There is a unique prime ideal $\mathfrak{q}' \in \mathrm{Spec}(\widehat{A}[X])$ lying over $\mathfrak{q}$. Consider the following commutative diagram

$$
\begin{array}{ccc}
B_\mathfrak{q} & \xrightarrow{\varphi} & \widehat{B_\mathfrak{q}} \\
\theta \downarrow & & \downarrow \eta \\
\widehat{A}[X]_{\mathfrak{q}'} & \xrightarrow{\psi} & \widehat{\widehat{A}[X]_{\mathfrak{q}'}}
\end{array}
$$

The morphism θ is regular. If we prove that ψ is regular, by Lemma 1.6.16 it follows that $\eta \circ \phi = \psi \circ \theta$ is regular and by Proposition 1.6.18 it follows that φ is regular. Therefore we may assume that $(A, \mathfrak{m})$ is a Noetherian complete local ring and that $\mathfrak{q}$ is a prime ideal of $B = A[X]$ lying over $\mathfrak{m}$. Let $\mathfrak{r} \subset \mathfrak{q}$ be a prime ideal of $A[X]$. We have to show that $\widehat{B_\mathfrak{q}} \otimes_B k(\mathfrak{r})$ is geometrically regular over $k(\mathfrak{r})$. Let $\mathfrak{p} := \mathfrak{r} \cap A$. Replacing A by $A/\mathfrak{p}$ and the prime ideals $\mathfrak{q}$ and $\mathfrak{r}$ by their images in $A/\mathfrak{p}[X]$, we may assume that A is a domain and $\mathfrak{r} \cap A = (0)$. There exists a regular local ring $R \subseteq A$ such that A is finite over R. Then it is enough to prove that $\widehat{B_\mathfrak{q}} \otimes_B k(\mathfrak{r})$ is geometrically regular over $k(\mathfrak{r})$, where A is a complete Noetherian local regular domain. Hence $\widehat{B_\mathfrak{q}} \otimes_B k(\mathfrak{r})$ is a regular ring and we are done if $\mathrm{char}(Q(A)) = 0$. In the case $\mathrm{char}(A) > 0$, we have $A = k[[X_1, \ldots, X_m]]$. Then we have two cases:

Case 1: $\mathfrak{r} = (0)$. We apply Lemma 1.6.36.
Case 2: $\mathfrak{r} \neq (0)$. In this case we apply Lemma 2.3.14.   □

Theorem 2.3.16 (Grothendieck [132, Th. 7.4.4]) *Let P be a property of Noetherian local rings satisfying the axioms $(\mathbf{A_1}) - (\mathbf{A_3})$ and $(\mathbf{A_5})$. Let A be a P-ring and B an A-algebra of finite type. Then B is a P-ring.*

Proof As in the proof of Theorem 2.3.15, we can reduce ourselves to the situation that $(A, \mathfrak{m})$ is a Noetherian local P-ring and $\mathfrak{q}$ is a maximal ideal of $A[X]$ lying over $\mathfrak{m}$ and we have to show that the fibres of the morphism $A[X]_\mathfrak{q} \to \widehat{A[X]_\mathfrak{q}}$ are geometrically P. Let us observe that there is a unique prime ideal $\mathfrak{q}' \in \mathrm{Spec}(\widehat{A}[X])$ lying over $\mathfrak{q}$. Consider the following commutative diagram

$$A[X]_{\mathfrak{q}} \xrightarrow{\varphi} \widehat{A[X]_{\mathfrak{q}}}$$

$$\theta \downarrow \qquad \qquad \downarrow \eta$$

$$\widehat{A[X]_{\mathfrak{q}'}} \xrightarrow{\psi} \widehat{A[X]_{\mathfrak{q}'}}$$

The morphism θ is a **P**-morphism, by applying Lemma 1.6.16 to the **P**-morphism $A \to \widehat{A}$. If we prove that ψ is a **P**-morphism, by Proposition 1.6.17 it follows that $\eta \circ \phi = \psi \circ \theta$ is a **P**-morphism, and by Proposition 1.6.18, it follows that φ is a **P**-morphism. But by Theorem 1.6.37 $\widehat{A}$ has geometrically regular formal fibres; hence ψ is a regular morphism. $\square$

Corollary 2.3.17 *Let* **P** *be a property verifying* $(\mathbf{A_1}) - (\mathbf{A_3})$ *and* $(\mathbf{A_5})$. *Let* A *be a Noetherian ring. The following are equivalent:*

(i) $A_{\mathfrak{p}}$ *is a* **P**-*ring, for any* $\mathfrak{p} \in \mathrm{Spec}(A)$.
(ii) $A_{\mathfrak{m}}$ *is a* **P**-*ring, for any* $\mathfrak{m} \in \mathrm{Max}(A)$.
(iii) *For any* A-*algebra of finite type* B *and any* $\mathfrak{q} \in \mathrm{Spec}(B)$, $B_{\mathfrak{q}}$ *is a* **P**-*ring.*

Proof $(i) \Leftrightarrow (ii)$ follows from Theorem 1.6.38.
$(i) \Leftrightarrow (iii)$ follows from Theorem 2.3.16. $\square$

Corollary 2.3.18 *Let* A *be a* **P**-*ring and* S *a multiplicatively closed system in* A. *Then* $S^{-1}A$ *is a* **P**-*ring.*

Proof Clear from Corollary 2.3.17, because the localizations in the prime ideals of $S^{-1}A$ are localizations in prime ideals of A. $\square$

Proposition 2.3.19 *Let* **P** *be a property of Noetherian local rings satisfying* $(\mathbf{A_1}) - (\mathbf{A_3})$, $(\mathbf{A_5})$ *and* $(\mathbf{A_6})$. *Let* A *be a semilocal* **P**-*ring. Then* A *is* **P**-*2.*

Proof Let $u : A \to \widehat{A}$ be the canonical completion morphism and let $g := \mathrm{Spec}(u) : \mathrm{Spec}(\widehat{A}) \to \mathrm{Spec}(A)$ be the morphism induced on the spectra. Then g is surjective and the topology on $\mathrm{Spec}(A)$ is the quotient topology [248, 6.H, Th. 7]. By Corollary 1.6.15 we know that $g^{-1}(\mathbf{P}(A)) = \mathbf{P}(\widehat{A})$. Since $\mathbf{P}(\widehat{A})$ is open in $\mathrm{Spec}(\widehat{A})$, we get that $\mathbf{P}(A)$ is open in $\mathrm{Spec}(A)$. From Theorem 2.3.16 it follows that A is **P**-2. $\square$

The above Theorem 2.3.16 provides, in conjunction with Theorem 2.3.12 and Lemma 1.2.9, a proof of the stability of Nagata rings to algebras of finite type (Theorem 2.3.20).

Theorem 2.3.20 (Nagata [266, Th. (36.5)]) *Let A be a Noetherian Nagata ring and B be an A-algebra of finite type. Then B is a Nagata ring.*

Proof We can assume that $B = A[X]$. Let C be a domain which is finitely generated over B. Then C is also finitely generated over A and consequently $\mathrm{Nor}(C)$ is open. By Theorem 2.3.16 we obtain that the formal fibres of B are geometrically reduced. Now Theorem 2.3.12 concludes the proof. $\square$

From the Zariski-Nagata theorem, it follows that there is a strong connection between the finiteness of the integral closure of a Noetherian local domain and the reducedness of its completion. We shall investigate a little more in this direction.

Proposition 2.3.21 *Let $(A, \mathfrak{m})$ be an analytically unramified Noetherian local ring and let $\mathfrak{p} \in \mathrm{Spec}(A)$. Then also $A_\mathfrak{p}$ is analytically unramified.*

Proof Let $B = \widehat{A}$ and let $\mathfrak{q} \in \mathrm{Spec}(B)$ be such that $\mathfrak{q} \cap A = \mathfrak{p}$. Then $B_\mathfrak{q}$ is a reduced Nagata ring; hence by Theorem 2.3.6 the ring $\widehat{B_\mathfrak{q}}$ is reduced. The morphism $\widehat{A_\mathfrak{p}} \to \widehat{B_\mathfrak{q}}$ is faithfully flat; consequently $\widehat{A_\mathfrak{p}}$ is reduced. $\square$

Lemma 2.3.22 (Nagata [266, Th. (36.8)]) *Let A be a Noetherian ring such $A_\mathfrak{m}$ is analytically unramified for any maximal ideal $\mathfrak{m}$. Then for any natural number n and for any prime ideal $\mathfrak{p}$ of $B := A[X_1, \ldots, X_n]$, the ring $B_\mathfrak{p}$ is analytically unramified.*

Proof Let $\mathfrak{p} \in \mathrm{Spec}(B)$ and $\mathfrak{q} = \mathfrak{p} \cap A$. Replacing A by $A_\mathfrak{q}$ and using Proposition 2.3.21, we can assume that A is a local ring, and it is analytically unramified. Therefore $C := \widehat{A}[X_1, \ldots, X_n]$ is a reduced Nagata ring. Let $\mathfrak{r} \in \mathrm{Spec}(C)$ be a prime ideal such that $\mathfrak{r} \cap B = \mathfrak{p}$. The canonical morphism $B_\mathfrak{p} \to C_\mathfrak{r}$ is faithfully flat; hence the morphism $\widehat{B_\mathfrak{p}} \to \widehat{C_\mathfrak{r}}$ is faithfully flat. Since $\widehat{C_\mathfrak{r}}$ is reduced, it follows that $\widehat{B_\mathfrak{p}}$ is reduced. $\square$

Lemma 2.3.23 *Let A be an integral domain, let a, b be a regular sequence in A and let $u : A[X] \to A[\frac{b}{a}]$, $u(X) = \frac{b}{a}$. Then $\ker(u) = (aX - b)$.*

Proof Obviously $aX - b \in \ker(u)$. Conversely, let $f = c_n X^n + c_{n-1} X^{n-1} + \ldots + c_0 \in \ker(u)$, that is, $c_n b^n + a(c_{n-1} b^{n-1} + \ldots + c_0 a^{n-1}) = 0$. This means that we have a relation $ad = c_n b^n$ with $d \in A$. But a, b is a regular sequence; hence $c_n = as, s \in A$. This implies that $f - s X^{n-1}(aX - b) \in \ker(u)$ and we continue by induction on n. $\square$

Theorem 2.3.24 (Rees [308]) *Let A be a semilocal Noetherian domain with field of fractions K. The following are equivalent:*

(i) A is analytically unramified.

(ii) Any A-algebra of finite type B such that $A \subseteq B \subseteq K$ is N–1.

Proof $(i) \Rightarrow (ii)$: Since $A \to \widehat{A}$ is faithfully flat, by Lemma 2.1.36 it is enough to show that $T := \widehat{A} \otimes_A B'$ is a finite $R := \widehat{A} \otimes_A B$-algebra. The ring $\widehat{A}$ is reduced; hence the rings R and T are reduced, being contained in $\widehat{A} \otimes_A K$. Let $\mathfrak{p}_1, \ldots, \mathfrak{p}_n$ be the minimal prime ideals of R and let $R_i := R/\mathfrak{p}_i$. Then Lemma 2.1.35 implies that $R' = (R_1)' \times \ldots \times (R_n)'$. But the rings R_i are finitely generated $\widehat{A}$- algebras, hence they are Japanese rings. This means that R' is a finite R-algebra. Since the extension $R \subseteq \widehat{A} \otimes_A B'$ is an integral extension, we get that $T = \widehat{A} \otimes_A B' \subseteq R'$. Therefore T is a finite R-module.

$(ii) \Rightarrow (i)$: Since A' is a finite A-algebra, we have $\widehat{A} \subseteq \widehat{A'} = \prod_{i=1}^{n} \widehat{A'_{\mathfrak{m}_i}}$, where $\mathfrak{m}_1, \ldots, \mathfrak{m}_n$ are the maximal ideals of A'. Replacing A by $A'_{\mathfrak{m}_i}$, we may assume that $(A, \mathfrak{m})$ is a Noetherian normal local domain. We proceed by induction on $r := \dim(A)$. If $r = 0$, A is a field and there is nothing to prove. If $r = 1$, A is a regular local ring, hence $\widehat{A}$ is regular and again there is nothing to prove. If $r \geq 2$, then $\text{depth}(A) \geq 2$ and there exists a regular sequence $a, b \in \mathfrak{m}$. By Lemma 2.3.23 we have $B := A[\frac{b}{a}] \cong A[X]/(aX - b)$. Then $B/\mathfrak{m}B = A[X]/(\mathfrak{m}, aX - b) = A/\mathfrak{m}[X]$; hence $\mathfrak{m}B$ is a prime ideal of B. If $C := A[X]_{\mathfrak{m}[X]}$, we have $\dim(C) = \dim(A)$. One can see that $B_{\mathfrak{m}B} = C/(aX - b)$; hence $\dim(B_{\mathfrak{m}B}) = \dim(A) - 1$. Moreover, one can easily see that $B_{\mathfrak{m}B}$ satisfies the assumptions of A, whence, by the induction hypothesis, $\widehat{B_{\mathfrak{m}B}}$ is reduced. Let $\mathfrak{n}$ denote the maximal ideal of C. Then $\mathfrak{n}^t \cap A = \mathfrak{m}^t$, for any $t \in \mathbb{N}$. Thus $\widehat{A} \subseteq \widehat{C}$. Since $aX - b$ is prime in C and $C/(aX - b)$ is analytically unramified, it follows by 2.3.1 that $\widehat{C}$ is reduced and consequently its subring $\widehat{A}$ is reduced.   □

Let us see a nice application of Theorem 2.3.6, involving the ring of restricted power series. For the notion of a Zariski ring, the reader can see [249, pag.62].

Lemma 2.3.25 *Let A be a Noetherian Nagata integral domain. Assume that A is a Zariski ring and let $\mathfrak{a} \subsetneq A$ be a non-zero ideal of definition of A. Then:*

(i) *There exists a semilocal principal ring B containing A and such that the radical topology of B induces on A the $\mathfrak{a}$-adic topology.*

(ii) *If moreover A is integrally closed, then the restricted power series $A_{\mathfrak{a}}\{T_1, \ldots, T_p\}$ is integrally closed, for any natural number p.*

Proof

(i) Let us denote by $F(\mathfrak{a})$ the set of elements $x \in A$ such that x divides 0 on $A/\mathfrak{a}^n$ for some $n \geq 1$. Then $F(\mathfrak{a})$ is the union of finitely many prime ideals of A, that is, the primes of $A(\mathfrak{a})$ (see Remark 2.1.18) and maybe some more, finitely many prime ideals (see also [257]). Therefore we can write $F(\mathfrak{a}) = \{\mathfrak{q}_1, \ldots, \mathfrak{q}_m\} \subseteq \text{Spec}(A)$. Let $C := \bigcap_{i=1}^{m} A_{\mathfrak{q}_i}$. Then C is a Nagata ring and $\mathfrak{a}^n C \cap A = \mathfrak{a}^n$, for any $n \in \mathbb{N}$.

Case 1. $\mathfrak{a} = xA$ is a principal ideal. Then $F(\mathfrak{a})$ is the set of maximal elements of $\mathrm{Ass}_A(A/xA)$. Since C', the integral closure of C, is finite over C, the $\mathfrak{a}$-adic to[ology of C' induces on C the $\mathfrak{a}$-adic topology. Moreover, one can easily see that $\mathfrak{a}^n C' = \mathfrak{a}^n B \cap C'$, where $B = \displaystyle\bigcap_{\mathfrak{p} \in \mathrm{Ass}(C'/xC')} C'_{\mathfrak{p}}$. Thus B is a semilocal principal ring with the desired properties.

Case 2. $\mathfrak{a}$ is not principal. Let $R := C[\mathfrak{a}T, T^{-1}]$ be the extended Rees ring. Then, if we put $t = T^{-1}$, it is well-known that $t^n R \cap C = \mathfrak{a}^n C$ and we can use the argument in Case 1 for R and the ideal tR.

(ii) By (i) there exists a semilocal principal ring B containing A and such that the radical topology of B induces on A the $\mathfrak{a}$-adic topology. To simplify things, let $T := \{T_1, \ldots, T_p\}$. Then $A_\mathfrak{a}\{T\}$ is a subring of $B_\mathfrak{n}\{T\}$, where $\mathfrak{n}$ is the Jacobson radical of B. Therefore $A_\mathfrak{a}\{T\} = A[[T]] \cap B_\mathfrak{n}\{T\}$. The ring $A[[T]]$ is integrally closed. The ring $B_\mathfrak{n}\{T\}$ is the completion of the regular ring $B[T]$ in the $\mathfrak{n}T$-adic topology, hence is a regular ring [131, Lemma 17.3.8.1]. Consequently $B_\mathfrak{n}\{T\}$ is integrally closed, whence $A_\mathfrak{a}\{T\}$ is integrally closed.   □

Theorem 2.3.26 (Marot [239, Prop. 7]) *Let A be a Nagata ring. Assume that A is a Zariski ring and let $\mathfrak{a} \subsetneq A$ be a non-zero ideal of definition of A. Then the morphism $A \to A_\mathfrak{a}\{T_1, \ldots, T_p\}$ is a normal morphism, for any $p \in \mathbb{N}$.*

Proof Let $T := \{T_1, \ldots, T_p\}$. The morphisms $A \to A[T]$ and $A[T] \to A_\mathfrak{a}\{T\}$ are flat; hence $A \to A_\mathfrak{a}\{T\}$ is flat. Let $\mathfrak{p} \in \mathrm{Spec}(A)$. Then

$$A_\mathfrak{a}\{T\} \otimes_A k(\mathfrak{p}) \cong (A/\mathfrak{p})_\mathfrak{a}\{T\} \otimes_{A/\mathfrak{p}} k(\mathfrak{p}),$$

so that we may assume that $\mathfrak{p} = (0)$ and A is an integral domain. Let $K := Q(A)$ be the field of fractions of A and L a field, finite extension of K. Let $D := A'_L$ be the integral closure of A in L. Then D is a finite A-algebra and

$$(A_\mathfrak{a}\{T\} \otimes_A K) \otimes_K L \cong (A_\mathfrak{a}\{T\} \otimes_A D) \otimes_D L \cong D_\mathfrak{a}\{T\} \otimes_D L.$$

But from Lemma 2.3.25, it follows that $D_\mathfrak{a}\{T\}$ is integrally closed; hence its fraction ring $D_\mathfrak{a}\{T\} \otimes_D L$ is integrally closed too.   □

We end the section with some applications to local rings whose completion are domains.

Lemma 2.3.27 *Let $(A, \mathfrak{m}, k)$ be a Noetherian local reduced ring and let $\mathfrak{p}_1, \ldots, \mathfrak{p}_n$ be the minimal prime ideals of A, for some $n \geq 2$. Let $r \in \{1, \ldots, n-1\}$ and let $\mathfrak{a} = \mathfrak{p}_1 \cap \ldots \cap \mathfrak{p}_r$ and $\mathfrak{b} = \mathfrak{p}_{r+1} \cap \ldots \cap \mathfrak{p}_n$. If $\mathfrak{a} + \mathfrak{b}$ is an $\mathfrak{m}$-primary ideal, then $\mathrm{depth}(A) \leq 1$.*

Proof Because A is reduced, we have $\mathfrak{a} \cap \mathfrak{b} = (0)$. Suppose that $\operatorname{depth}(A) > 1$. Consider the exact sequence

$$0 \to A = A/\mathfrak{a} \cap \mathfrak{b} \to A/\mathfrak{a} \oplus A/\mathfrak{b} \to A/(\mathfrak{a} + \mathfrak{b}) \to 0.$$

From this we get an exact sequence

$$0 \to \operatorname{Hom}_A(k, A) \to \operatorname{Hom}_A(k, A/\mathfrak{a} \oplus A/\mathfrak{b}) \to \operatorname{Hom}_A(k, A/(\mathfrak{a} + \mathfrak{b})) \to$$

$$\to \operatorname{Ext}^1_A(k, A) \to \operatorname{Ext}^1_A(k, A/\mathfrak{a} \oplus A/\mathfrak{b}) \to \operatorname{Ext}^1_A(k, A/(\mathfrak{a} + \mathfrak{b})).$$

Since $\mathfrak{a} + \mathfrak{b}$ is $\mathfrak{m}$-primary, it follows that $\operatorname{depth}(A/(\mathfrak{a} + \mathfrak{b})) = 0$ and this implies that $\operatorname{Hom}_A(k, A/(\mathfrak{a} + \mathfrak{b})) \neq (0)$. Because $\operatorname{depth}(A) > 1$ we obtain that

$$\operatorname{Hom}_A(k, A) = \operatorname{Ext}^1_A(k, A) = (0),$$

hence

$$\operatorname{Hom}_A(k, A/\mathfrak{a} \oplus A/\mathfrak{b}) \neq (0),$$

that is, $\operatorname{depth}(A/\mathfrak{a} \oplus A/\mathfrak{b}) = 0$. But A is a submodule of $A/\mathfrak{a} \oplus A/\mathfrak{b}$, so that $\operatorname{depth}(A) = 0$. Contradiction! $\qquad\square$

Theorem 2.3.28 (Ogoma [276, Th. 1]) *Let $(A, \mathfrak{m})$ be a Noetherian Nagata normal local domain. The following are equivalent:*

(i) $\widehat{A}$ is an integral domain.

(ii) $Q(A) \otimes_A \widehat{A}$ is locally an integral domain.

Proof $(i) \Rightarrow (ii)$ Obvious.
$(ii) \Rightarrow (i)$ Suppose that $\widehat{A}$ is not a domain. Let $\operatorname{Min}(\widehat{A}) = \{\mathfrak{q}_1, \ldots, \mathfrak{q}_n\}$. However, $\widehat{A}$ is reduced and this implies that $n \geq 2$, because $\bigcap_{i=1}^{n} \mathfrak{q}_i = (0)$. Set $\mathfrak{a} = \mathfrak{q}_1 + \mathfrak{q}_2 \cap \ldots \cap \mathfrak{q}_n$ and let $\mathfrak{q} \in \operatorname{Min}(\mathfrak{a})$. Consider $r \in \{2, \ldots, n\}$ such that $\mathfrak{q}_i \subseteq \mathfrak{q}$ for $i = 1, \ldots, r$ and $\mathfrak{q}_i \not\subseteq \mathfrak{q}$ for $i = r + 1, \ldots, n$. Then the ideal $\mathfrak{q}_1 \widehat{A}_{\mathfrak{q}} + \bigcap_{j=2}^{r} \mathfrak{q}_j \widehat{A}_{\mathfrak{q}}$ is $\mathfrak{q}\widehat{A}_{\mathfrak{q}}$-primary and $\operatorname{Min}(\widehat{A}_{\mathfrak{q}}) = \{\mathfrak{q}_1 \widehat{A}_{\mathfrak{q}}, \ldots, \mathfrak{q}_r \widehat{A}_{\mathfrak{q}}\}$. Let $\mathfrak{p} = \mathfrak{q} \cap A$.

Case 1.   Suppose that $\operatorname{ht}(\mathfrak{p}) \geq 2$. Since A is normal, it follows that $\operatorname{depth}(A_{\mathfrak{p}}) \geq 2$. Then

$$\operatorname{depth}(\widehat{A}_{\mathfrak{q}}) = \operatorname{depth}(A_{\mathfrak{p}}) + \operatorname{depth}(\widehat{A}_{\mathfrak{q}}/\mathfrak{p}\widehat{A}_{\mathfrak{q}}) \geq 2 + \operatorname{depth}(\widehat{A}_{\mathfrak{q}}/\mathfrak{p}\widehat{A}_{\mathfrak{q}}) \geq 2.$$

But this contradicts Lemma 2.3.27.

Case 2. Suppose that $\text{ht}(\mathfrak{p}) = 1$. Then $A_\mathfrak{p}$ is a discrete valuation ring and there are three possibilities:

(a) $\dim(\widehat{A}_\mathfrak{q}) > 1$. Then $\dim(\widehat{A}_\mathfrak{q}/\mathfrak{p}\widehat{A}_\mathfrak{q}) = \dim(\widehat{A}_\mathfrak{q}) - \dim(A_\mathfrak{p}) \geq 1$. But $\text{depth}(\widehat{A}_\mathfrak{q}/\mathfrak{p}\widehat{A}_\mathfrak{q}) = 0$ and this implies that $\widehat{A}_\mathfrak{q}/\mathfrak{p}\widehat{A}_\mathfrak{q}$ is not reduced. This contradicts the fact that A is a Nagata ring.

(b) $\dim(\widehat{A}_\mathfrak{q}) = 1$. In this case we have $\dim(\widehat{A}_\mathfrak{q}/\mathfrak{p}\widehat{A}_\mathfrak{q}) = \dim(\widehat{A}_\mathfrak{q}) - \dim(A_\mathfrak{p}) = 0$ and $\widehat{A}_\mathfrak{q}/\mathfrak{p}\widehat{A}_\mathfrak{q}$ is a reduced ring; hence $\widehat{A}_\mathfrak{q}/\mathfrak{p}\widehat{A}_\mathfrak{q}$ is a regular local ring and consequently also $\widehat{A}_\mathfrak{q}$ is regular. But then $n = 1$. Contradiction!

(c) $\mathfrak{p} = (0)$. Then $\widehat{A}_\mathfrak{q}$ is a localization of the generic formal fibre of A; hence is an integral domain. But this contradicts the fact that $r \geq 2$. $\square$

Corollary 2.3.29 (Ogoma [276, Cor. to Th. 1]) *Let $(A, \mathfrak{m})$ be a Nagata unibranched local domain. The following are equivalent:*

(i) $\widehat{A}$ is an integral domain.

(ii) $Q(A) \otimes_A \widehat{A}$ is locally an integral domain.

Proof Let A' be the integral closure of A. Then A' is a finite A-algebra and $\widehat{A'} \cong A' \otimes_A \widehat{A}$. Thus we have

$$\widehat{A'} \otimes_A Q(A) \cong A' \otimes_A \widehat{A} \otimes_A Q(A) \cong (A' \otimes_A Q(A)) \otimes_A \widehat{A} \cong Q(A) \otimes_A \widehat{A}.$$

Since the morphism $A \to \widehat{A}$ is flat, it follows that $\widehat{A} \to \widehat{A'}$ is injective. But A' is local because A is unibranched. This implies that we may assume that $A = A'$. Now we apply Theorem 2.3.28. $\square$

Excellent Rings and Regular Morphisms

3

Chapter 3, together with Chap. 5, is one of the most consistent parts of the book. The first section contains results about chain conditions in Noetherian rings, a topic that always proved to be quite subtle. Then we focus on criteria about regular morphisms and excellent rings. One can find here the famous André theorem about the localization of formal smoothness, theorem that is also the starting point for Chap. 4. There are many results about F-finite rings, rings that were long ago proved to be excellent. In order to include Scheja-Storch results about excellent rings, we chose to include the main features of the theory of universally finite module of differentials.

3.1   Chain Conditions

The study of chain conditions in commutative rings, even in Noetherian rings, is rather difficult. It goes back to Cohen, Nagata and Grothendieck. A big contribution was given by Ratliff [301–305]. We will make a detailed presentation of this topic, focusing on the Noetherian case.

Definition 3.1.1 Let A be a ring, $\mathfrak{p} \subsetneq \mathfrak{q}$ prime ideals of A. A chain of prime ideals $\mathfrak{p} = \mathfrak{p}_0 \subset \mathfrak{p}_1 \subset \ldots \subset \mathfrak{p}_{k-1} \subset \mathfrak{p}_k = \mathfrak{q}$ of A is called **saturated** if for any $i \in \{0, \ldots, k-1\}$ there is no prime ideal strictly between $\mathfrak{p}_i$ and $\mathfrak{p}_{i+1}$. A saturated chain of prime ideals such that $\mathfrak{p}_0 \in \text{Min}(A)$ and $\mathfrak{p}_k \in \text{Max}(A)$ is called a **maximal chain of prime ideals**.

Definition 3.1.2 A ring A is called **catenary** if for any prime ideals $\mathfrak{p} \subsetneq \mathfrak{q}$ of A, $\text{ht}(\mathfrak{q}/\mathfrak{p})$ is finite and is equal to the length of any saturated chain of prime ideals between $\mathfrak{p}$ and $\mathfrak{q}$.

© The Author(s), under exclusive license to Springer Nature Switzerland AG 2023
C. Ionescu, *Classes of Good Noetherian Rings*, Frontiers in Mathematics,
https://doi.org/10.1007/978-3-031-22292-4_3

Remark 3.1.3

(i) If A is a Noetherian ring, then $\mathrm{ht}(q/p)$ is always finite.
(ii) If A is a local Cohen-Macaulay ring and $\mathfrak{a}$ is an ideal of A, then $A/\mathfrak{a}$ is catenary [131, Cor. 16.5.12].
(iii) It is not easy to construct examples of non-catenary rings. For a long time, people thought that every Noetherian ring is catenary, until Nagata constructed the first counterexample [262]. We will see in example 3.1.8 such a ring constructed by S. Greco [124].

Lemma 3.1.4 *Let A be a Noetherian ring. The following are equivalent:*

(i) For any prime ideals $p \subseteq q$ of A, $\mathrm{ht}(q) = \mathrm{ht}(p) + \mathrm{ht}(q/p)$.
(ii) For any saturated chain of prime ideals $p \subsetneq q$ of A, $\mathrm{ht}(q) = \mathrm{ht}(p) + 1$.

Proof $(i) \Rightarrow (ii)$: Obvious.
$(ii) \Rightarrow (i)$: Considering a saturated chain $p \subsetneq p_1 \subsetneq \cdots \subsetneq p_n \subsetneq q$, the assertion follows by repeatedly applying the assumption.   □

Proposition 3.1.5 *Let A be a Noetherian ring. The following are equivalent:*

(i) A is catenary.
(ii) For any prime ideals $p_1 \subseteq p_2 \subseteq p_3$ of A, $\mathrm{ht}(p_3/p_1) = \mathrm{ht}(p_3/p_2) + \mathrm{ht}(p_2/p_1)$.

Proof $(i) \Rightarrow (ii)$: Let $s := \mathrm{ht}(p_3/p_2)$, $t := \mathrm{ht}(p_2/p_1)$ and let

$$p_1 = p'_0 \subset \cdots \subset p'_t = p_2$$

and

$$p_2 = p''_0 \subset \cdots \subset p''_s = p_3$$

be saturated chains. It follows that

$$p_1 = p'_0 \subset \cdots \subset p'_t = p_2 \subset p_1'' \subset \cdots \subset p''_t = p_3$$

is a saturated chain between p_1 and p_3. Since A is catenary, we get that $\mathrm{ht}(p_3/p_1) = s + t$.
$(ii) \Rightarrow (i)$: Let $p \subset q$ be two prime ideals of A and let $p = p'_0 \subset \cdots \subset p'_m = q$ and $p = p''_0 \subset \cdots \subset p''_n = q$, $n \leq m$ be two saturated chains. We show that $n = m$ by induction on n. If $n = 1$, it follows that the chain $p \subset q$ is saturated and consequently $m = 1$. Assume now that $n > 1$ and $n < m$. But $\mathrm{ht}(q/p) \geq m > n$ and $\mathrm{ht}(p''_n/p''_{n-1}) = 1$. It follows that $\mathrm{ht}(p''_{n-1}/p''_0) > n - 1$ and this contradicts the induction hypothesis.   □

Proposition 3.1.6 *Let A be a catenary ring, $\mathfrak{a}$ an ideal of A and S a multiplicatively closed system in A. Then $A/\mathfrak{a}$ and $S^{-1}A$ are catenary rings.*

Proof Follows from the description of the prime ideals in $A/\mathfrak{a}$ and $S^{-1}A$. □

Proposition 3.1.7 *Let A be a ring. Then the following are equivalent:*

(i) A is catenary;
(ii) $A_{\mathfrak{p}}$ is catenary for any prime ideal $\mathfrak{p} \in \mathrm{Spec}(A)$;
(iii) $A_{\mathfrak{m}}$ is catenary for any maximal ideal $\mathfrak{m} \in \mathrm{Max}(A)$.

Proof It follows at once from Proposition 3.1.6 and from the structure of prime ideals in the rings of fractions. □

Example 3.1.8 An example of a Noetherian semilocal domain that is not catenary (Greco [124, Lemma 1.4]).

Proof Let k_0 be a field and $k = k_0(Y_1, Y_2, \ldots)$ be the field of rational fractions in infinitely many indeterminates over k. Let $R := k[X, Y, Z]$ and consider in R the prime ideals $\tilde{\mathfrak{m}}_1 = (X, Y, Z)$, $\tilde{\mathfrak{m}}_2 = (X, 1 - YZ)$ and $\tilde{\mathfrak{n}} = (X + 1, Y, Z)$. Let us observe that $\mathrm{ht}(\tilde{\mathfrak{m}}_1) = \mathrm{ht}(\tilde{\mathfrak{n}}) = 3$ and $\mathrm{ht}(\tilde{\mathfrak{m}}_2) = 2$. Let $S := R \setminus (\tilde{\mathfrak{m}}_1 \cup \tilde{\mathfrak{m}}_2 \cup \tilde{\mathfrak{n}})$ and $C := S^{-1}R$. Thus C is a semilocal Noetherian domain having exactly three maximal ideals, that is, $\tilde{\mathfrak{m}}_1 C$, $\tilde{\mathfrak{m}}_2 C$ and $\tilde{\mathfrak{n}} C$. In the ring C, we consider the ideals $\tilde{\mathfrak{p}} := (Y, Z)C$ and $\tilde{\mathfrak{q}} := (X + 1 - XZ)C$ and let $\tilde{\mathfrak{a}} := \tilde{\mathfrak{p}} \cap \tilde{\mathfrak{q}}$. Let B be the $\tilde{\mathfrak{a}}$-adic completion of C. Set $\overline{\mathfrak{m}}_i = \tilde{\mathfrak{m}}_i B$ for $i = 1, 2$, $\overline{\mathfrak{n}} := \tilde{\mathfrak{n}} B$, $\overline{\mathfrak{p}} := \tilde{\mathfrak{p}} B$, $\overline{\mathfrak{q}} := \tilde{\mathfrak{q}} B$ and $\overline{\mathfrak{a}} := \tilde{\mathfrak{a}} B$. Then in $\mathrm{Spec}(B)$, we have $V(\overline{\mathfrak{a}}) = V(\overline{\mathfrak{p}}) \cup V(\overline{\mathfrak{q}})$ and $\overline{\mathfrak{n}} \in V(\overline{\mathfrak{p}}) \cap V(\overline{\mathfrak{q}})$. It follows that B is a regular semilocal Noetherian domain with maximal ideals $\overline{\mathfrak{m}}_1$, $\overline{\mathfrak{m}}_2$ and $\overline{\mathfrak{n}}$. Moreover $\mathrm{ht}(\overline{\mathfrak{m}}_1) = \mathrm{ht}(\overline{\mathfrak{n}}) = 3$, $\mathrm{ht}(\overline{\mathfrak{m}}_2) = \mathrm{ht}(\overline{\mathfrak{p}}) = 2$ and $\mathrm{ht}(\overline{\mathfrak{q}}) = 1$. Since $B/\overline{\mathfrak{m}}_1 = k = k_0(Y_1, Y_2, \ldots)$ and $B/\overline{\mathfrak{m}}_2 = k(Y) = k_0(Y, Y_1, Y_2, \ldots)$, there exists an isomorphism $f : B/\overline{\mathfrak{m}}_1 \to B/\overline{\mathfrak{m}}_2$. Let $p_i : B \to B/\overline{\mathfrak{m}}_i$, $i = 1, 2$ be the canonical surjections and $A := \{b \in B \mid f(p_1(b)) = p_2(b)\}$. Set $\mathfrak{m} := \overline{\mathfrak{m}}_1 \cap A = \overline{\mathfrak{m}}_2 \cap A$, $\mathfrak{n} := \overline{\mathfrak{n}} \cap A$, $\mathfrak{p} := \overline{\mathfrak{p}} \cap A$, $\mathfrak{q} := \overline{\mathfrak{q}} \cap A$, $\mathfrak{a} = \mathfrak{p} \cap \mathfrak{q}$. Then by Lemma 2.1.33, we know that A is a semilocal Noetherian domain, with maximal ideals $\mathrm{Max}(A) = \{\mathfrak{m}, \mathfrak{n}\}$ and such that $\mathrm{ht}(\mathfrak{m}) = \mathrm{ht}(\mathfrak{n}) = 3$, $\mathrm{ht}(\mathfrak{p}) = 2$. We will show that A is not catenary. We have $\tilde{\mathfrak{p}} \subseteq \tilde{\mathfrak{m}}_1 C \cap \tilde{\mathfrak{n}} C$ and $\tilde{\mathfrak{p}} \not\subseteq \tilde{\mathfrak{m}}_2 C$. Hence $\dim(B_{\overline{\mathfrak{m}}_1}/\overline{\mathfrak{a}} B_{\overline{\mathfrak{m}}_1}) = 1$. Moreover, $\tilde{\mathfrak{q}} \subseteq \tilde{\mathfrak{m}}_2 C \cap \tilde{\mathfrak{n}} C$ and $\tilde{\mathfrak{p}} \not\subseteq \tilde{\mathfrak{m}}_1 C$, hence $\dim(B/\overline{\mathfrak{q}}) = 2$. From Lemma 2.1.33 it follows that $B_{\mathfrak{m}}/\mathfrak{a} B_{\mathfrak{m}}$ is finite over $A_{\mathfrak{m}}/\mathfrak{a} A_{\mathfrak{m}}$, hence $\dim(A_{\mathfrak{m}}/\mathfrak{a} A_{\mathfrak{m}}) = 1$. As $B/\overline{\mathfrak{q}}$ is a $A/\mathfrak{q}$-module of finite type, it follows that $\dim(A/\mathfrak{q}) = 2$. Moreover $\dim(A) = 3$, hence $\mathrm{ht}(\mathfrak{q}) = 1$. But $\dim(A_{\mathfrak{m}}/\mathfrak{q} A_{\mathfrak{m}}) \leq \dim(A_{\mathfrak{m}}/\mathfrak{a} A_{\mathfrak{m}}) = 1$ and $\mathrm{ht}(\mathfrak{q} A_{\mathfrak{m}}) = 1$. Thus we have the local domain $A_{\mathfrak{m}}$, with $\dim(A_{\mathfrak{m}}) = 3$ and with the saturated chain $(0) \subset \mathfrak{q} \subset \mathfrak{m}$. Hence $A_{\mathfrak{m}}$ is not a catenary ring. By Proposition 3.1.6 it follows that A itself is not a catenary ring. □

Together with the notion of catenarity, there are other notions that show up in the study of Noetherian rings as well as in the study of algebraic varieties.

Definition 3.1.9 A ring A is called **equidimensional** if $\dim(A) < \infty$ and if for any minimal prime ideal $\mathfrak{p}$ of A we have $\dim(A/\mathfrak{p}) = \dim(A)$ and **equicodimensional** if for any maximal ideal $\mathfrak{m}$ of A we have $\mathrm{ht}(\mathfrak{m}) = \dim(A)$.

Example 3.1.10

  (i) Obviously any integral domain is equidimensional and any local ring is equicodimensional.

  (ii) Any local Cohen-Macaulay ring is equidimensional. Indeed, for any $\mathfrak{p} \in \mathrm{Ass}(A)$, we have $\dim(A/\mathfrak{p}) = \dim(A)$ [77, Th. 2.1.2].

  (iii) Let k be a field and $A := k[X, Y, Z]/(XY, XZ) = k[X, Y, Z]/((X) \cap (Y, Z)) = k[x, y, z]$. Then $\mathrm{Min}(A) = \{(x), (y, z)\}$ and $\dim(A/(x)) = 2$, $\dim(A/(y, z)) = 1$. Thus A is not equidimensional.

  (iv) Let A be a Noetherian ring, let $\mathfrak{p}$ and $\mathfrak{q}$ be two prime ideals of A such that $\mathrm{ht}(\mathfrak{p}) \neq \mathrm{ht}(\mathfrak{q})$ and let $S := A \setminus (\mathfrak{p} \cup \mathfrak{q})$. Then clearly $S^{-1}A$ is a semilocal ring that is not equicodimensional.

Lemma 3.1.11 *Let* $(A, \mathfrak{m})$ *be an equidimensional catenary local ring. Then, for any prime ideals* $\mathfrak{p} \subseteq \mathfrak{q}$ *of* A, *we have* $\mathrm{ht}(\mathfrak{q}) = \mathrm{ht}(\mathfrak{p}) + \mathrm{ht}(\mathfrak{q}/\mathfrak{p})$. *In particular, for any prime ideal* $\mathfrak{p}$ *of* A, *we have* $\dim(A) = \dim(A/\mathfrak{p}) + \mathrm{ht}(\mathfrak{p})$.

Proof Let $\mathfrak{p}_0$ be a minimal prime ideal of A contained in $\mathfrak{p}$. Then the catenarity of A implies by Proposition 3.1.5 that $\mathrm{ht}(\mathfrak{q}/\mathfrak{p}) = \mathrm{ht}(\mathfrak{q}/\mathfrak{p}_0) - \mathrm{ht}(\mathfrak{p}/\mathfrak{p}_0)$. On the other hand, $\mathrm{ht}(\mathfrak{p}/\mathfrak{p}_0) = \mathrm{ht}(\mathfrak{m}/\mathfrak{p}_0) - \mathrm{ht}(\mathfrak{m}/\mathfrak{p}) = \dim(A) - \mathrm{ht}(\mathfrak{m}/\mathfrak{p})$ and this doesn't depend on $\mathfrak{p}_0$, hence $\mathrm{ht}(\mathfrak{p}) = \mathrm{ht}(\mathfrak{p}/\mathfrak{p}_0)$. Similarly $\mathrm{ht}(\mathfrak{q}) = \mathrm{ht}(\mathfrak{q}/\mathfrak{p}_0)$, therefore $\mathrm{ht}(\mathfrak{q}) = \mathrm{ht}(\mathfrak{p}) + \mathrm{ht}(\mathfrak{q}/\mathfrak{p})$. $\qquad\square$

The above Lemma is not true in the non-local case, even if A is a catenary integral domain.

Example 3.1.12 An example of a Noetherian catenary domain A with a prime ideal $\mathfrak{p}$, such that $\dim(A/\mathfrak{p}) + \mathrm{ht}(\mathfrak{p}) < \dim(A)$.

Proof Let (D, tD) be a discrete valuation ring with parameter t and let $A := D[X]$. Consider the ideal $\mathfrak{p} = (tX - 1)A$, which is a maximal ideal of A. Then $\mathrm{ht}(\mathfrak{p}) = 1$ and $\dim(A/\mathfrak{p}) = 0$, while $\dim(A) = 2$. $\qquad\square$

If a ring A is catenary, the ring of polynomials over A is not necessarily catenary. We will give such an example later in Example 3.1.43. Therefore we need the following definition. We will restrict to Noetherian rings, as we will not deal with the non-Noetherian case.

Definition 3.1.13 A Noetherian ring A is called **universally catenary** if any A-algebra of finite type is catenary.

Universally catenary rings are one of the most useful classes of Noetherian rings in which chain conditions are satisfied. They will appear in the definition of excellent rings. One of our main aims in this section is to prove some important characterizations of universally catenary rings.

Lemma 3.1.14 *Let A be a universally catenary ring. Then any ring of fractions of A and any A-algebra of finite type are universally catenary.*

Proof Let S be a multiplicatively closed system in A. If B is a $S^{-1}A$- algebra of finite type, then $B = S^{-1}C$, where C is a A-algebra of finite type. By assumption C is catenary, hence by Proposition 3.1.6 we get that B is catenary too. The other statement follows from the definition of universally catenary rings. □

Lemma 3.1.15 *Let A be a Noetherian ring. The following are equivalent:*

 (i) *A is universally catenary;*
 (ii) *$A_{\mathfrak{p}}$ is universally catenary for any $\mathfrak{p} \in \text{Spec}(A)$;*
 (iii) *$A_{\mathfrak{m}}$ is universally catenary for any $\mathfrak{m} \in \text{Max}(A)$.*

Proof Using Lemma 3.1.14 we have only to prove $(iii) \Rightarrow (i)$. Let B be an A-algebra of finite type and $\mathfrak{q}$ be a prime ideal of B. Setting $\mathfrak{p} = \mathfrak{q} \cap A$, by assumption $A_{\mathfrak{p}}$ is universally catenary; hence $B_{\mathfrak{p}}$ is catenary. But $B_{\mathfrak{q}}$ is a ring of fractions of $B_{\mathfrak{p}}$, therefore is catenary. Now we apply Proposition 3.1.7. □

Remark 3.1.16 Let A be a Noetherian ring.

(i) A is catenary if and only if $A/\mathfrak{p}$ is catenary for any $\mathfrak{p} \in \text{Min}(A)$;
(ii) A is universally catenary if and only if $A/\mathfrak{p}$ is universally catenary for any $\mathfrak{p} \in \text{Min}(A)$.

Lemma 3.1.17 *A Noetherian ring A is universally catenary if and only if the polynomial ring $A[X_1, \ldots, X_n]$ is catenary for any $n \in \mathbb{N}$.*

Proof If A is universally catenary, then obviously $A[X_1, \ldots, X_n]$ is catenary. Conversely, let $B := A[x_1, \ldots, x_n]$ be an A-algebra of finite type. Then we have $B \simeq A[X_1 \ldots, X_n]/\mathfrak{a}$, where $\mathfrak{a}$ is an ideal in $A[X_1, \ldots, X_n]$. Now apply Proposition 3.1.6. □

Proposition 3.1.18 *Let A be a Cohen-Macaulay ring and $\mathfrak{a}$ an ideal of A. Then $A/\mathfrak{a}$ is universally catenary.*

Proof For any $n \in \mathbb{N}$ the ring $A[T_1, \ldots, T_n]$ is Cohen-Macaulay, so that the quotient ring $A/\mathfrak{a}[T_1, \ldots, T_n] \simeq A[T_1, \ldots, T_n]/\mathfrak{a}A[T_1, \ldots, T_n]$ is catenary. Now we apply Lemma 3.1.17. $\square$

Corollary 3.1.19 *Any complete semilocal ring is universally catenary.*

One of the most important properties of universally catenary rings is the so-called dimension formula. Recall that if $A \subseteq B$ is an extension of integral domains, by tr.$\deg_A B$ we mean tr.$\deg_{Q(A)} Q(B)$.

Theorem 3.1.20 (I.S. Cohen [84]) *Let A be a Noetherian domain and B a (not necessarily Noetherian) domain containing A. If $\mathfrak{q}$ is a prime ideal of B and $\mathfrak{p} := \mathfrak{q} \cap A$, then*

$$\mathrm{ht}(\mathfrak{q}) + \mathrm{tr.deg}_{k(\mathfrak{p})} k(\mathfrak{q}) \leq \mathrm{ht}(\mathfrak{p}) + \mathrm{tr.deg}_A B.$$

If B is a polynomial ring in finitely many variables over A, or if A is universally catenary and B is finitely generated over A, we have equality.

Proof Changing A with $A_\mathfrak{p}$ we may assume that A is a local ring with maximal ideal $\mathfrak{p}$. Let $s, t \in \mathbb{N}$ be such that $s \leq \mathrm{ht}(\mathfrak{q})$ and $t \leq \mathrm{tr.deg}_{k(\mathfrak{p})} k(\mathfrak{q})$. Then there exists a chain of prime ideals in B of the form $\mathfrak{q}_0 \subsetneq \mathfrak{q}_1 \subsetneq \cdots \subsetneq \mathfrak{q}_s = \mathfrak{q}$. For any $i = 1, \ldots, s$, pick an element $y_i \in \mathfrak{q}_i \setminus \mathfrak{q}_{i-1}$. Let also $z_1, \ldots, z_t \in B$ be such that their images $\bar{z}_1, \ldots, \bar{z}_t \in k(\mathfrak{q})$ are transcendental over $k(\mathfrak{p})$. Let $C := A[y_1, \ldots, y_s, z_1, \ldots, z_t]$ and $\mathfrak{r} := \mathfrak{q} \cap C$. Assume that the assertion is true for C. Then

$$\mathrm{ht}(\mathfrak{r}) + \mathrm{tr.deg}_{k(\mathfrak{p})} k(\mathfrak{r}) \leq \mathrm{ht}(\mathfrak{p}) + \mathrm{tr.deg}_A C \leq \mathrm{ht}(\mathfrak{p}) + \mathrm{tr.deg}_A B.$$

But by construction, we have $\mathrm{ht}(\mathfrak{r}) \geq s$ and $\mathrm{tr.deg}_{k(\mathfrak{p})} k(\mathfrak{r}) \geq t$ so that we obtain $s + t \leq \mathrm{ht}(\mathfrak{p}) + \mathrm{tr.deg}_A B$. Since $s \leq \mathrm{ht}(\mathfrak{q})$ and $t \leq \mathrm{tr.deg}_{k(\mathfrak{p})} k(\mathfrak{q})$ are arbitrary, by changing B with C, we may assume that B is a finitely generated A-algebra, say $B = A[x_1, \ldots, x_n]$. If the assertion is true for $A[x_1]$ then

$$\mathrm{ht}(\mathfrak{q}) + \mathrm{tr.deg}_{k(\mathfrak{q} \cap A[x_1])} k(\mathfrak{q}) \leq \mathrm{ht}(\mathfrak{q} \cap A[x_1]) + \mathrm{tr.deg}_{A[x_1]} B$$

and

$$\mathrm{ht}(\mathfrak{q} \cap A[x_1]) + \mathrm{tr.deg}_{k(\mathfrak{p})}(k(\mathfrak{q} \cap A[x_1])) \leq \mathrm{ht}(\mathfrak{p}) + \mathrm{tr.deg}_A A[x_1].$$

This shows that it is enough to prove the assertion for $n = 1$.

Case 1: x_1 is transcendental over A. Then either $\mathfrak{q} = \mathfrak{p}B$ or $\mathrm{ht}(\mathfrak{q}) = \mathrm{ht}(\mathfrak{p}B)+1 = \mathrm{ht}(\mathfrak{p})+1$. If $\mathfrak{q} = \mathfrak{p}B$, then $\mathrm{tr.deg}_{k(\mathfrak{p})}k(\mathfrak{q}) = 1$. In the other case, $\mathrm{tr.deg}_{k(\mathfrak{p})}k(\mathfrak{q}) = 0$. In both cases we are done. This also shows that equality holds if B is a polynomial ring in finitely many indeterminates over A.

Case 2: x_1 is not transcendental over A. Then $\mathrm{tr.deg}_A B = 0$. We can write $B = A[X]/\mathfrak{a}$, where $\mathfrak{a} \neq (0)$ is a prime ideal of $A[X]$. Since $A \subseteq B$, by inverting all non-zero elements of A we get that $\mathrm{ht}(\mathfrak{a}) = 1$. Let $\mathfrak{r} \in \mathrm{Spec}(A[X])$ such that $\mathfrak{q} = \mathfrak{r}/\mathfrak{a}$. Then $\mathfrak{r} \cap A = \mathfrak{p}$ and $k(\mathfrak{r}) = k(\mathfrak{q})$. By the transcendental case we have that

$$\mathrm{ht}(\mathfrak{r}) + \mathrm{tr.deg}_{k(\mathfrak{p})}k(\mathfrak{r}) = \mathrm{ht}(\mathfrak{p}) + \mathrm{tr.deg}_A A[X].$$

But $\mathrm{ht}(\mathfrak{q}) + 1 = \mathrm{ht}(\mathfrak{q}) + \mathrm{ht}(\mathfrak{a}) \leq \mathrm{ht}(\mathfrak{r})$, with equality by Lemma 3.1.11 if A is universally catenary, because in this case $A[X]$ is a catenary domain. It follows that $\mathrm{ht}(\mathfrak{q}) + \mathrm{tr.deg}_{k(\mathfrak{p})}k(\mathfrak{q}) \leq \mathrm{ht}(\mathfrak{p})$. $\qquad\square$

Definition 3.1.21 We say that a Noetherian integral domain A satisfies the **dimension formula** if for any finitely generated A-algebra B which is a domain and for any $\mathfrak{q} \in \mathrm{Spec}(B)$, if we put $\mathfrak{p} := \mathfrak{q} \cap A$, then

$$\mathrm{ht}(\mathfrak{q}) + \mathrm{tr.deg}_{k(\mathfrak{p})}k(\mathfrak{q}) = \mathrm{ht}(\mathfrak{p}) + \mathrm{tr.deg}_A B.$$

A Noetherian ring A satisfies the **dimension formula**, if $A/\mathfrak{p}$ satisfies the dimension formula, for any $\mathfrak{p} \in \mathrm{Min}(A)$.

Theorem 3.1.20 shows that:

Corollary 3.1.22 *Any universally catenary Noetherian ring A satisfies the dimension formula.*

Lemma 3.1.23 *Let A be a Noetherian ring.*

(i) *If A satisfies the dimension formula and S is a multiplicatively closed system in A, then $S^{-1}A$ satisfies the dimension formula.*

(ii) *A satisfies the dimension formula if and only if $A_{\mathfrak{m}}$ satisfies the dimension formula for any $\mathfrak{m} \in \mathrm{Max}(A)$.*

(iii) *If A is a domain that satisfies the dimension formula and B is a ring of fractions of a finitely generated integral domain over A, then B satisfies the dimension formula.*

Proof

(i) Follows from the properties of rings of fractions.
(ii) Follows from (i) and from the fact that any prime ideal is contained in a maximal one.
(iii) Suppose that $B = S^{-1}C$ where C is a domain, finitely generated over A. Then every finitely generated domain over C is also finitely generated over A. Now the conclusion follows by definition and applying (i). $\qquad\square$

Corollary 3.1.24 *Let $(A, \mathfrak{m})$ be a Noetherian universally catenary local domain of dimension n, B be a domain finite extension of A and $\mathfrak{n} \in \mathrm{Max}(B)$. Then $height(\mathfrak{n}) = n$. In particular B is equicodimensional.*

Proof We have $\mathrm{tr.deg}_A B = 0$ and $k(B_\mathfrak{n})$ is an algebraic extension of $Q(A)$. Moreover $\mathfrak{n} \cap A = \mathfrak{m}$. Now we apply Theorem 3.1.20. $\qquad\square$

Now we introduce another important class of Noetherian rings that, as we will see later, is much connected with the universally catenary rings.

Definition 3.1.25 A Noetherian semilocal ring A is called **formally equidimensional** or **quasi-unmixed** if $\widehat{A}$ is equidimensional, where $\widehat{A}$ is the completion of A in the radical topology. A is called **strictly formally equidimensional** or **unmixed** if $\widehat{A}$ is equidimensional and has no embedded primes, that is, $\mathrm{Ass}(\widehat{A}) = \mathrm{Min}(\widehat{A})$.

The following theorem gives a classical example of (quasi-)unmixed rings.

Theorem 3.1.26 (Nagata [262]) *Let A be a local domain which is a quotient of a Cohen-Macaulay ring. Then A is unmixed.*

Proof Let R be a Cohen-Macaulay ring and $\mathfrak{p} \in \mathrm{Spec}(R)$ be such that $A \simeq R/\mathfrak{p}$. Assume that $\mathrm{ht}(\mathfrak{p}) = r$ and applying [266, Th. 9.5] take elements $x_1, \ldots, x_r \in \mathfrak{p}$ such that if $\mathfrak{a}$ is the ideal generated by $x_1, \ldots, x_r$, we have $\mathrm{ht}(\mathfrak{a}) = r$. Since $\widehat{R}$ is Cohen-Macaulay, the unmixedness theorem [249, Th. 17.6] implies that $\mathrm{Ass}(\widehat{R}/\mathfrak{a}\widehat{R}) = \mathrm{Min}(\mathfrak{a}\widehat{R})$ and $\widehat{R}/\mathfrak{a}\widehat{R}$ is unmixed. Let $\mathfrak{q} \in \mathrm{Ass}(\widehat{R}/\mathfrak{p}\widehat{R})$. As $\mathfrak{p} \in \mathrm{Min}(\mathfrak{a})$ it follows that $\mathfrak{q} \in \mathrm{Ass}(\widehat{R}/\mathfrak{a}\widehat{R})$. Thus $\widehat{R}/\mathfrak{p}\widehat{R}$ is unmixed and the assertion is proved. $\qquad\square$

Lemma 3.1.27 *Let A be a formally equidimensional semilocal ring and let $\mathfrak{p} \in \mathrm{Spec}(A)$. Then $A/\mathfrak{p}$ is formally equidimensional and*

$$\dim(A) = \mathrm{ht}(\mathfrak{p}) + \dim(A/\mathfrak{p}).$$

Proof Applying [249, Th. 8.15] and due to the form of prime ideals in a direct product of rings, we may assume that A is a local ring. Let $q \in \text{Min}(\mathfrak{p}\widehat{A})$. Then

$$\text{ht}(q) = \text{ht}(\mathfrak{p}) + \text{ht}(q/\mathfrak{p}\widehat{A}) = \text{ht}(\mathfrak{p}).$$

Since $\widehat{A}$ is a local universally catenary and equidimensional ring, by Lemma 3.1.11 we have

$$\dim(\widehat{A}/q) = \dim(\widehat{A}) - \text{ht}(q) = \dim(A) - \text{ht}(\mathfrak{p}).$$

On the other hand, $\widehat{A/\mathfrak{p}} \cong \widehat{A}/\mathfrak{p}\widehat{A}$. Now from the above considerations, we get that $A/\mathfrak{p}$ is formally equidimensional and that

$$\dim(A) = \text{ht}(\mathfrak{p}) + \dim(A/\mathfrak{p}).$$

$\square$

The property proved in Lemma 3.1.27 for quasi-unmixed (formally equidimensional) rings is no more true for unmixed rings, as we shall see in the next example. In particular, this provides an example of a quasi-unmixed local ring which is not unmixed.

Example 3.1.28 There exists a Noetherian local ring A such that $\widehat{A}$ is a domain, with a prime ideal $\mathfrak{p}$ of A such that $A/\mathfrak{p}$ is not unmixed. In particular, $A/\mathfrak{p}$ is a quasi-unmixed local ring which is not unmixed.

Proof The example was constructed by Brodmann and Rotthaus [75, Cor. (15)] and answers an old question of Nagata [262, §5, Problem 2]. We sketch the construction, leaving to the reader the details. We use the second construction from Sect. 1.5. Let K_0 be a countable field and $n = 3$. Consider the polynomials $G_j \in K[X, Z_1, Z_2, Z_3]$, $j = 1, \ldots, 4$, as follows:

$$G_1(X, Z_1, Z_2, Z_3) = Z_2^3 - Z_3^2; \quad G_2(X, Z_1, Z_2, Z_3) = Z_2 X^2 - Z_1^3;$$

$$G_3(X, Z_1, Z_2, Z_3) = Z_2 Z_1 - X Z_3; \quad G_4(X, Z_1, Z_2, Z_3) = Z_2^2 X - Z_3 Z_1.$$

Then, performing the second construction in Sect. 1.5, we see that Theorem 1.5.18 gives us a local Noetherian domain $(A, \mathfrak{m})$ with a prime element x such that

$$\widehat{A} \simeq K[[x, \zeta_1, \zeta_2, \zeta_3, \zeta_4]]/(\zeta_2^3 - \zeta_3^2, \zeta_2 x^2 - \zeta_1^3, \zeta_2 \zeta_1 - x\zeta_3, \zeta_2^2 x - \zeta_3 \zeta_1)$$

and

$$\widehat{A}/x\widehat{A} \simeq K[[\zeta_1, \zeta_2, \zeta_3, \zeta_4]]/(\zeta_2^3 - \zeta_3^2, \zeta_1^2, \zeta_2\zeta_1, \zeta_3\zeta_1) \simeq$$

$$\simeq K[[\zeta_1, \zeta_2, \zeta_3, \zeta_4]]/(\zeta_2^3 - \zeta_3^2, \zeta_1) \cap (\zeta_1, \zeta_2, \zeta_3)^2.$$

We leave to the reader the task to check that $\widehat{A}$ is a domain and to compute $\mathrm{Ass}(\widehat{A}/x\widehat{A})$.
□

Lemma 3.1.29 *Let A be a semilocal ring and let $\{\mathfrak{q}_1, \dots, \mathfrak{q}_r\}$ be the minimal prime ideals of A. The following are equivalent:*

(i) A is formally equidimensional;
(ii) A is equidimensional and $A/\mathfrak{q}_i$ is formally equidimensional for any $i = 1, \dots, r$.

Proof $(i) \Rightarrow (ii)$: Follows from Lemma 3.1.27.
$(ii) \Rightarrow (i)$: Let $\mathfrak{q} \in \mathrm{Min}(\widehat{A})$. Then there is $i \in \{1, \dots, r\}$ such that $\mathfrak{q} \in \mathrm{Min}(\mathfrak{q}_i\widehat{A})$. It follows that

$$\dim(\widehat{A}/\mathfrak{q}) = \dim(\widehat{A}/\mathfrak{q}_i\widehat{A}) = \dim(A/\mathfrak{q}_i) = \dim(A) = \dim(\widehat{A}),$$

hence A is formally equidimensional.
□

Lemma 3.1.30 *Let A be a semilocal formally equidimensional ring, $\mathfrak{a}$ be an ideal of A and $\{\mathfrak{p}_1, \dots, \mathfrak{p}_t\}$ be the minimal prime over-ideals of $\mathfrak{a}$. Then $A/\mathfrak{a}$ is formally equidimensional if and only if $\mathrm{ht}(\mathfrak{p}_i) = \mathrm{ht}(\mathfrak{p}_j)$, for all $i, j \in \{1, \dots, t\}$.*

Proof From Lemma 3.1.27, we get that $A/\mathfrak{p}_i \simeq (A/\mathfrak{a})/(\mathfrak{p}_i/\mathfrak{a})$ is formally equidimensional for all i. From Lemma 3.1.29, it follows that $A/\mathfrak{a}$ is formally equidimensional if and only if $\dim((A/\mathfrak{a})/(\mathfrak{p}_i/\mathfrak{a})) = \dim(A/\mathfrak{a})$, $\forall i = 1, \dots, t$. This is equivalent to $\dim(A/\mathfrak{p}_i) = \dim(A/\mathfrak{p}_j)$, $\forall i, j = 1, \dots, t$. But as $\dim(A/\mathfrak{p}_i) = \dim(A) - \mathrm{ht}(\mathfrak{p}_i)$, we get that $A/\mathfrak{a}$ is formally equidimensional if and only if $\mathrm{ht}(\mathfrak{p}_i) = \mathrm{ht}(\mathfrak{p}_j)$, $\forall i, j = 1, \dots, t$. □

Corollary 3.1.31 *Let A be a semilocal Cohen-Macaulay ring and $\mathfrak{a}$ an ideal of A. Then $A/\mathfrak{a}$ is unmixed if and only if $\mathrm{ht}(\mathfrak{p}) = \mathrm{ht}(\mathfrak{a})$ for any $\mathfrak{p} \in \mathrm{Ass}(A/\mathfrak{a})$.*

Lemma 3.1.32 *Let $u : (A, \mathfrak{m}) \to (B, \mathfrak{n})$ be a local flat morphism of Noetherian local rings. If B is equidimensional and catenary, then A is equidimensional and catenary and $B/\mathfrak{p}B$ is equidimensional for any prime ideal $\mathfrak{p}$ of A.*

Proof Let $\mathfrak{p}$ be a minimal prime ideal of A and $\mathfrak{q}$ be a prime ideal of B, minimal over $\mathfrak{p}B$. Then $\mathfrak{q}$ is a minimal prime ideal of B, because u is flat. Thus

$$\dim(B/\mathfrak{q}) = \dim(B) - \mathrm{ht}(\mathfrak{q}) = \dim(B),$$

hence $B/\mathfrak{p}B$ is equidimensional and catenary. We have

$$\mathrm{ht}(\mathfrak{m}/\mathfrak{p}) = \mathrm{ht}(\mathfrak{n}/\mathfrak{p}B) - \mathrm{ht}(\mathfrak{n}/\mathfrak{m}B) = \dim(B) - \mathrm{ht}(\mathfrak{n}/\mathfrak{m}B)$$

and this doesn't depend on $\mathfrak{p} \in \mathrm{Min}(A)$. This shows that A is equidimensional. Now take $\mathfrak{p} \in \mathrm{Spec}(A)$ and $\mathfrak{q} \in \mathrm{Min}(\mathfrak{p}B)$. Then $\mathfrak{q} \cap A = \mathfrak{p}$ and $\mathrm{ht}(\mathfrak{q}) = \mathrm{ht}(\mathfrak{p})$. We have

$$\mathrm{ht}(\mathfrak{n}/\mathfrak{q}) = \mathrm{ht}(\mathfrak{n}) - \mathrm{ht}(\mathfrak{q}) = \mathrm{ht}(\mathfrak{n}) - \mathrm{ht}(\mathfrak{p})$$

and this depends only on $\mathfrak{p}$. Therefore $B/\mathfrak{p}B$ is equidimensional. Now let $\mathfrak{p}' \in \mathrm{Spec}(A)$, such that $\mathfrak{p}' \subset \mathfrak{p}$ with $\mathrm{ht}(\mathfrak{p}/\mathfrak{p}') = 1$. Also, let $\mathfrak{q}' \in \mathrm{Min}(\mathfrak{p}'B)$ be such that $\mathfrak{q}' \subset \mathfrak{q}$. Then the induced morphism $A/\mathfrak{p}' \to B/\mathfrak{p}'B$ is flat and $B/\mathfrak{p}'B$ is equidimensional. This implies that $\mathrm{ht}(\mathfrak{q}/\mathfrak{q}') = \mathrm{ht}(\mathfrak{q}/\mathfrak{p}'B) = \mathrm{ht}(\mathfrak{p}/\mathfrak{p}') = 1$. But B is equidimensional and catenary, hence

$$\mathrm{ht}(\mathfrak{q}/\mathfrak{q}') = \mathrm{ht}(\mathfrak{q}) - \mathrm{ht}(\mathfrak{q}') = \mathrm{ht}(\mathfrak{p}) - \mathrm{ht}(\mathfrak{p}'),$$

therefore $\mathrm{ht}(\mathfrak{p}) = \mathrm{ht}(\mathfrak{p}') + 1$ and A is catenary. $\qquad\square$

Lemma 3.1.33 *Let $(A, \mathfrak{m})$ be a local Noetherian formally equidimensional ring and $\mathfrak{p}$ a prime ideal of A. Then $A_{\mathfrak{p}}$ is formally equidimensional.*

Proof Let $\mathfrak{q} \in \mathrm{Spec}(\widehat{A})$ be such that $\mathfrak{q} \cap A = \mathfrak{p}$. Then the induced morphism $A_{\mathfrak{p}} \to B := \widehat{A}_{\mathfrak{q}}$ is flat, hence also $\widehat{A_{\mathfrak{p}}} \to \widehat{B}$ is flat. But B is a quotient of a regular ring, hence by Theorem 3.1.26 we know that $\widehat{B}$ is equidimensional. Now Lemma 3.1.32 says that $\widehat{A_{\mathfrak{p}}}$ is equidimensional. $\qquad\square$

Lemma 3.1.34 *Let $(A, \mathfrak{m})$ be a Noetherian local formally equidimensional ring, n a natural number and $\mathfrak{q} \in \mathrm{Spec}(A[X_1, \ldots, X_n])$. Then $A[X_1, \ldots, X_n]_{\mathfrak{q}}$ is formally equidimensional.*

Proof The morphism $A[X_1, \ldots, X_n] \to \widehat{A}[X_1, \ldots, X_n]$ is faithfully flat. Let $\mathfrak{p} \in \mathrm{Spec}(\widehat{A}[X_1, \ldots, X_n])$ be such that $\mathfrak{p} \cap A[X_1, \ldots, X_n] = \mathfrak{q}$. The induced morphism $B := A[X_1, \ldots, X_n]_{\mathfrak{q}} \to C := \widehat{A}[X_1, \ldots, X_n]_{\mathfrak{p}}$ is faithfully flat and then the morphism $\widehat{B} \to \widehat{C}$ is flat. The ring C is equidimensional and is a quotient of a regular ring, hence $\widehat{C}$ is equidimensional. By Lemma 3.1.32 it follows that $\widehat{B}$ is equidimensional. $\qquad\square$

Theorem 3.1.35 *Let A be a Noetherian local formally equidimensional domain. Then A is universally catenary.*

Proof Let $\mathfrak{p} \in \mathrm{Spec}(A)$. Then by Lemma 3.1.27, it follows that $A/\mathfrak{p}$ is formally equidimensional and for any prime ideal $\mathfrak{q} \supset \mathfrak{p}$, we have

$$\dim(A/\mathfrak{p}) = \mathrm{ht}(\mathfrak{q}/\mathfrak{p}) + \dim(A/\mathfrak{q}).$$

Consequently if we have $\mathfrak{p} \subset \mathfrak{q} \subset \mathfrak{r}$ prime ideals in A, we obtain

$$\mathrm{ht}(\mathfrak{r}/\mathfrak{p}) = \mathrm{ht}(\mathfrak{r}/\mathfrak{q}) + \mathrm{ht}(\mathfrak{q}/\mathfrak{p}),$$

so that Lemma 3.1.5 tells us that A is catenary. Let n be a natural number and $A_n :=$ $A[X_1, \ldots, X_n]$. Then for any $\mathfrak{p} \in \mathrm{Spec}(A_n)$, by Lemma 3.1.34 it follows that $(A_n)_{\mathfrak{p}}$ is formally equidimensional and the above proof shows that $(A_n)_{\mathfrak{p}}$ is catenary. By Lemma 3.1.14 and Lemma 3.1.17, we get the assertion.   □

Theorem 3.1.36 (Ratliff [302, Prop. 2.13]) *Let A be a Noetherian domain. The following are equivalent:*

 (i) A is universally catenary;
 (ii) A satisfies the dimension formula;
 (iii) $A_n := A[X_1, \ldots, X_n]$ is catenary, for sufficiently large n.

Proof $(i) \Rightarrow (iii)$: Obvious.
$(iii) \Rightarrow (i)$: Since $A_{n-1} \simeq A_n/(X_n)$, by Lemma 3.1.6 we get that A_n is catenary for any $n \in \mathbb{N}$.
$(i) \Rightarrow (ii)$: Is already proved in Corollary 3.1.22.
$(ii) \Rightarrow (i)$: Assume that for some $n \geq 0$, the ring A_n is not catenary. Then there exist prime ideals $\mathfrak{p}$, $\mathfrak{q}$ and $\mathfrak{r}$ in A_n with $\mathfrak{r} \subseteq \mathfrak{p} \subseteq \mathfrak{q}$ such that $\mathfrak{r}$ is a minimal prime ideal and $\mathrm{ht}(\mathfrak{q}/\mathfrak{r}) > \mathrm{ht}(\mathfrak{p}/\mathfrak{r}) + \mathrm{ht}(\mathfrak{q}/\mathfrak{p})$. This implies that there exist prime ideals $\mathfrak{q}$ and $\mathfrak{p}$ in A_n with $\mathfrak{p} \subseteq \mathfrak{q}$ and such that $d := \mathrm{ht}(\mathfrak{q}/\mathfrak{p}) < \mathrm{ht}(\mathfrak{q}) - \mathrm{ht}(\mathfrak{p})$. Let $s := \mathrm{ht}(\mathfrak{p})$ and $R := (A_n)_{\mathfrak{q}}$. Let also $\mathfrak{m} := \mathfrak{q}R = \mathfrak{q}A[X_1, \ldots, X_n]_{\mathfrak{q}}$ and $\mathfrak{n} := \mathfrak{p}R = \mathfrak{p}A[X_1, \ldots, X_n]_{\mathfrak{q}}$. Then $(R, \mathfrak{m})$ is a Noetherian local domain containing a prime ideal $\mathfrak{n}$ such that $\mathrm{ht}(\mathfrak{n}) = s$, $\dim(R/\mathfrak{n}) = d$ and $s + d < \dim(R)$. By [266, Th. 9.5] we can take elements $\alpha_1, \ldots, \alpha_s \in \mathfrak{n}$ such that, if we put $\mathfrak{a} := (\alpha_1, \ldots, \alpha_s)R$, we have $\mathrm{ht}(\mathfrak{a}) = s$ and $\mathfrak{n} \in \mathrm{Min}(\mathfrak{a})$. Then, applying [18, Lemma 4.4] we see that there exists $b \in \mathfrak{m}\backslash\mathfrak{n}$ such that $\mathfrak{a} : b^j R$ is the $\mathfrak{n}$-primary component of $\mathfrak{a}$ for any $j \geq 1$. Set $\mathfrak{a}^* := \mathfrak{a} + bR$, $y_i := \frac{\alpha_i}{b}$ for $i = 1, \ldots, s$, $D := R[y_1, \ldots, y_s]$ and $\mathfrak{b} := (y_1, \ldots, y_s)D$. If z is an element of D, then we can write $z = \frac{\alpha}{b^i}$ for i sufficiently large, with $\alpha \in \mathfrak{a}^{*i}$. As no power of b is in $\mathfrak{a}$, it follows that $\mathfrak{b} \subsetneq D$. We have

$$\mathfrak{a} : bR \subseteq \mathfrak{a}\mathfrak{a}^{*j} : b^{j+1}R \subseteq \mathfrak{a} : b^{j+1}R = \mathfrak{a} : bR, \ \forall j \geq 0.$$

Then $\mathfrak{b} \cap R = \mathfrak{a} : bR$. Therefore $\mathfrak{b} \cap R = \mathfrak{a} : bR$ is $\mathfrak{n}$-primary. Thus $D/\mathfrak{b} = R/\mathfrak{b} \cap R$ is a local ring of dimension d. It follows that $\mathfrak{r} := (\mathfrak{m} + \mathfrak{b})D \subseteq \mathrm{Max}(D)$ and $D_{\mathfrak{r}}/\mathfrak{b}D_{\mathfrak{r}} = R/\mathfrak{b} \cap R$. Since $\mathfrak{b}D_{\mathfrak{r}}$ is generated by s elements and $\dim(D_{\mathfrak{r}}/\mathfrak{b}D_{\mathfrak{r}}) = d$, we have that $\dim(D_{\mathfrak{r}}) \leq s + d$. Since $\mathfrak{r}D_{\mathfrak{r}} \cap R = \mathfrak{m}$ and $s + d < \mathrm{ht}(\mathfrak{m})$ it follows that R doesn't satisfy the dimension formula, hence A doesn't satisfy the dimension formula. $\qquad\square$

Corollary 3.1.37 *Let A be a Noetherian domain, $\mathfrak{p} \in \mathrm{Spec}(A)$ and B a domain finitely generated over $A/\mathfrak{p}$. If A satisfies the dimension formula, then B satisfies the dimension formula.*

Proof A is universally catenary, so also B is universally catenary. Now apply Theorem 3.1.36. $\qquad\square$

Theorem 3.1.38 (Mc Adam [256, Th. 1]) *Let A be a Noetherian ring and $\mathfrak{p}$ a prime ideal of A of height n. Let $H := \{\mathfrak{q} \in \mathrm{Spec}(A) \mid \mathfrak{q} \supset \mathfrak{p} \text{ and } \mathrm{ht}(\mathfrak{q}/\mathfrak{p}) = 1\}$. Then all but finitely many prime ideals $\mathfrak{q}$ of H have height $n + 1$.*

Proof By [190, Th. 153] there exist elements $a_1, \dots, a_n$ in A such that $\mathfrak{p} \in \mathrm{Min}(\mathfrak{a})$, where $\mathfrak{a} := (a_1, \dots, a_n)$. Let $\mathfrak{p}_2, \dots, \mathfrak{p}_t$ be the other minimal prime ideals of $\mathfrak{a}$. Let $G := \{\mathfrak{q} \in \mathrm{Spec}(A) \mid \mathfrak{q} \supset \mathfrak{p}, \ \mathrm{ht}(\mathfrak{q}/\mathfrak{p}) = 1 \text{ and } \mathrm{ht}(\mathfrak{q}) > n + 1\}$. Suppose that G is infinite. Then, since $\mathrm{ht}(\mathfrak{q}/\mathfrak{p}) = 1$ for any $\mathfrak{q} \in G$, it follows that $\bigcap_{\mathfrak{q} \in G} \mathfrak{q} = \mathfrak{p}$. This means that there exists $\mathfrak{q} \in G$ such that $\mathfrak{q} \not\supseteq \mathfrak{p}_j, \ \forall j \geq 2$. Thus $\mathrm{ht}(\mathfrak{q}) > n + 1$ and $\mathrm{ht}(\mathfrak{q}/\mathfrak{a}) = \mathrm{ht}(\mathfrak{q}/\mathfrak{p}) = 1$. But then [190, Th. 154] shows that $\mathrm{ht}(\mathfrak{q}) \leq n + 1$. Contradiction! $\qquad\square$

Proposition 3.1.39 (Ratliff [302, Prop. 2.16]) *Let $(A, \mathfrak{m})$ be a Noetherian local catenary domain of dimension n. Assume that there exists $\mathfrak{q} \in \mathrm{Min}(\widehat{A})$ such that $1 < d := \dim(\widehat{A}/\mathfrak{q}) < n$. For each $i = 1, 2, \dots, d - 1$, let*

$$\mathcal{P}_i := \{\mathfrak{p} \in \mathrm{Spec}(A) \mid \mathrm{ht}(\mathfrak{p}) = i, \ \exists \mathfrak{n} \in \mathrm{Min}(\mathfrak{p}\widehat{A}) \cap V(\mathfrak{q}), \ \dim(\widehat{A}/\mathfrak{n}) = d - i\}.$$

Then the sets $\mathcal{P}_1, \dots \mathcal{P}_{d-1}$ are not empty.

Proof We will prove that $\mathcal{P}_i \neq \emptyset$ for $i = 1, \dots, d - 1$ by induction on i. Let $a \in \mathfrak{m}, a \neq 0$ and let $\mathfrak{n} \in \mathrm{Min}(a\widehat{A} + \mathfrak{q})$. Then $\mathrm{ht}(\mathfrak{n}/\mathfrak{q}) = 1$ and $a \in \mathfrak{n} \cap A$, hence $\mathfrak{n} \cap A \neq (0)$. Thus, if we set

$$\mathcal{M} := \{\mathfrak{n} \in \mathrm{Spec}(\widehat{A}) \mid \mathfrak{q} \subset \mathfrak{n}, \ \mathrm{ht}(\mathfrak{n}/\mathfrak{q}) = 1 \text{ and } \mathfrak{n} \cap A \neq (0)\},$$

we have $\mathfrak{m} = \bigcup_{\mathfrak{n} \in \mathcal{M}} (\mathfrak{n} \cap A)$. Now $\mathrm{ht}(\mathfrak{m}\widehat{A}/\mathfrak{q}) = d > 1$, hence $\mathfrak{m}\widehat{A} \notin \mathcal{M}$. Thus $\mathfrak{m}\widehat{A}$ is not of the form $\mathfrak{n} \cap A$ with $\mathfrak{n} \in \mathcal{M}$. Then both $\mathcal{M}$ and $\{\mathfrak{n} \cap A \mid \mathfrak{n} \in \mathcal{M}\}$ are infinite sets. Applying Theorem 3.1.38 we get that the set $\mathcal{N} := \{\mathfrak{n} \in \mathcal{M} \mid \mathrm{ht}(\mathfrak{n}) = 1\}$ is infinite. Pick $\mathfrak{n} \in \mathcal{N}$ and let $\mathfrak{p} := \mathfrak{n} \cap A$. Then

$$0 < \mathrm{ht}(\mathfrak{p}) = \mathrm{ht}(\mathfrak{p}\widehat{A}) \leq \mathrm{ht}(\mathfrak{n}) = 1.$$

Thus $\mathrm{ht}(\mathfrak{p}) = 1$ and $\mathfrak{n} \in \mathrm{Min}(\mathfrak{p}\widehat{A})$. Since $\widehat{A}/\mathfrak{q}$ is a catenary domain it follows that

$$\dim(\widehat{A}/\mathfrak{n}) = \dim(\widehat{A}/\mathfrak{q}) - \mathrm{ht}(\mathfrak{n}/\mathfrak{q}) = d - 1.$$

This implies that $\mathfrak{p} \in \mathcal{P}_1$ and we are done for $i = 1$.

Let $i > 1$ and take $\mathfrak{p} \in \mathcal{P}_1$. Let $\overline{A} := A/\mathfrak{p}$ and $\overline{\mathfrak{n}} := \mathfrak{n}/\mathfrak{p}\widehat{A}$. Since A is a catenary domain, we have $\dim(\overline{A}) = n - 1$ and $\overline{\mathfrak{n}}$ is a minimal prime in $\widehat{\overline{A}} = \widehat{A}/\mathfrak{p}\widehat{A}$. Moreover

$$\dim(\widehat{\overline{A}}/\overline{\mathfrak{n}}) = \dim(\widehat{A}/\mathfrak{n}) = d - 1 < n - 1.$$

By induction there exists $\mathfrak{r}_i = \mathfrak{p}_i/\mathfrak{p} \in \mathrm{Spec}(\overline{A})$ such that $\mathrm{ht}(\mathfrak{r}_i) = i - 1$ and a minimal prime of $\mathfrak{r}_i \widehat{\overline{A}}$ say $\overline{\mathfrak{n}}_i$, such that $\overline{\mathfrak{n}} \subset \overline{\mathfrak{n}}_i$ and

$$\dim(\widehat{\overline{A}}/\overline{\mathfrak{n}}_i) = (d - 1) - (i - 1) = d - i.$$

Suppose that $\overline{\mathfrak{n}}_i = \mathfrak{n}_i/\mathfrak{p}\widehat{A}$. Then $\mathfrak{n}_i$ is a minimal prime of $\mathfrak{p}_i\widehat{A}$ containing $\mathfrak{n}$. Hence $\mathfrak{n}_i$ contains $\mathfrak{q}$ and

$$\dim(\widehat{A}/\mathfrak{n}_i) = \dim(\widehat{\overline{A}}/\overline{\mathfrak{n}}_i) = d - i.$$

Since A is catenary, we get $\mathrm{ht}(\mathfrak{p}_i) = \mathrm{ht}(\mathfrak{r}_i) + \mathrm{ht}(\mathfrak{p}) = i$. Therefore $\mathfrak{p}_i \in \mathcal{P}_i$. $\square$

Lemma 3.1.40 (Ratliff [302, Lemma. 2.17]) *Let* $(A, \mathfrak{m})$ *be a Noetherian local domain of dimension* $d > 1$ *with* $\mathrm{Ass}(\widehat{A}) = \{\mathfrak{q}_1, \ldots, \mathfrak{q}_s\}$. *Consider an irredundant primary decomposition* $(0) = \bigcap_{i=1}^{s} \mathfrak{r}_i$ *in* $\widehat{A}$, *such that* $\mathfrak{r}_i$ *is* $\mathfrak{q}_i$-*primary for any* $i = 1, \ldots, s$. *Assume that* $\mathrm{ht}(\mathfrak{q}_1) = 0$ *and* $\dim(\widehat{A}/\mathfrak{q}_1) = 1$. *Then there exists* $b, c \in \mathfrak{m}$ *and* $d \in \bigcap_{i=2}^{s} \mathfrak{r}_i \setminus \mathfrak{q}_1$ *such that:*

(i) $b - d \in \mathfrak{r}_1$;
(ii) $(b, d)\widehat{A} = (b, c)\widehat{A}$;
(iii) $\frac{c}{b}$ *is integral over* A *and* $\frac{c}{b} \notin A$.

Proof Because q_1 is minimal and the primary decomposition is irredundant, $\bigcap_{j=2}^{s} \mathfrak{r}_j \nsubseteq q_1$, so we can pick an element $d' \in \bigcap_{i=2}^{s} \mathfrak{r}_i \setminus q_1$. As $\dim(\widehat{A}/q_1) = 1$, the ideal $\mathfrak{r}_1 + d'\widehat{A}$ is $\widehat{\mathfrak{m}}$-primary and consequently $\mathfrak{b} := (\mathfrak{r}_1 + d'\widehat{A}) \cap A$ is an $\mathfrak{m}$-primary ideal. So, if we take an element $b \in \mathfrak{b}$, in $\widehat{A}$ we have $b = z + td'$, with $z \in \mathfrak{r}_1$ and $t \in \widehat{A}$. Then $td' \in \bigcap_{i=2}^{s} q_i$ and as b is not a zero divisor in $\widehat{A}$, we get that $td' \in q_1$. Let $d := td'$. Then $b - d \in \mathfrak{r}_1$ and we have (i). In order to prove the existence of c we will first show that:

(a) $d \notin b\widehat{A}$.

Indeed, if $d = rb$ with $r \in \widehat{A}$, then $b(1 - r) = b - d \in \mathfrak{r}_1$. Since b is not a zero divisor in $\widehat{A}$, we have $1 - r \in \mathfrak{r}_1$. But $\widehat{A}$ is a local ring, hence r is a unit in $\widehat{A}$. Consequently $b = \frac{d}{r}$ is a zero divisor in $\widehat{A}$. Contradiction!

(b) If $\mathfrak{c} \neq (0)$ is an ideal of A such that $\mathfrak{m} \notin \mathrm{Ass}_A(A/\mathfrak{c})$, then $d \in \mathfrak{c}\widehat{A}$.

It is easily seen that $\widehat{\mathfrak{m}} \notin \mathrm{Ass}(\widehat{A}/\mathfrak{c}\widehat{A})$. Indeed, since $\mathfrak{c} : \mathfrak{m} = \mathfrak{c}$, we obtain $\mathfrak{c}\widehat{A} : \mathfrak{m}\widehat{A} = (\mathfrak{c} : \mathfrak{m})\widehat{A} = \mathfrak{c}\widehat{A}$. Let $\mathfrak{p} \in \mathrm{Ass}(\widehat{A}/\mathfrak{c}\widehat{A})$. Then $\mathfrak{p} \neq \mathfrak{m}\widehat{A}$ and $\mathfrak{p} \neq q_1$, since $\mathfrak{p} \cap A \neq (0)$. We have $\mathfrak{p} \nsupseteq q_1$, because $\dim(\widehat{A}/q_1) = 1$, hence $\mathfrak{p} \nsupseteq \mathfrak{r}_1$. Pick an element $\alpha \in \mathfrak{r}_1 \setminus \mathfrak{p}$. If $\mathfrak{q}$ is the $\mathfrak{p}$-primary component of $\mathfrak{c}\widehat{A}$, then $\alpha d = 0 \in \mathfrak{q}$, hence $d \in \mathfrak{q}$. This shows that $d \in \mathfrak{c}\widehat{A}$.

From (a) and (b), it follows that $\mathfrak{m} \in \mathrm{Ass}_A(A/bA)$, so that we can write $bA = \mathfrak{a} \cap \mathfrak{q}$, where $\mathfrak{q}$ is an $\mathfrak{m}$-primary ideal. Then $b\widehat{A} = \mathfrak{c}\widehat{A} \cap \mathfrak{q}\widehat{A}$ and $d \in \mathfrak{c}\widehat{A} \setminus \mathfrak{q}\widehat{A}$. Since

$$(\mathfrak{c}\widehat{A} + \mathfrak{q}\widehat{A})/\mathfrak{q}\widehat{A} = ((\mathfrak{c} + \mathfrak{q})\widehat{A})/\mathfrak{q}\widehat{A} \cong (\mathfrak{c} + \mathfrak{q})/\mathfrak{q},$$

there exists $c \in \mathfrak{c}$ such that $c - d \in \mathfrak{q}\widehat{A}$. Then $c - d \in \mathfrak{c}\widehat{A} \cap \mathfrak{q}\widehat{A} = b\widehat{A}$. It follows that $(b, d)\widehat{A} = (b, c)\widehat{A}$ and (ii) is proved. Now $b - d \in q_1$ and $dq_1 = (0)$, hence $d^2 = db$. Since $c - d \in b\widehat{A}$, we get that $c^2 \in b(b, c)\widehat{A} \cap A = b(b, c)A$, hence $\frac{c}{b}$ is integral over A. Moreover, since $b\widehat{A} \subsetneq (b, d)\widehat{A} = (b, c)\widehat{A}$, it follows that $c \notin bA$ and the proof is complete. $\square$

Before proving the nice result of Ratliff characterizing universally catenary rings (Theorem 3.1.42), we insert a result that will be very useful in the proofs of the theorems about the lifting of **P**-rings in the general case, for example, 4.6.15 or 4.6.14.

Proposition 3.1.41 (Ratliff [302, Prop. 3.5], J. Nishimura and T. Nishimura [273, Prop. 1.8]) *Let $(A, \mathfrak{m})$ be a Noetherian local domain with integral closure A' and $\mathfrak{a}$ a non-zero ideal of A. The following are equivalent:*

(i) *A' has a maximal ideal of height 1;*
(ii) *the henselization $A^h = (A, \mathfrak{a})^h$ of A with respect to $\mathfrak{a}$ has a minimal prime ideal $\mathfrak{p}$ such that $\dim(A^h/\mathfrak{p}) = 1$;*

(iii) *the completion* $A^* = \widehat{(A, \mathfrak{a})}$ *of* A *in the* $\mathfrak{a}$-*adic topology has a minimal prime ideal* $\mathfrak{p}$
    such that $\dim(A^*/\mathfrak{p}) = 1$;
(iv) *the completion* $\widehat{A}$ *of* A *in the* $\mathfrak{m}$-*adic topology has a minimal prime ideal* $\mathfrak{p}$ *such that*
    $\dim(\widehat{A}/\mathfrak{p}) = 1$;

Proof Since the statement is clearly true if $\dim(A) = 1$, we may assume that $\dim(A) > 1$.
(i) $\Rightarrow$ *(ii)* : Let $\mathfrak{m}$ be a maximal ideal of A' of height 1. Then we can find a finite A-algebra
$B \subseteq A'$, which is a semilocal ring with maximal ideals $\mathfrak{n}_1, \ldots, \mathfrak{n}_t$ and such that one of its
maximal ideals, say $\mathfrak{n}_1$, has height 1. Then the $\mathfrak{a}B$-henselization of B is $B^h \simeq B \otimes_A A^h$
[359, ch. 15, Lemma 12.7] Note that

$$B/\mathfrak{a}B \simeq (B/\mathfrak{a}B)_{\mathfrak{n}_1} \oplus T^{-1}(B/\mathfrak{a}B),$$

where $T = B \setminus \bigcup_{j \geq 2} \mathfrak{n}_j$. Therefore, $B^h = (B_{\mathfrak{n}_1})^h \oplus (T^{-1}B)^h$, where $(B_{\mathfrak{n}_1})^h$ is the $\mathfrak{a}B_{\mathfrak{n}_1}$-
henselization of $B_{\mathfrak{n}_1}$ and $(T^{-1}B)^h$ is the henselization of $T^{-1}B$ with respect to $\mathfrak{a}T^{-1}B$.
This shows that B^h has a minimal prime ideal $\mathfrak{q}$ such that $\dim(B^h/\mathfrak{q}) = 1$, hence A^h has
a minimal prime ideal $\mathfrak{p}$ such that $\dim(A^h/\mathfrak{p}) = 1$.
(ii) $\Rightarrow$ *(iii)* : Since $A^* \simeq (A^h)^*$ and the morphism $A^h \to A^*$ is faithfully flat, the
assertion follows.
(iii) $\Rightarrow$ *(iv)* : The completion of A and A^* with respect to the topology of the maximal
ideal are isomorphic.
(iv) $\Rightarrow$ *(i)* : Let $\text{Ass}(\widehat{A}) = \{\mathfrak{p} = \mathfrak{p}_1, \mathfrak{p}_2, \ldots, \mathfrak{p}_m\}$. Let b and c be the elements given by
the Lemma 3.1.40 and let $D = A[\frac{c}{b}]$. Then, since D is a finite A-algebra, it is enough to
prove that there exists a maximal ideal of height 1 in D. Let R be the integral closure of $\widehat{A}$
and $S := A \setminus (0)$. Then

$$\widehat{D} = D \otimes_A \widehat{A} \subseteq S^{-1}A \otimes_A \widehat{A} = S^{-1}\widehat{A}.$$

Because the elements of S are not zero divisors in $\widehat{A}$, it follows that $\widehat{D} \subseteq Q(\widehat{A})$. The ring
extension $\widehat{A} \subseteq \widehat{D}$ is an integral one, hence

$$\widehat{A} \subseteq \widehat{D} \subseteq \widehat{A}[\frac{c}{b}] = \widehat{A}[\frac{d}{b}] \subseteq R.$$

Let $\mathfrak{q}_i := \mathfrak{p}_i Q(\widehat{A}) \cap D$, for $i = 1, \ldots, m$. Then $\text{Ass}(D) = \{\mathfrak{q}_1, \ldots, \mathfrak{q}_m\}$ and $\text{ht}(\mathfrak{q}_i) = $
$\text{ht}(\mathfrak{p}_i)$, $i = 1, \ldots, m$. Since $\widehat{D}$ is integral over $\widehat{A}$, it results that $\dim(\widehat{A}/\mathfrak{p}_i) = \dim(\widehat{D}/\mathfrak{p}_i)$,
for $i = 1, \ldots, m$. Let $(0) = \bigcap_{i=1}^{m} \mathfrak{r}_i$ be a primary decomposition of (0) in $\widehat{A}$ such that $\mathfrak{r}_i$
is $\mathfrak{p}_i$-primary for any $i = 1, \ldots, m$. Let $\overline{\mathfrak{p}}$ be a maximal ideal of $\widehat{D}$ containing $\mathfrak{q}_1$. From
Lemma 3.1.40 it follows that $d \in \bigcap_{i=2}^{m} \mathfrak{r}_i$ and that $d - b \in \mathfrak{r}_1$, that $\frac{d}{b} \in \mathfrak{q}_i$, for $i = 2, \ldots, m$

and that $\frac{d}{b} - 1 \in \mathfrak{q}_1$. Then $\frac{\frac{d}{b}-1}{1} \in \mathfrak{q}_1 \widehat{D}_{\overline{\mathfrak{p}}}$ hence $\frac{(\frac{d}{b})}{1}$ is invertible in $\mathfrak{q}_i \widehat{D}_{\overline{\mathfrak{p}}}$, for $i = 2, \ldots, m$ and this implies that $\mathfrak{q}_i \not\subseteq \overline{\mathfrak{p}}$, for $i = 2, \ldots, m$. It follows that $\mathfrak{q}_1$ is the only associated prime of (0) in $\widehat{D}$ contained in $\overline{\mathfrak{p}}$. Therefore

$$\mathrm{ht}(\overline{\mathfrak{p}}) = \dim(\widehat{D}/\mathfrak{q}_1) = \dim(\widehat{A}/\mathfrak{p}_1) = 1.$$

Now let $\mathfrak{p} = \overline{\mathfrak{p}} \cap D$. Then $\mathfrak{p}$ is a maximal ideal of D and $\mathfrak{p}\widehat{D} = \overline{\mathfrak{p}}$. It follows that $\mathrm{ht}(\mathfrak{p}) = 1$.

$\square$

Theorem 3.1.42 (Ratliff [302, Th. 3.1]) *Let $(A, \mathfrak{m})$ be a Noetherian local domain. The following are equivalent:*

  (i) A is formally equidimensional;
 (ii) A satisfies the dimension formula;
(iii) A is universally catenary.

Proof $(i) \Rightarrow (ii)$ and $(ii) \Leftrightarrow (iii)$: These implications were proved in Theorem 3.1.35 and Theorem 3.1.36.

$(ii) \Rightarrow (i)$: Suppose that A is not formally equidimensional. Then $d := \dim(A) > 1$. As by assumption A is a catenary local domain, any maximal chain of prime ideals in A has length equal to d. Because $\widehat{A}$ is not equidimensional, there exists $\mathfrak{q} \in \mathrm{Min}(\widehat{A})$ such that $t := \dim(\widehat{A}/\mathfrak{q}) < d$. Assume first that $t > 1$ and using the notations of Proposition 3.1.39, let $\mathfrak{p} \in \mathcal{P}_{t-1}$. Then $\mathrm{ht}(\mathfrak{p}) = t - 1$ and there exists $\mathfrak{q}^* \in \mathrm{Min}(\widehat{A}/\mathfrak{p}\widehat{A})$ such that $\dim((\widehat{A}/\mathfrak{p}\widehat{A})/\mathfrak{q}^*) = 1$. If we prove that $A/\mathfrak{p}$ does not satisfy the dimension formula, then by Corollary 3.1.37 it follows that A does not satisfy the dimension formula, contradicting (ii). Thus we may assume $t = 1$. Let $\mathrm{Ass}(\widehat{A}) = \{\mathfrak{q} = \mathfrak{q}_1, \mathfrak{q}_2, \ldots, \mathfrak{q}_s\}$ and let $(0) = \bigcap_{i=1}^{s} \mathfrak{r}_i$ be a primary decomposition of (0), where $\mathfrak{r}_i$ is $\mathfrak{q}_i$-primary, for $i = 1, \ldots, s$. From Lemma 3.1.40 we get elements $b, c \in \mathfrak{m}$ and $d \in \bigcap_{i=2}^{s} \mathfrak{r}_i \setminus \mathfrak{q}_1$ such that $b - d \in \mathfrak{r}_1, (b, d)\widehat{A} = (b, c)\widehat{A}$ and $\frac{c}{b} \in A' \setminus A$, where A' is the integral closure of A. Let $D := A[\frac{c}{b}]$. Let also $T^* := Q(\widehat{A})$ and $\widehat{A}'$ be the integral closure of $\widehat{A}$ in T^*. Since $A \subset A[\frac{c}{b}] = D \subseteq A'$ and $(b, d)\widehat{A} = (b, c)\widehat{A}$, we have that

$$\widehat{A} \subseteq \widehat{D} = \widehat{A}[\frac{c}{b}] = \widehat{A}[\frac{d}{b}] \subseteq \widehat{A}'.$$

Let $\mathfrak{Q}_i := \mathfrak{q}_i T^* \cap \widehat{D}$ for $i = 1, \ldots, s$. Then $\mathrm{Ass}(\widehat{D}) = \{\mathfrak{Q}_1, \ldots, \mathfrak{Q}_s\}$ and $\mathrm{ht}(\mathfrak{Q}_i) = \mathrm{ht}(\mathfrak{q}_i)$. Since $\widehat{D}$ is integral over $\widehat{A}$, it follows that $\dim(\widehat{D}/\mathfrak{Q}_i) = \dim(\widehat{A}/\mathfrak{q}_i)$. Let $\mathfrak{p}^* \in \mathrm{Max}(\widehat{D})$ be such that $\mathfrak{Q}_1 \subseteq \mathfrak{p}^*$. Since $d \in \bigcap_{i=2}^{s} \mathfrak{r}_i$ and $d - b \in \mathfrak{r}_1$, it follows that $\frac{d}{b} \in \bigcap_{i=2}^{s} \mathfrak{Q}_i$ and

$\frac{d}{b} - 1 \in \mathfrak{Q}_1$. If there is some $i \geq 2$ such that $\mathfrak{Q}_i \subseteq \mathfrak{p}^*$, then $\frac{d}{b} \in \mathfrak{Q}_1$ and consequently $1 \in \mathfrak{Q}_1$. Thus $\mathfrak{Q}_i \not\subseteq \mathfrak{p}^*$, for any $i \geq 2$. It follows that $\mathrm{ht}(\mathfrak{p}^*) = 1$. Let $\mathfrak{p} := \mathfrak{p}^* \cap D$. Then $\mathfrak{p} \in \mathrm{Max}(D)$ and $\mathfrak{p}^* = \mathfrak{p}\hat{D}$, hence $\mathrm{ht}(\mathfrak{p}) = 1$. Now clearly $\mathfrak{p} \cap A = \mathfrak{m}$. Hence $1 = \mathrm{ht}(\mathfrak{p}) < \mathrm{ht}(\mathfrak{m})$ and $\mathrm{tr.deg}_{A/\mathfrak{m}}(D/\mathfrak{p}) = \mathrm{tr.deg}_A D = 0$, that is, A does not satisfy the dimension formula.   □

We can give now the promised example of a catenary ring that is not universally catenary.

Example 3.1.43 Let K_0 be a countable field of any characteristic and let $n = 3$. Let $F_1(Z_1, Z_2, Z_3) = Z_1 Z_2$ and $F_2(Z_1, Z_2, Z_3) = Z_1 Z_3$. Then, by Corollary 1.5.9 we obtain a Noetherian local domain A such that $\dim(A) = 2$ and

$$\hat{A} \simeq K[[\zeta_1, \zeta_2, \zeta_3]]/(\zeta_1\zeta_2, \zeta_1\zeta_3) = K[[\zeta_1, \zeta_2, \zeta_3]]/(\zeta_1) \cap (\zeta_2, \zeta_3).$$

Thus A is catenary, being a local domain of dimension 2 and is not universally catenary by Theorem 3.1.42, since $\hat{A}$ is not equidimensional.

Example 3.1.44 Theorem 3.1.42 is not true in the case of local rings that are not domains. Indeed, if we consider the ring

$$A = K[[X, Y, Z]]/(XY, XZ) = K[[X, Y, Z]]/(X) \cap (Y, Z),$$

where K is any field, then A is a complete local ring, hence universally catenary, but is not equidimensional.

Theorem 3.1.42 generalizes to the non-local case as follows:

Theorem 3.1.45 (Ratliff [302, Th. 3.6]) *Let A be a Noetherian domain and for any natural number n let $A_n := A[X_1, \ldots, X_n]$. The following are equivalent:*

  (i) A is locally formally equidimensional;
 (ii) A satisfies the dimension formula;
(iii) A is universally catenary;
 (iv) $A_1 = A[X]$ is catenary.

Proof $(i) \Rightarrow (ii)$: Follows from Theorem 3.1.42 and Lemma 3.1.23.
$(ii) \Rightarrow (iii)$: Follows from Theorem 3.1.36.
$(iii) \Rightarrow (iv)$: Obvious.
$(iv) \Rightarrow (i)$: $A[X]$ is catenary, hence for any maximal ideal $\mathfrak{m}$ of A the localization $A_{\mathfrak{m}}$ is a catenary local domain. Suppose that $A_{\mathfrak{m}}$ is not formally equidimensional. By

Proposition 3.1.39 there exists a prime ideal $\mathfrak{p} \subset \mathfrak{m}$ such that if we put $R := A_{\mathfrak{m}}/\mathfrak{p}A_{\mathfrak{m}}$, there exists $\mathfrak{q} \in \mathrm{Min}(\widehat{R})$ with $\dim(\widehat{R}/\mathfrak{q}) = 1 < \dim(R)$. Using the notations from Lemma 3.1.40, we have that $D := R[\frac{c}{b}]$ does not satisfy the dimension formula with respect to R, therefore $R[X]$ is not catenary. But $R[X] = A[X]_{\mathfrak{m}}/\mathfrak{p}A[X]_{\mathfrak{m}}$. Contradiction!

□

Definition 3.1.46 A Noetherian local ring A is called **formally catenary**, if $A/\mathfrak{p}$ is formally equidimensional, for any prime ideal $\mathfrak{p}$ of A.

Theorem 3.1.47 ([132, Th. 7.1.8]) *Let A be a Noetherian local ring. The following are equivalent:*

(i) A is formally catenary;
(ii) $A/\mathfrak{p}$ is formally equidimensional, for any minimal prime ideal $\mathfrak{p}$ of A.
(iii) Any local equidimensional essentially of finite type A-algebra B is formally equidimensional.

Proof $(iii) \Rightarrow (i) \Rightarrow (ii)$: Obvious.

$(ii) \Rightarrow (i)$: Let $\mathfrak{q} \in \mathrm{Spec}(A)$. There exists $\mathfrak{p} \in \mathrm{Min}(A)$ such that $\mathfrak{p} \subseteq \mathfrak{q}$. Then $A/\mathfrak{q}$ is formally equidimensional and by Lemma 3.1.27 it follows that $A/\mathfrak{p}$ is formally equidimensional.

$(i) \Rightarrow (iii)$: From Lemma 3.1.29, it is enough to prove that $B/\mathfrak{p}$ is formally equidimensional, for any $\mathfrak{p} \in \mathrm{Min}(B)$. But $B/\mathfrak{p}$ is an A-algebra essentially of finite type, so we may assume that B is a domain. Let $\mathfrak{n}$ be the maximal ideal of B and $\mathfrak{q} = \mathfrak{n} \cap A$. Then B is an $A_{\mathfrak{q}}$-algebra essentially of finite type. Moreover, any quotient ring of $A_{\mathfrak{q}}$ which is a domain is also formally equidimensional. Hence we may assume that the structure morphism $u : A \to B$ is a local one. Let $\mathfrak{r} := \ker(u)$. Then $\mathfrak{r}$ is a prime ideal of A. By assumption, any quotient ring of $A/\mathfrak{r}$ which is a domain is formally equidimensional. Since B is an $A/\mathfrak{r}$-algebra essentially of finite type, it follows that we may also assume that u is injective. Thus A is a domain. Let $C := A[T_1, \ldots, T_n]$ be a polynomial ring over A and let $\mathfrak{p} \in \mathrm{Spec}(C)$ be such that $\mathfrak{p} \cap A = \mathfrak{m}$ and B is a quotient ring of $C_{\mathfrak{p}}$. Then, by Lemma 3.1.29 it is enough to prove that $C_{\mathfrak{p}}$ is formally equidimensional, that is, we may assume that $B = C_{\mathfrak{p}}$. Let $D := \widehat{A}[T_1, \ldots, T_n] = C \otimes_A \widehat{A}$. There exists a prime ideal $\mathfrak{p}' \in \mathrm{Spec}(D)$ such that $\mathfrak{p}' \cap \widehat{A} = \mathfrak{m}\widehat{A}$ and consequently $\mathfrak{p}' \cap A = \mathfrak{p}$. If $E := D_{\mathfrak{p}'}$, then the canonical morphism $B \to E$ is flat and local. Hence also the canonical morphism $\widehat{B} \to \widehat{E}$ is flat. We have $\mathrm{Min}(D) = \{\mathfrak{q}D \mid \mathfrak{q} \in \mathrm{Min}(\widehat{A})\}$ and $D/\mathfrak{q}D = (\widehat{A}/\mathfrak{q})[T_1, \ldots, T_n]$. Since $\widehat{A}$ is equidimensional, it follows that also D is equidimensional. But E is a quotient of a regular local ring, hence formally equidimensional by Theorem 3.1.26. Then, applying Lemma 3.1.32 we get that $\widehat{B}$ is equidimensional.

□

Corollary 3.1.48 *Let A be a Noetherian local ring. Then A is formally equidimensional if and only if A is formally catenary and equidimensional.*

Proof It follows by Lemma 3.1.27 and Theorem 3.1.47.   □

Corollary 3.1.49 *Let A be a Noetherian local ring. If A is formally catenary, then A^h is formally catenary.*

Proof We have $A^h = \varinjlim_{i \in I} A_i$, where A_i are essentially étale local A-algebras. Let $\mathfrak{p} \in \mathrm{Spec}(A^h)$ and $\mathfrak{q} = \mathfrak{p} \cap A$. Then $\mathfrak{q} A^h \subseteq \mathfrak{p}$, hence $A^h/\mathfrak{p}$ is a factor ring of $A^h/\mathfrak{q}A^h$. Therefore it is enough to show that $A^h/\mathfrak{q}A^h \cong (A/\mathfrak{q})^h$ is formally catenary, for every $\mathfrak{q} \in \mathrm{Spec}(A)$. Thus it is enough to show that if A is a formally catenary local domain, then A^h is formally equidimensional. But this follows at once because $\widehat{A^h} \cong \widehat{A}$.   □

Corollary 3.1.50 ([132, Prop. 7.1.10]) *Let A be a Noetherian local ring that is an homomorphic image of a Cohen-Macaulay ring. Then A is formally catenary.*

Proposition 3.1.51 *Let A be a Noetherian formally catenary local ring and B a Noetherian local ring which is an A-algebra essentially of finite type. Then B is formally catenary.*

Proof Let C be a B-algebra essentially of finite type. Then C is also an A-algebra essentially of finite type.   □

Proposition 3.1.52 ([132, Prop. 7.1.11]) *Let A be a Noetherian formally catenary local ring. Then A is universally catenary.*

Proof If $\mathfrak{p} \in \mathrm{Min}(A)$, then $A/\mathfrak{p}$ is formally equidimensional, hence catenary. It follows by Remark 3.1.16 that A is catenary. Let C be an A-algebra of finite type and $\mathfrak{p} \in \mathrm{Spec}(C)$. Then $C_{\mathfrak{p}}$ is catenary, hence C is catenary.   □

Example 3.1.53 Let A be a Noetherian local ring, $\dim(A) = 1$. Then obviously A and $\widehat{A}$ are equidimensional; hence A is formally catenary.

In order to generalize the characterizations 3.1.42 and 3.1.45 of universally catenary rings to the non-local and non-domain case, we need to introduce a new notion.

Definition 3.1.54 A Noetherian ring A is called **locally formally catenary** if $A_{\mathfrak{p}}$ is formally catenary, for any $\mathfrak{p} \in \mathrm{Spec}(A)$.

The following lemma follows immediately from the definitions.

Lemma 3.1.55 *Let A be a Noetherian ring. The following are equivalent:*

(i) A is locally formally catenary;
(ii) $A_{\mathfrak{m}}$ is formally catenary, for any $\mathfrak{m} \in \mathrm{Max}(A)$;
(iii) $A/\mathfrak{q}$ is locally formally equidimensional, for any $\mathfrak{q} \in \mathrm{Min}(A)$.

Now we are able to prove the general characterization of universally catenary rings.

Theorem 3.1.56 (Ratliff [303, Th. 2.6]) *Let A be a Noetherian ring. The following are equivalent:*

(i) A is universally catenary;
(ii) A is locally formally catenary;
(iii) A satisfies the dimension formula;
(iv) $A[X]$ is catenary.

Proof Let B be any ring and $n \in \mathbb{N}$. Then $\mathrm{Min}(B[X_1, \ldots, X_n]) = \{\mathfrak{p}B[X_1, \ldots, X_n] \mid \mathfrak{p} \in \mathrm{Min}(B)\}$ and $B[X_1, \ldots, X_n]/\mathfrak{p}B[X_1, \ldots, X_n] = (B/\mathfrak{p})[X_1, \ldots, X_n]$. Hence (i), (iii) and (iv) are true for B iff they are true for $B/\mathfrak{p}$, for any $\mathfrak{p} \in \mathrm{Min}(B)$. Also A is locally formally catenary iff $A/\mathfrak{p}$ is locally formally equidimensional, for any $\mathfrak{p} \in \mathrm{Min}(A)$. Now we apply Theorem 3.1.45. □

Corollary 3.1.57 *Let A be a Noetherian ring whose formal fibres are domains. Then A is universally catenary.*

We shall see now what happens to some of the chain conditions considered, when we pass to formal power series or completion.

Lemma 3.1.58 *Let A be a Noetherian ring, $B := A[[T]]$ the formal power series ring over A, $\mathfrak{n}$ a maximal ideal of B and $\mathfrak{m} = \mathfrak{n} \cap A$. Then $\widehat{A_{\mathfrak{m}}}[[T]] \simeq \widehat{B_{\mathfrak{n}}}$.*

Proof Since B is complete in the TB-adic topology, it follows that $T \in \mathfrak{n} = (\mathfrak{m}, T)B$ and that $\widehat{B_{\mathfrak{n}}}/T\widehat{B_{\mathfrak{n}}} \simeq \widehat{A_{\mathfrak{m}}}$. Let $D := \widehat{A_{\mathfrak{m}}}[[T]]$ and let $u : D \to \widehat{B_{\mathfrak{n}}}$ be the canonical morphism. Then $u \otimes 1_{D/TD}$ is an isomorphism; hence u is an isomorphism. □

Lemma 3.1.59 *Let A be a local formally equidimensional ring. Then $A[[T]]$ is formally equidimensional.*

Proof Applying Lemma 3.1.58, it is enough to show that $A[[T]]$ is equidimensional. But $\text{Min}(A[[T]]) = \{\mathfrak{p}[[T]] \mid \mathfrak{p} \in \text{Min}(A)\}$ and

$$\dim(A[[T]]/\mathfrak{p}[[T]]) = \dim(A/\mathfrak{p}[[T]]) = \dim(A) + 1.$$

□

Proposition 3.1.60 *Let A be a local formally equidimensional ring, $\mathfrak{a}$ an ideal of A and B the completion of A in the $\mathfrak{a}$-adic topology. Then B is formally equidimensional.*

Proof Applying [249, Th. 8.12], it is enough to show that $A[[T]]/(T - a)$ is formally equidimensional, for any $a \in A$. By Lemma 3.1.59 $A[[T]]$ is formally equidimensional and by Lemma 3.1.30 it is enough to show that all minimal prime ideals of $(T - a)A[[T]]$ have the same height. If $\mathfrak{p}$ is such an ideal, then $\text{ht}(\mathfrak{p}) \leq 1$ by Krull's principal ideal theorem. But $T - a$ is not a zero divisor in $A[[T]]$, hence $\text{ht}(\mathfrak{p}) = 1$.   □

Theorem 3.1.61 (Seydi [345, Th. 1.12]) *Let A be a Noetherian universally catenary ring. Then the formal power series ring $A[[T_1, \ldots, T_n]]$ is universally catenary.*

Proof We may assume $n = 1$. Let $\text{Min}(A) = \{\mathfrak{p}_1, \ldots, \mathfrak{p}_m\}$ and $\mathfrak{q}_i = \mathfrak{p}_i B$ for $i = 1, \ldots, m$, where $B := A[[T]]$. Then $\text{Min}(B) = \{\mathfrak{q}_1, \ldots, \mathfrak{q}_m\}$. Let $\mathfrak{n} \in \text{Max}(B)$ and let $\mathfrak{q} \in \text{Min}(B)$ such that $\mathfrak{q} \subseteq \mathfrak{n}$. From Theorem 3.1.56, it is enough to show that $C := \widehat{B_{\mathfrak{n}}}/\mathfrak{q}B_{\mathfrak{n}}$ is equidimensional. Let $\mathfrak{m} := \mathfrak{n} \cap A$ and assume that $\mathfrak{q} = \mathfrak{q}_1$. Then from 3.1.58 we get that $C \simeq \widehat{A_{\mathfrak{m}}/\mathfrak{p}_i A_{\mathfrak{m}}}[[T]]$. But from Theorem 3.1.56, it follows that $\widehat{A_{\mathfrak{m}}/\mathfrak{p}_i A_{\mathfrak{m}}}$ is equidimensional. From the correspondence between $\text{Min}(A_{\mathfrak{m}}/\mathfrak{p}_i A_{\mathfrak{m}})$ and $\text{Min}(C)$, it follows that C is equidimensional.   □

Corollary 3.1.62 *Let A be a Noetherian universally catenary ring, $\mathfrak{a}$ an ideal of A and B the completion of A in the $\mathfrak{a}$-adic topology. Then B is universally catenary.*

Proof We apply [249, Th. 8.12] and Theorem 3.1.61.   □

Lemma 3.1.63 *Let $(B_i, \phi_{ij})_{i \in I}$ be a directed system of Noetherian rings. Assume that:*

(i) *ϕ_{ij} is faithfully flat for any $i \leq j$;*

(ii) *$B := \varinjlim_{i \in I} B_i$ is a Noetherian ring;*

(iii) *B_i is catenary (resp. universally catenary) for any $i \in I$.*

Then B is catenary (resp. universally catenary).

Proof The directed system (B_i, ϕ_{ij}) induces canonically a new directed system $(B_i[T], \psi_{ij})$, where ψ_{ij} is the canonical extension of ϕ_{ij} to the polynomial ring. The system $(B_i[T], \psi_{ij})$ satisfies the same assumptions, so that it is enough to show that B is catenary. Let $\mathfrak{p}_0 \supset \mathfrak{p}_1 \supset \ldots \supset \mathfrak{p}_r$ be a saturated chain of prime ideals in B. If we put $\mathfrak{p}_{ki} := \mathfrak{p}_k \cap B_i$ for $k = 1, \ldots, r$ and $i \in I$, we obtain a chain of prime ideals $\mathfrak{p}_{0i} \supseteq \mathfrak{p}_{1i} \supseteq \ldots \supseteq \mathfrak{p}_{ri}$, that is a chain of prime ideals in B_i. It is enough to show that for sufficiently large i this chain is saturated. As $\mathfrak{p}_k$ is finitely generated, there exists $j \in I$ such that $\forall\, l \geq j$, $\mathfrak{p}_k = \mathfrak{p}_{kl} B_k$. Therefore, $\mathrm{ht}(\mathfrak{p}_k/\mathfrak{p}_{k+1}) = \mathrm{ht}(\mathfrak{p}_{kl}/\mathfrak{p}_{k+1,l})$, since B is flat over B_l. □

Proposition 3.1.64 *Let A be a Noetherian local ring. If A is universally catenary, then the henselization A^h is universally catenary.*

Proof We have $A^h = \varinjlim_{i \in I} A_i$, where A_i is an essentially étale local A-algebra, hence A_i is universally catenary and we apply Lemma 3.1.63. □

Example 3.1.65 (Greco [124, Lemma 1.4 and Lemma 1.5]) There exists a semilocal Noetherian domain A and an ideal $\mathfrak{a}$ of A such that:

(i) A is $\mathfrak{a}$-adically complete;
(ii) $A/\mathfrak{a}$ is universally catenary and has geometrically regular formal fibres;
(iii) A is not catenary.

Proof Let us consider the ring A from Example 3.1.8 keeping the same notations. We have proved that A is not catenary. On the other hand, $B/\mathfrak{a}$ has geometrically regular formal fibres and is a finite extension of $A/\mathfrak{a}$; hence also $A/\mathfrak{a}$ has geometrically regular formal fibres. The ring $(A/\mathfrak{a})_{\mathfrak{m}}$ is a reduced ring of dimension 1, hence universally catenary. Since $A_{\mathfrak{n}} = B_{\overline{\mathfrak{n}}}$, it follows that $(A/\mathfrak{a})_{\mathfrak{n}}$ is a factor ring of a regular ring; hence it is universally catenary. Finally, $A/\mathfrak{a}$ is universally catenary. □

However, if the ring A as in Example 3.1.65 is already catenary, it is even universally catenary, as it will be shown in Theorem 3.1.68.

Lemma 3.1.66 *Let A be a Noetherian ring and Z a non-empty, locally closed subset of $\mathrm{Spec}(A)$. Then there exists $\mathfrak{p} \in Z$ such that $\dim(A/\mathfrak{p}) \leq 1$.*

Proof Since Z is locally closed, we may assume that $Z = D(f) \cap V(\mathfrak{q})$, where $f \in A$, $\mathfrak{q} \in \mathrm{Spec}(A)$ and $f \notin \mathfrak{q}$. Then Z is homeomorphic to $\mathrm{Spec}((A/\mathfrak{q})_{\overline{f}})$, where $\overline{f}$ is the image of f in $A/\mathfrak{q}$. Let $\mathfrak{m} \in \mathrm{Max}((A/\mathfrak{q})_{\overline{f}})$ and let $\mathfrak{p} := \mathfrak{m} \cap A$. Let g be the image of f

in $A/\mathfrak{p}$. Then

$$A_f/\mathfrak{p}A_f = (A/\mathfrak{q})_{\overline{f}}/\mathfrak{m} = (A/\mathfrak{p})[g^{-1}]$$

is a field. It follows that all non-zero prime ideals of $A/\mathfrak{p}$ contain g; hence we have $\dim(A/\mathfrak{p}) \leq 1$. □

Lemma 3.1.67 *Let* $(A, \mathfrak{m})$ *be a Noetherian local ring of dimension* $d > 0$. *Assume that for any* $\mathfrak{p} \in \operatorname{Spec}(A)$ *such that* $\dim(A/\mathfrak{p}) = 1$, *we have* $\operatorname{ht}(\mathfrak{p}) = d - 1$. *Then* A *is equidimensional.*

Proof Let $\mathfrak{q} \in \operatorname{Min}(A) = \{\mathfrak{q}, \mathfrak{p}_1, \ldots, \mathfrak{p}_n\}$ and let $s \in \bigcap_{i=1}^{n} \mathfrak{p}_i \setminus \mathfrak{q}$. Set $F := D(s) \cap V(\mathfrak{q})$. Then F is a non-empty locally closed set. From Lemma 3.1.66 it follows that there exists $\mathfrak{p} \in F$ such that $\dim(A/\mathfrak{p}) = 1$. By assumption we have $\operatorname{ht}(\mathfrak{p}) = d-1 = \operatorname{ht}(\mathfrak{p}/\mathfrak{q})$, whence $\dim(A/\mathfrak{q}) = d$. □

Theorem 3.1.68 (Greco [124, Prop. 2.3]) *Let* A *be a Noetherian ring and* $\mathfrak{a}$ *an ideal of* A. *Assume that:*

  (i) A *is* $\mathfrak{a}$-*adically complete;*
 (ii) $A/\mathfrak{a}$ *is universally catenary;*
(iii) A *is catenary.*

Then A *is universally catenary.*

Proof We may assume that A is a domain and pick $\mathfrak{m} \in \operatorname{Max}(A)$. From Theorem 3.1.47 and Theorem 3.1.56, we have to prove that $B := \widehat{A_\mathfrak{m}}$ is equidimensional. Let $\mathfrak{P} \in \operatorname{Spec}(B)$ be such that $\dim(B/\mathfrak{P}) = 1$. Let also $d = \dim(B) = \operatorname{ht}(\mathfrak{m})$ and $\mathfrak{Q} = \mathfrak{P} \cap A$. We have two cases to consider.

Case 1: Assume first that $\mathfrak{a} \not\subseteq \mathfrak{p}$. Then $V(\mathfrak{a}B + \mathfrak{P}) = \mathfrak{m}B$, that is $\mathfrak{a}B + \mathfrak{P}$ is an ideal of definition of B and therefore the canonical map $A/(\mathfrak{a} + \mathfrak{p}) \to B/(\mathfrak{a}B + \mathfrak{P})$ is finite, actually even surjective. As $A/\mathfrak{p}$ is $\mathfrak{a}$-adically complete and $B/\mathfrak{P}$ is $\mathfrak{a}$-adically separated, it follows that $A/\mathfrak{p} \to B/\mathfrak{P}$ is a finite morphism. Hence $\dim(A/\mathfrak{p}) = \dim(B/\mathfrak{P}) = 1$. Since A is a catenary domain, we get that $d = \operatorname{ht}(\mathfrak{m}) = 1 + \operatorname{ht}(\mathfrak{p})$, hence by flatness $\operatorname{ht}(\mathfrak{P}) \geq \operatorname{ht}(\mathfrak{p}) = d - 1$. Now we apply Lemma 3.1.67.

Case 2: Assume that $\mathfrak{a} \subseteq \mathfrak{p}$. Let $\mathfrak{Q} \in \operatorname{Min}(\mathfrak{p}B)$ be such that $\mathfrak{Q} \subseteq \mathfrak{P}$. Then $\operatorname{ht}(\mathfrak{p}) = \operatorname{ht}(\mathfrak{Q})$. Since $(A/\mathfrak{p})_\mathfrak{m}$ is universally catenary, $B/\mathfrak{q}B$ is equidimensional by Theorem 3.1.56. Hence

$$\dim(B/\mathfrak{Q}) = \dim(B/\mathfrak{p}B) = \dim((A/\mathfrak{p})_\mathfrak{m}).$$

The rings $A_{\mathfrak{m}}$ and $B/\mathfrak{q}B$ are both catenary. Then

$$\mathrm{ht}(\mathfrak{P}) \geq \mathrm{ht}(\mathfrak{p}) + \mathrm{ht}(\mathfrak{P}/\mathfrak{Q}) = \mathrm{ht}(\mathfrak{p}) + \dim(B/\mathfrak{Q}) - \dim(B/\mathfrak{P}) =$$

$$= \mathrm{ht}(\mathfrak{p}) + \dim(A_{\mathfrak{m}}/\mathfrak{p}A_{\mathfrak{m}}) - 1 = d - 1.$$

Now we apply Lemma 3.1.67.   □

3.2 Regular Morphisms and André's Theorem

We start with some properties and characterizations of smooth and regular morphisms.

Lemma 3.2.1 *Let* $u : A \to B$ *be a morphism of Noetherian rings. The following are equivalent:*

(i) u is a regular morphism;
(ii) for any $\mathfrak{q} \in \mathrm{Spec}(B)$, the induced morphism $u_{\mathfrak{q}} : A_{\mathfrak{p}} \to B_{\mathfrak{q}}$ is $\mathfrak{q}B_{\mathfrak{q}}$-smooth, where $\mathfrak{p} := \mathfrak{q} \cap A$.

Proof $(i) \Rightarrow (ii)$: From the flatness of u we obtain the flatness of $u_{\mathfrak{q}}$. Since u is regular, it follows that $k(\mathfrak{p}) \to k(\mathfrak{p}) \otimes B_{\mathfrak{q}}$ is geometrically regular. Now we apply Theorem 1.4.20.
$(ii) \Rightarrow (i)$: From Theorem 1.4.20, we get that all the localizations of u at the prime ideals are flat; hence u is flat. Applying again Theorem 1.4.20, we obtain that the fibres of u are geometrically regular.   □

Proposition 3.2.2 *Let* $u : A \to B$ *be a morphism of Noetherian rings. Assume that:*

(i) A is a regular ring;
(ii) $H_i(B, k(\mathfrak{q}), k(\mathfrak{q})) = (0)$ for any $i \geq 3$ and for any $\mathfrak{q} \in \mathrm{Spec}(B)$.

Then $H_n(A, B, -) = (0)$ for any $n \geq 2$.

Proof Let $\mathfrak{q} \in \mathrm{Spec}(B)$. From the sequence $A \to B \to k(\mathfrak{q})$, for any $i \geq 2$ we get the associated Jacobi-Zariski exact sequence

$$(0) = H_{i+1}(B, k(\mathfrak{q}), k(\mathfrak{q})) \to H_i(A, B, k(\mathfrak{q})) \to H_i(A, k(\mathfrak{q}), k(\mathfrak{q})) = (0).$$

It follows that $H_i(A, B, k(\mathfrak{q})) = (0)$, for any $i \geq 2$. Now we apply [6, Supp. Prop. 29].
                                                                                            □

In the following, we will use the notions of formally projective module and formally left-invertible morphism. The reader can find the definitions and main properties of these notions in [131, 19.1.5 and 19.2.1].

Theorem 3.2.3 *Let $u : A \to B$ be a morphism of Noetherian rings, $\mathfrak{b}$ be an ideal of B and $C := B/\mathfrak{b}$. The following are equivalent:*

(i) B is $\mathfrak{b}$-smooth over A;
(ii) $\Omega_{B/A} \otimes_B C$ is a projective C-module and $H_1(A, B, C) = (0)$.

Proof Let R be a polynomial ring over A and let $\varphi : R \to B$ be a surjection such that $\ker(\varphi) = \mathfrak{b}$. Then we have an exact sequence

$$0 \to H_1(A, B, C) \to \mathfrak{b}/\mathfrak{b}^2 \xrightarrow{\delta \otimes 1_C} \Omega_{R/A} \otimes C \to \Omega_{B/A} \otimes_B C \to 0.$$

$(i) \Rightarrow (ii)$: Since u is $\mathfrak{b}$-smooth, $\Omega_{B/A}$ is a formally projective B-module [131, Th. 20.4.9]; hence $\Omega_{B/A} \otimes C$ is a projective C-module. But δ is formally left invertible [131, Prop. 20.7.2], hence $\delta \otimes 1_C$ is left invertible and therefore $H_1(A, B, C) = (0)$.
$(ii) \Rightarrow (i)$: The assumptions tells us that $\delta \otimes 1_C$ is injective. It follows that δ is formally left invertible; hence u is $\mathfrak{b}$-smooth. $\square$

Corollary 3.2.4 *Let $u : (A, \mathfrak{m}, K) \to (B, \mathfrak{n}, L)$ be a morphism of Noetherian local rings. The following are equivalent:*

(i) u is formally smooth (that is $\mathfrak{n}$-smooth);
(ii) $H_1(A, B, L) = (0)$;
(iii) $H_i(A, B, L) = (0)$, $\forall i \geq 1$.

Proof $(i) \Leftrightarrow (ii)$: Follows at once from Theorem 3.2.3.
$(i) \Rightarrow (iii)$: Let $C := B \otimes_A K$. Since u is flat, by [6, Prop. IV.54] it follows that for any $i \in \mathbb{N}$ we have

$$H_i(A, B, L) \cong H_i(K, C, L).$$

But K and C are regular rings, therefore, from Proposition 3.2.2 we get that $H_i(K, C, L) = (0)$, for any $i \geq 2$.
$(iii) \Rightarrow (ii)$: Clear. $\square$

Lemma 3.2.5 *Let $u : A \to B$ be a morphism of Noetherian rings. The following conditions are equivalent:*

(i) $H_1(A, B, B) = (0)$;
(ii) $H^1(A, B, E) \cong \text{Ext}^1_B(\Omega_{B/A}, E)$, *for any B-module E*;
(iii) $H_1(A, B, E) \cong \text{Tor}^B_1(\Omega_{B/A}, E)$, *for any B-module E*.

Proof $(i) \Rightarrow (ii)$ and $(i) \Rightarrow (iii)$ follow from [6, Lemme III.19].
$(iii) \Rightarrow (i)$ is obvious.
$(ii) \Rightarrow (i)$: From the assumption, it follows that $H^1(A, B, M) = 0$ for any injective B-module M. Let now I be the injective hull of $H_1(A, B, B)$. Then from [6, Lemme III.21] we get

$$(0) = H^1(A, B, I) \cong \text{Hom}_B(H_1(A, B, B), I).$$

This implies that $H_1(A, B, B) = (0)$.   □

Definition 3.2.6 (Granjean-Vale [119, Def.]) A morphism of Noetherian rings $u : A \to B$ is called **almost smooth** if $H_1(A, B, B) = (0)$.

Almost smooth morphisms will be studied in Sect. 6.2. The next result clarifies the word *almost* in Definition 3.2.6.

Theorem 3.2.7 *Let $u : A \to B$ be a morphism of Noetherian rings. The following are equivalent:*

(i) *u is smooth;*
(ii) $\Omega_{B/A}$ *is a projective B-module and $H_1(A, B, B) = 0$;*
(iii) $H^1(A, B, -) = 0$.

Proof $(i) \Leftrightarrow (ii)$: Follows from Theorem 3.2.3 applied for $\mathfrak{b} = 0$.
$(ii) \Rightarrow (iii)$: From Lemma 3.2.5 and the fact that $\Omega_{B/A}$ is a projective B-module, it follows that

$$H^1(A, B, E) \cong \text{Ext}^1(\Omega_{B/A}, E) = (0).$$

$(iii) \Rightarrow (ii)$: Since $H^1(A, B, -) = (0)$, it follows that

$$H^0(A, B, -) \cong \text{Der}_A(B, -) \cong \text{Hom}_B(\Omega_{B/A}, -)$$

is right-exact so that $\Omega_{B/A}$ is a projective B-module. Let E be any injective B-module. Then from the assumptions and [6, Lemme III.21], we obtain

$$(0) = H^1(A, B, E) \cong \operatorname{Hom}_B(H_1(A, B, B), E).$$

Hence $H_1(A, B, B) = (0)$. $\square$

Theorem 3.2.8 *Let $u : A \to B$ be a morphism of Noetherian rings. The following are equivalent:*

(i) u is regular;
(ii) $\Omega_{B/A}$ is a flat B-module and $H_1(A, B, B) = (0)$;
(iii) $H_1(A, B, -) = (0)$.
(iv) $H_i(A, B, -) = (0)$, for any $i \geq 1$.

Proof $(ii) \Rightarrow (iii)$: Since $\Omega_{B/A}$ is a flat B-module, from Lemma 3.2.5 it follows that for any B-module E we have

$$H_1(A, B, E) \cong \operatorname{Tor}_1(\Omega_{B/A}, E) = (0).$$

$(iii) \Rightarrow (ii)$: Since $H_1(A, B, B) = (0)$, from Lemma 3.2.5 it follows that

$$(0) = H_1(A, B, E) \cong \operatorname{Tor}_1^B(\Omega_{B/A}, E),$$

for any B-module E. This means that $\Omega_{B/A}$ is a flat B-module.
$(iv) \Rightarrow (iii)$: Obvious.
$(i) \Rightarrow (iv)$: Let $\mathfrak{q} \in \operatorname{Spec}(B)$. Then $A \to B_{\mathfrak{q}}$ is formally smooth; hence from Corollary 3.2.4 it results that $H_i(A, B_{\mathfrak{q}}, k(\mathfrak{q})) = (0)$, $\forall i \geq 1$. From [6, Supp. Prop. 29] it follows that $H_i(A, B, -) = 0$, $\forall i \geq 1$.
$(ii) \Rightarrow (i)$: Let $\mathfrak{q} \in \operatorname{Spec}(B)$ and $\mathfrak{p} := \mathfrak{q} \cap A$. Then from [6, Cor. IV.59 and Cor. V.27] we have

$$H_1(A_{\mathfrak{p}}, B_{\mathfrak{q}}, k(\mathfrak{q})) \cong H_1(A, B, k(\mathfrak{q})) = (0).$$

From Corollary 3.2.4 we get that the morphism of local rings $A_{\mathfrak{p}} \to B_{\mathfrak{q}}$ is formally smooth. Now we apply Lemma 3.2.1. $\square$

From Theorem 3.2.7, Theorem 3.2.8 and [100, Th. 19.2], we have:

Corollary 3.2.9 *Let $u : A \to B$ be a smooth morphism of Noetherian rings. Then u is regular. The converse is true if u is of finite type.*

Actually in the case of morphisms of finite type smoothness, regularity and formal smoothness coincide, as it is shown in the next result.

Corollary 3.2.10 *Let* $u : A \to (B, \mathfrak{n}, L)$ *be a morphism essentially of finite type of Noetherian rings, where moreover B is local and let $\mathfrak{b} \subseteq \mathfrak{n}$ be an ideal of B. The following are equivalent:*

(i) *u is smooth (that is (0)-smooth);*
(ii) *u is $\mathfrak{b}$-smooth;*
(iii) *u is $\mathfrak{n}$-smooth;*

Proof It is clear that we have only to show $(iii) \Rightarrow (i)$. By Corollary 3.2.4 we know that $H_1(A, B, L) = (0)$. From the universal coefficient exact sequence [230, Prop. 1.4.5]

$$\mathrm{Tor}_2^B(\Omega_{B/A}, L) \to H_1(A, B, B) \otimes_B L \to H_1(A, B, L) \to \mathrm{Tor}_1^B(\Omega_{B/A}, L) \to 0$$

it follows that $\mathrm{Tor}_1^B(\Omega_{B/A}, L) = (0)$; hence $\Omega_{B/A}$ is a projective B-module since B is Noetherian and $\Omega_{B/A}$ is finitely generated. The above exact sequence shows that $H_1(A, B, B) \otimes_B L = (0)$. By Nakayama's Lemma it follows that $H_1(A, B, B) = (0)$ and Theorem 3.2.7 concludes the proof.   □

Before proving André's theorem on the localization of formal smoothness, we will prove a nice recent result of Majadas [229] that will be successfully applied in several situations.

Theorem 3.2.11 (Majadas [229, Th. 1]) *Consider a commutative diagram of morphisms of Noetherian local rings:*

$$
\begin{array}{ccc}
(A, \mathfrak{m}_A, k_A) & \xrightarrow{\ u\ } & (B, \mathfrak{m}_B, k_B) \\
\downarrow{\scriptstyle f} & & \downarrow{\scriptstyle g} \\
(C, \mathfrak{m}_C, k_C) & \xrightarrow{\ v\ } & (D, \mathfrak{m}_D, k_D).
\end{array}
$$

Assume that:

(i) *$\mathrm{Tor}_i^A(C, B) = (0)$ for all $i > 0$;*
(ii) *the canonical morphism $H_1(A, B, k_D) \to H_1(C, D, k_D)$ is zero;*
(iii) *$(C \otimes_A B)_{\mathfrak{p}}$ is a Noetherian ring, where $\mathfrak{p} = \mathfrak{m}_D \cap (C \otimes_A B)$;*
(iv) *$\mathrm{fd}_{C \otimes_A B}(D) < \infty$.*

Then u is formally smooth.

Proof We consider the commutative diagram

$$
\begin{array}{ccccc}
C & \longrightarrow & C \otimes_A B & \longrightarrow & D \\
\downarrow{\scriptstyle 1_C} & & \downarrow{\scriptstyle 1_{C\otimes_A B}} & & \downarrow \\
C & \longrightarrow & C \otimes_A B & \longrightarrow & k_D.
\end{array}
$$

Applying the Jacobi-Zariski exact sequences [6, Th. V.1], we obtain the commutative diagram

$$
\begin{array}{ccc}
H_2(C \otimes_A B, D, k_D) & \xrightarrow{\ \delta\ } & H_1(C, C \otimes_A B, k_D) \\
\downarrow{\scriptstyle \gamma} & & \|{\scriptstyle =} \\
H_2(C \otimes_A B, k_D, k_D) & \longrightarrow & H_1(C, C \otimes_A B, k_D).
\end{array}
$$

On the other hand, applying [6, Cor. V.27], from the ring morphisms $C \otimes_A B \to D \to k_D$ we obtain a Jacobi-Zariski exact sequence

$$
H_2(C \otimes_A B, D, k_D) \xrightarrow{\gamma} H_2(C \otimes_A B, k_D, k_D) = H_2((C \otimes_A B)_{\mathfrak{p}}, k_D, k_D) \xrightarrow{\beta}
$$

$$
\xrightarrow{\beta} H_2(D, k_D, k_D).
$$

From [21, Théorème] it follows that β is injective, hence $\gamma = 0$. By assumption (i) and [6, Prop. IV.54] it follows that $H_1(A, B, k_D) \simeq H_1(C, C \otimes_A B, k_D)$. Thus from the commutative diagram

$$
\begin{array}{ccc}
H_1(A, B, k_D) & \xrightarrow{\ \simeq\ } & H_1(C, C \otimes_A B, k_D) \\
& \searrow{\scriptstyle 0} \qquad \swarrow{\scriptstyle \alpha} & \\
& H_1(C, D, k_D) &
\end{array}
$$

we get $\alpha = 0$. From the morphisms $C \to C \otimes_A B \to D$ we have a Jacobi-Zariski exact sequence

$$
H_2(C \otimes_A B, D, k_D) \xrightarrow{\delta} H_1(C, C \otimes_A B, k_D) \xrightarrow{\alpha} H_1(C, D, k_D).
$$

Therefore δ is surjective. This implies that $H_1(C, C \otimes_A B, k_D) = (0)$, therefore $H_1(A, B, k_D) = (0)$. Now Corollary 3.2.4 and [6, Prop. III.20] show that u is formally smooth.

□

Remark 3.2.12 In the above proof, we have used [21, Théorème]. More precisely, we have used the following consequence of this result:

Let $u : (A, \mathfrak{m}, K) \to (B, \mathfrak{n}, L)$ be a local morphism of Noetherian local rings such that $\mathrm{fd}_A(B) < \infty$. *Then the canonical morphism $H_2(A, K, L) \to H_2(B, L, L)$ is injective.*

The case $\mathrm{fd}_A(B) = 0$, which is the case u flat, was already proved by Avramov in [20, Th. 1.1]. The proof given in [21] exceeds the scientific limits of this book and unfortunately the unique source is the original paper [21], at least to our knowledge. The only reasonable choice was to cite this result from its original source. A proof of the case $\mathrm{fd}_A(B) = 0$ can be found in [231, Cor. 4.2.2]. It would be useful to give an elementary proof of the above statement.[1]

Corollary 3.2.13 (Majadas [229, Cor. 3]) *Consider a commutative diagram of morphisms of Noetherian local rings:*

$$
\begin{array}{ccc}
(A, \mathfrak{m}_A, k_A) & \xrightarrow{u} & (B, \mathfrak{m}_B, k_B) \\
\downarrow{\scriptstyle f} & & \downarrow{\scriptstyle g} \\
(C, \mathfrak{m}_C, k_C) & \xrightarrow{v} & (D, \mathfrak{m}_D, k_D)
\end{array}
$$

Assume that:

(i) $\mathrm{Tor}_i^A(C, B) = (0)$ *for all $i > 0$;*
(ii) *the canonical morphism $H_1(A, B, k_D) \to H_1(C, D, k_D)$ is zero;*
(iii) *D is a flat $C \otimes_A B$-module.*

Then u is formally smooth.

Proof It is clear that condition (iii) of Theorem 3.2.11 follows by flatness.

□

Corollary 3.2.14 (Majadas [229, Cor. 4]) *Consider a commutative diagram of morphisms of Noetherian local rings:*

---

[1] Very recently a new and simpler proof of [21, Théorème] was given in [3].

$$(A,\ \mathfrak{m}_A, k_A) \xrightarrow{u} (B,\ \mathfrak{m}_B, k_B)$$

$$\downarrow f \qquad\qquad \downarrow g$$

$$(C,\ \mathfrak{m}_C, k_C) \xrightarrow{v} (D,\ \mathfrak{m}_D, k_D)$$

Assume that:

(i) $\operatorname{Tor}_i^A(C, B) = (0)$ *for all* $i > 0$;
(ii) *for each* $\mathfrak{q} \in \operatorname{Spec}(B)$ *there exists* $\mathfrak{r} \in \operatorname{Spec}(D)$ *such that* $\mathfrak{r} \cap B = \mathfrak{q}$ *and the canonical morphism* $H_1(A, B, k(\mathfrak{r})) \to H_1(C, D, k(\mathfrak{r}))$ *is zero;*
(iii) D *is a flat* $C \otimes_A B$-*module.*

Then u *is a regular morphism.*

Proof By Lemma 3.2.1, we have to prove that for any $\mathfrak{q} \in \operatorname{Spec}(B)$ the morphism $A_\mathfrak{p} \to B_\mathfrak{q}$ is formally smooth, where $\mathfrak{p} = \mathfrak{q} \cap A$. Let $\mathfrak{r}$ be the prime ideal of D given by (ii) and set $\mathfrak{n} = \mathfrak{r} \cap C$. We apply Corollary 3.2.13 to the commutative square

$$\begin{array}{ccc} A_\mathfrak{p} & \longrightarrow & B_\mathfrak{q} \\ \downarrow & & \downarrow \\ C_\mathfrak{n} & \longrightarrow & D_\mathfrak{r} \end{array}$$

Indeed, conditions (i) and (iii) are true by localization, while applying [6, Cor. V.27] we obtain that the morphism

$$H_1(A_\mathfrak{p}, B_\mathfrak{q}, k(\mathfrak{r})) = H_1(A, B, k(\mathfrak{r})) \to H_1(C, D, k(\mathfrak{r})) = H_1(C_\mathfrak{n}, D_\mathfrak{r}, k(\mathfrak{r}))$$

is zero. Therefore $A_\mathfrak{p} \to B_\mathfrak{q}$ is formally smooth.   □

Corollary 3.2.15 (Majadas [229, Cor. 5]) *Consider a commutative diagram of morphisms of Noetherian local rings:*

$$(A,\ \mathfrak{m}_A, k_A) \xrightarrow{u} (B,\ \mathfrak{m}_B, k_B)$$

$$\downarrow f \qquad\qquad \downarrow g$$

$$(C,\ \mathfrak{m}_C, k_C) \xrightarrow{v} (D,\ \mathfrak{m}_D, k_D)$$

Assume that:

  (i) $\mathrm{Tor}_i^A(C, B) = (0)$ *for all* $i > 0$;
  (ii) v *is a formally smooth morphism;*
  (iii) $C \otimes_A B$ *is a Noetherian ring;*
  (iv) $\mathrm{fd}_{C \otimes_A B}(D) < \infty$.

Then u is formally smooth.

Proof By Corollary 3.2.4 it follows that $H_1(C, D, k_D) = (0)$; hence we can apply Theorem 3.2.11.
        $\square$

To illustrate Theorem 3.2.11, we note the following generalization of Proposition 1.6.42.

Proposition 3.2.16 (Majadas [229, Cor. 8]) *Let* $u : A \to B$ *be a flat local essentially of finite type morphism of local rings. If the formal fibres of B are geometrically regular, then the formal fibres of A are geometrically regular.*

Proof Consider the commutative diagram

$$
\begin{array}{ccc}
A & \xrightarrow{f} & \widehat{A} \\
{\scriptstyle u}\downarrow & & \downarrow{\scriptstyle \widehat{u}} \\
B & \xrightarrow{g} & \widehat{B}
\end{array}
$$

and try to apply Corollary 3.2.14. Since u is flat, also $\widehat{u}$ is flat, hence $\mathrm{Spec}(\widehat{u})$ is surjective. It is well-known [134, Lemma 7.9.3.1] that $\widehat{B}$ is isomorphic to the completion of $(B \otimes_A \widehat{A})_\mathfrak{n}$, where $\mathfrak{n}$ is a maximal ideal of $B \otimes_A \widehat{A}$ lying over the maximal ideals of A and B. This shows that the morphism $B \otimes_A \widehat{A} \to \widehat{B}$ is flat. Because the morphism $B \to \widehat{B}$ is regular, by Theorem 3.2.8 we know that $H_1(B, \widehat{B}, -) = (0)$. Now we can apply Corollary 3.2.14 to get that f is a regular morphism.
        $\square$

For the proof of André's theorem on the localization of formal smoothness, we need some preparations. The first result is a criterion of regularity for morphisms.

Theorem 3.2.17 (Radu [297, Prop. 3]) *Let* $u : A \to B$ *and* $v : B \to C$ *be morphisms of Noetherian local rings. Assume that:*

  (i) v *is formally smooth;*

(ii) v ∘ u is a regular morphism;

(iii) $\Omega_{B/A} \otimes_B C/\mathfrak{q}$ is a separated C-module, for any $\mathfrak{q} \in \mathrm{Spec}(C)$.

Then v is a regular morphism.

Proof ([64]) Let $\mathfrak{q} \in \mathrm{Spec}(C)$ and let $L := C_{\mathfrak{q}}/\mathfrak{q}C_{\mathfrak{q}}$. We have two exact sequences

$$H_1(A, C, L) \to H_1(B, C, L) \to \Omega_{B/A} \otimes_B L \xrightarrow{\delta} \Omega_{C/A} \otimes_B L$$

and

$$\Omega_{B/A} \otimes_B C \xrightarrow{v_0} \Omega_{C/A} \to \Omega_{C/B} \to 0.$$

Because v is formally smooth, v_0 is formally left-invertible, hence also the morphism $f := v_0 \otimes_C 1_{C/\mathfrak{q}} : \Omega_{B/A} \otimes_B C/\mathfrak{q} \to \Omega_{C/A} \otimes_B C/\mathfrak{q}$ is formally left-invertible. Let us denote by $E := \Omega_{B/A} \otimes_B C/\mathfrak{q}$, let G be an open submodule of E and $p : E \to E/G$ be the canonical surjection. Then there exists $h : \Omega_{C/A} \otimes_B C/\mathfrak{q} \to G$ such that $h \circ f = p$. Let $x \in \ker(f)$. Then $p(x) = h(f(x)) = 0$, that is, $x \in G$. Since G was an arbitrary open submodule of E and E is separated, we get that $x = 0$, so that f is injective. Then also the morphism δ is injective. Since $v \circ u$ is a regular morphism, we get that $H_1(A, C, L) = 0$ and consequently $H_1(B, C, L) = 0$. From Corollary 3.2.4 we obtain that $C_{\mathfrak{q}}$ is formally smooth over B; hence v is regular.   □

Definition 3.2.18 Let A be a Noetherian ring. An A-module E is called **well-separated** if for any Noetherian local A-algebra B and any finitely generated B-module F, the B-module $E \otimes_A F$ is separated in the topology of the maximal ideal of B.

Remark 3.2.19

(1) It is easy to see that projective modules, as well as finitely generated modules, are well-separated.
(2) Well-separated modules have a strong relation with excellent rings of positive characteristic. This connection will be discussed later.

Lemma 3.2.20 *Let k be a field of characteristic $p > 0$ and let $(k_i)_{i \in I}$ be a directed family of subfields of k such that $\bigcap_{i \in I} k_i = k^p$. If $X = (X_1, \ldots, X_m)$ is a finite family of indeterminates, then*

$$\bigcap_{i \in I} k_i((X)) = k^p((X)).$$

Proof It is obvious that

$$k^p((X)) \subseteq \bigcap_{i \in I} k_i((X)).$$

Conversely, let $0 \neq f \in \bigcap_{i \in I} k_i((X))$. Let $i \in I$ and $g \in k_i[[X]]$ such that $f \cdot g \in k_i[[X]]$.
Replacing, if necessary, g by g^p we may assume that $g \in k[[X]]$. Then, for any index $j \geq i$ we have

$$g \in k_i[[X]] \cap k_j((X)) = k_j[[X]].$$

But $\bigcap_{i \in I} k_i[[X]] = k^p[[X]]$, hence $f \in k^p((X))$.   □

Theorem 3.2.21 (André [8, App., Corollaire]) *Let k be a field of characteristic $p > 0$
and let $A = k[[X]] := k[[X_1, \ldots, X_n]]$. Then Ω_A is well-separated.*

Proof Denote by J a p-basis of k. For any finite subset I of J, let $k_I := k^p(J \setminus I)$. Then
$[k : k_I] < \infty$, I is a p-basis of k over k_I and $\bigcap_{I \subseteq J,\ I \text{ finite}} k_I = k^p$. Let also $A_I := k_I[[X^p]]$.
Then we have

$$A^p \subseteq A_I \subseteq A,$$

$$\bigcap_I A_I = A^p = k^p[[X^p]].$$

For any A-module F there is a canonical morphism

$$i_F : \Omega_A \otimes_A F \to \prod_I (\Omega_{A/A_I} \otimes_A F).$$

Because $I \cup X$ is a p-basis of A over A_I, it results that Ω_{A/A_I} are free finitely generated
A-modules [131, Cor. 21.2.25]. Hence, by Remark 3.2.19 it is enough to show that i_F is
injective for any A-module F. This means that, using a standard argument, it is enough to
show that $i_{A/\mathfrak{q}}$ is injective for any $\mathfrak{q} \in \mathrm{Spec}(A)$.
Case 1. $\mathfrak{q} = (0)$. Let $K := Q(A)$ and $K_I := Q(A_I)$. We have a commutative diagram

$$
\begin{array}{ccccc}
\Omega_A & \xrightarrow{\ f\ } & \Omega_A \otimes_A K & \xrightarrow{\ \cong\ } & \Omega_K \\
{\scriptstyle i_A}\downarrow & & {\scriptstyle i_K}\downarrow & & \downarrow{\scriptstyle h} \\
\prod_I \Omega_{A/A_I} & \longrightarrow & \prod_I (\Omega_{A/A_I} \otimes_A K) & \xrightarrow{\ \cong\ } & \prod_I \Omega_{K/K_I}.
\end{array}
$$

From Lemma 3.2.20 we have $\bigcap_I K_I = K^p$ and by [131, Th. 21.8.3] we obtain that h is injective. Because $\mathbb{F}_p$ is a perfect field and A is a regular local ring, it follows that the morphism $\mathbb{F}_p \to A$ is a regular morphism. From Theorem 3.2.8 we get that Ω_A is a flat, hence torsion-free A-module, therefore f is injective. Consequently i_A is injective.

Case 2. $\mathfrak{q} \neq (0)$. We prove the assertion by induction on n. Let $f \in \mathfrak{q}$, $f \neq 0$. Replacing, if necessary, f by f^p, we can assume that $f \in A^p$. Using a suitable automorphism, we can also assume that $f(0, \ldots, 0, X_n) \neq 0$ [56, ch. VII, §3, n. 7, Lemme 3]. Let us denote $Y := X_n$, $C := k[[X_1, \ldots, X_{n-1}]]$, $C_I := k_I[[X_1^p, \ldots, X_{n-1}^p]]$. Then there exists a polynomial $g \in D := C[Y]$ associated with f [56, ch. VII, §3, n. 8, Prop. 6]. Changing f with g we get $f \in C^p[Y] = D^p$ and $f \in \mathfrak{q} \cap D$. Then $A/fA \cong D/fD := T$. Set $F := A/\mathfrak{q}$. Then

$$\Omega_A \otimes_A F \cong \Omega_T \otimes_T F \cong \Omega_D \otimes_D F.$$

Set $D_I := C_I[Y^p]$. Then we have

$$A_I/fA_I \cong D_I/fD_I := T_I.$$

Hence

$$\Omega_{A/A_I} \otimes_A F \cong \Omega_{T/T_I} \otimes_T F \cong \Omega_{D/D_I} \otimes_D F \cong \Omega_{D/C_I} \otimes_D F.$$

Consequently we get the commutative diagram

$$
\begin{array}{ccccccccc}
0 & \longrightarrow & \Omega_C \otimes_C F & \longrightarrow & \Omega_D \otimes_D F & \longrightarrow & \Omega_{D/C} \otimes_C F & \longrightarrow & 0 \\
& & \downarrow{\scriptstyle u} & & \downarrow{\scriptstyle i_{A/\mathfrak{q}}} & & \downarrow{\scriptstyle w} & & \\
0 & \longrightarrow & \prod_I (\Omega_{C/C_I} \otimes_C F) & \longrightarrow & \prod_I (\Omega_{D/C_I} \otimes_D F) & \longrightarrow & \prod_I (\Omega_{D/C} \otimes_D F) & \longrightarrow & 0.
\end{array}
$$

Because D is smooth over C, the rows of this diagram are exact. By induction u is injective and w is obviously injective. It follows that also $i_{A/\mathfrak{q}}$ is injective. $\square$

We will prove now a famous result, known as the *localization of formal smoothness*. This will solve affirmatively the problem of localization 4.2.1 for regular morphisms. Note that condition (i) of 3.2.22 is exactly condition (i) of 4.2.1 for regular morphisms.

Theorem 3.2.22 (André [8, Théorème]) *Let* $u : A \to B$ *be a local morphism of Noetherian local rings. Suppose that:*

(i) *A has geometrically regular formal fibres;*
(ii) *u is formally smooth.*

Then u is a regular morphism.

Proof ([64]) We have the commutative diagram

$$
\begin{array}{ccc}
A & \overset{u}{\longrightarrow} & B \\
{\scriptstyle f}\downarrow & & \downarrow{\scriptstyle g} \\
\widehat{A} & \overset{\widehat{u}}{\longrightarrow} & \widehat{B}
\end{array}
$$

where $\widehat{A}$ and $\widehat{B}$ are the completions of A resp. B in the topology of the respective maximal ideal, u and consequently $\widehat{u}$ is formally smooth and f is regular. If we prove that $\widehat{u}$ is regular, then $\widehat{u} f = gu$ is regular and since g is faithfully flat, from Proposition 1.6.18 it will follow that u is a regular morphism. This means that we can assume that A and B are complete local rings. Let $\mathfrak{p} \in \mathrm{Spec}(B)$. Then the fibre of u in $\mathfrak{p}$ is the fibre of $u \otimes_A A/\mathfrak{p} : A/\mathfrak{p} \to B/\mathfrak{p}B$ in the zero ideal. So we can assume that moreover A is a domain and we have only to show that the fibre of u in (0) is geometrically regular. By [231, Prop. 6.1.8] there exists a commutative diagram

$$
\begin{array}{ccc}
C & \overset{v}{\longrightarrow} & D \\
{\scriptstyle f}\downarrow & & \downarrow{\scriptstyle g} \\
A & \overset{u}{\longrightarrow} & B
\end{array}
$$

where C and D are Noetherian local complete regular rings, f is finite and injective and $B \cong A \otimes_C D$. If the fibre of v in (0) is geometrically regular, it follows at once that the fibre of u in (0) is geometrically regular. This means that we can moreover assume that A and B are regular local rings. Let $p = \mathrm{char}(A)$.

Case 1. $p = 0$. Then $B \otimes_A Q(A)$ is a regular ring, hence geometrically regular over $Q(A)$.

Case 2. $p \neq 0$. Then p is a prime number, because A is a local domain. Since A is a regular local ring containing a field, from Cohen's structure theorem we get that $A \cong k[[X_1, \ldots . X_n]]$, where k is the residue field of A. By Theorem 3.2.21 it follows that Ω_A is well-separated. Now we apply Theorem 3.2.17 to the morphisms $\mathbb{F}_p \to A \to B$ and we get that u is regular.   □

Remark 3.2.23 This famous theorem, which has many important applications, was proved for the first time by André [8], using André-Quillen homology. The above proof was given by Brezuleanu and Radu in [64]. It uses André-Quillen homology in the proof of Theorem 3.2.17. On the other hand, Theorem 3.2.17 can be proved in an elementary way (see [297]). Another proof of Theorem 3.2.22, without using homological tools, was given by Seydi [353, Th. 5.1].

We present an application of the above results due to Majadas and Rodicio.

Proposition 3.2.24 (Majadas and Rodicio [230, Prop. 6.8]) *Let $u : A \to B$ be a smooth morphism of Noetherian rings. If the formal fibres of A are geometrically regular, the formal fibres of B are geometrically regular.*

Proof Let $\mathfrak{q} \in \mathrm{Spec}(B)$ and $\mathfrak{p} = \mathfrak{q} \cap A$. We have the canonical morphisms

$$A_{\mathfrak{p}} \xrightarrow{\phi} B_{\mathfrak{q}} \xrightarrow{\psi} \widehat{B_{\mathfrak{q}}}.$$

By assumption ϕ is regular, hence formally smooth because it is local. Therefore $\psi \circ \phi$ is formally smooth. By Theorem 3.2.22 $\psi \circ \phi$ is regular. The $B_{\mathfrak{q}}$-module $\Omega_{B_{\mathfrak{q}}/A_{\mathfrak{p}}}$ is projective, hence well-separated. From Theorem 3.2.17 we obtain that ψ is regular. $\square$

Now we want to know more about the connections between excellent rings and the good separation of the module of differentials. We start with some easy consequences of Theorem 3.2.17.

Proposition 3.2.25 *Let $(A, \mathfrak{m}, k)$ be a Noetherian regular local ring. Assume that either A contains a field or $\mathrm{char}(k) = p > 0$ and $p \notin \mathfrak{m} \smallsetminus \mathfrak{m}^2$. If $\Omega_A \otimes_A (\widehat{A}/\mathfrak{p})$ is separated for any $\mathfrak{p} \in \mathrm{Spec}(\widehat{A})$, then A has geometrically regular formal fibres.*

Proof If A contains a field let S be the prime field contained in A. If A doesn't contain a field let $S = \mathbb{Z}_{(p)}$. However, in both cases the formal fibres of S are geometrically regular. As A is regular, the canonical morphism $u : S \to A$ is formally smooth; hence the composition $S \xrightarrow{u} A \xrightarrow{v} \widehat{A}$ is formally smooth and André's theorem 3.2.22 implies that $v \circ u$ is a regular morphism. By Theorem 3.2.17 we get that v is a regular morphism. $\square$

Theorem 3.2.26 (André [8, App. Théorème]) *Let A be a regular local ring of characteristic $p > 0$. Then A has geometrically regular formal fibres if and only if Ω_A is well-separated.*

Proof Assume first that the formal fibres of A are geometrically regular. Let $(C, \mathfrak{n})$ be a Noetherian local A-algebra, E a finitely generated C-module and $\mathfrak{p} := \mathfrak{n} \cap A$. Then

$\Omega_A \otimes_A E \simeq \Omega_A \otimes_A A_{\mathfrak{p}} \otimes_{A_{\mathfrak{p}}} E$. Hence we may assume that $(A, \mathfrak{p})$ is a local ring and that $A \to C$ is a local morphism. Let $L := k(\mathfrak{p})$. Then $\widehat{A} \simeq L[[X_1, \ldots, X_n]]$ and by Theorem 3.2.21 we get that $\Omega_{\widehat{A}}$ is well-separated. Now the canonical morphism $A \to \widehat{A}$ is regular; hence by Theorem 3.2.8 we have $H_1(A, \widehat{A}, E) = (0)$. From the Jacobi-Zariski exact sequence

$$H_1(A, \widehat{A}, E) \to \Omega_A \otimes E \to \Omega_{\widehat{A}} \otimes E$$

it follows that the canonical morphism $\Omega_A \otimes_A \widehat{E} \to \Omega_{\widehat{A}} \otimes_{\widehat{A}} \widehat{E}$ is injective, therefore $\Omega_A \otimes_A \widehat{E}$ is separated and Ω_A is well-separated. The converse follows from Proposition 3.2.25. $\qquad\square$

Lemma 3.2.27 *Let $(A, \mathfrak{m})$ be a Noetherian local ring and let*

$$0 \to M \to N \to P \to 0$$

be an exact sequence of A-modules.

(i) *If A is a complete local ring, M is finitely generated and N is $\mathfrak{m}$-adically separated, then P is $\mathfrak{m}$-adically separated;*
(ii) *If P is flat and $\mathfrak{m}$-adically separated, then N is $\mathfrak{m}$-adically separated.*

Proof Considering the completions in the $\mathfrak{m}$-adic topology, we have the commutative diagram

$$
\begin{array}{ccccccccc}
0 & \longrightarrow & M & \longrightarrow & N & \longrightarrow & P & \longrightarrow & 0 \\
& & \downarrow{\scriptstyle\alpha} & & \downarrow{\scriptstyle\beta} & & \downarrow{\scriptstyle\gamma} & & \\
0 & \longrightarrow & \widehat{M} & \longrightarrow & \widehat{N} & \longrightarrow & \widehat{P} & \longrightarrow & 0
\end{array}
$$

(i) The topology of M is induced by the topology of N by Chevalley's Lemma [266, Theorem (30.1)]; hence the bottom line is exact too. Moreover α is an isomorphism and β is injective. Then γ is injective so that P is separated.
(ii) Since P is flat, we have $\mathfrak{m}^n N \cap M = \mathfrak{m}^n M$ for any $n \geq 1$. This implies that the bottom line is exact. But α and γ are injective, hence β is injective. $\qquad\square$

Lemma 3.2.28 *Let A be a ring and E an A-module. Assume that for any Noetherian complete local A-algebra $(B, \mathfrak{m})$ and for any finitely generated B-module F, the B-module $E \otimes_A F$ is separated in the $\mathfrak{m}$-adic topology. Then E is well-separated.*

Proof Let $(C, \mathfrak{n})$ be a Noetherian local A-algebra and G a finitely generated C-module. We consider the completions in the $\mathfrak{n}$-adic topologies. Since $\widehat{C}$ is a faithfully flat C-module, the canonical morphism $E \otimes_A G \to E \otimes_A G \otimes_C \widehat{C}$ is injective. Then $E \otimes_A G$ is separated; hence E is well-separated.   □

Proposition 3.2.29 *Let $(A, \mathfrak{m}, k)$ be a Noetherian local ring of characteristic $p > 0$, where p is prime. If A has geometrically regular formal fibres, then Ω_A is well-separated.*

Proof We can write $\widehat{A} = C/\mathfrak{a}$, where C is a complete regular local ring of characteristic p. Then for any $\widehat{A}$-module E we have an exact sequence

$$\mathfrak{a}/\mathfrak{a}^2 \otimes_{\widehat{A}} E \to \Omega_C \otimes_C E \to \Omega_{\widehat{A}} \otimes_{\widehat{A}} E \to 0.$$

Let D be a Noetherian local complete A-algebra and E a finitely generated D-module. Then from the above exact sequence we obtain an exact sequence of D-modules

$$0 \to F \to \Omega_C \otimes_C E \to \Omega_{\widehat{A}} \otimes_{\widehat{A}} E \to 0,$$

where F is a finitely generated D-module. Since $\Omega_C \otimes_C E$ is a separated D-module, it follows from Lemma 3.2.27 that $\Omega_{\widehat{A}} \otimes_{\widehat{A}} E$ is a separated D-module, hence $\Omega_{\widehat{A}}$ is well-separated. Since the fibres of A are geometrically regular, from Theorem 3.2.8 we have $H_1(A, \widehat{A}, E) = (0)$ and therefore we have the exact sequence

$$0 \to \Omega_A \otimes_A E \to \Omega_{\widehat{A}} \otimes_{\widehat{A}} E \to \Omega_{\widehat{A}/A} \otimes_{\widehat{A}} E \to 0.$$

Applying Lemmas 3.2.27 and 3.2.28, we obtain that Ω_A is well-separated.   □

Example 3.2.30

(i) It is easy to see that for rings containing a field of characteristic 0, the above results are no more true. Indeed, let $A = \mathbb{Q}[[X]]$. Then A is a regular local ring with geometrically regular formal fibres (Th.1.6.37) and $\widehat{\Omega}_A$ is a free A-module of finite rank. Since $\mathrm{tr.deg}_{\mathbb{Q}}(\mathbb{Q}((X))) = \infty$, it follows that Ω_A is not a finitely generated A-module. Hence Ω_A is not separated.

(ii) Let p be a prime number and let $\mathbb{Z}_p = \widehat{\mathbb{Z}_{(p)}}$ be the ring of p-adic numbers. Obviously $\mathbb{Z}_p$ is a regular local ring with geometrically regular formal fibres (Theorem 1.6.37). Since $\mathbb{Z}_{(p)} \to \mathbb{Z}_p$ is étale, we have $\widehat{\Omega}_{\mathbb{Z}_p} = 0$. Since $Q(\mathbb{Z}_p)$ is not countable, it follows that $\mathrm{tr.deg}_{\mathbb{Q}}(Q(\mathbb{Z}_p)) = \infty$. Thus $\Omega_{\mathbb{Z}_p}$ is not a finitely generated $\mathbb{Z}_p$-

module, hence is not separated. This shows that also in the unequal characteristic case Proposition 3.2.29 is not true.

3.3 Excellent Rings

The class of excellent rings was first considered by A. Grothendieck [132, Def. 7.8.2]. Let us quote his motivation for introducing this notion:

Soulignons en outre que dans l'étude de ces problèmes nous nous sommes systématique-ment préoccupés de savoir si la réponse affirmative à l'un d'eux est stable pour les deux opérations les plus importantes de l'Algèbre Commutative: la localisation et le passage à une algèbre de type fini. Les résultats obtenus dans l'étude de ces problèmes amènent à dégager la définition d'une classe d'anneaux noethériens dont le comportement à cet égard est le meilleur possible [132, p. 214].

Definition 3.3.1 A Noetherian ring A is called **quasi-excellent** if:

 (i) A has geometrically regular formal fibres;
(ii) A is Reg-2.

A quasi-excellent universally catenary ring is called **excellent**.

Lemma 3.3.2 *Let A be a (quasi)-excellent ring. Then any ring of fractions of A and any A-algebra of finite type are (quasi)-excellent rings.*

Proof This is a consequence of Lemma 1.2.9, Theorem 2.3.15, Corollary 2.3.18 and Lemma 3.1.14.   □

Example 3.3.3

 (i) Fields and algebras essentially of finite type over fields are excellent rings.
(ii) Any complete semilocal ring is excellent. (cf. Theorem 1.2.44, Theorem 1.6.37 and Corollary 3.1.19)
(iii) If D is a Dedekind domain with fraction field of characteristic zero, then D is excellent.
     Indeed, if $\mathfrak{p} \in \mathrm{Spec}(A)$, then $A_\mathfrak{p}$ is a DVR, hence also $\widehat{A_\mathfrak{p}}$ is a DVR, therefore the formal fibres of A are geometrically regular. The remaining properties follow from Example 1.2.43 and Proposition 3.1.18.
(iv) Examples of rings which are not quasi-excellent can be obtained from the examples presented in the preceding sections. In Example 1.5.20 a ring whose regular locus is not open is constructed. Consequently that ring is not quasi-excellent.

(iv) There are rings that are locally excellent, but not excellent. For example, if A is the ring constructed in Example 1.2.4, then for any prime ideal $\mathfrak{p}$ of A, the localization $A_\mathfrak{p}$ is essentially of finite type over a field, hence excellent, while A itself is not excellent. It must be noted that universal catenarity and condition (i) of Definition 3.3.1 are of local nature, while condition (ii) is not.

Example 3.3.4 An excellent ring which is not finite dimensional (Tanaka [366]).

Proof We will show that the classical example of Nagata [266, Ex. 1] of an infinite dimensional Noetherian ring is actually excellent. Let us briefly remind the construction: let k be any field and $B := k[X_1, \ldots, X_i, \ldots] = \bigcup_{i=0}^{\infty} k[X_1, \ldots, X_i]$. Let $\mathfrak{m}_i :=$ $(X_{2^i}, \ldots, X_{2^{i+1}-1})$. Then $\mathfrak{m}_i$ is a prime ideal of B and the set $S := B \setminus \bigcup_{i=0}^{\infty} \mathfrak{m}_i$ is a multiplicative system of B. Let $A := S^{-1}B$. Then A is an infinite dimensional Noetherian ring whose maximal ideals are $\mathfrak{m}_i A$, $i \geq 1$ and for each $i \geq 1$, the ring $A_{\mathfrak{m}_i}$ is an algebra essentially of finite type over a field. Therefore A is an universally catenary(even regular) ring with geometrically regular formal fibres. It remains to show that any finitely generated A-algebra D has open regular locus. We can write $D = C/\mathfrak{a}$, where $C := A[Y_1, \ldots, Y_m]$ and $\mathfrak{a} = (f_1, \ldots, f_r)$ is an ideal of C. The generators of $\mathfrak{a}$ are of the form

$$f_i = \sum \frac{p_{j_1,\ldots,j_m}}{q_{j_1,\ldots,j_m}} Y_1^{j_1} \cdot \ldots \cdot Y_m^{j_m}, \quad \text{with } p_{j_1,\ldots,j_m}, q_{j_1,\ldots,j_m} \in B.$$

Clearly only finitely many variables $X_1, \ldots, X_t$ occur in all the elements p_l, q_l, so that, enlarging t if necessary, we may assume that there is i such that $t = 2^{i+1} - 1$. Then let $\overline{B} := k[X_1, \ldots, X_t]$, $\overline{S} := \overline{B} \setminus \bigcup_{j=1}^{i} \mathfrak{m}_j$, $\overline{A} := \overline{S}^{-1}\overline{B}$, $\overline{C} := \overline{A}[Y_1, \ldots, Y_m]$, $\overline{\mathfrak{a}} :=$ $(f_1, \ldots, f_r)\overline{C}$ and $\overline{D} := \overline{C}/\overline{\mathfrak{a}}$. Let $\mathfrak{q} \in \mathrm{Spec}(D)$ and $\overline{\mathfrak{q}} := \mathfrak{q} \cap \overline{D}$. There is a commutative diagram

$$
\begin{array}{ccccccccc}
\overline{B} & \longrightarrow & \overline{A} & \longrightarrow & \overline{C} & \longrightarrow & \overline{D} & \longrightarrow & \overline{D}_{\overline{\mathfrak{q}}} \\
\downarrow{\alpha} & & \downarrow{\beta} & & \downarrow{\gamma} & & \downarrow{\delta} & & \downarrow{\delta_\mathfrak{q}} \\
B & \longrightarrow & A & \longrightarrow & C & \longrightarrow & D & \longrightarrow & D_\mathfrak{q}
\end{array}
$$

The morphism α is flat and smooth; hence β is flat and smooth. This implies that γ and δ are flat and smooth. By Theorem 3.2.7 and Theorem 3.2.8, it follows that $\delta_\mathfrak{q}$ is formally smooth. This means that $\overline{D}_{\overline{\mathfrak{q}}}$ is a regular local ring iff $D_\mathfrak{q}$ is a regular local ring, in other words

$$\mathrm{Spec}(\delta)^{-1}(\mathrm{Reg}(\overline{D})) = \mathrm{Reg}(D).$$

But $\overline{D}$ is an algebra of finite type over a field, hence $\mathrm{Reg}(\overline{D})$ is open, therefore $\mathrm{Reg}(D)$ is open too. □

Theorem 3.3.5 (Greco [122, Th. 3.1]) *Let $u : A \to B$ be a morphism of finite type such that* $\ker(u)$ *is a nilpotent ideal of A. Then:*

(i) *A is quasi-excellent if and only if B is quasi-excellent;*
(ii) *A is excellent if and only if A is universally catenary and B is quasi-excellent.*

Proof

(i) Follows from Lemma 3.3.2, Proposition 1.2.11 and Proposition 1.6.42.
(ii) Follows immediately from (i) and from the definitions. □

For semilocal rings, one has only to check the properties of the formal fibres in order to decide the quasi-excellence:

Proposition 3.3.6 *Let A be a semilocal ring with geometrically regular formal fibres. Then A is quasi-excellent.*

Proof It is a special case of Proposition 2.3.19. □

Proposition 3.3.7 *Let A be a Noetherian quasi-excellent ring. Then A is a Nagata ring.*

Proof It is a consequence of Theorem 2.3.12. □

The converse of Proposition 3.3.7 is not true. The first example of a Nagata local ring that is not quasi-excellent was constructed by Rotthaus [325]. We shall present such an example using the tools of Sect. 1.5. We need a preparatory result.

Lemma 3.3.8 *Let R be a local domain with geometrically regular formal fibres and $x \in R$ be a prime element. Let $\mathfrak{p}$ be a prime ideal of $\widehat{R}$ such that $x \notin \mathfrak{p}$ and let $B := \widehat{R}/\mathfrak{p}$. If $Q(B/xB)$ is a separable field extension of $Q(R/xR)$, then $Q(B)$ is separable over $Q(R)$.*

Proof Let $D := R_{xR}$ and $E := \widehat{A}_{x\widehat{A}}$. From the exact sequence

$$0 \to E \xrightarrow{x} E \to E/xE \to 0,$$

where the map on the left is multiplication by x, we obtain by [6, Lemme III.22] an exact sequence

$$H_2(D, E, E/xE) \to H_1(D, E, E) \xrightarrow{x} H_1(D, E, E) \to H_1(D, E, E/xE).$$

By [6, Prop. IV.54] we have

$$H_i(D, E, E/xE) \cong H_i(D/xD, E/xE, E/xE), \text{ for } i = 1, 2.$$

Because E/xE is separable over D/xD, by [6, Prop. VII.4, VII.13] we know that

$$H_i(D/xD, E/xE, E/xE) = (0), \text{ for } i \geq 1.$$

Hence $H_1(D, E, E/xE) = H_2(D, E, E/xE) = (0)$. Therefore

$$H_1(Q(R), Q(\widehat{A}), Q(\widehat{A})) = H_1(D, E, E) = \bigcap_{j=1}^{\infty} x^j H_1(D, E, E).$$

This means that it suffices to show that $H_1(D, E, E)$ is x-adically separated. Let $\mathfrak{q} = \mathfrak{p} + x\widehat{R}$ which is a prime ideal of $\widehat{R}$. We have $\mathfrak{q} \cap R = xR$; hence there are canonical morphisms $D = R_{xR} \xrightarrow{\phi} \widehat{R}_{\mathfrak{q}} \xrightarrow{\psi} E = \widehat{A}_{x\widehat{A}}$. We observe that ψ is surjective and that ϕ is regular by Theorem 3.2.22, because R has geometrically regular formal fibres. Then we have an exact sequence

$$(0) = H_1(D, \widehat{R}_{\mathfrak{q}}, E) \to H_1(D, E, E) \to H_1(\widehat{R}_{\mathfrak{q}}, E, E)$$

and using [6, Prop. IV.55] we see that $H_1(D, E, E)$ is a finitely generated E-module, therefore it is separated [18, Cor. 10.19].   □

Example 3.3.9 (Ogoma [275, Ex. 1], cf. [271, Ex. 4.4])   There exists a Nagata regular local ring A, such that $\dim(A) = 2$, $\operatorname{char}(A) = p > 0$ and the generic formal fibre of A is not geometrically normal. Consequently A is not quasi-excellent.

Proof Let $p > 2$ be a prime number and K_0 be a countable field of characteristic p and take $n = 2$. Let $G(X, Z_1, Z_2) := X + Z_1^2 + Z_2^p \in K_0[X, Z_1, Z_2]$. We take $\varepsilon_i = p \cdot i$ for $i \in \mathbb{N}$ and we perform the second construction from Sect. 1.5, keeping the notations from that section. Then by Corollary 1.5.19, we obtain a regular local ring $(A, \mathfrak{m})$ with

$\dim(A) = 2$ and a prime element x such that $\widehat{A} \cong K[[x, \zeta_1, \zeta_2]]/(x + \zeta_1^2 + \zeta_2^p)$. By Lemma 3.3.8 applied to $A_{xA} = R_{xR}$ and to $B = \widehat{A}$, we have only to show that $Q(\widehat{A}/x\widehat{A})$ is separable over $Q(A/xA)$, or in other words that $Q(K[[\bar{\zeta}_1, \bar{\zeta}_2]]/(\bar{\zeta}_1^2 + \bar{\zeta}_2^p))$ is separable over $K(\bar{Z}_1, \bar{Z}_2)$. But this is the same as showing that

$$K[[\bar{\zeta}_1, \bar{\zeta}_2]]/(\bar{\zeta}_1^2 + \bar{\zeta}_2^p) \otimes_{K[\bar{Z}_1, \bar{Z}_2]} K^{1/p}[\bar{Z}_1^{1/p}, \bar{Z}_2^{1/p}]$$

is reduced. Let us notice that $K[[\bar{\zeta}_1, \bar{\zeta}_2]] = K[[\bar{Z}_1, \bar{Z}_2]]$ and that

$$K[[\bar{Z}_1, \bar{Z}_2]] \otimes_{K[\bar{Z}_1, \bar{Z}_2]} K^{1/p}[\bar{Z}_1^{1/p}, \bar{Z}_2^{1/p}] \cong K[[\bar{Z}_1^{1/p}, \bar{Z}_2^{1/p}]][K^{1/p}] =$$

$$= \bigcup_j K[[\bar{Z}_1^{1/p}, \bar{Z}_2^{1/p}]][a_{11}^{1/p}, a_{21}^{1/p}, \ldots, a_{1j}^{1/p}, a_{2j}^{1/p}] =$$

$$= \bigcup_j K(a_{11}^{1/p}, a_{21}^{1/p}, \ldots, a_{1j}^{1/p}, a_{2j}^{1/p})[[\bar{Z}_1^{1/p}, \bar{Z}_2^{1/p}]].$$

By a direct limit argument, it is enough to show that

$$K(a_{11}^{1/p}, a_{21}^{1/p}, \ldots, a_{1j}^{1/p}, a_{2j}^{1/p})[[\bar{Z}_1^{1/p}, \bar{Z}_2^{1/p}]]/(\bar{\zeta}_1^2 + \bar{\zeta}_2^p)$$

is a reduced ring, for any $j \geq 1$. Now let us denote

$$H_j := K(a_{11}^{1/p}, a_{21}^{1/p}, \ldots, a_{1j}^{1/p}, a_{2j}^{1/p})[[\bar{Z}_1^{1/p}, \bar{Z}_2^{1/p}]]$$

and consider the derivation $D = \frac{\partial}{\partial a_l} \in \mathrm{Der}(H_j)$, for some $l > j$. Then it is easy to see that $D(\bar{\zeta}_1^2 + \bar{\zeta}_2^p) = q_l^{\epsilon_l} \bar{\zeta}_1$. On the other hand, in H_j we have

$$(x + \zeta_1^2 + \zeta_2^p, q_l^{p^l} \zeta_1) = (\zeta_1, x + \zeta_2^p) \cap (x + \zeta_1^2 + \zeta_2^p, q_l)$$

and the ideals on the right side are prime ideals of height 2. Therefore the ideal $(x + \zeta_1^2 + \zeta_2^p)H_j$ has height 1. But H_j is a regular ring; hence locally factorial and it follows easily that $(x + \zeta_1^2 + \zeta_2^p)H_j$ is a radical ideal. $\qquad \square$

Let us point out Rotthaus' example as well.

Example 3.3.10 (Rotthaus [325], cf. [271, Ex. 4.5]) There exists a Nagata regular local ring A, such that $\dim(A) = 3$ and A is not quasi-excellent.

Proof Let $p > 2$ be a prime number and K_0 be a countable field of characteristic 0 or of characteristic p and take $n = 3$. Let $G(X, Z_1, Z_2, Z_3) := X + Z_1^2 + Z_2^p \in$

$K_0[X, Z_1, Z_2, Z_3]$. In the characteristic zero case, in the definition of G we can take $p > 1$ any natural number. We take $\varepsilon_i = p \cdot i$ in the positive characteristic case, and we perform the second construction from Sect. 1.5, keeping the notations from that section. By Theorem 1.5.18, we obtain a Noetherian regular local ring A of dimension 3, having a prime element $x \in A$ such that

$$\widehat{A} \cong K[[x, \zeta_1, \zeta_2, \zeta_3]]/(x + \zeta_1^2 + \zeta_2^p)$$

and

$$\widehat{A}/x\widehat{A} \cong K[[\zeta_1, \zeta_2, \zeta_3]]/(\zeta_1^2 + \zeta_2^p).$$

The same proof as in Example 3.3.9 shows that A is the required example.   □

However, in small dimensions the above phenomena cannot occur.

Proposition 3.3.11 *Let $(A, \mathfrak{m}, k)$ be a Noetherian local ring with geometrically normal formal fibres.*

(i) *If $\dim(A) \leq 2$, then A is quasi-excellent.*
(ii) *If $\dim(A) \leq 3$, then A is Reg-2.*

Proof

(i) Let $\mathfrak{p} \in \mathrm{Spec}(A)$ and $k(\mathfrak{p}) \subseteq K$ be a finite field extension. If $\mathfrak{p} = \mathfrak{m}$, then

$$\dim(\widehat{A} \otimes_A K) = \dim(\widehat{A} \otimes_A k(\mathfrak{m})) = 0$$

hence $\widehat{A} \otimes_A K$ is a regular ring. If $0 < \dim(A/\mathfrak{p}) \leq 2$, then

$$\dim(\widehat{A} \otimes_A K) = \dim(\widehat{A} \otimes_A k(\mathfrak{p})) \leq \dim(A) - 1 + \mathrm{ht}(\mathfrak{p}) \leq 1$$

and is a normal ring, hence regular.
(ii) The non-trivial case is $\dim(A) = 3$. Any finite A-algebra has geometrically normal formal fibres; hence by Proposition 1.2.11 we can suppose that A and consequently $\widehat{A}$ are normal rings, and it is sufficient to prove that $\mathrm{Reg}(A)$ is open. We may restrict to the case $\mathrm{Reg}(A) \subsetneq \mathrm{Spec}(A)$. Since $\widehat{A}$ is normal, in $\mathrm{Sing}(\widehat{A})$ we have only $\mathfrak{m}$ and maybe finitely many prime ideals of height 2, say $\mathfrak{q}_1, \ldots, \mathfrak{q}_r$. If $f : \mathrm{Spec}(\widehat{A}) \to \mathrm{Spec}(A)$ is the canonical map, then $f^{-1}(\mathrm{Reg}(\widehat{A}) = \mathrm{Spec}(\widehat{A}) \setminus \{\mathfrak{m}, \mathfrak{q}_1, \ldots, \mathfrak{q}_r\}$. But $\dim(A) = 3$ implies that $\{\mathfrak{m}, \mathfrak{q}_i\}$ is a closed set.   □

Just along the same lines one can prove:

Proposition 3.3.12 *Let A be a Noetherian local ring with geometrically reduced formal fibres.*

(i) If $\dim(A) \le 1$, then A is quasi-excellent.
(ii) If $\dim(A) \le 2$, then A is Reg-2.

Proof We leave the details to the reader. □

Corollary 3.3.13 *A Nagata ring A with $\dim(A) \le 1$ is an excellent ring.*

Proof By Proposition 3.3.11 the formal fibres of A are geometrically regular and A is Reg-2. Moreover any one-dimensional ring is universally catenary. □

Let us point out another example, similar to Examples 3.3.9 and 3.3.10.

Example 3.3.14 ([173, Cor. 12]) There exists a local ring A, such that $\dim(A) = 3$, the formal fibres of A are geometrically normal, A is Reg-2, but A is not quasi-excellent.

Proof We perform the first construction from Sect. 1.5 in the following situation: let $K_0 = \mathbb{Q}, n = 3, r = 1, F_1(Z_1, Z_2, Z_3) = Z_1^2 + Z_2^2 - Z_3^2$. Then by Theorem 1.5.8 we obtain a Noetherian local domain A with $\dim(A) = 3$ and such that $\widehat{A} \cong K[[\zeta_1, \zeta_2, \zeta_3]]/(\zeta_1^2 + \zeta_2^2 - \zeta_3^2)$; hence the formal fibre of A in (0) is geometrically normal and not geometrically regular. By Proposition 3.3.11 the ring A is Reg-2. □

The following consequence of Proposition 3.2.24 provides us with other examples of excellent rings.

Proposition 3.3.15 (Majadas and Rodicio [230, Prop. 6.7]) *Let k be a field and A a Noetherian smooth k-algebra. Then A is an excellent ring.*

Proof By Proposition 3.2.24 we get that A has geometrically regular formal fibres. If B is an A-algebra of finite type, then by [248, Rem. 1, p.221] it follows that $\mathrm{Reg}(B)$ is open. Therefore A is quasi-excellent. But A is regular being smooth over a field, hence is universally catenary. □

In the following, we will give some examples of excellent rings using tools similar to the well-known Jacobian criterion. These are mainly due to the Japanese mathematician Hideyuki Matsumura (1930–1995).

Notation 3.3.16 Let A be a ring and $\mathfrak{p} \in \mathrm{Spec}(A)$. Let $D_1, \ldots, D_r \in \mathrm{Der}(A)$ and $x_1, \ldots, x_t \in A$. Then we denote

$$J(x_1, \ldots, x_t; D_1, \ldots, D_r)(\mathfrak{p}) := (D_i x_j)(\mathrm{mod}\ \mathfrak{p})_{i,j} \in M_{t \times r}(A/\mathfrak{p}).$$

Definition 3.3.17 (Matsumura [252]) Let $(A, \mathfrak{m}, K)$ be an equicharacteristic local ring. A subfield k of A is a **quasi-coefficient field** of A if K is 0-étale over k. If $(A, \mathfrak{m}, k)$ is a local ring of mixed characteristic such that $\mathrm{char}(k) = p$, a Noetherian local subring (R, pR, K) of A is called a **quasi-coefficient ring** if K is 0-étale over k.

Remark 3.3.18 Remember that a field extension $k \subseteq K$ is étale if and only if it is separable and $\Omega_{K/k} = (0)$ [248, (38.E), p. 273].

One can find the main properties and facts about the existence of quasi-coefficient fields and quasi-coefficient rings in [249] and [252]. In particular, the existence of the fields k and K in the next theorem follows from [249, Th. 29.6].

Theorem 3.3.19 (Nomura [274]) *Let $(A, \mathfrak{m})$ be a regular local ring of dimension n containing a field, with a regular system of parameters $x_1, \ldots, x_n$. Let k be a quasi-coefficient field of A and let K be a coefficient field of $\widehat{A}$ containing k. Then:*

(i) *$\widehat{A} \simeq K[[x_1, \ldots, x_n]]$ and using this representation $\mathrm{Der}_k(\widehat{A}) = \mathrm{Der}_K(\widehat{A})$ is a free $\widehat{A}$-module with basis $\{\frac{\partial}{\partial x_1}, \ldots, \frac{\partial}{\partial x_n}\}$.*
(ii) *The following conditions are equivalent:*
    (a) *$\frac{\partial}{\partial x_i}(A) \subseteq A$, for any $i = 1, \ldots, n$.*
    (b) *there exist derivations $D_1, \ldots, D_n \in \mathrm{Der}_k(A)$ and elements $a_1, \ldots, a_n \in A$ such that $D_i a_j = \delta_{ij}$, $1 \leq i, j \leq n$.*
    (c) *there exist derivations $D_1, \ldots, D_n \in \mathrm{Der}_k(A)$ and elements $a_1, \ldots, a_n \in A$ such that $\det(D_i a_j) \notin \mathfrak{m}$.*
    (d) *$\mathrm{Der}_k(A)$ is a free A-module of rank n.*
    (e) *$\mathrm{rk}_A(\mathrm{Der}_k(A)) = n$.*

Proof

(i) The field extension $k \subseteq K$ is 0-étale, hence for any derivation $D \in \mathrm{Der}(A)$ such that $D = 0$ on k, it follows that $D = 0$ also on K. Let $D \in \mathrm{Der}_K(\widehat{A})$ and set $y_i := Dx_i$. For any $f \in \widehat{A} = K[[x_1, \ldots, x_n]]$ we have

$$D(f) = \sum_{i=1}^{n} (\frac{\partial f}{\partial x_i}) Dx_i = \sum_{i=1}^{n} (\frac{\partial f}{\partial x_i}) y_i,$$

therefore $D = \sum_{i=1}^{n} y_i \frac{\partial}{\partial x_i}$. Conversely, if we take $y_1, \ldots, y_n$ we can construct

a derivation $D = \sum_{i=1}^{n} y_i \frac{\partial}{\partial x_i}$. Hence $\mathrm{Der}_K(\widehat{A})$ is the free $\widehat{A}$-module with basis

$\frac{\partial}{\partial x_1}, \ldots, \frac{\partial}{\partial x_n}$.

(ii) The implications $a) \Rightarrow b) \Rightarrow c)$ and $d) \Rightarrow e)$ are obvious.

$c) \Rightarrow d)$: The derivations $D_1, \ldots, D_n$ are linearly independent over A and over $\widehat{A}$ by assumption. From (i) it follows that any $D \in \mathrm{Der}_k(A)$ can be written $D = \sum_{i=1}^{n} c_i D_i$ with $c_i \in Q(\widehat{A})$. Since $Da_j = \sum_{i=1}^{n} c_i D_i a_j$, solving these equations we get that $c_i \in A$. Therefore $D_1, \ldots, D_n$ is a basis of $\mathrm{Der}_k(A)$.

$e) \Rightarrow a)$: If $D_1, \ldots, D_n$ are linearly independent over A, there exist elements $a_1, \ldots, a_n \in A$ such that $\det(D_i a_j) \neq 0$. Then $D_1, \ldots, D_n$ are linearly independent over $\widehat{A}$ too. Then we can write $\frac{\partial}{\partial x_i} = \sum_{j=1}^{n} c_{ij} D_j$, with $c_{ij} \in Q(\widehat{A})$. Hence we get

$\delta_{ih} = \sum_{j=1}^{n} c_{ij} D_j x_h$. Then the matrix (c_{ij}) is the inverse of $(D_j x_h)$ and $c_{ij} \in Q(A)$. It follows that $(\frac{\partial}{\partial x_i})(A) \subseteq Q(A) \cap \widehat{A} = A$. $\square$

Definition 3.3.20 (Matsumura [251, Definition]) Let A be a regular ring and $\mathfrak{p}$ be a prime ideal of A of height r. We say that the **weak Jacobian condition** holds at $\mathfrak{p}$ if there exist derivations $D_1, \ldots, D_r \in \mathrm{Der}(A)$ and elements $x_1, \ldots, x_r \in \mathfrak{p}$ such that $\det(D_i x_j) \notin \mathfrak{p}$. We say that the **weak Jacobian condition** holds in A, if it holds at any prime ideal $\mathfrak{p} \in \mathrm{Spec}(A)$. If in the above definitions we consider $D_1, \ldots, D_r \in \mathrm{Der}_k(A)$, where k is a subfield of A, we say that the **weak Jacobian condition holds over** k.

Lemma 3.3.21 *Let A be a regular ring and $\mathfrak{p}$ a prime ideal of A of height r. The following are equivalent:*

(i) the weak Jacobian condition holds at $\mathfrak{p}$;

(ii) for any $\mathfrak{q} \in \mathrm{Spec}(A)$ with $\mathrm{ht}(\mathfrak{q}) = s \leq r$ such that $\mathfrak{q} \subseteq \mathfrak{p}$ and such that $A_{\mathfrak{p}}/\mathfrak{q}A_{\mathfrak{p}}$ is a regular local ring, there exist derivations $D_1, \ldots, D_s \in \mathrm{Der}(A)$ and elements $y_1, \ldots, y_s \in \mathfrak{q}$ such that $\det(D_i y_j) \notin \mathfrak{p}$.

Proof It is enough to show $(i) \Rightarrow (ii)$. By assumption there exist derivations $D_1, \ldots, D_r \in \mathrm{Der}(A)$ and elements $x_1, \ldots, x_r \in \mathfrak{p}$ such that $\det(D_i x_j) \notin \mathfrak{p}$. It follows that the images of $x_1, \ldots, x_r$ in $\mathfrak{p}A_{\mathfrak{p}}/\mathfrak{p}^2 A_{\mathfrak{p}}$ are linearly independent over $k(\mathfrak{p})$, whence $x_1, \ldots, x_r$ is a regular system of parameters of $A_{\mathfrak{p}}$. If $A_{\mathfrak{p}}/\mathfrak{q}A_{\mathfrak{p}}$ is a regular ring, we can take elements $y_1, \ldots, y_s \in \mathfrak{q}$ which are a minimal system of generators of $\mathfrak{q}A_{\mathfrak{p}}$ and

$y_{s+1} \ldots, y_r \in \mathfrak{p}$ such that $y_1, \ldots, y_r$ is a regular system of parameters of $A_\mathfrak{p}$. Since $\det(D_i x_j) \notin \mathfrak{p}$ it follows that $\det(D_i y_j) \notin \mathfrak{p}$. $\qquad\square$

Theorem 3.3.22 (Matsumura [251, Th. 2], [249, Th. 30.7]) *Let $(A, \mathfrak{m})$ be a Noetherian local domain of dimension n containing $\mathbb{Q}$ and let $k \subseteq A$ be a subfield such that* tr.deg$_k(A/\mathfrak{m}) = r < \infty$. *Then* Der$_k(A)$ *is a finitely generated A-submodule of A^{n+r} and*

$$\mathrm{rk}_A(\mathrm{Der}_k(A)) \leq \dim(A) + \mathrm{tr.deg}_k(A/\mathfrak{m}) = n + r.$$

Proof Let L be a quasi-coefficient field of A containing k and let K be a coefficient field of $\widehat{A}$ containing L [249, Th. 28.3]. Let $u_1, \ldots, u_r$ be a transcendence basis of L over k and let $x_1, \ldots, x_n$ be a system of parameters of A. Let

$$\phi : \mathrm{Der}_k(A) \to A^{n+r}, \quad \phi(D) = (Du_1, \ldots, Du_r, Dx_1, \ldots, Dx_n).$$

Then ϕ is an A-linear map and we shall prove that ϕ is injective. Suppose that $\phi(D) = 0$, hence $Du_1 = \cdots = Du_r = Dx_1 = \cdots = Dx_n = 0$. D has a continuous extension D' to $\widehat{A}$ and $D' = 0$ on $B := K[[x_1, \ldots, x_n]]$. It is known that $\widehat{A}$ is a finite B-module. Let $a \in A$. Then a is integral over B and if $f \in B[X]$ is such that $f(a) = 0$ and is of minimal degree, then $f'(a) \neq 0$. Then $0 = D(f(a)) = f'(a)Da$. As $Da \in A$ it follows that Da is a non-zero divisor in $\widehat{A}$, therefore $Da = 0$. Hence $D = 0$. It follows that ϕ is injective and the theorem is proved. $\qquad\square$

Proposition 3.3.23 (Matsumura [251, Th. 6], [249, Th. 30.8]) *Let $(A, \mathfrak{m})$ be a regular local ring containing a field of characteristic zero and let k be a quasi-coefficient field of A. The following conditions are equivalent:*

   (i) *the weak Jacobian condition holds at $\mathfrak{m}$ over k.*
  (ii) $\mathrm{rk}(\mathrm{Der}_k(A)) = \dim(A)$.
 (iii) *the weak Jacobian condition holds in A over k.*

If these conditions hold, then for any prime ideal $\mathfrak{p}$ of A, every derivation $D \in \mathrm{Der}_k(A/\mathfrak{p})$ is induced by an element of Der$_k(A)$ *and*

$$\mathrm{rk}(\mathrm{Der}_k(A/\mathfrak{p})) = \dim(A/\mathfrak{p}).$$

Proof $(i) \Leftrightarrow (ii)$: Follows from Theorem 3.3.19.
$(iii) \Rightarrow (i)$: Obvious.

$(i) \Rightarrow (iii)$: Let K be a coefficient field of $\widehat{A}$ containing k [249, Th. 28.3]. Then, by Theorem 3.3.19, if $x_1, \ldots, x_n$ is a regular system of parameters of A, the derivations $\frac{\partial}{\partial x_1}, \ldots, \frac{\partial}{\partial x_n} \in \mathrm{Der}_k(A)$ and form a basis of this A-module. Let $\mathfrak{p} \in \mathrm{Spec}(A)$ and let $\phi : A \to A/\mathfrak{p}$ be the canonical morphism. Let $D' \in \mathrm{Der}_k(A/\mathfrak{p})$. The fact that D' is induced by $D \in \mathrm{Der}_k(A)$ means that there exists $D \in \mathrm{Der}_k(A)$ such that $\phi \circ D = D' \circ \phi$. If we have such a D', then $D' \circ \phi \in \mathrm{Der}_k(A, A/\mathfrak{p})$ and this has a unique extension to an element of $\mathrm{Der}_k(\widehat{A}, \widehat{A}/\mathfrak{p}\widehat{A})$ so that it is uniquely determined by its values on $x_1, \ldots, x_n$. Hence, if we choose $b_1, \ldots, b_n \in A$ such that $D'(\phi(x_i)) = \phi(b_i)$ and we put $D := \sum\limits_{i=1}^{n} b_i \frac{\partial}{\partial x_i}$, then D' is induced by D. Now $\mathrm{Der}_k(A, A/\mathfrak{p})$ is a free $A/\mathfrak{p}$-module with basis $\{\phi \circ \frac{\partial}{\partial x_i},\ i = 1, \ldots, n\}$ and $\mathrm{Der}_k(A/\mathfrak{p})$ can be identified with the submodule $N := \{\delta \in \mathrm{Der}_k(A, A/\mathfrak{p}) \mid \delta(f) = 0, \forall f \in \mathfrak{p}\}$. Hence, if $\mathfrak{p} = (f_1, \ldots, f_t)$ and $\mathrm{ht}(\mathfrak{p}) = r$ then

$$\mathrm{rk}(\mathrm{Der}_k(A/\mathfrak{p})) = n - \mathrm{rk}(J(f_1, \ldots, f_t; \frac{\partial}{\partial x_1}, \ldots, \frac{\partial}{\partial x_n})(\mathfrak{p})).$$

From [249, Th. 30.4] we get that

$$n - \mathrm{rk}(J(f_1, \ldots, f_t; \frac{\partial}{\partial x_1}, \ldots, \frac{\partial}{\partial x_n})(\mathfrak{p})) \geq n - r$$

and from Theorem 3.3.22 we get that

$$\mathrm{rk}(\mathrm{Der}_k(A/\mathfrak{p})) \leq \dim(A/\mathfrak{p}) - r,$$

so that

$$\mathrm{rk}(J(f_1, \ldots, f_t; \frac{\partial}{\partial x_1}, \ldots, \frac{\partial}{\partial x_n})(\mathfrak{p})) = r$$

and

$$\mathrm{rk}(\mathrm{Der}_k(A/\mathfrak{p})) = \dim(A/\mathfrak{p}).$$

$\square$

Theorem 3.3.24 (Mizutani [249, Th. 32.6]) *Let A be a regular ring such that the weak Jacobian condition holds in $A[X_1, \ldots, X_n]$ for any $n \in \mathbb{N}$. Then A is an excellent ring.*

Proof Let us denote $A_n := A[X_1, \ldots, X_n]$ for any natural number n and let C be an integral domain which is a finite A-module. Then C can be written as $C = A_n/\mathfrak{q}$, where $n \in \mathbb{N}$ and $\mathfrak{q} \in \mathrm{Spec}(A_n)$. Let $\mathfrak{m}$ be a maximal ideal of C and let $\mathfrak{n}$ be the corresponding maximal ideal of A_n. Let $E := (A_n)_\mathfrak{n}$. It is enough to show that for every $\mathfrak{p} \in \mathrm{Spec}(\widehat{E})$ with $\mathfrak{p} \cap E = \mathfrak{q}E$, the ring $\widehat{E}_\mathfrak{p}/\mathfrak{q}\widehat{E}_\mathfrak{p}$ is a regular local ring. Let $r := \mathrm{ht}(\mathfrak{q})$. Then $\mathrm{ht}(\mathfrak{q}\widehat{E}_\mathfrak{p}) = r$.

There exist derivations $D_1, \ldots, D_r \in \mathrm{Der}(A_n)$ and elements $x_1, \ldots, x_r \in \mathfrak{q}$ such that $\det(D_i x_j) \notin \mathfrak{q}$. But one can extend canonically the derivations $D_1, \ldots, D_r$ to E and then to $\hat{E}$. As $\mathfrak{p} \cap A_n = \mathfrak{q}$, it follows that $\det(D_i x_j) \notin \mathfrak{p}$, so that $\widehat{E}_\mathfrak{p}/\mathfrak{q}\widehat{E}_\mathfrak{p}$ is a regular local ring [249, Th. 30.4]. We have proved that the formal fibres of A are geometrically regular.

Let us now prove that A is Reg-2. Let $B := A_n/\mathfrak{a}$ be an A-algebra of finite type. It is enough to prove that, for any $\mathfrak{p}' \in \mathrm{Spec}(B)$, the set $\mathrm{Reg}(B/\mathfrak{p}')$ is open in $\mathrm{Spec}(B/\mathfrak{p}')$. Let $\mathfrak{p} \in \mathrm{Spec}(A_n)$ be the prime ideal corresponding to $\mathfrak{p}'$. Then $B/\mathfrak{p}' = A_n/\mathfrak{p}$. So, it is enough to show that $\mathrm{Reg}(A_n/\mathfrak{p})$ is open in $\mathrm{Spec}(A_n/\mathfrak{p})$. Let then $\mathfrak{q}' \in \mathrm{Reg}(A_n/\mathfrak{p})$ and let $\mathfrak{q} \in \mathrm{Spec}(A_n)$ be the prime ideal corresponding to $\mathfrak{q}'$. Since $(A_n)_\mathfrak{q}/\mathfrak{p}$ is a regular local ring and A_n satisfies the weak Jacobian condition, it follows that there exist derivations $D_1, \ldots, D_r \in \mathrm{Der}(A_n)$ and elements $x_1, \ldots, x_r \in \mathfrak{p}$, such that $d := \det(D_i x_j) \notin \mathfrak{p}$, where $\mathrm{ht}(\mathfrak{p}) = r$. Hence, if $\mathfrak{r} \in \mathrm{Spec}(A_n)$ is such that $\mathfrak{r} \supseteq \mathfrak{p}$ and $d \notin \mathfrak{r}$, the ring $(A_n)_\mathfrak{r}/\mathfrak{p}$ is regular. It follows that, if $\bar{d}$ is the image of d in $A_n/\mathfrak{p}$, then the open set $D(\bar{d})$ is contained in $\mathrm{Reg}(A_n/\mathfrak{p})$; hence $\mathrm{Reg}(A_n/\mathfrak{p})$ is open.   □

Remark 3.3.25 It can be seen from the proof that it is enough to assume that the weak Jacobian condition holds at every prime ideal $\mathfrak{p}$ of $A[X_1, \ldots, X_n]$ such that $A[X_1, \ldots, X_n]/\mathfrak{p}$ is a finite A-module.

Theorem 3.3.26 (Matsumura [251, Th. 7]) *Let A be a regular ring containing $\mathbb{Q}$ such that the weak Jacobian condition holds in A. Then the weak Jacobian condition holds in $A[X]$ too.*

Proof Clearly $A[X]$ is regular. Let $\mathfrak{q} \in \mathrm{Spec}(A[X])$ and $\mathfrak{p} = \mathfrak{q} \cap A$. Then $A[X]_\mathfrak{q}$ is a ring of fractions of $A_\mathfrak{p}[X]$. Moreover, $A_\mathfrak{p}[X]/\mathfrak{p}A_\mathfrak{p}[X] = k(\mathfrak{p})[X]$. Hence either $\mathfrak{q} = \mathfrak{p}A[X]$ and $\mathrm{ht}(\mathfrak{q}) = \mathrm{ht}(\mathfrak{p})$ or $\mathrm{ht}(\mathfrak{q}) = \mathrm{ht}(\mathfrak{p}) + 1$ and $\mathfrak{q}A_\mathfrak{p}[X] = (\mathfrak{p}, f)$ where f is a monic polynomial with coefficients in $A_\mathfrak{p}[X]$ such that $f' \notin \mathfrak{q}A_\mathfrak{p}[X]$. Let now $D_1, \ldots, D_n \in \mathrm{Der}(A)$ and $x_1, \ldots, x_n \in \mathfrak{p}$ where $n = \mathrm{ht}(\mathfrak{p})$, such that $\det(D_i x_j) \notin \mathfrak{p}$. We extend the derivations $D_1, \ldots, D_n$ to derivations $D'_1, \ldots, D'_n \in \mathrm{Der}(A[X])$, by putting $D'_i(X) = 0$ for $i = 1, \ldots, n$. If $\mathrm{ht}(\mathfrak{q}) = \mathrm{ht}(\mathfrak{p})$, we have $\det((D'_i x_j)_{1 \le i, j \le n}) \notin \mathfrak{q}$. If $\mathrm{ht}(\mathfrak{q}) = \mathrm{ht}(\mathfrak{p}) + 1$, we have $f'(X) \notin \mathfrak{q}$, hence taking $D'_{n+1} = \frac{\partial}{\partial X}$ and $x_{n+1} = f$, we get that $\det((D'_i x_j)_{1 \le i, j \le n+1}) \notin \mathfrak{q}$.   □

Corollary 3.3.27 *Let A be a regular ring containing $\mathbb{Q}$ such that the weak Jacobian condition holds in A. Then A is excellent.*

Proof It follows from Theorem 3.3.24 and Theorem 3.3.26.   □

Corollary 3.3.28 *Let k be a field of characteristic zero and $n, m \in \mathbb{N}$. Then the ring $k[X_1, \ldots, X_n][[Y_1, \ldots, Y_m]]$ is excellent.*

Proof It is easy to see that $\mathrm{rk}_k(\mathrm{Der}_k(k[X_1, \ldots, X_n][[Y_1, \ldots, Y_m]])) = n + m$. $\qquad\square$

Corollary 3.3.29 *Let $k = \mathbb{R}$ or $k = \mathbb{C}$ (or more generally, any field of characteristic 0 with a multiplicative valuation) and $n \in \mathbb{N}$. Then the ring $k\langle\langle X_1, \ldots, X_n\rangle\rangle$ of convergent power series over k (cf. [266, §45, pag.191]) is excellent.*

Proof This follows by Corollary 3.3.27 and the relations $\frac{\partial X_i}{\partial X_j} = \delta_{ij}$. $\qquad\square$

Corollary 3.3.30 *Let k be a field of characteristic 0 with a complete non-Archimedean absolute value that is non-trivial. Then the ring of restricted power series in n variables $k\{X_1, \ldots, X_n\}$ is an excellent ring.*

We will try to apply Theorem 3.3.24 to rings containing a field of characteristic $p > 0$. As usual this situation is more complicated. We shall again make use of the notion of p-basis that we've already used.

Lemma 3.3.31 *Let A be a ring of characteristic $p > 0$, B a subring of A containing A^p and M an A-module. Assume that there exists a p-basis $X = \{x_i\}_{i \in I}$ of A over B. Then for any map $f : X \to M$, there exists a unique $D \in \mathrm{Der}_B(A, M)$ such that $D(x_i) = f(x_i)$ for any $i \in I$.*

Proof Let $a \in A$. We have an unique representation

$$a = \sum_{0 \le j_1, \ldots, j_n \le p-1} u^p_{j_1, \ldots, j_n} x^{j_1}_{s_1} \ldots x^{j_n}_{s_n}, \text{ with } u^p_{j_1, \ldots, j_n} \in B.$$

Then we put

$$D(a) = \sum j_1 u^p_{j_1, \ldots, j_n} x^{j_1-1}_{s_1} \ldots x^{j_n}_{s_n} f(x_{s_1}) + \ldots$$

$$\ldots + \sum j_n u^p_{j_1, \ldots, j_n} x^{j_1-1}_{s_1} \ldots x^{j_n-1}_{s_n} f(x_{s_n}).$$

One can see at once that this formula gives a unique B-derivation of A in M with the required property. $\qquad\square$

Corollary 3.3.32 *Let A be a ring of characteristic $p > 0$ having a p-basis $\{x_i\}_{i \in I}$. Then for each $i \in I$ there exists a unique derivation $d_i \in \mathrm{Der}(A)$ such that $d_i(x_j) = \delta_{ij}$.*

Proposition 3.3.33 *Let A be a reduced ring of characteristic $p > 0$ and S a multiplicatively closed system of A. Assume that A has a p-basis $X = \{x_i\}_{i \in I}$. Then $S^{-1}X := \{\frac{x_i}{1}\}_{i \in I}$ is a p-basis of $S^{-1}A$.*

Proof Let $\frac{a}{s} \in S^{-1}A$. Then there exists $q \in S$ such that

$$q^p a s^{p-1} = \sum_k b_k^p x_1^{s_{1,k}} \dots x_n^{s_{n,k}},$$

hence

$$\frac{a}{s} = \frac{a s^{p-1} q^p}{s^p q^p} = \sum_k \frac{b_k^p}{s^p q^p} \frac{x_1^{s_{1,k}}}{1} \dots \frac{x_n^{s_{n,k}}}{1}.$$

It follows that $\frac{a}{s} \in ((S^{-1})^p A^p)[S^{-1}X]$. Assume now that $\sum_k \frac{b_k^p}{c_k^p} x_1^{s_{1,k}} \dots x_n^{s_{n,k}} = 0$. Then

$$\frac{1}{q^p} \left(\sum_k a_k^p x_1^{s_{1,k}} \dots x_n^{s_{n,k}} \right) = 0,$$

hence there exists an element $t \in S$ such that

$$\sum_k t^p a_k^p x_1^{s_{1,k}} \dots x_n^{s_{n,k}} = 0.$$

But $t^p \neq 0$, because A is reduced, whence $t^p a_k^p = 0$. □

Proposition 3.3.34 (Kimura and Niitsuma [197, Th. 3.1]) *Let $(A, \mathfrak{m}, k)$ be a Noetherian local reduced ring of characteristic $p > 0$ having a p-basis Γ. Then A is a regular local ring and $\Gamma = \Delta \cup \{z_1 + v_1, \dots, z_r + v_r\}$, where Δ is a system of representatives of a p-basis of k, $\{z_1, \dots, z_r\}$ is a minimal system of generators of $\mathfrak{m}$ and $v_1, \dots, v_r \in A^p[\Delta]$.*

Proof We set $\mathfrak{m}^{[p]} := \{x^p \mid x \in \mathfrak{m}\}$. By Kunz's Theorem [248, (42.B) Th. 107] A is a regular local ring. Since $A = A^p[\Gamma]$, it follows that $k = k^p(\overline{\Gamma})$, where $\overline{\Gamma}$ is the set of residue classes modulo $\mathfrak{m}$ of the elements of Γ. Therefore we may choose a subset $\Delta \subseteq \Gamma$ such that $\overline{\Delta}$ is a p-basis of k. Then $\Lambda := \Gamma - \Delta$ is a p-basis of A over $A^p[\Delta]$. Since $A = A^p[\Delta] + \mathfrak{m}$, we may assume that $\Lambda \subseteq \mathfrak{m}$. Because $A = A^p[\Delta][\Lambda]$, we obtain that $\mathfrak{m} = \mathfrak{n} + \Lambda A$, where $\mathfrak{n}$ is the maximal ideal $\mathfrak{m}^{[p]} A^p[\Delta]$ of $A^p[\Delta]$. Then $\mathfrak{m} = \mathfrak{m}^2 + \Lambda A$ and from Nakayama's Lemma, we get $\mathfrak{m} = \Lambda A$. Then we can choose a minimal system of generators of $\mathfrak{m}$ contained in Λ. Let $\{z_1, \dots, z_r\} \subseteq \Lambda$ be such a system. Suppose that $\{z_1, \dots, z_r\} \subsetneq \Lambda$. Then there exists $b \in \Lambda$ such that $b \neq z_i$, for all $i = 1, \dots, r$. But $b \in \mathfrak{m}$, hence we have $b = \sum_{i=1}^r \gamma_i z_i$, with $\gamma_1, \dots, \gamma_r \in A$. On the other hand, for any $i = 1, \dots, r$ we have $\gamma_i = \sum_{(e_\lambda)} \alpha_{ie(\lambda)} \prod_\lambda b_\lambda^{e_\lambda}$, where $\alpha_{ie(\lambda)} \in A^p[\Delta, z_1, \dots, z_r]$ and $b_\lambda \in \Lambda \setminus \{z_1, \dots, z_r\}$, $0 \leq e_\lambda \leq p-1$. From these relations and the p-independence of the elements of $\Lambda \setminus \{z_1, \ldots, z_r\}$ over $k^p(\Delta, z_1, \ldots, z_r)$, it follows that we have $\sum_{i=1}^{r} \beta_i z_i = 1$, with $\beta_1, \ldots, \beta_r \in A$. But this is a contradiction, hence $\{z_1, \ldots, z_r\} = \Lambda$. $\qquad\square$

Example 3.3.35 Not any regular local ring of characteristic $p > 0$ has a p-basis. Let us consider a field k of characteristic $p > 0$ such that $[k : k^p] = \infty$ and $A = k[[X_1 \ldots, X_m]]$. Then A does not have a p-basis.

Proof We assume for simplicity that $m = 1$ and that A has a p-basis $\Gamma = \{\gamma_i \mid i \in I\}$. If $\Gamma = \{\gamma_1, \ldots, \gamma_s\}$ is finite, then k will be a finite k^p-vector space, generated by $\prod_{i=1}^{s} \gamma_i^{r_i}(0)$, $0 \le r_i < p$. Contradiction!
If $\Gamma = \{\gamma_i \mid i \in I\}$ is infinite, pick a countable subset $\{\gamma_i \mid i \in \mathbb{N}\} \subseteq \Gamma$ and let $z = \sum_{i=0}^{\infty} \gamma_i X^{ip}$. Then we must have

$$z = \sum_{k} a_k^p \gamma_{j_1}^{s_{k,1}} \ldots \gamma_{j_r}^{s_{k,r}}, \text{ with } 0 \le s_{k,j} < p.$$

Suppose that $\gamma_0 \in \{\gamma_{j_1}, \ldots, \gamma_{j_r}\}$ and let $D' : \Gamma \to A$ such that $D'(\gamma_0) = 1$ and $D'(\gamma_r) = 0$, for any $r > 0$. Then D' can be extended to a map $D : A \to A$ and we have

$$D(z) = D\Big(\sum_{k} a_k^p \gamma_{j_1}^{s_{k,1}} \ldots \gamma_{j_r}^{s_{k,r}} \Big) = 0,$$

and

$$D\Big(\sum_{i=0}^{\infty} \gamma_i X^{ip} \Big) = \sum_{i=0}^{\infty} i p \gamma_i X^{ip-1} D(X) + \sum_{i=0}^{\infty} X^{ip} D(\gamma_i) = 1.$$

Contradiction! $\qquad\square$

Proposition 3.3.36 *Let A be a reduced Noetherian ring of characteristic $p > 0$ having a p-basis and let $\mathfrak{p}$ be a prime ideal of A of height r. Then there exist $D_1, \ldots, D_r \in \mathrm{Der}(A)$ and $x_1, \ldots, x_r \in \mathfrak{p}$ such that $\det(D_i x_j) \notin \mathfrak{p}$.*

Proof We already know that A is a regular ring. Let X be a p-basis of A and $X_{\mathfrak{p}} := \{\frac{x}{1} \mid x \in X\}$. Then by Proposition 3.3.33 and Proposition 3.3.34, we have that $X_{\mathfrak{p}}$ is a p-basis of $A_{\mathfrak{p}}$ and moreover that $X_{\mathfrak{p}} = T_1 \cup \{z_1 + v_1, \ldots, z_r + v_r\}$, where T is a system of representatives of a p-basis of $k(\mathfrak{p})$, $v_1, \ldots, v_r \in A_{\mathfrak{p}}[T^p]$ and $z_1, \ldots, z_r$ is a regular system of parameters of $A_{\mathfrak{p}}$. There exist $a_1, \ldots, a_r \in X$ such that $\frac{a_i}{1} = z_i + v_i$ for $i = 1, \ldots, r$. Therefore there exist derivations $D_1, \ldots, D_r \in \mathrm{Der}(A)$ such that $D_i(a_i) = 1$ and $D_i \mid_{X \setminus \{a_i\}} = 0$. Let D_i' be the extension of D_i to $A_{\mathfrak{p}}$ for $i = 1, \ldots, r$. Then $D_i'(z_j) =$

$D_i'(z_j + v_j) = \delta_{ij}$. Let $\mathfrak{m} = \mathfrak{p}A_\mathfrak{p}$. Since $z_i \in \mathfrak{m}$, there exist $x_1, \ldots, x_r \in \mathfrak{p}$ and $y \in A \setminus \mathfrak{p}$ such that $\frac{x_i}{y} = z_i$ for $i = 1, \ldots, r$. Then

$$1 = \det(D_i'(z_j)) = \frac{1}{y^{2r}} \det(y D_i'(x_j) - x_j D_i'(y)),$$

hence $y^r = \det(D_i'(x_j) + c)$, for some $c \in \mathfrak{m}$. It follows that $\det(D_i'(x_j)) \notin \mathfrak{m}$ and $\frac{\det(D_i(x_j))}{1} = \det(D_i'(x_j))$, hence $\det(D_i(x_j)) \notin \mathfrak{p}$. $\square$

Theorem 3.3.37 (Furuya [114, Prop. 7]) *Let A be a reduced Noetherian ring of characteristic $p > 0$. If A has a p-basis, then A is excellent.*

Proof We know that A is a regular ring. Since A has a p-basis, it follows at once that $A[X_1, \ldots, X_n]$ has a p-basis, for every $n \in \mathbb{N}$. By 3.3.36, the weak Jacobian condition holds in $A[X_1, \ldots, X_n]$, $\forall n \in \mathbb{N}$. Now we apply 3.3.24. $\square$

Corollary 3.3.38 *Let k be a field of characteristic $p > 0$ such that $[k : k^p] < \infty$ and $n, m \in \mathbb{N}$. Then the ring $k[X_1, \ldots, X_n][[Y_1, \ldots, Y_m]]$ is excellent.*

Proof It is easy to see that if Γ is a finite p-basis of k, then the set $\Gamma \cup \{X_1, \ldots, X_n, Y_1, \ldots, Y_m\}$ is a p-basis of $k[X_1, \ldots, X_n][[Y_1, \ldots, Y_m]]$. Now we apply Theorem 3.3.37. $\square$

Example 3.3.39 Let k be a field of characteristic $p > 0$ such that $[k : k^p] = \infty$. We will see later (Theorem 3.6.58) that $A = k[X_1, \ldots, X_n][[Y_1, \ldots, Y_m]]$ is excellent. However, A does not have a p-basis as it follows from example 3.3.35, so we cannot apply Theorem 3.3.37 in this case.

3.4 Criteria of Excellence Using the Frobenius Morphism

Throughout this section, we fix a prime number $p > 0$. We are mainly interested in rings of characteristic p, where by a ring of characteristic p we mean a ring containing the prime field with p elements. As it is well-known, for such rings the map $F_A : A \to A$, defined by $F_A(a) = a^p$, is a ring morphism. This morphism plays a crucial role in the study of rings of characteristic p.

Notation 3.4.1 Let A be ring of characteristic $p > 0$. The morphism $F_A : A \to A$ defined by $F_A(a) = a^p$ is called the **Frobenius morphism** of A. We will denote $F_A^r = F_A \circ \ldots \circ F_A$, the composition of F_A with itself r times. We will also denote by $A^{(p^r)}$ the A-algebra A with the structure given by the r-th power of the Frobenius morphism $F_A^r : A \to A$. For $r = 1$ we will simply denote $A^{(p^1)} = A^{(p)}$.

Lemma 3.4.2 *Let A be a ring of characteristic p. Then A is reduced if and only if the Frobenius morphism F_A is injective.*

Proof If $F_A(a) = a^p = 0$, then the reducedness of A shows that $a = 0$; hence F_A is injective. Conversely, assume that F_A is injective and let $a \in A$ be such that $a^r = 0$. Set $e := \min\{t \in \mathbb{N} \mid a^{p^t} = 0\}$. If $e \geq 1$, then $a^{p^{t-1}} \neq 0$ and $F(a^{p^{t-1}}) = 0$, contradiction. Therefore $e = 0$; hence $a = 0$.   □

If $u : A \to B$ is a ring morphism between rings of characteristic p, then we have a commutative diagram

$$
\begin{array}{ccc}
A & \xrightarrow{\ u\ } & B \\
\downarrow{\scriptstyle F_A^r} & & \downarrow{\scriptstyle F_A^r \otimes 1_B} \\
A^{(p^r)} & \longrightarrow A^{(p^r)} \otimes_A B \xrightarrow{\ \omega_{B/A}^{[r]}\ } & B^{(p^r)}
\end{array}
$$

where F_A is the Frobenius morphism of A and

$$\omega_{B/A}^{[r]}(a \otimes b) = u(a) \cdot b^{p^r}, \ \forall a \in A, \ b \in B.$$

Definition 3.4.3 The morphism $\omega_{B/A}^{[r]}$ is called the r-th **André-Radu morphism** associated to u and the ring $A^{(p^r)} \otimes_A B$ is called the r-th **André-Radu ring** associated to u.

Notation 3.4.4 Let us also denote by A^{perf} (or $A^{(p^\infty)}$) the limit of the inductive system of A-algebras

$$A^{(p)} \xrightarrow{F_A} A^{(p^2)} \xrightarrow{F_A} \ldots \xrightarrow{F_A} A^{(p^r)} \xrightarrow{F_A} A^{(p^{r+1})} \xrightarrow{F_A} \ldots$$

Remark 3.4.5

(i) $\mathrm{Spec}(F_A) = 1_{\mathrm{Spec}(A)}$, hence $\mathrm{Spec}(\omega_{B/A}^{[r]})$ is bijective, for all $r \geq 1$. Therefore B and $A^{(p^r)} \otimes_A B$ have practically the same spectrum.

(ii) Since $(u^{(p^r)} : A^{(p^r)} \to B^{(p^r)})_{r \geq 1}$ is a morphism between the inductive systems $(A^{(p^r)})_{r \geq 1}$ and $(B^{(p^r)})_{r \geq 1}$, we obtain a morphism between the inductive limits

$$u^{perf} = \varinjlim_r u^{(p^r)} : A^{perf} \to B^{perf}.$$

(iii) The morphism $F_{A^{perf}} : A^{perf} \to A^{perf}$ is bijective, that is, A^{perf} is a perfect ring.

(iv) Let $G_A : A \to A^{perf}$ and $G_B : B \to B^{perf}$ be the canonical morphisms. Then $G_B \circ u = u^{perf} \circ G_A$. It follows that we have a canonical morphism of A-algebras

$$\omega_{B/A}^{perf} : A^{perf} \otimes_A B \to B^{perf}.$$

More precisely $\omega_{B/A}^{perf}(a \otimes b) = ab^{p^r}$ for $a \in A^{(p^r)}$ and $b \in B$. It is clear that we have

$$\omega_{B/A}^{perf} = \varinjlim_r \omega_{B/A}^{[r]}.$$

(v) It follows from (i) that if one of the morphisms F_A, $\omega_{B/A}^{[r]}$ or $\omega_{B/A}^{perf}$ is flat, then it is faithfully flat.

(vi) We have the relations

$$u^{(p^r)} = \omega_{B/A}^{[r]} \circ (1_{A^{(p^r)}} \otimes_A u)$$

and

$$\omega_{B/A}^{[r+1]} = (\omega_{B/A})^{(p^r)} \circ (1_{A^{(p^{r+1})}} \otimes_{A^{(p^r)}} \omega_{B/A}^{[r]}).$$

(vii) If C is an A-algebra and $D := C \otimes_A B$ it follows that

$$\omega_{D/C}^{[r]} = 1_{C^{(p^r)}} \otimes_{A^{(p^r)}} \omega_{B/A}^{[r]}$$

and

$$\omega_{D/C}^{perf} = 1_{C^{perf}} \otimes_{A^{perf}} \omega_{B/A}^{perf}.$$

(viii) If $\omega_{B/A}$ is flat, it follows by induction that $\omega_{B/A}^{[r]}$ is flat for all $r \geq 1$; hence also $\omega_{B/A}^{perf}$ is flat.

Notation 3.4.6 Let $u : A \to B$ be a morphism of Noetherian rings of characteristic $p > 0$. For any $A^{(p)}$-module W we get a morphism

$$\omega_{B/A} \otimes_{A^{(p)}} 1_W : A^{(p)} \otimes_A B \otimes_{A^{(p)}} W \to B^{(p)} \otimes_{A^{(p)}} W.$$

We set $E(W) := \ker(\omega_{B/A} \otimes_{A^{(p)}} 1_W)$.

Remark 3.4.7

(i) With the notations from 3.4.6, it is easy to check that we obtain an additive functor E from the category of $A^{(p)}$-modules to the category of abelian groups.
(ii) The functor E commutes with direct limits.
(iii) If u is flat, then E is left-exact.

Lemma 3.4.8 (André [11, Lemme 48]) *Let $u : A \to B$ be a reduced morphism of rings of characteristic p. Then $E = 0$.*

Proof The functor E is left-exact due to the fact that u is flat and moreover commutes with inductive limits; hence, since A is Noetherian, it is easy to see that it is enough to show that $E(k(\mathfrak{p})) = (0)$, for any $\mathfrak{p} \in \mathrm{Spec}(A)$. But then it is clear that we may assume that A is a domain and $\mathfrak{p} = (0)$. Now consider the natural morphisms $g : Q(A)^{(p)} \otimes_A B \to Q(A) \otimes_A B$ and $h : Q(A) \otimes_A B \to Q(A)^{(p)} \otimes_A B$. Clearly h is injective. But then

$$E(Q(A)) = \ker(h \circ g) = \ker(F_{Q(A)^{(p)} \otimes_A B}).$$

Since u is reduced, $F_{Q(A)^{(p)} \otimes_A B}$ is injective and we get the assertion. $\square$

Lemma 3.4.9 (Dumitrescu [94, Lemma 1]) *Let $u : A \to B$ be a morphism of Noetherian rings of characteristic p.*

(i) If $\omega_{B/A}$ is flat, then u is flat.
(ii) If $\omega_{B/A}$ is injective and $B/A[B^p]$ is a flat A-module, then u is flat.

Proof We may assume that A and B are local rings. Let $\mathfrak{m}$ be the maximal ideal of A and let $A_n := A/\mathfrak{m}^n$ and $B_n := B/\mathfrak{m}^n B$. By the Local Flatness Criterion [248, Th. 49, (20.C)], it is enough to show that $u_n := u \otimes 1_{A/\mathfrak{m}^n} : A_n \to B_n$ is flat for any $n \in \mathbb{N}$. Let us denote for short $\omega := \omega_{B/A}$ and $\omega_n := \omega_{B_n/A_n}$. Then we can write $u = \omega \circ (u \otimes_A 1_{A^{(p)}})$, hence $u_n = \omega_n \circ (u_n \otimes_{A_n} 1_{A_n^{(p)}})$. Because F_{A_n} factorizes through the canonical map $A_n \to A_{n-1}$ we obtain

$$u_n \otimes_{A_n} 1_{A_n^{(p)}} = (u_n \otimes_{A_n} A_{n-1}) \otimes_{A_{n-1}} 1_{A_n^{(p)}} = u_{n-1} \otimes_{A_{n-1}} 1_{A_n^{(p)}}.$$

(i) Notice that $\omega_n = \omega \otimes_{A^{(p)}} 1_{A_n^{(p)}}$. Since u_1 and ω_n, for any $n \geq 1$ are flat morphisms, the assertion follows by induction on n.
(ii) Since $A^{(p)} \otimes_A B \simeq A[B^p]$ and $B/A[B^p]$ is a flat A-module, it follows that ω_n is

injective. Moreover $B_n/A_n[B_n^p]$ is a flat A_n-module. Also we have the exact sequence of $A_n^{(p)}$-modules

$$0 \to A_n^{(p)} \otimes_{A_n} B_n = A_n^{(p)} \otimes_{A_{n-1}} B_{n-1} \to B_n^{(p)} = B_n \to B_n/A_n[B_n^p] \to 0.$$

Since $A_n^{(p)} \otimes_{A_n} B_n$ and $B_n/A_n[B_n^p]$ are flat, we can proceed by induction on n, as in the proof of (i).   □

Let us now see a characterization of reduced morphisms between rings of characteristic p.

Theorem 3.4.10 (Dumitrescu [94, Th. 3]) *Let $u : A \to B$ be a morphism of Noetherian rings of characteristic p. The following are equivalent:*

(i)   *u is a reduced morphism;*
(ii) *u is flat and for each $\mathfrak{p} \in \mathrm{Spec}(A)$, if we put $\overline{A} := A/\mathfrak{p}$ and $\overline{B} := B/\mathfrak{p}B$, the morphism $\omega_{\overline{B}/\overline{A}}$ is injective;*
(iii) *u is flat and for every A-algebra $\overline{A}$, if we put $\overline{B} := B \otimes_A \overline{A}$, the morphism $\omega_{\overline{B}/\overline{A}}$ is injective;*
(iv) *u is flat, $\omega_{B/A}$ is injective and $\mathrm{Im}(\omega_{B/A}) = A[B^p]$ is a pure A-submodule of B;*
(v)   *$\omega_{B/A}$ is injective and $B/A[B^p]$ is a flat A-module.*

Proof $(i) \Rightarrow (iv)$: We consider the functor defined in 3.4.6. Then from Lemma 3.4.8 we get that $E = 0$.
$(iv) \Rightarrow (v)$: Since B is flat and $A[B^p]$ is pure, it follows that $B/A[B^p]$ is a flat A-module [209, Cor. 4.86].
$(v) \Rightarrow (iii)$: From Lemma 3.4.9 it follows that u is flat. If $\overline{A}$ is an A-algebra and $\overline{B} = \overline{A} \otimes_A B$, we have $\omega_{\overline{B}/\overline{A}} = \omega_{B/A} \otimes_{A^{(p)}} 1_{\overline{A}^{(p)}}$. Hence $\omega_{\overline{B}/\overline{A}}$ is injective.
$(iii) \Rightarrow (ii)$: Obvious.
$(ii) \Rightarrow (i)$: Let $\mathfrak{p} \in \mathrm{Spec}(A)$ and set $\overline{A} = A/\mathfrak{p}$, $\overline{B} = B/\mathfrak{p}B$, $L = k(\mathfrak{p})$ and $\tilde{B} = L \otimes_A B$. Then we obtain the injective morphisms $\omega_{\overline{B}/\overline{A}}$ and $\omega_{\tilde{B}/L} = \omega_{\overline{B}/\overline{A}} \otimes_{\overline{A}^{(p)}} 1_{L^{(p)}}$. If we put $C := L^{(p)} \otimes_L \tilde{B}$, we have that also $(F_L \otimes_L 1_{\tilde{B}}) \circ \omega_{\tilde{B}/L} = F_C$ is injective. It follows that C is reduced, hence the fibre of u in $\mathfrak{p}$ is geometrically reduced.   □

Corollary 3.4.11 *Let A be a reduced semilocal Noetherian ring of characteristic p. The following are equivalent:*

(i)   *A is a Nagata ring;*
(ii) *$A^{(p)} \otimes_A \widehat{A}$ is reduced and $\widehat{A}/A[\widehat{A}^p]$ is a flat A-module.*

Proof The canonical morphism $g : \widehat{A} \rightarrow B := A^{(p)} \otimes_A \widehat{A}$ is injective because A is reduced. Since $g \circ \omega_{\widehat{A}/A} = F_B$ we get that $\omega_{\widehat{A}/A}$ is injective if and only if B is reduced. Now we apply Theorem 3.4.10. $\qquad\qquad\qquad\qquad\qquad\qquad\qquad\qquad\qquad\qquad\qquad\qquad\qquad$ $\square$

In order to prove the characterization 3.4.15 of regular local rings given by Rodicio and the André-Radu characterization 3.4.19 of regular morphisms between rings of characteristic p, we need some preparations.

Lemma 3.4.12 (Majadas [229, Lemma 10]) *Let A be a Noetherian ring, $\mathfrak{a}$ be an ideal of A and M a finitely generated A-module. Let also $f : A \rightarrow A$ be a ring morphism such that $f(\mathfrak{a}) \subseteq \mathfrak{a}^2$, $g : M \rightarrow M$ be a morphism such that $g(am) = f(a)g(m)$ and $n > 0$ be a natural number. Then there exists an integer $s > 0$ such that the map*

$$\mathrm{Tor}_n^A(M, A/\mathfrak{a}) \rightarrow \mathrm{Tor}_n^A(M, A/\mathfrak{a})$$

induced by f^s and g^s is the zero map.

Proof Let us consider a projective resolution of M with finitely generated A-modules

$$\cdots \longrightarrow F_2 \xrightarrow{d_2} F_1 \xrightarrow{d_1} F_0 \longrightarrow M \longrightarrow 0.$$

Then $\mathrm{Tor}_n^A(M, A/\mathfrak{a})$ is the homology of the complex

$$F_{n+1}/\mathfrak{a}F_{n+1} \xrightarrow{\delta_{n+1}} F_n/\mathfrak{a}F_n \xrightarrow{\delta_n} F_{n-1}/\mathfrak{a}F_{n-1},$$

where δ_n and δ_{n+1} are induced by d_n and d_{n+1}, that is

$$\mathrm{Tor}_n^A(M, A/\mathfrak{a}) = \ker(\delta_n)/\mathrm{Im}(\delta_{n+1}) = \ker\left(\frac{F_n/\mathfrak{a}F_n}{\mathrm{Im}(\delta_{n+1})} \rightarrow F_{n-1}/\mathfrak{a}F_{n-1}\right).$$

From the exact sequence

$$0 \rightarrow \mathrm{Im}(d_{n+1}) \rightarrow F_n \rightarrow \mathrm{Im}(d_n) \rightarrow 0$$

by tensorizing with $A/\mathfrak{a}$ we obtain that

$$\frac{F_n/\mathfrak{a}F_n}{\mathrm{Im}(\delta_{n+1})} = \frac{\mathrm{Im}(d_n)}{\mathfrak{a} \cdot \mathrm{Im}(d_n)}.$$

Therefore we have

$$\operatorname{Tor}_n^A(M, A/\mathfrak{a}) = \ker\left(\frac{\operatorname{Im}(d_n)}{\mathfrak{a} \cdot \operatorname{Im}(d_n)} \to F_{n-1}/\mathfrak{a}F_{n-1}\right) = \frac{\operatorname{Im}(d_n) \cap \mathfrak{a}F_{n-1}}{\mathfrak{a} \cdot \operatorname{Im}(d_n)}.$$

By the Artin-Rees lemma [249, Th. 8.5], there exists $t > 0$ such that

$$\operatorname{Im}(d_n) \cap \mathfrak{a}^t \cdot \operatorname{Im}(d_{n-1}) \subseteq \mathfrak{a} \cdot \operatorname{Im}(d_n).$$

Thus let $s \in \mathbb{N}$ be such that $f^s(\mathfrak{a}) \subseteq \mathfrak{a}^t$. Then f^s and g^s give the zero map on $\operatorname{Tor}_n^A(M, A/\mathfrak{a})$.   □

Proposition 3.4.13 (Majadas [229, Prop. 11]) *Let $(A, \mathfrak{m}, k)$ be a Noetherian local ring of characteristic $p > 0$. For any integer $n \geq 0$, there exists an integer $s > 0$ such that the s-th power of the Frobenius morphisms F_A^s and F_k^s induces the zero map of functors*

$$H_n(A, k, -) \to H_n(A, k, -).$$

Proof We have to prove that for any k-module M, the map induced by F_A^s and F_k^s on $H_n(A, k, M) \to H_n(A, k, M)$ is zero, where in the left part M is considered as a k-module by restriction of scalars via F_k^s. This is obviously true for $n = 0$, since $H_0(A, k, M) = \Omega_{k/A} \otimes_k M = (0)$. Assume that $n > 0$. By Lemma 3.4.12, for each $i = 1, \ldots, n$ we find s_i such that $F_k^{s_i}$ induces the zero map

$$\operatorname{Tor}_i^A(k, k) \to \operatorname{Tor}_i^A(k, k).$$

Applying [6, Prop. X.12] we get that for $s \geq s_1 + \ldots + s_n$ the map

$$H_n(A, k, M) \to H_n(A, k, M)$$

induced by F_A^s and F_k^s is zero.   □

Corollary 3.4.14 *Let $u : (A, \mathfrak{m}, k) \to (B, \mathfrak{n}, l)$ be a morphism of Noetherian local rings of characteristic p. Then, for any integer $n \geq 0$, there exists an integer $s > 0$ such that for any l-module M, the Frobenius morphisms F_A^s and F_B^s induce the zero map*

$$H_n(A, B, M) \to H_n(A, B, M).$$

Proof Consider the commutative diagram with exact rows

$$
\begin{array}{ccccc}
H_{n+1}(B, l, M) & \xrightarrow{\delta} & H_n(A, B, M) & \xrightarrow{\epsilon} & H_n(A, l, M) \\
\downarrow{\beta_i} & & \downarrow{\alpha_i} & & \downarrow{\gamma_i} \\
H_{n+1}(B, l, M) & \xrightarrow{\delta} & H_n(A, B, M) & \xrightarrow{\epsilon} & H_n(A, l, M)
\end{array}
$$

where the vertical maps are induced by F_A^i, F_B^i and F_l^i. Applying Proposition 3.4.13, for $i >> 0$ we have $\beta_i = \gamma_i = 0$. Then $\mathrm{Im}(\alpha_i) \subseteq \ker(\epsilon) = \mathrm{Im}(\delta)$ and $\mathrm{Im}(\delta) \subseteq \ker(\alpha_i)$, hence $\alpha_{2i} = \alpha_i^2 = 0$. □

Theorem 3.4.15 (Rodicio [319, Th. 2]) *Let* $(A, \mathfrak{m}, k)$ *be a Noetherian local ring of characteristic* $p > 0$. *The following are equivalent:*

(i) *A is a regular local ring;*
(ii) $\mathrm{fd}_A(A^{(p^n)}) < \infty$, *for any* $n > 0$;
(iii) *there exists* $n > 0$ *such that* $\mathrm{fd}_A(A^{(p^n)}) < \infty$.

Proof ([229]) $(i) \Rightarrow (ii)$: This follows from the well-known characterization of regular rings of characteristic p given by Kunz [199, Th. 2.1].
$(ii) \Rightarrow (iii)$: Obvious.
$(iii) \Rightarrow (i)$: We apply Corollary 3.4.14 to the commutative square

$$
\begin{array}{ccc}
\mathbb{F}_p & \longrightarrow & A \\
F_{\mathbb{F}_p}^s \downarrow & & \downarrow F_A^s \\
\mathbb{F}_p & \longrightarrow & A
\end{array}
$$

so that, changing if necessary F_A with F_A^s, we may assume that F_A induces the zero map $H_1(\mathbb{F}_p, A, M) \to H_1(\mathbb{F}_p, A, M)$. Now we apply Corollary 3.2.13. □

Lemma 3.4.16 *Let A be a ring, B and C two A-algebras such that* $\mathrm{Tor}_j^A(C, B) = 0$ *for any* $j > 0$ *and let M be a C-module. Then*

$$
\mathrm{Tor}_n^C(C \otimes_A B, M) \simeq \mathrm{Tor}_n^A(B, M), \text{ for any } n \geq 0.
$$

Proof This follows from the change of rings spectral sequence [324, Th. 10.71]

$$E^2_{p,q} = \text{Tor}^C_p(\text{Tor}^A_q(C, B), M) \underset{p}{\Rightarrow} (\text{Tor}^A_n(B, M)).$$

$\square$

Corollary 3.4.17 *Let A be a Noetherian ring of characteristic $p > 0$. The following are equivalent:*

 (i) A is a regular ring;
(ii) there exists $n > 0$ such that $\text{fd}_A(A^{(p^n)}) < \infty$.

Proof $(i) \Rightarrow (ii)$: Obvious.
$(ii) \Rightarrow (i)$: Let $\mathfrak{q} \in \text{Spec}(A)$ and E an $A_\mathfrak{q}$-module. Then from Lemma 3.4.16 or by [324, Prop. 7.17], we obtain $\text{Tor}^{A_\mathfrak{q}}_n(A^{(p^n)}_\mathfrak{q}, E) \simeq \left(\text{Tor}^A_n(A^{(p^n)}, E)\right)_\mathfrak{q}$. $\square$

Theorem 3.4.18 (Majadas [229, Th. 13]) *Let $u : (A, \mathfrak{m}, k) \to (B, \mathfrak{n}, l)$ be a morphism of Noetherian local rings of characteristic $p > 0$. If there exists an integer j such that the André-Radu morphism $\omega^{[j]}_{\widehat{B}/\widehat{A}}$ is flat, then u is formally smooth.*

Proof By [131, Prop. 19.3.6] we may assume that $\omega^{[j]}_{B/A}$ is flat, hence by Remark 3.4.5, viii) also $\omega^{[2j]}_{B/A}$ is flat. By Corollary 3.4.14 there exists $s \in \mathbb{N}$ such that F^{js}_A and F^{js}_B induce the zero map $H_1(A, B, M) \to H_1(A, B, M)$ for any l-module M. Therefore, replacing F_A and F_B by a suitable power of them, we may assume that $\omega_{B/A}$ is flat and the Frobenius morphisms of A and B induce the zero map $H_1(A, B, M) \to H_1(A, B, M)$ for any l-module M. From Lemma 3.4.9 we obtain that u is flat and Corollary 3.2.13 says that u is formally smooth. $\square$

The next result is a very interesting characterization of regular morphisms in the case of rings containing a field of characteristic $p > 0$. This result was obtained by M. André and N. Radu in several papers, with different proofs [14, 15, 299]. We present here the most elementary proof obtained by Dumitrescu [93] for one implication and by Majadas [229] for the other implication.

Theorem 3.4.19 (André [14] and Radu [299]) *Let $u : A \to B$ be a morphism of Noetherian rings of characteristic $p > 0$. Then u is a regular morphism if and only if the André-Radu morphism $\omega_{B/A}$ is flat.*

Proof ([93, 229]) We may assume that $(A, \mathfrak{m}, k)$ and $(B, \mathfrak{n}, L)$ are local rings, because flatness and regularity of morphisms are local properties.

Suppose that u is regular and assume first that A is a field, that is, B is a geometrically regular A-algebra. By Theorem 1.4.19, the rings $B \otimes_A A^{(p)}$ and $B^{(p)}$ are Noetherian regular local rings. As in the proof of Theorem 1.4.19, we see that $\omega_{B/A}$ is flat. More precisely, we write $A^{(p)}$ as the direct limit of its finite subextensions $(K_i)_i$. For such a subextension, the induced morphism $K_i \otimes_A B \to B^{(p)}$ is an integral local morphism between two regular local rings, hence by [249, Th. 23.1] is flat. Since $\omega_{B/A}$ is the direct limit of these morphisms, it is flat.

Assume now that A is not a field. Let $B' := B/\mathfrak{m}B$, $\overline{B} := A^{(p)} \otimes_A B$ and let $\overline{\mathfrak{n}}$ be the maximal ideal of $\overline{B}$. Then $\overline{\mathfrak{n}} = \omega_{B/A}^{-1}(\mathfrak{n})$. Let also $\overline{B}' := \overline{B}/\mathfrak{m}\overline{B}$. Then $\overline{B}' = k^{(p)} \otimes_k B'$ and $\omega_{B/A} \otimes 1_{A'} = \omega' : \overline{B}' \to \overline{B}'^{(p)}$. As k is a field, ω' is flat. Because u is flat, we get $\mathfrak{m} \otimes_A B = \mathfrak{m}B$. Then $\mathfrak{m}\overline{B} \otimes_{\overline{B}} B \simeq \mathfrak{m}B$. We have the surjective morphisms

$$\mathfrak{m} \otimes_A B \simeq \mathfrak{m} \otimes_A \overline{B} \otimes_{\overline{B}} B \overset{u}{\to} \mathfrak{m}\overline{B} \otimes_{\overline{B}} B \overset{v}{\to} \mathfrak{m}\overline{B} \cdot B = \mathfrak{m}B,$$

and $v \circ u$ is an isomorphism. It follows that u and v are isomorphisms. Then, by the Local Flatness Criterion [248, Th. 49, (20.C)], the canonical morphism $\omega_n : \overline{B}/\mathfrak{m}^n\overline{B} \to B^{(p)}/\mathfrak{m}^n B^{(p)}$ is flat, for any $n \in \mathbb{N}$. If $\overline{B}$ is Noetherian, then $\omega_{B/A}$ is flat. Assume that $\overline{B}$ is not Noetherian. Let $\mathfrak{q}$ be an ideal of $\overline{B}$ which is maximal among the non-finitely generated ideals and let $\mathfrak{p} := \mathfrak{q} \cap A^{(p)}$. Then $\mathfrak{q}$ and consequently $\mathfrak{p}$ are prime ideals. If we change $A \to B$ by $A/\mathfrak{p} \to \overline{B}/\mathfrak{p}\overline{B} \to B/\mathfrak{p}B$ we may assume that A is an integral domain and that $\mathfrak{q} \cap A = (0)$. Since A is not a field we can take a non-zero element $a \in A$ which is not invertible in A. Then we consider the morphisms $A_x \to \overline{B}_x \to B_x$. Since $\overline{B}_x = A_x^{(p)} \otimes_{A_x} B_x$ and $\dim(B_x) < \dim(B)$, arguing by induction on $\dim(B)$ we may assume that $\overline{B}_x$ is Noetherian. Then $(\omega_{B/A})_x$ is flat, therefore $(\mathrm{Tor}_1^{\overline{B}}(\overline{B}/\mathfrak{q}, B))_x = 0$. We consider the sequence of morphisms $A/xA \to \overline{B}/x\overline{B} \to B/xB$. As $\dim(B/xB) < \dim(B)$, it follows that $\omega_{B/A}/x\omega_{B/A}$ is flat. Since u is flat, $\mathrm{Tor}_1^A(A/xA, B) = 0$ if and only if $xA \otimes_A B \to B$ is an isomorphism. Then, arguing as above, $x\overline{B} \otimes_{\overline{B}} B \simeq xB$, hence $\mathrm{Tor}_1^{\overline{B}}(\overline{B}/x\overline{B}, B) = 0$. From the Local Flatness Criterion [248, Th. 49, (20.C)] it follows that $\mathrm{Tor}_1^{\overline{B}}(E, B) = 0$, for any $\overline{B}/x\overline{B}$-module E. Let $\mathfrak{q}' := \mathfrak{q} + xB$. From the exact sequence

$$0 \to x\overline{B} \to \mathfrak{q}' \to \mathfrak{q}'/x\overline{B} \to 0,$$

we get $\mathrm{Tor}_1^{\overline{B}}(\mathfrak{q}', B) = 0$. From the exact sequence

$$0 \to \mathfrak{q}' \to \overline{B} \to \overline{B}/\mathfrak{q}' \to 0,$$

we get $\mathrm{Tor}_2^{\overline{B}}(\overline{B}/\mathfrak{q}', B) = 0$. Finally from the exact sequence

$$0 \to \overline{B}/\mathfrak{q} \overset{x}{\to} \overline{B}/\mathfrak{q} \to \overline{B}/\mathfrak{q}' \to 0$$

we obtain that the multiplication by x on $\operatorname{Tor}_1^{\overline{B}}(\overline{B}/\mathfrak{q}, B) = 0$ is injective. This, together with $(\operatorname{Tor}_1^{\overline{B}}(\overline{B}/\mathfrak{q}, B))_x = 0$, implies that $\operatorname{Tor}_1^{\overline{B}}(\overline{B}/\mathfrak{q}, B) = 0$. Since $\overline{B}/\mathfrak{q}$ is Noetherian, from the fact that $\operatorname{Tor}_1^{\overline{B}}(E, B) = 0$ for any $\overline{B}/x\overline{B}$-module E and from the Local Flatness Criterion, it follows that

$$\omega' = \omega(\operatorname{mod} \mathfrak{q}) : \overline{B}/\mathfrak{q} \to B/\mathfrak{q}B$$

is faithfully flat. From $\operatorname{Tor}_1^{\overline{B}}(\overline{B}/\mathfrak{q}, B) = 0$ we obtain that

$$\mathfrak{q}/\mathfrak{q}^2 \otimes_{\overline{B}/\mathfrak{q}} B/\mathfrak{q}B \simeq \mathfrak{q}B/\mathfrak{q}^2B,$$

hence $\mathfrak{q}/\mathfrak{q}^2$ is a finitely generated $\overline{B}/\mathfrak{q}$-module. Then there exists a finitely generated ideal $\mathfrak{a} \subset \overline{B}, \mathfrak{a} \subseteq \mathfrak{q}$ such that $\mathfrak{a} + \mathfrak{q}^2 = \mathfrak{q}$. Since $\overline{B}_x$ is Noetherian, enlarging $\mathfrak{a}$ we may assume that $\mathfrak{a}\overline{B}_x = \mathfrak{q}\overline{B}_x$. As $\overline{B}/x\overline{B}$ is Noetherian, we may furthermore assume that $\mathfrak{a} + x\overline{B} = \mathfrak{q} + x\overline{B}$. Then, since $x \notin \mathfrak{q}$, we get that $\mathfrak{q} \subseteq \mathfrak{a} + x\mathfrak{q}$. It follows that $\mathfrak{q} \subseteq \mathfrak{a} + x^j\mathfrak{q}, \forall j \in \mathbb{N}$. Let $s, t \in \mathfrak{q}$. There exists $j \in \mathbb{N}$ such that $x^j t \in \mathfrak{a}$. Let $s = a + x^j b, a \in \mathfrak{a}, b \in \mathfrak{q}$. Then $st = at + x^j bt \in \mathfrak{a}$, hence $\mathfrak{q}^2 \subseteq \mathfrak{a}$. It follows that $\mathfrak{a} = \mathfrak{q}$.

Conversely, suppose that $\omega_{B/A}$ is flat. By Lemma 3.4.9 we get that u is flat. To show that u is regular we have to show that for any prime ideal $\mathfrak{q}$ of B, the induced morphism $A_{\mathfrak{q} \cap A} \to B_\mathfrak{q}$ is formally smooth. This follows from Theorem 3.4.18. $\square$

Corollary 3.4.20 *Let $u : A \to B$ be a morphism of Noetherian local rings of characteristic $p > 0$. The following are equivalent:*

(i) u is regular;
(ii) $\omega_{B/A}$ is flat;
(iii) u is formally smooth and $A^{(p)} \otimes_A B$ is a Noetherian ring.

Proof $(i) \Leftrightarrow (ii)$: This is Theorem 3.4.19.
$(ii) \Rightarrow (iii)$: Obvious.
$(iii) \Rightarrow (ii)$: Let k be the residue field of A. Since u is formally smooth it follows that $1_{k^{(p)}} \otimes \omega_{B/A} : k^{(p)} \otimes_A B \to (k^{(p)} \otimes_A B)^{(p)}$ is flat. Because u is flat, we get easily by [249, Appl. 3, (20.G)] that $\omega_{B/A}$ is flat. $\square$

Corollary 3.4.21 *Let A be a Noetherian local rings of characteristic $p > 0$. The following are equivalent:*

(i) A is quasi-excellent;
(ii) $\omega_{\hat{A}/A}$ is flat;
(iii) $A^{(p)} \otimes_A \hat{A}$ is a Noetherian ring.

Corollary 3.4.22 ([183]) *Let $u : A \to B$ be a morphism of Noetherian rings of characteristic $p > 0$. Assume that B has a p-basis over A and that the André-Radu morphism $\omega_{B/A}$ is injective. Then u is smooth.*

Proof Let $\{x_i\}_{i \in I}$ be a p-basis of B over A. Because $\omega_{B/A}$ is injective, we have that $A[B^p] \cong A^{(p)} \otimes_A B$ and clearly $\{x_i\}_{i \in I}$ is a p-basis of B over $A^{(p)} \otimes_A B$. Thus $\omega_{B/A}$ is flat and by Theorem 3.4.19 it follows that u is regular. Since B has a p-basis over A, it follows that $\Omega_{B/A}$ is free, hence by Theorem 3.2.7 we get that u is smooth. $\square$

Corollary 3.4.23 ([183]) *Let $u : A \to B$ be an injective morphism of Noetherian rings of characteristic $p > 0$. Assume that B has a p-basis over A and that $A^{(p)} \otimes_A B$ is reduced. Then u is smooth.*

Proof Since $A^{(p)} \otimes_A B$ is reduced, it follows that $F_{A^{(p)} \otimes_A B}$ is injective. But we have $F_{A^{(p)} \otimes_A B} = (F_A \otimes_A B)^{(p)} \circ \omega_{B/A}$. It follows that $\omega_{B/A}$ is injective and we apply Corollary 3.4.22. $\square$

Now we want to prove a generalization of Theorem 3.4.19.

Lemma 3.4.24 (Dumitrescu [95, Lemma 2.3]) *Let $u : A \to B$ be a ring morphism such that F_A is flat. The following are equivalent:*

(i) $\omega_{B/A}$ *is flat;*
(ii) $\omega_{B/A}^{perf}$ *is flat;*
(iii) $F_{A^{perf} \otimes_A B}$ *is flat;*
(iv) $F_{A^{(p)} \otimes_A B}$ *is flat.*

Proof $(i) \Rightarrow (ii)$: It is already observed in Remark 3.4.5, vii).
$(iii) \Rightarrow (iv)$: We have the following commutative diagram:

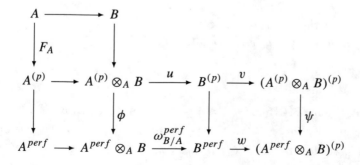

Since F_A is flat, we get that ϕ, hence also ψ are flat. By assumption $F_{A^{perf} \otimes_A B} = v \circ u$ is flat. Thus $\psi \circ v \circ u = w \circ \omega_{B/A}^{perf} \circ \phi$ is flat. It follows that $F_{A^{(p)} \otimes_A B} = v \circ u$ is flat.

$(ii) \Rightarrow (iii)$: We have a similar commutative diagram as above for p^∞. Since F_A is flat, it follows that $F_A^{perf} := \varinjlim_t F_A^t$, $(F_A \otimes_A 1_B)^{perf}$ and $F_A^{perf} \otimes_A 1_B$ are flat. Then $F_B^{perf} = \omega_{B/A}^{perf} \circ (F_A^{perf} \otimes_A 1_B)$ is flat. But flatness in all these cases means faithful flatness. Hence

$$(F_A \otimes_A 1_B)^{perf} \circ \omega_{B/A}^{perf} \circ (F_A^{perf} \otimes_A 1_B) = (F_{A^{perf} \otimes_A B}) \circ (F_A^{perf} \otimes_A 1_B)$$

is flat. It follows that $F_{A^{perf} \otimes_A B}$ is flat.

$(iv) \Rightarrow (i)$: Set $\omega := \omega_{B/A}$. We have the following commutative diagram:

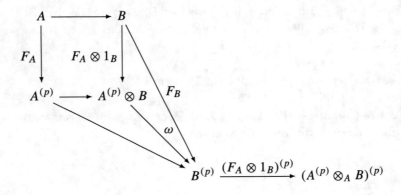

Since $F_{A^{(p)} \otimes_A B} = (F_A \otimes_A 1_B)^{(p)} \circ \omega$ is flat and $(F_A \otimes_A 1_B)^{(p)}$ is faithfully flat, it follows that $\omega = \omega_{B/A}$ is flat. □

Lemma 3.4.25 *Let* $u : A \to B$ *be a ring morphism,* $v : A \to M$ *be an A-algebra and* $N = M \otimes_A B$. *Assume that* $\mathrm{fd}(\omega_{B/A}) < \infty$. *Then:*

(i) If u *or* v *is flat, then* $\mathrm{fd}(\omega_{N/M}) < \infty$.
(ii) If $\mathfrak{q} \in \mathrm{Spec}(A)$ *and* $\mathfrak{p} = \mathfrak{q} \cap A$, *then* $\mathrm{fd}(\omega_{B_\mathfrak{q}/A_\mathfrak{p}}) < \infty$.

Proof

(i) Applying Lemma 3.4.16 for $A := A^{(p)}, C := A^{(p)} \otimes_A B, B := M^{(p)}$ and $E := B^{(p)}$ we get that $\mathrm{Tor}_j^{A^{(p)} \otimes_A B}(M^{(p)} \otimes_A B, B^{(p)}) = 0$, $\forall j > 0$. If we apply again Lemma 3.4.16 for $A := A^{(p)} \otimes_A B, C := M^{(p)} \otimes_A B := M^{(p)} \otimes_M N, B := B^{(p)}$

and E an arbitrary B-module, we get the assertion. Indeed, we have

$$M^{(p)} \otimes_A B \otimes_{A^{(p)} \otimes_A B} B^{(p)} \simeq M^{(p)} \otimes_{A^{(p)}} A^{(p)} \otimes_A B \otimes_{A^{(p)} \otimes_A B} B^{(p)} \simeq$$

$$\simeq M^{(p)} \otimes_{A^{(p)}} B^{(p)} \simeq N^{(p)},$$

hence

$$0 = \mathrm{Tor}_n^{A^{(p)} \otimes_A B}(B^{(p)}, E) \simeq \mathrm{Tor}_n^{M^{(p)} \otimes_M N}(N^{(p)}, E),$$

for each $n > \mathrm{fd}(\omega_{B/A})$.

(ii) We have $\omega_{B_q/A_p} = \omega_{B/A} \otimes_B B_q$ and B_q is flat over B, so we can apply again Lemma 3.4.16.   □

Lemma 3.4.26 *Let k be a field of characteristic $p > 0$ and B be a k-algebra such that $\mathrm{fd}_B(\omega_{B/k}) < \infty$. Then $\omega_{B/k}$ is flat.*

Proof Let $(k_i)_{i \in I}$ be the family of subfields of $k^{(p)}$ that are finite extensions of k. Then $k^{(p)} = \bigcup_{i \in I} k_i$. We have a diagram

$$
\begin{array}{ccc}
k & \longrightarrow & B \\
\downarrow & & \downarrow{\scriptstyle \eta_i} \\
k_i & \longrightarrow & B_i = k_i \otimes_k B \\
\downarrow{\scriptstyle \alpha_i} & & \downarrow{\scriptstyle \beta_i} \\
k^{(p)} & \longrightarrow & k^{(p)} \otimes_k B \xrightarrow{\omega_{B/k}} B^{(p)} \xrightarrow{\delta_i} B_i^{(p)}
\end{array}
$$

The morphisms η_i, β_i and $\delta_i = (F_k \otimes_k B)^{(p)}$ are flat and $F_{B_i} = \delta_i \circ \omega_{B/k} \circ \beta_i$, hence by Lemma 3.4.16 we obtain that $\mathrm{fd}(F_{B_i}) < \infty$. Now from Theorem 3.4.17, we get that F_{B_i} is flat; hence the canonical morphism $\omega_i : B_i \to B^{(p)}$ is flat. Therefore $\omega_{B/k}$ is flat, being the colimit of the morphisms w_i.   □

Now we can prove a generalization of Theorem 3.4.19.

Theorem 3.4.27 (Dumitrescu [95, Th. 2.13]) *Let $u : A \to B$ be a morphism of Noetherian rings of characteristic $p > 0$. The following are equivalent:*

   (i) u is regular;
   (ii) u and $\omega_{B/A}^{[\infty]}$ are flat;
   (iii) u is flat and $\mathrm{fd}(\omega_{B/A}) < \infty$.

Proof $(i) \Rightarrow (ii)$: Follows from Remark 3.4.5, (vii).
$(ii) \Rightarrow (i)$: We have to show that the fibres of u are geometrically regular. By Remark 3.4.5, (vii) we may assume that A is a field. Now apply Lemma 3.4.24.
$(i) \Rightarrow (iii)$: Follows at once from Theorem 3.4.19.
$(iii) \Rightarrow (i)$: Again we have to show that the fibres of u are geometrically regular. By Lemma 3.4.25 we may assume that A is a field. Now apply Lemma 3.4.26.   □

Corollary 3.4.28 *Let A be a Noetherian local ring of characteristic $p > 0$. Then A is quasi-excellent if and only if $\mathrm{fd}_A(\omega_{\widehat{A}/A}) < \infty$.*

3.5   F-Finite Rings

This section is dedicated to F-finite rings. This is an important class of rings, which is extremely close to the class of excellent rings. This class of rings occurs in the study of singularities in positive characteristic. In this section, $p > 0$ will be a fixed prime number and all rings will be supposed to be of characteristic p, that is to contain a field of characteristic p.

Definition 3.5.1 A Noetherian ring A of characteristic p is called **F-finite** if the Frobenius morphism of A is finite, that is, A is a finite A-module via F_A.

Example 3.5.2 A perfect field k is obviously F-finite. Moreover, if X is a variable, also $k[X]$ and $k[[X]]$ are F-finite. If I is an infinite set, $k[X_i; i \in I]$ is not F-finite.

   The next results will provide more examples of F-finite rings.

Lemma 3.5.3 *Let A be a F-finite ring, $\mathfrak{a}$ be an ideal of A and S a multiplicatively closed system in A. Then $A/\mathfrak{a}$ and $S^{-1}A$ are F-finite rings.*

Proof Let $\{x_1, \ldots, x_n\}$ be a system of generators of the A^p-module A. Then $A = \sum_{i=1}^{n} A^p x_i$, hence $A/\mathfrak{a} = \sum_{i=1}^{n}(A/\mathfrak{a})^p x_i$. Also, since $\frac{a}{w} = \frac{aw^{p-1}}{w^p}$ we have $(S^{-1}A)^p = S^{-1}(A^p)$, therefore $S^{-1}A = \sum_{i=1}^{n}(S^{-1}A)^p x_i$.   □

Corollary 3.5.4 *Let A be a F-finite ring and B an A-algebra of finite type. Then B is F-finite.*

Proof Using Lemma 3.5.3 it is enough to show that $B = A[T]$ is F-finite. But if $A = \sum_{i=1}^{n} A^p x_i$, then clearly $A[T] = \sum_{1 \le i \le n, 1 \le j \le p-1} A^p[T^p]x_i T^j$. ☐

The next lemma is a well-known general result and does not need any characteristic assumption.

Lemma 3.5.5 *Let $u : A \to B$ be a ring morphism and $\mathfrak{a}$ a finitely generated nilpotent ideal of B. If $B/\mathfrak{a}$ is finite over A, then u is finite.*

Proof We consider the exact sequence

$$0 \to \mathfrak{a}^i/\mathfrak{a}^{i+1} \to B/\mathfrak{a}^i \to B/\mathfrak{a}^{i+1} \to 0.$$

Since $\mathfrak{a}^i/\mathfrak{a}^{i+1}$ is a finite B-module for all $i \in \mathbb{N}$, we get by induction that $B/\mathfrak{a}^i$ is a finite A-module for all $i \in \mathbb{N}$. As $\mathfrak{a}$ is nilpotent we get the conclusion. ☐

Lemma 3.5.6 *Let A be a ring and $\mathfrak{a}$ an ideal of B such that B is $\mathfrak{a}$-adically complete. If $B/\mathfrak{a}$ is F-finite, then B is F-finite.*

Proof Since $B^{(p)}/\mathfrak{a}^{(p)} \to B/\mathfrak{a}$ is finite, by Lemma 3.5.5 we get that $B/\mathfrak{a}^{(p)}B \to B$ is finite. As $\bigcap_{i=0}^{\infty} \mathfrak{a}^i = 0$, we have that $\bigcap_{i=0}^{\infty} (\mathfrak{a}^{(p)})^i = 0$ and $B^{(p)}$ is $\mathfrak{a}^{(p)}$-adically complete. By [249, Th. 8.4] it follows that $B^{(p)} \to B$ is finite. ☐

Corollary 3.5.7 *Let $(A, \mathfrak{m}, k)$ be a complete local ring of characteristic $p > 0$. Then A is F-finite if and only if k is F-finite.*

Proof Follows from Lemmas 3.5.3 and 3.5.6. ☐

Corollary 3.5.8 *Let A be a Noetherian ring and $\mathfrak{a}$ an ideal of A. If $A/\mathfrak{a}$ is F-finite, then $(A, \widehat{\mathfrak{a}})$, the completion of A in the $\mathfrak{a}$-adic topology, is F-finite.*

Corollary 3.5.9 *If A is F-finite, then $A[[T]]$ is F-finite.*

Proof We apply Corollary 3.5.8 to the ideal (T) of the polynomial ring $A[T]$. ☐

The next proposition shows that, when working with F-finite rings, we may seldom assume that the ring is reduced.

Proposition 3.5.10 *Let A be a Noetherian ring of characteristic $p > 0$. Then A is F-finite if and only if A_{red} is F-finite.*

Proof If A is F-finite, by Lemma 3.5.3 it follows that A_{red} is F-finite. Conversely, let $\mathfrak{a} = (v_1, \ldots, v_s)$ be an ideal of A such that $A/\mathfrak{a}$ is F-finite. Then we can write $A/\mathfrak{a} = \sum_{i=1}^{n} (A/\mathfrak{a})^p \bar{u}_i$, with $\bar{u}_i \in A/\mathfrak{a}$. Let $T := A^p u_1 + \cdots + A^p u_n$, where u_i are representatives for $\bar{u}_i$. Then $A = T + Av_1 + \cdots + Av_h$. If we substitute the same formula for each copy of A occurring in a term of the form Av_j on the right, we find that $A = T + \sum_{i,j} A^p u_i v_j + \sum_{s,t} Av_s v_t$. It follows that $\{u_1, \ldots, u_n, u_1 v_1, \ldots, u_n v_t\}$ is a system of generators of $A/\mathfrak{a}^2$ over $(A/\mathfrak{a}^2)^p$; hence $A/\mathfrak{a}^2$ is F-finite. Going on we obtain that $A/\mathfrak{a}^{2k}$ is F-finite for any $k \in \mathbb{N}$. Taking $\mathfrak{a} = N(A)$, the nilradical of A, we get that A is F-finite. $\square$

Proposition 3.5.11 *Let A be a Noetherian ring of characteristic $p > 0$. The following are equivalent:*

  (i) *A is F-finite;*
  (ii) *$F_A^t : A \to A$ is a finite morphism, for any $t \geq 1$;*
  (iii) *there exists $t \geq 1$ such that F_A^t is a finite morphism.*

Proof $(i) \Rightarrow (ii) \Rightarrow (ii)$: Obvious.
$(iii) \Rightarrow (i)$: Let $B := A_{red}$. Then also $F_B^t : B \to B$ is finite. Setting $q := p^t$, we have $B \subseteq B^{\frac{1}{p}} \subseteq B^{\frac{1}{q}}$ and since $B^{\frac{1}{q}}$ is a Noetherian B-module, it follows that also $B^{\frac{1}{p}}$ is a Noetherian, hence finitely generated, B-module. Now we apply 3.5.10. $\square$

Our next aim now is to prove Theorem 3.5.30, result proved by Gabber, that shows a very important and useful property of F-finite rings. The author gave in [116] only a very quick sketch of the proof. Here we follow closely the detailed proof, as it was presented by L. Ma and T. Polstra in [226, Th. 12.5].

Notation 3.5.12 From now on, until the end of the proof of Theorem 3.5.30, A will be a Noetherian F-finite ring.

Remark 3.5.13 Since A is F-finite, it makes sense to consider a finite system of generators of A over A^p, say $a_1, \ldots, a_s$. Let

$$A_n := A[Z_1, \ldots, Z_s]/(Z_1^{p^n} - a_1, \ldots, Z_s^{p^n} - a_s).$$

and let

$$\phi_n : A_n \to A_{n-1},$$

$$\phi_n(a) = F_A(a) = a^p, \ \forall a \in A,$$

$$\phi_n(Z_j) = Z_j, \ 1 \leq j \leq s.$$

Lemma 3.5.14 *With the above notations, ϕ_n is a well-defined surjective morphism and* $(A_n, \phi_n)_n$ *is an inverse system of ring morphisms.*

Proof It is easy to see that ϕ_n is well-defined and that these morphisms build an inverse system. Let $\alpha \in A$. Since $A = A^p[a_1, \ldots, a_s]$, there exists a polynomial $H \in A^p[T_1, \ldots, T_s]$ with $\alpha = H(a_1, \ldots, a_s)$. If we look at α as an element in A_{n-1} we can write

$$\alpha = H(a_1, \ldots, a_s) = \sum_{j_1, \ldots, j_s = 0}^{p-1} c_{j_1 \ldots j_s}^p a_1^{j_1} \ldots a_s^{j_s}, \quad c_{j_1 \ldots j_s} \in A.$$

Then we see that

$$\phi_n\left(\sum_{j_1, \ldots, j_s = 0}^{p-1} c_{j_1 \ldots j_s} (Z_1^{p^{n-1}})^{j_1} \ldots (Z_s^{p^{n-1}})^{j_s} \right) = \alpha.$$

Since $\phi_n(Z_j) = Z_j$ for $j = 1, \ldots, s$ it follows that ϕ_n is surjective. □

Remark 3.5.15 Let us note that the morphisms of the inverse system are actually

$$\phi_{n+m,n} : A_{n+m} \to A_n,$$

$$\phi_{n+m,n}(a) = F_A^m(a) = a^{p^m}, \quad \phi_{n+m,n}(Z_j) = Z_j, \ j = 1, \ldots, s.$$

Notation 3.5.16 We put $A_\infty := \varprojlim_{n \in \mathbb{N}} A_n$. We also denote $\mathfrak{a}_{n+m,n} := \ker(\phi_{n+m,n})$.

The proof of the next Lemma is straightforward.

Lemma 3.5.17 A_n *is a free A-module with basis* $\{Z_1^{j_1} \ldots Z_s^{j_s} \mid 0 \le j_1, \ldots, j_s < p^n\}$ *and a free A_n^p-module with basis* $\{Z_1^{j_1} \ldots Z_s^{j_s} \mid 0 \le j_1, \ldots, j_s \le p - 1\}$.

Remark 3.5.18 The ring A_n is not a regular ring, even if A_n is a free A_n^p-module. This comes from the fact that A_n is not reduced, so that one cannot apply Kunz's theorem [248, (42.B), Th. 107].

Notation 3.5.19 In order to simplify the notations, we shall denote by (i) an s-uple $i_1, \ldots, i_s$. If $(i) = i_1, \ldots, i_s$ and $(j) = j_1, \ldots, j_s$ are two s-uples, by $p^l \mid (i) - (j)$ we mean that $i_r \equiv j_r \pmod{p^l}$ for all $r = 1, \ldots, s$.

Lemma 3.5.20 $\ker(\phi_n) = \{x \in A_n \mid x^p = 0\}$.

Proof Pick an element $x \in A_n$. Then we can write

$$x = \sum_{i_1,\ldots,i_s=0}^{p^n-1} b_{i_1\ldots i_s} Z_1^{i_1} \ldots Z_s^{i_s}, \; b_{i_1\ldots i_s} \in A,$$

hence we have

$$x^p = \sum_{i_1,\ldots,i_s=0}^{p^n-1} b_{i_1\ldots i_s}^p Z_1^{pi_1} \ldots Z_s^{pi_s} =$$

$$= \sum_{j_1,\ldots,j_s=0}^{p^{n-1}-1} \left(\sum_{p^{n-1}|(i)-(j)} b_{i_1\ldots i_s}^p Z_1^{p(i_1-j_1)} \ldots Z_s^{p(i_s-j_s)} \right) Z_1^{pj_1} \ldots Z_s^{pj_s} =$$

$$= \sum_{j_1,\ldots,j_s=0}^{p^{n-1}-1} \left(\sum_{p^{n-1}|(i)-(j)} b_{i_1\ldots i_s}^p a_1^{\frac{i_1-j_1}{p^{n-1}}} \ldots a_s^{\frac{i_s-j_s}{p^{n-1}}} \right) Z_1^{pj_1} \ldots Z_s^{pj_s}.$$

Therefore, applying Lemma 3.5.17 we see that $x^p = 0$ if and only if for each $(j) = j_1, \ldots, j_s$ we have

$$\sum_{p^{n-1}|(i)-(j)} b_{i_1\ldots i_s}^p a_1^{\frac{i_1-j_1}{p^{n-1}}} \ldots a_s^{\frac{i_s-j_s}{p^{n-1}}} = 0.$$

Applying once more Lemma 3.5.17, this is equivalent to the fact that in A_{n-1}

$$\sum_{j_1,\ldots,j_s=0}^{p^{n-1}-1} \left(\sum_{p^{n-1}|(i)-(j)} b_{i_1\ldots i_s}^p a_1^{\frac{i_1-j_1}{p^{n-1}}} \ldots a_s^{\frac{i_s-j_s}{p^{n-1}}} \right) Z_1^{j_1} \ldots Z_s^{j_s} = 0.$$

At the same time in A_{n-1} we have

$$\sum_{j_1,\ldots,j_s=0}^{p^{n-1}-1} \left(\sum_{p^{n-1}|(i)-(j)} b_{i_1\ldots i_s}^p a_1^{\frac{i_1-j_1}{p^{n-1}}} \ldots a_s^{\frac{i_s-j_s}{p^{n-1}}} \right) Z_1^{j_1} \ldots Z_s^{j_s} =$$

$$= \sum_{j_1,\ldots,j_s=0}^{p^n-1} b_{i_1}^p \ldots b_{i_s}^p Z_1^{i_1} \ldots Z_s^{i_s} = \phi_n(x).$$

Thus we have proved that $\phi_n(x) = 0$ iff $x^p = 0$. $\square$

From Lemma 3.5.20 it follows immediately that

Corollary 3.5.21 A_∞ *is a reduced ring.*

What we need to prove and does not seem straightforward at all is that A_∞ is a Noetherian ring.

Lemma 3.5.22 $\mathfrak{a}_{n+m,n} = (\mathfrak{a}_{n+m,0})^{[p^n]} = (y^{p^n} \mid y \in \mathfrak{a}_{n+m,0})$, *for all* $n \geq 0$ *and* $m \geq 1$.

Proof From Lemma 3.5.20 it results at once that $(\mathfrak{a}_{n+m,0})^{[p^n]} \subseteq \mathfrak{a}_{n+m,n}$. Conversely, let $r \in \mathfrak{a}_{n+m,n}$. Then we can write

$$r = \sum_{i_1,\ldots,i_s=0}^{p^{n+m}-1} r_{i_1\ldots i_s} Z_1^{i_1} \ldots Z_s^{i_s} =$$

$$= \sum_{j_1,\ldots,j_s=0}^{p^n-1} \left(\sum_{p^n \mid (i)-(j)} r_{i_1\ldots i_s} Z_1^{i_1-j_1} \ldots Z_s^{i_s-j_s} \right) Z_1^{j_1} \ldots Z_s^{j_s}, \ r_{i_1\ldots i_s} \in A.$$

Because $\{a_1,\ldots,a_s\}$ is a system of generators of A over A^p, we see that $\{a_1^{k_1} \ldots a_s^{k_s} \mid 0 \leq k_1,\ldots,k_s < p^n\}$ is a system of generators of A over A^{p^n}. Therefore in A_{n+m} we have

$$r_{i_1\ldots i_s} = \sum_{k_1,\ldots,k_s=0}^{p^n-1} b_{i_1\ldots i_s,k_1\ldots k_s}^{p^n} a_1^{k_1} \ldots a_s^{k_s} =$$

$$= \sum_{k_1,\ldots,k_s=0}^{p^n-1} b_{i_1\ldots i_s,k_1\ldots k_s}^{p^n} Z_1^{k_1 p^{n+m}} \ldots Z_s^{k_s p^{n+m}},$$

with $b_{i_1\ldots i_s,k_1\ldots k_s} \in A$. Then

$$r = \sum_{j_1,\ldots,j_s=0}^{p^n-1} \left(\sum_{p^n \mid (i)-(j)} \left(\sum_{k_1,\ldots,k_s=0}^{p^n-1} b_{i_1\ldots i_s,k_1\ldots k_s} Z_1^{k_1 p^m} \ldots Z_s^{k_s p^m} \right) \right.$$

$$\left. \cdot Z_1^{\frac{i_1-j_1}{p^n}} \ldots Z_s^{\frac{i_s-j_s}{p^n}} \right)^{p^n} \cdot Z_1^{j_1} \ldots Z_s^{j_s}$$

Hence, in order to show that $r \in \mathfrak{a}_{n+m,0}^{[p^n]}$, it is enough to show that for all s-uples $j_1, \ldots, j_s$ we have

$$c_{j_1 \ldots j_s} = \sum_{p^n|(i)-(j)} \left(\sum_{k_1,\ldots,k_s=0}^{p^n-1} b_{i_1\ldots i_s,k_1\ldots k_s} Z_1^{k_1 p^m} \ldots Z_s^{k_s p^m} \right) \cdot Z_1^{\frac{i_1-j_1}{p^n}} \ldots Z_s^{\frac{i_s-j_s}{p^n}} \in \mathfrak{a}_{n+m,0}.$$

But in $A_0 = A$ we have $Z_i = a_i$, for $i = 1, \ldots, s$ hence the image of $c_{j_1 \ldots j_s}$ in A_0 is

$$c_{j_1 \ldots j_s} = \sum_{p^n|(i)-(j)} \left(\sum_{k_1,\ldots,k_s=0}^{p^n-1} b_{i_1\ldots i_s,k_1\ldots k_s}^{p^{n+m}} a_1^{k_1 p^m} \ldots a_s^{k_s p^m} \right) a_1^{\frac{i_1-j_1}{p^n}} \ldots a_s^{\frac{i_s-j_s}{p^n}}.$$

But $\phi_{n+m,n}(r) = 0$, so that in A_n we have

$$\sum_{j_1,\ldots,j_s=0}^{p^n-1} \left(\sum_{p^n|(i)-(j)} r_{i_1\ldots i_s}^{p^m} Z_1^{i_1-j_1} \ldots Z_s^{i_s-j_s} \right) Z_1^{j_1} \ldots Z_s^{j_s} =$$

$$= \sum_{j_1,\ldots,j_s=0}^{p^n-1} \left(\sum_{p^n|(i)-(j)} r_{i_1\ldots i_s}^{p^m} a_1^{\frac{i_1-j_1}{p^n}} \ldots a_s^{\frac{i_s-j_s}{p^n}} \right) Z_1^{j_1} \ldots Z_s^{j_s} = 0.$$

By Lemma 3.5.17 we obtain that for every $j_1, \ldots, j_s$ we have

$$0 = \sum_{p^n|(i)-(j)} r_{i_1\ldots i_s}^{p^m} a_1^{\frac{i_1-j_1}{p^n}} \ldots a_s^{\frac{i_s-j_s}{p^n}} =$$

$$= \sum_{p^n|(i)-(j)} \left(\sum_{k_1,\ldots,k_s=0}^{p^n-1} b_{i_1\ldots i_s,k_1\ldots k_s} a_1^{k_1 p^m} \ldots a_s^{k_s p^m} \right) a_1^{\frac{i_1-j_1}{p^n}} \ldots a_s^{\frac{i_s-j_s}{p^n}} = c_{j_1\ldots j_s}. \qquad \square$$

Notation 3.5.23 We put $\mathfrak{c}_n := \ker(A_\infty \to A_n)$ and $\mathfrak{c} := \mathfrak{c}_0 = \ker(A_\infty \to A)$.

Remark 3.5.24 It is clear that $\mathfrak{c} = \mathfrak{c}_0 \supseteq \mathfrak{c}_1 \supseteq \ldots \supseteq \mathfrak{c}_n \supseteq \ldots$. On the other hand, the sequence of ideals $(\mathfrak{c}_l)_{l \geq 0}$ is not a filtration and consequently it doesn't define a ring topology on A_∞; therefore we need to modify a little bit this sequence.

Lemma 3.5.25 $\mathfrak{c}_n \subseteq \bigcap_{m \geq 0} (\mathfrak{c}^n + \mathfrak{c}_m)$ *for any natural number n.*

Proof Since c_n is an ideal of $A_\infty = \varprojlim_{i \in \mathbb{N}} A_i$, let us remember that an element x_* of c_n is actually a thread, that is, $x_* = (x_i)_{i \in \mathbb{N}} \in \prod_{l=0}^{\infty} A_l$ such that $\phi_{j+1,j}(x_{j+1}) = x_j$ for any $j \in \mathbb{N}$. If $x_* \in c_n = \ker(A_\infty \to A_n)$, this means that $x_n = 0$ and consequently $x_m = 0$, for $m \geq n$. Thus $x_m \in \ker(\phi_{m,n})$. But then Lemma 3.5.22 shows that $x_m = \sum r_{im} y_{im}^{p^n}$, where $r_{im} \in A_m$ and $y_{im} \in a_{m,0}$. The inverse system defining A_∞ has surjective transition maps, therefore r_{im} and y_{im} are images of some threads r_{i*} and resp. y_{i*} from A_∞ and moreover $y_{i*} \in c_0 = c$ by construction. Looking at the m-th components we see that $x_* - \sum r_{i*} y_{i*}^{p^n} \in c_m$. Thus $c_n \subseteq c^{[p^n]} + c_m$, hence $c_n \subseteq \bigcap_{m \geq 0} (c^{[p^n]} + c_m)$ and since $\bigcap_{m \geq 0} (c^{[p^n]} + c_m) \subseteq \bigcap_{m \geq 0} (c^n + c_m)$, we are done. $\qquad\square$

Notation 3.5.26 Set $b_n := \bigcap_{l \geq 0} (c^n + c_l)$.

Lemma 3.5.27 *With the above notations we have:*

(i) $c = b_1 \supseteq b_2 \supseteq \ldots \supseteq b_s \supseteq \ldots$;
(ii) $b_k \cdot b_l \subseteq b_{k+l}$ *for any natural numbers* k, l;
(iii) *the sequence of ideals* $(b_l)_{l \geq 0}$ *defines a linear topology on the ring* A_∞.

Proof Straightforward. $\qquad\square$

Lemma 3.5.28 *The ring* A_∞ *is complete in the topology given by the filtration* $\mathcal{F} :=$ $(b_l)_{l \geq 0}$.

Proof From Lemma 3.5.25 we know that $c_n \subseteq b_n$, for any $n \in \mathbb{N}$. Therefore we have a commutative diagram

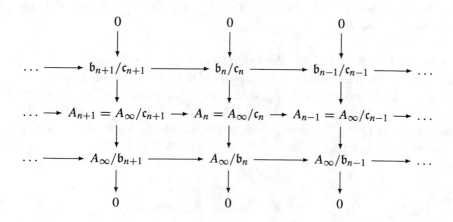

The projective limit of the middle row is $A_\infty = \varprojlim A_n$ and the projective limit of the bottom row is the completion of A_∞ in the topology defined by the filtration $\mathcal{F} = (\mathfrak{b}_l)_{l \geq 0}$. Thus it is enough to show that the projective limit of the top row is zero, whence it is enough to show that for any natural number $n \geq 1$, there exists $k \gg 0$ such that $\mathfrak{b}_k \subseteq \mathfrak{c}_n$. Let us consider a thread $y_* = (y_n)_n \in \mathfrak{b}_1 = \mathfrak{c}$. This means that $y_1 = 0$ and moreover $y_{n+1} \in \mathfrak{a}_{n+1,0}$ for any $n \geq 0$. By Lemma 3.5.22 we know that $\mathfrak{a}_{n+1,n} = (\mathfrak{a}_{n,0})^{[p^n]}$; hence, since $\mathfrak{a}_{n,0}$ is finitely generated, we can find $k \gg 0$ depending on n, such that $(\mathfrak{a}_{n,0})^k \subseteq (\mathfrak{a}_{n,0})^{[p^n]} = \mathfrak{a}_{n+1,n}$. Thus let us take a thread $x_* = (x_l)_{l \in \mathbb{N}} \in \mathfrak{b}_k = \bigcap_{m \geq 0} (\mathfrak{c}^k + \mathfrak{c}_m) \subseteq \mathfrak{c}^k + \mathfrak{c}_{n+1}$. The elements on the $n+1$-th place of the threads from $\mathfrak{c}^k$ are belonging to $(\mathfrak{a}_{n+1,0})^k$ while the elements on the $n+1$-th place of the threads from $\mathfrak{c}_{n+1}$ are 0. Therefore $x_{n+1} \in (\mathfrak{a}_{n+1,0})^k$. By the choice of k it follows that $x_{n+1} \in \mathfrak{a}_{n+1,n}$. Thus $x_n = 0$, that is $x_* \in \mathfrak{c}_n$ and finally $\mathfrak{b}_k \subseteq \mathfrak{c}_n$. $\square$

We can finally prove the most important property of the ring A_∞.

Proposition 3.5.29 A_∞ *is a Noetherian ring.*

Proof Using [58, Ch. III, §2, n. 9, Prop. 12] and Lemma 3.5.28, it is enough to show that $\mathrm{gr}_{\mathcal{F}}(A_\infty) = A_\infty/\mathfrak{b}_1 \oplus \mathfrak{b}_1/\mathfrak{b}_2 \oplus \ldots \oplus \mathfrak{b}_n/\mathfrak{b}_{n+1} \oplus \ldots$ is a Noetherian ring. It is clear that $A_\infty/\mathfrak{b}_1 \cong A_\infty/\mathfrak{c} \cong A$ is a Noetherian ring. We also remark that $\mathfrak{b}_1/\mathfrak{b}_2$ is a finitely generated ideal, as it is an ideal of the ring $A_\infty/\mathfrak{b}_2$ that is a factor ring of the Noetherian ring $A_\infty/\mathfrak{c}_2 \cong A_2$. Thus, in order to show that $\mathrm{gr}_{\mathcal{F}}(A_\infty)$ is Noetherian, it suffices to show that for any natural number $n \geq 1$ we have $\mathfrak{b}_n/\mathfrak{b}_{n+1} = (\mathfrak{b}_1/\mathfrak{b}_2)^n$ or what is the same that $\mathfrak{b}_n = \mathfrak{b}_1^n + \mathfrak{b}_{n+1}$. Clearly $\mathfrak{b}_1^n + \mathfrak{b}_{n+1} \subseteq \mathfrak{b}_n$. To show the other inclusion, note that from Lemma 3.5.25 we know that $\mathfrak{c}_{n+1} \subseteq \mathfrak{b}_{n+1}$, whence it is enough to show that $\mathfrak{b}_n/\mathfrak{c}_{n+1} \subseteq (\mathfrak{b}_1^n + \mathfrak{b}_{n+1})/\mathfrak{c}_{n+1}$. But $\mathfrak{b}_n/\mathfrak{c}_{n+1} = \left(\bigcap_{m \geq 0} (\mathfrak{c}^n + \mathfrak{c}_m) \right)/\mathfrak{c}_{n+1}$, which is obviously generated by $\mathfrak{c}^n = \mathfrak{b}_1^n$. $\square$

Theorem 3.5.30 (Gabber [116, Rem. 13.6]) *Let A be a Noetherian F-finite ring. Then A is isomorphic to a quotient of a regular ring.*

Proof From Corollary 3.5.21 and Proposition 3.5.29 it results that A_∞ is a Noetherian reduced ring. Using Kunz's theorem [248, (42.B), Th. 107], it is enough to show that A_∞ is flat over A_∞^p. Let us denote by Z_{i*} the constant thread having Z_i on all places. By Lemma 3.5.17 the family $\{Z_1^{j_1} \ldots Z_s^{j_s} \mid 0 \leq j_1, \ldots, j_s \leq p - 1\}$ is a basis of A_n over A_n^p, for any natural number n. But inverse limits commute with finite direct sums, hence A_∞ is a free A_∞^p-module having as basis the family of threads $\{Z_{1*}^{j_1} \ldots Z_{s*}^{j_s} \mid 0 \leq j_1, \ldots, j_s \leq p - 1\}$. $\square$

Corollary 3.5.31 *Any F-finite ring is universally catenary.*

Corollary 3.5.31 can be proved without using Theorem 3.5.30. The interested reader can see [248, Th. 108].

Proposition 3.5.32 *Let $(A, \mathfrak{m}, k)$ be a Noetherian local F-finite ring. Then:*

(i) *The canonical morphism $(A^p)\widehat{\ } \otimes_{A^p} A \to \widehat{A}$ is an isomorphism;*
(ii) *$(\widehat{A})^p \simeq (A^p)\widehat{\ }$;*
(iii) *$\Omega_A \otimes_A \widehat{A} \simeq \Omega_{\widehat{A}}$;*
(iv) *If A is reduced, then $\widehat{A}$ is reduced;*

Proof

(i) Let $B := A^p$. Since the canonical morphism $B \to A$ is finite and injective, B is a topological subspace of A. Then $\widehat{B} \subseteq \widehat{A}$. The $\mathfrak{m}$-adic topology on A coincide with the deduced topology of the B-module A; hence $\widehat{A} \simeq \widehat{B} \otimes_B A$ and the natural morphism $\widehat{B} \to \widehat{A}$ is injective and finite.
(ii) The Frobenius morphism $F_A : A \to C = A^p$ is surjective; hence also $\widehat{F_A} : \widehat{A} \to \widehat{C}$ is surjective. Let $x \in A$. Then $\widehat{F_A}(x) = x^p$, therefore, by continuity, we have $\widehat{F_A}(y) = y^p$ for any $y \in \widehat{A}$. Then $\widehat{C} \simeq (\widehat{A})^p$.
(iii) From the properties of the module of differentials and from (i) we get $\Omega_A = \Omega_{A/A^p}$ and $\Omega_{A/A^p} \otimes_A \widehat{A} = \Omega_{\widehat{A}/\widehat{A^p}} = \Omega_{\widehat{A}}$.
(iv) Since A is reduced, the Frobenius morphism F_A is injective. Then $\widehat{F_A}$ is injective; hence $\widehat{A}$ is reduced. $\square$

Lemma 3.5.33 *Let K be a field of characteristic $p > 0$ such that Ω_K is finitely generated. Then K is F-finite.*

Proof Indeed, in this case K has a finite p-basis [131, Th. 21.4.5]. $\square$

Theorem 3.5.34 (Kunz [199, Cor. 2.6] and Seydi [350, Th. 3.3]) *Let $(A, \mathfrak{m}, k)$ be a Noetherian local ring. The following are equivalent:*

(i) *A is F-finite;*
(ii) *A is excellent and k is F-finite;*
(iii) *A has geometrically regular formal fibres and k is F-finite;*
(iv) *A is a Nagata ring and k is F-finite;*
(v) *The generic formal fibres of A are geometrically reduced and k is F-finite.*

Proof (i) $\Rightarrow$ (ii) : From Corollary 3.5.31 A is universally catenary. Let B be a finite reduced A-algebra. Then from Corollary 3.5.4 and Proposition 3.5.32 we get that $\widehat{B}$ is reduced and that $\widehat{B} \simeq \widehat{B}^p \otimes_{B^p} B$. Let $g : \mathrm{Spec}(\widehat{B}) \to \mathrm{Spec}(B)$ be the canonical map. Let $\mathfrak{q} \in \mathrm{Spec}(\widehat{B})$ and $\mathfrak{p} = \mathfrak{q} \cap B$. Then $B_\mathfrak{p} \to \widehat{B}_\mathfrak{q}$ is a local flat morphism. If $\mathfrak{q} \in \mathrm{Reg}(\widehat{B})$, then obviously $\mathfrak{p} \in \mathrm{Reg}(B)$. Conversely, assume that $\mathfrak{p} \in \mathrm{Reg}(B)$. Then by Kunz's theorem, the morphism $(B_\mathfrak{p})^p \to B_\mathfrak{p}$ is flat. From the isomorphism above, we obtain an isomorphism of $(\widehat{B}_\mathfrak{q})^p$-modules $\widehat{B}_\mathfrak{q} \simeq (\widehat{B}_\mathfrak{q})^p \otimes_{(B_\mathfrak{p})^p} B_\mathfrak{p}$. Hence, the morphism obtained by base change $(\widehat{B}_\mathfrak{q})^p \to \widehat{B}_\mathfrak{q}$ is flat too. Again by Kunz's theorem, it follows that $\mathfrak{q} \in \mathrm{Reg}(\widehat{B})$. We have proved that $g^{-1}(\mathrm{Reg}(B)) = \mathrm{Reg}(\widehat{B})$. Therefore, by Theorem 1.6.23 the formal fibres of A are geometrically regular. Since A is local, it follows from Proposition 3.3.6 that A is excellent. The fact that k is F-finite follows from Lemma 3.5.3.

$(ii) \Rightarrow (iii) \Rightarrow (iv) \Rightarrow (v)$: Obvious.

$(v) \Rightarrow (i)$: Let $\mathfrak{p} \in \mathrm{Min}(A)$ and $L := k(\mathfrak{p})$. Then $L \otimes_A \widehat{A}$ is a reduced ring; hence $Q(L \otimes_A \widehat{A}) = L_1 \times \cdots \times L_n$, where $L_1, \ldots, L_n$ are fields, separable extensions of L. For any $i = 1, \ldots, n$ we have an exact sequence

$$0 \to \Omega_L \otimes_L L_i \to \Omega_{L_i} \to \Omega_{L/L_i} \to 0.$$

Since k is F-finite, by Corollary 3.5.7 it results that $\widehat{A}$ is F-finite and then L_i is F-finite. From the exact sequences above and Lemma 3.5.33, we get that also L is F-finite. As $A/\mathfrak{p}$ is a Nagata ring, the integral closure of $A/\mathfrak{p}$ in $L^{\frac{1}{p}}$ is a finite $A/\mathfrak{p}$-module; hence $A/\mathfrak{p}$ is F-finite. Therefore, if $\mathfrak{a} = N(A)$, then $A/\mathfrak{a}$ is F-finite. From Proposition 3.5.10 it follows that A is F-finite.   □

Corollary 3.5.35 (Kunz [200, Th. 2.5]) *Let A be a Noetherian F-finite ring of characteristic p. Then A is an excellent ring.*

Proof By Theorem 3.5.34 and Corollary 3.5.31 it follows that A has geometrically regular formal fibres and is universally catenary. Moreover, if B is an A-algebra of finite type, then B is F-finite. But $\mathrm{Reg}(B) = \{\mathfrak{q} \in \mathrm{Spec}(B) \,|\, B_\mathfrak{q}$ is a free $B_\mathfrak{q}^p$ − module$\}$ and this set is well-known to be open since B is F-finite.   □

Corollary 3.5.36 *Let k be a field of characteristic $p > 0$ which is complete with respect to a non-trivial non-Archimedean absolute value. If k is F-finite, then the ring of restricted power series $k\{X_1, \ldots, X_n\}$ is an excellent ring.*

Remark 3.5.37 The general case of Corollary 3.5.36 in characteristic $p > 0$ was proved by Kiehl [195, Th. 3.3]. The proof is difficult and we will not include it. One can find more details about the proof in [85, Th. 1.1.3].

The connection between excellent and F-finite rings is even stronger, as we now show.

Proposition 3.5.38 (Datta and Smith [88, Th. 2.4]) *Let A be a Noetherian domain whose fraction field K is F-finite. Then A is F-finite if and only if A is excellent.*

Proof If A is F-finite, it follows from Corollary 3.5.35 that A is excellent. Conversely, if A is excellent, then A^p is excellent, because it is isomorphic to A via the Frobenius morphism. Note that the ring of fractions of A^p is K^p. Since K is F-finite, the integral closure C of A^p in K is a finite A^p-module. But the elements of A are integral over A^p; hence A is an A^p- submodule of the Noetherian A^p-module C. Thus A is a Noetherian, hence finitely generated A^p-module.   □

Theorem 3.5.39 (Datta and Smith [88, Cor. 2.6]) *Let A be a reduced Noetherian ring whose total quotient ring K is F-finite. Then A is F-finite if and only if A is excellent.*

Proof We only have to show that if A is excellent, then it is F-finite. Let $\mathfrak{q}_1, \ldots, \mathfrak{q}_n$ be the minimal prime ideals of A and let $K_i := Q(A/\mathfrak{q}_i)$, so that K_i^p is the fraction field of $A^p/\mathfrak{q}_i^p$. We have the commutative diagram of A^p-modules

$$
\begin{array}{ccccc}
A^p & \xrightarrow{\gamma} A^p/\mathfrak{q}_1^p \times \ldots \times A^p/\mathfrak{q}_n^p & \hookrightarrow & K_1^p \times \ldots \times K_n^p \simeq K^p \\
\big\uparrow & \quad\big\uparrow{\scriptstyle\theta} & & \big\uparrow \\
A & \longrightarrow A/\mathfrak{q}_1 \times \ldots \times A/\mathfrak{q}_n & \hookrightarrow & K_1 \times \ldots \times K_n \simeq K.
\end{array}
$$

The horizontal maps are injections because A is reduced. Since A is excellent, the quotients $A/\mathfrak{q}_i$ are excellent too. The fields K_i are F-finite, since K is F-finite. Thus Proposition 3.5.38 implies that $A/\mathfrak{q}_i$ is F-finite for any $i = 1, \ldots, n$. Then θ is a finite map and clearly also γ is finite. Therefore $A/\mathfrak{q}_1 \times \ldots \times A/\mathfrak{q}_n$ is a finitely generated A^p-module; hence its submodule A is finitely generated too. This means that A is F-finite.   □

The next Lemma is valid in general, not only for rings of positive prime characteristic.

Lemma 3.5.40 *Let $u : (A, \mathfrak{m}, k) \to (B, \mathfrak{n}, K)$ be a formally smooth morphism of Noetherian local rings. Assume that:*

(i) $H_1(A, B, B)$ is a finitely generated B-module;
(ii) $\Omega_{B/A}$ is a flat B-module.

Then u is a regular morphism.

Proof Consider a polynomial ring $R := A[X_i, i \in I]$ over A and a surjective morphism of A-algebras $v : R \to B$. Then we get the following Jacobi-Zariski exact sequences associated to $A \to R \to B = R/\mathfrak{a}$

$$0 \to H_1(A, B, B) \to \mathfrak{a}/\mathfrak{a}^2 \to \Omega_{R/A} \otimes_R B \to \Omega_{B/A} \to 0$$

and

$$0 \to H_1(A, B, K) \to \mathfrak{a}/\mathfrak{a}^2 \otimes_B K \to \Omega_{R/A} \otimes_R K \to \Omega_{B/A} \otimes_B K \to 0.$$

Because u is formally smooth, by Corollary 3.2.4 it follows that $H_1(A, B, K) = (0)$. Since $\Omega_{R/A} \otimes_R B$ and $\Omega_{B/A}$ are flat B-modules, it follows that $H_1(A, B, B) \otimes_B K = H_1(A, B, K) = (0)$, hence by Nakayama's lemma $H_1(A, B, B) = (0)$. From Theorem 3.2.8 it results that u is regular. $\qquad\square$

Theorem 3.5.41 (Brezuleanu and Radu [64, Th. 3.4]) *Let A be a Noetherian ring of characteristic $p > 0$. Then A is F-finite if and only if Ω_A is a finitely generated A-module.*

Proof Clearly if A is F-finite, $\Omega_A = \Omega_{A/A^p}$ is a finitely generated A-module. Conversely, assume that Ω_A is finitely generated. Suppose first that $(A, \mathfrak{m}, k)$ is a local ring. Then Ω_k is a finitely generated k-vector space; hence by [131, Th. 21.4.5] k is F-finite. By Corollary 3.5.7 $\widehat{A}$ is F-finite and by the first implication, $\Omega_{\widehat{A}}$ is a finitely generated $\widehat{A}$-module and then also $\Omega_{\widehat{A}/A}$ is finitely generated. Since $A \to \widehat{A}$ is formally étale, $\Omega_{\widehat{A}/A} = (0)$. Hence $\Omega_{\widehat{A}/A} = (0)$. Let C be a regular local ring containing $\mathbb{F}_p$ such that $\widehat{A} = C/\mathfrak{a}$ for some ideal $\mathfrak{a}$ of C. Since the morphism $\mathbb{F}_p \to C$ is regular, by Theorem 3.2.8 we have $H_1(\mathbb{F}_p, C, -) = 0$. From the Jacobi-Zariski exact sequence associated to $\mathbb{F}_p \to C \to \widehat{A}$, we obtain an injective morphism $H_1(\mathbb{F}_p, \widehat{A}, \widehat{A}) \hookrightarrow \mathfrak{a}/\mathfrak{a}^2$. Hence the $\widehat{A}$-module $H_1(\mathbb{F}_p, \widehat{A}, \widehat{A})$ is finitely generated. From the exact sequence

$$H_1(\mathbb{F}_p, \widehat{A}, \widehat{A}) \to H_1(A, \widehat{A}, \widehat{A}) \to \Omega_A \otimes_A \widehat{A}$$

it follows that also the $\widehat{A}$-module $H_1(A, \widehat{A}, \widehat{A})$ is finitely generated. From Lemma 3.5.40 it follows that A has geometrically regular formal fibres. If A is not necessarily local, we obtain that all the localizations of A have geometrically regular formal fibres, and it follows that A itself has geometrically regular formal fibres.

Let us prove now that A is Reg $-$ 2. Let B be a finite A-algebra which is a domain. Then Ω_B is a finitely generated B-module. Let $\mathfrak{q} \in \mathrm{Spec}(B)$ and $C := B_\mathfrak{q}$. It follows that C is a Nagata ring, being a local ring with geometrically regular formal fibres. Then C is a regular local ring if and only if $\Omega_C = \Omega_B \otimes_B C$ is a formally projective, or equivalently free C-module [61, Th. 15.5]. Since Ω_B is finitely generated, the set $\{\mathfrak{p} \in \mathrm{Spec}(B) \mid (\Omega_B)_\mathfrak{p} \text{ is free}\}$ is open in $\mathrm{Spec}(B)$ and consequently A is Reg-2. Hence, by Theorem 2.3.12, A is a Nagata ring. Let $\mathfrak{p} \in \mathrm{Min}(A)$, $B := A/\mathfrak{p}$ and $L := Q(B)$. Then

$\Omega_L = \Omega_B \otimes_B L$ is a finitely generated L-vector space; hence L is F-finite. As $B^p \simeq B$, it follows that B^p is Japanese. Then B is F-finite. Now if $\mathfrak{a} = N(A)$ it follows that $A/\mathfrak{a}$ is F-finite; hence by Proposition 3.5.10 we get that A is F-finite. □

Lemma 3.5.42 *Let A be a Noetherian F-finite ring of characteristic $p > 0$, $\mathfrak{a}$ be an ideal of A and $\widehat{A}$ the completion of A in the $\mathfrak{a}$-adic topology. Then the canonical completion morphism $A \to \widehat{A}$ is étale.*

Proof Since A is F-finite, the A-module Ω_A is finitely generated. By Corollary 3.5.8 $\widehat{A}$ is F-finite; hence also $\Omega_{\widehat{A}}$ is a finitely generated $\widehat{A}$-module and then the $\widehat{A}$-module $\Omega_{\widehat{A}/A}$ is finitely generated too. But $A \to \widehat{A}$ is formally étale, hence $\widehat{\Omega_{\widehat{A}/A}} = (0)$, therefore $\Omega_{\widehat{A}/A} = (0)$. On the other hand, by Corollary 3.5.35 the morphism $A \to \widehat{A}$ is regular, hence $H_1(A, B, B) = (0)$. Applying Theorem 3.2.7 we get that u is smooth, hence étale. □

Theorem 3.5.43 *Let $(A, \mathfrak{m}, k)$ be a Noetherian local ring of characteristic $p > 0$. The following are equivalent:*

(i) A is F-finite;
(ii) The canonical morphism $A \to \widehat{A}$ is étale and k is F-finite.

Proof $(i) \Rightarrow (ii)$: Follows from Lemmas 3.5.42 and 3.5.3.
$(ii) \Rightarrow (i)$: Since k is F-finite, by Lemma 3.5.6 it results that $\widehat{A}$ is F-finite and by Theorem 3.5.41 it follows that $\Omega_{\widehat{A}}$ is finitely generated. As $A \to \widehat{A}$ is an étale morphism we get that $\Omega_A \otimes_A \widehat{A} \simeq \Omega_{\widehat{A}}$; hence $\Omega_A \otimes_A \widehat{A}$ is a finitely generated $\widehat{A}$-module. By flatness, Ω_A is finitely generated and we apply Theorem 3.5.41. □

In the following, we shall give more properties and characterizations of F-finite rings. In view of the previous results, they can be considered as characterizations of excellent rings. Let us begin with the case of fields.

Lemma 3.5.44 (Tanimoto [367, Lemma 2.1]) *Let k be a field. The following are equivalent:*

(i) $\mathrm{char}(k) = p > 0$ and k is F-finite;
(ii) $k[[X_1, \ldots, X_n]]$ is smooth over k for any $n \geq 1$;
(iii) there exists $n \geq 1$ such that $k[[X_1, \ldots, X_n]]$ is smooth over k;
(iv) $\Omega_{k[[X_1,\ldots,X_n]]/k}$ is (X)-adically separated for any $n \geq 1$;
(v) there exists $n \geq 1$ such that $\Omega_{k[[X_1,\ldots,X_n]]/k}$ is (X)-adically separated.

Proof $(i) \Rightarrow (ii)$: Set $X := \{X_1, \ldots, X_n\}$. Since k is F-finite it follows by Corollary 3.5.9 that also $k[[X]]$ is F-finite; hence $\Omega_{k[[X]]/k}$ is finitely generated, as a quotient of $\Omega_{k[[X]]}$. Therefore $\Omega_{k[[X]]/k} \simeq (\widehat{\Omega_{k[[X]]/k}})$, which is a free $k[[X]]$-module [131, Cor. 21.9.3]. Moreover the morphism $k \to k[[X]] = k[[X_1, \ldots, X_n]]$ is regular, hence $H_1(k, k[[X]], k[[X]]) = (0)$. From Theorem 3.2.7 it follows that $k[[X_1, \ldots, X_n]]$ is smooth over k.

$(ii) \Rightarrow (iii)$: Obvious.

$(iii) \Rightarrow (i)$: Set again $X := \{X_1, \ldots, X_n\}$. The Jacobi-Zariski exact sequence associated to the morphisms $k \to k[X] \to k[[X]]$ is:

$$\Omega_{k[X]/k} \otimes_{k[X]} k[[X]] \xrightarrow{\phi} \Omega_{k[[X]]/k} \to \Omega_{k[[X]]/k[X]} \to 0.$$

Since $k[[X]]$ is a local ring smooth over k, it follows from Theorem 3.2.7 that $\Omega_{k[[X]]/k}$ is free. But

$$\Omega_{k[[X]]/k} \otimes_{k[[X]]} k[[X]]/(X) \simeq \bigoplus_{i=1}^{n} k[[X]]/(X) dX_i.$$

It follows that $\Omega_{k[[X]]/k} \simeq \bigoplus_{i=1}^{n} k[[X]] dX_i$. Thus ϕ is an isomorphism and $\Omega_{k[[X]]/k[X]} = (0)$, hence $\Omega_{k((X))/k(X)} = (0)$. But it is well-known that $\operatorname{tr.deg}_{k(X)} k((X)) = \infty$ [399, Rem. 4, p. 220]. It follows that $\operatorname{char}(k) = p > 0$ and $k((X))^p[k(X)] = k((X))$. Therefore k is F-finite.

$(ii) \Rightarrow (iv)$: By Theorem 3.2.7 we have that $\Omega_{k[[X]]/k}$ is projective, hence free.

$(iv) \Rightarrow (v)$: Obvious.

$(v) \Rightarrow (i)$: Since $\Omega_{k[[X]]/k}$ is (X)-adically separated, we have that $\Omega_{k[[X]]/k} \subseteq \widehat{\Omega_{k[[X]]/k}}$. But $\widehat{\Omega_{k[[X]]/k}} \simeq \bigoplus_{i=1}^{n} k[[X]] dX_i$ is a free finitely generated $k[[X]]$-module with basis $\{dX_1, \ldots, dX_n\}$. It follows that also $\Omega_{k[[X]]/k}$ is a free $k[[X]]$-module with basis $\{dX_1, \ldots, dX_n\}$. Let $C := k[X]$. We have an exact sequence

$$\Omega_{C/k} \otimes_C k[[X]] \simeq \Omega_{k[[X]]/k} \to \Omega_{k[[X]]/C} \to 0.$$

It follows that $\Omega_{k[[X]]/C} = (0)$, hence $\Omega_{k((X))/k(X)} = (0)$. But as we already noticed $\operatorname{tr.deg}_{k(X)} k((X)) = \infty$, whence $\operatorname{char}(k) = p > 0$ and $k((X))^p[k(X)] = k((X))$. Therefore k is F-finite.   □

Theorem 3.5.45 (Tanimoto [367, Th. 2.2]) *Let A be a Noetherian ring containing a field k. The following are equivalent:*

  (i) *$A[[X_1, \ldots, X_n]]$ is smooth over A for any $n \geq 1$;*
  (ii) *there exists $n \geq 1$ such that $A[[X_1, \ldots, X_n]]$ is smooth over A;*
  (iii) *$\operatorname{char}(k) = p > 0$ and A is F-finite.*

Proof $(ii) \Rightarrow (iii)$: In order to simplify things, we denote $X := \{X_1, \ldots, X_n\}$ and $B = A[[X]] = A[[X_1, \ldots, X_n]]$. Then $\Omega_{B/A}$ is a projective B-module, so that it is (X)-adically separated. Hence $\Omega_{B/A} \subseteq \widehat{(\Omega_{B/A})}$, where completion means (X)-adic completion. Let $d := d_{B/A}$ be the universal B-derivation. Then $\widehat{(\Omega_{B/A})}$ is a free $\widehat{B} = B$-module with basis $\{dX_1, \ldots, dX_n\} \subseteq \Omega_{B/A}$. It follows that $\Omega_{B/A}$ is a free B-module with basis $\{dX_1, \ldots, dX_n\}$. Let $\mathfrak{a} = (X_2, \ldots, X_n)B$ and let $E := B/\mathfrak{a} = A[[X_1]]$. Then $\delta : \mathfrak{a}/\mathfrak{a}^2 \to \Omega_{B/A} \otimes_B E$ is left-invertible. It follows that E is smooth over A, so that we may assume that $n = 1$. Let $\mathfrak{m} \in \mathrm{Max}(A)$ and let $K := A/\mathfrak{m}$. Then $K[[X]]$ is smooth over K and by Lemma 3.5.44 it follows that $\mathrm{char}(K) = \mathrm{char}(k) = p > 0$ and K is F-finite. We may assume that A is a domain. Indeed, if we know that $A/\mathfrak{p}$ is F-finite for any minimal prime ideal $\mathfrak{p}$ of A, then, since $A_{red} \subseteq \bigoplus_{\mathfrak{p} \in \mathrm{Min}(A)} A/\mathfrak{p}$, we will get that A_{red} is finite over A^p and finally by Lemma 3.5.5 it will follow that A is F-finite. Let $J := \{\mathfrak{q} \in \mathrm{Spec}(A) \mid A/\mathfrak{q} \text{ is not } F - \text{finite}\}$. Suppose that $J \neq \emptyset$ and let $\mathfrak{p}$ be a maximal element of J. Then $\mathfrak{p}$ is not a maximal ideal of A as it follows from Lemma 3.5.44. Let $R := A/\mathfrak{p}$ and $C := R[[X]]$. Let also $a \in R$ be a non-zero and non-invertible element. If we put $R^* := \widehat{(R, aR)}$ the completion of R in the aR-adic topology, then from the maximality of $\mathfrak{p}$ we get that $R/aR \simeq R^*/aR^*$ is F-finite. By Theorem 3.5.34 it follows that R/aR is a Nagata ring, hence by Theorem 2.2.5 we get that R^* is a Nagata ring. By [249, Th. 8.12] we know that $R^* \simeq C/(X-a)C$. The natural morphism $\delta : (X-a)C/(X-a)^2C \to \Omega_{C/A} \otimes_C R^*$ is an isomorphism; therefore R^* is smooth over R. Let $S := 1 + aR$. Then $S^{-1}R \to R^*$ is smooth and faithfully flat; hence it is a reduced morphism. Then by Proposition 2.2.9 we obtain that $S^{-1}R$ is a Nagata ring. Let $\mathfrak{n} \in \mathrm{Max}(R)$ be such that $\mathfrak{n} \cap S = \emptyset$. Then $R_\mathfrak{n}$ is a Nagata ring and $R_\mathfrak{n}/\mathfrak{n}R_\mathfrak{n}$ is F-finite. Hence $R_\mathfrak{n}$ is F-finite and $Q(R)$ is F-finite. We can write $R = \varinjlim_i R_i$, where R_i are the finitely generated R^p-algebras contained in R. Then $R^p[[X]][R] = \varinjlim_i R_i[[X]]$. The morphism $R \to C$ is smooth. Consider the exact sequence

$$\Omega_{R[X]/R} \otimes_{R[X]} C \to \Omega_{C/R} \to \Omega_{C/R[X]} \to 0.$$

The C-module $\Omega_{C/R} = \Omega_{R[[X]]/R}$ is free and finitely generated. It follows that $\Omega_{R[[X]]/R[X]} = (0)$. Then

$$Q(R[[X]]) = Q(R[[X]])^p (Q(R[X])) = Q(R^p[[X]][R]).$$

Let $h \in (R)_a[[X]]$. There exist $i \in I$ and elements $f, g \in R_i[[X]]$ such that $h = \frac{g}{f}$. Let b be the coefficient of the lowest degree in f and $a = b^p \neq 0$. Then $h \in (R_i)_a[[X]]$. Let $S := 1 + aR$ and $T := S^p$. The ring $(S^{-1}R)^p = T^{-1}R^p$ is a Nagata ring and $Q(R) = Q(T^{-1}R)$ is F-finite. It follows that $S^{-1}R$ is finite over $(T^{-1}R)^p$; hence there exists $j \geq i$ such that $T^{-1}R_j = T^{-1}R = S^{-1}R$. Then

$$h \in (R_j)_a[[X]] \cap (T^{-1}R_j)[[X]] = R_j[[X]] \subseteq R^p[[X]][R],$$

therefore $R[[X]] = R^p[[X]][R]$. It follows that R is F-finite, contradiction! Hence $J = \emptyset$.
$(iii) \Rightarrow (i)$: Let us denote $X := \{X_1, \ldots, X_n\}$. From Corollary 3.5.4 and Theorem 3.5.34, we get that also $A[X]$ is F-finite and excellent and by Lemma 3.5.42 the morphism $A[X] \to A[[X]]$ is étale. $\square$

The above results can be extended to the case of rings not containing a field. Let us begin with a lemma.

Lemma 3.5.46 *Let A be a Noetherian ring with $\mathrm{char}(A) = n > 0$ and let B be a Noetherian flat A-algebra. Assume that for any prime divisor p of n, the morphism $A/pA \to B/pB$ is unramified (resp. regular, resp. étale). Then $A \to B$ is unramified (resp. regular, resp. étale).*

Proof The regularity of the morphism $A \to B$ follows easily using Theorem 1.4.20. Let $n = p_1 \ldots p_t$ be the decomposition of n into prime elements and let $E := \Omega_{B/A}$. If $t = 1$, then n is prime and $p_1 A = 0$. It follows that $\Omega_{B/A} = 0$. Assume $t > 1$ and let $p := p_1$ and $q := p_2 \ldots p_t$. Then we have an exact sequence

$$0 \to pB \to B \to B/pB \to 0,$$

therefore we get an exact sequence

$$\Omega_{B/A} \otimes_B pB \to \Omega_{B/A} \to \Omega_{B/A} \otimes_B B/pB \to 0.$$

But from the induction hypothesis, we have

$$\Omega_{B/A} \otimes pB \simeq (\Omega_{B/A} \otimes_B B/qB) \otimes_B pB = \Omega_{(B/qB)/(A/qA)} \otimes_A pB = 0.$$

On the other hand, by assumption we have $\Omega_{B/A} \otimes_A (A/pA) = 0$. Hence from the above exact sequence, we obtain $\Omega_{B/A} = 0$. $\square$

Theorem 3.5.47 (Tanimoto [368, Th. 3.2]) *Let A be a Noetherian ring. The following conditions are equivalent:*

(i) *$A[[X_1, \ldots, X_n]]$ is smooth over A for any $n \geq 1$;*
(ii) *there exists $n \geq 1$ such that $A[[X_1, \ldots, X_n]]$ is smooth over A;*
(iii) *$\mathrm{char}(A) = n > 0$ and A/pA is F-finite for any prime divisor p of n.*

Proof $(i) \Rightarrow (ii)$: Obvious.

$(ii) \Rightarrow (iii)$: Assume that $\mathrm{char}(A) = 0$. Then there exists a prime ideal $\mathfrak{p}$ of A such that $\mathrm{char}(A/\mathfrak{p}) = 0$. Therefore we may assume that A is a domain with $\mathrm{char}(A) = 0$. Let $X := \{X_1, \ldots, X_n\}$. We have the exact sequence

$$\Omega_{A[X]/A} \otimes_{A[X]} A[[X]] \to \Omega_{A[[X]]/A} \to \Omega_{A[[X]]/A[X]} \to 0.$$

By assumption, $\Omega_{A[[X]]/A}$ is a free $A[[X]]$-module of rank n and also $\Omega_{A[X]/A} \otimes_{A[X]} A[[X]]$ is free of rank n. It follows that $\Omega_{A[[X]]/A[X]} = 0$. Since $\mathrm{char}(A) = 0$, this means that $Q(A[[X]])$ is algebraic over $Q(A[X])$. This is a contradiction [399, Rem. 4, p. 220], so that $\mathrm{char}(A) > 0$ and we apply Theorem 3.5.45.

$(iii) \Rightarrow (i)$: Follows immediately from Theorem 3.5.45 and Lemma 3.5.46. □

Corollary 3.5.48 *Let A be a Noetherian ring. Assume that there exists $n \geq 1$ such that $A \to A[[X_1, \ldots X_n]]$ is smooth. Then A is an excellent ring.*

Proof Follows at once from Theorem 3.5.47 and Theorem 3.5.34. □

Corollary 3.5.49 *Let A be a Noetherian ring. The following are equivalent:*

(i) *$A[[X_1, \ldots, X_n]]$ is smooth over A for any $n \geq 1$;*
(ii) *there exists $n \geq 1$ such that $A[[X_1, \ldots, X_n]]$ is smooth over A;*
(iii) *there exists $n \geq 1$ such that $A[[X_1, \ldots, X_n]]$ has geometrically regular formal fibres and $A_{\mathfrak{m}}/pA_{\mathfrak{m}}$ is F-finite for any $\mathfrak{m} \in \mathrm{Max}(A)$, where p is a prime number such that $\mathrm{char}(A_{\mathfrak{m}}) = p^t$;*
(iv) *A has geometrically reduced formal fibres, $\mathrm{char}(A) > 0$ and there exists an ideal $\mathfrak{a} \subseteq J(A)$ and an integer $n \geq 1$ such that $A[[X_1, \ldots, X_n]]$ is $\mathfrak{a}A[[X_1, \ldots, X_n]]$-smooth over A;*
(v) *$\mathrm{char}(A) = n > 0$ and A/pA is F-finite for any prime divisor p of n;*
(vi) *$\mathrm{char}(A) > 0$ and Ω_A is a finitely generated A-module;*
(vii) *A is a Nagata ring and $A_{\mathfrak{m}}/pA_{\mathfrak{m}}$ is F-finite for any $\mathfrak{m} \in \mathrm{Max}(A)$, where $\mathrm{char}(A_{\mathfrak{m}}) = p^t$, for some prime number p.*

Proof First let us shortly denote $X := \{X_1, \ldots, X_n\}$.

$(i) \Leftrightarrow (ii) \Leftrightarrow (v)$: This is Theorem 3.5.47.

$(iii) \Rightarrow (i)$: Let $\mathfrak{m} \in \mathrm{Max}(A)$ and let $\mathfrak{n} := (\mathfrak{m}, X)A[[X]]$. If we put $C := A_{\mathfrak{m}}[X]$ and $D := A_{\mathfrak{m}}[[X]]$, we have the ring morphisms $C \xrightarrow{\phi} A[[X]]_{\mathfrak{n}} \xrightarrow{\psi} D$. The morphism ψ is $(\mathfrak{m}, X)$-smooth; hence by Theorem 3.2.22 it is a regular morphism, because $A[[X]]_{\mathfrak{n}}$ has geometrically regular formal fibres. We have the exact sequence

$$0 = H_1(A[[X]]_{\mathfrak{n}}, D, D) \to \Omega_{A[[X]]_{\mathfrak{n}}/C} \otimes_{A[[X]]_{\mathfrak{n}}} D \to \Omega_{D/C}.$$

As in the proof of Theorem 3.5.47, $(ii) \Rightarrow (iii)$ we get that $\Omega_{D/C} = 0$. Therefore we obtain that $\Omega_{A[[X]]_n/C} = 0$ for any maximal ideal n of $A[[X]]$, so that $\Omega_{A[[X]]/A[X]} = 0$. From the assumption and Theorem 3.5.34, it follows that A has geometrically regular formal fibres. Then the morphism $A[X] \to A[[X]]$ is regular, hence étale. Consequently $A \to A[[X]]$ is smooth.

$(v) \Rightarrow (vi)$: Let p be a prime divisor of $\mathrm{char}(A)$ and $B := A/pA$. Then Ω_B is a finitely generated B-module. Now we proceed as in the proof of Lemma 3.5.46. Namely consider the exact sequence

$$0 \to pA \to A \to A/pA \to 0$$

and tensorize with Ω_A. We obtain the exact sequence

$$\Omega_A \otimes pA \to \Omega_A \to \Omega_A \otimes A/pA \to 0.$$

But the right-hand module is $\Omega_{A/pA}$ and the left hand-one is $\Omega_{A/pA} \otimes pA$. The conclusion follows.

$(vi) \Rightarrow (v)$: Follows from Theorem 3.5.41.

$(v) \Rightarrow (vii)$: Follows from Theorem 3.5.34.

$(vii) \Rightarrow (v)$: Let $q \in \mathrm{Spec}(A)$ and let $\mathrm{char}(A_q) = p^m$, for some prime number p. Then by assumption $k(q)$ is F-finite. The ring A/q is a Nagata ring; hence $(A/q)^p$ is also a Nagata ring. It follows that A/q is F-finite.

$(v) \Rightarrow (iii)$: Follows from the fact that also $A[[X]]$ satisfies the assumptions of (v).

$(iv) \Rightarrow (vii)$: The morphism $A/\mathfrak{a} \to A[[X]]/\mathfrak{a}A[[X]] = A/\mathfrak{a}[[X]]$ is smooth. From Corollary 3.5.48 we get that $A/\mathfrak{a}$ is a Nagata ring. Let A^* be the $\mathfrak{a}$-adic completion of A. Then A^* is a Nagata ring. As by Proposition 1.6.40 the canonical morphism $A \to A^*$ is faithfully flat and reduced, it follows by Proposition 2.2.9 that A is a Nagata ring. Let $m \in \mathrm{Max}(A)$. Then $A/m \to A[[X]]/mA[[X]]$ is smooth. From Lemma 3.5.44 it results that A/m is F-finite.

$(i) \Rightarrow (iv)$: So far we have proved that (i), (ii), (iii), (v), (vi) and (vii) are equivalent. This means that A has geometrically reduced formal fibres, that $\mathrm{char}(A) > 0$ and that $A \to A[[X]]$ is smooth, that is, we can take $\mathfrak{a} = (0)$.   □

We close this section with an example. From Lemma 3.5.3 we see that any localization of a F-finite ring is F-finite. Is it true that the property of a Noetherian ring of being F-finite is a local one? In other words, if a Noetherian ring A is such that all of its localizations in prime ideals are F-finite, does it follow that A itself is F-finite? The next example shows that this is not the case, thus answering a question of Datta and Murayama.

Example 3.5.50 ([99]) There exists a locally F-finite Noetherian domain that is not F-finite.

Proof Let K be an algebraically closed field of characteristic $p > 2$ and X an indeterminate. Consider the ring

$$A := K\left[X, \frac{1}{\sqrt{(X+a)^3} + \sqrt{b^3}}; \ a, b \in K, \ b \neq 0\right].$$

More precisely, for each square radical above, we *choose* one of its two values. Hence the denominator $\sqrt{(X+a)^3} - \sqrt{b^3}$ is not in our list. Note that

$$\sqrt{(X+a)^3} = \frac{1}{\sqrt{(X+a)^3} + \sqrt{b^3}}((X+a)^3 - b^3) + \sqrt{b^3} \in A.$$

It follows that A is a fraction ring of $K\left[X, \sqrt{(X+a)^3}; \ a \in K\right]$, which in turn is an integral extension of $K[X]$. Thus $\dim(A) \leq 1$. □

Claim 3.5.51 The factor ring $A/(X, \sqrt{X^3})$ is isomorphic to K, while the factor ring $A/(X)$ is isomorphic to $K[Y]/(Y^2)$, where Y is an indeterminate.

Proof Indeed, from Kneser's theorem [191, Th. 5.1] we obtain that

$$\sqrt{X+c} \notin K(X, \sqrt{X+a}, \ a \in K, a \neq c)$$

for every $c \in K$ (this also follows by adapting the well-known argument used to prove that $\sqrt{p_n} \notin \mathbb{Q}(\sqrt{p_1}, \ldots, \sqrt{p_{n-1}})$, when $p_1, \ldots, p_n$ are distinct primes). Consequently, we get a ring isomorphism

$$\frac{K[X, T_a, \ a \in K]}{(T_a^2 - (X+a)^3, a \in K)} \simeq K\left[X, \sqrt{(X+a)^3}, \ a \in K\right]$$

sending each indeterminate T_a into $\sqrt{(X+a)^3}$, which extends to an isomorphism

$$\frac{K\left[X, T_a, (T_a + \sqrt{b^3})^{-1}, \ a, b \in K, b \neq 0\right]}{(T_a^2 - (X+a)^3, a \in K)} \simeq A.$$

It follows that

$$A/(X) \simeq \frac{K\left[T_a, (T_a + \sqrt{b^3})^{-1}, \ a, b \in K, b \neq 0\right]}{(T_a^2 - a^3, a \in K)}$$

$$\simeq \frac{K\left[T_a, (T_a + \sqrt{b^3})^{-1}, \ a, b \in K, b \neq 0\right]}{(T_0^2, T_a - \sqrt{a^3}, a \in K, a \neq 0)}$$

$$\simeq \frac{K\left[T_0, (\sqrt{a^3} + \sqrt{b^3})^{-1}, (T_0 + \sqrt{b^3})^{-1}, \ a, b \in K - \{0\}\right]}{(T_0^2)} \simeq \frac{K[T_0]}{(T_0^2)}$$

so

$$A/(X, \sqrt{X^3}) \simeq K.$$

Note that $\sqrt{a^3} + \sqrt{b^3}$ is non-zero for $a, b \in K - \{0\}$, by our initial one-value-choice for $\sqrt{b^3}$. Also note that $T_0 + \sqrt{b^3}$ is a unit modulo T_0^2.   □

Claim 3.5.52 The non-zero prime ideals of A are $(X + a, \sqrt{(X + a)^3})A$ with $a \in K$. In particular, A is a Noetherian domain of dimension one.

Proof By Claim 3.5.51, $(X, \sqrt{X^3})$ is the only prime ideal of A containing X. One can see that every non-zero prime ideal of A contains some $X + a$, so the assertion follows from Claim 3.5.51, performing a translation in K. The final assertion follows from Cohen's Theorem [249, Th. 3.4].   □

Claim 3.5.53 A_m is F-finite for each maximal ideal m of A.

Proof Performing a translation in K, it suffices to work with $m = (X, \sqrt{X^3})$. As noted above, A is a fraction ring of

$$B := K\left[X, \sqrt{(X + a)^3}; \ a \in K\right].$$

For $a \in K - \{0\}$, we have that $X + a$ is a unit of A_m so

$$\sqrt{(X + a)^3} = \frac{(\sqrt{(X + a)^3})^p}{(X + a)^{pk}} (X + a)^{(p-1)k} \in (A_m)^p[X, \sqrt{X^3}]$$

where k is the integer $3(p - 1)/2$. Hence

$$B \subseteq (A_m)^p[X, \sqrt{X^3}] \subseteq A_m$$

Let $n = mA_m \cap B$. Since A is a fraction ring of B and $1/b^p \in (A_m)^p$ for each $b \in B - n$, we get

$$A_m = B_n \subseteq (A_m)^p[X, \sqrt{X^3}], \quad \text{therefore } A_m = (A_m)^p[X, \sqrt{X^3}].$$

and Claim 3.5.53 is proved.   □

Suppose now that A_q is regular for some maximal ideal q of A. By the above translation argument, it follows that A_m is regular for each maximal ideal m of A. Then A is normal, so $\sqrt{X} \in A$. Hence X divides $\sqrt{X^3}$ in A, which contradicts Claim 3.5.51. This means that $\text{Reg}(A) = \{(0)\}$; hence it is not open in $\text{Spec}(A)$. By Theorem 3.5.35 it follows that A is not F-finite. $\qquad\qquad\qquad\qquad\qquad\qquad\qquad\qquad\qquad\qquad\qquad\qquad\qquad\qquad\square$

Remark 3.5.54 The ring A constructed above is not excellent, because $\text{Reg}(A)$ is not open. On the other hand, for any prime ideal p of A, the localization A_p is F-finite, hence excellent. Thus this ring is another example of a locally excellent ring that is not excellent, as we noticed in Example 3.3.3,iv). A Noetherian ring A that is locally F-finite is universally catenary and has geometrically regular formal fibres. Moreover, if A is reduced, then A has F-finite total quotient ring; hence by Theorem 3.5.39 we get that A is F-finite if and only if A is excellent, that is, if and only if A satisfies condition Reg-2.

Remark 3.5.55 One can construct a similar example in the case $p = 2$ as follows [99, Rem. 10]. Let K be an algebraically closed field of characteristic two and X an indeterminate. Consider the ring

$$B := K\left[X, \sqrt[3]{(X+a)^4}, \frac{1}{\sqrt[3]{(X+a)^8} + \sqrt[3]{(X+a)^4}\sqrt[3]{b^4} + \sqrt[3]{b^8}}; \ a, b \in K, b \neq 0\right].$$

For each cube radical above, we *choose* one of its three values. By Kneser's theorem [191, Th. 5.1], it follows that

$$\left[K(X, \sqrt[3]{X+a_1}, ..., \sqrt[3]{X+a_n}) : K(X)\right] = 3^n$$

whenever $a_1,...,a_n$ are distinct elements of K, so we get a ring isomorphism

$$\frac{K\left[X, T_a, (T_a^2 + T_a\sqrt[3]{b^4} + \sqrt[3]{b^8})^{-1}; \ a, b \in K, b \neq 0\right]}{(T_a^3 - (X+a)^4, \ a \in K)} \simeq B$$

sending each indeterminate T_a into $\sqrt[3]{(X+a)^4}$. It easily follows that $B/(X) \simeq K[T_0]/(T_0^3)$ and $B/(X, \sqrt[3]{X^4}) \simeq K$. Now all arguments used above can be adapted to show that B is locally F-finite but not F-finite. The reader is invited to do it.

3.6   Universally Finite Module of Differentials and Excellent Rings

We begin with an example.

Example 3.6.1 There exists a complete regular local ring with module of differentials not finitely generated.

Proof Let k be a field of characteristic zero and let $A = k[[X]]$ be the formal power series ring in one variable X over k. Then $\mathrm{Der}_k(A) \cong \mathrm{Hom}_A(\Omega_{A/k}, A)$ is a finitely generated free A-module of rank one, with basis $\frac{d}{dX}$. On the other hand, $\mathrm{rk}_A \Omega_{A/k} = \mathrm{tr.deg}_k k((X)) = \infty$, so that $\Omega_{A/k}$ is not finitely generated.   □

Obviously one can consider finitely many variables in the above example, instead of only one. However, since the finite generation of the module $\Omega_{A/k}$ would much simplify many things, it would be useful to find such a module that could replace $\Omega_{A/k}$. This idea, which generated what we may call the *Scheja-Storch theory of the universally finite module of differentials* [340], celebrating the German mathematicians Günther Scheja (1932–2014) and Uwe Storch (1940–2017), is investigated in this section, together with its applications in our frame. Our approach is mainly inspired by Matsumura's notes [247] of the lectures given at the Istituto Matematico del Politecnico di Torino and by Ch. I, §8, of a preliminary (and unfortunately final) version (March 2002) of an unpublished manuscript of late Prof. E. Kunz (1933–2021), *Algebraic Differential Calculus*, which presumably was intended to be a new and slightly extended version of his book [202]. And of course by the original celebrated paper by Scheja and Storch [340].

We will not assume for the moment that all rings are Noetherian.

Definition 3.6.2 Let $u : A \to B$ be a ring morphism and M a B-module. A derivation $d \in \mathrm{Der}_A(B, M)$ is called **finite** if BdB is a finite B-module.

Remark 3.6.3 If B is Noetherian, a derivation $d : B \to M$ is always finite if M is a finitely generated B-module.

Definition 3.6.4 Let $u : A \to B$ be a ring morphism. A **universally finite A-derivation** of B is a finite A-derivation $d : B \to M_0$ such that for any finite A-derivation $d' : B \to M$, there exists a uniquely determined B-linear map $f : M_0 \to M$ such that $d' = f \circ d_0$.

Remark 3.6.5

(i) It is not true that the universally finite derivation always exists.
(ii) If the universally finite derivation $d : B \to M$ exists, then M is finitely generated. Moreover, in this case, the universally finite derivation is unique up to canonical isomorphism.

Definition 3.6.6 If the universally finite derivation $d : B \to M_0$ exists, the B-module M_0 is called the **universally finite module of differentials** of B over A and d_0 is called the **universally finite derivation** of B over A.

Notation 3.6.7 If the universally finite module of differentials of B over A exists, it will be denoted $\Omega^f_{B/A}$ and the universally finite derivation will be denoted $d^f_{B/A} : B \to \Omega^f_{B/A}$.

Proposition 3.6.8 *Let* $u : A \to B$ *be a ring morphism. Let* $\{U_i\}_{i\in I}$ *be the family of* B-*submodules of* $\Omega_{B/A}$ *such that* $\Omega_{B/A}/U_i$ *is finitely generated and let* $U := \bigcap_{i\in I} U_i$.
Then $\Omega^f_{B/A}$ *exists if and only if* $\Omega_{B/A}/U$ *is finitely generated. If this is the case,* $\Omega^f_{B/A} \simeq \Omega_{B/A}/U$ *and the universally finite derivation* $d^f_{B/A} : B \to \Omega^f_{B/A}$ *is the map induced by the universal derivation* $d_{B/A}$.

Proof If $\Omega^f_{B/A}$ exists, then $\Omega^f_{B/A} = \Omega_{B/A}/E$ for some B-submodule $E \subseteq \Omega_{B/A}$, such that E belongs to the family $(U_i)_{i\in I}$, hence $U \subseteq E$. Let now $i \in I$. Then there is an induced surjection $\Omega_{B/A}/E \to \Omega_{B/A}/U_i$, hence $E \subseteq \bigcap_{i\in I} U_i = U$.

Conversely, suppose that $\Omega_{B/A}/U$ is finitely generated and let d be the composition $B \xrightarrow{d_{B/A}} \Omega_{B/A} \xrightarrow{\pi} \Omega_{B/A}/U$, where π is the canonical surjection. Let $\delta : B \to N$ be a finite derivation. Then there exists a B-linear map $\eta : \Omega_{B/A} \to N$ such that $\delta = \eta \circ d_{B/A}$. Then η factors through $\Omega_{B/A}/U$; hence d is the universally finite derivation of B over A. □

Remark 3.6.9 Let $u : A \to B$ be a ring morphism.

(i) If $\Omega_{B/A}$ is a finitely generated B-module, then $\Omega^f_{B/A}$ exists and

$$\Omega^f_{B/A} \cong \Omega_{B/A}.$$

(ii) If $\Omega^f_{B/A}$ exists, then for any finitely generated B-module N we have

$$\mathrm{Der}_A(B, N) \cong \mathrm{Hom}_B(\Omega^f_{B/A}, N).$$

Lemma 3.6.10 *Let* $u : A \to K$ *be a ring morphism such that* K *is a field. Then* $\Omega^f_{K/A}$ *exists if and only if* $\dim_K \Omega_{K/A} < \infty$.

Proof Suppose that $\dim_K \Omega_{K/A} = \infty$. Let $n \in \mathbb{N}$. There exists a subspace $U_n \subseteq \Omega_{K/A}$ such that $\dim_K \Omega_{K/A}/U_n = n$. Let $u_n : \Omega_{K/A} \to \Omega_{K/A}/U_n$ be the canonical map and let $\delta_n = u_n \circ d_{K/A}$. Then $\delta_n \in \mathrm{Der}_A(K, \Omega_{K/A}/U_n)$ is such that $\dim(K\delta_n(K)) = n$. It follows that $\Omega^f_{K/A}$ does not exists. The converse is clear. □

Example 3.6.11 Let k be a field of characteristic zero and $K = k((X))$. We saw already that $\dim_K \Omega_{K/k} = \infty$. From Lemma 3.6.10 it follows that $\Omega^f_{K/k}$ does not exists.

Corollary 3.6.12 *Let $u : A \to B$ and $v : B \to C$ be ring morphisms. Assume that B is a Noetherian ring and v is finite. If $\Omega^f_{B/A}$ exists, then $\Omega^f_{C/A}$ exists and $\Omega^f_{C/A} \cong \Omega_{C/A}/C \cdot U$ where U is the B-submodule of $\Omega_{B/A}$ considered in Proposition 3.6.8.*

Proof Since $\Omega^f_{B/A}$ exists, from Proposition 3.6.8 it follows that $\Omega_{C/A}/C \cdot U$ is a finitely generated B-module; hence it is a finitely generated C-module. The canonical map $\Omega_{B/A} \to \Omega_{C/A}$ induces a map $\Omega^f_{B/A} = \Omega_{B/A}/U \to \Omega_{C/A}/C \cdot U$. Let $d : C \to M$ be a finite A-derivation of C. Then, since B is Noetherian and v is finite, M is also a finitely generated B-module and also BdB is finitely generated. The composition map $B \to C \to M$ is an A-derivation; hence there exists a B-linear map $\phi : \Omega^f_{B/A} \to M$ such that $d = \phi \circ d^f_{B/A}$. Let $\psi : \Omega_{C/A} \to M$ be the C-linear map such that $d^f_{B/A} = \psi \circ d_{C/A}$. Then $\psi(C \cdot U) = 0$ and ψ induces a C-linear map $\rho : \Omega_{C/A}/C \cdot U \to M$. $\square$

Proposition 3.6.13 *Let $u : A \to B$ and $v : A \to C$ be ring morphisms and let $\psi : B \to C$ be a morphism of A-algebras. If $\Omega^f_{C/A}$ exists, then $\Omega^f_{C/B}$ exists and we have an exact sequence*

$$\Omega_{B/A} \otimes_B C \to \Omega^f_{C/A} \to \Omega^f_{C/B} \to 0.$$

Proof Let $d : C \to M$ be a finite B-derivation of C into a C-module M. Then d is also an A-derivation; hence there exists a C-linear map $\phi : \Omega^f_{C/A} \to M$ such that $d = \phi \circ d^f_{C/A}$. For any element $b \in B$, we have $\phi(d^f_{C/A}(b)) = 0$; hence a C-linear map $\overline{\phi} : \Omega^f_{C/A}/U \to M$ is induced, where U is the C-submodule of $\Omega^f_{C/A}$ generated by the set $\{d^f_{C/A}(b), b \in B\}$. If $\pi : \Omega^f_{C/A} \to \Omega^f_{C/A}/U$ is the canonical map, then $\pi \circ d^f_{C/A}$ is a B-derivation and it is clearly the universally finite B-derivation of C. Thus $\Omega^f_{C/B} \cong \Omega^f_{C/A}/U$ and this means exactly the exactness of the above sequence. $\square$

Proposition 3.6.14 *Let $u : A \to B$ be a ring morphism such that $\Omega^f_{B/A}$ exists, let $\mathfrak{a}$ be an ideal of B and let $C = B/\mathfrak{a}$. Then $\Omega^f_{C/A}$ exists and we have an exact sequence of C-modules*

$$\mathfrak{a}/\mathfrak{a}^2 \xrightarrow{\delta} \Omega^f_{B/A}/\mathfrak{a}\Omega^f_{B/A} \to \Omega^f_{C/A} \to 0.$$

Proof Let $\tilde{d} : \mathfrak{a} \to \Omega^f_{B/A}$ be the restriction of $d^f_{B/A}$ to $\mathfrak{a}$ and $\delta : \mathfrak{a}/\mathfrak{a}^2 \to \Omega^f_{B/A}$ be the map induced by $\tilde{d}$. Let $\overline{d} : C \to M$ be a finite A-derivation and let $d = \overline{d} \circ u : B \to M$. Since d is finite, there exists a B-linear map $\phi : \Omega^f_{B/A} \to M$ such that $\phi \circ d^f_{B/A} = d$. Then $\phi(\mathfrak{a}\Omega^f_{B/A} + B \cdot d^f_{B/A}(\mathfrak{a})) = 0$, hence ϕ induces a unique C-linear map $\psi : \Omega^f_{B/A}/(\mathfrak{a}\Omega^f_{B/A} + B \cdot d^f_{B/A}(\mathfrak{a})) \to M$ such that $\overline{d} = \psi \circ d^f_{C/A}$, where $d^f_{C/A}$ is the derivation induced by $d^f_{B/A}$. It follows that $d^f_{C/A}$ is universally finite and consequently $\Omega^f_{C/A} = \Omega^f_{B/A}/(\mathfrak{a}\Omega^f_{B/A} + B \cdot d^f_{B/A}(\mathfrak{a}))$. $\qquad\qquad\qquad\qquad\qquad\qquad\qquad\qquad\qquad\qquad$ □

Corollary 3.6.15 *Let $u : A \to B$ be a ring morphism such that $\Omega^f_{B/A}$ exists, let $\mathfrak{m}$ be a maximal ideal of B and $K := B/\mathfrak{m}$. Then $\dim_K \Omega_{K/A} < \infty$.*

Proof Follows from Proposition 3.6.14 and Lemma 3.6.10. $\qquad\qquad\qquad\qquad\qquad$ □

Theorem 3.6.16 *Let $u : A \to B$ be a ring morphism such that B is a Noetherian semilocal ring and let $\mathfrak{m} = J(B)$ be the Jacobson radical of B. The following are equivalent:*

(i) $\Omega^f_{B/A}$ exists;

(ii) $\Omega_{B/A}/ \bigcap\limits_{i=0}^{\infty} \mathfrak{m}^i\Omega_{B/A}$ is a finitely generated B-module.

If this is the case, then $\Omega^f_{B/A} = \Omega_{B/A}/ \bigcap\limits_{i=0}^{\infty} \mathfrak{m}^i\Omega_{B/A}$.

Proof Let $U_\alpha \subseteq \Omega_{B/A}$ be a submodule such that $M := \Omega_{B/A}/U_\alpha$ is finitely generated and let U denote the intersection of all these U_α. By Krull's intersection theorem, we get that $\bigcap\limits_{i\geq 0} \mathfrak{m}^i M = (0)$ hence $\bigcap\limits_{i=0}^{\infty} \mathfrak{m}^i\Omega_{B/A} \subseteq U$.

$(ii) \Rightarrow (i)$: Since $\Omega_{B/A}/ \bigcap\limits_{i=0}^{\infty} \mathfrak{m}^i\Omega_{B/A}$ is finitely generated, it follows that $U = \bigcap\limits_{i=0}^{\infty} \mathfrak{m}^i\Omega_{B/A}$. The assertion follows from Proposition 3.6.8.

$(i) \Rightarrow (ii)$: Let $\mathrm{Max}(A) = \{\mathfrak{m}_1, \ldots, \mathfrak{m}_r\}$. Then by the Chinese remainder theorem, $B/\mathfrak{m} \cong \prod\limits_{i=1}^{r} B/\mathfrak{m}_i$. Let $k_i := B/\mathfrak{m}_i$. By Corollary 3.6.15 $\dim_{k_i}(\Omega_{k_i/A}) < \infty$, hence $\Omega_{B/\mathfrak{m}/A} \cong \prod\limits_{i=1}^{r} \Omega_{k_i/A}$ is a finitely generated $B/\mathfrak{m}$-module. From the exact sequence

$$\mathfrak{m}/\mathfrak{m}^2 \to \Omega_{B/A}/\mathfrak{m}\Omega_{B/A} \to \Omega_{B/\mathfrak{m}/A} \to 0$$

it follows that $\Omega_{B/A}/\mathfrak{m}\Omega_{B/A}$ is a finitely generated B-module; hence $\Omega_{B/A}/\mathfrak{m}^i\Omega_{B/A}$ is a finitely generated B-module for any $i \in \mathbb{N}$. The canonical map $\Omega_{B/A} \to \Omega^f_{B/A}$ induces a surjection $\alpha : \Omega_{B/A}/ \bigcap\limits_{i=0}^{\infty} \mathfrak{m}^i\Omega_{B/A} \to \Omega^f_{B/A}$. We want to show that α is an isomorphism.

Let $\Omega := \Omega_{B/A} / \bigcap_{i=0}^{\infty} \mathfrak{m}^i \Omega_{B/A}$ and take a non-zero element $x \in \Omega$. Then there exists $n \in \mathbb{N}$ such that $x \notin \mathfrak{m}^n \Omega$. We have a commutative diagram

$$\begin{array}{ccccc} \Omega_{B/A} & \longrightarrow & \Omega & \overset{\alpha}{\longrightarrow} & \Omega^f_{B/A} \\ & & \downarrow & & \downarrow \\ & & \Omega_{B/\mathfrak{m}^{n+1}/A} & \cong & \Omega^f_{(B/\mathfrak{m}^{n+1})/A} \end{array}$$

Since

$$\Omega_{(B/\mathfrak{m}^{n+1})/A} = \Omega_{B/A} / (\mathfrak{m}^{n+1}\Omega_{B/A} + B \cdot d^f_{B/A}\mathfrak{m}^{n+1})$$

and

$$B \cdot d^f_{B/A}\mathfrak{m}^{n+1} \subseteq \mathfrak{m}^{n+1}\Omega_{B/A},$$

it follows that the image of x in $\Omega^f_{(B/\mathfrak{m}^{n+1})/A}$ is not zero. $\square$

Corollary 3.6.17 *Let $u : A \to (B, \mathfrak{n}, L)$ be a ring morphism such that B is a Noetherian local ring for which $\Omega^f_{B/A}$ exists. Let $\mathfrak{m} = \mathfrak{n} \cap A$ and $K = A_\mathfrak{m}/\mathfrak{m}A_\mathfrak{m}$.*

(i) If the extension $K \subseteq L$ is separable, the canonical sequence

$$0 \to \mathfrak{n}/(\mathfrak{n}^2 + \mathfrak{m}B) \to \Omega^f_{B/A}/\mathfrak{n}\Omega^f_{B/A} \to \Omega_{L/K} \to 0$$

is exact.

(ii) If $\mathrm{char}(L) = p > 0$ and $\mathfrak{q} = \mathfrak{n} \cap A[B^p]$, there is a canonical exact sequence

$$0 \to \mathfrak{n}/(\mathfrak{n}^2 + \mathfrak{q}B) \to \Omega^f_{B/A}/\mathfrak{n}\Omega^f_{B/A} \to \Omega_{L/K} \to 0.$$

Proof From Theorem 3.6.16 we have that $\Omega^f_{B/A}/\mathfrak{n}\Omega^f_{B/A} \cong \Omega_{B/A}/\mathfrak{n}\Omega_{B/A}$. Hence the sequences are derived from the corresponding Jacobi-Zariski exact sequences. $\square$

Corollary 3.6.18 *Let A be a ring, let B and C be two Noetherian semilocal A-algebras with Jacobson radicals $\mathfrak{m}$ and $\mathfrak{n}$, respectively, and let $\phi : B \to C$ be an A-algebra morphism such that $\phi(\mathfrak{m}) \subseteq \mathfrak{n}$. If $\Omega^f_{B/A}$ and $\Omega^f_{C/A}$ exist, then ϕ induces a canonical map $\phi^* : \Omega^f_{B/A} \to \Omega^f_{C/A}$ such that $\phi^* \circ d^f_{B/A} = d^f_{C/A} \circ \phi$.*

Proof Follows from Theorem 3.6.16 and the fact that the canonical map $\tilde{\phi} : \Omega_{B/A} \to \Omega_{C/A}$, induced by ϕ to the differential modules, has the property that

$$\tilde{\phi}\left(\bigcap_{i=0}^{\infty} \mathfrak{m}^i \Omega_{B/A}\right) \subseteq \bigcap_{i=0}^{\infty} \mathfrak{n}^i \Omega_{C/A}.$$

$\square$

Corollary 3.6.19 *Let $u : A \to B$ be a ring morphism such that $l_B(B) < \infty$. Then $\Omega_{B/A}^f$ exists if and only if $\Omega_{B/A}$ is finitely generated.*

Proof By assumption B is semilocal and if $\mathfrak{m}$ is the Jacobson radical of B, then there exists $i \in \mathbb{N}$ such that $\mathfrak{m}^i = (0)$. Hence $\bigcap_{i=0}^{\infty} \mathfrak{m}^i \Omega_{B/A} = (0)$ and we apply Theorem 3.6.16.

$\square$

Corollary 3.6.20 *Let $u : A \to (B, \mathfrak{n}, L)$ be a ring morphism such that B is a Noetherian local ring. Let $\mathfrak{m} = \mathfrak{n} \cap A$ and $K = A_{\mathfrak{m}}/\mathfrak{m}A_{\mathfrak{m}}$. Assume that:*

(i) the extension $K \subseteq L$ is separable;
(ii) $\Omega_{B/A}^f$ exists.

Then $\mu(\Omega_{B/A}^f) = \mathrm{edim}(B/\mathfrak{m}B) + \mathrm{tr.deg}_K L.$

Proof Let $C := B/(\mathfrak{n}^2 + \mathfrak{m}B)$. By Proposition 3.6.14 we get that $\mu(\Omega_{B/A}^f) = \mu(\Omega_{C/A}^f)$. Since C is of finite length, by Corollary 3.6.19 we have $\Omega_{C/A}^f \cong \Omega_{C/A}$. Now we apply [202, Th. 6.5].

$\square$

Corollary 3.6.21 *Let k be a field of characteristic 0 and $(A, \mathfrak{m}, K)$ be a Noetherian local k-algebra such that $\Omega_{A/k}^f$ exists. Then*

$$\mu(\Omega_{A/k}^f) = \mathrm{edim}(A) + \mathrm{tr.deg}_k(K).$$

Remark 3.6.22 Let $u : A \to B$ be a ring morphism, M a B-module and $d \in \mathrm{Der}_A(B, M)$. We consider an ideal $\mathfrak{a}$ of B and on B and M we consider the $\mathfrak{a}$-adic topology. Then it is easy to see that d is continuous. If moreover B is Noetherian and M is a finitely generated B-module, we extend d by continuity to $\hat{d} : \hat{B} \to \hat{M}$. It follows at once that $\hat{d}$ is an A-derivation.

Proposition 3.6.23 *Let $u : A \to B$ be a ring morphism, where B is a Noetherian ring. Let $\mathfrak{a}$ be an ideal of B and let $\hat{B}$ be the completion of B in the $\mathfrak{a}$-adic topology. If $\Omega_{B/A}$ is* *finitely generated, then the canonical map* $\widehat{d}_{B/A} : \widehat{B} \to \widehat{\Omega_{B/A}}$ *induced by completion from the universal derivation* $d_{B/A}$ *of B over A is the universally finite derivation of* $\widehat{B}$ *over A and*

$$\Omega^f_{\widehat{B}/A} \cong \widehat{\Omega_{B/A}} \cong \widehat{B} \otimes_B \Omega_{B/A}.$$

Proof Clearly $\widehat{\Omega_{B/A}}$ is a finite $\widehat{B}$-module. Let $\delta : \widehat{B} \to M$ be an arbitrary A-derivation into a finite $\widehat{B}$-module M. Then there exists a B-linear map $\phi : \Omega_{B/A} \to M$ such that $d = \phi \circ d_{B/A}$ where d is the composition of the canonical map $B \to \widehat{B}$ with δ. Since $\widehat{B} \otimes_B \Omega_{B/A}$ is generated as a $\widehat{B}$-module by the elements $\{d_{B/A}(b), b \in B\}$, there is only one $\widehat{B}$-linear map with the above property. It follows that $\Omega^f_{\widehat{B}/A}$ exists and coincides with $\widehat{\Omega_{B/A}}$. $\square$

Theorem 3.6.24 *Let* $u : A \to B$ *be a ring morphism. Assume that B is a Noetherian complete semilocal ring with maximal ideals* $\mathfrak{m}_1, \ldots, \mathfrak{m}_r$ *and denote* $k_i := B/\mathfrak{m}_i$ *for* $i = 1, \ldots, r$. *Then* $\Omega^f_{B/A}$ *exists if and only if* $\dim_{k_i}(\Omega_{k_i/A}) < \infty$ *for* $i = 1, \ldots, r$.

Proof If $\Omega^f_{B/A}$ exists, then by Corollary 3.6.15 we get $\dim_{k_i}(\Omega_{k_i/A}) < \infty$, $i = 1, \ldots, r$. Conversely, since B is complete, by [249, Th. 8.15] we have $B \cong B_1 \times \ldots \times B_r$, where B_i is the completion of B in the $\mathfrak{m}_i$-adic topology. Let $\mathfrak{m} = \mathfrak{m}_1 \cap \ldots \cap \mathfrak{m}_r$ be the Jacobson radical of B. The exact sequence

$$\mathfrak{m}_i/\mathfrak{m}_i^2 \to \Omega_{B_i/A}/\mathfrak{m}_i\Omega_{B_i/A} \to \Omega_{k_i/A} \to 0$$

shows that $\dim_{k_i}(\Omega_{B_i/A}/\mathfrak{m}_i\Omega_{B_i/A}) < \infty$, hence

$$\Omega_{B/A}/\mathfrak{m}\Omega_{B/A} \cong \prod_{i=1}^{r} \Omega_{B_i/A}/\mathfrak{m}_i\Omega_{B_i/A}$$

is a finitely generated $B/\mathfrak{m}$-module. Let $M := \Omega_{B/A}/ \bigcap_{i=0}^{\infty} \mathfrak{m}^i\Omega_{B/A}$ be the separated module of $\Omega_{B/A}$. Since B is complete, M is finitely generated. By Theorem 3.6.16, we deduce that $\Omega^f_{B/A}$ exists. $\square$

Corollary 3.6.25 *Let* $u : A \to B$ *be a ring morphism. Assume that B is a Noetherian semilocal ring,* $\mathfrak{m}$ *is the Jacobson radical of B and* $\widehat{B}$ *is the completion of B in the* $\mathfrak{m}$-*adic* *topology. If $\Omega^f_{B/A}$ exists, then also $\Omega^f_{\widehat{B}/A}$ exists and*

$$\Omega^f_{\widehat{B}/A} \cong \widehat{B} \otimes_B \Omega^f_{B/A}.$$

Proof The existence of $\Omega^f_{\widehat{B}/A}$ follows from Theorem 3.6.24 and Corollary 3.6.15. Moreover, we have $\Omega^f_{\widehat{B}/A} = \Omega_{\widehat{B}/A} / \bigcap_{i=0}^{\infty} \mathfrak{m}^i \Omega_{\widehat{B}/A}$. But this $\widehat{B}$-module is finitely generated and complete, hence we have

$$\Omega^f_{\widehat{B}/A} \cong \varprojlim \Omega_{\widehat{B}/A} / \mathfrak{m}^n \Omega_{\widehat{B}/A} \cong \varprojlim \Omega^f_{B/A} / \mathfrak{m}^n \Omega^f_{B/a} \cong \widehat{B} \otimes_{\widehat{B}} \Omega^f_{B/A}.$$

$\square$

Example 3.6.26 Let A be a Noetherian ring, let $B = A[X_1, \ldots, X_n]$ be the polynomial ring in n variables over A and let $\mathfrak{a} := (X_1, \ldots, X_n)$. Then the completion of B in the $\mathfrak{a}$-adic topology is the ring of formal power series $\widehat{B} = A[[X_1, \ldots, X_n]]$. By Proposition 3.6.23 it follows that $\Omega^f_{\widehat{B}/A}$ exists and that

$$\Omega^f_{A[[X_1, \ldots, X_n]]/A} = \Omega^f_{\widehat{B}/A} = \widehat{B} \otimes_B \Omega_{B/A} = \sum_{i=1}^{n} \widehat{B} dX_i.$$

Moreover $d^f_{\widehat{B}/A}(f) = \sum_{i=1}^{n} \frac{\partial f}{\partial X_i} dX_i$.

Proposition 3.6.27 *Let $u : A \to B$ be a morphism of Noetherian rings such that $\Omega^f_{B/A}$ exists. Then, for any maximal ideal $\mathfrak{m}$ of B, $\Omega^f_{B_\mathfrak{m}/A}$ exists and moreover $\Omega^f_{B_\mathfrak{m}/A} = (\Omega^f_{B/A})_\mathfrak{m}$.*

Proof Let $d^f_{B/A} : B \to \Omega^f_{B/A}$ be the universal finite derivation. We prove that $d_\mathfrak{m} : B_\mathfrak{m} \to (\Omega^f_{B/A})_\mathfrak{m}$ satisfies the universal property of $d^f_{B_\mathfrak{m}/A}$. For this, let $\delta : B_\mathfrak{m} \to N = B_\mathfrak{m} \delta B_\mathfrak{m}$ be a finite derivation. For any natural number i, we have $B_\mathfrak{m} / \mathfrak{m}^i B_\mathfrak{m} \cong B/\mathfrak{m}^i$ and $N/\mathfrak{m}^i N$ is a finitely generated $B/\mathfrak{m}^i$- module; hence it is a finitely generated B-module. Let $\delta_i : B \to N/\mathfrak{m}^i N$ be the map obtained as the composition

$$B \to B_\mathfrak{m} \xrightarrow{\delta} N \to N/\mathfrak{m}^i N.$$

Then, for any natural number i, there exists a linear map

$$\phi_i : \Omega^f_{B/A} \to N/\mathfrak{m}^i N$$

such that $\phi_i \circ d^f_{B/A} = \delta_i$. The map ϕ_i induces by localization the map

$$\psi_i = (\phi_i)_\mathfrak{m} : (\Omega^f_{B/A})_\mathfrak{m} \to N/\mathfrak{m}^i N.$$

Let

$$\theta : (\Omega^f_{B/A})_\mathfrak{m} \to \prod_{i=1}^{\infty} N/\mathfrak{m}^i N$$

be the map given by

$$\theta((d^f_{B/A}(b))_\mathfrak{m}) = (\psi_i((d^f_{B/A}(b))_\mathfrak{m}))_i.$$

Let also $\mu : N \to \prod_{i=1}^{\infty} N/\mathfrak{m}^i N$ be the map induced by the canonical surjections $N \to N/\mathfrak{m}^i N$. Since $\bigcap_{i=1}^{\infty} \mathfrak{m}^i N = (0)$, it results that θ is injective. We have a commutative diagram

$$
\begin{array}{ccc}
B_\mathfrak{m} & \xrightarrow{(d^f_{B/A})_\mathfrak{m}} & (\Omega^f_{B/A})_\mathfrak{m} \\
\downarrow{\scriptstyle\delta} & & \downarrow{\scriptstyle\theta} \\
N & \xrightarrow{\mu} & \prod_{i=1}^{\infty} N/\mathfrak{m}^i N
\end{array}
$$

Therefore $\theta((\Omega^f_{B/A})_\mathfrak{m}) \subset \mu(N)$ and θ induces a $B_\mathfrak{m}$-linear map $h : (\Omega^f_{B/A})_\mathfrak{m} \to N$ such that $\delta = h \circ (d^f_{B/A})_\mathfrak{m}$. $\qquad\square$

Example 3.6.28

(i) If $\Omega^f_{B/A}$ exists, it is not true that for any prime ideal $\mathfrak{p}$ of A, the module $\Omega^f_{B_\mathfrak{p}/A}$ exists. Indeed, let $A = k$ be a field, $B = k[[X]]$ and $\mathfrak{p} = (0)$. Then from Example 3.6.26 we know that $\Omega^f_{B/A}$ exists, while from Example 3.6.11 it follows that $\Omega^f_{B_{(0)}/A} = \Omega^f_{k((X))/k}$ does not exist.

(ii) If $\Omega^f_{B/A}$ exists, it does not follow that $\Omega^f_{B[X]/A}$ exists. Indeed, let $A = k$ be a field and $B = k[[Y]]$. Then $\Omega^f_{B/A}$ exists by Example 3.6.26. Let $C = B[X] = k[[Y]][X]$ and let $D := C/(XY - 1)$. If $\Omega^f_{C/A}$ exists, then by Proposition 3.6.14 it follows that $\Omega^f_{D/A}$ exists too. But $D = C/(XY - 1) = k[[Y]][\frac{1}{Y}] = k((Y))$.

Our purpose now is to prove a Jacobian criterion of regularity in characteristic zero in terms of the universally finite module of differentials (Theorem 3.6.37). This criterion was proved by Scheja and Storch [340] and roughly speaking, characterizes the regular prime ideals of a k-algebra A for which $\Omega^f_{A/k}$ exists, as exactly the primes for which the localization of $\Omega^f_{A/k}$ is free. We need some preparations.

Lemma 3.6.29 *Let k be a field of characteristic zero and $(A, \mathfrak{m}, K)$ be a Noetherian local k-algebra. If $\Omega^f_{A/k}$ exists, then $\dim_K \Omega^f_{K/k} = \mathrm{tr.deg}_k(K) < \infty$.*

Proof If $\Omega^f_{A/k}$ exists, then by Proposition 3.6.14 it follows that $\Omega^f_{K/k}$ exists too, hence $\dim_K(\Omega^f_{K/k}) < \infty$. If $x_1, \cdots, x_s \in K$ are algebraically independent over k, then $d^f_{K/k}(x_1), \cdots, d^f_{K/k}(x_s)$ are linearly independent over K. $\qquad\square$

Lemma 3.6.30 *Let $(A, \mathfrak{m})$ be a Noetherian complete local ring of dimension n and k a coefficient field of A. Assume that $\mathrm{char}(k) = 0$ and that $\Omega^f_{A/k}$ exists. Then A is a regular local ring if and only if $\Omega^f_{A/k}$ is a free A-module.*

Proof If A is regular, then $A \cong k[[X_1, \cdots, X_n]]$. We have seen in Example 3.6.26 that $\Omega^f_{A/k}$ is a free A-module. Conversely, assume that $\Omega^f_{A/k}$ is free. By Corollary 3.6.20 we have $\mathrm{edim}(A) = \mu(\Omega^f_{A/k})$. Let $s = \mathrm{edim}(A)$ and let $\{x_1, \ldots, x_s\}$ be a minimal system of generators of $\mathfrak{m}$. Suppose that $s > n$. Then there exists a system of parameters $y_1, \ldots, y_n \in \mathfrak{m}$, such that $y_i \equiv x_i(\mathrm{mod}\ \mathfrak{m}^2)$ for any index $i = 1, \ldots, n$. Hence $\mathfrak{m} = (y_1, \ldots, y_n, x_{n+1}, \ldots, x_s)$. Thus we may assume that $\{x_1, \ldots, x_n\}$ is a system of parameters. It follows that $A = k[[x_1, \ldots, x_n, x_{n+1}, \ldots, x_s]]$ and $x_{n+1}, \ldots, x_s$ are algebraically dependent over $k[[x_1, \ldots, x_n]]$. Let $f(T) = \sum_i \phi_i(x_1, \ldots, x_n)T^i$ be the minimal polynomial of x_s. Then

$$f(x_s) = \sum_i \phi_i(x_1, \ldots, x_n)x_s^i = 0,$$

hence

$$d^f_{A/k}(f(x_s)) = \sum_{h=1}^n \left(\sum_i \frac{\partial \phi_i}{\partial x_h}x_s^i\right)d^f_{A/k}(x_h) + f'(x_s)d^f_{A/k}(x_s) = 0.$$

But from Corollary 3.6.20 we see that $\{d^f_{A/k}(x_1), \ldots, d^f_{A/k}(x_n), d^f_{A/k}(x_{n+1}), \ldots,$ $d^f_{A/k}(x_s)\}$ is a minimal system of generators of $\Omega^f_{A/k}$. Since this module is free, this is a basis of $\Omega^f_{A/k}$ over A; hence $f'(x_s) = 0$. But this contradicts the minimality of f. □

Lemma 3.6.31 *Let k be a field of characteristic zero and $(A, \mathfrak{m})$ be a Noetherian complete local ring containing k. Let K be a coefficient field of A containing k and let $a_1, \ldots, a_t \in K$ be algebraically independent elements over k. Assume that $\Omega^f_{A/k}$ exists.*

(i) *If A is an integral domain, then $d^f_{A/k}(a_1), \ldots, d^f_{A/k}(a_t) \in \Omega^f_{A/k}$ are linearly independent over A;*

(ii) *Let $\mathfrak{p}$ be a prime ideal of A and let $b_1, \ldots, b_t$ be elements in A such that*
$$\sum_{i=1}^t b_i d^f_{A/k}(a_i) \in \mathfrak{p}\Omega^f_{A/k}.\ Then\ b_1, \ldots, b_t \in \mathfrak{p}.$$

Proof

(i) Let $x_1, \ldots, x_n$ be a system of parameters of A and let $\{a_i\}_{i \in I}$ be a transcendence basis of K over k containing $a_1, \ldots, a_t$. Then A is finite over $B := K[[x_1, \ldots, x_n]]$ and we can write $K = k(a_i; i \in I)(u_j; j \in J)$, where u_j is algebraic over $k(a_i; i \in I)$ for any $j \in J$. For any $s = 1, \ldots, t$ we define $D_s : B \to B$, by

$$D_s(a_i) = \delta_{si}, \ \forall i \in I,$$

$$D_s(x_l) = 0, \ l = 1, \ldots, n.$$

Then $D_s \in \mathrm{Der}_k(B)$. We can extend D_s to a map $\tilde{D}_s : A \to Q(A)$ and then clearly $\tilde{D}_s \in \mathrm{Der}_K(A, Q(A))$. Since A is a finite B-module, it follows that also $A\tilde{D}_s(A)$ is a finite A-module. Thus, for any $s = 1, \ldots, t$ we have the following commutative diagram:

$$
\begin{array}{ccc}
A & \xrightarrow{\ d^f_{A/k}\ } & \Omega^f_{A/k} \\
\ \downarrow{\scriptstyle \tilde{D}_s} & \swarrow{\scriptstyle \phi_s} & \\
A\tilde{D}_s(A) & &
\end{array}
$$

Suppose that $\sum_{i=1}^{t} \lambda_i d^f_{A/k}(a_i) = 0$, with $\lambda_1, \ldots, \lambda_s \in A$. Then, for any $s = 1, \ldots, t$ we have

$$0 = \phi_s\left(\sum_{i=1}^{t} \lambda_i d^f_{A/k}(a_i)\right) = \sum_{i=1}^{t} \lambda_i \phi_s\left(d^f_{A/k}(a_i)\right).$$

We get

$$\sum_{i=1}^{t} \lambda_i \tilde{D}_s(a_i) = \lambda_s = 0.$$

Thus $d^f_{A/k}(a_1), \ldots, d^f_{A/k}(a_t)$ are linearly independent over A.

(ii) Since $k \subseteq K \subseteq C := A/\mathfrak{p}$, we know from Proposition 3.6.14 that $\Omega^f_{C/k}$ exists and from (i) that $d^f_{C/k}(\bar{a}_1), \ldots, d^f_{C/k}(\bar{a}_s)$ are linearly independent over $A/\mathfrak{p}$. On the other hand, from Proposition 3.6.14 we have

$$\Omega^f_{C/k} = \Omega^f_{A/k}/(\mathfrak{p}\Omega^f_{A/k} + Ad^f_{A/k}(\mathfrak{p}))$$

and

$$d^f_{C/k}(\bar{a}_i) = d^f_{A/k}(a_i) + (\mathfrak{p}\Omega^f_{A/k} + Ad^f_{A/k}(\mathfrak{p})).$$

Now (ii) follows from (i) applied for $A/\mathfrak{p}$. $\qquad\square$

Definition 3.6.32 Let $(A, \mathfrak{m}, K)$ be a complete Noetherian local ring containing a field k such that $[K : k] < \infty$ and let $\mathfrak{p} \in \operatorname{Spec}(A)$. A system of elements $E = \{f_1, \ldots, f_r\} \subseteq \mathfrak{m}$ such that the images $\{\bar{f}_1, \ldots, \bar{f}_r\}$ in $A/\mathfrak{p}$ are a system of parameters of $A/\mathfrak{p}$ is called a **reference system** of A in $\mathfrak{p}$.

Lemma 3.6.33 *Let $(A, \mathfrak{m})$ be a Noetherian complete local ring, k a coefficient field of A and $\{f_1, \ldots, f_r\}$ a reference system of A in a prime ideal $\mathfrak{p}$. Then:*

(i) $k((f_1, \ldots, f_r)) \subseteq A_\mathfrak{p}$;
(ii) the field extension $k((f_1, \ldots, f_r)) \subseteq A_\mathfrak{p}/\mathfrak{p}A_\mathfrak{p}$ is finite.

Proof

(i) Set $K := k((f_1, \ldots, f_r))$. We have the following commutative diagram:

$$k[[f_1,\ldots,f_r]] \xrightarrow{\psi} k[[\overline{f}_1,\ldots,\overline{f}_r]]$$

$$A \longrightarrow A/\mathfrak{p}$$

Since by [249, Th. 14.5] the elements $\overline{f}_1,\ldots,\overline{f}_r$ are analytically independent over k, it follows that also $f_1,\ldots,f_r$ are analytically independent over k; hence ψ is bijective. Then $\mathfrak{p} \cap k[[f_1,\ldots,f_r]] = (0)$ and this implies that $K \subseteq A_\mathfrak{p}$.

(ii) As k is isomorphic to the residue field of A and $A/\mathfrak{p}$, we get that $A/\mathfrak{p}$ is a finite $k[[\overline{f}_1,\ldots,\overline{f}_r]]$-module, hence $A_\mathfrak{p}/\mathfrak{p}A_\mathfrak{p} = Q(A/\mathfrak{p})$ is a finite algebraic extension of K.   □

Lemma 3.6.34 *Let $(A,\mathfrak{m})$ be a Noetherian complete local ring and k a coefficient field of A. Let $\mathfrak{p}$ be a prime ideal of A, $\{f_1,\ldots,f_r\}$ a reference system of A in $\mathfrak{p}$ and $K := k((f_1,\ldots,f_r))$. Then $\Omega^f_{A_\mathfrak{p}/K}$ exists and*

$$\Omega^f_{A_\mathfrak{p}/K} = (\Omega^f_{A/k})_\mathfrak{p}/\left(\sum_{i=1}^r (d^f_{A/k})_\mathfrak{p}(f_i)A_\mathfrak{p}\right).$$

Proof We have to check that the canonical map

$$\delta_K : A_\mathfrak{p} \xrightarrow{(d^f_{A/k})_\mathfrak{p}} (\Omega^f_{A/k})_\mathfrak{p} \xrightarrow{\pi} (\Omega^f_{A/k})_\mathfrak{p}/\left(\sum_{i=1}^r (d^f_{A/k})_\mathfrak{p}(f_i)A_\mathfrak{p}\right)$$

is the universally finite K-derivation of $A_\mathfrak{p}$. Since $\delta_K(f_i) = 0$ for any $i = 1,\ldots,r$ and $\delta_K(a) = 0$, for any $a \in k$, it follows that δ_K is a K-derivation. Let now M be a finitely generated $A_\mathfrak{p}$-module and let $D \in \mathrm{Der}_K(A_\mathfrak{p}, M)$. If $i : A \to A_\mathfrak{p}$ is the canonical morphism, it is enough to prove that $(D \circ i)(A)$ generates a finite A-module, so that we can factorize $D \circ i$ through $\Omega^f_{A/k}$, hence through $(\Omega^f_{A/k})_\mathfrak{p}$ by localization, as in the following commutative diagrams:

$$A \xrightarrow{D \circ i} M \qquad\qquad A_\mathfrak{p} \xrightarrow{D} M$$
$$d^f_{A/k} \downarrow \quad \nearrow v \qquad\qquad (d^f_{A/k})_\mathfrak{p} \downarrow \quad \nearrow v_\mathfrak{p}$$
$$\Omega^f_{A/k} \qquad\qquad\qquad (\Omega^f_{A/k})_\mathfrak{p}$$

Indeed, in such a case we have $v_{\mathfrak{p}}((d^f_{A/k})_{\mathfrak{p}}(f_i)) = D(f_i) = 0$, hence $v_{\mathfrak{p}}$ factorizes, giving the following commutative diagram:

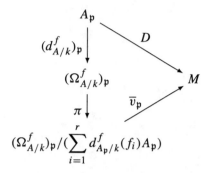

Set $\delta_K = \pi \circ (d^f_{A/k})_{\mathfrak{p}}$. This diagram realizes the factorization of D through δ_K, and it remains to prove that $\mathrm{Im}(D \circ i) = (D \circ i(A))A$ is a finite A-module. Let $\{g_1, \ldots, g_s\}$ be a system of generators of $\mathfrak{p}$. Then the ideal $\mathfrak{a} := (f_1, \ldots, f_r, g_1, \ldots, g_s)$ is $\mathfrak{m}$-primary, that is, there exists $t \in \mathbb{N}$, such that $\mathfrak{m}^t \subseteq \mathfrak{a}$. It follows that A is finite over $k[[f_1, \ldots, f_r, g_1, \ldots, g_s]]$. Let $B := k[[X_1, \ldots, X_r, Y_1, \ldots, Y_s]]$ and let $\phi : B \to A$ be the morphism of A-algebras given by

$$\phi(X_i) = f_i, \ i = 1, \ldots, r$$

$$\phi(Y_j) = g_j, \ j = 1, \ldots, s.$$

Since ϕ is finite, by Corollary 3.6.12 it induces a morphism $\Omega^f(\phi) : \Omega^f_{B/k} \to \Omega^f_{A/k}$. Moreover $\phi^{-1}(\mathfrak{p}) = \sum_{j=1}^{s} BY_j := \mathfrak{q}$ so that ϕ induces by localization a morphism $\psi :$ $B_{\mathfrak{q}} \to A_{\mathfrak{p}}$. Let us denote $E := (\Omega^f_{B/k})_{\mathfrak{q}}/(\sum_{i=1}^{r} B_{\mathfrak{q}}(d^f_{B/k})_{\mathfrak{q}}(X_i))$ and remember that $K := k((X_1, \ldots, X_r)) = k((f_1, \ldots, f_r))$. We have a morphism

$$\lambda : E \to (\Omega^f_{A/k})_{\mathfrak{p}}/\left(\sum_{i=1}^{r} A_{\mathfrak{p}}(d^f_{A/k})_{\mathfrak{p}}(f_i)\right),$$

which is naturally induced by $\Omega^f(\phi)$. Since $\Omega^f_{B/k}$ is a free B-module with basis $\{d^f_{B/k}(X_i), d^f_{B/k}(Y_j), \ 1 \leq i \leq r, \ 1 \leq j \leq s\}$, it follows that E is a free B-module with basis $\{(d^f_{B/k})_{\mathfrak{q}}(Y_j),\ 1 \le j \le s\}$. We can define a $B_{\mathfrak{q}}$-linear map

$$u : E = \bigoplus_{j=1}^{s} d^f_{B/k}(Y_j)B_{\mathfrak{q}} \to M$$

by

$$u\big(d^f_{B/k}(Y_j)\big) = D(g_j),\ j = 1, \ldots, s.$$

The diagram

$$
\begin{array}{ccc}
B_{\mathfrak{q}} & \xrightarrow{\ \psi\ } & A_{\mathfrak{p}} \\
{\scriptstyle (d^f_{B/k})_{\mathfrak{q}}}\big\downarrow & & \big\downarrow{\scriptstyle D} \\
E & \xrightarrow{\ u\ } & M
\end{array}
$$

is commutative. Indeed, $D \circ \psi = u \circ (\tilde{d}_{B/k})_{\mathfrak{q}}$ on K. Moreover, we have $D(\psi(Y_j)) = u((d^f_{B/k})_{\mathfrak{q}}((Y_j))$ for any $j = 1, \ldots, s$. As $Y_1, \ldots, Y_j$ is a system of parameters of $B_{\mathfrak{q}}$ and M is a separated $B_{\mathfrak{q}}$-module, it follows that $D \circ \psi = u \circ (d^f_{B/k})_{\mathfrak{q}}$. We consider the following commutative diagram:

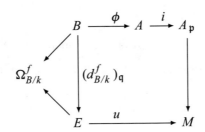

Let $w_1, \ldots, w_n$ be a system of generators of the B-module A. Then we have $A = \sum_{i=1}^{n} w_i \phi(B)$, hence

$$(D \circ i)(A) = (D \circ i)(\sum_{i=1}^{n} w_i \phi(B)) = \sum_{i=1}^{n}(D \circ i)\phi(B)w_j + \sum_{i=1}^{n} \phi(B)(D \circ i)(w_j),$$

where

$$(D \circ i \circ \phi)(B) = u \circ (d^f_{B/k})_\mathfrak{q}(B) = u\left(\bigoplus_{l=1}^{s} d^f_{B/k}(B)Y_l B\right) =$$

$$= \sum_{l=1}^{s} u \circ d^f_{B/k}(B)Y_l = \sum_{l=1}^{s} D(g_l)B.$$

It follows that $(D \circ i(A))A = \sum_{l=1}^{s} D(g_l)A + \sum_{j=1}^{n} D(i(w_j))A$, hence $(D \circ i)(A)A$ is a finite A-module. $\qquad\square$

We need more preparatory results.

Lemma 3.6.35 *Let $(A, \mathfrak{m}, K)$ be a complete Noetherian local domain containing a field of characteristic zero k. Assume that $[K : k] < \infty$, that $\Omega^f_{A/k}$ exists and let $\{f_1, \ldots, f_r\}$ be a system of parameters of A. Then:*

(i) $d^f_{A/k}(f_1), \ldots, d^f_{A/k}(f_r)$ are linearly independent over A;

(ii) $\Omega^f_{A/k}/\sum_{i=1}^{r} d^f_{A/k}(f_i)A$ is a torsion A-module;

(iii) $\mathrm{rk}(\Omega^f_{A/k}) = \dim(A)$.

Proof

(i) Let $B = k[[f_1, \ldots, f_r]]$ and assume that $\sum_{j=1}^{r} \lambda_j d^f_{A/k}(f_j) = 0$, where $\lambda_1, \ldots, \lambda_r \in A$. Let $\delta_i \in \mathrm{Der}_k(B)$ be the partial derivatives. These derivations can be extended to derivations $D_i \in \mathrm{Der}_k(Q(A))$ and moreover, since $D_i(B) \subseteq B$ and A is finite over B, there exists a non-zero $a \in A$, such that $(aD_i)(A) \subseteq A$. Hence $aD_i \in \mathrm{Der}_k(A)$, for $i = 1, \ldots, r$ and $(aD_i)(f_j) = a\delta_{ij}$, where δ_{ij} is the Kronecker delta. For any $i = 1, \ldots, r$ we consider the commutative diagram where u_i is given by the Definition 3.6.4, that is, by the universal property of the universally finite module of differentials. From $\sum_{j=1}^{r} \lambda_j d_{A/k}^f(f_j) = 0$ it follows that

$$0 = u_i \left(\sum_{j=1}^{r} \lambda_j d_{A/k}^f(f_j) \right) = \sum_{j=1}^{r} \lambda_j u_i \left(d_{A/k}^f(f_j) \right).$$

Hence $\sum_{j=1}^{r} \lambda_j a D_i(f_j) = 0$ and consequently $a\lambda_j = 0$ for $j = 1, \ldots, r$. It results that $\lambda_i = 0$, because $a \neq 0$ and A is a domain.

(ii) From the exact sequence

$$\Omega_{B/k} \otimes_B A \to \Omega_{A/k}^f \to \Omega_{A/B}^f \to 0$$

we get that $\Omega_{A/B}^f \cong \Omega_{A/k}^f / (\sum_{i=1}^{r} d_{A/k}^f(f_i)A)$. On the other hand, since A is finite over B we have that $\Omega_{A/B}^f = \Omega_{A/B}$. But the field extension $Q(B) \subseteq Q(A)$ is a finite algebraic one and then

$$\Omega_{A/B} \otimes_A Q(A) = \Omega_{Q(A)/B} = \Omega_{Q(A)/Q(B)} = 0.$$

It follows that $\Omega_{A/B} \cong \Omega_{A/k}^f / \left(\sum_{i=1}^{r} d_{A/k}^f(f_i)A \right)$ is a torsion module.

(iii) Follows from (i) and (ii). $\square$

Lemma 3.6.36 *Let $(A, \mathfrak{m}, K)$ be a complete Noetherian local ring containing a field k of characteristic zero. Assume that K is an algebraic extension of k and that $\Omega_{A/k}^f$ exists. Let $\mathfrak{p}$ be a prime ideal of A and let $f_1, \ldots, f_r \in A$ be a reference system for $\mathfrak{p}$. Then $\{d_{A/k}^f(f_1), \ldots, d_{A/k}^f(f_r)\}$ are a part of a minimal system of generators of $(\Omega_{A/k}^f)_\mathfrak{p}$.*

Proof We have to prove that the images $\overline{d_{A/k}^f(f_i)} \in (\Omega_{A/k}^f)_\mathfrak{p}/\mathfrak{p}(\Omega_{A/k}^f)_\mathfrak{p}$ are linearly independent over $k(\mathfrak{p}) = A_\mathfrak{p}/\mathfrak{p}A_\mathfrak{p}$. If we have a relation

$$\sum_{i=1}^{r} a_i d_{A/k}^f(f_i) \in \mathfrak{p}(\Omega_{A/k}^f)_\mathfrak{p} \text{ with } a_1, \ldots. a_r \in A_\mathfrak{p},$$

there exists $s \in A \setminus \mathfrak{p}$ such that

$$s\left(\sum_{i=1}^{r} a_i d^f_{A/k}(f_i)\right) \in \mathfrak{p}\Omega^f_{A/k}.$$

On the other hand, from Proposition 3.6.14 it follows that

$$\Omega^f_{(A/\mathfrak{p})/k} \cong \Omega^f_{A/k}/(\mathfrak{p}\Omega^f_{A/k} + d^f_{A/k}(\mathfrak{p})A).$$

Then, taking classes modulo $\mathfrak{p}\Omega^f_{A/k} + d^f_{A/k}(\mathfrak{p})A$ we get that $\sum_{i=1}^{r} \overline{sa_i} d^f_{A/k}(f_i) = 0$
in $\Omega^f_{(A/\mathfrak{p})/k}$. From Lemma 3.6.35 we have that $d^f_{A/k}(f_1), \ldots, d^f_{A/k}(f_r)$ are linearly independent in $\Omega^f_{(A/\mathfrak{p})/k}$. Hence $\overline{sa_i} = 0$, for $i = 1, \ldots, r$ or in other words $sa_1, \ldots, sa_r \in \mathfrak{p}$. Therefore $a_1, \ldots, a_r \in \mathfrak{p}A_\mathfrak{p}$.   □

Now we can prove the promised Jacobian criterion.

Theorem 3.6.37 (Scheja-Storch Regularity Criterion (Scheja and Storch [340, Th. 8.7])) *Let A be a Noetherian ring containing a field k of characteristic zero. Assume that $\Omega^f_{A/k}$ exists and let $\mathfrak{p} \in \mathrm{Spec}(A)$. The following are equivalent:*

(i) $A_\mathfrak{p}$ is a regular local ring;
(ii) $(\Omega^f_{A/k})_\mathfrak{p}$ is a free $A_\mathfrak{p}$-module.

Proof Step 1: *We can assume that A is a local ring.* Indeed, by Proposition 3.6.27 if $\mathfrak{m}$ is a maximal ideal of A containing $\mathfrak{p}$, then $A_\mathfrak{m}$ satisfies the same conditions as A does and $(\Omega^f_{A_\mathfrak{m}/k})_\mathfrak{p} = ((\Omega^f_{A/k})_\mathfrak{m})_\mathfrak{p} = (\Omega^f_{A/k})_\mathfrak{p}$.
Step 2: *We may reduce ourselves to the case A is a complete local ring.* Thus assume that the assertion is true for a complete Noetherian local ring. Let us first prove that, in this case, if $\mathfrak{p}$ is any prime ideal of A, then the completion $\widehat{A/\mathfrak{p}}$ is reduced. Set $B := A/\mathfrak{p}$. Then $\Omega^f_{\widehat{B}/k}$ exists. In order to prove that $\widehat{B}$ is reduced, we have to prove that for any $\mathfrak{q} \in \mathrm{Ass}(\widehat{B})$, the ring $\widehat{B}_\mathfrak{q}$ is a regular local ring. Therefore by assumption it is enough to show that $(\Omega^f_{\widehat{B}/k})_\mathfrak{q}$ is free over $\widehat{B}_\mathfrak{q}$. But

$$(\Omega^f_{\widehat{B}/k})_\mathfrak{q} \cong \Omega^f_{\widehat{B}/k} \otimes_B \widehat{B} \otimes_{\widehat{B}} \widehat{B}_\mathfrak{q} \cong \Omega^f_{\widehat{B}/k} \otimes_B \widehat{B}_\mathfrak{q}.$$

By flatness $\mathfrak{q} \cap B = (0)$, hence all non-zero elements of B are invertible in $\widehat{B}_\mathfrak{q}$. This means that $\widehat{B}_\mathfrak{q}$ is a $Q(B)$-module, hence

$$\Omega^f_{\widehat{B}/k} \otimes_B \widehat{B}_\mathfrak{q} \cong (\Omega^f_{\widehat{B}/k} \otimes_B Q(B)) \otimes_{Q(B)} \widehat{B}_\mathfrak{q}.$$

Since $\Omega^f_{B/k} \otimes_B Q(B)$ is a $Q(B)$-vector space, it follows that

$$(\Omega^f_{\widehat{B}/k})_{\mathfrak{q}} \cong \Omega^f_{B/k} \otimes_B Q(B) \otimes_{Q(B)} \widehat{B}_{\mathfrak{q}}$$

is a free $\widehat{B}_{\mathfrak{q}}$-module. Thus the ring $\widehat{B}$ is reduced. Therefore, for any $\mathfrak{p} \in \mathrm{Spec}(A)$ and for any $\mathfrak{n} \in \mathrm{Min}(\mathfrak{p}\widehat{A})$, we have $\mathfrak{p}\widehat{A}_{\mathfrak{n}} = \mathfrak{n}\widehat{A}_{\mathfrak{n}}$. We have

$$\mathfrak{n}\widehat{A}_{\mathfrak{n}}/\mathfrak{n}^2\widehat{A}_{\mathfrak{n}} = \mathfrak{p}\widehat{A}_{\mathfrak{n}}/\mathfrak{p}^2\widehat{A}_{\mathfrak{n}} = (\mathfrak{p}/\mathfrak{p}^2) \otimes_A \widehat{A} \otimes_{\widehat{A}} \widehat{A}_{\mathfrak{n}} = \mathfrak{p}/\mathfrak{p}^2 \otimes_A \widehat{A}_{\mathfrak{p}} =$$

$$= \mathfrak{p}/\mathfrak{p}^2 \otimes_A A_{\mathfrak{p}} \otimes_{A_{\mathfrak{p}}} \widehat{A}_{\mathfrak{n}} = \mathfrak{p}A_{\mathfrak{p}}/\mathfrak{p}^2 A_{\mathfrak{p}} \otimes_{A_{\mathfrak{p}}} \widehat{A}_{\mathfrak{n}} =$$

$$= \mathfrak{p}A_{\mathfrak{p}}/\mathfrak{p}^2 A_{\mathfrak{p}} \otimes_{k(\mathfrak{p})} k(\mathfrak{p}) \otimes_{A_{\mathfrak{p}}} \widehat{A}_{\mathfrak{n}}.$$

But

$$k(\mathfrak{n}) \cong \widehat{A}_{\mathfrak{n}}/\mathfrak{n}\widehat{A}_{\mathfrak{n}} \cong \widehat{A}_{\mathfrak{n}}/\mathfrak{p}\widehat{A}_{\mathfrak{n}} \cong A_{\mathfrak{p}}/\mathfrak{p}A_{\mathfrak{p}} \otimes_{A_{\mathfrak{p}}} \widehat{A}_{\mathfrak{n}} \cong k(\mathfrak{p}) \otimes_{A_{\mathfrak{p}}} \widehat{A}_{\mathfrak{n}}.$$

It follows that $\mathrm{edim}(A_{\mathfrak{p}}) = \mathrm{edim}(\widehat{A}_{\mathfrak{n}})$. Since by faithful flatness we have also $\dim(A_{\mathfrak{p}}) = \dim(\widehat{A}_{\mathfrak{n}})$, we get that $A_{\mathfrak{p}}$ is regular if and only if $\widehat{A}_{\mathfrak{n}}$ is regular. On the other hand, since $A_{\mathfrak{p}} \to \widehat{A}_{\mathfrak{n}}$ is faithfully flat, by Proposition 3.6.23 and [248, (4.E), ii)] it results that $(\Omega^f_{A/k})_{\mathfrak{p}}$ is free over $A_{\mathfrak{p}}$ if and only if $(\Omega^f_{A/k})_{\mathfrak{n}}$ is free over $\widehat{A}_{\mathfrak{n}}$. Finally, it follows that we may assume A complete.

Step 3: *We may furthermore assume that k is a coefficient field of A.* Indeed, let $\tilde{k}$ be a coefficient field of A containing k, let $\{a_1, \ldots, a_t\}$ be a transcendence basis of $\tilde{k}$ over k and let $L := k(a_1, \ldots, a_t)$. It is enough to show that we can replace k by L, because any L-derivation is also a $\tilde{k}$-derivation. Since by Proposition 3.6.13 it follows that $\Omega^f_{A/L}$ exists and moreover $\Omega^f_{A/L} \cong \Omega^f_{A/k}/(\sum_{i=1}^t Ad^f_{A/k}(a_i))$, actually we have to prove the following two assertions:

(a) If $(\Omega^f_{A/k})_{\mathfrak{p}}$ is free over $A_{\mathfrak{p}}$, then also $(\Omega^f_{A/L})_{\mathfrak{p}}$ is free over $A_{\mathfrak{p}}$;
(b) If $(\Omega^f_{A/L})_{\mathfrak{p}}$ is free over $A_{\mathfrak{p}}$ and $A_{\mathfrak{p}}$ is regular, then $(\Omega^f_{A/k})_{\mathfrak{p}}$ is free over $A_{\mathfrak{p}}$.

Proof of (a): Since

$$\Omega^f_{A/L} \cong \Omega^f_{A/k}/\left(\sum_{i=1}^t Ad^f_{A/k}(a_i)\right),$$

by localization we get

$$(\Omega_{A/L}^f)_{\mathfrak{p}} \cong (\Omega_{A/k}^f)_{\mathfrak{p}}/\Big(\sum_{i=1}^{t} A_{\mathfrak{p}}d_{A/k}^f(a_i)\Big).$$

From Lemma 3.6.31, (ii) it results that $d_{A/k}^f(a_1), \ldots, d_{A/k}^f(a_t)$ is a part of a minimal system of generators of $(\Omega_{A/k}^f)_{\mathfrak{p}}$ over $A_{\mathfrak{p}}$. This proves a).

Proof of (b): Since $(\Omega_{A/k}^f)_{\mathfrak{p}}$ is free, we have

$$(\Omega_{A/k}^f)_{\mathfrak{p}} \cong (\Omega_{A/L}^f)_{\mathfrak{p}} \oplus \Big(\sum_{i=1}^{t} A_{\mathfrak{p}}d_{A/k}^f(a_i)\Big).$$

Therefore it is enough to show that $d_{A/k}^f(a_1), \ldots, d_{A/k}^f(a_t)$ are linearly independent over $A_{\mathfrak{p}}$. Note first that from Lemma 3.6.31, (ii) applied to all minimal prime ideals of A, we obtain that if we have a relation $\sum_{i=1}^{t} b_i d_{A/k}^f(a_i) = 0$ with $b_1, \ldots, b_t \in A$, it follows that $b_1, \ldots, b_t$ are nilpotent elements of A. So, if $\sum_{i=1}^{t} b_i d_{A/k}^f(a_i) = 0$ with $b_1, \ldots, b_t \in A_{\mathfrak{p}}$, it follows that $b_1, \ldots, b_t$ are nilpotent elements of $A_{\mathfrak{p}}$. But $A_{\mathfrak{p}}$ is a domain, being a regular local ring, hence $b_1 = \ldots = b_t = 0$. This proves the assertion b).

So finally we may assume that $(A, \mathfrak{m})$ is a complete local ring containing a coefficient field k of characteristic 0 and such that $\Omega_{A/k}^f$ exists.

$(ii) \Rightarrow (i)$: If $\mathfrak{p} = \mathfrak{m}$, the assertion follows by Lemma 3.6.30. Assume now that $\mathfrak{p} \subsetneq \mathfrak{m}$ and let $K := k((f_1, \ldots, f_r))$ where $\{f_1, \ldots, f_r\}$ is a reference system for $\mathfrak{p}$. By Lemma 3.6.33 the field extension $K \subseteq k(\mathfrak{p})$ is an algebraic one. From Lemma 3.6.36 it follows that since $(\Omega_{A/k}^f)_{\mathfrak{p}}$ is $A_{\mathfrak{p}}$-free, then $\Omega_{A_{\mathfrak{p}}/K}^f$ is $A_{\mathfrak{p}}$-free. Finally from Lemma 3.6.30 it follows that $A_{\mathfrak{p}}$ is a regular local ring.

$(i) \Rightarrow (ii)$: If $\mathfrak{p} = \mathfrak{m}$, the assertion follows again by Lemma 3.6.30. Assume now that $\mathfrak{p} \subsetneq \mathfrak{m}$. From the regularity of $A_{\mathfrak{p}}$ it follows that $\Omega_{A_{\mathfrak{p}}/K}^f$ is free, hence

$$(\Omega_{A/k}^f)_{\mathfrak{p}} \cong (\Omega_{A_{\mathfrak{p}}/K}^f) \oplus \Big(\sum_{i=1}^{r} A_{\mathfrak{p}}d_{A/k}^f(f_i)\Big).$$

Moreover, from Corollary 3.6.20 we get that $\mu(\Omega_{A_{\mathfrak{p}}/K}^f) = \dim(A_{\mathfrak{p}})$. It follows that

$$\mu\big((\Omega_{A/k}^f)_{\mathfrak{p}}\big) \leq \operatorname{ht}(\mathfrak{p}) + r = \operatorname{ht}(\mathfrak{p}) + \dim(A/\mathfrak{p}).$$

Because $A_\mathfrak{p}$ is a regular local ring, it follows that $\mathfrak{p}$ contains only one minimal prime, say $\mathfrak{q}$. Since A is complete, it results that $A/\mathfrak{q}$ is catenary, hence

$$\operatorname{ht}(\mathfrak{p}) + \dim(A/\mathfrak{p}) = \dim(A/\mathfrak{q}).$$

We have also

$$\operatorname{rk}_{A_\mathfrak{p}}\left((\Omega^f_{A/k})_\mathfrak{p}\right) = \operatorname{rk}_{A_\mathfrak{q}}\left((\Omega^f_{A/k})_\mathfrak{q}\right).$$

But $A_\mathfrak{q}$ is a field being a regular local ring of dimension 0, hence $(\Omega^f_{A/k})_\mathfrak{q}$ is free over $A_\mathfrak{q}$, so that

$$\operatorname{rk}_{A_\mathfrak{q}}\left((\Omega^f_{A/k})_\mathfrak{q}\right) = \mu\left((\Omega^f_{A/k})_\mathfrak{q}\right).$$

From Lemma 3.6.36 we have

$$\mu\left((\Omega^f_{A/k})_\mathfrak{q}\right) \geq \dim(A/\mathfrak{q}).$$

Finally we obtain

$$\operatorname{rk}_{A_\mathfrak{p}}\left((\Omega^f_{A/k})_\mathfrak{p}\right) \geq \dim(A/\mathfrak{q}),$$

hence

$$\operatorname{rk}_{A_\mathfrak{p}}\left((\Omega^f_{A/k})_\mathfrak{p}\right) = \mu\left((\Omega^f_{A/k})_\mathfrak{p}\right),$$

so that $(\Omega^f_{A/k})_\mathfrak{p}$ is a free $A_\mathfrak{p}$-module. $\square$

Theorem 3.6.38 *Let A be a Noetherian local domain containing a field k of characteristic zero. If $\Omega^f_{A/k}$ exists, then $\widehat{A}$ is reduced and equidimensional.*

Proof In order to prove that $\widehat{A}$ is reduced we need to show that if $\mathfrak{q} \in \operatorname{Ass}(\widehat{A})$, then $\widehat{A}_\mathfrak{q}$ is a regular local ring. Applying Theorem 3.6.37 it is enough to show that $(\Omega^f_{\widehat{A}/k})_\mathfrak{q}$ is free over $\widehat{A}_\mathfrak{q}$. But

$$(\Omega^f_{\widehat{A}/k})_\mathfrak{q} \cong \Omega^f_{\widehat{A}/k} \otimes_A \widehat{A} \otimes_{\widehat{A}} \widehat{A}_\mathfrak{q} \cong \Omega^f_{A/k} \otimes_A \widehat{A}_\mathfrak{q}.$$

By flatness, $q \cap A = (0)$, hence all elements of A are invertible in $\widehat{A}_q$. This means that $\widehat{A}_q$ is a $Q(A)$-module, hence

$$\Omega^f_{A/k} \otimes_A \widehat{A}_q \cong (\Omega^f_{A/k} \otimes_A Q(A)) \otimes_{Q(A)} \widehat{A}_q.$$

Since $\Omega^f_{A/k} \otimes_A Q(A)$ is a $Q(A)$-vector space, it follows that

$$(\Omega^f_{\widehat{A}/k})_q \cong \Omega^f_{A/k} \otimes_A Q(A) \otimes_{Q(A)} \widehat{A}_q$$

is a free $\widehat{A}_q$-module. Consequently $\widehat{A}$ is reduced. In order to show that $\widehat{A}$ is equidimensional, let $q \in \mathrm{Min}(\widehat{A})$ and $\tilde{k}$ a coefficient field of $\widehat{A}$. Let $f_1, \ldots, f_r$ be a reference system of $\widehat{A}$ with respect to q and $K := k((f_1, \ldots, f_r))$. Then by Lemma 3.6.34 we have that

$$\Omega^f_{\widehat{A}_q/K} \cong (\Omega^f_{\widehat{A}/k})_q / (\sum_{i=1}^{r} d^f_{\widehat{A}/k}(f_i)\widehat{A}_q).$$

But $\widehat{A}_q$ is a field, which is algebraic over K by Lemma 3.6.33. Then $\Omega^f_{\widehat{A}_q/K} = (0)$, hence

$$(\Omega^f_{\widehat{A}/\tilde{k}})_q = \sum_{i=1}^{r} \widehat{A}_q d^f_{A/k}(f_i).$$

By Lemma 3.6.35 the elements $d^f_{A/k}(f_1), \ldots, d^f_{A/k}(f_r) \in (\Omega^f_{\widehat{A}/\tilde{k}})_q$ are linearly independent over $\widehat{A}_q$. It follows that for any minimal prime ideal q of $\widehat{A}$ we have

$$\mathrm{rk}_{\widehat{A}}(\Omega^f_{\widehat{A}/\tilde{k}}) = \mathrm{rk}_{\widehat{A}_q}(\Omega^f_{\widehat{A}/\tilde{k}})_q = r = \dim(\widehat{A}/q).$$

$\square$

Now we are ready to prove the main result of this section.

Theorem 3.6.39 (Scheja and Storch [340, Prop. 8.10]) *Let A be a Noetherian ring containing a field k of characteristic zero. If $\Omega^f_{A/k}$ exists, then A is an excellent ring.*

Proof Let $\mathfrak{p} \in \mathrm{Spec}(A)$ and $B := A/\mathfrak{p}$. Then B satisfies the same conditions that A does. Since from the Regularity Criterion 3.6.37 we have

$$\mathrm{Reg}(A) = \{\mathfrak{p} \in \mathrm{Spec}(A) \mid (\Omega^f_{A/k})_\mathfrak{p} \text{ is free}\}$$

and $\Omega^f_{A/k}$ is a finite A-module, it follows that $\text{Reg}(A)$ is open. As $\text{char}(A) = 0$, we get that A is Reg-2. From Theorem 3.6.38 and Theorem 3.1.35, we obtain that A is universally catenary. In order to show that the formal fibres of A are geometrically regular, as $\text{char}(A) = 0$, it is enough to show that, for any $\mathfrak{m} \in \text{Max}(A)$ and for any $\mathfrak{q} \in \text{Spec}(\widehat{A_\mathfrak{m}})$ such that $\mathfrak{q} \cap A_\mathfrak{m} = \mathfrak{p}$, the ring $(\widehat{A_\mathfrak{m}})_\mathfrak{q}/\mathfrak{p}(\widehat{A_\mathfrak{m}})_\mathfrak{q}$ is regular. But also $A_\mathfrak{m}$ satisfies the same assumptions that A does; hence it is enough to prove that for any $\mathfrak{q} \in \text{Spec}(\widehat{A})$, if $\mathfrak{p} = \mathfrak{q} \cap A$, then the ring $\widehat{A}_\mathfrak{q}/\mathfrak{p}\widehat{A}_\mathfrak{q}$ is regular. But $\widehat{A}_\mathfrak{q}/\mathfrak{p}\widehat{A}_\mathfrak{q} = \widehat{(A/\mathfrak{p})}_\mathfrak{q}$ and $A/\mathfrak{p}$ satisfies the same assumptions as A, so that, changing A with $A/\mathfrak{p}$ it is enough to prove that if A is a domain and $\mathfrak{q}$ is a prime ideal of $\widehat{A}$ such that $\mathfrak{q} \cap A = (0)$, then $\widehat{A}_\mathfrak{q}$ is a regular local ring. By Theorem 3.6.37 it is enough to prove that $(\Omega^f_{\widehat{A}/k})_\mathfrak{q}$ is a free $\widehat{A}_\mathfrak{q}$-module. We have

$$(\Omega^f_{\widehat{A}/k})_\mathfrak{q} \cong \Omega^f_{A/k} \otimes_A Q(A) \otimes_{Q(A)} \widehat{A}_\mathfrak{q}.$$

But $\Omega^f_{A/k} \otimes_A Q(A)$ is a free $Q(A)$-vector space and the map $Q(A) \to \widehat{A}_\mathfrak{q}$ is faithfully flat. It follows that $(\Omega^f_{\widehat{A}/k})_\mathfrak{q}$ is a free $\widehat{A}_\mathfrak{q}$-module and we are done.   □

In order to prove similar results in positive characteristic, we need more preparations. First let us recall the following notion [131, §20.6].

Definition 3.6.40 ([131, Def. 20.6.1]) Consider $A \to B \to C$ ring morphisms and the associated Jacobi-Zariski exact sequence

$$\Omega_{B/A} \otimes_B C \xrightarrow{\delta} \Omega_{C/A} \to \Omega_{C/B} \to 0.$$

Then $\ker(\delta)$ is called the **module of imperfection** of the B-algebra C over A and is denoted by $\gamma_{C/B/A}$. If $A = \mathbb{Z}$ is the ring of integers, then we put $\gamma_{C/B} := \gamma_{C/B/\mathbb{Z}}$.

The definition above is given in the general situation. We will need here only the case when all ring morphisms are actually field extensions.

Remark 3.6.41

(i) If B and C are algebras over the prime field $\mathbb{P}$, then $\gamma_{C/B/\mathbb{P}} = \gamma_{C/B}$.

(ii) If $k \subseteq K \subseteq L$ are field extensions and $k \subseteq K$ is separable, then $\gamma_{L/K/k} = H_1(K, L, L)$, the André-Quillen homology module, because $H_2(k, K, L) = (0)$. In particular $\gamma_{L/K} = H_1(K, L, L)$.

(iii) [131, 21.6.1] If $k_0 \subseteq k \subseteq K \subseteq L$ are field extensions, then we have an exact sequence

$$0 \to \gamma_{K/k/k_0} \otimes_K L \xrightarrow{v} \gamma_{L/k/k_0} \xrightarrow{u} \gamma_{L/K/k_0} \xrightarrow{s} \gamma_{L/K/k} \to 0.$$

Definition 3.6.42 ([131, 21.6.1]) Let $k_0 \subseteq k \subseteq K \subseteq L$ be field extensions. With the above notations, we say that k is k_0-**admissible** for L over K if the morphism s in the above sequence is bijective. If k_0 is the prime field we simply say that k is **admissible** for L over K.

We shall need the following result from [131].

Lemma 3.6.43 ([131, Cor. 21.7.2]) *Let $k \subseteq K \subseteq L$ be field extensions such that L is finitely generated over K. Then:*

(i) $\Omega_{L/K}$ and $\gamma_{L/K/k}$ are finite dimensional L-vector spaces;
(ii) $\mathrm{rk}_L(\Omega_{L/K}) - \mathrm{rk}_L(\gamma_{L/K/k}) \geq \mathrm{tr.deg}_K L;$
(iii) $\mathrm{rk}_L(\Omega_{L/K}) - \mathrm{rk}_L(\gamma_{L/K/k}) = \mathrm{tr.deg}_K L$ if and only if k is admissible for L over K.

Lemma 3.6.44 *Let $k \subseteq K$ be a field extension and $A = K[[X_1, \ldots, X_n]]$. The following are equivalent:*

(i) $\Omega^f_{A/k}$ exists;
(ii) $\dim_K(\Omega_{K/k}) < \infty;$
(iii) $\mathrm{char}(k) = 0$ and $\mathrm{tr.deg}_k K < \infty$ or $\mathrm{char}(k) = p > 0$ and $[K : k(K^p)] < \infty$.

If this is the case, $\Omega^f_{A/k}$ is a free A-module and $\mathrm{rk}(\Omega^f_{A/k}) = n + \dim_K(\Omega_{K/k})$.

Proof $(i) \Rightarrow (ii)$: Follows from Corollary 3.6.15.
$(ii) \Rightarrow (iii) \Rightarrow (i)$: Follow from well-known facts about differentials. $\qquad \square$

Proposition 3.6.45 *Let k be a field and $(A, \mathfrak{m}, K)$ be a Noetherian local k-algebra. Assume that:*

(i) $\Omega^f_{A/L}$ exists for any cofinite subfield $L \subseteq k$;
(ii) $\mathrm{rk}_K(\gamma_{K/k}) < \infty$.

Then there exists a subfield $N \subseteq k$ and a quasi-coefficient field M of A containing N such that $\Omega^f_{A/M}$ exists and the conditions (i) and (ii) hold for N instead of k.

Proof Assume first that $\mathrm{char}(k) = 0$. From [249, Th. 28.3] it follows that A has a quasi-coefficient field containing k. Now we apply Proposition 3.6.13.
Consider now the case $\mathrm{char}(k) = p > 0$. Let B be a p-basis of k. From (ii) it follows that there exists a finite subset F of B such that, if we put $G := B \setminus F$ and $N = \mathbb{F}_p(G)$, the field extension $N \subseteq K$ is separable. By [249, Th. 28.3] it follows that A has a quasi-coefficient field M containing N. Let $k' := k^p(N) = k^p(G)$. Since $[k : k'] < \infty$, we

get that $\Omega^f_{A/k'}$ exists. Since $\Omega^f_{A/N} = \Omega^f_{A/k'}$, by Proposition 3.6.13 we obtain that $\Omega^f_{A/M}$ exists. Let $k_1 \subseteq N$ be such that $[N : k_1] < \infty$. Then we have $[k^p(N) : k^p(k_1)] < \infty$, hence $[k : k^p(k_1)] < \infty$. This implies that $\Omega^f_{A/k^p(k_1)} = \Omega^f_{A/k_1}$ exists. $\qquad\square$

For 3.6.46–3.6.50, we fix the following setting: let K be a field of characteristic $p > 0$ and k be a subfield such that $\dim_K \Omega_{K/k} < \infty$. We consider $A = K[[X_1, \ldots, X_n]]$ and let $\mathfrak{p}$ be a prime ideal of A. Put $B = A/\mathfrak{p}$. If $d = \dim(B)$, there exists a k-subalgebra $B_1 = K[[T_1, \ldots, T_d]]$ such that B is a finite B_1-algebra. Let also $B_0 = k(K^p)[[T_1^p, \ldots, T_d^p]]$.

Proposition 3.6.46 *With the above notations $\Omega^f_{B/k}$ exists and $\Omega^f_{B/k} \cong \Omega_{B/B_0}$.*

Proof By Lemma 3.6.44 we know that $\Omega^f_{B/k}$ exists. Since $\dim_K \Omega_{K/k} < \infty$ it results that B is finite over B_0, hence Ω_{B/B_0} is a finite B-module. Let N be an arbitrary finite B-module. Then N is separated in the $\mathfrak{m}_B$-adic topology and $T_i \in \mathfrak{m}_B$, hence any k-derivation of B into N vanishes on B_0. This means that

$$\mathrm{Hom}_B(\Omega^f_{B/k}, N) \cong \mathrm{Der}_k(B, N) \cong \mathrm{Hom}_B(\Omega_{B/B_0}, N).$$

This implies that $\Omega^f_{B/k} \cong \Omega_{B/B_0}$. $\qquad\square$

Remark 3.6.47 If $[K : k] < \infty$, then we can replace B_0 in the above proof by $k[[T_1^p, \ldots, T_d^p]]$.

Corollary 3.6.48 *Let $M = Q(B)$ and $M_0 = Q(B_0)$. Then*

$$\mathrm{rk}_B(\Omega^f_{B/k}) = \dim_M(\Omega_{M/M_0}).$$

Corollary 3.6.49 *Let $M_1 := Q(B_1)$. The following are equivalent:*

(i) M_0 is admissible for M over M_1;
(ii) $\mathrm{rk}_B(\Omega^f_{B/k}) = \dim(B) + \dim_K(\Omega_{K/k})$.

Proof Let F be a p-basis of K over k. Then $F \cup \{T_1, \ldots, T_d\}$ is a p-basis of B_1 over B_0. Thus $\dim_{M_1}(\Omega_{M_1/M_0}) = d + \dim_K(\Omega_{K/k})$. We have the exact sequence

$$0 \to \gamma_{M/M_1/M_0} \to \Omega_{M_1/M_0} \otimes_{M_1} M \to \Omega_{M/M_0} \to \Omega_{M/M_1} \to 0.$$

Since $[M : M_1] < \infty$, the condition (i) is equivalent to the equality

$$\dim_{M_1}(\Omega_{M_1/M_0}) = \dim_M(\Omega_{M/M_0}).$$

But Corollary 3.6.48 shows that this is equivalent to condition (ii).   □

Proposition 3.6.50 (Yamauchi [395, Prop. 1.8]) *Assume that* $\mathrm{rk}_K (\gamma_{K/k}) < \infty$. *Then there exists a cofinite subfield* k' *of* k *such that for any cofinite subfield* $k'' \subseteq k'$, *we have*

$$\mathrm{rk}_B \Omega^f_{B/k''} = \dim(B) + \dim_K \Omega_{K/k''}.$$

Proof As above set $M = Q(B)$ and $M_1 = Q(B_1)$. Consider $(k_\alpha)_{\alpha \in I}$ a downward directed family of cofinite subfields of k with $\bigcap_{\alpha \in I} k_\alpha = k^p$. Since $\dim_K \gamma_{K/k} < \infty$, from [131, Th. 21.8.3] we obtain $\bigcap_{\alpha \in I} k_\alpha(K^p) = K^p$. Let $M_\alpha = k_\alpha(K^p)((T_1^p, \ldots, T_d^p))$. Then from [248, (30.E), Prop.] it follows that $\bigcap_{\alpha \in I} M_\alpha = K^p((T_1^p, \ldots, T_d^p)) = M_1^p$. Since $[M : M_1] < \infty$ we get $\dim_M(\gamma_{M/M_1}) < \infty$ and again by [131, Th. 21.8.3] there exists $\alpha \in I$ such that M_α is admissible for M over M_1. If k'' is a cofinite subfield of k_α, then $M'' = k''(K^p)((T_1^p, \ldots, T_d^p))$ is also admissible for M over M_1 by [131, Cor.21.6.5]. By Corollary 3.6.49 we have

$$\mathrm{rk}_B \Omega^f_{B/k''} = \dim(B) + \dim_K \Omega_{K/k''}$$

and we can take $k' = k_\alpha$.   □

Lemma 3.6.51 (Yamauchi [395, Lemma 1.9]) *Let* k *be a field and* A *be a Noetherian local* k-*algebra which is a domain, such that* $\Omega^f_{A/k}$ *exists and* $\widehat{A}$ *is reduced. Then for any* $\mathfrak{p} \in \mathrm{Ass}(\widehat{A})$ *we have*

$$\mathrm{rk}_A \Omega^f_{A/k} = \mathrm{rk}_{\widehat{A}/\mathfrak{p}} \Omega^f_{(\widehat{A}/\mathfrak{p})/k}.$$

Proof From Proposition 3.6.13 we obtain by tensorizing with $k(\mathfrak{p}) = A_\mathfrak{p}/\mathfrak{p}A_\mathfrak{p}$ an exact sequence

$$\mathfrak{p}/\mathfrak{p}^2 \otimes_{\widehat{A}} k(\mathfrak{p}) \to \Omega^f_{\widehat{A}/k} \otimes_{\widehat{A}} k(\mathfrak{p}) \to \Omega^f_{(\widehat{A}/\mathfrak{p})/k} \otimes_{\widehat{A}/\mathfrak{p}} k(\mathfrak{p}) \to 0.$$

But $\widehat{A}$ is reduced and $\mathfrak{p} \in \mathrm{Ass}(\widehat{A})$, hence $\widehat{A}_\mathfrak{p}$ is a field, so that $k(\mathfrak{p}) = \widehat{A}_\mathfrak{p}$. Then

$$\Omega^f_{\widehat{A}/k} \otimes_{\widehat{A}} k(\mathfrak{p}) \cong \Omega^f_{(\widehat{A}/\mathfrak{p})/k} \otimes_{\widehat{A}/\mathfrak{p}} k(\mathfrak{p}).$$

It follows that

$$\mathrm{rk}_A \Omega^f_{A/k} = \mathrm{rk}_{k(\mathfrak{p})} \Omega^f_{\widehat{A}/k} \otimes_{\widehat{A}} k(\mathfrak{p}) = \mathrm{rk}_{\widehat{A}/\mathfrak{p}} \Omega^f_{(\widehat{A}/\mathfrak{p})/k}.$$

□

Theorem 3.6.52 (Yamauchi [395, Th. 1.10, Cor. 1.11]) *Let A be a Noetherian local domain of characteristic $p > 0$, let K be a quasi-coefficient field of A and k a subfield of K. Assume that:*

(i) for any cofinite subfield k_1 of k the module Ω^f_{A/k_1} exists;
(ii) $\mathrm{rk}_K(\gamma_{K/k}) < \infty$;
(iii) $\widehat{A}$ is reduced.

Then there exists a cofinite subfield k' of k such that for any cofinite subfield k'' of k' we have $\mathrm{rk}_A \Omega^f_{A/k''} = \dim(A) + \mathrm{rk}_K \Omega_{K/k''}$ and A is formally equidimensional.

Proof Let K^* be a coefficient field of $\widehat{A}$ containing K. For any subfield $K' \subseteq K$ we have

$$\gamma_{K/K'} \otimes_K K^* \cong \gamma_{K^*/K'} \text{ and } \Omega_{K/K'} \otimes_K K^* \cong \Omega_{K^*/K'}.$$

In particular $\mathrm{rk}_{K^*}(\gamma_{K^*/k}) < \infty$. Let k_1 be a cofinite subfield of K and let $\mathfrak{p}$ be a prime ideal in $\mathrm{Ass}(\widehat{A})$ such that $\dim(\widehat{A}) = \dim(\widehat{A}/\mathfrak{p})$. Then by Lemma 3.6.51 we have

$$\mathrm{rk}_A(\Omega^f_{A/k_1}) = \mathrm{rk}_{\widehat{A}/\mathfrak{p}}(\Omega^f_{(\widehat{A}/\mathfrak{p})/k_1}).$$

The assertion follows now from Proposition 3.6.50.

In order to prove the second assertion, let $\mathrm{Ass}(\widehat{A}) = \{\mathfrak{p}_1, \dots, \mathfrak{p}_r\}$. From the first part it follows that there exists a cofinite subfield k' of k such that

$$\mathrm{rk}_{\widehat{A}/\mathfrak{p}_i}(\Omega^f_{(\widehat{A}/\mathfrak{p}_i)/k'}) = \dim(\widehat{A}/\mathfrak{p}_i) + \mathrm{rk}_K(\Omega_{K/k'}), \ \forall i = 1, \dots, r.$$

But by Lemma 3.6.51 we see that

$$\mathrm{rk}_A(\Omega^f_{A/k'}) = \mathrm{rk}_{\widehat{A}/\mathfrak{p}_i}(\Omega^f_{(\widehat{A}/\mathfrak{p}_i)/k'}).$$

Hence $\dim(\widehat{A}/\mathfrak{p}_i)$ does not depend on i and we are done. □

Proposition 3.6.53 (Yamauchi [395, Prop. 1.13]) *Let $k \subseteq K$ be a field extension and $A := K[[X_1, \dots, X_n]]$. Take a radical ideal $\mathfrak{a}$ of A and a prime ideal $\tilde{\mathfrak{p}}$ of A containing*

$\mathfrak{a}$. *Put* $B := A/\mathfrak{a}$ *and* $\mathfrak{p} := \tilde{\mathfrak{p}}/\mathfrak{a}$. *Moreover assume that* $\Omega^f_{B/k}$ *exists and that for any* $\mathfrak{q} \in \mathrm{Ass}(B)$ *we have* $\mathrm{rk}_{B/\mathfrak{q}}(\Omega^f_{(B/\mathfrak{q})/k}) = \dim(B/\mathfrak{q}) + \dim_K(\Omega_{K/k})$. *Then:*

(i) *If* $(\Omega^f_{B/k})_\mathfrak{p}$ *is a free* $B_\mathfrak{p}$*-module, then* $B_\mathfrak{p}$ *is a regular local ring;*

(ii) *Conversely, if* $B_\mathfrak{p}$ *is a regular local ring and* $\mathrm{rk}_{B/\mathfrak{p}}(\Omega^f_{(B/\mathfrak{p})/k}) = \dim(B/\mathfrak{p}) + \mathrm{rk}_K(\Omega_{K/k})$, *then* $(\Omega^f_{B/k})_\mathfrak{p}$ *is a free* $B_\mathfrak{p}$*-module.*

Proof

(i) Let $\tilde{\mathfrak{q}} \in \mathrm{Ass}_A(A/\mathfrak{a})$ be such that $\tilde{\mathfrak{q}} \subseteq \tilde{\mathfrak{p}}$ and $\mathrm{ht}(\mathfrak{a}A_{\tilde{\mathfrak{p}}}) = \mathrm{ht}(\tilde{\mathfrak{q}}A_{\tilde{\mathfrak{p}}}) = \mathrm{ht}(\tilde{\mathfrak{q}})$. Then $\mathfrak{q} := \tilde{\mathfrak{q}}/\mathfrak{a} \in \mathrm{Min}(B)$ and $\mathfrak{q} \subseteq \mathfrak{p}$. Since B is reduced we have

$$\Omega^f_{B/k} \otimes_B k(\mathfrak{q}) \cong \Omega^f_{(B/\mathfrak{q})/k} \otimes_{B/\mathfrak{p}} k(\mathfrak{q}).$$

Let us consider the following commutative diagram with exact rows:

$$
\begin{array}{ccccccc}
\mathfrak{a}/\mathfrak{a}^2 \otimes_B k(\mathfrak{p}) & \xrightarrow{\delta} & \Omega^f_{A/k} \otimes_A k(\mathfrak{p}) & \longrightarrow & \Omega^f_{B/k} \otimes_B k(\mathfrak{p}) & \longrightarrow & 0 \\
{\scriptstyle i}\downarrow & & \| & & & & \\
\tilde{\mathfrak{p}}/\tilde{\mathfrak{p}}^2 \otimes_{A/\tilde{\mathfrak{p}}} k(\tilde{\mathfrak{p}}) & \xrightarrow{\tilde{\delta}} & \Omega^f_{A/k} \otimes_A k(\tilde{\mathfrak{p}}) & \longrightarrow & \Omega^f_{(B/\mathfrak{p})/k} \otimes_{B/\mathfrak{p}} k(\mathfrak{p}) & \longrightarrow & 0
\end{array}
$$

Since $(\Omega^f_{B/k})_\mathfrak{p}$ is free, we have

$$\mathrm{rk}_{k(\mathfrak{p})}\left(\Omega^f_{B/k} \otimes_B k(\mathfrak{p})\right) = \mathrm{rk}_{k(\mathfrak{q})}\left(\Omega^f_{B/k} \otimes_B k(\mathfrak{q})\right) = \mathrm{rk}_{B/\mathfrak{q}}(\Omega^f_{(B/\mathfrak{q})/k}) =$$

$$= \dim(B/\mathfrak{q}) + \dim_K(\Omega_{K/k}).$$

It follows that

$$\mathrm{rk}_{k(\mathfrak{p})}(\mathrm{Im}(i)) \geq \mathrm{rk}_{k(\mathfrak{p})}(\mathrm{Im}(\delta)) =$$

$$= \dim(A) + \dim_K(\Omega_{K/k}) - (\dim(B/\mathfrak{q}) + \dim_K(\Omega_{K/k})) =$$

$$= \mathrm{ht}(\tilde{\mathfrak{q}}) = \mathrm{ht}(\mathfrak{a}A_{\tilde{\mathfrak{p}}}).$$

But the inequality $\mathrm{rk}_{k(\mathfrak{p})}(\mathrm{Im}(i)) \leq \mathrm{ht}(\mathfrak{a}A_{\tilde{\mathfrak{p}}})$ is always true, hence $\mathrm{rk}_{k(\mathfrak{p})}(\mathrm{Im}(i)) = \mathrm{ht}(\mathfrak{a}A_{\tilde{\mathfrak{p}}})$. This means that i is injective, so that $\mathfrak{a}A_{\tilde{\mathfrak{p}}}$ is generated by a subset of a regular system of parameters of $A_{\tilde{\mathfrak{p}}}$. In other words, $B_\mathfrak{p}$ is a regular local ring.

(ii) The assumption implies that $\tilde{\delta}$ is injective. Since $B_{\mathfrak{p}}$ is a regular local ring, we get that i is injective and consequently also δ is injective. Moreover

$$\mathrm{rk}_{k(\mathfrak{p})}(\Omega^f_{B/k} \otimes_B k(\mathfrak{p})) = \dim(A) + \dim_K(\Omega_{K/k}) - \mathrm{ht}(\mathfrak{a} A_{\tilde{\mathfrak{p}}}).$$

On the other hand, from the regularity of $B_{\mathfrak{p}}$ and the exact sequence

$$\mathfrak{a}/\mathfrak{a}^2 \otimes B_{\mathfrak{p}} \to \Omega^f_{A/k} \otimes_A B_{\mathfrak{p}} \to (\Omega^f_{B/k})_{\mathfrak{p}} \to 0$$

we get

$$\mathrm{rk}\big((\Omega^f_{B/k})_{\mathfrak{p}}\big) \geq \dim(A) + \dim_K(\Omega_{K/k}) - \mathrm{ht}(\mathfrak{a} A_{\tilde{\mathfrak{p}}}).$$

Since we always have $\mathrm{rk}\big((\Omega^F_{B/k} \otimes_B k(\mathfrak{p})\big) \geq \mathrm{rk}\big((\Omega^f_{B/k})_{\mathfrak{p}}\big)$, it follows that

$$\mathrm{rk}\big((\Omega^f_{B/k}) \otimes_B k(\mathfrak{p})\big) = \mathrm{rk}\big((\Omega^f_{B/k})_{\mathfrak{p}}\big),$$

hence $(\Omega^f_{B/k})_{\mathfrak{p}}$ is free. $\square$

Corollary 3.6.54 (Yamauchi [395, Th. 1.14]) *Let A be a Noetherian local domain containing a field k and let $\mathfrak{p} \in \mathrm{Spec}(A)$. Assume that:*

(i) A has a quasi-coefficient field K containing k;
(ii) $\widehat{A}$ is reduced;
(iii) $\Omega^f_{A/E}$ exists for any cofinite subfield E of k;
(iv) $\mathrm{rk}_A(\Omega^f_{A/k}) = \dim(A) + \dim_K(\Omega_{K/k})$;
(v) $(\Omega^f_{A/k})_{\mathfrak{p}}$ is a free $A_{\mathfrak{p}}$-module.

Then $A_{\mathfrak{p}}$ is a regular local ring.

Proof Let $\mathfrak{q} \in \mathrm{Spec}(\widehat{A})$ be such that $\mathfrak{q} \cap A = \mathfrak{p}$. Then

$$(\Omega^f_{\widehat{A}/k})_{\mathfrak{q}} \cong (\Omega^f_{A/k})_{\mathfrak{p}} \otimes_{A_{\mathfrak{p}}} \widehat{A}_{\mathfrak{q}}$$

is a free $\widehat{A}_{\mathfrak{q}}$-module. Let L be a coefficient field of $\widehat{A}$ containing K. Since by Theorem 3.6.52 the ring A is formally equidimensional, for any $\mathfrak{r} \in \mathrm{Ass}(\widehat{A})$ we have that

$$\mathrm{rk}_{\widehat{A}/\mathfrak{r}}(\Omega^f_{(\widehat{A}/\mathfrak{r})/k}) = \mathrm{rk}_A(\Omega^f_{A/k}) = \dim(A) + \dim_K(\Omega_{K/k}) =$$

$$= \dim(\widehat{A}/\mathfrak{r}) + \dim_L \Omega_{L/k}.$$

By Proposition 3.6.53 we get that $\widehat{A}_{\mathfrak{q}}$ is regular; hence by flatness $A_{\mathfrak{p}}$ is regular. $\square$

Example 3.6.55 (Yamauchi [395])

(i) Let k be a field of characteristic $p > 0$ and $A = k[X]/(X^p)$. Then A is a 0-dimensional local ring and A is not regular. But $\Omega^f_{A/k}$ exists and is a free A-module of rank 1. If $\mathfrak{m}$ is the maximal ideal of A, then $\dim_{A/\mathfrak{m}}(\Omega^f_{(A/\mathfrak{m})/k}) = 0$.

(ii) Let k be a field of characteristic $p > 0$ such that $k \neq k^p$ and let $a \in k \setminus k^p$. Let $A := k[[X, Y]]$ and $f = aX^p + Y^p$. Then f is an irreducible element of A and if we denote $B := A/fA$, then B is a local domain, $\dim(B) = 1$ and B is not a regular local ring. On the other hand $\Omega^f_{B/k}$ is a free B-module of rank 2.

Examples (i) and (ii) above show that the Jacobian criterion of regularity of Scheja and Storch 3.6.37 is not true in full generality in positive characteristic.

(iii) Let k be a field of characteristic $p > 0$ such that $[k : k^p] = \infty$. Let $A = k^p[[T]][k]$. Take $u \in k[[T]], u \notin A$ and set $a = u^p$. Then $B := A[X]/(X^p - a)$ is a local domain of dimension 1. Also $\widehat{B} = \widehat{A}[X]/(X - u)^p$ is not reduced. It is easy to see that for any cofinite subfield k' of k such that $k^p \subseteq k'$ the module $\Omega^f_{B/k'}$ exists and

$$\Omega^f_{B/k'} = (\Omega_{k/k'} \otimes_k B) \oplus B dT \oplus B dX,$$ hence $\mathrm{rk}_B(\Omega^f_{B/k'}) = \dim_k(\Omega_{k/k'}) + 2$. It follows that assumption (iv) in Theorem 3.6.54 is essential in positive characteristic.

We want to prove extensions of Theorem 3.6.39 to the positive characteristic case.

Theorem 3.6.56 (Yamauchi [395, Th. 2.1]) *Let k be a field of characteristic $p > 0$ and let A be a Noetherian k-algebra. Assume that:*

(i) *A has geometrically reduced formal fibres;*
(ii) *$\Omega^f_{A/k'}$ exists for any cofinite subfield k' of k;*
(iii) *$\dim_{k(\mathfrak{m})} \gamma_{k(\mathfrak{m})/k} < \infty$ for any maximal ideal $\mathfrak{m}$ of A.*

Then A has geometrically regular formal fibres and is universally catenary. If moreover k is a perfect field, then A is excellent.

Proof To prove that A has geometrically regular formal fibres, we can assume that $(A, \mathfrak{m})$ is a local ring. Let B be a finite A-algebra and $\mathfrak{q}$ be a prime ideal of B such that $\mathfrak{q} \cap B = (0)$. Let $\mathfrak{n}^*$ be a maximal ideal of $\widehat{B}$ containing $\mathfrak{q}$ and let $\mathfrak{n} = \mathfrak{n}^* \cap B$. Then $\mathfrak{n}$ is a maximal ideal of B and $\widehat{B}_{\mathfrak{n}^*} = \widehat{B_\mathfrak{n}}$. Since B is finite over A, it follows that also B satisfies the assumptions (i), (ii) and (iii). So, replacing A by $B_\mathfrak{n}$ we have to prove that if A is a local domain with the properties (i), (ii) and (iii), the generic formal fibre of A is regular. For this, let $\mathfrak{q}$ be a prime ideal of $\widehat{A}$ such that $\mathfrak{q} \cap A = (0)$. By (ii), (iii) and Proposition 3.6.45, we may also assume that k is contained in a quasi-coefficient field of A. By Theorem 3.6.52 there exists a cofinite subfield k' of k such that

$$\mathrm{rk}_A(\Omega^f_{A/k'}) = \dim(A) + \dim_{k(\mathfrak{m})}(\Omega_{k(\mathfrak{m})/k}).$$

Since $\widehat{A}$ is reduced and $\Omega^f_{\widehat{A}_{\mathfrak{q}}/k'} = \Omega^f_{A/k'} \otimes_A Q(A) \otimes_{Q(A)} \widehat{A}_{\mathfrak{q}}$ is a free $\widehat{A}_{\mathfrak{q}}$-module, by Lemma 3.6.51, Theorem 3.6.52 and Proposition 3.6.53 it follows that $\widehat{A}_{\mathfrak{q}}$ is a regular local ring.

To prove that A is universally catenary, we may also assume that A is a Noetherian local domain. We can also assume, as above, that k is contained in a quasi-coefficient field of A. Since $\widehat{A}$ is reduced, Theorem 3.6.52 implies that A is formally equidimensional; hence by Theorem 3.1.42 it follows that A is universally catenary.

Finally let us assume that k is a perfect field. Notice that we cannot assume anymore that A is local, because the property Reg-2 is not a local one. We have to prove that for any finite A-algebra B, the regular locus of B is open. By Theorem 1.2.16 it suffices to prove that for any prime ideal $\mathfrak{p}$ of B the set $\mathrm{Reg}(B/\mathfrak{p})$ contains a non-empty open subset. Since $B/\mathfrak{p}$ satisfies the same assumptions as A, we have to prove that if A is moreover a domain, then $\mathrm{Reg}(A)$ contains a non-empty open subset. Let $\mathfrak{m}$ be a maximal ideal of A. Then $\widehat{A_{\mathfrak{m}}}$ is reduced. Since k is perfect, by [249, Th. 28.3] it follows that $A_{\mathfrak{m}}$ has a quasi-coefficient field containing k. By Theorem 3.6.52 we get

$$\mathrm{rk}(\Omega^f_{A_{\mathfrak{m}}/k}) = \dim(A_{\mathfrak{m}}) + \dim(\Omega_{k(\mathfrak{m})/k}).$$

Let $U := \{\mathfrak{p} \in \mathrm{Spec}(A) \mid (\Omega^f_{A/k})_{\mathfrak{p}}$ is free$\}$. Let $\mathfrak{p} \in U$ and let $\mathfrak{m} \in \mathrm{Max}(A)$ such that $\mathfrak{m} \supseteq \mathfrak{p}$. Then $(\Omega^f_{A/k})_{\mathfrak{p}} = (\Omega^f_{A/k} \otimes_A A_{\mathfrak{m}})_{\mathfrak{p}} = (\Omega^f_{A_{\mathfrak{m}}/k})_{\mathfrak{p}}$ is a free $A_{\mathfrak{p}}$-module and

$$\mathrm{rk}(\Omega^f_{A_{\mathfrak{m}}/k}) = \dim(A_{\mathfrak{m}}) + \dim(\Omega_{k(\mathfrak{m})/k}).$$

By Corollary 3.6.54 we obtain that $\mathfrak{p} \in \mathrm{Reg}(A)$. Hence $U \subseteq \mathrm{Reg}(A)$ and clearly U is open.   □

Finally we prove a result belonging to Matusmura, which is somehow similar to Theorem 3.6.56 and is the positive characteristic version of Corollary 3.3.27. The present proof belongs to Yamauchi [395].

Theorem 3.6.57 (Matsumura [251, Th. 15]) *Let A be a regular ring containing a field k of characteristic $p > 0$. Assume that:*

(i) *A has geometrically reduced formal fibres;*
(ii) *$\Omega^f_{A/k'}$ exists for any cofinite subfield k' of k;*
(iii) *$\dim_{k(\mathfrak{m})} \Upsilon_{k(\mathfrak{m})/k} < \infty$ for any maximal ideal $\mathfrak{m}$ of A.*

Then the weak Jacobian criterion holds in A and A is an excellent ring.

Proof ([395]) Let us denote for simplicity $A_n = A[X_1, \ldots, X_n]$. Let $\mathfrak{p}$ be a prime ideal of A_n such that $B := A_n/\mathfrak{p}$ is a finite A-algebra. We will prove that the weak Jacobian condition holds at $\mathfrak{p}$. Let $\mathfrak{n}$ be a maximal ideal of A_n containing $\mathfrak{p}$ and let $\mathfrak{m} := \mathfrak{n} \cap A$. Since B is finite over A the ideal $\mathfrak{m}$ is a maximal ideal of A. Replacing A by $A_{\mathfrak{m}}$ and A_n by $(A_n)_{\mathfrak{m}}$ we may assume that $(A, \mathfrak{m})$ is a regular local ring. Let P be a p-basis of k and let $\{F_\alpha\}_{\alpha \in I}$ be the family of finite subsets of P. Put $k_\alpha := k^p(P_\alpha)$ and $k'_\alpha = \mathbb{F}_p(P_\alpha)$, where $P_\alpha := P \setminus F_\alpha$. Then $\{k_\alpha\}_{\alpha \in I}$ and $\{k'_\alpha\}_{\alpha \in I}$ are downward directed families of cofinite subfields of k and respectively $\mathbb{F}_p(P)$ such that $\bigcap_{\alpha \in I} k_\alpha = k^p$, respectively such that $\bigcap_{\alpha \in I} k'_\alpha = (\mathbb{F}_p(P))^p$. By (ii) it follows that Ω^f_{A/k_α} exists for any $\alpha \in I$ and that $\Omega^f_{A/k_\alpha} = \Omega^f_{A/k'_\alpha}$. Moreover Ω^f_{B/k_α} exists for any $\alpha \in I$ and $\Omega^f_{B/k_\alpha} = \Omega^f_{B/k'_\alpha}$. By (iii) there exists $\alpha \in I$ such that the field extension $k'_\alpha \subseteq k(\mathfrak{m})$ is separable; hence A has a quasi-coefficient field K_A containing k'_α by [249, Th. 28.3]. Changing k with k'_α we may assume that k is contained in K_A. In a similar way, since $[k(\mathfrak{n}) : k(\mathfrak{m})] < \infty$ and consequently $\dim_{k(\mathfrak{n})}(\gamma_{k(\mathfrak{n})/k}) < \infty$, we may assume that k is contained in a quasi-coefficient field of $B_{\mathfrak{n}}$. Because $\dim_{k(\mathfrak{n})}(\gamma_{k(\mathfrak{n})/k(\mathfrak{m})}) < \infty$, by [131, Th. 21.8.3] there exists $\beta \in I$ such that k_β is admissible for $k(\mathfrak{n})$ over $k(\mathfrak{m})$. We have the exact sequence

$$0 \to \gamma_{k(\mathfrak{n})/k(\mathfrak{m})/k_\beta} \to \Omega_{k(\mathfrak{m})/k_\beta} \otimes_{k(\mathfrak{m})} k(\mathfrak{n}) \to \Omega_{k(\mathfrak{n})/k_\beta} \to \Omega_{k(\mathfrak{n})/k(\mathfrak{m})} \to 0.$$

Since Ω^f_{A/k'_β} and $\Omega^f_{B_{\mathfrak{n}}/k_\beta}$ exist, it follows that

$$\dim_{k(\mathfrak{m})} \Omega_{k(\mathfrak{m})/k_\beta} < \infty \text{ and } \dim_{k(\mathfrak{n})} \Omega_{k(\mathfrak{n})/k_\beta} < \infty.$$

By Lemma 3.6.43 we have

$$\dim_{k(\mathfrak{m})} \Omega_{k(\mathfrak{m})/k_\beta} - \dim_{k(\mathfrak{n})} \Omega_{k(\mathfrak{n})/k_\beta} =$$

$$= \dim_{k(\mathfrak{n})} \Omega_{k(\mathfrak{n})/k(\mathfrak{m})} - \dim_{k(\mathfrak{n})} \gamma_{k(\mathfrak{n})/k(\mathfrak{m})/k_\beta} = 0.$$

On the other hand, since $\widehat{B_{\mathfrak{n}}}$ is reduced, by Theorem 3.6.52 there exists $\gamma \in I$ such that

$$\mathrm{rk}_{B_{\mathfrak{n}}} \Omega^f_{B_{\mathfrak{n}}/k_\gamma} = \dim(B_{\mathfrak{n}}) + \dim_{k(\mathfrak{n})} \Omega_{k(\mathfrak{n})/k_\gamma}.$$

We can assume that $k = k_\beta = k_\gamma$. Let $D := (\Omega^f_{A/k} \otimes_A B) \oplus (\bigoplus_{i=1}^{n} B dX_i)$. There are natural B-morphisms

$$\mathfrak{p}/\mathfrak{p}^2 \to \Omega_{A_n/k} \otimes_{A_n} B = (\Omega_{A/k} \otimes_A B) \oplus (\bigoplus_{i=1}^{n} B dX_i)$$

and

$$\Omega_{A/k} \to \Omega^f_{A/k}.$$

Hence we get a B-morphism $\delta : \mathfrak{p}/\mathfrak{p}^2 \to D$. We have also another B-morphism $u : D \to \Omega^f_{B/k}$ defined by the natural map $\Omega^f_{A/k} \otimes_A B \to \Omega^f_{B/k}$ and $dX_i \to dx_i$ where $x_i = X_i(\bmod\ \mathfrak{p})$. Finally we get a sequence of B-morphisms

$$\mathfrak{p}/\mathfrak{p}^2 \overset{\delta}{\to} D \overset{u}{\to} \Omega^f_{B/k} \to 0.$$

We will prove that this sequence is exact. For this let M be an arbitrary finitely generated B-module. Then M is also a finitely generated A-module and we have

$$\mathrm{Hom}_B(D, M) \cong \mathrm{Hom}_B(\Omega^f_{A/k} \otimes_A B, M) \oplus \mathrm{Hom}_B(\overset{n}{\underset{i=1}{\oplus}} B dX_i, M) \cong$$

$$\cong \mathrm{Der}_k(A, M) \oplus \mathrm{Hom}_{A_n}(\overset{n}{\underset{i=1}{\oplus}} A_n dX_i, M) \cong \mathrm{Der}_k(A_n, M).$$

This implies that the following sequence is exact

$$0 \to \mathrm{Hom}_B(\Omega^f_{B/k}, M) \to \mathrm{Hom}_B(D, M) \to \mathrm{Hom}_B(\mathfrak{p}/\mathfrak{p}^2, M)$$

and from this we obtain that the initial sequence is exact. Tensorizing with $k(\mathfrak{p}) = Q(B)$ we get the exact sequence

$$\mathfrak{p}/\mathfrak{p}^s \otimes_B k(\mathfrak{p}) \overset{\Delta}{\to} (\Omega^f_{A/k} \otimes_A k(\mathfrak{p})) \oplus (\overset{n}{\underset{i=1}{\oplus}} k(\mathfrak{p}) dX_i) \to \Omega^f_{B/k} \otimes k(\mathfrak{p}) \to 0.$$

Since $\widehat{A}$ is a complete regular local ring, by Lemma 3.6.44 it results that $\Omega^f_{\widehat{A}/k} = \Omega^f_{A/k} \otimes_A \widehat{A}$ is a free $\widehat{A}$-module and

$$\mathrm{rk}_{\widehat{A}}(\Omega^f_{\widehat{A}/k}) = \dim(\widehat{A}) + \dim(\Omega_{k(\mathfrak{m})/k}).$$

Now by [248, Prop. (4.E)] we have that $\Omega^f_{A/k}$ is a free A-module and

$$\mathrm{rk}(\mathrm{Im}(\Delta)) = \mathrm{rk}(\Omega^f_{A/k}) + n - \mathrm{rk}(\Omega^f_{B/k}) =$$

$$= \dim(A) + \dim \Omega_{k(\mathfrak{m})/k} + n - (\dim \Omega_{k(\mathfrak{n})/k} + \dim(B_n)) - \mathrm{ht}(\mathfrak{p}).$$

This means that the weak Jacobian condition holds at $\mathfrak{p}$. By Theorem 3.3.24 and Remark 3.3.25 it follows that A is excellent. $\square$

Corollary 3.6.58 (Valabrega [384, Th. 4]) *Let k be a field and n, m be natural numbers. Then $k[X_1, \ldots, X_n][[Y_1, \ldots, Y_m]]$ is excellent.*

Proof ([251]) It follows from Corollary 3.3.28 and Theorem 3.6.57, observing that if a Noetherian local ring A satisfies the assumptions (i), (ii) and (iii) in Theorem 3.6.57, then also $A[[Y]]$ does. Indeed (i) holds by Corollary 2.2.6, (ii) holds by Proposition 3.6.23 and (iii) obviously continues to hold since every maximal ideal of $A[[Y]]$ contains Y. $\qquad \square$

Example 3.6.59 Let k be a field of characteristic $p > 0$ such that $[k : k^p] = \infty$ and $A = k^p[[T]][k]$. Then A is not excellent but $\Omega^f_{A/k}$ is a free A-module of rank 1. Thus Theorem 3.6.39 doesn't hold in full generality in positive characteristic.

Corollary 3.6.60 (Valabrega [384, Cor. 5]) *Let k be a field and A be a k-algebra of finite type. Then for any ideal $\mathfrak{a}$ of A, the $\mathfrak{a}$-adic completion of A is an excellent ring.*

Proof Suppose that $A \cong k[X_1, \ldots, X_n]/\mathfrak{b}$, where $\mathfrak{b} = (f_1, \ldots, f_m)$ is an ideal in $k[X_1, \ldots, X_n]$. Then the completion of A in the $\mathfrak{a}$-adic topology is a factor ring of $k[X_1, \ldots, X_n][[Y_1, \ldots, Y_m]]$. We apply Corollary 3.6.58. $\qquad \square$

Corollary 3.6.61 (Valabrega [384, Prop. 7]) *Let $(A, \mathfrak{m})$ be a Noetherian complete local ring containing a field and let $X := (X_1, \ldots, X_n)$ be variables over A. Then the restricted power series ring $A\{X_1, \ldots, X_n\}$ is an excellent ring, considering on A the $\mathfrak{m}$-adic topology.*

Proof The ring A contains a coefficient field k. By Cohen's structure theorem [56, ch. IX, §2 , no. 5, Th. 3], we know that A is an homomorphic image of a formal power series ring over k, say $A \cong k[[Y_1, \ldots, Y_m]]/\mathfrak{a}$. Let $Y := (Y_1, \ldots, Y_m)$. Then $A\{X\} \cong k[[Y]]\{X\}/\mathfrak{a}\{X\}$, where $\mathfrak{a}\{X\}$ is the set of restricted power series with coefficients in $\mathfrak{a}$. This means that it is enough to show that $k[[Y]]\{X\}$ is an excellent ring. But $k[[Y]]\{X\}$ is the completion of $k[X_1, \ldots, X_n, Y_1, \ldots, Y_m] := k[X, Y]$ in the (Y)-adic topology; hence it is the (Y)-adic closure of $k[X, Y]$ in $k[[X, Y]]$. But this is exactly $k[X][[Y]]$. $\qquad \square$

3.7 Inductive Limits of P-Rings and Applications

We want to study the behaviour of **P**-rings to inductive limits. The main results of this section belong to Marot [240] and Doretti [91]. In the following theorem, **P** is one of the properties: reduced, normal, regular, (R_k), (S_k), Cohen-Macaulay, Gorenstein, complete intersection.

Theorem 3.7.1 (Marot [240, Prop. 1.1]) *Let $(A_i, \varphi_{ij})_{i \in I}$ and $(B_i, \psi_{ij})_{i \in I}$ be inductive systems of Noetherian rings and $(u_i : A_i \to B_i)_{i \in I}$ be a morphism of inductive systems.*

Let $A = \varinjlim A_i$, $B = \varinjlim B_i$ and $u = \varinjlim u_i : A \to B$. Let P be a property of Noetherian rings satisfying (A_2), (A_3) and (A_5). Assume that:

(i) *A and B are Noetherian rings;*
(ii) *for any $i \leq j$, the morphisms φ_{ij} and ψ_{ij} are flat;*
(iii) *for any index $i \in I$, the morphism u_i is a P-morphism.*

Then u is a P-morphism.

Proof Using Lemma 1.6.12 we may assume that A_i and B_i are local rings for any $i \in I$ and that the morphisms φ_{ij} and ψ_{ij} are local ones. In particular the morphisms ϕ_{ij} and ψ_{ij} are faithfully flat. Let $\mathfrak{p}$ be a prime ideal of A and $\mathfrak{p}_i := \mathfrak{p} \cap A_i$. Denote by $k := k(\mathfrak{p})$ and $k_i := k(\mathfrak{p}_i)$ the residue fields in $\mathfrak{p}$ and $\mathfrak{p}_i$, respectively. Because A is Noetherian, there exists $i_0 \in I$, such that $\mathfrak{p} = \mathfrak{p}_i A$ for any $i \geq i_0$. It follows that the inductive systems induced on $(A_i/\mathfrak{p}_i, \varphi_{ij})_{i \in I}$ and $(B_i/\mathfrak{p}_i B_i, \psi_{ij})_{i \in I}$ satisfy the conditions (i), (ii) and (iii) for $i \geq i_0$. This shows that we may assume that A is an integral domain with $\mathfrak{p} = (0)$ and that A_i is a domain and $\mathfrak{p}_i = (0)$ for any $i \in I$. Therefore $k_i = Q(A_i)$, the field of fractions of A_i. From (ii) we get that the canonical morphism $h_{ij} : B_i \otimes_{A_i} k_i \to B_j \otimes_{A_j} k_j$ is flat for $i_0 \leq i \leq j$. Let K be a field, finite extension of k. By Lemma 1.3.13 there exists $i_1 \in I$, that we may assume greater than i_0 and a finite field extension $k_{i_1} \subseteq K_{i_1}$ such that $K = k \otimes_{k_{i_1}} K_{i_1}$. For any $i \geq i_1$, the extension $k_i \subseteq K_i = k_i \otimes_{k_{i_1}} K_{i_1}$ is finite. We have $K = \varinjlim_i K_i$ and we have an inductive system of Noetherian rings $(B_i \otimes_{A_i} K_i)_{i \in I}$ such that $B \otimes_A K = \varinjlim_i B_i \otimes_{A_i} K_i$. For $j \geq i \geq i_1$ the morphisms of this system are flat. Indeed, since $K_j = k_j \otimes_{k_i} K_i$, the morphism $g_{ij} : B_i \otimes_{A_i} K_i \to B_j \otimes_{A_j} K_j$ is induced by the flat morphism h_{ij}, via the flat base change $k_i \to K_i$. Because u is flat, by Lemma 1.6.12 we have to show that $B \otimes_A K$ has the property P, knowing that for any $i \geq i_1$, the ring $B_i \otimes_{A_i} K_i$ has the property P. For this we apply Proposition 1.3.10, Corollary 1.3.11 and Proposition 1.3.12. $\square$

Proposition 3.7.2 (Marot [240, Prop. 4.1]) *Let $(A_i, \varphi_{ij})_{i \in I}$ be an inductive system of Noetherian rings and let $A := \varinjlim A_i$. Let P be a property of Noetherian rings satisfying (A_1), (A_2), (A_3) and (A_5) and assume that:*

(i) *A is a Noetherian ring;*
(ii) *φ_{ij} is flat for any $i \leq j$;*
(iii) *for any $\mathfrak{m} \in \operatorname{Max}(A)$, the canonical morphism $\varinjlim_{i \in I}(\widehat{A_i})_{\mathfrak{m}_i} \to \widehat{A}_{\mathfrak{m}}$ is regular, where*

   $\mathfrak{m}_i := \varphi_i^{-1}(\mathfrak{m}),;$
(iv) *A_i is a P-ring, for any $i \in I$.*

Then A is a P-ring.

Proof Let $C_i = (A_i)_{\mathfrak{m}_i}$ for any $i \in I$ and let $C = A_{\mathfrak{m}}$. By Corollary 1.3.18 the ring $\varinjlim_{i \in I} \widehat{C}_i$ is a Noetherian local ring and $\widehat{C} \cong (\varinjlim \widehat{C}_i)\widehat{}$. On the other hand, from (i) and (ii) it follows that the inductive systems $(C_i, \varphi_{ij})_{i \in I}$ and $(\widehat{C}_i, \widehat{\varphi}_{ij})_{i \in I}$ satisfy the conditions of Theorem 3.7.1. For any index i, the morphism $C_i \to \widehat{C}_i$ is a **P**-morphism; hence by Theorem 3.7.1 the canonical morphism $\varphi : C \to \varinjlim \widehat{C}_i$ is a **P**-morphism. The canonical completion morphism $C \to \widehat{C}$ factorizes as $C \xrightarrow{\varphi} \varinjlim \widehat{C}_i \xrightarrow{\psi} \widehat{C}$. Since from (iii) it follows that ψ is regular, we obtain that $\psi \circ \varphi$ is a **P**-morphism, that is, A is a **P**-ring. $\qquad\square$

Proposition 3.7.3 (Marot [240, Prop. 3.1]) *Let $(A_i, \varphi_{ij})_{i \in I}$ be an inductive system of complete Noetherian local rings and let $A := \varinjlim_{i \in} A_i$. Let k_i be the residue field of A_i for any $i \in I$ and let k be the residue field of A. Assume that:*

(i) A is a Noetherian ring;
(ii) for any pair of indices $i \leq j$ the morphism φ_{ij} is flat and the field extension $k_i \subseteq k_j$ is separable.

Then A is an excellent Henselian ring.

Proof By [133, Prop. 18.6.14] and Lemma 3.1.63 the ring A is Henselian and universally catenary. Let $\mathfrak{m}$ be the maximal ideal of A and for any index i let $\mathfrak{m}_i := \varphi_i^{-1}(\mathfrak{m})$. Since A is Noetherian, there exists $i_0 \in I$ such that $\mathfrak{m} = \mathfrak{m}_i A$ for any $i \geq i_0$. Therefore, for $i \geq i_0$ we have $\widehat{A} \otimes_{A_i} k_i = k$ and condition (ii) implies that the field extension $k_i \subseteq k$ is separable. Thus $A_i \to \widehat{A}$ is formally smooth for any $i \geq i_0$. By Theorem 3.2.22 the morphism $A_i \to \widehat{A}$ is regular. Thus Theorem 3.7.1 implies that the morphism $A \to \widehat{A}$ is regular. $\qquad\square$

Corollary 3.7.4 (Marot [240, Cor. 4.2]) *Let k be a field of characteristic zero and let $(A_i, \varphi_{ij})_{i \in I}$ be an inductive system of Noetherian k-algebras. Let **P** be a property of Noetherian rings satisfying $(\mathbf{A_1})$, $(\mathbf{A_2})$, $(\mathbf{A_3})$ and $(\mathbf{A_5})$. Assume that:*

(i) $A := \varinjlim A_i$ is a Noetherian ring;
(ii) φ_{ij} is flat for any $i \leq j$;
*(iii) A_i is a **P**-ring for any $i \in I$.*

*Then A is a **P**-ring.*

Proof Using Proposition 3.7.3 we see that the condition (iii) in Proposition 3.7.2 is satisfied. $\qquad\square$

Corollary 3.7.5 (Marot [240, Cor. 4.3]) *Let* $(A_i, \varphi_{ij})_{i \in I}$ *be an inductive system of Noetherian local rings and let* k_i *be the residue field of* A_i. *Let* **P** *be a property of Noetherian rings satisfying* **(A₁)**, **(A₂)**, **(A₃)** *and* **(A₅)**. *Assume that:*

(i) $A := \varinjlim A_i$ *is a Noetherian ring;*
(ii) φ_{ij} *is a local flat morphism for any* $i \leq j$;
(iii) φ_{ij} *induces a separable field extension* $k_i \subseteq k_j$, *for any* $i \leq j$;
(iv) A_i *is a* **P***-ring for any* $i \in I$.

Then A *is a* **P***-ring.*

Proof Follows from Proposition 3.7.2 and Proposition 3.7.3.   □

Corollary 3.7.6 (Marot [240, Cor. 4.4]) *Let* $(A_i, \varphi_{ij})_{i \in I}$ *be an inductive system of Noetherian local rings and let* k_i *be the residue field of* A_i. *Assume that:*

(i) $A := \varinjlim_i A_i$ *is a Noetherian ring;*

(ii) φ_{ij} *is a local flat morphism for any* $i \leq j$;
c) φ_{ij} *induces a separable field extension* $k_i \subseteq k_j$, *for any* $i \leq j$;
d) A_i *is a Nagata(resp. excellent) ring for any* $i \in I$.

Then A *is a Nagata(resp. excellent) ring.*

Proof Follows from Corollary 3.7.5, Theorem 2.3.6 and Lemma 3.1.63.   □

For the properties Cohen-Macaulay, Gorenstein and Complete intersection we will prove later (4.2.20) a stronger result.

The above result is not valid in the global case as we will show in the next example.

Example 3.7.7 Let k be a field. There exists an inductive system $(A_i, \phi_{ij})_{i \in \mathbb{N}}$ of semilocal Noetherian k-algebras that are integral domains such that:

(i) the ring $A := \varinjlim A_i$ is a Noetherian domain of dimension 1;
(ii) for any indices $i \leq j$, the morphisms ϕ_{ij} are faithfully flat reduced morphisms that are not normal morphisms;
(iii) for any index $i \in \mathbb{N}$, the ring A_i is excellent;
(iv) the ring A is not a Nagata ring;
(v) for any maximal ideal $\mathfrak{m}$ of A, the localization $A_{\mathfrak{m}}$ is an excellent ring.

Proof We consider the example constructed in Example 1.2.4. We can describe that example as follows: let k be any field and let $B_i := k[X_i, Y_i]/(Y_i^2 - X_i^3)$ for any

$i \geq 1$. Let us define inductively $R_1 := B_1 = k[X_1, Y_1]/(Y_1^2 - X_1^3)$ and furthermore $R_n := R_{n-1} \otimes_k B_n = k[X_1, Y_1, \ldots, X_n, Y_n]/(Y_1^2 - X_1^3, \ldots, Y_n^2 - X_n^3)$. Also let $\phi_{n,n+1} : R_n \to R_{n+1}$ be the canonical induced morphism. If we consider the maximal ideal $\mathfrak{m}_i = (X_i, Y_i)B_i$ and $S_n := R_n \setminus \bigcup_{i=1}^{n} \mathfrak{m}_i R_n$, we easily see that we have an inductive system $(A_m := S_m^{-1} R_m, \phi_{m,n})_{m \in \mathbb{N}}$. Let $A := \varinjlim A_n$. It is easy to see that A is exactly the ring constructed in Example 1.2.4. Thus (i), (iii) and the faithfully flatness of (ii) are clear. Since $\mathrm{Nor}(A)$ is not open, as it was proved in Example 1.2.4, from Theorem 2.3.12 it follows that A is not a Nagata ring. If $\mathfrak{m}$ is a maximal ideal of A, taking account of the description of the maximal ideals of A in Example 1.2.4, we see at once that $A_{\mathfrak{m}}$ is an excellent ring. Therefore (iv) and (v) are true. In order to show that the morphisms ϕ_{ij} are reduced and not normal, let $\mathfrak{p} \in \mathrm{Spec}(A_n)$, for some natural number n. Then we have

$$k(\mathfrak{p}) \otimes_{A_n} A_{n+1} \cong k(\mathfrak{p}) \otimes_{A_n} \left(A_n[X_{n+1}, Y_{n+1}]/(Y_{n+1}^2 - X_{n+1}^3) \right) \cong$$

$$\cong k(\mathfrak{p})[X_{n+1}, Y_{n+1}]/(Y_{n+1}^2 - X_{n+1}^3),$$

which is a reduced ring that is not normal. □

Later, in Example 6.4.11, we will see another example of the same type.

Notation 3.7.8 Let A be a Noetherian ring, S a multiplicatively closed system in A and $\mathfrak{a}$ an ideal of A such that $\mathfrak{a} \cap S = \emptyset$ and f an element of A. According to [135, Ch. 0, §7.6] we denote:

(i) $A\{S^{-1}\} := \widehat{(S^{-1}A, \mathfrak{a}S^{-1}A)}$, the completion of $S^{-1}A$ in the $\mathfrak{a}S^{-1}A$-adic topology [135, Not. 7.6.1];

(ii) $A_{\{f\}} := A\{S_f^{-1}\}$, where $S_f := \{f^n, n \geq 0\}$ is the multiplicative system of the powers of f [135, Not. 7.6.15]

c) $A_{\{S\}} := \varinjlim_{f \in S} A_{\{f\}}$ [135, Not. 7.6.15].

For the most important properties of the rings considered above we refer to [135, Ch. 0, §7.6].

Proposition 3.7.9 *Let A be a Noetherian ring, S a multiplicatively closed system in A and $\mathfrak{a}$ an ideal of A such that $\mathfrak{a} \cap S = \emptyset$. Then:*

(i) *$A_{\{S\}}$ is a Zariski ring for the $\mathfrak{a}A_{\{S\}}$-adic topology;*

(ii) *$A_{\{S\}}$ is a Noetherian ring;*

(iii) *$\widehat{(A_{\{S\}}, \mathfrak{a}A_{\{S\}})} \cong A\{S^{-1}\}$.*

Proof Follows from 1.3.18 and [127, Ch. 0, §7.6].   □

We will recall from [215] some properties of ring epimorphisms that will be needed. It is well-known that in the category of commutative rings, any surjection is an epimorphism, while the converse is not true. For example, any ring of fractions of A is epimorphic over A. First let us consider some notations.

Notation 3.7.10 With a ring morphism $f : A \to B$ we consider the following canonical morphisms:

$i_1 : B \to B \otimes_A B,\ i_1(b) = b \otimes 1$;
$i_2 : B \to B \otimes_A B,\ i_2(b) = 1 \otimes b$;
$p : B \otimes_A B \to B,\ p(x \otimes y) = xy$.

Lemma 3.7.11 Let $f : A \to B$ be a ring morphism. Then f is an epimorphism if and only if p is an isomorphism.

Proof Assume that f is an epimorphism. For any element $a \in A$ we have

$$i_1 \circ f(a) = a \otimes 1 = 1 \otimes a = i_2 \circ f(a),$$

hence $i_1 \circ f = i_s \circ f$. Therefore $i_1 = i_2$ or in other words $b \otimes 1 = 1 \otimes b$ for any element $b \in B$. But it is known that $\ker(p)$ is generated by $\{b \otimes 1 - 1 \otimes b \mid p \in B\}$, that is, p is injective. Conversely, consider two ring morphisms $\alpha, \beta : B \to X$ such that $\alpha \circ f = \beta \circ f$. It is easy to see that they produce a ring morphism $\lambda : B \otimes_A B \to X$, defined by $\lambda(b_1 \otimes b_2) = \alpha(b_1) \cdot \beta(b_2)$. Then for any element $b \in B$ we have

$$\alpha(b) = \alpha(b) \cdot \beta(1) = \lambda(b \otimes 1) = \lambda(1 \otimes b) = \alpha(1) \cdot \beta(b) = \beta(b),$$

that is $\alpha = \beta$.   □

Lemma 3.7.12 Let $f : A \to B$ and $g : B \to C$ be ring morphisms.

(i) if f and g are epimorphisms, then $g \circ f$ is an epimorphism;
(ii) if f is an epimorphism and $h : A \to C$ is a ring morphism, then the induced morphism $1_C \otimes h : C \to C \otimes_A B$ is an epimorphism.

Proof Straightforward.   □

Lemma 3.7.13 *Let $f : A \rightarrow B$ be a faithfully flat ring epimorphism. Then f is an isomorphism.*

Proof Let $K = \ker(f)$ and $L = \mathrm{Coker}(f)$. From Lemma 3.7.11 we get that the canonical map $A \otimes_A B \rightarrow B \otimes_A B$ is an isomorphism. Then $K \otimes_A B = L \otimes_A B = (0)$. By faithful flatness we obtain $K = L = (0)$. $\square$

Corollary 3.7.14 *Let $f : A \rightarrow B$ be a ring epimorphism.*

(i) if A is a field, then f is an isomorphism;

(ii) if A and B are local rings and f is a local flat morphism, then f is an isomorphism.

Proposition 3.7.15 *Let $f : A \rightarrow B$ be a ring morphism. The following are equivalent:*

(i) f is an epimorphism;
(ii) (a) $\mathrm{Spec}(f)$ is injective;
   (b) for any $\mathfrak{q} \in \mathrm{Spec}(B)$, we have $k(\mathfrak{q} \cap A) = k(\mathfrak{q})$;
   (c) the ideal $\mathfrak{a} = \ker(p)$ is finitely generated;
   (d) $\Omega_{B/A} = (0)$.

Proof $(i) \Rightarrow (ii)$: (a) Let $\mathfrak{p} \in \mathrm{Spec}(A)$. Then either $B_\mathfrak{p}/\mathfrak{p}B_\mathfrak{p}$ is a field and by Corollary 3.7.14 there is an isomorphism $B_\mathfrak{p}/\mathfrak{p}B_\mathfrak{p} \cong A_\mathfrak{p}/\mathfrak{p}A_\mathfrak{p}$ or $B_\mathfrak{p}/\mathfrak{p}B_\mathfrak{p} = (0)$. But $(\mathrm{Spec}(f))^{-1}(\mathfrak{p}) = \mathrm{Spec}(B_\mathfrak{p}/\mathfrak{p}B_\mathfrak{p})$, hence $\mathrm{Spec}(f)$ is injective. (b) follows from Corollary 3.7.14, (i). From Lemma 3.7.11 we obtain $\mathfrak{a} = \ker(p) = (0)$. This implies (c) and (d).

$(ii) \Rightarrow (i)$: Let C be a reduced ring and let $g, h : B \rightarrow C$ be two ring morphisms such that $g \circ f = h \circ f$. Condition a) tells us that $\mathrm{Spec}(g) = \mathrm{Spec}(h)$. Now let $Z := \mathrm{Spec}(C)$, $Y := \mathrm{Spec}(h)(Z) = \mathrm{Spec}(g)(Z)$ and $X := \mathrm{Spec}(f)(Y)$. Consider the diagram:

$$
\begin{array}{ccccc}
A & \xrightarrow{\ f\ } & B & \overset{g}{\underset{h}{\rightrightarrows}} & C \\
\downarrow{\scriptstyle a} & & \downarrow{\scriptstyle b} & & \downarrow{\scriptstyle c} \\
\prod_{\mathfrak{p}\in X} k(\mathfrak{p}) & \xrightarrow{\ f'\ } & \prod_{\mathfrak{q}\in Y} k(\mathfrak{q}) & \overset{g'}{\underset{h'}{\rightrightarrows}} & \prod_{\mathfrak{m}\in Z} k(\mathfrak{m})
\end{array}
$$

We have $g' \circ f' = h' \circ f'$. From (a) and (b) it follows that f' is an isomorphism; hence $g' = h'$ and then $c \circ h = c \circ g$. Since C is reduced, the morphism c is injective, hence $g = h$. Let now $\phi : B \otimes_A B \rightarrow (B \otimes_A B)_{red}$ be the canonical surjection. From what we have proved, it results that $\phi \circ i_1 = \phi \circ i_2$. Let $x \in B$. Then $1 \otimes x - x \otimes 1 \in \ker(\phi)$.

Since $\mathfrak{a}$ is generated by such elements, we get that $\mathfrak{a} \subseteq \ker(\phi)$ is a nilideal. As $\mathfrak{a}$ is finitely generated, it is a nilpotent ideal. But then $\mathfrak{a} = \mathfrak{a}^2 = \ldots = \mathfrak{a}^s = (0)$. Now we apply Lemma 3.7.11.   □

Proposition 3.7.16 *Let* $f : A \to B$ *be a flat epimorphism and* $\mathfrak{b}$ *be an ideal of* B. *Then* $(\mathfrak{b} \cap A)B = \mathfrak{b}$.

Proof Let $\mathfrak{a} := \mathfrak{b} \cap A$. Then

$$B/\mathfrak{b} \cong B/\mathfrak{b} \otimes_B B \cong B/\mathfrak{b} \otimes_B (B \otimes_A B) \cong B/\mathfrak{b} \otimes_A B.$$

Thus from the injection $A/\mathfrak{a} \to B/\mathfrak{b}$, by tensorizing with B, we obtain the injection $B/\mathfrak{a}B \to B/\mathfrak{b}$. As this morphism is clearly surjective, the conclusion follows.   □

Corollary 3.7.17 *Let* $f : A \to B$ *be a flat epimorphism. Then* $\mathrm{Spec}(f)$ *is a homeomorphism between* $\mathrm{Spec}(B)$ *and its image.*

Proof Let $V(\mathfrak{b})$ be a closed set in $\mathrm{Spec}(B)$ and $\mathfrak{a} = \mathfrak{b} \cap A$. By Proposition 3.7.16 it follows that

$$\mathrm{Spec}(f)(V(\mathfrak{b})) = \mathrm{Spec}(f)(V(\mathfrak{a}B)) = V(\mathfrak{a}) \cap \mathrm{Spec}(f)(\mathrm{Spec}(B))$$

which is a closed subset of $\mathrm{Spec}(B)$.   □

Corollary 3.7.18 *Let* $f : A \to B$ *be a flat epimorphism. If* A *is Noetherian, then* B *is Noetherian.*

Proof Let $\mathfrak{b}$ be an ideal of B and let $\mathfrak{a} := \mathfrak{b} \cap A$. Then there exists a finite system of generators of $\mathfrak{a}$, say $a_1, \ldots, a_n \in A$. But by Proposition 3.7.16 it results that $\mathfrak{b} = \mathfrak{a}B = (a_1, \ldots, a_n)B$, whence B is Noetherian.   □

Proposition 3.7.19 *Let* $f : A \to B$ *be a flat epimorphism. If* A *is a Nagata ring, then also* B *is a Nagata ring.*

Proof By Corollary 3.7.18 the ring B is Noetherian. Let $\mathfrak{q} \in \mathrm{Spec}(B)$ and $\mathfrak{p} = \mathfrak{q} \cap A$. Then $A_\mathfrak{p} \cong B_\mathfrak{q}$. From Theorem 2.3.12 we get that $B_\mathfrak{q}$ has geometrically reduced formal fibres. Since by Corollary 3.7.17 $\mathrm{Spec}(B)$ is homeomorphic to a subset of $\mathrm{Spec}(A)$ and A is Nor-2 by Theorem 2.3.12, we easily deduce that B is Nor-2. Now we apply Theorem 2.3.12 once more.   □

We shall use the above considerations about ring epimorphisms later too. Now we need some of them for our study of inductive limits.

Proposition 3.7.20 (Marot [240, Prop. 4.8]) *Let $(A_i, \varphi_{ij})_{i \in I}$ be an inductive system of Noetherian rings and for any index $i \in I$ let $\mathfrak{a}_i$ be an ideal of A_i, such that $\varphi_{ij}(\mathfrak{a}_i) \subseteq \mathfrak{a}_j, \forall i \leq j$. Let $B_i = \widehat{(A_i, \mathfrak{a}_i)}$ be the completion of A_i in the $\mathfrak{a}_i$-adic topology and let $(B_i, \psi_{ij})_{i \in I}$ be the inductive system constructed in Prop. 1.3.15. Assume that:*

(i) φ_{ij} is a flat epimorphism $\forall i \leq j$;
(ii) there exists an index $i_0 \in I$ such that A_{i_0} is a Nagata ring.

Then $B := \varinjlim_{i \in I} B_i$ is a Nagata ring.

Proof Since φ_{ij} is a flat epimorphism, it follows that the canonical morphisms $\phi_i : A_i \to A := \varinjlim A_i$ are flat epimorphisms; hence by Lemma 3.7.18 the ring A is Noetherian. Let $B := \varinjlim B_i$ and $\psi_i : B_i \to B$ be the canonical morphisms. Let $\mathfrak{m} \in \mathrm{Max}(B)$ and $\mathfrak{m}_i := \psi_i^{-1}(\mathfrak{m})$. We consider the inductive system $((B_i)_{\mathfrak{m}_i}, \psi_{ij})_i$. This inductive system verifies the conditions of Corollary 3.7.6. Indeed, by Proposition 1.3.17 it follows that $(B_i)_{\mathfrak{m}_i}$ is Noetherian and ψ_{ij} is flat, while the residue fields of $(B_i)_{\mathfrak{m}_i}$ and $(B_j)_{\mathfrak{m}_j}$ are the same for any $i, j \in I$. So, in order to apply Corollary 3.7.5 we have to show that $(B_i)_{\mathfrak{m}_i}$ is a Nagata ring. By Lemma 2.2.9 it is enough to show that the canonical morphism $\psi : B \to C := \varprojlim_n (\varinjlim_i A_i/\mathfrak{a}_i^n)$, as defined in Proposition 1.3.15, is faithfully flat and reduced and that C is a Nagata ring. By Proposition 1.3.15 we see that the morphism ψ is faithfully flat. Since φ_{ij} are flat epimorphisms, by Proposition 3.7.19 the rings A_i for $i \geq i_0$ and A are Nagata rings. By Marot's theorem 2.2.5, it follows that B_i for $i \geq i_0$ and C are Nagata rings. Thus $(B_i)_{\mathfrak{m}_i}$ is a Nagata ring. Now by Corollary 3.7.6 we get that $B_\mathfrak{m}$ has geometrically reduced formal fibres. Since $C = \widehat{(B, \mathfrak{a})}$, it follows that the morphism $\psi : B \to C$ is reduced and Proposition 2.2.9 shows that B is a Nagata ring.   □

Corollary 3.7.21 (Marot [240, Prop. 4.7]) *Let A be a Noetherian ring, S a multiplicatively closed system in A and $\mathfrak{a}$ an ideal of A such that $\mathfrak{a} \cap S = \emptyset$. If A is a Nagata ring, then $A\{S^{-1}\}$ and $A_{\{S\}}$ are Nagata rings.*

Proof By Theorem 2.2.5 it follows that $A\{S^{-1}\}$ is a Nagata ring. The second assertion follows from Proposition 3.7.20 and the definition of $A_{\{S\}}$ (Notation 3.7.8).   □

Proposition 3.7.22 (Doretti [91, Prop. 2.4]) *Let $(A_i, \varphi_{ij})_{i \in I}$ be an inductive system of Noetherian local rings and $A := \varinjlim A_i$. Let **P** be a property of Noetherian rings satisfying $(\mathbf{A_1})$, $(\mathbf{A_2})$, $(\mathbf{A_3})$ and $(\mathbf{A_5})$. Assume that:*

(i) A is a Noetherian ring;
(ii) $\varphi_i : A_i \to A$ is a faithfully flat reduced morphism for any index $i \in I$;
*(iii) A_i is a **P**-ring, for any index $i \in I$.*

Then A is a P-ring.

Proof For any index $i \in I$, let $u_i : A_i \to \widehat{A_i}$ be the canonical completion morphism. By assumption, u_i is a **P**-morphism. We have an inductive system of Noetherian local rings $(\widehat{A_i}, \widehat{\varphi_{ij}})_{i \in I}$ such that $u_j \circ \varphi_{ij} = \widehat{\varphi_{ij}} \circ u_i$ for any pair of indices $i \leq j$. Let us denote $A^* := \varinjlim \widehat{A_i}$. Since the limit morphism $A^* \to \widehat{A}$ is faithfully flat, it results that A^* is a Noetherian ring and by Theorem 3.7.1 that $f : A \to A^*$ is a **P**-morphism. Since by Corollary 1.3.18 we have $\widehat{A} \cong \widehat{(A^*)}$, we obtain that the canonical completion morphism $A \to \widehat{A}$ factorizes as $A \overset{f}{\to} A^* \overset{g}{\to} \widehat{A}$. Thus it suffices to prove that g is a **P**-morphism. Let $\mathfrak{m}_i$ be the maximal ideal of A_i and $\mathfrak{m} = \varinjlim \mathfrak{m}_i$ be the maximal ideal of A. Since A is Noetherian, there exists $i_0 \in I$ such that for any $i \geq i_0$ we have $\mathfrak{m}_i A = \mathfrak{m}$; hence the morphism $g_i : \widehat{A_i} \to \widehat{A}$ is a **P**-morphism. The closed fibre of g_i is

$$\widehat{A} \otimes_{\widehat{A_i}} \widehat{A_i}/\widehat{\mathfrak{m}_i} \cong \widehat{A}/(\mathfrak{m}_i A)\widehat{} \cong \widehat{A}/\mathfrak{m}\widehat{A} \cong A/\mathfrak{m}.$$

Let k and k_i be the residue fields of A resp. A_i. Let L_i be a finite extension of k_i. By (ii) we get that $k \otimes_{k_i} L_i$ is reduced and hence regular, being zero-dimensional. It follows that the closed fibre of g_i is geometrically regular. By Theorem 3.2.22 the morphism g_i is a regular morphism, and by Theorem 3.7.1 it follows that g is a regular morphism.   □

Corollary 3.7.23 (Doretti [91, Cor. 2.5]) *Let $(A_i, \varphi_{ij})_{i \in I}$ be an inductive system of Noetherian local rings and let $A := \varinjlim A_i$. Assume that:*

  (i) *A is a Noetherian ring;*
 (ii) *$\varphi_i : A_i \to A$ is a faithfully flat reduced morphism for any index $i \in I$;*
(iii) *A_i is a Nagata (resp. quasi-excellent) ring, for any index $i \in I$.*

Then A is a Nagata (resp. quasi-excellent) ring.

Finally let us notice a result regarding the openness of loci.

Proposition 3.7.24 (Doretti [91, Prop. 2.9]) *Let $(A_i, \varphi_{ij})_{i \in I}$ be an inductive system of Noetherian rings and let $A := \varinjlim A_i$. Let **P** be a property of Noetherian rings satisfying $(\mathbf{A_1})$, $(\mathbf{A_2})$, $(\mathbf{A_3})$ and $(\mathbf{A_5})$. Assume that:*

  (i) *A is a Noetherian ring;*
 (ii) *$\varphi_i : A_i \to A$ is a **P**-morphism for any index $i \in I$;*
(iii) *A_i is a **P**-2 ring, for any index $i \in I$.*

Then A is a P-2 ring.

Proof Let B be a finitely generated A-algebra. Since B is Noetherian, by Lemma 1.3.13 there exists $i \in I$ and a finitely generated A_i-algebra B_i such that $B \cong B_i \otimes_{A_i} B$. By assumption B_i is **P**-1, hence $\mathbf{P}(B_i)$ is open. Since φ_i is a **P**-morphism, also $\psi_i : B_i \to B_i \otimes_{A_i} A \cong B$ is a **P**-morphism. Then $\mathbf{P}(B) = \mathrm{Spec}(\psi_i)^{-1}(\mathbf{P}(B_i))$ is open. □

Localization and Lifting Theorems

4

Introducing the notions of quasi-excellent rings and more generally of **P**-rings, Grothendieck left some unsolved problems. Among them, two questions seem to be the most important ones: the localization problem for **P**-morphisms and the lifting problem for **P**-rings. Probably the most spectacular result is Corollary 4.6.21 stating that power series ring over quasi-excellent rings remains quasi-excellent. The results presented in this chapter need a serious geometric preparation, because the proofs use an important result of Gabber, known as the weak local uniformization theorem. Because Flenner's form of the second Theorem of Bertini is used in the proof, we decided to present this result that is important in many places in Commutative Algebra and Algebraic Geometry. This chapter uses notions and results from Algebraic Geometry. They are collected in Sect. 4.1.

4.1   Some Geometric Preliminaries

This section contains some notions and results from Algebraic Geometry that are used in the following sections. For most of the results, no proofs are provided, but precise references are given. We assume that the reader is familiar with the basic notions of Algebraic Geometry such as (affine) scheme, Noetherian or locally Noetherian scheme and affine, projective or quasi-projective variety. We refer for these notions to [100, 127, 131–135, 142, 201, 359].

Definition 4.1.1 ([135, Def. 2.1.8]) A scheme X is called an **integral scheme** if the topological space X is irreducible and X is a reduced scheme (i.e. $\mathcal{O}_{X,x}$ is a reduced ring for any $x \in X$).

The notion of (quasi)-excellent ring is globalized to the case of schemes.

Definition 4.1.2 ([132, Def. 7.8.5]) A Noetherian scheme X is **(quasi)-excellent** if X has an open affine covering $X = \bigcup_{i \in I} X_i$ such that $X_i = \mathrm{Spec}(A_i)$, where A_i is a (quasi)-excellent ring for any $i \in I$.

We also need some notions about scheme morphisms.

Definition 4.1.3 ([135, Def. 5.2.1]) A scheme morphism $f : X \to Y$ is called **separated** if the diagonal morphism $\Delta_f : X \times_Y X \to X$ is a closed immersion.

Remark 4.1.4 ([135, Prop. 5.2.2]) Any morphism of affine schemes is separated. This comes essentially from the fact that if $u : A \to B$ is a ring morphism, then the induced morphism $B \otimes_A B \to B$, $b_1 \otimes b_2 \mapsto b_1 \cdot b_2$ is surjective.

Definition 4.1.5 ([135, Def. 6.1.1]) A scheme morphism $f : X \to Y$ is called **quasi-compact** if for any open quasi-compact subset U of Y, the set $f^{-1}(U)$ is a quasi-compact subset of X.

Definition 4.1.6 ([135, Def. 6.2.1]) Let $f : X \to Y$ be a scheme morphism, $x \in X$ and $y = f(x)$. We say that f is a morphism **of finite type in** x if there exists an affine open neighbourhood V of y and an affine open neighbourhood U of x such that $f(U) \subseteq V$ and the coordinate ring of U is an algebra of finite type over the coordinate ring of V. We say that f is a morphism **locally of finite type** if it is a morphism of finite type in every point of X.

Proposition 4.1.7 ([135, Prop. 6.2.5]) *Let* $f : \mathrm{Spec}(B) \to \mathrm{Spec}(A)$ *be a morphism of affine schemes. Then* f *is a morphism locally of finite type if and only if B is an A-algebra of finite type.*

Definition 4.1.8 ([135, Def. 6.3.2]) A scheme morphism $f : X \to Y$ is called of **finite type** if it is locally of finite type and quasi-compact.

Definition 4.1.9 ([135, Def. 2.3.3] or [359, Def. 29.8.1]) A scheme morphism $f : X \to Y$ is called **dominant** if $\mathrm{Im}(f)$ is a dense subset of Y.

Remark 4.1.10

(i) The notion of dominant morphism involves only the topological structure of the schemes.

(ii) If $f : Y = \mathrm{Spec}(B) \to X = \mathrm{Spec}(A)$ is a morphism of affine schemes and $u : A \to B$ is the associated ring morphism, then f is dominant if and only if $\ker(u)$ is contained in the nilradical $N(A)$. Indeed, for any ideal $\mathfrak{b}$ of B, $V(u^{-1}(\mathfrak{b})) = \overline{f(V(\mathfrak{b}))}$.

Thus, f is dominant iff $V(\ker(u)) = \overline{f(X)} = Y$ iff $\ker(u) \subseteq N(A)$. In particular if A is reduced, then u must be injective.

Definition 4.1.11 ([135, Ch. 0, 2.1.1]) If X is a topological space, a point $x \in X$ is called a **maximal point** of X if x is the generic point of an irreducible component of X.

Remark 4.1.12 If $X = \mathrm{Spec}(A)$ is an affine scheme, a maximal point of X corresponds exactly to a minimal prime ideal of A.

Definition 4.1.13 ([171, Exp. II, Def. 1.1.2]) A scheme morphism $f : Y \to X$ is called **maximally dominant** if every maximal point of Y maps to a maximal point of X.

Remark 4.1.14

(i) If $f : Y = \mathrm{Spec}(B) \to X = \mathrm{Spec}(A)$ is a morphism of affine schemes and $u : A \to B$ is the associated ring morphism, then f is maximally dominant if and only if for any minimal prime ideal $\mathfrak{q}$ of B, the prime ideal $\mathfrak{q} \cap A = u^{-1}(\mathfrak{q})$ is a minimal prime ideal of A.

(ii) If f is flat, then it is maximally dominant as it follows from [135, Prop. 3.9.3 ii) and Prop. 3.9.5 ii)].

Definition 4.1.15 ([135, Def. 3.8.1]) A scheme morphism $f : X \to Y$ is called **universally closed** if for any base change $T \to Y$, the morphism $X \times_Y T \to T$ is a closed map of the underlying topological spaces.

Remark 4.1.16 A universally closed morphism is quasi-compact [359, Lemma 29.40.9].

Definition 4.1.17 ([135, Def. 2.3.4 and Rem. 2.3.4.1]) A scheme morphism $f : X \to Y$ is called **birational** if the restriction of f to the set of maximal points of X is a bijection between the sets of maximal points of X and Y and for any maximal point x of X the morphism $\mathcal{O}_{Y,f(x)} \to \mathcal{O}_{X,x}$ is an isomorphism.

Definition 4.1.18 ([142, Def. p. 100]) A scheme morphism $f : X \to Y$ is called **proper** if f is separated, of finite type and universally closed.

Remark 4.1.19 A closed immersion is a proper morphism [359, Lemma 29.40.6].

We will also need the notion of projective morphism.

Definition 4.1.20 ([142, Def. 4.8]) A morphisms of schemes $f : X \to Y$ is called **projective** if it factors as a closed immersion $i : X \to \mathbf{P}_Y^n$ for some $n \in \mathbb{N}$, followed by the projection $\mathbf{P}_Y^n \to Y$.

The dream of everyone working in Algebraic Geometry is to work with a non-singular object. Equivalently, the dream of every Commutative Algebraist would be to work with a regular ring. Since this is not always possible, people tried to find something non-singular as close as possible to the initial object. Sometimes this is easy: if we consider, for example, the ring $\mathbb{Z}[X^2, X^3] \cong \mathbb{Z}[T, U]/(T^2 - U^3)$, this is not a regular ring, but its integral closure, which is $\mathbb{Z}[X]$, is a regular ring that is finite over it. Unfortunately this is far from being always possible, at least not in this easy way. The next definitions formalize this wish.

Definition 4.1.21 A morphism of algebraic varieties $f : Y \to X$ is called a **modification** if it is proper and birational.

Definition 4.1.22 A **resolution of singularities** of a variety X is a modification $f : Y \to X$, where Y is a regular scheme.

The most important and famous result obtained so far is the following:

Theorem 4.1.23 (Hironaka [160]) *If X is a variety over a field of characteristic zero, there exists a resolution of singularities of X.*

Grothendieck has shown that there is a connection between the resolution of singularities and excellence.

Proposition 4.1.24 (Grothendieck [132, Prop. 7.9.5]) *Let X be a locally Noetherian scheme such that for any integral scheme Y which is finite over X we can resolve the singularities of Y. Then X is quasi-excellent.*

Grothendieck also conjectured that the converse is true, but this was proved only recently by Temkin.

Theorem 4.1.25 (Temkin [374, Thm. 1.1]) *Let X be a scheme of characteristic zero. Then X is quasi-excellent if and only if any integral scheme of finite type over X has a resolution of singularities.*

The proof is quite complicated. Many people tried to avoid the restriction about the characteristic zero in the above results. Unfortunately, so far there does not exist a result similar to 4.1.23 in the general case. One of the most satisfactory results was obtained by giving up the birationality request.

Definition 4.1.26 ([171, Prop. 1.1.7]) Let X be a Noetherian scheme. A morphism of finite-type, maximally dominant $f : X \to Y$ is called **generically finite** if for any maximal point $x \in X$, the field extension $k(f(x)) \subseteq k(x)$ is finite.

Definition 4.1.27 A morphism of algebraic varieties $f : Y \to X$ is called an **alteration** if it is surjective, proper and generically finite.

Theorem 4.1.28 (de Jong [189]) *If X is a variety over a field k, there exists an alteration $f : Y \to X$, where Y is a non-singular variety.*

Remark 4.1.29

  (i) Every modification is an alteration.
 (ii) Any alteration $f : Y \to X$ can be factored as $Y \xrightarrow{g} Z \xrightarrow{h} X$, where $g : Y \to Z$ is a modification and $h : Z \to X$ is a finite morphism.
(iii) A morphism of projective varieties is a modification iff it is birational iff it is an isomorphism almost everywhere; it is an alteration if it is finite almost everywhere.

To generalize Theorem 4.1.28 for schemes, we need to generalize the notion of alteration.

Definition 4.1.30 ([171, Exp. II, Def. 2.3.1 and Th. 3.2.1]) Let Y be a Noetherian integral scheme. A finite family of scheme maps $\{f_i : X_i \to Y\}_{1 \le i \le m}$ is called an **alteration covering** of Y, if there exists a proper surjective morphism $f : V \to Y$, an open covering $V = \bigcup_{i=1}^{m} V_i$ with a family of scheme maps $\{g_i : V_i \to X_i\}_{1 \le i \le m}$ such that for each $i = 1, \ldots, m$ the following diagram is commutative

$$
\begin{array}{ccc}
V_i & \longrightarrow & V \\
\downarrow{g_i} & & \downarrow{f} \\
X_i & \xrightarrow[f_i]{} & Y
\end{array}
$$

where $V_i \to V$ is the natural immersion. If for each $i = 1, \ldots, m$ the scheme X_i is a regular integral scheme, then we say that $\{f_i : X_i \to Y\}_{1 \le i \le m}$ is a **regular alteration covering** of Y.

We will use the following important result, known as Gabber's weak local uniformization theorem [171].

Theorem 4.1.31 (Gabber, see [171, Exp. II, Th. 4.3.1]) *Let Y be a quasi-excellent Noetherian integral scheme. Then there exists a regular alteration covering of Y.*

This result will be used as a device to replace in the general situation the well-known resolution of singularities 4.1.23, which unfortunately is valid only in characteristic zero so far.

A problem similar to the resolution of singularities was considered by changing the requirement that the resolving scheme is regular, to the one of being Cohen-Macaulay. The first to consider this device, called Macaulayfication, was Faltings [105].

Definition 4.1.32 A locally Noetherian scheme X is called **CM-quasi-excellent** if the formal fibres of the local rings $\mathcal{O}_{X,x}$ are Cohen-Macaulay and every integral closed subscheme $Z \subseteq X$ has a non-empty Cohen-Macaulay open subscheme.

Example 4.1.33 Any quasi-excellent scheme is CM-quasi-excellent. Any Cohen-Macaulay scheme is CM-excellent.

Theorem 4.1.34 (Česnavičius [80, Th. 5.3]) *For every Noetherian CM-quasi-excellent scheme, X there exist a Cohen-Macaulay scheme Y and a modification $f : Y \to X$ that is an isomorphism over the Cohen-Macaulay locus $CM(X) \subset X$.*

Without the condition about the Cohen-Macaulay locus, the result was obtained by Kawasaki [192]. But this improvement allowed Česnavičius to obtain several interesting consequences that are beyond our interest.

Definition 4.1.35 Let **P** be a property of Noetherian local rings. If X is a locally Noetherian reduced scheme, we say that a scheme Y is a **P**-resolution of X if Y has the property **P**, and there exists a modification $f : Y \to X$. The morphism f is called a **P-resolvent morphism**.

Example 4.1.36

   (i) For the property **P**=regular, we have Hironaka's result above 4.1.23.
   (ii) If A is a Nagata ring, then A has a **P**-resolution, for **P**=normal, namely, the normalization of A.
   (iii) If A is a Japanese ring of dimension 1, then the normalization of A is a resolution of singularities, that is, a **P**-resolution, for **P**=regular.
   (iv) If A is a Japanese ring of dimension 2, then the normalization of A is a **P**-resolution, for **P**=Cohen-Macaulay.

There is a connection, as we mentioned earlier, between the existence of **P**-resolutions and the good properties of a ring.

Proposition 4.1.37 ([132], [388, Prop. 3.1]) *Let X be a locally Noetherian scheme such that any X-scheme integral and finite over X has a **P**-resolution, where **P** is a property* *satisfying the properties* $(\mathbf{A_1})$, $(\mathbf{A_2})$, $(\mathbf{A_3})$, $(\mathbf{A_5})$ *and* $(\mathbf{NC})$*. Then X is a* $\mathbf{P} - 2$ *scheme, that is, X has an open covering with affine schemes that are spectra of* $\mathbf{P}$*-2 rings.*

Proof If we can **P**-resolve X, by [128, Prop. 5.4.2] we can **P**-resolve any scheme induced to an open subset of X and any scheme of the form $\mathcal{O}_{X,x}$. Moreover, there exists an open subset U of X such that $\overline{U} = X$ and $U \subseteq \mathbf{P}(X)$. $\qquad\square$

Lemma 4.1.38 *Let A be a Noetherian local reduced ring and let* **P** *be a property satisfying the properties* $(\mathbf{A_1})$, $(\mathbf{A_2})$, $(\mathbf{A_3})$, $(\mathbf{A_4})$ *and* $(\mathbf{A_6})$*. If any finite integral A-algebra B has a* **P**-*resolution, then A is a* **P**-*ring.*

Proof We can assume that A is a domain, and we have to prove that the fibre in (0) of the completion morphism $A \to \widehat{A}$ has the property **P**. Let Y be a **P**-resolution of A. We have the commutative diagram of canonical scheme morphisms

$$
\begin{array}{ccc}
Y & \xrightarrow{\;\;f\;\;} & X = \mathrm{Spec}(A) \\
\big\uparrow{\scriptstyle g'} & & \big\uparrow{\scriptstyle g} \\
Y' = Y \times_X X' & \xrightarrow{\;\;f'\;\;} & X' = \mathrm{Spec}(\widehat{A}).
\end{array}
$$

Since f' is a finite-type morphism, the scheme Y' is Noetherian. Let $\mathfrak{m}'$ be the closed point of X', let $\mathfrak{m}$ be the closed point of X and let $y' \in (f')^{-1}(\mathfrak{m}')$. Then $g^{-1}(\mathfrak{m}) = \mathfrak{m}'$ and $k(\mathfrak{m}) = k(\mathfrak{m}')$. Thus $(g')^{-1}(f^{-1}(\mathfrak{m})) = (f')^{-1}(\mathfrak{m}')$ and the restriction $f^{-1}(\mathfrak{m}) \to (f')^{-1}(\mathfrak{m}')$ is an isomorphism of fibres. Let $y = g'(y')$ and $B = \mathcal{O}_{Y,y}$. If $\mathfrak{n}$ is the maximal ideal of B, there is only one maximal ideal $\mathfrak{n}'$ of $B' := B \otimes_A \widehat{A}$ lying over $\mathfrak{n}$ and $\mathcal{O}_{Y',y'} = B'_{\mathfrak{n}'}$. By [132, Lemma 7.9.3.1], we have that $\widehat{B} \cong \widehat{B'_{\mathfrak{n}'}}$. But B has the property **P**; hence $\mathcal{O}_{Y',y'}$ has the property **P**. This implies that the open set $\mathbf{P}(Y')$ contains $(f')^{-1}(\mathfrak{m}')$. But then, for any point $y \in Y'$, the closure of y contains at least one point y' since f' is closed. Hence $\mathcal{O}_{Y',y}$ has **P**. $\qquad\square$

Theorem 4.1.39 (Valabrega [388, Th. 3.2]) *Let X be a locally Noetherian scheme such that any X-scheme integral and finite over X has a* **P**-*resolution, where* **P** *is a property satisfying the properties* $(\mathbf{A_1})$, $(\mathbf{A_2})$, $(\mathbf{A_3})$, $(\mathbf{A_4})$ *and* $(\mathbf{A_6})$*. Then for any* $x \in X$*, the canonical completion morphism* $\mathcal{O}_{X,x} \to \widehat{\mathcal{O}_{X,x}}$ *is a* **P**-*morphism.*

Proof It follows easily from Lemma 4.1.38. $\qquad\square$

4.2   Localization Theorems

Let us consider a property **P** of Noetherian local rings. In his monumental work *Eléments de géometrie algébrique* [127]-[135], A. Grothendieck, who initiated the study of **P**-rings, considered the following problems:

Problem 4.2.1 (The Localization Problem for P-morphisms [133, Rem. 7.5.4])   Consider a local and flat morphism of Noetherian local rings $u : (A, \mathfrak{m}, k) \rightarrow (B, \mathfrak{n}, K)$. Assume that:

(i) A is a **P**-ring.
(ii) $u \otimes 1_k : k \rightarrow B \otimes_A k$ is a **P**-morphism.
    Does it follow that u is a **P**-morphism?

Problem 4.2.2 (The Lifting Problem for P-rings [133, Rem. 7.4.8])   Let A be a Noetherian ring and $\mathfrak{a}$ an ideal of A such that A is complete in the $\mathfrak{a}$-adic topology. If $A/\mathfrak{a}$ is a **P**-ring, does it follow that A is a **P**-ring?

A weaker form of Problem 4.2.2 is the following problem.

Problem 4.2.3 (The Stability to Completion for P-rings [133, Rem. 7.4.8])   Let A be a Noetherian ring, let $\mathfrak{a}$ be an ideal of A and let B be the completion of A in the $\mathfrak{a}$-adic topology. If A is a **P**-ring, does it follow that B is a **P**-ring?

Remark 4.2.4

(i) Consider the assumptions in Problem 4.2.1. Condition (ii) means that the fibre of u at the closed point is a **P**-morphism. The conclusion means that all the fibres of u are **P**-morphisms. In other words, does the property of being a geometrically **P** algebra *localize* from the fibre at the closed point to all the fibres?
(ii) It is clear that a positive answer to Problem 4.2.2 for a property **P** implies that also Problem 4.2.3 has a positive answer for **P**.

The aim of this chapter is to study the status of these problems. This section deals with Problem 4.2.1. The first general result was Theorem 3.2.22, given by M. André for **P**=regular. We start this section with a positive answer for Problem 4.2.1 for the property **P**=reduced. The first proof of this result was given by Nishimura [269]. His proof used in an essential way André's theorem of the localization of the formal smoothness (3.2.22). Here we shall give a direct proof that does not use the Theorem of André.

Theorem 4.2.5 (Nishimura [269, Prop. 2.4]) *Let* $u : (A, \mathfrak{m}, k) \to (B, \mathfrak{n}, K)$ *be a local flat morphism of Noetherian local rings. Assume that:*

(i) *A has geometrically reduced formal fibres.*
(ii) *$B \otimes_A k$ is a geometrically reduced k-algebra.*
    Then u is a reduced morphism.

Proof ([177]) First of all let us remark that, by Theorem 2.3.6, the assumption (i) means that A is a Nagata ring. Let B^* be the $\mathfrak{m}B$-adic completion of B. We have the canonical commutative diagram:

$$
\begin{array}{ccc}
A & \xrightarrow{\ u\ } & B \\
{\scriptstyle f}\downarrow & & \downarrow{\scriptstyle g} \\
\widehat{A} & \xrightarrow{\ v\ } & B^*
\end{array}
$$

By assumption it follows that f is reduced. If we prove that v is reduced, from Proposition 1.6.17 and Proposition 1.6.18, it will follow that u is reduced. Hence we can suppose that A is a complete local ring. We have to prove that for any prime ideal $\mathfrak{p}$ of A the $k(\mathfrak{p})$-algebra $B \otimes_A k(\mathfrak{p})$ is geometrically reduced. By Noetherian induction, we may assume that for any prime ideal $\mathfrak{q}$ strictly containing $\mathfrak{p}$, the $k(\mathfrak{q})$-algebra $B \otimes_A k(\mathfrak{q})$ is geometrically reduced. Replacing A with $A/\mathfrak{p}$, we may assume that A is a domain, that for any non-zero prime ideal $\mathfrak{q}$ of A the $k(\mathfrak{q})$-algebra $B \otimes_A k(\mathfrak{q})$ is geometrically reduced, and we have to show that the fibre of u in (0) is geometrically reduced. Let L be a field, finite extension of $K = Q(A)$. Let C be a finite extension of A having L as field of fractions. Then C is a semilocal ring, and setting $D := C \otimes_A B$, we have $B \otimes_A L \cong D \otimes_C L$. Let us observe that every local ring of $B \otimes_A L$ is a local ring of $D_{\mathfrak{q}} \otimes_{C_{\mathfrak{p}}} L$ for some prime ideals $\mathfrak{p}$ of C and $\mathfrak{q}$ of D. Moreover,

$$B \otimes_A k(\mathfrak{p}) \cong B \otimes_A C \otimes_C k(\mathfrak{p}) \cong D \otimes_C k(\mathfrak{p}).$$

Thus $D_{\mathfrak{q}} \otimes_{C_{\mathfrak{p}}} k(\mathfrak{p})$ is a localization of $B \otimes_A k(\mathfrak{p})$. Hence we may replace A by $C_{\mathfrak{p}}$, B by $D_{\mathfrak{q}}$ and K by L, and we have only to show that the local rings of $B \otimes_A K$ are reduced. Taking C to be the integral closure of A in L in the above consideration, we may also assume that A is normal. Finally we can assume that A is a local normal domain having geometrically regular formal fibres, and we have only to show that $B \otimes_A Q(A)$ is a reduced ring. For this take a prime ideal $\mathfrak{P} \in \mathrm{Spec}(B)$ such that $\mathfrak{P} \cap A = (0)$ and $\mathrm{depth}(B_{\mathfrak{P}}) = 0$. Let $x \in A$ be a non-zero and non-invertible element and take $\mathfrak{Q} \in \mathrm{Min}(\mathfrak{P} + xB)$. Then $\mathrm{depth}(B_{\mathfrak{Q}}) = 1$ because by flatness x is also B-regular. Let $\mathfrak{q} := \mathfrak{Q} \cap A$. Then clearly we have depth$(A_q) = 1$. Consequently,

$$\text{depth}(B_\mathfrak{Q}/qB_\mathfrak{Q}) = \text{depth}(B_\mathfrak{Q}) - \text{depth}(A_q) = 0.$$

Since the fibre of u in q is reduced, from Theorem 1.1.36 it follows that $B_\mathfrak{Q}/qB_\mathfrak{Q}$ is regular. But as depth$(A_q) = 1$ and A is normal, from Theorem 1.1.43 it follows that A_q is regular. Hence $B_\mathfrak{Q}$ is regular and consequently also $B_\mathfrak{P}$ is regular, being a localization of $B_\mathfrak{Q}$.   □

The following theorem gives a complete and uniform answer to Problem 4.2.2 for many properties **P**. Essentially the proof goes back to Marot [244, Th. 2.1], who used Hironaka's resolution of singularities 4.1.23, thus restricting to the characteristic zero case. Instead, Murayama used Gabber's weak local uniformization Theorem 4.1.31 result that was not available when Marot wrote [244]. We need some more topological preparations.

Definition 4.2.6 Let X be a topological space and let $x, y \in X$. We say that x is a **specialization** of y or that y is a **generalization** of x, if $x \in \overline{\{y\}}$.

Remark 4.2.7 In the affine case $X = \text{Spec}(A)$, for two points $x = q$ and $y = \mathfrak{p}$ in X, one can see that y is a specialization of x iff $q \subseteq \mathfrak{p}$.

Lemma 4.2.8 *Let $f : X \to Y$ be a continuous map of topological spaces. Let $y, y' \in Y$ be such that y is a specialization of y' (i.e. $y \in \overline{\{y'\}}$), and let $x' \in X$ be such that $f(x') = y'$. If f is closed, there exists x a specialization of x' in X, such that $f(x) = y$.*

Proof Let $T := \overline{\{x'\}} \subset X$. Then $f(T)$ is closed in Y and $y' \in f(T)$, hence $y \in f(T)$. Thus $y = f(x), x \in T$ and x is a specialization of x'.   □

Lemma 4.2.9 ([132, Lemma 7.5.1.1]) *Let **P** be a property of Noetherian local rings satisfying $(\mathbf{A_7})$ and $u : (A, \mathfrak{m}, k) \to (B, \mathfrak{n}, K)$ be a flat local morphism of Noetherian local rings. If A is a regular local ring and $B \otimes_A k$ has the property **P**, then B has the property **P**.*

Proof We argue by induction on $n = \dim(A)$. If $n = 0$, then $A = k$, and there is nothing to prove. Let $n > 0$ and take $t \in \mathfrak{m} \setminus \mathfrak{m}^2$. Then A/tA is a regular local ring of dimension $n - 1$. Moreover,

$$B/tB \otimes_{A/tA} k \cong B \otimes_A k,$$

hence by induction B/tB has the property **P**. The element t is B-regular, because u is flat. Hence by $(\mathbf{A_7})$ the ring B has the property **P**.   □

Theorem 4.2.10 (Murayama [260, Cor. 3.6], see also Marot [244, Th. 2.2]) *Let u : $(A, \mathfrak{m}, k) \to (B, \mathfrak{n}, K)$ be a local flat morphism of Noetherian local rings and let P be a property of Noetherian local rings satisfying $(\mathbf{A_1})$, $(\mathbf{A_2})$, $(\mathbf{A_3})$, $(\mathbf{A_5})$ and $(\mathbf{A_7})$. Assume that:*

(i) A is a P-ring.
(ii) $B \otimes_A k$ is a geometrically P k-algebra.
   Then u is a P-morphism.

Proof As in the proof of Theorem 4.2.5, we may assume that A is a complete local domain with field of fractions K, that the fibres of u in all non-zero prime ideals are geometrically **P**, and we have to show that the fibre of u in (0) is geometrically **P**, that is, we have to show that if L is a finite field extension of K, the local rings of $B \otimes_A L$ have the property **P**. Let C be a finite extension of A with field of fractions L. Hence C is a semilocal ring, and if we set $D := B \otimes_A C$, we have $B \otimes_A L \cong D \otimes_C L$. Every local ring of $B \otimes_A L$ is a local ring of $D_{\mathfrak{q}} \otimes_{C_{\mathfrak{p}}} L$ for some prime ideals $\mathfrak{p}$ of C and $\mathfrak{q}$ of D. Moreover we have

$$B \otimes_A k(\mathfrak{p}) \cong B \otimes_A C \otimes_C k(\mathfrak{p}) \cong D \otimes_C k(\mathfrak{p}).$$

Hence $D_{\mathfrak{q}} \otimes_{C_{\mathfrak{p}}} L$ is a localization of $B \otimes_A k(\mathfrak{p})$. Thus replacing A by $C_{\mathfrak{p}}$, B by $D_{\mathfrak{q}}$ and K by L, we may assume that A is a quasi-excellent local domain with field of fractions K, and we have to show that the local rings of $B \otimes_A K$ have the property **P**. By Gabber's weak local uniformization theorem 4.1.31, since $X := \mathrm{Spec}(A)$ is quasi-excellent, there exists a finite alteration covering $\{f_i : Y_i \to X\}_{1 \leq i \leq m}$ of X such that Y_i is regular and integral for every $i = 1, \ldots, m$. Since X is irreducible, by [171, Exp. II, Th. 3.2.1], there exists a proper surjective morphism $\pi : V \to X$ and a Zariski open covering $V = \bigcup_{i=1}^{m} V_i$ together with a family of morphisms $h_i : V_i \to Y_i$, $i = 1, \ldots, m$ such that the following diagram commutes for every $i = 1, \ldots, m$:

$$
\begin{array}{ccc}
V_i & \hookrightarrow & V \\
\downarrow{\scriptstyle h_i} & & \downarrow{\scriptstyle \pi} \\
Y_i & \xrightarrow{f_i} & X
\end{array}
$$

Setting $X' := \mathrm{Spec}(B)$ and changing the base along the canonical morphism $\theta := \mathrm{Spec}(u) : X' \to X$, we get the commutative diagrams with Cartesian squares, for every

$i = 1, \ldots, m$:

$$
\begin{array}{ccc}
V'_i & \hookrightarrow & V' \\
h'_i \downarrow & & \downarrow \pi' \\
Y'_i & \xrightarrow{f'_i} & X' \\
g_i \downarrow & & \downarrow \theta \\
Y_i & \xrightarrow{f_i} & X.
\end{array}
$$

Let $\eta' \in X'$ be a point lying over the generic point of X. We have to show that $\mathcal{O}_{X',\eta'}$ has the property **P**. There exists a point $\xi' \in V'$ such that $\pi'(\xi') = \eta'$, because π' is surjective. Note that $\mathfrak{n}$ is a specialization of η' in $X' = \mathrm{Spec}(B)$. By Lemma 4.2.8 there exists a specialization v' of η' in X' such that $\pi'(v') = \mathfrak{n}$. Since $V' = \bigcup_{i=1}^{m} V'_i$ is a Zariski open covering, there exists an index $i_0 \in \{1, \ldots, m\}$ such that $v' \in V'_{i_0}$. Since open sets are stable by generalization [248, Rem. (6.C)], it follows that $\xi' \in V'_{i_0}$. Let us look again at the commutative diagram

$$
\begin{array}{ccc}
Y'_{i_0} & \xrightarrow{f'_{i_0}} & X' \\
g_{i_0} \downarrow & & \downarrow \theta \\
Y_{i_0} & \xrightarrow{f_{i_0}} & X.
\end{array}
$$

The morphism f_{i_0} is maximally dominant. Thus, changing the base along the morphism $\mathrm{Spec}(K) \to X$, localizing at the generic point of Y_{i_0} and taking global sections give the commutative co-Cartesian diagram

$$
\begin{array}{ccc}
K & \longrightarrow & L_{i_0} \\
\downarrow & & \downarrow \\
B \otimes_A K & \longrightarrow & B \otimes_A L_{i_0}
\end{array}
$$

where L_{i_0} is the function field of Y_{i_0}. Thus, from the top faithfully flat morphism, by base change and localization, we obtain a faithfully flat morphism

$$
\mathcal{O}_{X',\eta'} \cong (B \otimes_A K)_{\eta'} \to (B \otimes_A L_{i_0})_{h'_{i_0}(\xi')} \cong \mathcal{O}_{Y'_{i_0},h'_{i_0}(\xi')}.
$$

Hence, if we prove that $\mathcal{O}_{Y'_{i_0},h'_{i_0}(\xi')}$ has the property **P**, it will follow that $\mathcal{O}_{X',\eta'}$ has the property **P** and we are done. Let $y' := h'_{i_0}(v')$ and $y := g_{i_0}(y')$. Consider the commutative diagram

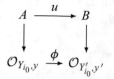

Since f_{i_0} is of finite type, the residue field extension $k \subseteq k(y)$ is finitely generated; hence the closed fibre of the flat morphism ϕ above is a geometrically $\mathbf{P}$ algebra over $k(y)$. On the other hand, $\mathcal{O}_{Y_{i_0},y}$ is a regular local ring by construction. From Lemma 4.2.9 we deduce now that $\mathcal{O}_{Y'_{i_0},y'}$ has the property $\mathbf{P}$. By continuity the specialization v' of ξ' maps to the specialization $y' = h'_{i_0}(v')$ of $h'_{i_0}(\xi')$. It follows that $\mathcal{O}_{Y'_{i_0},h'_{i_0}(\xi')}$ has the property $\mathbf{P}$. $\square$

Remark 4.2.11

(i) Theorem 4.2.10 provides a positive answer to the Localization Problem 4.2.1, as well as a uniform proof for the following properties of Noetherian local rings: regular, complete intersection, Gorenstein, Cohen-Macaulay, (S_n), (R_n) and (S_{n+1}), normal, reduced. As one can easily see, the same proof works for all these properties.

(ii) For the property $\mathbf{P} = $ regular, as it is well-known and we already mentioned, the result was proved by André [8] (see 3.2.22). In his proof he used André-Quillen homology. The above theorem gives a new proof of this result.

(iii) For the property $\mathbf{P} = $ complete intersection, the result was first proved by Tabaâ [363] using André-Quillen homology. Another proof was given by Avramov and Foxby [27].

(iv) For the property $\mathbf{P} = $ Gorenstein, the result was proved by Hall and Sharp [140, Th. 3.3] using dualizing complexes. Another proof was given by Avramov and Foxby [27].

(v) For the property $\mathbf{P} = $ Cohen-Macaulay, a proof quite similar to the above one, inspired by Marot's proof in [244], was given in [62] using Faltings's Macaulayfication [105]. Because at that time the general results of Kawasaki [192] or Česnavičius [80] didn't exist, the authors of [62] stated the result only in low dimensions. The first complete and different proof was given by Avramov and Foxby [27]. One can see this proof in Sect. 6.1.

(vi) The property (S_n) of Serre satisfies the assumptions of Theorem 4.2.10. In [62, Prop. 3.3] it was shown that one can also obtain such a result from a positive result for the property Cohen-Macaulay.

(vii) For the property (R_n) and (S_{n+1}), the first proof was given in [62]. Before this, J. Nishimura proved that the localization problem is true for $\mathbf{P} = $ normal and $\mathbf{P} = $ reduced [269]. All these proofs rely on André's Theorem 3.2.22.

(viii) From Theorem 4.2.10 and Lemma 1.1.47, it follows that if the localization problem holds for **P**, it holds also for $\mathbf{P}'_k$. This was first proved by Cangemi and Imbesi [78].

Remark 4.2.12 The proof of Theorem 4.2.10 uses arguments belonging to Algebraic Geometry. It would be nice to have a uniform proof involving only Commutative Algebra arguments.

Lemma 4.2.13 ([174, Lemme 2.1]) *Let* $(A, \mathfrak{m})$ *be a Noetherian catenary and equidimensional local ring, let* k *be a natural number and let* $x \in \mathfrak{m}$ *be a non-zero-divisor. If* A/xA *has the property* (R_k), *then* A *has the property* (R_k).

Proof Let $\mathfrak{p}$ be a prime ideal of A of height $r \leq k$. If $x \in \mathfrak{p}$, then $\mathrm{ht}(\mathfrak{p}/xA) = k - 1$; hence $(A/xA)_\mathfrak{p}$ is a regular local ring. Now by $(\mathbf{A_7})$ for regular local rings, it follows that $A_\mathfrak{p}$ is regular. If $x \notin \mathfrak{p}$, let $\mathfrak{q}$ be a minimal prime over-ideal of $\mathfrak{p} + xA$. Then $\mathrm{ht}(\mathfrak{q}/\mathfrak{p}) = 1$. Since A is catenary and equidimensional, it follows that also $A_\mathfrak{p}$ is a local catenary and equidimensional ring; hence $\mathrm{ht}(\mathfrak{q}) = \mathrm{ht}(\mathfrak{p}) + 1 = r + 1$. It follows that $\mathrm{ht}(\mathfrak{q}/xA) = k$; hence $A_\mathfrak{q}/xA_\mathfrak{q}$ is regular. Thus $A_\mathfrak{q}$ and consequently $A_\mathfrak{p}$ are regular local rings.   □

Remark 4.2.14 It is easy to see that if we consider a property **P** satisfying conditions $(\mathbf{A_1})$, $(\mathbf{A_2})$, $(\mathbf{A_3})$ and $(\mathbf{A_4})$, then with the same proof as above, one can show that in the set-up of Lemma 4.2.13, if A/xA has the property $\mathbf{P_k}$, then A has the property $\mathbf{P_k}$.

Now we can prove a result about the localization problem for the properties of type $\mathbf{P_k}$.

Theorem 4.2.15 (Cangemi and Imbesi [78]) *Let* $u : (A, \mathfrak{m}, k) \to (B, \mathfrak{n}, K)$ *be a local flat morphism of Noetherian local rings, let* **P** *be a property of Noetherian local rings satisfying* $(\mathbf{A_1})$, $(\mathbf{A_2})$, $(\mathbf{A_3})$, $(\mathbf{A_4})$ *and* $(\mathbf{A_7})$ *and let* n *be a natural number. Assume that:*

- *(i)* A *is a* $\mathbf{P_n}$-*ring.*
- *(ii) The* k-*algebra* $B \otimes_A k$ *is a geometrically* $\mathbf{P_n}$ k-*algebra.*
- *(iii)* B *is a formally equidimensional ring.*
- *(iv) The localization problem holds for* **P**.

  Then u *is a* $\mathbf{P_n}$-*morphism.*

Proof Let B^* be the $\mathfrak{m}B$-adic completion of B, and consider the commutative diagram

$$
\begin{array}{ccc}
A & \xrightarrow{\;u\;} & B \\
{\scriptstyle f}\downarrow & & \downarrow{\scriptstyle g} \\
\widehat{A} & \xrightarrow{\;v\;} & B^*
\end{array}
$$

It suffices to prove that v is a $\mathbf{P}_n$-morphism, that is, applying Lemma 3.1.60, we can assume that A has geometrically regular formal fibres. By Noetherian induction we can suppose that A is a domain, that for any $\mathfrak{q} \in \mathrm{Spec}(A)$ such that $\mathfrak{q} \neq (0)$ the fibre of u in $\mathfrak{q}$ is geometrically $\mathbf{P}_n$, and we have only to prove that $B \otimes_A Q(A)$ is a k-algebra geometrically $\mathbf{P}_n$. Take $\mathfrak{q} \in \mathrm{Spec}(B)$ such that $\mathfrak{q} \cap A = (0)$ and $\mathrm{ht}(\mathfrak{q}) \leq n$. Let also $x \in \mathfrak{m}$, $x \neq 0$ and $\mathfrak{q}' \in \mathrm{Min}(\mathfrak{q} + xB)$. Since by Lemma 3.1.29 and Proposition 3.1.52 we have that B is catenary and equidimensional, it follows that $\mathrm{ht}(\mathfrak{q}'/\mathfrak{q}) \leq 1$; hence $\mathrm{ht}(\mathfrak{q}') \leq n + 1$. Put $\mathfrak{p}' = \mathfrak{q}' \cap A$. We have

$$\dim(B_{\mathfrak{q}'}/\mathfrak{p}' B_{\mathfrak{q}'}) = \dim(B_{\mathfrak{q}'}) - \dim(A_{\mathfrak{p}'}) \leq n + 1 - 1 = n$$

Since $A_{\mathfrak{p}'}$ is a $\mathbf{P}$-ring and the localization problem holds for $\mathbf{P}$, it follows that the morphism $A_{\mathfrak{p}'} \to B_{\mathfrak{q}'}$ is a $\mathbf{P}$-morphism, and consequently its localization $Q(A) \to B_{\mathfrak{q}}$ is a $\mathbf{P}$-morphism too. Thus $B_{\mathfrak{q}}$ has the property $\mathbf{P}$ and then $B \otimes_A Q(A)$ has the property $\mathbf{P}_n$. Let now L be a finite purely inseparable extension of $Q(A)$ and let C be a finite A-algebra with field of fractions L. Since $Q(A) \subseteq L$ is purely inseparable, $B \otimes_A C$ is a finite free local B-algebra. Then $B \otimes_A C$ is formally equidimensional, and by the case already proved, we get that $B \otimes_A L$ has the property $\mathbf{P}_n$. $\square$

Corollary 4.2.16 ([174]) *Let $u : (A, \mathfrak{m}, k) \to (B, \mathfrak{n}, K)$ be a local flat morphism of Noetherian local rings and n a natural number. Suppose that:*

(i) *A has geometrically (R_n) formal fibres.*
(ii) *$B/\mathfrak{m}B$ is a geometrically (R_n) k-algebra.*
(iii) *B is a formally equidimensional ring.*
     Then u is an (R_n) morphism.

In [132, Ex. 7.5.3] besides Serre property (R_n), Grothendieck considers the property $\mathbf{P}=$ normal and (R_n) for $n \geq 0$. Taking account of Theorem 1.1.43, one sees immediately that for $n = 0, 1$ this property is exactly the property of being normal, while for $n \geq 2$ the property $\mathbf{P}=$ normal and (R_n) means $\mathbf{P}= (R_n) + (S_2)$. For the property $\mathbf{P}=$normal and (R_n), we can prove only a partial but significant result.

Theorem 4.2.17 ([62]) *Let $u : (A, \mathfrak{m}, k) \to (B, \mathfrak{n}, K)$ be a local flat morphism of Noetherian local rings and let $n \geq 2$ be a natural number. Suppose that:*

(i) *A has geometrically normal and (R_n) formal fibres.*
(ii) *$B/\mathfrak{m}B$ is a geometrically normal and (R_n) k-algebra.*
(iii) *B is a universally catenary ring.*
     Then u is a normal and (R_n) morphism.

Proof From Theorem 4.2.10 it follows that u is a normal morphism. Let B^* be the $\mathfrak{m}B$-adic completion of B, and consider the commutative diagram

$$
\begin{array}{ccc}
A & \xrightarrow{\ u\ } & B \\
{\scriptstyle f}\downarrow & & \downarrow{\scriptstyle g} \\
\widehat{A} & \xrightarrow{\ v\ } & B^*
\end{array}
$$

The morphism f is (R_n). Then, as in the proof of Theorem 4.2.10, since by Theorem 3.1.61 it follows that B^* is universally catenary, we see that it suffices to prove that the morphism v is (R_n). Hence we can assume that A has geometrically regular formal fibres. By Noetherian induction we can assume that A is a domain, that for any non-zero prime ideal $\mathfrak{q}$ of A the fibre of u in $\mathfrak{q}$ is geometrically (R_n), and we have only to prove that $B \otimes_A Q(A)$ is (R_n). Replacing A with its normalization, we can moreover assume that A is normal. Since u is normal, it follows that also B is normal. But B is a local ring, consequently a domain. Let now $\mathfrak{q}$ be a prime ideal B such that $\mathfrak{q} \cap A = (0)$ and $\mathrm{ht}(\mathfrak{q}) \leq n$. Let also $x \in \mathfrak{m}$, $x \neq 0$ and $\mathfrak{q}' \in \mathrm{Min}(\mathfrak{q} + xB)$. Since B is a catenary domain and $\mathrm{ht}(\mathfrak{q}'/\mathfrak{q}) \leq 1$, it follows that $\mathrm{ht}(\mathfrak{q}') \leq n + 1$. Put $\mathfrak{p}' = \mathfrak{q}' \cap A$. We have

$$
\dim(B_{\mathfrak{q}'}/\mathfrak{p}'B_{\mathfrak{q}'}) = \dim(B_{\mathfrak{q}'}) - \dim(A_{\mathfrak{p}'}) \leq n + 1 - 1 = n.
$$

Since $A_{\mathfrak{p}'}$ has geometrically regular formal fibres, from Theorem 3.2.22 or from Theorem 4.2.10, it follows that the morphism $A_{\mathfrak{p}'} \to B_{\mathfrak{q}'}$ is regular, and consequently also its localization $Q(A) \to B_{\mathfrak{q}}$ is regular. Therefore $B_{\mathfrak{q}}$ is a regular ring.   □

Question 4.2.18 Is condition (iii) necessary in Theorem 4.2.15 or in Theorem 4.2.17?

As an application we prove the promised result about inductive limits for rings having Cohen-Macaulay, Gorenstein or complete intersection formal fibres (see the comment before 3.7.7).

Lemma 4.2.19 *Let $(A_i, \varphi_{ij})_{i \in I}$ be an inductive system of Noetherian complete local rings and let $A := \varinjlim A_i$. Let k_i be the residue field of A_i for any index $i \in I$ and k be the residue field of A. Assume that:*

(i) A is a Noetherian ring.
(ii) φ_{ij} is a local flat morphism, $\forall i \leq j$.
    Then $A \to \widehat{A}$ is a complete intersection morphism.

Proof Since A is Noetherian, there exists $i_0 \in I$ such that for any $i \geq i_0$ we have $\widehat{A} \otimes_{A_i} k_i \cong k$; hence the morphism $k_i \to k_i \otimes_{A_i} \widehat{A}$ is complete intersection. By Theorem 4.2.10 the morphism $A_i \to \widehat{A}$ is a complete intersection morphism. Now we apply Theorem 3.7.1.   □

Theorem 4.2.20 (Doretti [91, Th. 1.7]) *Let $(A_i, \varphi_{ij})_{i \in I}$ be an inductive system of Noetherian rings and $A := \varinjlim A_i$. Assume that:*

(i) A is a Noetherian ring.

(ii) φ_{ij} is a flat morphism, $\forall i \leq j$.

(iii) A_i has geometrically Cohen-Macaulay (resp. Gorenstein, resp. complete intersection) formal fibres for any index $i \in I$.

Then A has geometrically Cohen-Macaulay (resp. Gorenstein, resp. complete intersection) formal fibres.

Proof Let $\mathfrak{m} \in \operatorname{Max}(A)$ and $\mathfrak{m}_i := \varphi_i^{-1}(\mathfrak{m})$. Also let $B := A_{\mathfrak{m}}$ and $B_i = (A_i)_{\mathfrak{m}_i}$. By Corollary 1.3.18 we have the factorization $B \xrightarrow{f} \varinjlim \widehat{B_i} := C \xrightarrow{g} \widehat{B} \cong \widehat{C}$. By assumption (iii) and Theorem 3.7.1, it follows that f is a Cohen-Macaulay (resp. Gorenstein, resp. complete intersection) morphism. By Lemma 4.2.19 it results that g is a Cohen-Macaulay (resp. Gorenstein, resp. complete intersection) morphism. Thus $B \to \widehat{B}$ is a Cohen-Macaulay (resp. Gorenstein, resp. complete intersection) morphism.   □

4.3 Lifting Theorems in the Semilocal Case

In this section we will present an answer to Problem 4.2.2 in the case of semilocal rings. We start with some preparations.

Lemma 4.3.1 *Let A be a Noetherian ring, $\mathfrak{b}$ an ideal of A and $A \supseteq \mathfrak{a}_1 \supseteq \mathfrak{a}_2 \supseteq \ldots a$ descending sequence of ideals of A. Suppose that:*

(i) A is $\mathfrak{b}$-adically complete.

(ii) There exists $r > 0$ such that $\mathfrak{a}_m \not\subseteq \mathfrak{b}^r$ for any $m > 0$.

(iii) For any $n > 0$, there exists $t(n)$ such that for $m \geq t(n)$ we have

$$\mathfrak{a}_{t(n)} + \mathfrak{b}^n = \mathfrak{a}_m + \mathfrak{b}^n.$$

Then $\bigcap\limits_{i \in \mathbb{N}} \mathfrak{a}_i \neq (0)$.

Proof (see [266, Proof of Th. 30.1]) We can suppose that $t(n) < t(n+1)$ for any $n \in \mathbb{N}$. From (ii) we get that $\mathfrak{a}_{t(r)} \not\subseteq \mathfrak{b}^r$; hence we can choose $x_r \in \mathfrak{a}_{t(r)} \setminus \mathfrak{b}$. Then from condition (iii), we obtain $\mathfrak{a}_{t(r)} + \mathfrak{b} = \mathfrak{a}_{t(r+1)} + \mathfrak{b}^r$. This means that there exists $x_{r+1} \in \mathfrak{a}_{t(r+1)} \setminus \mathfrak{b}^r$ such that $x_{r+1} - x_r \in \mathfrak{b}^r$. In this way we construct by induction the sequence $(x_n)_n$ such that $x_n \in \mathfrak{a}_{t(n)} \setminus \mathfrak{b}^r$ and $x_{n+1} - x_n \in \mathfrak{b}^n$, for $n \geq r$. Since A is $\mathfrak{b}$-adically complete, there exists $x = \lim_n x_n$. Since $x_{m+n} \in \mathfrak{b}_{t(m)}$ for any n, it follows that $x \in \mathfrak{a}_{t(m)}$. But $m \leq t(m)$; hence $x \in \bigcap_n \mathfrak{a}_n$. Since $x_{n+r} - x_r \in \mathfrak{b}^r$ for any natural number n, it results that $x - x_r \in \mathfrak{b}^r$. But $x_r \notin \mathfrak{b}^r$; hence $x \neq 0$. $\qquad\square$

Lemma 4.3.2 (Nishimura [269, Lemma 1.2]) *Let A be a Noetherian ring, $\mathfrak{c}$ be an ideal of A and B be an A-algebra. Let $\mathfrak{b}$ be an ideal of B and set $\mathfrak{a} := \mathfrak{b} \cap A$. For any natural number n, let $\mathfrak{b}_n := \mathfrak{b} + \mathfrak{c}^n B$ and $\mathfrak{a}_n := \mathfrak{b}_n \cap A$. Suppose that:*

(i) A is $\mathfrak{c}$-adically complete.
(ii) For any $n > 0$ we have $\mathfrak{a}_n B = \mathfrak{b}_n$ and the induced morphism $A/\mathfrak{a}_n \to B/\mathfrak{b}_n$ is faithfully flat.
(iii) $\bigcap\limits_{n \in \mathbb{N}} (\mathfrak{a} B + \mathfrak{c}^n B) = \mathfrak{a} B$ and $\bigcap\limits_{n \in \mathbb{N}} \mathfrak{b}_n = \mathfrak{b}$.
Then $\mathfrak{a} B = \mathfrak{b}$.

Proof Since $\mathfrak{a} B \cap A = \mathfrak{a}$, we may assume that $\mathfrak{a} = (0)$. Then $\mathfrak{b} \cap A = (0)$, hence the first relation in condition (iii) becomes $\bigcap\limits_{n \in \mathbb{N}} \mathfrak{c}^n B = (0)$. We have to show that $\mathfrak{b} = (0)$. Assume the contrary. Then there exists $r > 0$ such that $\mathfrak{b} \not\subseteq \mathfrak{c}^r B$. If there exists $m \in \mathbb{N}$ such that $\mathfrak{a}_m \subseteq \mathfrak{c}^r$, then $\mathfrak{a}_t \subseteq \mathfrak{c}^r$ for any $t \geq m$. Hence $\mathfrak{a}_t B \subseteq \mathfrak{c}^r B$ and by condition (ii) this implies that $\mathfrak{b}_t \subseteq \mathfrak{c}^r B$. Condition (iii) gives $\mathfrak{b} \subseteq \mathfrak{c}^r B$ and this is not possible. Thus $\mathfrak{a}_m \not\subseteq \mathfrak{c}^r$ for any m. Let n be fixed and let $m \geq n$. Then

$$(\mathfrak{a}_m + \mathfrak{c}^n) B = \mathfrak{a}_m B + \mathfrak{c}^n B = \mathfrak{b}_m + \mathfrak{c}^n B = \mathfrak{b} + \mathfrak{c}^n B = \mathfrak{b}_n = \mathfrak{a}_n B.$$

But

$$\mathfrak{a}_n = \mathfrak{a}_n B \cap A = (\mathfrak{a}_m + \mathfrak{c}^n) B \cap A = \mathfrak{a}_m + \mathfrak{c}^n.$$

Thus we can apply Lemma 4.3.1 to get $\bigcap\limits_{n \in \mathbb{N}} \mathfrak{a}_n \neq (0)$. On the other hand

$$(0) = \mathfrak{a} = \mathfrak{b} \cap A = \left(\bigcap\limits_{n \in \mathbb{N}} \mathfrak{b}_n\right) \cap A = \bigcap\limits_{n \in \mathbb{N}} \mathfrak{a}_n.$$

Contradiction! $\qquad\square$

Corollary 4.3.3 (Rotthaus [328, Lemma 2]) *Let A be a semilocal ring, $\mathfrak{a}$ be an ideal of A such that A is $\mathfrak{a}$-adically complete, $\mathfrak{b}$ a non-zero ideal of $\widehat{A}$ and let $\mathfrak{b}_n := (\mathfrak{b} + \mathfrak{a}^n \widehat{A}) \cap A$.*

If there exists $m \in \mathbb{N}$ such that $\mathfrak{b}_n \widehat{A} = \mathfrak{b} + \mathfrak{a}^n \widehat{A}$ for any $n \geq m$, then $(\mathfrak{b} \cap A)\widehat{A} = \mathfrak{b}$. In particular $\mathfrak{b} \cap A \neq (0)$.

Proof We apply 4.3.2 for $B = \widehat{A}$, the completion of A in the topology of the Jacobson radical. $\qquad\square$

Corollary 4.3.4 *Let A be a semilocal ring, $\mathfrak{a}$ be an ideal of A such that A is $\mathfrak{a}$-adically complete and $\mathfrak{b}$ a non-zero ideal of $\widehat{A}$. If $\sqrt{\mathfrak{b} + \mathfrak{a}\widehat{A}} = \mathfrak{m}_1 \cap \ldots \cap \mathfrak{m}_t$, where $\mathfrak{m}_1, \ldots, \mathfrak{m}_t$ are maximal ideals of $\widehat{A}$, then $(\mathfrak{b} \cap A)\widehat{A} = \mathfrak{b}$.*

Proof For any $n \in \mathbb{N}$, we have the relation

$$\sqrt{\mathfrak{b} + \mathfrak{a}^n \widehat{A}} = \sqrt{\mathfrak{b} + \mathfrak{a}\widehat{A}},$$

hence

$$\sqrt{\mathfrak{b} + \mathfrak{a}^n \widehat{A}} = \mathfrak{q}_{1,n} \cap \cdots \cap \mathfrak{q}_{l,n},$$

where $\mathfrak{q}_{j,n}$ is a $\mathfrak{m}_j$-primary ideal, for $j = 1, \ldots, l$. Then

$$\mathfrak{b}_n = (\mathfrak{q}_{1,n} \cap A) \cap \cdots \cap (\mathfrak{q}_{l,n} \cap A).$$

We get

$$\mathfrak{b}\widehat{A} = \mathfrak{q}_{1,n} \cap \cdots \cap \mathfrak{q}_{l,n}.$$

Now we apply Corollary 4.3.3. $\qquad\square$

Remember that if **P** is a property of local rings and A is a ring, then $i(\mathbf{P}(A))$ is the radical ideal of the Zariski closure of **NP**(A)(see 1.2.5).

Lemma 4.3.5 *Let P be a property satisfying the conditions $(\mathbf{A}_2)$, $(\mathbf{A}_3)$ and $(\mathbf{A}_5)$, let $u : A \to B$ be a faithfully flat P-morphism and let $f := \mathrm{Spec}(u)$ be the induced morphism. Then:*

(i) $i(\mathbf{P}(A)) = i(\mathbf{P}(B)) \cap A$.
(ii) If moreover any ring having P is reduced and $\mathbf{P}(A)$ is open, then $i(\mathbf{P}(A))B = i(\mathbf{P}(B))$.

Proof Since u is a **P**-morphism, from Corollary 1.6.14 it follows that A has the property **P** if and only if B has the property **P**. Hence we have to consider only the case when A does not have the property **P**, that is, $\mathbf{P}(A) \neq \mathrm{Spec}(A)$.

(i) Let $x \in i(\mathbf{P}(A))$ and let $\mathfrak{q} \in \mathbf{NP}(B)$. Then $\mathfrak{p} := \mathfrak{q} \cap A \in \mathbf{NP}(A)$; hence $x \notin \mathfrak{p}$, that is, $x \in i(\mathbf{P}(B)) \cap A$. Conversely, let $x \in i(\mathbf{P}(B)) \cap A$ and let $\mathfrak{p} \in \mathbf{NP}(A)$. If $\mathfrak{q} \in \mathrm{Min}(\mathfrak{p}B)$, then obviously $\mathfrak{q} \in \mathbf{NP}(B)$. Thus $x \notin \mathfrak{q}$, hence $x \notin \mathfrak{p}$, that is, $x \in i(\mathbf{P}(A))$.

(ii) Let $i(\mathbf{P}(A)) = \mathfrak{p}_1 \cap \ldots \cap \mathfrak{p}_n$. where $\mathrm{Min}(i(\mathbf{P}(A))) = \{\mathfrak{p}_1, \ldots, \mathfrak{p}_n\}$. Since $\mathbf{P}(A)$ is open, it follows that $\mathfrak{p}_i \in \mathbf{NP}(B)$. Let $i \in \{1, \ldots, n\}$ and put $\mathfrak{p} = \mathfrak{p}_i$. Set $S := (A/\mathfrak{p}) \setminus (0)$. Then $S^{-1}(B/\mathfrak{p}B)$ is a geometrically $\mathbf{P}$ algebra over $Q(A/\mathfrak{p})$. In particular, $B/\mathfrak{p}B$ is a reduced ring; hence $\mathfrak{p}B = \sqrt{\mathfrak{p}B} = \mathfrak{q}_1 \cap \cdots \cap \mathfrak{q}_r$, where $\mathfrak{q}_i \in \mathrm{Min}(\mathfrak{p}B)$. But $\mathfrak{q}_i \cap A = \mathfrak{p}$; hence $\mathfrak{q}_i \in \mathbf{NP}(B)$. Thus $i(\mathbf{P}(B)) \subseteq i(\mathbf{P}(A))B$. The other inclusion follows from (i). $\qquad \square$

Now we can prove the lifting theorem in the semilocal case. This result was first proved by Rotthaus [327] for the property $\mathbf{P} = $ regular and then by Nishimura [269] for $\mathbf{P} = $ normal and $\mathbf{P} = $ reduced and extended in [62] for other properties.

Theorem 4.3.6 (Rotthaus [327, Th. 3], Nishimura [269, Theorem], cf. [62, Th. 2.3]) *Let $\mathbf{P}$ be a property of Noetherian local rings that satisfies the conditions $(A_1), (A_2), (A_3), (A_5)$ and (A_6). Let A be a semilocal ring and let $\mathfrak{a}$ be an ideal of A. Suppose that:*

(i) A is complete in the $\mathfrak{a}$-adic topology.
(ii) A is a Nagata ring.
(iii) The localization property is valid for $\mathbf{P}$ in the case of complete local rings.
(iv) $A/\mathfrak{a}$ is a $\mathbf{P}$-ring.
    Then A is a $\mathbf{P}$-ring.

Proof First of all let us consider a finite A-algebra E which is a domain and let $\mathfrak{q}$ be a prime ideal of $\widehat{E}$ such that $\mathfrak{q} \supseteq \mathfrak{a}\widehat{E}$. As usual, by $\widehat{E}$ we mean the completion of the semilocal ring E in the radical-adic topology. Set

$$\mathfrak{p} := \mathfrak{q} \cap E, \quad C := \widehat{E_\mathfrak{p}} \text{ and } D := (\widehat{E}_\mathfrak{q})\widehat{}.$$

If $\mathfrak{q} \in \mathrm{Max}(\widehat{E})$, clearly $C = D$. Suppose now that $\mathfrak{q}$ is not a maximal ideal of $\widehat{E}$. Since $E/\mathfrak{a}E$ is a $\mathbf{P}$-ring, it follows that $E/\mathfrak{a}E \to \widehat{E}/\mathfrak{a}\widehat{E}$ is a $\mathbf{P}$-morphism; hence also $E/\mathfrak{p} \to \widehat{E}/\mathfrak{p}\widehat{E}$ is a $\mathbf{P}$-morphism. Set $S := (E/\mathfrak{p}) \setminus 0$. Then, by Lemma 1.6.11, $S^{-1}(E/\mathfrak{p}) \to S^{-1}(\widehat{E}/\mathfrak{p}\widehat{E})$ is a $\mathbf{P}$-morphism too. It follows that $C/\mathfrak{p}C \to D/\mathfrak{p}D$ is a $\mathbf{P}$-morphism. Because C and D are complete local rings, we get from assumption (iii) that $C \to D$ is a $\mathbf{P}$-morphism. By Lemma 4.3.5 we have that $i(\mathbf{P}(C))D = i(\mathbf{P}(D))$. In order to prove the theorem, by Theorem 1.6.23 it is enough to prove that for any finite A-algebra E that is a domain and for any $\mathfrak{Q} \in \mathbf{NP}(\widehat{E})$, we have $\mathfrak{Q} \cap E \neq (0)$. So, replacing A with E, we can suppose that A is a domain, and we must prove that for every $\mathfrak{Q} \in \mathbf{NP}(\widehat{A})$, we have $\mathfrak{Q} \cap A \neq (0)$. Let

$$B := \widehat{A}, \quad F := V(\mathfrak{a}B) \setminus \mathrm{Max}(B), \quad G := \{\mathfrak{p} \in \mathbf{NP}(B) \mid V(\mathfrak{p}) \cap F \neq \emptyset\}.$$

Hence

$$G = \emptyset \Leftrightarrow \mathbf{NP}(B) = \mathrm{Max}(B).$$

Let $\mathfrak{P} \in F$ and $\mathfrak{p} := \mathfrak{P} \cap A$, so that $\mathfrak{a} \subseteq \mathfrak{p}$. We have the canonical commutative diagram

Since B is a complete local domain, the morphism φ is regular. From this and from the above considerations, we get the relations

$$i(\mathbf{P}(C))D = i(\mathbf{P}(D)),$$

$$i(\mathbf{P}(B_{\mathfrak{P}})) = i(\mathbf{P}(D)) \cap B_{\mathfrak{P}},$$

$$i(\mathbf{P}(B_{\mathfrak{P}}))D = i(\mathbf{P}(D)).$$

We put

$$\mathfrak{c} := \bigcap_{\mathfrak{P} \in F} (i(\mathbf{P})(D)) \cap B = \bigcap_{\mathfrak{P} \in F} i(\mathbf{P}(B_{\mathfrak{P}})) \cap B.$$

Suppose that $\mathfrak{c} = (0)$. Then B is reduced, so $\mathbf{P}(B) \neq \emptyset$; hence by $(\mathbf{A_6})$, since $\mathbf{P}(B)$ is open, $i(\mathbf{P}(B)) \neq 0$. But this contradicts the fact that $i(\mathbf{P}(B)) \subseteq \mathfrak{c}$. Obviously $\mathfrak{c}$ is a radical ideal. Now we show that if $\mathfrak{P} \in F$, then $\mathfrak{c}D = i(\mathbf{P}(D))$. Indeed,

$$\mathfrak{c}D \subseteq i(\mathbf{P}(B_{\mathfrak{P}}))D = i(\mathbf{P}(D)).$$

If $\mathfrak{c} = B$, then $\mathfrak{c}D = D$. If $\mathfrak{c} \neq B$, we have $\mathfrak{c} = \mathfrak{P}_1 \cap \cdots \cap \mathfrak{P}_r$, with $\mathfrak{P}_i \in \mathrm{Min}(\mathfrak{c})$. Because $\mathbf{P}(B)$ is open, it follows that $\mathfrak{P}_i \in \mathbf{NP}(B)$. We obtain

$$\mathfrak{c}B_{\mathfrak{P}} = \mathfrak{P}_1 B_{\mathfrak{P}} \cap \cdots \cap \mathfrak{P}_r B_{\mathfrak{P}} \supseteq i(\mathbf{P}(B_{\mathfrak{P}})).$$

But then $\mathfrak{c}D = i(\mathbf{P}(D))$, because $i(\mathbf{P}(B_{\mathfrak{P}}))D = i(\mathbf{P}(D))$.

Let now $\mathfrak{P} \in \mathrm{Spec}(B)$, $\mathfrak{Q}$ a $\mathfrak{P}$-primary ideal in B and $\mathfrak{q} := \mathfrak{Q} \cap A$. We first want to show that $\mathfrak{q}C = \mathfrak{Q}D \cap C$ and that $\mathfrak{q}C$ is a $\mathfrak{p}C$-primary ideal. Indeed, $\mathfrak{Q}B_{\mathfrak{P}}$ is a $\mathfrak{P}B_{\mathfrak{P}}$-primary ideal; hence $\mathfrak{Q}B_{\mathfrak{P}} \cap A_{\mathfrak{P}}$ is $\mathfrak{p}A_{\mathfrak{p}}$-primary. Then $\mathfrak{Q}B_{\mathfrak{P}} \cap A$ is $\mathfrak{p}$-primary. But

$$\mathfrak{Q}B_{\mathfrak{P}} \cap A_{\mathfrak{p}} = \mathfrak{Q} \cap A = \mathfrak{q}$$

and also this last ideal is $\mathfrak{p}$-primary. Consequently $\mathfrak{Q}B_{\mathfrak{P}} \cap A_{\mathfrak{p}} = \mathfrak{q}A_{\mathfrak{p}}$. Since $C = \widehat{A_{\mathfrak{p}}}$, the ideal $\mathfrak{q}C$ is a $\mathfrak{p}C$-primary ideal. In the same way we prove that $\mathfrak{Q}D \cap C$ is $\mathfrak{p}C$-primary. But

$$\mathfrak{q}C \cap A = (\mathfrak{Q}D \cap C) \cap A$$

and the assertion is proved. Now we prove that

$$((\mathfrak{c} + \mathfrak{a}^n B) \cap A)B = \mathfrak{c} + \mathfrak{a}^n B, \ \forall n \in \mathbb{N}.$$

The relation is clear if $\mathfrak{c} = B$; hence we may assume $\mathfrak{c} \neq B$. Set $\mathfrak{c}_n := \mathfrak{c} + \mathfrak{a}^n B$. Using a primary decomposition of $\mathfrak{c}_n$, it is enough to prove that if $\mathfrak{Q}$ is a $\mathfrak{P}$-primary ideal of B and $\mathfrak{c}_n \subseteq \mathfrak{Q}$, then, setting $\mathfrak{q} := \mathfrak{Q} \cap A$, we have $\mathfrak{c}_n \subseteq \mathfrak{q}B$.
Indeed, if $\mathfrak{P} \in \mathrm{Max}(B)$, then $\mathfrak{q}B = \mathfrak{Q}$. If $\mathfrak{P} \notin \mathrm{Max}(B)$, then $\mathfrak{P} \in F$. Since $\mathfrak{a}^n \subseteq \mathfrak{q}$, it is enough to prove that $\mathfrak{c} \subseteq \mathfrak{q}B$. From $\mathfrak{c} \subseteq \mathfrak{Q}$ and $\mathfrak{c}D = i(\mathbf{P}(D))$, we get that

$$i(\mathbf{P}(C))D = i(\mathbf{P}(D)) \subseteq \mathfrak{Q}D.$$

So $i(\mathbf{P}(C)) \subseteq \mathfrak{q}C$, consequently $i(\mathbf{P}(D)) \subseteq \mathfrak{q}D$. Let $\mathfrak{r} \in F$ be such that $\mathfrak{r} \cap A = \mathfrak{p}$. Then

$$\mathfrak{c} \subseteq i(\mathbf{P}(B_{\mathfrak{P}})) \subseteq \mathfrak{q}\widehat{B_{\mathfrak{r}}}.$$

Consider a reduced primary decomposition of $\mathfrak{q}B = \mathfrak{Q}_1 \cap \cdots \cap \mathfrak{Q}_m$, where $\mathfrak{Q}_i$ is a $\mathfrak{P}_i$-primary ideal. Then, since $\mathrm{Ass}_B(B/\mathfrak{p}B) = \mathrm{Ass}_B(B/\mathfrak{q}B)$ and B is flat over A, it follows that $\mathfrak{P}_i \cap A = \mathfrak{p}$. Hence $\mathfrak{P}_i \in F$. Then

$$\mathfrak{c} \subseteq \mathfrak{q}\widehat{B_{\mathfrak{P}_i}} = \mathfrak{Q}\widehat{B_{\mathfrak{P}_i}}, \ i = 1, \ldots, m.$$

It follows that $\mathfrak{c} \subseteq \mathfrak{Q}_i B_{\mathfrak{P}_i} \cap B = \mathfrak{Q}_i$, that is, $\mathfrak{c} \subseteq \mathfrak{q}B$. Now let $\mathfrak{Q} \in \mathbf{NP}(B)$. If there is $\mathfrak{P} \in \mathrm{Spec}(B) \setminus \mathrm{Max}(B)$ such that $\mathfrak{Q} + \mathfrak{a}B \subseteq \mathfrak{P}$, then $\mathfrak{Q}B_{\mathfrak{P}} \in \mathbf{NP}(B_{\mathfrak{P}})$. It follows that $i(\mathbf{P}(B_{\mathfrak{P}})) \subseteq \mathfrak{Q}B_{\mathfrak{P}}$, that is, $\mathfrak{c} \subseteq \mathfrak{Q}$. But $\mathfrak{c} \cap A \neq (0)$, so that $\mathfrak{Q} \cap A \neq (0)$. If $\sqrt{\mathfrak{Q} + \mathfrak{a}B}$ is an intersection of maximal ideals, we apply Corollary 4.3.4. $\square$

Corollary 4.3.7 ([62]) *Let P a property of Noetherian local rings that satisfies the conditions* $(\mathbf{A_1})$, $(\mathbf{A_2})$, $(\mathbf{A_3})$, $(\mathbf{A_5})$, $(\mathbf{A_6})$ *and* $(\mathbf{A_7})$. *Let A be a semilocal Nagata ring and let $\mathfrak{a}$ be an ideal of A. Suppose that:*

 (i) *A is complete in the $\mathfrak{a}$-adic topology.*
(ii) *$A/\mathfrak{a}$ is a P-ring.*
     Then A is a P-ring.

Proof We apply Theorem 4.2.10 and Theorem 4.3.6.   □

Remark 4.3.8 Corollary 4.3.7 can be applied for the properties **P** = regular, complete intersection, Gorenstein, Cohen-Macaulay, (S_n), (R_n) and (S_{n+1}), normal, reduced.

What happens with the property (R_n) of Serre? In this case we have to apply Theorem 4.2.16.

Corollary 4.3.9 ([174]) *Let A be a semilocal Nagata ring, n a natural number and $\mathfrak{a}$ be an ideal of A. Suppose that:*

(i) A is complete in the $\mathfrak{a}$-adic topology.
(ii) $A/\mathfrak{a}$ is an (R_n)-ring.
(iii) A is a universally catenary ring.
   Then A is an (R_n)-ring.

Proof If moreover A is a domain, then $\widehat{A}$ is equidimensional. It follows by Lemma 3.1.33 that $A_\mathfrak{p}$ is formally equidimensional, for any $\mathfrak{p} \in \operatorname{Spec}(A)$. By Lemma 3.1.60, the $\mathfrak{a}A_\mathfrak{p}$-adic completion of $A_\mathfrak{p}$ is formally equidimensional. Now we can apply Theorem 4.3.6 and Corollary 4.2.16.   □

Remark 4.3.10 The condition that A is a Nagata ring in Theorem 4.3.6 can be replaced by the condition that any ring having the property **P** is reduced. The proof is obviously the same.

Question 4.3.11 Is the condition (iii) necessary in Corollary 4.3.9? Similarly, is the condition (ii) necessary in Theorem 4.3.6?

   As an application of the lifting property for quasi-excellent rings in the semilocal case, we will give an interesting characterization of quasi-excellent local rings belonging to André [9]. We need some preparations involving André-Quillen homology.

Definition 4.3.12 Let $u : A \to B$ be a morphism of Noetherian rings. We say that u is **H-finite** if for any Noetherian B-algebra C and any finitely generated C-module E, the C-module $H_1(A, B, E)$ is finitely generated.

Lemma 4.3.13 *If $u : A \to B$ and $v : B \to C$ are H-finite morphisms, then $v \circ u$ is H-finite.*

Proof We have an exact sequence

$$H_1(A, B, E) \to H_1(A, C, E) \to H_1(B, C, E).$$

The first and the third modules are finitely generated. It follows immediately that the module in the middle is finitely generated. $\square$

Remark 4.3.14

(i) It follows from Theorem 3.2.8 that any regular morphism is H-finite.
(ii) Any surjective morphism u is H-finite, since

$$H_1(A, B, E) \simeq \mathfrak{a}/\mathfrak{a}^2 \otimes_B E,$$

where $\mathfrak{a} = \ker(u)$.

Lemma 4.3.15 (cf. [6, Lemme II.47]) *Let $u : A \to B$ be a ring morphism, N a non-zero finitely generated B-module and $\mathfrak{a}$ an ideal of A. Assume that $aN = (0)$ for any $a \in \mathfrak{a}$ and $aN = N$ for any $a \notin \mathfrak{a}$. Then there exists $\mathfrak{m} \in \mathrm{Max}(B)$ such that $\mathfrak{m} \cap A = \mathfrak{a}$.*

Proof Since $N \neq (0)$, there exists $\mathfrak{m} \in \mathrm{Max}(B)$ such that $N_\mathfrak{m} \neq (0)$. If $a \notin \mathfrak{a}$, then $aN_\mathfrak{m} = N_\mathfrak{m}$. Applying Nakayama Lemma to $N_\mathfrak{m}$, we get that $u(a) \notin \mathfrak{m}B_\mathfrak{m}$, so that $u(a) \notin \mathfrak{m}$. Suppose that $a \in \mathfrak{a}$. Then $aN = (0)$. Hence $u(a) \in \mathfrak{m}$; otherwise, from Nakayama Lemma, it would follow that $N_\mathfrak{m} = (0)$. Thus $\mathfrak{m} \cap A = \mathfrak{a}$. $\square$

Proposition 4.3.16 (André [6, Prop. II.54]) *Let $u : A \to B$ be a morphism of Noetherian rings. Let $F : \mathrm{Mod}(A) \to \mathrm{Mod}(B)$ be an additive semiexact functor that commutes with direct limits and maps finitely generated A-modules into finitely generated B-modules. Assume moreover that for any $a \in A$ and any A-module M, if $\phi_a : M \to M$ is the multiplication by a on M, then $F(\phi_a)$ is the multiplication by $u(a)$ on $F(M)$. Assume that $F(A/\mathfrak{a}) = (0)$ for any ideal $\mathfrak{a}$ of A such that there exists $\mathfrak{m} \in \mathrm{Max}(B)$ with $\mathfrak{m} \cap A = \mathfrak{a}$. Then $F = 0$.*

Proof Since F commutes with direct limits, it is enough to show that F vanishes on finitely generated A-modules, and since it is semiexact, it suffices to show that it vanishes on cyclic A-modules, that is, we need to show that $F(A/\mathfrak{a}) = (0)$ for any ideal $\mathfrak{a}$ of A. If $\mathfrak{a} \cap A = \mathfrak{m}$ for some $\mathfrak{m} \in \mathrm{Max}(A)$, then $F(A/\mathfrak{a}) = (0)$ by assumption. Assume that $\mathcal{A} = \{\mathfrak{a} \mid F(A/\mathfrak{a}) \neq (0)\} \neq \emptyset$. Because A is Noetherian, there exists a maximal element $\mathfrak{a} \in \mathcal{A}$. Let $a \in A$ be such that $a \notin \mathfrak{a}$. We have an exact sequence

$$0 \to A/(\mathfrak{a} : aA) \xrightarrow{\phi} A/\mathfrak{a} \xrightarrow{\psi} A/(\mathfrak{a} + aA) \to 0,$$

where ϕ is the multiplication by a. Since F is semiexact, we obtain an exact sequence

$$F(A/(\mathfrak{a} : aA)) \xrightarrow{F(\phi)} F(A/\mathfrak{a}) \xrightarrow{F(\psi)} F(A/(\mathfrak{a} + aA)).$$

But $F(A/\mathfrak{a})$ is a non-zero finitely generated B-module and $F(A/(\mathfrak{a}+aA)) = (0)$, because $\mathfrak{a}$ is maximal in $\mathcal{A}$. Hence $F(\phi)$ is surjective and $F(A/(\mathfrak{a} : aA)) \neq (0)$. It results that $\mathfrak{a} = \mathfrak{a} : aA$. If $a \notin \mathfrak{a}$, then $aF(A/\mathfrak{a}) = F(A/\mathfrak{a})$, because $F(\phi)$ is the multiplication by $u(a)$. If $a \in \mathfrak{a}$, then for the same reason $aF(A/\mathfrak{a}) = (0)$. From Lemma 4.3.15 it results that there exists $\mathfrak{m} \in \text{Max}(B)$ such that $\mathfrak{m} \cap A = \mathfrak{a}$. Contradiction! $\qquad\square$

Proposition 4.3.17 *Let* $u : (A, \mathfrak{m}, k) \to (B, \mathfrak{n}, K)$ *be a local morphism of Noetherian local rings. Then* u *is a regular morphism if and only if* u *is formally smooth and H-finite.*

Proof If u is regular, then it is obviously formally smooth and H-finite. Conversely, for a B-module E, set $F(E) := H_1(A, B, E)$. It follows from Proposition 4.3.16 and Corollary 3.2.4 that $F(E) = 0$. Now we apply Theorem 3.2.8. $\qquad\square$

Corollary 4.3.18 *Let* $u : A \to B$ *be a morphism of Noetherian local rings such that:*

(i) $u = p_n \circ p_{n-1} \circ \cdots \circ p_1$, *where* $p_1, \ldots, p_n$ *are morphisms of Noetherian local rings.*
(ii) p_i *is a regular morphism for any odd* i.
(iii) p_i *is a surjective morphism for any even* i.
(iv) u *is formally smooth.*
    Then u *is regular.*

Theorem 4.3.19 (André [9, Th. 1]) *Let* A *be a Noetherian local ring. Then* A *is quasi-excellent if and only if the morphism* $A \to A[[X]]$ *is regular.*

Proof If A is quasi-excellent, then by Theorem 2.3.15 and Proposition 1.6.40, it follows that the morphism $A \to A[[X]]$ is regular. To prove the converse, we argue by induction on $n := \dim(A)$. If $n = 0$, then A is Artinian; hence quasi-excellent. Suppose that $n > 0$. Pick $x \in A$ such that $\dim(A/xA) = n - 1$. Since the morphism $A \to A[[X]]$ is regular, it follows that also the morphism $A/xA \to A/xA[[X]]$ is regular; hence by induction A/xA is quasi-excellent. Let $B = \widehat{(A, xA)} \simeq A[[X]]/(X - x)$ be the completion of A in the x-adic topology. Then $B/xB \simeq A/xA$ is quasi-excellent. Since B is xB-adically complete, from Corollary 4.3.7, it follows that B is quasi-excellent. Note that the local rings A and B have the same completions in the topologies given by the maximal ideals, that is, $\widehat{A} \simeq \widehat{B}$. From the canonical morphism $A \to B$, we get a surjection

$$p_2 : A[[X]] \to B, \quad p(X) = x.$$

Now we have:

$$A \xrightarrow{p_1} A[[X]] \xrightarrow{p_2} B \xrightarrow{p_3} \widehat{A} = \widehat{B}.$$

Applying Corollary 4.3.18, we get that A is quasi-excellent. $\qquad\square$

Corollary 4.3.20 *Let $(A, \mathfrak{m}, k)$ be a Noetherian local ring. The following are equivalent:*

(i) *A is a quasi-excellent ring.*
(ii) *$A[[X_1, \ldots, X_n]]$ is quasi-excellent for any $n \in \mathbb{N}$.*
(iii) *The canonical morphism $A \to A[[X_1, \ldots, X_n]]$ is regular for any $n \in \mathbb{N}$.*
(iv) *The canonical morphism $A \to A[[X_1, \ldots, X_n]]$ is regular for $n = \dim_k(\mathfrak{m}/\mathfrak{m}^2) = \mathrm{edim}(A)$.*
(v) *The canonical morphism $A \to A[[X]]$ is regular.*

Proof $(i) \Rightarrow (ii)$: Follows from Theorem 4.3.6.

$(ii) \Rightarrow (i)$: Obvious.

$(i) \Rightarrow (iii)$: The morphism $A \to A[[X_1, \ldots, X_n]]$ is formally smooth; hence Theorem 4.2.10 (or Theorem 3.2.22) shows that it is regular.

$(iii) \Rightarrow (iv)$ and $(iii) \Rightarrow (v)$: Clear.

$(v) \Rightarrow (i)$: Follows from Th. 4.3.19.

$(iv) \Rightarrow (i)$: Pick a minimal system of generators $x_1, \ldots, x_n$ of $\mathfrak{m}$, and consider the ring morphisms

$$A \xrightarrow{u} C := A[[X_1, \ldots, X_n]] \xrightarrow{v} \widehat{A}$$

where u is the canonical inclusion morphism and $v(X_i) = x_i$, for $i = 1, \ldots, n$. Then v is a surjective morphism and $\ker(v) = (X_1 - x_1, \ldots, X_n - x_n) := \mathfrak{a}$. Then for any $\widehat{A}$-module E, we have the Jacobi-Zariski exact sequence below, where the left module is (0) by assumption:

$$(0) = H_1(A, C, E) \to H_1(A, \widehat{A}, E) \to \mathfrak{a}/\mathfrak{a}^2 \otimes_C E \xrightarrow{\delta} \Omega_{C/A} \otimes_{\widehat{A}} E$$

It follows that $\ker \delta = H_1(A, \widehat{A}, E)$. But since $A \to \widehat{A}$ is formally smooth by [131, Cor. 20.7.9], we get that δ is formally left-invertible. Because $\mathfrak{a}/\mathfrak{a}^2$ is separated, it follows easily that δ is injective. Now we apply Th. 3.2.8. □

We want to present a similar result in the special case of rings of positive prime characteristic. We begin with a lemma about preserving flat dimension after a faithfully flat morphism, which is actually valid in any characteristic.

Lemma 4.3.21 *Let $u : A \to B$ and $v : B \to C$ be ring morphisms such that v is faithfully flat. Then $\mathrm{fd}_A(B) = \mathrm{fd}_A(C)$.*

Proof Let E be an A-module. Then we have a change of ring spectral sequence [324, Th. 10.71]

$$E^2_{p,q} = \operatorname{Tor}^B_p(C, \operatorname{Tor}^A_q(B, E)) \underset{p}{\Rightarrow} \operatorname{Tor}^A_n(C, E).$$

Therefore $C \otimes_B \operatorname{Tor}^A_n(B, E) \simeq \operatorname{Tor}^A_n(C, E)$. Since v is faithfully flat, it follows that $\operatorname{Tor}^A_n(C, E) = (0)$ if and only if $\operatorname{Tor}^A_n(B, E) = (0)$, that is, $\operatorname{fd}_A(B) = \operatorname{fd}_A(C)$. $\square$

Remark 4.3.22 Let A be a Noetherian ring of characteristic $p > 0$ and let $X := (X_1, \ldots, X_n)$ be indeterminates over A. We can write $A^{(p)} = \varinjlim_i A_i$, where A_i are the finite A-subalgebras of $A^{(p)}$. If $\alpha_i : A_i \to A^{(p)}$ is the inclusion, we denote by

$$\bar{\alpha}_i := \alpha_i \otimes_A 1_{A[[X]]} : A_i[[X]] \to A^{(p)} \otimes_A A[[X]]$$

and

$$\beta_i := \omega_{A[[X]]/A} \circ \bar{\alpha}_i : A_i[[X]] \to A[[X]]^{(p)}.$$

(i) It is clear that β_i extends α_i and that $\beta_i(X_j) = X^p_j$ for any $i \in I$ and any $j = 1, \ldots, n$.

(ii) It follows from (i) that β_i are injective morphisms, so that also the morphism $\varinjlim_i \beta_i = \omega_{A[[X]]/A}$ is injective.

(iii) We have

$$\operatorname{Im}(\omega_{A[[X]]/A}) = (A^p[[X^p_1, \ldots, X^p_n]][A]])^{(p)} \subseteq A[[X]]^{(p)}.$$

(iv) We have

$$A^p[[X^p]][A] \subseteq A[[X^p]] \subseteq A[[X]]$$

and the morphism $A[[X]]$ has the p-basis $\{X_1, \ldots, X_n\}$ over $A[[X^p]]$. Applying (iii) and Lemma 4.3.21 we get

$$\operatorname{fd}(\omega_{A[[X]]/A}) = \operatorname{fd}_{A^p[[X^p]][A]}(A[[X^p]]) = \operatorname{fd}_{A^p[[X]][A]}(A[[X]]).$$

(v) We notice that $A[[X]]$ is the completion of the ring $A^p[[X]][A]$ in the (X)-adic topology and that $(A[[X]])^p \subseteq A^p[[X]][A]$; hence the morphism $\alpha : A^p[[X]][A] \to A[[X]]$ is flat if and only if $A^p[[X]][A]$ is a Noetherian ring.

(vi) From Theorem 3.4.27 and Theorem 3.4.19, we see that $\mathrm{fd}_{A^P[[X]][A]}(A[[X]])$ can be 0 or ∞.

Theorem 4.3.23 (Dumitrescu [95, Th. 3.3]) *Let A be a Noetherian ring of characteristic $p > 0$. The following are equivalent:*

(i) *A has geometrically regular formal fibres.*
(ii) *The morphism $A \to A[[X_1, \ldots, X_n]]$ is regular for each $n \in \mathbb{N}$.*
(iii) *The inclusion morphism $A^P[[X_1, \ldots, X_n]][A] \to A[[X_1, \ldots, X_n]]$ is flat for each $n \in \mathbb{N}$.*
(iv) *The inclusion morphism $A^P[[X_1, \ldots, X_n]][A] \to A[[X_1, \ldots, X_n]]$ has finite flat dimension for each $n \in \mathbb{N}$.*
(v) *$A^P[[X_1, \ldots, X_n]][A]$ is a Noetherian ring for each $n \in \mathbb{N}$.*

Proof $(ii) \Leftrightarrow (iii) \Leftrightarrow (iv) \Leftrightarrow (v)$; Follows from Theorem 3.4.27 and Remark 4.3.22.
$(i) \Rightarrow (ii)$: Follows from Proposition 1.6.40.
$(iv) \Rightarrow (i)$: Let $\mathfrak{m}$ be a maximal ideal of A and $a_1, \ldots, a_n$ be a system of generators of $\mathfrak{m}$. Then the completion of A in the $\mathfrak{m}$-adic topology is isomorphic to the completion of the local ring $A_\mathfrak{m}$ in the topology of the maximal ideal. Thus

$$\widehat{A_\mathfrak{m}} \simeq A[[X_1, \ldots, X_n]]/(X_1 - a_1, \ldots, X_n - a_n).$$

Since $A^{(p)} \otimes_A A[[X_1, \ldots, X_n]]/(X_1 - a_1, \ldots, X_n - a_n)$ is isomorphic to a factor ring of $A^P[[X_1, \ldots, X_n]][A]$, it follows that $(A_\mathfrak{m})^{(p)} \otimes_{A_\mathfrak{m}} \widehat{A_\mathfrak{m}}$ is Noetherian. Thus from Corollary 3.4.21, we obtain that $A_\mathfrak{m}$ has geometrically regular formal fibres. Because $\mathfrak{m}$ is arbitrary in $\mathrm{Max}(A)$, it follows from Corollary 1.6.39 that A has geometrically regular formal fibres. $\qquad\square$

In the case of local rings, Theorem 4.3.23 specifies as follows.

Corollary 4.3.24 (Dumitrescu [95, Cor. 3.4]) *Let A be a Noetherian local ring of characteristic $p > 0$ and X an indeterminate. The following are equivalent:*

(i) *A is quasi-excellent.*
(ii) *The morphism $A \to A[[X]]$ is regular.*
(iii) *The morphism $A^P[[X]][A] \to A[[X]]$ is flat.*
(iv) *The morphism $A^P[[X]][A] \to A[[X]]$ has finite flat dimension.*
(v) *$A^P[[X]][A]$ is a Noetherian ring.*
(vi) *$A^{(p)} \otimes_A \widehat{A}$ is a Noetherian ring.*

4.4   Lifting of the Property Reg-2

First let us remind the definition of constructible set of a topological space.

Definition 4.4.1 ([248, Def. (6.B)]) A subset C of a Noetherian topological space X is called **constructible** if C is a finite union of sets which are intersections of a closed set with an open set.

Lemma 4.4.2 *Let A be a Noetherian ring and $\mathfrak{a}$ be an ideal of A such that $A/\mathfrak{a}$ is Reg-2. Let B be an A-algebra of finite type and $Z \subseteq V(\mathfrak{a}B)$ be a closed subset of $\operatorname{Spec}(B)$. Then $\operatorname{Reg}(B) \cap Z$ is a constructible set.*

Proof We shall prove the assertion by Noetherian induction, so we suppose that for any closed subset $U \subsetneq Z$, the set $\operatorname{Reg}(B) \cap U$ is a constructible set. We may assume that $Z = V(\mathfrak{q})$, where $\mathfrak{q}$ is a prime ideal of B such that $\mathfrak{a}B \subsetneq \mathfrak{q}$. Since $A/\mathfrak{a}$ is Reg-2, it follows that $\operatorname{Reg}(B/\mathfrak{q})$ is a non-empty open set of $\operatorname{Spec}(B/\mathfrak{q})$. If $\mathfrak{q} \notin \operatorname{Reg}(B)$, then $\operatorname{Reg}(B) \cap Z = \operatorname{Reg}(B) \cap V(\mathfrak{q}) = \varnothing$; hence $\operatorname{Reg}(B) \cap Z$ is constructible. If $\mathfrak{q} \in \operatorname{Reg}(B)$, then the ideal $\mathfrak{q}B_{\mathfrak{q}}$ is generated by a regular sequence. But then, since $\operatorname{Reg}(B/\mathfrak{q})$ is open, it follows that there exists $f \in B \setminus \mathfrak{q}$ such that $\mathfrak{q}B_f$ is generated by a B_f-regular sequence and $B_f/\mathfrak{q}B_f$ is a regular ring. Thus, if $\mathfrak{p}$ is a prime ideal of B such that $\mathfrak{p} \in V(\mathfrak{q})$ and $\mathfrak{p} \notin V(\mathfrak{q} + fB)$, then $B_{\mathfrak{p}}$ is a regular local ring. We have the decomposition

$$\operatorname{Reg}(B) \cap V(\mathfrak{q}) = \left[\operatorname{Reg}(B) \cap V(\mathfrak{q} + fB)\right] \cup \left[\operatorname{Reg}(B) \cap (V(\mathfrak{q}) \setminus V(\mathfrak{q} + fB))\right].$$

By Noetherian induction $\operatorname{Reg}(B) \cap V(\mathfrak{q} + fB)$ is a constructible set. On the other hand, we have $\operatorname{Reg}(B) \cap (V(\mathfrak{q}) \setminus V(\mathfrak{q} + fB)) = V(\mathfrak{q}) \setminus V(\mathfrak{q} + fB)$ which is open in $V(\mathfrak{q})$. This shows that $\operatorname{Reg}(B) \cap V(\mathfrak{q})$ is constructible.   □

Lemma 4.4.3 *Let A be a Noetherian ring, $\mathfrak{a}$ be an ideal of A and $\pi : X \to \operatorname{Spec}(A)$ be a scheme morphism of finite type. If $A/\mathfrak{a}$ is a Reg-2 ring, then $\operatorname{Reg}(X) \cap \pi^{-1}(V(\mathfrak{a}))$ is an open subset of $\pi^{-1}(V(\mathfrak{a}))$.*

Proof We may assume that $X = \operatorname{Spec}(B)$ is an affine scheme, where B is an A-algebra of finite type. Let $\mathfrak{p} \in \operatorname{Reg}(B) \cap \pi^{-1}(V(\mathfrak{a}))$, that is, $\mathfrak{p} \cap A \supseteq \mathfrak{a}$, and let $\mathfrak{q} \in \pi^{-1}(V(\mathfrak{a}))$ be such that $\mathfrak{q} \subseteq \mathfrak{p}$. Then $B_{\mathfrak{p}}$ is regular and consequently $B_{\mathfrak{q}}$ is regular. Thus $\operatorname{Reg}(B) \cap \pi^{-1}(V(\mathfrak{a}))$ is stable by generalization and by Lemma 4.4.2 it is constructible. Now we apply [248, Lemma (6.G)] to get the conclusion.   □

The main ingredient of the proof is the following result.

Lemma 4.4.4 *Let A be a Noetherian domain, $\mathfrak{q}$ a prime ideal of A and $\mathfrak{a}$ an ideal of A. Suppose that $A_\mathfrak{q}$ has geometrically regular formal fibres and that $A/\mathfrak{a}$ is Reg-2. Then there exist an element $b \in A \setminus \mathfrak{q}$ and an alteration covering of A_b $\{\phi_{b,i} : X_{b,i} \to \operatorname{Spec}(A_b)\}_i$ such that for any i*

$$\phi_{b,i}^{-1}(\operatorname{Spec}(A_b/\mathfrak{a}A_b)) \subseteq \operatorname{Reg}(X_{b,i}).$$

Proof The localization $A_\mathfrak{q}$ is a local quasi-excellent domain; hence by Gabber's Weak Uniformization Theorem 4.1.31, there exists a regular alteration covering $\{\phi_i : X_i \to \operatorname{Spec}(A_\mathfrak{q})\}_{1 \le i \le m}$ and a proper morphism $f : V \to \operatorname{Spec}(A_\mathfrak{q})$ with $V := \bigcup_{i=1}^{m} V_i$ and a family of maps $\{\psi_i : V_i \to X_i\}_{1 \le i \le m}$ with the properties from the Definition 4.1.30, that is, we have a commutative diagram of scheme morphisms.

$$
\begin{array}{ccc}
V_i & \longrightarrow & V \\
\psi_i \downarrow & & \downarrow f \\
X_i & \xrightarrow{\phi_i} & \operatorname{Spec}(A_\mathfrak{q})
\end{array}
$$

By Chow's Lemma [142, Ch. II, Ex. 4.10], we may assume that the morphism f is projective. Therefore we have a commutative diagram

$$
\begin{array}{ccc}
V & \xrightarrow{\;\theta\;} & \mathbb{P}^r_\mathbb{Z} \otimes_\mathbb{Z} \operatorname{Spec}(A_\mathfrak{q}) \\
& \searrow{\scriptstyle f} \quad \swarrow & \\
& \operatorname{Spec}(A_\mathfrak{q}) &
\end{array}
$$

where θ is a closed immersion. This means that V can be identified with a closed subset of $\mathbb{P}^r_{A_\mathfrak{q}} = \operatorname{Proj}(A_\mathfrak{q}[T_0, \ldots, T_r])$, the projective space of dimension r over $A_\mathfrak{q}$. The ideal of V is generated by finitely many homogeneous polynomials $p_1, \ldots, p_t$ with coefficients in $A_\mathfrak{q}$. If we take an element $s \in A \setminus \mathfrak{q}$ which is a common divisor of all denominators of the coefficients of the polynomials $p_1, \ldots, p_t$, we can consider that these polynomials belong to $A_s[T_0, \ldots, T_r]$. Therefore we have a closed subscheme $V_s \subseteq \mathbb{P}^r_{A_s}$ and a commutative diagram

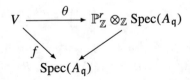

$$
\begin{array}{ccc}
V \simeq V_s \times_{\operatorname{Spec}(A_s)} \operatorname{Spec}(A_\mathfrak{q}) & \longrightarrow & V_s \\
f \downarrow & & \downarrow f_s \\
\operatorname{Spec}(A_\mathfrak{q}) & \longrightarrow & \operatorname{Spec}(A_s)
\end{array}
$$

The set $Z_i := V \setminus V_i$ is a closed Zariski subset of V. Since $V \subset \mathbb{P}^r_{A_\mathfrak{q}}$, we see that Z_i is the zero-set of finitely many homogeneous polynomials $q_{1,i}, \ldots, q_{l,i}$. If we take s

such that it is divisible by all the coefficients of all the polynomials $q_{i,j}$, we see that the same polynomials, regarded as elements of $A_s[T_1, \ldots, T_r]$, define a Zariski-closed subset $Z_{i,s} \subseteq V_s$ such that the following diagram is commutative

$$
\begin{array}{ccc}
Z_i \simeq Z_{i,s} \times_{\mathrm{Spec}(A_s)} \mathrm{Spec}(A_{\mathfrak{q}}) & \longrightarrow & V_s \\
f \downarrow & & \downarrow f_s \\
\mathrm{Spec}(A_{\mathfrak{q}}) & \longrightarrow & \mathrm{Spec}(A_s)
\end{array}
$$

and setting $V_{i,s} := V_s \setminus Z_{i,s}$, we get the commutative diagram

$$
\begin{array}{ccc}
V_i \simeq V_{i,s} \times_{\mathrm{Spec}(A_s)} \mathrm{Spec}(A_{\mathfrak{q}}) & \longrightarrow & V_{i,s} \\
\psi_i \downarrow & & \downarrow f_s \\
\mathrm{Spec}(A_{\mathfrak{q}}) & \longrightarrow & \mathrm{Spec}(A_s)
\end{array}
$$

Since ψ_i is a morphism of finite type, one can easily see that in this way we can finally find an element $c \in A \setminus \mathfrak{q}$ and an alteration covering $\{\phi_{c,i} : X_{c,i} \to \mathrm{Spec}(A_c)\}_{1 \le i \le m}$ such that $\phi_i = \phi_{c,i} \otimes_{A_c} A_{\mathfrak{q}}$, $i = 1, \ldots, m$. Let $\phi_{c,i}(s)$ be a generalization of $\mathfrak{q}A_c$. Then $s \in X_i$; hence the ring $\mathcal{O}_{X_{c,i},s}$ is a regular local ring. Let $Z_c := \mathrm{Spec}(A_c/\mathfrak{a}A_c)$. This is a closed subset of $\mathrm{Spec}(A_c)$. From Lemma 4.4.3 it follows that $W_{c,i} := \phi_{c,i}^{-1}(Z_c) \setminus \mathrm{Reg}(X_{c,i})$ is closed in $\phi_{c,i}^{-1}(Z_c)$; hence $W_{c,i}$ is a closed subset of $X_{c,i}$. By Chevalley's Constructibility Theorem [248, Th. 6, (6.E)] $\phi_{c,i}(W_{c,i})$ is a constructible subset of $\mathrm{Spec}(A_c)$. Hence the closure $\overline{\phi_{c,i}(W_{c,i})}$ is the set of the points of $\mathrm{Spec}(A_c)$ that are specializations of a point from $\phi_{c,i}(W_{c,i})$. Suppose that $\mathfrak{q} \in \overline{\phi_{c,i}(W_{c,i})}$. Then there exists $s \in W_{c,i}$ such that $\mathfrak{q} \in \{\phi_{c,i}(s)\}$, and we have seen above that $\mathcal{O}_{X_{c,i},s}$ is a regular local ring. But this contradicts the definition of $W_{c,i}$. This means that $\mathfrak{q} \notin \overline{\phi_{c,i}(W_{c,i})}$. Choose a non-zero element $b \in A$ such that $\mathrm{Spec}(A_b) \subset \mathrm{Spec}(A_c)$ and $\mathrm{Spec}(A_b) \cap \overline{\phi_{c,i}(W_{c,i})} = \emptyset$, for $i = 1, \ldots, m$. Consider the Cartesian commutative diagram

$$
\begin{array}{ccc}
X_{b,i} & \xrightarrow{\phi_{b,i}} & \mathrm{Spec}(A_b) \\
\downarrow & & \downarrow \\
X_{c,i} & \xrightarrow{\phi_{c,i}} & \mathrm{Spec}(A_c)
\end{array}
$$

Then $W_{c,i} \cap X_{b,i} = \emptyset$. Let $Z_b := \mathrm{Spec}(A_b/\mathfrak{a}A_b)$ and $W_{b,i} := \phi_{b,i}^{-1}(Z_b) \setminus \mathrm{Reg}(X_{b,i})$. Since $Z_b = Z_c \cap \mathrm{Spec}(A_b)$, we obtain

$$
W_{b,i} = (\phi_{c,i}^{-1}(Z_c) \cap X_{b,i}) \setminus \mathrm{Reg}(X_{b,i}) = (\phi_{c,i}^{-1}(Z_c) \cap X_{b,i}) \setminus \mathrm{Reg}(X_{c,i}) =
$$

$$
= W_{c,i} \cap X_{b,i} = \emptyset.
$$

This proves the claim of the Lemma. $\qquad \square$

Lemma 4.4.5 *Let Y be a Noetherian integral scheme, let $\{\phi_i : X_i \to Y\}_{1 \leq i \leq m}$ be an alteration covering of Y and let $y_1, \ldots, y_l$ points in Y such that $y_{j+1} \in \overline{\{y_j\}}$ for $j = 1, \ldots, l-1$. Then there exist an index $i \in \{1, \ldots, m\}$ and a sequence of points $x_1, \ldots, x_l \in X_i$ such that $\phi_i(x_j) = y_j$ for $j = 1, \ldots, l$ and $x_{j+1} \in \overline{\{x_j\}}$ for $j = 1, \ldots, l-1$.*

Proof For any $i = 1, \ldots, m$, there is a commutative diagram

$$
\begin{array}{ccc}
V_i & \longrightarrow & V \\
\psi_i \downarrow & & \downarrow f \\
X_i & \xrightarrow{\phi_i} & Y
\end{array}
$$

where $f : V \to Y$ is a proper surjective morphism of schemes and $V = \bigcup_{i=1}^{m} V_i$ is a Zariski open covering. First we construct a sequence $v_1, \ldots, v_l \in V$ such that $f(v_j) = y_j$ for $j = 1, \ldots, l$ and $v_{j+1} \in \overline{\{v_j\}}$ for $j = 1, \ldots, l-1$. Start with a point $v_1 \in V$ such that $f(v_1) = y_1$. Suppose that we already found a sequence $v_1, \ldots, v_t \in V$ such that $f(v_j) = y_j$, $j = 1, \ldots, t$ and $v_{j+1} \in \overline{\{v_j\}}$, $j = 1, \ldots, t-1$. Since f is proper hence closed, $f(\overline{\{v_t\}}) = \overline{\{y_t\}}$, therefore $y_{t+1} \in f(\overline{\{v_t\}})$ and there exists $v_{t+1} \in V$ such that $f(v_{t+1}) = y_{t+1}$ and $v_{t+1} \in \overline{\{v_t\}}$. There exists some index i such that $v_l \in V_i$, and since V_i is closed under generalization, it follows that $v_1, \ldots, v_l \in V_i$. Then the sequence $x_j := \psi_i(v_j)$, $j = 1, \ldots, l$ is the desired sequence. $\qquad\square$

The following theorem was first proved by Brodmann and Rotthaus in the case that A is universally catenary and contains a field of characteristic zero, using Hironaka's resolution of singularities [73, Th. 2]. The present general form was given by Gabber [117] and presented in all details by Kurano and Shimomoto [205, Th. 4.1].

Theorem 4.4.6 (Brodmann-Rotthaus-Gabber [73, Th. 2], [117], [205, Th. 4.1]) *Let A be a Noetherian ring and $\mathfrak{a}$ be an ideal contained in the Jacobson radical of A. Assume that $A/\mathfrak{a}$ is quasi-excellent and A has geometrically regular formal fibres. Then A is quasi-excellent, i.e. A is Reg-2.*

Proof We have to show that any finite A-algebra B has open regular locus. By Nagata's Criterion 1.2.16, it is enough to prove that $\mathrm{Reg}(B)$ contains a non-empty open set. Since B is finite over A, it follows that $\mathfrak{a}B$ is contained in the Jacobson radical of B and that $B/\mathfrak{a}B$ is quasi-excellent. Since B has geometrically regular formal fibres, the localization $B_{\mathfrak{q}}$ is quasi-excellent for any prime ideal $\mathfrak{q}$ of B. By Lemma 4.4.4 for any such prime $\mathfrak{q}$, there exists an element $b_{\mathfrak{q}} \in B \setminus \mathfrak{q}$ and an alteration covering

$$
\{\phi_{b_{\mathfrak{q}},i} : X_{b_{\mathfrak{q}},i} \to \mathrm{Spec}(B_{b_{\mathfrak{q}}})\}_{1 \leq i \leq m}
$$

such that for any $i = 1, \ldots, m$ we have

$$\phi_{b_q,i}^{-1}(\operatorname{Spec}(B_{b_q}/\mathfrak{a}B_{b_q})) \subseteq \operatorname{Reg}(X_{b_q,i}).$$

Because $\operatorname{Spec}(B)$ is quasi-compact [56, Ch. II, §4, n. 3, Cor. 7], we can choose finitely many elements $b_1, \ldots, b_s \in \{b_q \mid q \in \operatorname{Spec}(B)\}$ such that $\operatorname{Spec}(B) = \operatorname{Spec}(B_{b_1}) \cup \ldots \cup \operatorname{Spec}(B_{b_s})$. Since B_{b_j} is quasi-excellent for any $j = 1, \ldots, s$, applying Lemma 4.4.4, there exists a finite family of alteration coverings

$$\{\phi_{b_j,i} : X_{b_j,i} \to \operatorname{Spec}(B_{b_j})\}_i$$

such that

$$\phi_{b_j,i}^{-1}(\operatorname{Spec}(B_{b_j}/\mathfrak{a}B_{b_j})) \subseteq \operatorname{Reg}(X_{b_j,i}). \tag{4.1}$$

By generic flatness [248, Th. 52, (22.A)], there exists a non-zero element $c \in B$ such that the map

$$\phi_{b_j,i}^{-1}(\operatorname{Spec}(B_{b_jc})) \to \operatorname{Spec}(B_{b_jc})$$

is flat for all i and j. It is enough to show that

$$D(c) = \{\mathfrak{p} \in \operatorname{Spec}(B) \mid c \notin \mathfrak{p}\} \subseteq \operatorname{Reg}(B).$$

Let $\mathfrak{p} \in \operatorname{Spec}(B)$ be such that $c \notin \mathfrak{p}$, and take a maximal ideal $\mathfrak{m}$ of B containing $\mathfrak{p}$. Since $\mathfrak{a}B$ is contained in the Jacobson radical of B, we have $\mathfrak{a}B \subseteq \mathfrak{m}$. By the way the elements $b_1, \ldots, b_s$ are chosen, there exists j such that $\mathfrak{p}$ and $\mathfrak{m}$ are in $\operatorname{Spec}(B_{b_j})$. By Lemma 4.4.5 there exist elements $x_1, x_2 \in X_{b_j,i}$ for some i, such that $x_2 \in \overline{\{x_1\}}$, $\phi_{b_j,i}(x_1) = \mathfrak{p}$ and $\phi_{b_j,i}(x_2) = \mathfrak{m}$. By the above relation (4.1) and the fact that $\mathfrak{a}B \subseteq \mathfrak{m}$, it follows that $\mathcal{O}_{X_{b_j,i},x_2}$ is a regular local ring. But x_1 is a generalization of x_2; hence also $\mathcal{O}_{X_{b_j,i},x_1}$ is a regular local ring. Since $c \notin \mathfrak{p}$, the morphism $B_{\mathfrak{p}} \to \mathcal{O}_{X_{b_j,i},x_1}$ is flat; hence $B_{\mathfrak{p}}$ is regular, concluding the proof. □

Theorem 4.4.7 (Chiriacescu [83, Th. 1.6]) *Let A be a Noetherian ring, $\mathfrak{a}$ be an ideal of A contained in the Jacobson radical and $\widehat{A}$ the completion of A in the $\mathfrak{a}$-adic topology. Assume that:*

(i) The morphism $A \to \widehat{A}$ is reduced.
(ii) $A/\mathfrak{a}$ is a Nagata ring.
   Then A is a Nagata ring.

Proof Let $\mathfrak{p} \in \mathrm{Spec}(A)$. We need to show that $B := A/\mathfrak{p}$ is Japanese. If $\mathfrak{a} \subseteq \mathfrak{p}$, then B is a factor ring of $A/\mathfrak{a}$; hence is a Japanese ring. Assume that $\mathfrak{a} \nsubseteq \mathfrak{p}$. Let L be a field, finite extension of $Q(B)$, and let C be a finite A-algebra such that $Q(C) = L$. We have $(0) \neq \mathfrak{a}C \subseteq J(C)$. Let D be the completion of C in the $\mathfrak{a}C$-adic topology. Then the morphism $C \to D$ is reduced; hence D is a reduced ring. Since $D/\mathfrak{a}D \cong C/\mathfrak{a}C$ is a Nagata ring, it follows by Theorem 2.2.5 that D is a Nagata ring. Thus D', the integral closure of D, is a finite D-module. Since the morphism $C \to D$ is faithfully flat, by Lemma 2.1.36 it follows that C' is finite over C. $\qquad\square$

As a corollary we obtain the following counterpart of Theorem 4.4.6 for the property P=normal. This result was initially obtained by Brezuleanu and Rotthaus [71, Satz 1], with a more difficult proof.

Corollary 4.4.8 (Brezuleanu–Rotthaus [69, Satz 1]) *Let A be a Noetherian ring and $\mathfrak{a} \subseteq J(A)$ be an ideal of A. Assume that:*

(i) A has geometrically normal formal fibres.
(ii) $A/\mathfrak{a}$ is $\mathrm{Nor} - 2$.
    Then A is $\mathrm{Nor} - 2$.

Proof Follows from Lemma 2.1.36 and Theorem 4.4.7. $\qquad\square$

4.5   The Second Theorem of Bertini for Local Rings

This section deals with a famous result known as the Theorem of Bertini for local rings, obtained by H. Flenner [111] answering a Conjecture of Grothendieck [136, Exp. XIII, Conj. 2.6]. It has many applications in Commutative Algebra and Algebraic Geometry. The result will be used in the next section. We begin with some facts that will be necessary in the proof of the main theorem.

Notation 4.5.1 Consider a Noetherian ring A, a finitely generated A-module M and a prime ideal $\mathfrak{p} \in \mathrm{Spec}(A)$. We denote by $\mu_{\mathfrak{p}}(M) := \mu_{A_{\mathfrak{p}}}(M_{\mathfrak{p}})$ the minimal number of generators of the $A_{\mathfrak{p}}$-module $M_{\mathfrak{p}}$.

Definition 4.5.2 (Swan [360]) Let A be a Noetherian ring, M a finitely generated A-module, m an element of M and $\mathfrak{p}$ a prime ideal of A. We say that m is **basic** in $\mathfrak{p}$ if $\mu_{\mathfrak{p}}(M) - \mu_{\mathfrak{p}}(M/Am) = 1$.

Remark 4.5.3 It is easy to see that m is basic in $\mathfrak{p}$ iff $m \in M_{\mathfrak{p}} \setminus \mathfrak{p}M_{\mathfrak{p}}$, in other words if it belongs to a minimal system of generators of the $A_{\mathfrak{p}}$- module $M_{\mathfrak{p}}$.

Definition 4.5.4 Let A be a Noetherian ring A, let $\mathfrak{p}$ be a prime ideal of A and let M be a finitely generated A-module. If $k \geq 1$ is a natural number, we say that the elements $m_1, \ldots, m_s \in M$ are **k-times basic** in $\mathfrak{p}$, if $\mu_{\mathfrak{p}}(M) - \mu_{\mathfrak{p}}(M/ \sum_{i=1}^{s} Am_i) \geq k$.

If A is a commutative ring, in 4.5.5– 4.5.9, we shall denote for simplicity $A[X] :=$ $A[X_1, \ldots, X_n]$, where $n \geq 1$ is a natural number. Moreover, if M is an A-module, $M[X]$ will mean $M \otimes_A A[X]$.

Lemma 4.5.5 *Let A be a Noetherian ring and M a finitely generated A-module. Consider elements $m_1, \ldots, m_n \in M$ and polynomials $g_1, \ldots, g_s \in A[X]$ such that the element $\sum_{i=1}^{n} m_i X_i \in M[X]$ is basic in all prime ideals $\mathfrak{p} \in D(g_1, \ldots, g_s)$. Let $\alpha = (\alpha_1, \ldots, \alpha_n) \in$ A^n and $\mathfrak{q} \in D(g_1(\alpha), \ldots, g_s(\alpha))$. Then the element $\sum_{i=1}^{n} \alpha_i m_i \in M$ is basic in $\mathfrak{q}$.*

Proof Since $\mathfrak{q} \in D(g_1(\alpha), \ldots, g_s(\alpha))$, there exists $1 \leq i \leq s$, such that $g_i(\alpha) \notin \mathfrak{q}$. Then $g_i \notin \mathfrak{p} := \mathfrak{q}A[X] + \sum_{i=1}^{n} (X_i - \alpha_i)A[X]$. This means that $\mathfrak{p} \in D(g_1, \ldots, g_s)$; hence the class $\overline{\sum_{i=1}^{n} m_i X_i} \in (M[X]/\mathfrak{p}M[X])_{\mathfrak{q}}$ is not 0. Consider the canonical isomorphism $u :$ $(M[X]/\mathfrak{p}(M[X]))_{\mathfrak{q}} \cong (M/\mathfrak{q}M)_{\mathfrak{q}}$. Then $\sum_{i=1}^{n} \alpha_i m_i = u(\overline{\sum_{i=1}^{n} m_i X_i}) \neq 0$. $\qquad\square$

Lemma 4.5.6 (Flenner [111, Lemma 1.2]) *Let A be a Noetherian ring and M a finitely generated A-module. Consider the elements $m_1, \ldots, m_n \in M$ and let $M' := \sum_{i=1}^{n} Am_i$ be the submodule of M generated by these elements. Let $U \subseteq \operatorname{Spec}(A)$ be an open subset and*

$$U' = \phi^{-1}(U) = \{\mathfrak{q} \in \operatorname{Spec}(A[X]) \mid \mathfrak{q} \cap A \in U\},$$

where $\phi : \operatorname{Spec}(A[X]) \to \operatorname{Spec}(A)$ is the canonical map. Assume that for all $\mathfrak{p} \in U$ we have

$$\mu_{\mathfrak{p}}(M) - \mu_{\mathfrak{p}}(M/M') \geq \dim(U \cap V(\mathfrak{p})) - t,$$

for some $t \geq 0$. Then there exists a closed subset $V = V(\mathfrak{a}) \subseteq \operatorname{Spec}(A[X])$ for some homogeneous ideal $\mathfrak{a}$ of $A[X]$ such that:

(i) $\dim(V) \leq n + t$.
(ii) $\sum_{i=1}^{n} m_i X_i$ is basic at all prime ideals $\mathfrak{p} \in U' \cap (\operatorname{Spec}(A[X]) \setminus V)$.

Proof Since $U = \bigcup\limits_{i=1}^{k} U_i$, where $U_1, \ldots, U_k$ are affine open subsets of $\mathrm{Spec}(A)$, we may assume that $U = \mathrm{Spec}(A)$. Then $U' = \mathrm{Spec}(A[X])$, and for any $\mathfrak{p} \in \mathrm{Spec}(A)$, we have $\mu_{\mathfrak{p}}(M) - \mu_{\mathfrak{p}}(M/M') \geq \dim(A/\mathfrak{p}) - t$. We may assume that A is reduced, because $\mathrm{Spec}(A) = \mathrm{Spec}(A_{red})$. By Noetherian induction we may moreover assume that M is a free A-module and M' is a direct summand of M with basis $m_1, \ldots, m_r$. Thus $r \geq \dim(A)$. Let $s > r$ and $m_s = \sum\limits_{l=1}^{r} a_{ls} m_l$. Then

$$\sum_{i=1}^{n} m_i X_i = \sum_{l=1}^{r} m_l \left(X_l + \sum_{s=r+1}^{n} a_{ls} X_s \right) := h.$$

Thus, if $\mathfrak{q} \in \mathrm{Spec}(A[X])$, the polynomial h is basic in $\mathfrak{q}$ if and only if there exists $l \in \{1, \ldots, r\}$ such that $X_l + \sum\limits_{s=r+1}^{n} a_{ls} X_s \notin \mathfrak{q}$. Let

$$F_1 := X_1 + \sum_{s=r+1}^{n} a_{1s} X_s, \ldots, F_r := X_r + \sum_{s=r+1}^{n} a_{rs} X_s$$

and let $\mathfrak{a} := (F_1, \ldots, F_r) \subseteq A[X]$. It follows that $\sum\limits_{i=1}^{n} m_i X_i$ is basic in all prime ideals which are not in $V(\mathfrak{a})$. We have

$$\dim(V(\mathfrak{a})) = \dim(A[X]/\mathfrak{a}) \leq \dim(A[X_1, \ldots, X_n]) - \mathrm{ht}(\mathfrak{a}) =$$

$$= \dim(A) + n - r \leq n + t.$$
$\square$

Lemma 4.5.7 *Let A be a Noetherian ring, S a subring of A not necessarily with unit and $\mathfrak{p}_1, \ldots, \mathfrak{p}_r$ prime ideals of A such that the image of S in $A/\mathfrak{p}_i$ is infinite for any $i = 1, \ldots, r$. Then, for any elements $\alpha_1, \ldots, \alpha_s \in S$, we have*

$$S \neq \bigcup_{i=1}^{s} \bigcup_{j=1}^{r} (\alpha_i + \mathfrak{p}_j \cap S).$$

Proof Set $\mathfrak{q}_i = \mathfrak{p}_i \cap S$, for $i = 1, \ldots, r$. We may assume that $\mathfrak{q}_i \not\subseteq \mathfrak{q}_j$, for $i \neq j$. Let $\gamma_{12} \in \mathfrak{q}_2 \setminus \mathfrak{q}_1, \gamma_{13} \in \mathfrak{q}_3 \setminus \mathfrak{q}_1, \ldots, \gamma_{1r} \in \mathfrak{q}_r \setminus \mathfrak{q}_1$. Then

$$\gamma_1 := \gamma_{12} \cdot \gamma_{13} \cdot \ldots \cdot \gamma_{1r} \in \mathfrak{q}_2 \cap \ldots \cap \mathfrak{q}_r \setminus \mathfrak{q}_1,$$

because S is multiplicatively closed and $\mathfrak{p}_1, \ldots, \mathfrak{p}_r$ are prime ideals. Thus for each $i = 1, \ldots, r$, we can find $\gamma_i \in \bigcap_{j \neq i} \mathfrak{q}_j \setminus \mathfrak{q}_i$. Since for any index $i \in \{1, \ldots, r\}$ the image of S in $A/\mathfrak{p}_i$ is infinite, we can find elements $\beta_1, \ldots, \beta_r \in S$ such that for any $i = 1, \ldots, r$ we have $\gamma_i \beta_i - \alpha_j \notin \mathfrak{p}_i$, for $j = 1, \ldots, s$. Then $\alpha := \sum_{i=1}^{r} \gamma_i \beta_i \in S$ and $\alpha \notin \bigcup_{i=1}^{s} \bigcup_{j=1}^{r} (\alpha_i + \mathfrak{p}_j \cap S)$. $\qquad \square$

Lemma 4.5.8 (Flenner [111, Lemma 1.3]) *Let A be a Noetherian ring and $U \subseteq \mathrm{Spec}(A)$ be an open subset. Let $\phi : \mathrm{Spec}(A[X]) \to \mathrm{Spec}(A)$ be the canonical map, $U' = \phi^{-1}(U)$ and V a closed set in $\mathrm{Spec}(A[X])$ such that $\dim(V \cap U') \leq n + t$, for some $t \geq 0$. Let S be a subring of A not necessarily with unit such that the image of S in $A/\mathfrak{p}$ is infinite for every $\mathfrak{p} \in U$. Then:*

(i) There exists $\alpha = (\alpha_1, \ldots, \alpha_n) \in S^n$ such that

$$\dim(V \cap U' \cap V(X_1 - \alpha_1, \ldots, X_n - \alpha_n)) \leq t.$$

(ii) If $x_0, \ldots, x_n \in A$ and $G \subseteq D(x_0, \ldots, x_n) \cap U$ is a finite set of prime ideals, then we can choose $\alpha = (\alpha_1, \ldots, \alpha_n) \in S^n$ such that $x_0 + \sum_{i=1}^{n} \alpha_i x_i \notin \bigcup_{\mathfrak{p} \in G} \mathfrak{p}$.

(iii) If $V = V(\mathfrak{a})$, where $\mathfrak{a}$ is an homogeneous ideal of $A[X]$, then there exist $\alpha_2, \ldots, \alpha_n \in S$ such that

$$\dim(V \cap U' \cap V(X_1 - 1, X_2 - \alpha_2, \ldots, X_n - \alpha_n)) \leq t.$$

Moreover, if $x_1, \ldots, x_n \in A$ and $G \subseteq D(x_1, \ldots, x_n) \cap U$ is a finite set of prime ideals, then $\alpha \in S^n$ can be chosen such that $\sum_{i=1}^{n} \alpha_i x_i \notin \bigcup_{\mathfrak{p} \in G} \mathfrak{p}$.

Proof We assume that $n = 1$, because the general case follows easily by induction on n.
(i) and (ii): Let $\{\mathfrak{q}_1, \ldots, \mathfrak{q}_r\}$ be the prime ideals of $A[X]$ that are minimal in the set $V \cap U'$. Thus $\dim(V \cap U') = \max\{\dim(A/\mathfrak{q}_1), \ldots, \dim(A/\mathfrak{q}_r)\} \leq t + 1$. We have to find $\alpha \in S$ such that $X - \alpha \notin \bigcup_{i=1}^{r} \mathfrak{q}_i$. In order to fulfil also the condition (ii), we have to find $\alpha \in S$ such that $x_0 + \alpha x_1 \notin \mathfrak{p}$, for any $\mathfrak{p} \in G$. Therefore $\alpha \in S$ must be chosen in such a way that α is not in a finite number of sets of the form $\alpha_\mathfrak{p} + (\mathfrak{p} \cap S)$, $\mathfrak{p} \in U$. If $X - \alpha \notin \bigcup_{i=1}^{r} \mathfrak{q}_i, \forall \alpha \in S$ there is nothing to prove. If there is $i \in \{1, \ldots, n\}$ and $\alpha_i \in S$ such that $X - \alpha_i \in \mathfrak{q}_i$, then α must not be in the set $\alpha_i + (\mathfrak{q}_i \cap A \cap S)$, in order that $X - \alpha \notin \mathfrak{q}_i$. Now we apply Lemma 4.5.7.
(iii) Since all the ideals $\mathfrak{q}_1, \ldots, \mathfrak{q}_r$ are homogeneous, in the above proof we can choose $\alpha = 1$. $\qquad \square$

Proposition 4.5.9 (Flenner [111, Satz 1.5]) *Let A be a Noetherian local ring, M a finitely generated A-module and U ⊊ Spec(A) an open set. Let S be a subring of A not necessarily with unit such that for any $\mathfrak{p} \in U$ the image of S in $A/\mathfrak{p}$ is infinite. Let also $m_1, \ldots, m_n \in M$ and $M' := \sum_{i=1}^{n} A m_i$ be the submodule of M generated by these elements. Assume that for any prime ideal $\mathfrak{p}$ of U we have*

$$\mu_{\mathfrak{p}}(M) - \mu_{\mathfrak{p}}(M/M') \geq \dim(A/\mathfrak{p}).$$

Then:
(i) There exist elements $\alpha_2, \ldots, \alpha_n \in S$ such that $m_1 + \alpha_2 m_2 + \ldots + \alpha_n m_n$ is basic at all prime ideals $\mathfrak{p} \in U$ with $\dim(A/\mathfrak{p}) > 0$.
(ii) If $x_1, \ldots, x_n$ are elements of A and $G \subseteq U \cap D(x_1, \ldots, x_n)$ is a finite set of prime ideals, then we can choose the elements $\alpha_2, \ldots, \alpha_n \in S$ such that we also have $x_1 + \alpha_2 x_2 + \ldots + \alpha_n x_n \notin \mathfrak{p}$, for any $\mathfrak{p} \in G$.

Proof Since U is a proper open subset of Spec(A), from the assumption it follows that for any prime ideal $\mathfrak{p}$ from U we have

$$\mu_{\mathfrak{p}}(M) - \mu_{\mathfrak{p}}(M/M') \geq \dim(U \cap V(\mathfrak{p})) + 1.$$

Let $\phi : \text{Spec}(A[X]) \rightarrow \text{Spec}(A)$ be the canonical map and $U' = \phi^{-1}(U)$. By Lemma 4.5.6 there exists a closed set $V = V(\mathfrak{b}) \subseteq \text{Spec}(A[X])$, where $\mathfrak{b}$ is an homogeneous ideal of $A[X]$, such that $\sum_{i=1}^{n} m_i X_i$ is basic at all prime ideals $\mathfrak{p} \in U' \cap$ (Spec$(A[X] \setminus V)$. Moreover $\dim(V \cap U') \leq n - 1$. From Lemma 4.5.8 it results that for a given finite set $G \subseteq U \cap D(x_0, \ldots, x_n)$ there exist elements $\alpha_2, \ldots, \alpha_n \in S$ such that $x_1 + \alpha_2 x_2 + \ldots + \alpha_n x_n \notin \mathfrak{p}$ for any $\mathfrak{p} \in G$ and

$$\dim \left(V \cap U' \cap V(X_1 - 1, X_2 - \alpha_2, \ldots, X_n - \alpha_n) \right) = 0.$$

Let $\mathfrak{p} \in U$ be such that $\dim(A/\mathfrak{p}) > 0$. Then $\mathfrak{q} := \mathfrak{p}A[X] + (X_1 - 1, X_2 - \alpha_2, \ldots, X_n - \alpha_n) \in U'$ and $\dim(A[X]/\mathfrak{q}) > 0$. Therefore

$$\mathfrak{q} \notin V \cap U' \cap V(X_1 - 1, X_2 - \alpha_2, \ldots, X_n - \alpha_n),$$

hence

$$\mathfrak{q} \in (\text{Spec}(A[X_1, \ldots, X_n]) \setminus V) \cap U'.$$

It follows that $\sum_{i=1}^{n} m_i X_i$ is basic at $\mathfrak{q}$; hence $m_1 + \sum_{i=2}^{n} \alpha_i m_i \in M$ is basic at $\mathfrak{p}$. □

Notation 4.5.10 In order to make things easier, if $\mathfrak{p}$ is a prime ideal of a ring A, we denote $\mathfrak{p}^{(2)} := \mathfrak{p}^2 A_\mathfrak{p} \cap A$. Later on, in Definition 6.2.26, we will give a more general definition. Comparing Definition 6.2.26 with this notation, the reader will notice that in fact $\mathfrak{p}^{(2)}$ is exactly what is known to be the second symbolic power of $\mathfrak{p}$.

Lemma 4.5.11 *Let A be a Noetherian ring, $\mathfrak{p}$ a prime ideal of A and x an element of A. Assume that there exist an A-module M and a derivation $D \in \mathrm{Der}_A(A, M)$ such that $D(x)$ is basic in $\mathfrak{p}$. Then $x \notin \mathfrak{p}^{(2)}$.*

Proof Localizing at $\mathfrak{p}$ we may assume that A is a local ring with maximal ideal $\mathfrak{p}$. Then $\mathfrak{p}^{(2)} = \mathfrak{p}^2$; hence if $x \in \mathfrak{p}^2$, we would have a relation $x = a_1 b_1 + \ldots + a_n b_n$, for some elements $a_i, b_i \in \mathfrak{p}$. But then $D(x) = \sum_{i=1}^{n} a_i D(b_i) + \sum_{i=1}^{n} b_i D(a_i) \in \mathfrak{p}M$. Contradiction.

$\square$

Before going further, it is useful to explain why the condition $x \notin \mathfrak{p}^{(2)}$ from Lemma 4.5.11 is important. Consider a Noetherian regular ring A and a prime ideal $\mathfrak{p} \in \mathrm{Spec}(A)$. Then $A_\mathfrak{p}$ is a regular local ring. If x is an element of $\mathfrak{p}$, we want to know when the ring $A_\mathfrak{p}/xA_\mathfrak{p}$ is a regular local ring. It is known that this happens if and only if $x \in \mathfrak{p}A_\mathfrak{p} \setminus \mathfrak{p}^2 A_\mathfrak{p}$ [249, Th. 14.2]. But $\mathfrak{p}^2 A_\mathfrak{p} \cap A = \mathfrak{p}^{(2)}$; therefore $A_\mathfrak{p}/xA_\mathfrak{p}$ is a regular local ring if and only if $x \notin \mathfrak{p}^{(2)}$.

The next two propositions give an estimation of the minimal number of generators of the universally finite module of differentials in some special cases, estimations that will be needed in the main result.

Proposition 4.5.12 *Let $(A, \mathfrak{m}, k)$ be a complete local ring containing a field and $\mathfrak{p}$ a prime ideal of A. Then*

$$\mu_\mathfrak{p}(\Omega_{A/k}^f) \geq \dim(A/\mathfrak{p}).$$

Proof There exists a Noetherian complete regular local ring $R = k[[X_1, \ldots, X_n]]$ and an ideal $\mathfrak{a}$ of R such that $A = R/\mathfrak{a}$, where $n = \mathrm{edim}(A)$. Let $\mathfrak{q}$ be a prime ideal of R such that $\mathfrak{q} \supseteq \mathfrak{a}$ and let $\mathfrak{p} = \mathfrak{q}/\mathfrak{a}$. By Example 3.6.26 and Proposition 3.6.14, the module $\Omega_{A/k}^f$ exists, and there is an exact sequence

$$\mathfrak{q}/\mathfrak{q}^2 \to \Omega_{R/k}^f/\mathfrak{q}\Omega_{R/k}^f \to \Omega_{(R/\mathfrak{q})/k}^f \to 0.$$

Then

$$\mu_{\mathfrak{p}}(\Omega^f_{A/k}) = \mu_{\mathfrak{q}}(\Omega^f_{(R/\mathfrak{q})/k}) \geq \mu_{\mathfrak{q}}(\Omega^f_{R/k}/\mathfrak{q}\Omega^f_{R/k}) - \mu_{\mathfrak{q}}(\mathfrak{q}/\mathfrak{q}^2) =$$

$$= \dim_{k(\mathfrak{q})}\left(\Omega^f_{R/k} \otimes k(\mathfrak{q})\right) - \mathrm{ht}(\mathfrak{q}) =$$

$$= \dim(R) - \mathrm{ht}(\mathfrak{q}) = \dim(R/\mathfrak{q}) = \dim(A/\mathfrak{p}). \qquad \square$$

In the mixed characteristic case, we have the following similar result:

Proposition 4.5.13 *Let* $(A, \mathfrak{m}, k)$ *be a Noetherian local ring such that* $\mathrm{char}(A) = 0$ *and* $\mathrm{char}(k) = p > 0$. *Let* W *be a Cohen p-ring of* $\widehat{A}$ *and* $\mathfrak{p} \in \mathrm{Spec}(\widehat{A})$. *Then*

$$\mu_{\mathfrak{p}}(\Omega^f_{\widehat{A}/W}) \geq \dim(A/\mathfrak{p}) - 1.$$

Proof Assume that $\widehat{A} = R/\mathfrak{a}$, where R is a regular local ring. Then $\widehat{A}/\mathfrak{p} = R/\mathfrak{q}$, with $\mathfrak{q} \in \mathrm{Spec}(R)$. By Proposition 3.6.14 we have an exact sequence

$$\mathfrak{q}/\mathfrak{q}^2 \to \Omega^f_{R/W}/\mathfrak{q}\Omega^f_{R/W} \to \Omega^f_{(R/\mathfrak{q})/W} \to 0.$$

It follows that

$$\mu_{\mathfrak{q}}(\Omega^f_{(R/\mathfrak{q})/W}) \geq \mu_{\mathfrak{q}}(\Omega^f_{R/W}/\mathfrak{q}\Omega^f_{R/W}) - \mu_{\mathfrak{q}}(\mathfrak{q}/\mathfrak{q}^2).$$

Then

$$\mu_{\mathfrak{q}}(\Omega^f_{(R/\mathfrak{q})/W}) \geq \dim(R) - 1 - \mathrm{ht}(\mathfrak{q}) = \dim(R/\mathfrak{q}) - 1,$$

hence

$$\mu_{\mathfrak{p}}(\Omega^f_{(\widehat{A}/\mathfrak{p})/W}) \geq \dim(\widehat{A}/\mathfrak{p}) - 1$$

and consequently

$$\mu_{\mathfrak{p}}(\Omega^f_{\widehat{A}/W}) \geq \dim(\widehat{A}/\mathfrak{p}) - 1. \qquad \square$$

The next theorem is somehow out of the main stream of the book. We could only cite the result without proof, but we decided to include its proof because it is one of the most interesting results in Commutative Algebra that moreover shows the strong connections with Algebraic Geometry. As already mentioned it is usually known as **the second Bertini theorem for local rings**, and it was proved by H. Flenner [111]. A gap in the mixed characteristic case was afterward noticed and fixed by V. Trivedi [377, 378].

Theorem 4.5.14 (Flenner [111, Satz 2.1], Trivedi [377, Th. 1], [378]) *Let $(A, \mathfrak{m}, k)$ be a Noetherian local ring, $\mathfrak{a}$ an ideal of A and $G \subseteq D(\mathfrak{a})$ a finite set of prime ideals. Then there exists an element $x \in \mathfrak{a}$ such that:*

(i) $x \notin \mathfrak{p}^{(2)}$ *for any prime ideal* $\mathfrak{p} \in D(\mathfrak{a})$.
(ii) $x \notin \bigcup_{\mathfrak{q} \in G} \mathfrak{q}$.

Proof *Case 1) A contains a field.* Let $K = k \subseteq \widehat{A}$ be a coefficient field of $\widehat{A}$. Let $y_1, \ldots, y_s$ be a system of generators of $\mathfrak{m}$ and let $z_1, \ldots, z_l$ be a system of generators of $\mathfrak{a}$. By Example 3.6.26 and Proposition 3.6.14, the $\widehat{A}$-module $\Omega^f_{\widehat{A}/K}$ exists and is an homomorphic image of the free $\widehat{A}$-module with basis $dy_1, \ldots, dy_s$, where $d :=$ $d^f_{K[[y_1,\ldots,y_s]]/K}$ is the universally finite derivation. Thus $\Omega^f_{\widehat{A}/K}$ is generated by the images of $dy_1, \ldots, dy_s$. We consider the system of elements $T := \{dz_j, d(y_i z_j), i = 1, \ldots, s, j = 1, \ldots, l\}$. Let us first remark that we have the well-known relations

$$d(y_i z_j) = z_j dy_i + y_i dz_j, \ 1 \le i \le s, \ 1 \le j \le l.$$

Let $\mathfrak{p} \in G$. Since $\mathfrak{a} \not\subseteq \mathfrak{p}$, there exists j_0 such that $z_{j_0} \notin \mathfrak{p}$. Then in $(\Omega^f_{\widehat{A}/K})_\mathfrak{p}$ the above relation becomes essentially

$$\alpha_i d(y_i z_{j_0}) = dy_i + \beta_i y_i dz_{j_0}, \ 1 \le i \le s, \ \alpha_i, \beta_i \in A_\mathfrak{p}.$$

From this it follows at once that T is a system of generators of $(\Omega^f_{\widehat{A}/K})_\mathfrak{p}$, for any prime ideal $\mathfrak{p} \in D(\mathfrak{a}\widehat{A})$. Summarizing, we have found a system of generators of $\mathfrak{a}$ that we denote $\{x_1, \ldots, x_n\}$, such that the set $T = \{dx_1, \ldots, dx_n\}$ is a system of generators of $(\Omega^f_{\widehat{A}/K})_\mathfrak{p}$, for any prime ideal $\mathfrak{p} \in D(\mathfrak{a}\widehat{A})$. Now we apply Proposition 4.5.9 for $M = D_K(\widehat{A})$, $m_i = dx_i, i = 1, \ldots, n, \ t = 0$ and $U = D(\mathfrak{a})$. Moreover we set $S = \mathbb{Q}$ if A contains $\mathbb{Q}$ and $S = A^{[p]} = \{x^p \mid x \in A\}$ if A contains the prime field $\mathbb{F}_p$ for some prime number p. In the second case, let $\mathfrak{p} \in \mathrm{Spec}(\widehat{A})$ be such that $\mathfrak{p} \not\supseteq \widehat{\mathfrak{a}}$. Then $\mathfrak{q} := \mathfrak{p} \cap A$ is a prime ideal of A with $\dim(A/\mathfrak{q}) > 0$, hence $A/\mathfrak{q}$ is an infinite integral domain. Then $S \cdot A/\mathfrak{q} = (A/\mathfrak{q})^{[p]}$, which is infinite. Thus, using Proposition 4.5.12, we can apply Prop. 4.5.9, and it follows that there exist elements $\alpha_2, \ldots, \alpha_n \in S$ such that $dx_1 + \alpha_2 dx_2 + \ldots + \alpha_n dx_n$ is basic in all prime ideals $\mathfrak{p} \in D(\mathfrak{a}\widehat{A})$. But $d\alpha_i = 0$; hence if we set $x := x_1 + \alpha_2 x_2 + \ldots + \alpha_n x_n$, we have $dx = dx_1 + \alpha_2 dx_2 + \ldots + \alpha_n dx_n$. From Lemma 4.5.11 we get that $x \notin \mathfrak{p}^{(2)}$ for any prime ideal $\mathfrak{p} \in D(\mathfrak{a}\widehat{A})$. Let $\mathfrak{q} \in D(\mathfrak{a})$. There exists $\mathfrak{p} \in D(\mathfrak{a}\widehat{A})$ such that $\mathfrak{p} \cap A = \mathfrak{q}$.

Moreover $q^{(2)} \subseteq p^{(2)}$; hence $x \notin q^{(2)}$, for any $q \in D(\mathfrak{a})$. Let now $q \in G$. There exists $p \in \text{Spec}(\widehat{A})$ with $p \cap A = q$. Now we apply again Proposition 4.5.9.

Case (2) A *does not contain a field*; hence $\text{char}(k) = p > 0$. Let

$$G' = G \cap D(\mathfrak{a}) \cap V(pA) \quad \text{and} \quad G'' = G \cap D(\mathfrak{a}) \cap D(pA).$$

Thus $G = G' \cup G''$. Let $B = A/pA$. Then B contains a field of characteristic p, hence denoting by $\bar{}$ the classes in B, by case 1 we obtain that there exists $\bar{x}_1 \in \bar{\mathfrak{a}}$ such that $\bar{x}_1 \notin q^{(2)}$ for any $q \in D(\bar{\mathfrak{a}})$ and $\bar{x}_1 \notin q$ for any $q \in G' = D(\bar{\mathfrak{a}}) \cap G$. Let $x_1 \in \mathfrak{a}$ be a preimage of $\bar{x}_1$. Then $x_1 \notin q^{(2)}$ for any $q \in D(\mathfrak{a}) \cap V(pA)$ and $x_1 \notin q$ for any $q \in G'$.

Case 2-a) $\text{char}(A) = p^n, n > 0$ and $\text{char}(k) = p > 0$. Then $D(\mathfrak{a}) = D(\mathfrak{a}) \cap V(pA)$ and $G = G'$. Thus x_1 is the needed element.

Case 2-b) $\text{char}(A) = 0$ and $\text{char}(k) = p > 0$. Let W be a Cohen p-ring such that there exists a surjective morphism $W[[T_1, \ldots, T_n]] \to \widehat{A}$ and let $\tilde{d} := d^f_{\widehat{A}/W} : \widehat{A} \to \Omega^f_{\widehat{A}/W}$ be the universal finite derivation. Note that W is a complete discrete valuation ring of characteristic 0 with maximal ideal pW. Let $\{y_1 = x_1, y_2, \ldots, y_k\}$ be a system of generators of $\mathfrak{m}$ and $\{g_1, \ldots, g_l\}$ be a system of generators of $\mathfrak{a}$. We denote

$$\{x_1, \ldots, x_n\} := \{x_1, x_1 g_1, \ldots, x_1 g_l, y_2, y_2 g_1, \ldots, y_2 g_l, \ldots, y_k, y_k g_1, \ldots, y_k g_l\}.$$

Then, using the same argument as in *Case 1*), we have $(\Omega^f_{\widehat{A}/W})_p = \sum_{i=1}^n \widehat{A}_p \tilde{d} x_i$, for any prime ideal $p \in D(\mathfrak{a})$. Let $M := \widehat{A} \oplus \Omega^f_{\widehat{A}/W}$, let $m_i := (x_i, \tilde{d} x_i)$ for $i = 1, \ldots, n$ and let M' be the submodule of M generated by $m_1, \ldots, m_n$. Let $p \in D(\mathfrak{a})$. There exists $j \in \{1, \ldots, k\}$ such that $y_j \notin p$. From the relation

$$(0, y_j \tilde{d} g_i) = (y_j g_i, \tilde{d}(y_j g_i)) - g_i(y_j, \tilde{d} y_j)$$

it follows that $(0, \tilde{d} g_i) = y_j^{-1}(0, y_j \tilde{d} g_i) \in M'_p$, for all $i = 1, \ldots, l$. Then $M_p = M'_p$, for all prime ideals $p \in D(\mathfrak{a})$. Therefore, applying Prop. 4.5.13, for every prime ideal $p \in D(\mathfrak{a})$ we have

$$\mu_p(M) - \mu_p(M/M') = \mu_p(M) = 1 + \mu_p(\Omega^f_{\widehat{A}/W}) \geq \dim(A/p).$$

Let $U = D(\mathfrak{a}\widehat{A}) \cap D(p\widehat{A})$ and $S = p^2 \mathbb{Z}$. Then for any $p \in U$, the image of S in A/p is infinite. Applying Prop. 4.5.9 it follows that there exist elements $\alpha_2, \ldots, \alpha_n \in S$ such that:

(a) $m = m_1 + m_2 \alpha_2 + \ldots + m_n \alpha_n$ is basic for $\Omega^f_{\widehat{A}/W}$ in all prime ideals $q \in U$ such that $\dim(\widehat{A}/q) > 0$, that is, for any $q \in U$.

(b) $x = x_1 + \alpha_2 x_2 + \ldots + \alpha_n x_n \notin q$, for any prime ideal $q \in G''$.

Since $m = (x, \tilde{d}x) \in \hat{A} \oplus \Omega^f_{A/W}$ is basic in any prime ideal $\mathfrak{p} \in U$, it follows that for any prime ideal $\mathfrak{p} \in U$, we have

$$(x, \tilde{d}x) \notin \mathfrak{p}\hat{A}_\mathfrak{p} \oplus \mathfrak{p}(\Omega^f_{A/W})_\mathfrak{p}.$$

Therefore, for any prime ideal $\mathfrak{p} \in U$ either $x \notin \mathfrak{p}\hat{A}_\mathfrak{p}$ or $\tilde{d}x \notin \mathfrak{p}(\Omega^f_{A/W})_\mathfrak{p}$. Thus either $x \notin \mathfrak{p}$ or, by Lemma 4.5.11, $x \notin \mathfrak{p}^{(2)}$. Since $\bar{x} = \bar{x}_1$, we also have that $x \notin \bigcup_{\mathfrak{p} \in G' \cup G''} \mathfrak{p}$.   □

Let us see some nice applications of the Theorem of Bertini.

Corollary 4.5.15 *Let $(A, \mathfrak{m})$ be a Noetherian local ring, $\mathfrak{a} \subseteq \mathfrak{m}$ be an ideal of A and G a finite subset of $D(\mathfrak{a})$. Then there exists an element $x \in \mathfrak{a}$ such that*

$$\mathrm{Reg}(A) \cap V(xA) \cap D(\mathfrak{a}) \subseteq \mathrm{Reg}(A/xA) \cap D(\mathfrak{a}).$$

Moreover one can choose the element x such that it does not belong to any of the prime ideals $\mathfrak{p} \in G$.

Proof We apply Theorem 4.5.14. Pick $x \in \mathfrak{a}$ such that $x \notin \mathfrak{p}^{(2)}$ for any $\mathfrak{p} \in D(\mathfrak{a})$. Let $\mathfrak{p} \in \mathrm{Reg}(A) \cap V(xA) \cap D(\mathfrak{a})$. Then $A_\mathfrak{p}$ is regular, and since $x \notin \mathfrak{p}^{(2)} = \mathfrak{p}^2 A_\mathfrak{p} \cap A$, it follows that $(A/xA)_\mathfrak{p}$ is regular, that is, $\mathfrak{p} \in \mathrm{Reg}(A/xA)$.   □

Proposition 4.5.16 *Let $(A, \mathfrak{m})$ be a Noetherian normal local domain. Then there exists an element $x \in \mathfrak{m}$ such that A/xA is reduced.*

Proof From Theorem 4.5.14 it follows that there exists an element $x \in \mathfrak{m}$ such that $\mathrm{Reg}(A) \cap V(xA) \subseteq \mathrm{Reg}(A/xA)$. Let $\mathfrak{q} \in \mathrm{Spec}(A/xA)$ be such that $\mathrm{depth}((A/xA)_\mathfrak{q}) = 0$. Then $\mathrm{depth}(A_\mathfrak{q}) = 1$; hence $A_\mathfrak{q}$ is a regular local ring because A is normal. Then $(A/xA)_\mathfrak{q}$ is regular; therefore A/xA is reduced.   □

Corollary 4.5.17 *Let $(A, \mathfrak{m}_1, \ldots, \mathfrak{m}_r)$ be a semilocal Noetherian normal domain. Then there exists an element $x \in \mathfrak{m}_1 \cdot \ldots \cdot \mathfrak{m}_r$ such that A/xA is reduced.*

Proof From Proposition 4.5.16, it follows that for any $i = 1, \ldots, r$ there exists an element $x_i \in \mathfrak{m}_i$ such that $A_{\mathfrak{m}_i}/x_i A_{\mathfrak{m}_i}$ is reduced. Let $x = x_1 \cdot \ldots \cdot x_r$. If $\mathfrak{q} \in \mathrm{Spec}(A/xA)$ is such that $\mathrm{depth}(A_\mathfrak{q}/xA_\mathfrak{q}) = 0$, then $\mathrm{depth}(A_\mathfrak{q}) = \mathrm{ht}(\mathfrak{q}) = 1$ and $A_\mathfrak{q}$ is regular because A is normal. If $x_i \in \mathfrak{q} \subseteq \mathfrak{m}_i$, then $\mathfrak{q} \in \mathrm{Ass}(A/x_i A)$; hence $\mathrm{depth}((A/x_i A)_\mathfrak{q}) = 0$. It follows that $(A/x_i A)_\mathfrak{q}$ is regular and consequently also $A_\mathfrak{q}$ is regular. Thus A/xA is reduced.   □

The next corollary will be used in the next section in the proof of Theorem 4.6.15.

Corollary 4.5.18 *Let A be a Noetherian normal domain, $\mathfrak{p}_1, \ldots, \mathfrak{p}_r$ be prime ideals of A and $\mathfrak{a} = \mathfrak{p}_1 \cap \ldots \cap \mathfrak{p}_r$. Then there exists an element $x \in \mathfrak{a}, x \neq 0$ and an element $a \in A \setminus \bigcup_{i=1}^{r} \mathfrak{p}_i$, such that $(A/xA)_a$ is reduced.*

Proof Let $S = A \setminus \bigcup_{i=1}^{r} \mathfrak{p}_i$ and $B = S^{-1}A$. By Corollary 4.5.17, there exists an element $x \in \mathfrak{a}$, such that $S^{-1}(A/xA)$ is reduced. Now the conclusion follows.   □

We end this section with other nice consequences of Th. 4.5.14. We need a Lemma.

Lemma 4.5.19 *Let $(A, \mathfrak{m})$ be a Noetherian local ring, U an open subset of $\mathrm{Spec}(A)$ and n a natural number. Assume that $U \subseteq S_n(A)$ and that $S_{n+1}(A)$ is open. Then the set $P := \{\mathfrak{p} \in U \mid \mathrm{depth}(A_\mathfrak{p}) = n, \dim(A_\mathfrak{p}) > n\}$ is finite.*

Proof Let $A_{n+1} := (\mathrm{Spec}(A) \setminus S_{n+1}(A)) \cap U$. By assumption the set A_{n+1} is closed in U. Let $\mathfrak{p}$ be a prime ideal in A_{n+1}. Then $A_\mathfrak{p}$ has the property (S_n) and not the property (S_{n+1}), that is,

$$\min\{n, \mathrm{ht}(\mathfrak{p})\} \leq \mathrm{depth}(A_\mathfrak{p}) < \min\{\mathrm{ht}(\mathfrak{p}), n+1\},$$

hence $\dim(A_\mathfrak{p}) > n$. Let $\mathfrak{p}_1, \ldots, \mathfrak{p}_r$ be the minimal elements of the set A_{n+1} and let $\mathfrak{a} = \mathfrak{p}_1 \cap \ldots \cap \mathfrak{p}_r$. There exist elements $x_1 \ldots, x_n \in \mathfrak{a}$ that are a regular sequence of length n in any prime ideal $\mathfrak{p} \in U \cap A_{n+1}$. Let $\mathfrak{p} \in P$. Then $\mathfrak{p} \in A_{n+1}$; hence $x_1, \ldots, x_n \in \mathfrak{p}$. But $x_1, \ldots, x_n$ is a regular sequence in $A_\mathfrak{p}$; therefore $\mathfrak{p} \in \mathrm{Ass}_A(A/(x_1, \ldots, x_n))$. This shows that $P \subseteq \mathrm{Ass}_A(A/(x_1, \ldots, x_n))$, which is a finite set.   □

Proposition 4.5.20 *Let $(A, \mathfrak{m})$ be a Noetherian local ring such that $\mathrm{Reg}(A)$ is open, let $\mathfrak{a} \subseteq \mathfrak{m}$ be an ideal of A and let $U \subseteq D(\mathfrak{a})$ be an open set. Assume that for any prime ideal $\mathfrak{p}$ in U, the ring $A_\mathfrak{p}$ satisfies the Serre properties (R_k) and (S_l) for some $k, l \in \mathbb{N}$. Then there exists an element $x \in A$ such that $(A/xA)_\mathfrak{p}$ satisfies the Serre properties (R_k) and (S_l), for any prime ideal $\mathfrak{p} \in U \cap V(xA)$.*

Proof Let P_1 be the finite set of minimal prime ideals of the closed set $\mathrm{Sing}(A)$. From Lemma 4.5.19 we get that the set $P_2 := \{\mathfrak{p} \in U \mid \mathrm{depth}(A_\mathfrak{p}) = r, \dim(A_\mathfrak{p}) > r\}$ is finite. From Corollary 4.5.15 it follows that we can find an element $x \in \mathfrak{a}$ such that

$$\mathrm{Reg}(A) \cap V(x) \cap U \subseteq \mathrm{Reg}(A/xA) \cap U$$

and such that x does not belong to any of the prime ideals from the sets P_1, P_2 and the minimal prime ideals of U.   □

Corollary 4.5.21 *Let $(A, \mathfrak{m})$ be a Noetherian local ring such that $\mathrm{Reg}(A)$ is open and let $\mathfrak{a} \subseteq \mathfrak{m}$ be an ideal of A. Then there exist elements $x, y \in \mathfrak{a}$ such that*

$$D(\mathfrak{a}) \cap \mathrm{Nor}(A) \cap V(x) \subseteq D(\mathfrak{a}) \cap \mathrm{Nor}(A/xA)$$

and

$$D(\mathfrak{a}) \cap \mathrm{Red}(A) \cap V(y) \subseteq D(\mathfrak{a}) \cap \mathrm{Red}(A/yA)$$

Proof It follows at once from Proposition 4.5.20 applied for $U = D(\mathfrak{a})$, taking account of the Serre characterization of normal and reduced Noetherian rings from Theorem 1.1.36 and Theorem 1.1.43. □

4.6   Lifting in the General Case

In the general case, that is, the case of not necessarily semilocal rings, good properties of the formal fibres are not preserved by completion.

Example 4.6.1 (J. Nishimura [269, Ex. 5.3]) There exists a Noetherian ring B having geometrically regular formal fibres such that the formal fibres of the formal power series ring $B[[T]]$ are not even geometrically reduced.

The proof below belongs to Dumitrescu [96, Th. 2.2]. We start with a lemma:

Lemma 4.6.2 (Dumitrescu [96, Lemma 2.1]) *Let A be a Noetherian domain of prime characteristic $p > 0$, A' be the integral closure of A and $K = Q(A)$ be its field of fractions. Assume that the formal fibres of $A[[X]]$ are geometrically reduced. Then:*

- *(i) $Q(A[[X]]) \subseteq K((X))$ is a separable field extension.*
- *(ii) $K[[X]] \cap Q(A[[X]])^{p^{-1}} \subseteq Q(A[[X]])$.*
- *(iii) If E is a countably generated A-subalgebra of K such that $E^p \subseteq A$, then there exists a non-zero element $a \in A$ such that $E \subseteq A_a$.*
- *(iv) If A' is a countably generated A-algebra and $A'^p \subseteq A$, then there exists a non-zero element $a \in A$ such that $A_a = A'_a$.*

Proof (i) Because

$$A[[X]]_{(X)}/(X^i)A[[X]]_{(X)} \cong A[X]_{(X)}/(X^i)A[X]_{(X)} \cong$$

$$\cong K[X]_{(X)}/(X^i)K[X]_{(X)}$$

we see that $K[[X]]$ is the (X)-adic completion of $A[[X]]_{(X)}$. But then by Proposition 1.6.40, the morphism $A[[X]]_{(X)} \rightarrow K[[X]]$ is reduced; therefore its generic fibre $Q(A[[X]]) \subseteq K((X))$ is reduced, that is, a separable field extension.

(ii) From (i) it follows that

$$K((X)) \cap Q(A[[X]])^{p^{-1}} = Q(A[[X]]),$$

therefore

$$K[[X]] \cap Q(A[[X]])^{p^{-1}} \subseteq Q(A[[X]]).$$

(iii) Let $(e_n)_{n \in \mathbb{N}}$, $e_0 \neq 0$ be a system of generators of the A-algebra E and let $f = \sum_{i=0}^{\infty} e_i X^i$. Then $f^p \in A[[X]]$. From (ii) it follows that $f = \frac{g}{h}$, where $g, h \in A[[X]]$ are non-zero. Because $e_0 \neq 0$, we can assume that $a := h(0) \neq 0$. By induction on i we easily see that $a^{i+1} e_i \in A$ for any $i \geq 0$, hence $E \subseteq A_a$.

(iv) Follows from (iii). $\square$

Now we can prove the existence of the claimed example.

Proof Consider the ring in Example 1.2.4, in the case the field k is of characteristic 2. More precisely let $\{X_i, Y_i, \ i \geq 1\}$ be variables and $A_i := k[X_i^2, X_i^3]$, $\mathfrak{p}_i = (X_i^2, X_i^3)A_i$. Let us remind that $B = S^{-1}(\bigotimes_{i=1}^{\infty} A_i)$, where S is the complement of the union of all ideals $\mathfrak{p}_i$, is a Noetherian ring of dimension 1, and it is easy to see that B has geometrically regular formal fibres. If $B[[X]]$ would have geometrically reduced formal fibres, then, for any $i \in \mathbb{N}$, the ring $B'_{\mathfrak{p}_i}$ is finite over $B_{\mathfrak{p}_i}$. Therefore there exists a finite set F_i such that $B'_{\mathfrak{p}_i}[F_i] = B_{\mathfrak{p}_i}$. Then $F = \bigcup_{i=1}^{\infty} F_i$ is a countable set with the property that $B' = B[F]$. Now Lemma 4.6.2 tells us that there exists a non-zero element $b \in B$ such that $B_b = B'_b$. But the Jacobson radical of B is (0); hence there exists an index i such that $b \notin \mathfrak{p}_i$. This implies that $B_{\mathfrak{p}_i}$ is integrally closed. Contradiction! $\square$

In particular, if a ring A has geometrically regular formal fibres and B is the completion of A in the topology of an ideal $\mathfrak{a}$, it does not follow that the formal fibres of B are geometrically regular, not even geometrically reduced. We will prove now the best result in this direction. We need some preparations, the most important and difficult one being Lemma 4.6.13, known as Rotthaus' Hilfssatz. In order to state and prove this lemma, we have to fix some notations that will be used throughout this section.

Notation 4.6.3 Let A be a ring and $\Gamma := \{\gamma \subseteq \mathrm{Max}(A) \mid \gamma \text{ finite}\}$, the family of finite sets of maximal ideals of A. We will use the following notations:

$$\Gamma(\gamma_0) := \{\gamma \in \Gamma \mid \gamma \supset \gamma_0\}, \text{ for a fixed } \gamma_0 \in \Gamma \tag{4.2}$$

$$S_\gamma := A \setminus \bigcup_{\mathfrak{m} \in \gamma} \mathfrak{m} \text{ and } A_\gamma := S_\gamma^{-1}A, \text{ for any } \gamma \in \Gamma. \tag{4.3}$$

Notation 4.6.4 For a fixed element $x \in A$ and any A-algebra B, we denote $B^* := \widehat{(B, xB)}$, the completion of B in the xB-adic topology.

Remark 4.6.5 With the above notations, if $\gamma \subset \gamma'$, then $A_\gamma^* = (S_\gamma^{-1}A_{\gamma'}^*)^*$.

Definition 4.6.6 With the above notations, assume that for any $\gamma \in \Gamma(\gamma_0)$ we have a prime ideal $\mathfrak{p}_\gamma^* \in \mathrm{Spec}(A_\gamma^*)$. The set $\{\mathfrak{p}_\gamma^*\}_{\gamma \in \Gamma(\gamma_0)}$ is called a **prime ideals sequence** if for any $\gamma, \gamma' \in \Gamma(\gamma_0)$ such that $\gamma \subset \gamma'$, we have $\mathfrak{p}_{\gamma'}^* = \mathfrak{p}_\gamma^* \cap A_{\gamma'}^*$.

Definition 4.6.7 A prime ideals sequence $\{\mathfrak{p}_\gamma^*\}_{\gamma \in \Gamma(\gamma_0)}$ is called **good** if there exists $\gamma_1 \in \Gamma(\gamma_0)$, such that for any $\gamma, \gamma' \in \Gamma(\gamma_1)$ with $\gamma \subset \gamma'$ we have $\mathfrak{p}_\gamma^* = \mathfrak{p}_{\gamma'}^* A_\gamma^*$.

Notation 4.6.8 If $\{\mathfrak{p}_\gamma^*\}_{\gamma \in \Gamma(\gamma_0)}$ is a prime ideals sequence, we denote

$$B_\gamma := A_\gamma^*/\mathfrak{p}_\gamma^*, \quad \overline{B}_\gamma \text{ the integral closure of } B_\gamma \tag{4.4}$$

and

$$C = \bigcap_{\gamma \in \Gamma(\gamma_0)} \overline{B}_\gamma. \tag{4.5}$$

Notation 4.6.9 With notations as above, set

$$\Delta_\gamma(x) := \{\mathfrak{q} \cap A \mid \mathfrak{q} \in \mathrm{Ass}(\overline{B}_\gamma/x\overline{B}_\gamma)\} \tag{4.6}$$

and

$$\Delta(x) := \bigcup_{\gamma \in \Gamma(\gamma_0)} \Delta_\gamma(x). \tag{4.7}$$

Remark 4.6.10 If A is a Nagata ring, as it will be the case in Rotthaus' Hilfssatz, in the above notation, we have

$$\Delta_\gamma(x) = \{\mathfrak{q} \cap A \mid \mathfrak{q} \in \mathrm{Min}(\overline{B}_\gamma/x\overline{B}_\gamma)\}.$$

This follows from the fact that in this case $\overline{B}_\gamma$ is a Noetherian normal ring; hence principal ideals have no embedded primes.

Definition 4.6.11 A prime ideals sequence is called **bounded** if $\Delta(x)$ is a finite set. A prime ideals sequence is called **simple** if it is bounded and good.

Now we can proceed to prove the main Lemma, usually known as Rotthaus' Hilfssatz. This lemma was first proved by Rotthaus [328] and then improved by J. Nishimura and T. Nishimura [273]. First we need a simple but useful fact. We must remark that the local ring A in the next Lemma is not necessarily Noetherian. Thus we use the Bourbaki definition of associated primes [56, Ch. 4, §1, Def. 1].

Lemma 4.6.12 *Let* $(A, \mathfrak{m})$ *be a local ring contained in a Noetherian normal domain* V. *Assume that there exists a non-zero element* $y \in \mathfrak{m}$ *such that* $yV \cap A = yA$ *and* $\mathfrak{m} \in \mathrm{Ass}(A/yA)$. *Then* A *is a DVR.*

Proof By assumption we can find $z \in A$ such that $z \notin yA$ and $z \in yA : \mathfrak{m}$. Let $\alpha := \frac{z}{y} \in Q(A)$. Then it is enough to show that $\alpha \cdot \mathfrak{m} = A$. Assume not. Then $\alpha \cdot \mathfrak{m} \subset \mathfrak{m}$; hence $\alpha \cdot \mathfrak{m}V \subseteq \mathfrak{m}V$. But V is a normal domain, so that one can easily see that the above relation implies that $\alpha \in V$. Thus $z \in yV \cap A = yA$. This contradicts the choice of z. $\qquad\square$

Lemma 4.6.13 (Rotthaus' Hilfssatz [328, Hilfssatz], [273, Th. 1.9]) *Let* A *be a Noetherian ring,* $x \in A$ *and* $\gamma_0 \in \Gamma$. *Assume that* A *is a Nagata ring which is* xA-*adically complete and that* $\{\mathfrak{p}_\gamma^*\}_{\gamma \in \Gamma(\gamma_0)}$ *is a simple prime ideal sequence. For any* $\gamma \in \Gamma(\gamma_0)$, *let* $\mathfrak{p}_\gamma = \mathfrak{p}_\gamma^* \cap A$. *Then:*

(i) *There exists* $\gamma_2 \in \Gamma(\gamma_0)$ *such that* $\mathfrak{p}_\gamma^* \in \mathrm{Ass}(A_\gamma^*/\mathfrak{p}_\gamma A_\gamma^*)$ *for any* $\gamma \in \Gamma(\gamma_2)$.
(ii) *If* $\mathrm{ht}(\mathfrak{p}_\gamma^*) > 0$ *for any* $\gamma \in \Gamma(\gamma_0)$, *then* $\mathfrak{p}_\gamma^* \cap A \neq (0)$.

Proof ([273]) It is clear that we may assume that A is an integral domain, that $x \in A$ is a non-zero element and $\mathfrak{p}_\gamma = (0)$. Moreover, since the sequence is simple, hence good, we may assume that there exists $\gamma_1 \in \Gamma(\gamma_0)$ such that for any $\gamma, \gamma' \in \Gamma(\gamma_1)$ with $\gamma \subset \gamma'$, we have $\mathfrak{p}_\gamma^* = \mathfrak{p}_{\gamma'}^* A_\gamma$. Thus, it is enough to find $\gamma_2 \supset \gamma_1$ such that $\mathrm{ht}(\mathfrak{p}_\gamma^*) = 0$ for any $\gamma \in \Gamma(\gamma_2)$. For any $\gamma \supset \gamma'$ we have

$$B_\gamma = A_\gamma^*/\mathfrak{p}_\gamma^* = A_\gamma^*/\mathfrak{p}_{\gamma'}^* A_\gamma^* = (S_\gamma^{-1}B_{\gamma'})^* \tag{4.8}$$

hence the morphism $B_{\gamma'} \to B_\gamma$ is flat. It follows that

$$B_\gamma \subseteq \overline{B}_{\gamma'} \otimes_{B_{\gamma'}} B_\gamma \subseteq Q(B_\gamma). \tag{4.9}$$

Since B_γ is a Nagata ring, it follows that for any $\gamma' \supset \gamma$, the morphism

$$\overline{B}_{\gamma'} \otimes_{B_{\gamma'}} B_\gamma \to \overline{B}_\gamma$$

is finite and injective. Now $\overline{B}_{\gamma'} \otimes_{B_{\gamma'}} B_\gamma = (S_\gamma^{-1}\overline{B}_{\gamma'})^*$ and $\overline{B}_\gamma$ is the integral closure of $\overline{B}_{\gamma'} \otimes_{B_{\gamma'}} B_\gamma$. Let $\overline{q} \in \mathrm{Min}(\overline{B}_\gamma/x\overline{B}_\gamma)$. Then by [257, Prop. 3.5 and Prop. 3.17]

$$\overline{q} \cap (\overline{B}_{\gamma'} \otimes_{B_{\gamma'}} B_\gamma) \in A_a(x(\overline{B}_{\gamma'} \otimes_{B_{\gamma'}} B_\gamma)) = A_a(x\,S_\gamma^{-1}(\overline{B}_{\gamma'})) \subseteq$$

$$\subseteq \mathrm{Ass}(S_\gamma^{-1}(\overline{B}_{\gamma'}/x\overline{B}_{\gamma'})) = \mathrm{Min}(S_\gamma^{-1}(\overline{B}_{\gamma'}/x\overline{B}_{\gamma'})).$$

This means that

$$\overline{q} \cap \overline{B}_{\gamma'} \in \mathrm{Min}(\overline{B}_{\gamma'}/x\overline{B}_{\gamma'}) \text{ for any } \overline{q} \in \mathrm{Min}(\overline{B}_\gamma/x\overline{B}_\gamma). \tag{4.10}$$

Thus we have a canonical surjective map

$$\mathrm{Min}(\overline{B}_\gamma/x\overline{B}_\gamma) \to \mathrm{Min}(S_\gamma^{-1}(\overline{B}_{\gamma'}/x\overline{B}_{\gamma'})). \tag{4.11}$$

Since $\overline{B}_\gamma$ is a Krull domain, for any $\gamma \in \Gamma(\gamma_1)$ the set of essential valuations v of $\overline{B}_\gamma$ with $v(x) \neq 0$ corresponding to $\mathrm{Min}(\overline{B}_\gamma/x\overline{B}_\gamma)$ is finite. As $\Delta(x)$ is finite, there exists $\gamma_2 \in \Gamma(\gamma_1)$ such that for any $\gamma, \gamma' \in \Gamma(\gamma_2)$ with $\gamma' \supset \gamma$, the canonical map

$$\mathrm{Min}(\overline{B}_\gamma/x\overline{B}_\gamma) \to \mathrm{Min}(\overline{B}_{\gamma'}/x\overline{B}_{\gamma'})$$

is surjective. Let $\{w_1, \ldots, w_r\}$ be the set of essential valuations w of $\overline{B}_{\gamma_2}$ such that $w(x) \neq 0$ and let $W_1, \ldots, W_r$ be their associated DVRs. Then $v_{\gamma,i} = w_i \mid_{Q(\overline{B}_\gamma)}$ and $V_{\gamma,i} = W_i \cap Q(\overline{B}_\gamma)$ are exactly the essential valuations v and their associated DVRs in $\overline{B}_\gamma$ with $v(x) \neq 0$. Therefore

$$\mathrm{Min}(\overline{B}_\gamma/x\overline{B}_\gamma) = \{\overline{q}_{\gamma,1}, \ldots, \overline{q}_{\gamma,r}\} \text{ and } (\overline{B}_\gamma)_{\overline{q}_{\gamma,i}} = V_{\gamma,i}, \ i = 1, \ldots, r.$$

Let $\nabla(x) = \{\tilde{q} \in \mathrm{Spec}(C) \mid \exists \, i \in \{1, \ldots, r\} \text{ such that } \tilde{q} = \overline{q}_{\gamma,i} \cap C\}$. Let us denote $\nabla(x) = \{\tilde{q}_1, \ldots, \tilde{q}_s\}$. Then we have:

$$\tilde{q}_i \cap A \in \Delta(x) \text{ for any } i = 1, \ldots, s; \tag{4.12}$$

$$x\overline{B}_\gamma = x\overline{B}_{\gamma_2} \cap B_\gamma, \forall \gamma \in \Gamma(\gamma_2); \tag{4.13}$$

$$xC = x\overline{B}_\gamma \cap C; \tag{4.14}$$

$$xC_{\tilde{q}_i} = x(\overline{B}_\gamma)_{\tilde{q}_i} \cap C_{\tilde{q}_i}, \text{ for any } i = 1, \ldots, s; \tag{4.15}$$

$$\sqrt{xC} = \tilde{q}_1 \cap \ldots \cap \tilde{q}_s. \tag{4.16}$$

Now let $\tilde{q} \in V(x)$ and $\overline{q}_{\gamma_2} \in \mathrm{Min}(\overline{B}_{\gamma_2}/x\overline{B}_{\gamma_2})$ be such that $\tilde{q} = \overline{q}_{\gamma_2} \cap C$, and put $\overline{q}_\gamma = \overline{q}_{\gamma_2} \cap \overline{B}_\gamma$, for any $\gamma \in \Gamma(\gamma_2)$. We have the following diagram:

$$
\begin{array}{ccccc}
A/q & \longrightarrow & S_\gamma^{-1}(A/q) & \longrightarrow & Q(A/q) \\
\downarrow & & \downarrow & & \downarrow \\
C/\tilde{q} & \longrightarrow & \overline{B}_\gamma/\overline{q}_\gamma & \longrightarrow & Q(\overline{B}_{\gamma_2}/\overline{q}_{\gamma_2}).
\end{array}
$$

Let D be the integral closure of A/Q in $Q(\overline{B}_{\gamma_2}/\overline{q}_{\gamma_2})$. Since the field extension $Q(A/q) \subseteq Q(\overline{B}_{\gamma_2}/\overline{q}_{\gamma_2})$ is finite and A is a Nagata ring, it follows that D is a finite A/q-module. Because $S_\gamma^{-1}(A/q) \subseteq \overline{B}_\gamma/\overline{q}_\gamma$ is an integral extension, we have $\overline{B}_\gamma/\overline{q}_\gamma \subseteq S_\gamma^{-1}D$ and finally

$$C/\tilde{q} \subseteq \bigcap_{\gamma \in \Gamma(\gamma_1)} S_\gamma^{-1}D = D.$$

It follows that $C/\tilde{q}$ is a finite A-module, for any prime ideal $\tilde{q} \in V(x)$. Even if the ring C is not necessarily Noetherian, we can consider, as noted before, the set $\mathrm{Ass}(C/xC)$ as defined in [56, Ch. 4, §1, Def. 1]. By Lemma 4.6.12 and the relation (4.14), it follows that $C_{\tilde{q}}$ is a DVR for any $\tilde{q} \in \mathrm{Ass}(C/xC) \subseteq V(x)$. Let $\mathrm{Ass}(C/xC) = \{\tilde{q}_1, \ldots, \tilde{q}_t\}$. Then $xC = \tilde{q}_1^{(e_1)} \cap \ldots \cap \tilde{q}_t^{(e_t)}$. Thus it is enough to show that for any $\tilde{q} \in \mathrm{Ass}(C/xC)$, the ring $C/\tilde{q}^{(e)}$ is Noetherian for any $e \in \mathbb{N}$. Let $U := C \setminus \bigcup_{i=1}^t \tilde{q}_i$ and let v_i be the valuation associated with the discrete valuation ring $C_{\tilde{q}_i}$. We may assume for simplicity that $\tilde{q} = \tilde{q}_1$. Since $U^{-1}C$ is a semilocal Dedekind domain, there exists $c_1' = \frac{c}{u} \in U^{-1}C$ such that $v_1(c_1') = e_1 - 1$ and $v_j(c_1') \geq e_j$ for $j = 2, \ldots, t$. Let $\mathfrak{c} := \bigcap \{\tilde{q}' \mid \tilde{q}' \in V(x) \setminus \mathrm{Ass}(C/xC)\}$. Pick $c_2 \in \mathfrak{c} \setminus \bigcup_{i=1}^t \tilde{q}_i$ and let $c = c_1 \cdot c_2^v$, for sufficiently large v. Let $\xi := \frac{x}{c}$. Then from (4.16) we have

$$\tilde{q} \subseteq \bigcap \{\xi \overline{B}_\gamma \cap \overline{B}_\gamma\} = \xi C \cap C,$$

hence

$$\xi C \cap C = \tilde{q}$$

and

$$\xi^n C \cap C = \tilde{\mathfrak{q}}^{(n)} \text{ for any } n \geq 1.$$

As in the proof of Lemma 2.1.14, we can show that for any $n \geq 1$ we have a finite injective morphism

$$C/\tilde{\mathfrak{q}}^{(n)} \subseteq C[\xi]/\xi^n C[\xi]$$

and it suffices to show that $C[\xi]/\xi^n C[\xi]$ is Noetherian. But $C/\tilde{\mathfrak{q}} \cong C[\xi]/\xi C[\xi]$ is a Noetherian ring and by Lemma 2.1.13 it follows that $C[\xi]/\xi^n C[\xi]$ is Noetherian for any $n \geq 1$. Thus we obtain that C/xC is Noetherian and a finite A-module. By ([248, (28.P), Lemma]), it follows that C is a finite A-module. Let now $\mathfrak{p} \in \text{Min}(xA)$. Since C is finite over A, there exists $\tilde{\mathfrak{q}} \in \text{Ass}(C/xC)$ such that $\tilde{\mathfrak{q}} \cap \mathfrak{p}$. By (4.12) it follows that $\mathfrak{p} \in \Delta(x)$. Hence for any $\gamma \in \Gamma(\gamma_2)$, we have

$$1 = \text{ht}(\mathfrak{p}) \geq \text{ht}(\mathfrak{p}_\gamma^*) + 1.$$

Therefore $\text{ht}(\mathfrak{p}_\gamma^*) = 0$. $\qquad\square$

Proposition 4.6.14 (J. Nishimura-T. Nishimura [273, Prop. 1.18]) *Let P be a property of Noetherian local rings and $\mathfrak{m}$ a maximal ideal of A. We pick an element $x \in A$ and a prime ideal $\mathfrak{p}_\mathfrak{m}^*$ that is minimal in $\text{NP}(A_\mathfrak{m})$. For any $\gamma \in \Gamma(\mathfrak{m})$, let $\mathfrak{p}_\gamma^* = \mathfrak{p}_\mathfrak{m}^* \cap A_\gamma^*$. Assume that:*

 (i) *P satisfies the conditions $\mathbf{A}_3$, $\mathbf{A}_5$ and $\mathbf{A}_6$.*
 (ii) *A/xA is a Nagata P-ring.*
 (iii) *The sequence $S := \{\mathfrak{p}_\gamma^*\}_{\gamma \in \Gamma(\mathfrak{m})}$ is bounded.*
     Then the sequence S is good.

Proof For any γ and γ' in $\Gamma(\mathfrak{m})$ such that $\gamma' \supset \gamma$, we have

$$\mathfrak{p}_{\gamma'}^* A_\gamma^* = \mathfrak{p}_{\gamma,1}^* \cap \ldots \cap \mathfrak{p}_{\gamma,m}^*,$$

where $\mathfrak{p}_{\gamma,1}^* = \mathfrak{p}_\gamma^*, \ldots, \mathfrak{p}_{\gamma,m}^* \in \text{Min}(NP(A_\gamma^*))$. For $\mathfrak{q}_\gamma^* \in \text{Min}(\mathfrak{p}_{\gamma,i}^* + xA_\gamma^*)$, let us denote $\mathfrak{q}_{\gamma\gamma'}^* := \mathfrak{q}_\gamma^* \cap A_{\gamma'}^*$. Since

$$((A_{\gamma'}^*/\mathfrak{p}_{\gamma'}^*)_{\mathfrak{q}_{\gamma\gamma'}^*})^* \cong ((A_\gamma^*/\mathfrak{p}_\gamma^* A_\gamma^*)_{\mathfrak{q}_\gamma^*})^*,$$

by Proposition 3.1.41, we get that $q^*_{\gamma\gamma'} \cap A \in \Delta_{\gamma'}(x)$. Let $S := A \setminus \bigcup_{q \in \Delta(x)} q$. Then $q^*_\gamma \cap S = \emptyset$; hence

$$(\mathfrak{p}^*_{\gamma,i}, x)(S^{-1}A^*_\gamma) \subset q^*_\gamma (S^{-1}A^*_\gamma) \neq S^{-1}A^*_\gamma.$$

On the other hand, since by Theorem 4.3.6 the ring $S^{-1}A^*$ is a semilocal **P**-ring, we have

$$\bigcap_{\mathfrak{p}^* \in \mathbf{NP}(S^{-1}A^*)} \mathfrak{p}^* = \mathfrak{p}^*_1 \cap \ldots \cap \mathfrak{p}^*_n, \text{ where } \mathfrak{p}^*_i \in \mathrm{Min}(\mathbf{NP}(S^{-1}A)).$$

Let $\gamma_0 \in \Gamma(\mathfrak{m})$ be such that for any $q \in \Delta(x)$, there exists at least one maximal ideal $\mathfrak{m}' \in \gamma_0$ containing q. Since A^*_γ is a Nagata **P**-ring, the morphism $S^{-1}A^*_\gamma \to S^{-1}A^*$ is a reduced **P**-morphism for any $\gamma \in \Gamma(\gamma_0)$. Thus we get

$$\mathfrak{p}^*_{\gamma,i} S^{-1} A^* = \bigcap_{j \in \Lambda_{\gamma,i}} \mathfrak{p}^*_j, \text{ with } \Lambda_{\gamma,i} \subset \{1, \ldots, n\},$$

$$\mathfrak{p}^*_{\gamma'} S^{-1} A^* = \bigcap_{h \in \Lambda_{\gamma'}} \mathfrak{p}^*_h = \bigcap_{i=1}^{m} \bigcap_{j \in \Lambda_{\gamma,i}} \mathfrak{p}^*_j.$$

It follows that, denoting by $|L|$ the cardinality of a set L,

$$|\Lambda_{\gamma'}| = |\Lambda_{\gamma,1}| + \ldots + |\Lambda_{\gamma,m}| \leq n,$$

$$\text{where } |\Lambda_{\gamma,i}| > 0, \ i = 1, \ldots, m.$$

Then there exists $\gamma_1 \in \Gamma(\gamma_0)$ such that $\mathfrak{p}^*_{\gamma'} A^*_\gamma = \mathfrak{p}^*_\gamma$, for any $\gamma, \gamma' \in \Gamma(\gamma_1)$, with $\gamma' \supset \gamma$; therefore the sequence is good.   □

The next result, a lifting property for rings with geometrically normal formal fibres, was proved by Brezuleanu and Rotthaus [69] with some additional assumptions. The present general form was given by J. Nishimura and T. Nishimura [273].

Theorem 4.6.15 (Brezuleanu-Rotthaus [69, Satz 2], J. Nishimura-T. Nishimura [273, Th. A]) *Let A be a Noetherian ring and $\mathfrak{a}$ be an ideal of A such that A is $\mathfrak{a}$-adically complete. If $A/\mathfrak{a}$ is a Nagata ring with geometrically normal formal fibres, then A is a Nagata ring with geometrically normal formal fibres.*

Proof ([273]) By Marot's Theorem 2.2.5, we know that A is a Nagata ring; hence it is enough to show that A has geometrically normal formal fibres. Assume that this is not true. By Noetherian induction we may assume that $A/\mathfrak{b}$ has geometrically normal formal fibres for any ideal $\mathfrak{b} \neq 0$. Again by Noetherian induction, we may assume that for any ideal $\mathfrak{a}'$ strictly containing $\mathfrak{a}$, the completion of A in the $\mathfrak{a}'$-adic topology has geometrically normal formal fibres. In particular, since the $\mathfrak{a}$-adic topology coincides with the $\sqrt{\mathfrak{a}}$-adic topology, we may assume that $\mathfrak{a}$ is a radical ideal. Since the formal fibres of A are not all geometrically normal, there are prime ideals $\mathfrak{p} \supset \mathfrak{q}$ such that the generic fibre of the morphism

$$A_\mathfrak{p}/\mathfrak{q}A_\mathfrak{p} \to \widehat{A_\mathfrak{p}}/\mathfrak{q}\widehat{A_\mathfrak{p}}$$

is not geometrically normal. But the above map is a normal morphism if $\mathfrak{q} \neq (0)$; hence we may assume that A is an integral domain with field of fractions K. Let L be a field, finite extension of K, and let B be a finite A-algebra with fraction field L. We have to show that for any $\mathfrak{n} \in \mathrm{Max}(B)$, the ring $L \otimes_B \widehat{B_\mathfrak{n}}$ is normal. By the above assumptions, we may assume that B is a normal domain with $\mathfrak{b} = \sqrt{\mathfrak{a}B} = \mathfrak{q}_1 \cap \ldots \cap \mathfrak{q}_s$, where $\mathfrak{q}_1, \ldots, \mathfrak{q}_s$ are prime ideals, such that B is $\mathfrak{b}$-adically complete and $B/\mathfrak{c}$ has geometrically normal formal fibres for any ideal $\mathfrak{c} \neq 0$ of B. By Corollary 4.5.18 there exist elements $y \in \mathfrak{b}$ and $b \in B \setminus \bigcup_{i=1}^{s} \mathfrak{q}_i$ such that the ring $(B/yB)_b$ is reduced. For a ring C, we denote by C^* the yC-adic completion of C. The ring $B_\mathfrak{n}/yB_\mathfrak{n}$ has geometrically normal formal fibres; hence applying Cor. 4.3.7, it follows that $(B_\mathfrak{n})^*$ has geometrically normal formal fibres. This implies that the morphism $(B_\mathfrak{n})^* \to \widehat{B_\mathfrak{n}}$ is a normal morphism; therefore it is enough to show that for any $\mathfrak{n} \in \mathrm{Max}(B)$ the ring $L \otimes_B (B_\mathfrak{n})^*$ is normal. Thus, considering a maximal ideal $\mathfrak{n}$ of B and a prime ideal $\mathfrak{q}_\mathfrak{n}^*$ that is minimal in $\mathbf{NNor}(B_\mathfrak{n}^*)$, we need to show that $\mathfrak{q}_\mathfrak{n}^* \cap B \neq (0)$. Let us denote, as above, $\Gamma := \{\delta \subseteq \mathrm{Max}(B) \mid \delta \text{ is a finite set}\}$. If $\gamma \in \Gamma(\mathfrak{n})$ and $\mathfrak{q}_\gamma^* := \mathfrak{q}_\mathfrak{n}^* \cap B_\gamma^*$, we have $\mathrm{ht}(\mathfrak{q}_\gamma^*) > 0$. Therefore, applying Proposition 4.6.14, we need only to show that $\Delta(y)$ is a finite set. For $\gamma \in \Gamma$, set

$$\mathfrak{c}_\gamma^* := \bigcap_{\mathfrak{q}^* \in \mathbf{NNor}(B_\gamma^*)} \mathfrak{q}^* = \mathfrak{q}_{\gamma 1}^* \cap \ldots \cap \mathfrak{q}_{\gamma t_\gamma}^*, \text{ where } \mathfrak{q}_\gamma^* = \mathfrak{q}_{\gamma 1}^*$$

and let $\bar{C}_{\gamma k}$ be the integral closure of $C_{\gamma k} = B_\gamma^*/\mathfrak{q}_{\gamma k}^*$, for $1 \leq k \leq t_\gamma$. Then by [248, (17.I), Th. 39], we have $\mathrm{depth}(B_\gamma^*)_{\mathfrak{q}_{\gamma k}^*} = 1$. Let

$$\Delta_\gamma'(y) := \bigcup_{k=1}^{t_\gamma} \{\mathfrak{q} \mid \exists \, \bar{\mathfrak{q}} \in \mathrm{Min}(\bar{C}_{\gamma k}/y\bar{C}_{\gamma k}) \text{ such that } \mathfrak{q} = \bar{\mathfrak{q}} \cap B\}.$$

Clearly

$$\Delta_\gamma(y) = \{\mathfrak{q} \mid \exists \, \bar{\mathfrak{q}} \in \mathrm{Min}(\bar{C}_{\gamma 1}/y\bar{C}_{\gamma 1}) \text{ such that } \mathfrak{q} = \bar{\mathfrak{q}} \cap B\} \subseteq \Delta_\gamma'(y). \qquad (4.17)$$

We will first show that for any $\gamma \in \Gamma(n)$

$$\Delta_\gamma'(y) = \bigcup_{m \in \gamma} \Delta_{\{m\}}'(y). \tag{4.18}$$

Let $m \in \gamma$. Since by Corollary 4.3.7 the ring B_γ^* has geometrically normal formal fibres, it follows that the morphism

$$B_\gamma^* \xrightarrow{\rho_{\gamma,m}} B_m^* \xrightarrow{g} \widehat{B_m} = \widehat{B_\gamma^*}$$

is a normal morphism. Since g is faithfully flat, by Proposition 1.6.18 we get that the natural map $\rho_{\gamma,m} : B_\gamma^* \to B_m^*$ is a normal morphism. It follows that

$$q_{\gamma k}^* B_m^* = \bigcap_{h \in \Lambda_{\gamma km}} q_{mh}^*, \text{ with } \Lambda_{\gamma km} \subset \{1, \dots, t_m\}.$$

Then for any $q \subseteq m$ such that $q = \bar{q} \cap B$, for some $\bar{q} \in \mathrm{Min}(\bar{C}_{\gamma k}/y\bar{C}_{\gamma k})$, we have the diagram:

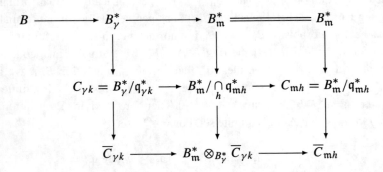

Thus, by Proposition 3.1.41 there exists $\tilde{q} \in \mathrm{Min}(\overline{C}_{mh}/y\overline{C}_{mh})$ for some $h \in \Delta_{\gamma km}$, such that $\bar{q} = \tilde{q} \cap \overline{C}_{\gamma k}$. Again by Proposition 3.1.41, it follows that for any $h \in \Lambda_{\gamma km}$ and for any $\tilde{q} \in \mathrm{Min}(\overline{C}_{mh}/y\overline{C}_{mh})$, we have $\bar{q} = \tilde{q} \cap \overline{C}_{\gamma k} \in \mathrm{Min}(\overline{C}_{\gamma k}/y\overline{C}_{\gamma k})$. This proves (4.18).

   Case 1. Suppose first that $b \in m$. Let D be the completion of B in the $(b + bB)$-adic topology and let

$$c^\# = \bigcap_{q \in \mathbf{NNor}(D)} q = q_1^\# \cap \dots \cap q_p^\#.$$

We have the diagram:

$$
\begin{array}{ccc}
B & \longrightarrow & D \\
\downarrow & & \downarrow \theta_{\mathrm{m}} \\
B_{\mathrm{m}}^{*} & \xrightarrow{\rho_{\mathrm{m}}} & D_{\mathrm{m}}
\end{array}
$$

By the choice of b, we have $\mathfrak{a} \subsetneq (\mathfrak{b}, bB) \cap A$; hence the formal fibres of D are geometrically normal. Then the morphisms $D \to \widehat{D} = \widehat{B_{\mathrm{m}}^{*}}$ is a normal morphism; hence ρ_{m} and θ_{m} are normal morphisms too. It follows that $\mathfrak{c}_{\mathrm{m}}^{*} D_{\mathrm{m}} = \mathfrak{c}^{\#} D_{\mathrm{m}}$. As in the above proof of (4.18), we get that

$$
\Delta'_{\{\mathrm{m}\}}(y) \subseteq \mathrm{Ass}(D/(\mathfrak{c}^{\#}, y)) \cap B. \tag{4.19}
$$

Case 2. Suppose now that $b \notin \mathfrak{m}$. Since the normal locus of B is open, we can find an element $z \in B$ such that $\{y, z\}$ is a regular sequence and $(B/yB)_{bz}$ is a normal ring. We will show that

$$
\Delta'_{\{\mathrm{m}\}} \subseteq \mathrm{Ass}\big(B/(y, z)B\big). \tag{4.20}
$$

Let $\tilde{\mathfrak{q}} \in \mathrm{Min}(\overline{C}_{\mathrm{m}h}/y\overline{C}_{\mathrm{m}h})$, for some $h \in \{1, \ldots, t_{\mathrm{m}}\}$. Let

$$
\mathfrak{q} = \tilde{\mathfrak{q}} \cap B, \quad \mathfrak{q}^{*} = \overline{\mathfrak{q}} \cap B_{\mathrm{m}}^{*} \text{ and } \mathfrak{q}' = \overline{\mathfrak{q}} \cap C_{\mathrm{m}h}.
$$

Then there exists $\mathfrak{p} \in \mathrm{Min}\big((C_{\mathrm{m}h})_{\mathfrak{q}'}^{*}\big)$ such that

$$
\dim\big((C_{\mathrm{m}h})_{\mathfrak{q}'}^{*}/\mathfrak{p}\big) = 1.
$$

Since $\mathrm{depth}\big((B_{\gamma}^{*})_{\mathfrak{q}_{\gamma h}^{*}}\big) = 1$, we get that $\mathrm{depth}(B_{\mathfrak{q}}) \leq 2$. Suppose that $z \notin \mathfrak{q}$. Then, if we set $\mathfrak{q}^{*} := \mathfrak{q}B_{\mathrm{m}}^{*}$, we see that $(B/yB)_{\mathfrak{q}} = (B_{\mathrm{m}}^{*}/yB_{\mathrm{m}}^{*})_{\mathfrak{q}^{*}}$ is a normal ring and $(B_{\mathrm{m}}^{*})_{\mathfrak{q}^{*}}$ is normal too. In particular $(B_{\mathrm{m}}^{*})_{\mathfrak{q}_{\mathrm{m}h}^{*}}$ is normal, and this contradicts the choice of $\mathfrak{q}_{\mathrm{m}h}^{*}$. Hence $z \in \mathfrak{q}$ and this proves (4.20).

From (4.18), (4.19) and (4.20) it follows that $\Delta(y)$ is a finite set. Let $\mathfrak{q}_{\gamma}^{*} := \mathfrak{q}_{\mathrm{n}}^{*} \cap B_{\gamma}^{*}$. Then $\mathrm{ht}(\mathfrak{q}_{\gamma}^{*}) = 0$ and from Lemma 4.6.13 it results that $\mathfrak{q}_{\mathrm{n}}^{*} \cap B \neq (0)$.   □

The next theorem is the most important lifting result. The first form was obtained by Rotthaus [328] in the case of finite dimensional universally catenary rings containing a field of characteristic zero. The proof used the resolution of singularities of Hironaka. Later the result was generalized by J. Nishimura and T. Nishimura [273] to any Noetherian ring containing a field of characteristic zero. The final general result was obtained by Gabber [117] using the weak local uniformization instead of resolution of singularities, and the details of the proof were presented by Kurano and Shimomoto [205].

Theorem 4.6.16 (Gabber [117], Nishimura-Nishimura [273], Rotthaus [328]) *Let A be a Noetherian ring and $\mathfrak{a}$ be an ideal of A. Assume that A is $\mathfrak{a}$-adically complete and $A/\mathfrak{a}$ is quasi-excellent. Then A is quasi-excellent.*

Proof ([205, Th. 5.1]) Suppose that A is not quasi-excellent. Let $\mathfrak{a} = (a_1, \ldots, a_n)$ and put $\mathfrak{a}_0 := (0)$ and $\mathfrak{a}_s := (a_1, \ldots, a_s)$, $s = 1, \ldots, n$. Since A is $\mathfrak{a}$-adically complete, it follows by [341, Rem. 2.2.9] that A is $\mathfrak{a}_i$-adically complete for any $i = 0, \ldots, n$. Moreover $A/\mathfrak{a}_i$ is $\mathfrak{a}_{i+1}/\mathfrak{a}_i$-adically complete for any $i = 0, \ldots, n-1$. Note that $A/\mathfrak{a}_n$ is quasi-excellent and that $A/\mathfrak{a}_0$ is not quasi-excellent. Hence there exists i such that $A/\mathfrak{a}_{i+1}$ is quasi-excellent and $A/\mathfrak{a}_i$ is not quasi-excellent. Let us also remark that $\mathfrak{a}_{i+1}/\mathfrak{a}_i$ is a principal ideal of the ring $A/\mathfrak{a}_i$. Therefore we may assume that

$$\mathfrak{a} = (x) \neq (0) \text{ is a principal ideal of } A. \tag{4.21}$$

Let $\mathcal{F} := \{\mathfrak{b} \mid A/\mathfrak{b} \text{ is not quasi-excellent}\}$. The family $\mathcal{F}$ is not empty because it contains (0), whence there exists a maximal element $\mathfrak{b}_0 \in \mathcal{F}$. Replacing $A/\mathfrak{b}_0$ by A, we may assume that

$$A/\mathfrak{b} \text{ is quasi-excellent for any ideal } \mathfrak{b} \neq (0). \tag{4.22}$$

By Theorem 4.4.6 not all the formal fibres of A are geometrically regular. This means that there are prime ideals $\mathfrak{p} \supset \mathfrak{q}$ of A such that the generic fibre of the morphism $A_\mathfrak{p}/\mathfrak{q}A_\mathfrak{p} \to \widehat{A_\mathfrak{p}}/\mathfrak{q}\widehat{A_\mathfrak{p}}$ is not geometrically regular. But (4.22) shows that $\mathfrak{q} = (0)$, that is, A is an integral domain. By Marot's Theorem 2.2.5 the ring A is Nagata, hence the integral closure A' is a finite A-algebra. By the Theorem of Greco 1.6.42, it results that A' is not quasi-excellent since A is not quasi-excellent. But A' is xA'-adically complete and satisfies the assumptions (4.22). Therefore, replacing A with A', we may assume that

$$A \text{ is a Nagata normal domain.} \tag{4.23}$$

By Theorem 4.6.15 the formal fibres of A are geometrically normal. Therefore, for any prime ideal $\mathfrak{p}$ of A, the ring $\widehat{A_\mathfrak{p}}$ is a normal local domain. By Ratliff's Theorem 3.1.42, we get that moreover A is universally catenary. From Th. 4.6.15 it follows that not all of the formal fibres of A are geometrically regular, and by Cor. 1.6.39 we see that there exists a maximal ideal $\mathfrak{m}$ of A such that the morphism $A_\mathfrak{m} \to \widehat{A_\mathfrak{m}}$ is not regular. But (4.22) says that all the non-zero fibres of this morphism are geometrically regular, whence we have to look at the generic fibre. Let $Q(A)$ be the field of fractions of A and L a finite algebraic extension of $Q(A)$. Let B be the integral closure of A in L, which is a finite A-module since A is a Nagata ring. Let $\mathcal{L} := \{\mathfrak{n} \in \mathrm{Max}(B) \mid \mathfrak{n} \cap A = \mathfrak{m}\}$. We have a commutative diagram

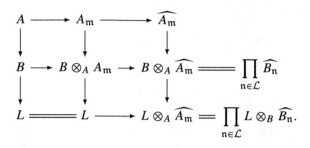

As A/xA is quasi-excellent, it follows that $B_{\mathfrak{n}}/xB_{\mathfrak{n}}$ is quasi-excellent. Let $(B_{\mathfrak{n}})^*$ be the x-adic completion of $B_{\mathfrak{n}}$. By the Local Lifting Theorem 4.3.6, it results that $(B_{\mathfrak{n}})^*$ is quasi-excellent. This means that the morphism $(B_{\mathfrak{n}})^* \to \widehat{(B_{\mathfrak{n}})^*}$ is regular and that $\mathbf{Reg}((B_{\mathfrak{n}})^*)$ is an open subset of $\mathrm{Spec}((B_{\mathfrak{n}})^*)$. Therefore $L \otimes_B (B_{\mathfrak{n}})^*$ is regular if and only if $L \otimes_B \widehat{B_{\mathfrak{n}}}$ is regular. Let

$$\mathcal{G}_L := \{\mathfrak{q}_{\mathfrak{n}}^* \mid \mathfrak{q}_{\mathfrak{n}}^* \text{ is a prime ideal of } \mathrm{Sing}((B_{\mathfrak{n}})^*) \text{ such that } \mathfrak{q}_{\mathfrak{n}}^* \cap B = (0),$$

$$\text{where } \mathfrak{n} \text{ is a maximal ideal of } B\}.$$

Since the formal fibres of A are not geometrically regular, there exists a field L, finite algebraic extension of $Q(A)$, such that $\mathcal{G}_L \neq \emptyset$. Let

$$h_0 := \min\{\mathrm{ht}(\mathfrak{q}) \mid \mathfrak{q} \in \mathcal{G}_L, \text{ for some } L\}. \tag{4.24}$$

Since A has geometrically normal formal fibres, it follows that also B has geometrically normal formal fibres. Therefore $B_{\mathfrak{n}}$ and $(B_{\mathfrak{n}})^*$ have geometrically normal formal fibres, and since $\widehat{B_{\mathfrak{n}}} = \widehat{(B_{\mathfrak{n}})^*}$ it follows that $B_{\mathfrak{n}}$, $(B_{\mathfrak{n}})^*$ and $\widehat{B_{\mathfrak{n}}}$ are normal local domains; hence $h_0 \geq 2$. ∎

Claim 4.6.17 Let $\mathfrak{p}$ be a prime ideal of A with $\mathrm{ht}(\mathfrak{p}) \leq h_0$. Then $A_{\mathfrak{p}}$ is excellent.

Proof If $\mathrm{ht}(\mathfrak{p}) = 0$, since A is a domain, $A_{\mathfrak{p}}$ is a field. Assume that $0 < \mathrm{ht}(\mathfrak{p}) \leq h_0$. We already know that A is universally catenary, and using the previous reductions, it is enough to show that the generic fibre of the morphism $A_{\mathfrak{p}} \to \widehat{A_{\mathfrak{p}}}$ is geometrically regular. Let L be a finite extension of $Q(A)$ and B_L be the integral closure of A in L. We have the following commutative diagram:

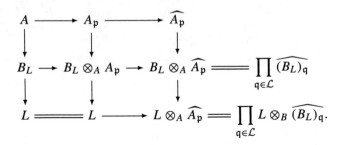

where $\mathcal{L} := \{q \in \mathrm{Spec}(B_L) \mid q \cap A = p\}$. The ring A is normal; hence by the going-down theorem, we have $0 < \mathrm{ht}(q) = \mathrm{ht}(p) \leq h_0$. Let n be a maximal ideal of B_L containing q and let $q^* \in \mathrm{Min}(q(B_L)_n^*)$. The flatness of the map $(B_L)_n \to (B_L)_n^*$ implies that $q^* \cap B_L = q$. Denote by $T := ((\widehat{B_L})_n^*)_{q^*}$ and consider the commutative diagram:

$$
\begin{array}{ccccc}
B_L \longrightarrow & (B_L)_n & \longrightarrow & (B_L)_q & \longrightarrow & \widehat{(B_L)_q} \\
& \downarrow & & \downarrow{\alpha} & & \downarrow{\gamma} \\
& (B_L)_n^* & \longrightarrow & ((B_L)_n^*)_{q^*} & \overset{\beta}{\longrightarrow} & T = ((\widehat{B_L})_n^*)_{q^*}
\end{array}
$$

The morphism α is a local flat one with zero-dimensional closed fibre, whence $\mathrm{ht}(q) = \mathrm{ht}(q^*)$. We have seen that the morphism β is regular and that $\mathrm{Sing}((B_L)_n^*)$ is a closed subset of $\mathrm{Spec}((B_L)_n^*)$. Let c_n^* be the radical ideal of $(B_n)^*$ defining the singular locus. Then $c_n^* T$ is the defining radical ideal of $\mathrm{Sing}(T)$, because β is a regular morphism. There are prime ideals $q_{n,1}^*, \ldots, q_{n,s}^*$ of $(B_L)_n^*$ such that $q_{n,i}^* \not\subseteq q_{n,j}^*$ if $i \neq j$ and $c_n^* = \bigcap_{i=1}^{s} q_{n,i}^*$.

There are three possible cases:

Case 1. $q_{n,i}^* \not\subseteq q^*$, for $i = 1, \ldots, s$. Then $c_n^* T = T$, hence $c_n^* T \cap B_L = B_L \neq (0)$.

Case 2. There exists $1 \leq i \leq s$ such that $q_{n,i}^* = q^*$. Then $c_n^* T = q^* T$. Thus

$$
c_n^* T \cap B_L = q^* T \cap B_L = \left(q^* T \cap (B_L)_n^*\right) \cap B_L = q^* \cap B_L = q \neq (0).
$$

Case 3. There is $t \in \{1, \ldots, s\}$ such that $q_{n,i}^* \subsetneq q^*$ for $1 \leq i \leq t$ and $q_{n,j}^* \not\subseteq q^*$ for $t+1 \leq j \leq s$. Then

$$
c_n^* T = q_{n,1}^* T \cap \ldots \cap q_{n,t}^* T.
$$

Moreover,

$$\mathrm{ht}(q_{n,i}^*) < \mathrm{ht}(q^*) = \mathrm{ht}(q) = \mathfrak{p} \le h_0 \text{ for any } 1 \le i \le t.$$

Since $q_{n,i}^* \in \mathrm{Sing}((B_L)_n^*)$, by the minimality of h_0 (4.24), we have

$$q_{n,i}^* \cap B_L \ne (0), \text{ for } i = 1, \dots, t.$$

Since B_L is a domain, we get $c_n^* T \cap B_L \ne (0)$.

Thus we always have $c_n^* T \cap B_L \ne (0)$. Pick an element $b \in c_n^* T \cap B_L, b \ne 0$. As $\mathrm{Sing}(T) = V(c_n^* T)$, it follows that $T \otimes_{B_L} (B_L)_b$ is a regular ring. The morphism

$$\widehat{(B_L)_q} \otimes_{B_L} (B_L)_b \to T \otimes_{B_L} (B_L)_b$$

is faithfully flat; hence $\widehat{(B_L)_q} \otimes_{B_L} (B_L)_b$ is regular. Therefore its fraction ring $\widehat{(B_L)_q} \otimes_{B_L} L$ is regular, and Claim 4.6.17 is proved. □

Let L be a finite algebraic extension of $Q(A)$ and B be the integral closure of A in L. Let n be a maximal ideal in B and q_n^* a prime ideal that is minimal in $\mathrm{Sing}(B_n^*)$ such that $q_n^* \cap B = (0)$ and $\mathrm{ht}(q_n^*) = h_0$. Let us remark that L, n and q_n^* exist by (4.24). We want to apply Rotthaus' Hilfssatz 4.6.13 and 4.6.14. Therefore we use the same notations:

$$\Gamma = \Gamma(n) = \{\gamma \subset \mathrm{Max}(B), n \in \gamma, \gamma \text{ is finite}\},$$

$$B_\gamma := S_\gamma^{-1} B, \text{ where } S_\gamma := B \setminus \bigcup_{\mathfrak{a} \in \gamma} \mathfrak{a},$$

$$B_\gamma^* \text{ is the } x B_\gamma - \text{adic completion of } B_\gamma,$$

$$q_\gamma^* := q_n \cap B_\gamma^*, \text{ for each } \gamma \in \Gamma,$$

$$\Delta_\gamma(x) := \{\bar{q} \cap B \mid \bar{q} \in \mathrm{Min}(\overline{(B_\gamma^*/q_\gamma^*)}/x\overline{(B_\gamma^*/q_\gamma^*)})\},$$

$$\Delta(x) := \bigcup_{\gamma \in \Gamma} \Delta_\gamma(x).$$

Let us remark that x is contained in all the maximal ideals of B, because B is complete in the x-adic topology [18, Prop. 10.15,iv)]. From the isomorphism $B_\gamma/x B_\gamma \cong B_\gamma^*/x B_\gamma^*$, it follows that there is a bijection between the set of maximal ideals of B_γ^* and γ. Let $n^* := n B_\gamma^*$ be the maximal ideal of B_γ^* corresponding to n.

Let $C_\gamma := B_\gamma^*/q_\gamma^*$. The ring C_γ is a Nagata ring, because A is Nagata. Therefore $\overline{C_\gamma}$ is a finite C_γ-algebra. Since $q_n^* \cap B = (0)$, we have that $x \notin q_\gamma^*$. Let $\bar{q} \in \mathrm{Min}(\overline{C_\gamma}/x\overline{C_\gamma})$. Since A is universally catenary, also C_γ is universally catenary. Then, applying Theorem 3.1.36

and Corollary 3.1.24, we see that $\bar{q} \cap C_\gamma \in \mathrm{Min}(C_\gamma/xC_\gamma)$. This means that

$$\Delta_\gamma(x) = \{\tilde{q} \cap B \mid \tilde{q} \in \mathrm{Min}(B_\gamma^*/(q_\gamma^* + xB_\gamma^*))\}. \tag{4.25}$$

Claim 4.6.18 For every $\gamma \in \Gamma(\mathfrak{n})$, $\mathrm{ht}(q_\gamma^*) = h_0$.

Proof We consider the morphisms

$$(B_\gamma^*)_{\mathfrak{n}^*} \xrightarrow{f} B_\mathfrak{n}^* \xrightarrow{g} \widehat{B_\mathfrak{n}} = \widehat{(B_\gamma^*)}_{\mathfrak{n}^*}.$$

Since Theorem 4.3.6 implies that B_γ^* has geometrically regular formal fibres, it results that gf is a regular morphism, and as g is faithfully flat, it follows that f is regular. Because $q_\gamma^*(B_\gamma^*)_{\mathfrak{n}^*} = q_\mathfrak{n}^* \cap (B_\gamma^*)_{\mathfrak{n}^*}$, we obtain that $q_\gamma^*(B_\gamma^*)_{\mathfrak{n}^*}$ is a minimal prime ideal of $\mathrm{Sing}((B_\gamma^*)_{\mathfrak{n}^*})$. Because $q_\mathfrak{n}^* \in \mathrm{Min}(q_\gamma^* B_\mathfrak{n}^*)$, it follows that

$$\mathrm{ht}(q_\gamma^*) = \mathrm{ht}(q_\gamma^*(B_\gamma^*)_{\mathfrak{n}^*}) = \mathrm{ht}(q_\mathfrak{n}^*) = h_0.$$

We have proved Claim 4.6.18.   □

Claim 4.6.19 For any $q \in \Delta(x)$, $\mathrm{ht}(q) = h_0 + 1$.

Proof Let $\mathfrak{r} \in \gamma$ and put $\mathfrak{r}^* := \mathfrak{r}B_\gamma^*$. As the formal fibres of B are geometrically normal, the ring $\widehat{B_\mathfrak{r}}$ is a local normal domain. The morphism $(B_\gamma^*)_{\mathfrak{r}^*} \to \widehat{(B_\gamma^*)}_{\mathfrak{r}^*} = \widehat{B_\mathfrak{r}}$ is faithfully flat; hence $(B_\gamma^*)_{\mathfrak{r}^*}$ is normal. Thus the ring B_γ^* is normal. There exists a prime ideal $\tilde{q} \in \mathrm{Min}(B_\gamma^*/(q_\gamma^* + xB_\gamma^*))$ such that $q = \tilde{q} \cap B$. By the flatness of the morphism $B_\gamma \to B_\gamma^*$ and because $qB_\gamma^* = \tilde{q}$, it follows that

$$\mathrm{ht}(\tilde{q}) = \dim\left((B_\gamma^*)_{\tilde{q}}\right) = \dim\left((B_\gamma)_q\right) = \mathrm{ht}(q). \tag{4.26}$$

The ring B_γ^* is a normal ring; hence is a direct product of finitely many normal domains. Moreover B_γ^* is universally catenary. Then $\mathrm{ht}(\tilde{q}) = \mathrm{ht}(q_\gamma^*) + 1$. From this equality, (4.26) and (4.6.18), we have that $\mathrm{ht}(q) = h_0 + 1$ and this proves Claim 4.6.19.   □

Let us continue the proof of Theorem 4.6.16. Let q be a prime ideal of B with $\mathrm{ht}(q) \le h_0$. The ring A is normal; hence $\mathrm{ht}(q \cap A) = \mathrm{ht}(q) \le h_0$. Therefore by Claim 4.6.17 the ring $A_{q \cap A}$ is excellent and consequently also B_q is excellent. By Lemma 4.4.4 there exists an element $b_q \in B \setminus q$ and an alteration covering

$$\{X_{b_q,i} \xrightarrow{\phi_{b_q,i}} \mathrm{Spec}(B_{b_q})\}_{1 \le i \le l} \tag{4.27}$$

such that for any $i = 1, \ldots, l$

$$\phi_{b_q,i}^{-1}\left(\mathrm{Spec}(B_{b_q}/x\,B_{b_q})\right) \subseteq \mathrm{Reg}(X_{b_q,i}). \tag{4.28}$$

Let us denote

$$\Omega := \bigcup_q \mathrm{Spec}(B_q) \subseteq \mathrm{Spec}(B), \quad \Lambda := \mathrm{Spec}(B) \setminus \Omega.$$

Then Ω is an open subset of $\mathrm{Spec}(B)$ containing all the prime ideals of B of height at most h_0. This implies that Λ contains only finitely many prime ideals of height $h_0 + 1$. Thus, to show that $\Delta(x)$ is finite, it is enough to prove that $\Delta(x) \subseteq \Lambda$. Assume this is not true, that is there exists $q \in \Delta(x) \cap \Omega$. There is some prime ideal $\tilde{q} \in \mathrm{Min}(B_\gamma^*/(q_\gamma^* + xB_\gamma^*))$ with $\tilde{q} \cap B = q$, for some $\gamma \in \Gamma(\mathfrak{n})$. On the other hand, since $q \in \Omega$, there is some prime ideal $\mathfrak{r}$ with $\mathrm{ht}(\mathfrak{r}) \le h_0$, such that $q \in \mathrm{Spec}(B_{b_\mathfrak{r}})$. By the definition of the alteration covering 4.1.30, there is a proper surjective generically finite morphism $\pi :$ $V \to \mathrm{Spec}(B_{b_\mathfrak{r}})$, together with an open covering $V = \bigcup\limits_{i=1}^{l} V_i$ and morphisms $\psi_i : V_i \to X_{b_\mathfrak{r},i}$, for $1 \le i \le l$ such that all the diagrams below commute:

$$
\begin{array}{ccc}
V_i & \longrightarrow & V \\
\psi_i \downarrow & & \downarrow \pi \\
X_{b_\mathfrak{r},i} & \xrightarrow{\ \phi_{b_\mathfrak{r},i}\ } & \mathrm{Spec}(B_\mathfrak{r})
\end{array}
$$

We have the following diagram:

$$
\begin{array}{ccccc}
V_i' = V_i \times_{\mathrm{Spec}(B)} \mathrm{Spec}(B_\gamma^*) & \subseteq & V' = V \times_{\mathrm{Spec}(B)} \mathrm{Spec}(B_\gamma^*) & & \\
\downarrow \psi_i' & & \downarrow f' & & \\
X_{b_\mathfrak{r},i}' = X_{b_\mathfrak{r},i} \times_{\mathrm{Spec}(B)} \mathrm{Spec}(B_\gamma^*) \xrightarrow{\ \phi_{b_\mathfrak{r},i}'\ } & \mathrm{Spec}((B_\gamma^*)_{b_\mathfrak{r}}) & \longrightarrow & \mathrm{Spec}(B_\gamma^*) \\
\downarrow \epsilon & & \downarrow \theta & & \downarrow \\
X_{b_\mathfrak{r},i} \xrightarrow{\qquad \phi_{b_\mathfrak{r},i} \qquad} & \mathrm{Spec}(B_{b_\mathfrak{r}}) & \longrightarrow & \mathrm{Spec}(B)
\end{array}
$$

Then $\tilde{q}$ and q_γ^* are belonging to $\mathrm{Spec}((B_\gamma^*)_{b_\mathfrak{r}})$. Since f' is proper and surjective, by Lemma 4.2.8 there exist $\xi_1, \xi_2 \in V'$ such that $f'(\xi_1) = \tilde{q}$, $f'(\xi_2) = q_\gamma^*$ and ξ_1 is a specialization of ξ_2. There exists some index $i \in \{1, \ldots, l\}$ such that $\xi_1 \in V_i'$ and then also $\xi_2 \in V_i'$, because V_i' is closed under generalization. Let $\eta_1 = \psi_i'(\xi_1)$, $\eta_2 = \psi_i'(\xi_2)$ and $\zeta_1 = \epsilon(\psi_i'(\xi_1)) \in X_{b_\mathfrak{r},i}$. Since $x \in q = \phi_{b_\mathfrak{r},i}(\eta_1)$ by (4.28) it results that $\mathcal{O}_{X_{b_\mathfrak{r},i},\eta_1}$ is a regular local ring. On the other hand, since the ring morphism $B \to B_\gamma^*$ is flat, the morphism $\mathcal{O}_{X_{b_\tau,i},\zeta_1} \to \mathcal{O}_{X'_{b_\tau,i},\eta_1}$ is a faithfully flat local morphism. The maximal ideal of $\mathcal{O}_{X_{b_\tau,i},\zeta_1}$ generates the maximal ideal of $\mathcal{O}_{X'_{b_\tau,i},\eta_1}$; hence the ring $\mathcal{O}_{X'_{b_\tau,i},\eta_1}$ is a regular local ring. Therefore also the ring $\mathcal{O}_{X'_{b_\tau,i},\eta_2}$ is a regular local ring, since η_2 is a generalization of η_1. Since $\mathfrak{q}_\gamma^* \cap B = \theta(\mathfrak{q}_\gamma^*) = (0)$, it obviously follows that the morphism $Q(B) \to \mathcal{O}_{X_{b_\tau,i},\epsilon(\eta_2)}$ is flat; hence also the morphism $(B_\gamma^*)_{\mathfrak{q}_\gamma^*} \to \mathcal{O}_{X'_{b_\tau,i},\eta_2}$ is flat. Therefore $(B_\gamma^*)_{\mathfrak{q}_\gamma^*}$ is a regular local ring. But this contradicts the choice of $\mathfrak{q}_n^* \in \mathrm{Sing}(B_n^*)$. Thus we have proved that $\Delta(x) \subseteq \Lambda$ and in particular that $\Delta(x)$ is a finite set. Together with Claim 4.6.18, this shows that we can apply Rotthaus' Hilfssatz 4.6.13 and 4.6.14 to conclude.   □

Corollary 4.6.20 *Let A be a Noetherian quasi-excellent ring, $\mathfrak{a}$ be an ideal of A and $\widehat{A}$ the completion of A in the $\mathfrak{a}$-adic topology. Then $\widehat{A}$ is quasi-excellent.*

Corollary 4.6.21 *Let A be a Noetherian quasi-excellent ring. Then the formal power series ring $A[[X_1, \ldots, X_n]]$ is a quasi-excellent ring for any natural number n.*

Structure of Regular Morphisms

This chapter is probably the most technical of all chapters of the book. The main aim of Chap. 5 is to present the important and very difficult Popescu's Theorem 5.2.56 showing that every regular morphism is a filtered direct limit of smooth morphisms of finite type. The proof is quite complicated. It would be nice to find any simplification of it. Together with Rotthaus' result 5.3.13, this will provide the identification of the class of excellent Henselian local rings with the class of local rings with Artin approximation property.

5.1 Rings with Artin Approximation Property

Rings with the approximation property refer to Noetherian local rings $(A, \mathfrak{m}, k)$ such that every finite system of polynomial equations in finitely many variables over A that has a solution in $\widehat{A}$ has also one in A. There are many applications and properties of Artin approximation that are not presented in this book. The interested reader can look at the survey papers [323] or [146].

Definition 5.1.1 A Noetherian local ring $(A, \mathfrak{m}, k)$ is called a **ring with Artin approximation property** if for any $m, s \in \mathbb{N}$, for any finite system of polynomials $f_1, \ldots, f_s \in A[X_1, \ldots, X_m]$, if there exist elements $u_1, \ldots u_m \in \widehat{A}$, such that $f_i(u_1, \ldots, u_m) = 0$, $i = 1, \ldots, s$, then there exist elements $v_1, \ldots, v_m \in A$ such that $f_i(v_1, \ldots, v_m) = 0$, $i = 1, \ldots, s$.

The study of rings with Artin approximation property initiated with two famous papers of M. Artin [16, 17], in which the author proved that some extremely important classes of rings have Artin approximation property: convergent power series rings over a non-trivial valued field containing $\mathbb{Q}$, the Henselization of a local ring essentially of finite type over a field and excellent Dedekind rings.

Notation 5.1.2 We shall abbreviate **ring with Artin approximation property** by **ring with AP(approximation property)** or **AP ring**.

Proposition 5.1.3 *Any local ring with Artin approximation property is a Henselian local ring.*

Proof Let $f \in A[Y]$ be such that $f(0) \in \mathfrak{m}$ and $f'(0) \notin \mathfrak{m}$. Since $\widehat{A}$ is a Henselian ring, it follows that f has a unique root $\alpha \in \mathfrak{m}\widehat{A}$. But since A has AP, we see $\alpha \in \mathfrak{m}$.   □

The following remark will be useful.

Lemma 5.1.4 *Let $(A, \mathfrak{m})$ be a local ring. Then A is an AP ring if and only if for any finitely generated A-subalgebra B of $\widehat{A}$ there exists an A-algebra retract $\psi : B \to A$.*

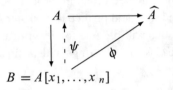

$$B = A[x_1, \ldots, x_n]$$

Proof Assume that A is an AP ring and that $B = A[X_1, \ldots, X_n]/(f_1, \ldots, f_m) = A[x_1, \ldots, x_n]$. Then $x_1, \ldots, x_n$ is a solution in $\widehat{A}$ of the system $f_1 = \ldots = f_m = 0$; hence there is a solution $y_1, \ldots, y_n$ in A. Then the map $A[X_1, \ldots, X_n] \to A$, defined by $X_i \mapsto y_i$, induces the required retraction $\psi : A[x_1, \ldots, x_n] \to A$.

Conversely, let $f_1 = \ldots = f_m = 0$ be a system of polynomial equations in $X_1, \ldots, X_n$ over A having a solution $s_1, \ldots, s_n$ in $\widehat{A}$. Then there exists an A-algebra retraction $\psi : B = A[s_1, \ldots, s_n] \to A$. Let $x_i = \psi(s_i)$, $i = 1, \ldots, n$. Since $f_i(s_1, \ldots, s_n) = 0$, $i = 1, \ldots, m$ and ψ is a ring morphism, it follows that $f_i(x_1, \ldots, x_n) = 0$, for $i = 1, \ldots, m$.   □

Notation 5.1.5 Let A be any ring, $m, s \in \mathbb{N}$ and $f := (f_1, \ldots, f_s)$ be a system of s polynomials in $A[X_1, \ldots, X_m]$. For any $r \leq m$, we shall denote by $J_{f,r}(X)$ the ideal of $A[X_1, \ldots, X_m]$ generated by the $r \times r$-minors of the matrix $(\partial f_i/\partial X_j)_{1 \leq i \leq s, 1 \leq j \leq m}$. For an element $a = (a_1, \ldots, a_m) \in A^m$, we put $J_{f,r}(a) = \det((\partial f_i/\partial X_j)(a)) \in A$.

We will need the following result, usually known as **Tougeron's implicit function theorem for Henselian rings** [187, Ch. III, Prop. 7.2]:

Theorem 5.1.6 *Let $(A, \mathfrak{m})$ be a Henselian local ring, let $F = (f_1, \ldots, f_r)$ be a system of polynomials in $A[X_1, \ldots, X_n]$ and let $J = (\frac{\partial f_i}{\partial X_j})_{i,j}$. Let $a \in A^n$ be such that $F(a) \equiv 0 \,(\text{mod } \mathfrak{m} \cdot e^2)$, where e is one of the $r \times r$-minors of $J(a)$. Then there exists an element $b \in A^n$ such that $F(b) = 0$ and $b \equiv a \,(\text{mod } \mathfrak{m} \cdot e)$.*

Proof We may assume that $e = (\frac{\partial f_i}{\partial X_j})_{j=1,\ldots,r}$. By completing $F = (f_1, \ldots, f_r)$ to $(f_1, \ldots, f_r, X_{r+1}, \ldots, X_n)$, we may assume that $r = n$ and $e = \det J(a)$. The Taylor expansion gives

$$F(a + eT) = F(a) + eJ(a)T + e^2 G(T),$$

where G is a vector polynomials, each beginning with terms of degree at least 2. Let J^* be the adjoint matrix to J. Thus $J^* J = J J^* = \det(J) I_n$. We can write $F(a) = e^2 \cdot c$, where $c = (c_1, \ldots, c_n)$, $c_i \in \mathfrak{m}$. Substituting $e^2 \cdot I_n = e \cdot J(0) J^*(0)$, we get $F(a + eT) = e \cdot J(0) H(T)$, where $H(T) = J^*(0) \cdot c + T \cdot J^*(0) G(T)$. Then we can find $t = (t_1, \ldots, t_n)$, where $t_i \in \mathfrak{m}$, such that $H(t) = 0$. Let $b = a + et$. Then $b \equiv a \,(\text{mod } \mathfrak{m}e)$ and $F(b) = 0$. $\square$

Lemma 5.1.7 *Let $(A, \mathfrak{m}, k)$ be a Henselian local ring, let $A[Y] := A[Y_1, \ldots, Y_n]$ be the polynomial ring in n variables over A and let $f := (f_1, \ldots, f_r) \in A[Y]$, for some $r \leq n$. Let $z \in \widehat{A}^n$ be such that $f(z) = 0$ and $J_{f,r}(z) \notin \mathfrak{m}\widehat{A}$. Then there exists $y \in A^n$, such that $f(y) = 0$ and $y - z \in \mathfrak{m}\widehat{A}^n$.*

Proof Let $u \in A^n$ be such that $u - z \in \mathfrak{m}\widehat{A}^n$. This is possible since $\widehat{A}/\mathfrak{m}\widehat{A} \simeq A/\mathfrak{m}$. Then $f(z) \in \mathfrak{m}\widehat{A}$ and $J_{f,r}(z) \notin \mathfrak{m}\widehat{A}$. By Theorem 5.1.6 it follows that there is $y \in A^n$ such that $f(y) = 0$ and $y - u \in \mathfrak{m}$. $\square$

The reason for using the word *approximation* in the definition of AP rings can be deduced from the following property:

Lemma 5.1.8 *Let $(A, \mathfrak{m}, k)$ be a Noetherian local ring. The following conditions are equivalent:*

(i) A is an AP ring.

(ii) For any natural numbers m and s, for any system of polynomials $f := (f_1, \ldots, f_s) \in A[X_1, \ldots, X_m]$, for any solution $u = (u_1, \ldots, u_m) \in \widehat{A}^m$ of f and for every $l \in \mathbb{N}$ such that $l \neq 0$, there exists a solution $v = (v_1, \ldots, v_m) \in A^m$ of f, such that $v_i \equiv u_i \,(\text{mod } \mathfrak{m}^l)$, for every $i = 1, \ldots, m$.

Proof $(i) \Rightarrow (ii)$: Take elements $w_1, \ldots, w_m \in A$ such that $w_i \equiv u_i \,(\text{mod } \mathfrak{m}^l)$, for $i = 1, \ldots, m$. Then there exist elements $\lambda_{i,j} \in \mathfrak{m}^l$ and $z_{i,j} \in A$, for $i = 1, \ldots, s$ and

$j = 1, \ldots, m$ such that

$$w_j - v_j = \sum_{i=1}^{s} \lambda_{i,j} z_{i,j}, \quad j = 1, \ldots, m.$$

Let $g_j := w_j - Y_j - \sum_{i=1}^{s} \lambda_i Z_{i,j} \in A[Y_1, \ldots, Y_m, Z_{1,1}, \ldots, Z_{s,m}], \; j = 1, \ldots, m$. Then the system of polynomial equations

$$\begin{cases} f_1 = \ldots = f_s = 0 \\ g_1 = \ldots = g_m = 0 \end{cases}$$

has the solution $(w_1, \ldots, w_m, z_{1,1}, \ldots, z_{s,m})$ in $\widehat{A}$. It follows that the system has a solution $(v_1, \ldots, v_m, t_{1,1}, \ldots, t_{s,m})$ in A. The conclusion follows.
$(ii) \Rightarrow (i)$: Obvious. $\square$

The above lemma says that $(A, \mathfrak{m})$ is an AP ring if for every finite system of polynomial equations over A, the set of the solutions of f in A is dense with respect to the $\mathfrak{m}$-adic topology in the set of the solutions of f in $\widehat{A}$.

Proposition 5.1.9 *Let $(A, \mathfrak{m}, k)$ be a Noetherian local AP ring. Let $f := (f_1, \ldots, f_n)$ be a system of polynomials in $A[Y] := A[Y_1, \ldots, Y_m]$ and let $l \in \mathbb{N}$. Suppose that the set of solutions of the system $f = 0$ belonging to $\mathfrak{m}^l$ is finite. Then there is no other solution of the system $f = 0$ in $\mathfrak{m}^l \widehat{A}$.*

Proof Let $y^1, \ldots, y^s$ be the solutions of the system in $\mathfrak{m}^l$, and suppose that there exists another solution $z \in \mathfrak{m}^l \widehat{A}$. It follows that there exists $t > l$ such that $y^i - z \notin \mathfrak{m}^t$, $i = 1, \ldots, s$. Since A has AP, from Lemma 5.1.8 we obtain that there is a solution $y \in A$ such that $y \equiv z \bmod \mathfrak{m}^t$. Then $y - y^i \notin \mathfrak{m}^t \widehat{A}^m$, $i = 1, \ldots, s$. This means that y is a solution of the system $f = 0$, different from $y^1, \ldots, y^s$. $\square$

Corollary 5.1.10 *Let $(A, \mathfrak{m}, k)$ be a Noetherian local AP ring. Then:*

(i) A is reduced if and only if $\widehat{A}$ is reduced.
(ii) If A is a domain, then A is algebraically closed in $\widehat{A}$.

Proof

(i) If A is reduced, let $n \in \mathbb{N}$ and consider the polynomial $f(Y) = Y^n \in A[Y]$. Then f has only one solution in A, namely, $y = 0$. From Proposition 5.1.9 it follows that f

cannot have any other solution in $\widehat{A}$, so that $\widehat{A}$ is reduced. The converse is clear and generally true.

(ii) Let $f \in A[Y]$ be such that $\deg(f) = s$. Then f has at most s solutions in A. By Proposition 5.1.9, f cannot have other roots in $\widehat{A}$. This means that A is algebraically closed in $\widehat{A}$. $\qquad\qquad\qquad\qquad\qquad\qquad\qquad\qquad\qquad\qquad\qquad\qquad\qquad\qquad\qquad\square$

Proposition 5.1.11 *Let $(A, \mathfrak{m}, k)$ be a Noetherian local AP ring with the approximation property and B a finite A-algebra which is a local ring. Then B is an AP ring.*

Proof Let $\{e_1, \ldots, e_s\}$ be a system of generators of the A-module B, and consider the A-linear map

$$\varphi : A^s \to B, \quad \varphi(a_1, \ldots, a_s) := \sum_{i=1}^{s} a_i e_i.$$

Let $u_j := (u_{j1}, \ldots, u_{js})$, $j = 1, \ldots, t$ be a system of generators of $\ker(\varphi)$. Consider $f := (f_1, \ldots, f_r)$ a system of polynomials in $A[Y] = A[Y_1, \ldots, Y_n]$. Note first that $\widehat{B} \cong \widehat{A} \otimes_A B$, because of the finiteness of B as an A-module. If $\widehat{y} := (\widehat{y}_1, \ldots, \widehat{y}_n) \in \widehat{B}^n$, then

$$\widehat{y}_j = \sum_{k=1}^{s} \widehat{y}_{jk} e_k, \quad \widehat{y}_{jk} \in \widehat{A}, \ j = 1, \ldots, n$$

and we have

$$f_i(\widehat{y}) = \sum_{\alpha=1}^{s} f_{i\alpha} \widehat{y}_{jk} e_\alpha, \quad f_{i\alpha} \in A[Y_{j\alpha}], \ i = 1, \ldots, r.$$

Obviously $\widehat{y} \in \widehat{B}$ is a solution of f if and only if there are $\widehat{z}_{i\alpha} \in \widehat{A}$, $1 \leq i \leq r$ and $1 \leq \alpha \leq s$ such that

$$f_{i\alpha}((\widehat{y}_{jk})) = \sum_{\beta=1}^{t} \widehat{z}_{i\beta} u_{\beta\alpha}.$$

Suppose that $\widehat{y}$ is a solution of f in $\widehat{B}$. Then $(\widehat{y}_{jk}), (\widehat{z}_{i\beta})$ is a solution in $\widehat{A}$ of the polynomial system

$$f_{i\alpha}((Y_{jk})) = \sum_{\beta=1}^{t} Z_{i\beta} u_{\beta\alpha}, \ i = 1, \ldots, r, \ \alpha = 1, \ldots, s.$$

This system must have a solution (y_{jk}), $(z_{i\beta})$ in A, since A has AP. Obviously, $(y_j) = \sum_{k=1}^{s} y_{jk}e_k$, is a solution of f in B. $\square$

Proposition 5.1.12 *Let $(A, \mathfrak{m}, k)$ be a Noetherian local AP ring.*

(i) A is an integral domain if and only if $\widehat{A}$ is an integral domain.
(ii) If $\mathfrak{p} \in \mathrm{Spec}(A)$, then $\mathfrak{p}\widehat{A} \in \mathrm{Spec}(\widehat{A})$.

Proof

(i) If $\widehat{A}$ is an integral domain, then clearly A is an integral domain, being a subring of $\widehat{A}$. If $\widehat{A}$ is not an integral domain, there are non-zero elements $x, y \in \widehat{A}$ such that $xy = 0$. Let $\alpha \in \mathbb{N}$ be such that $x, y \notin \mathfrak{m}^\alpha \widehat{A}$. From Lemma 5.1.7 it follows that there exist $x_1, y_1 \in A$, such that $x_1 \cdot y_1 = 0$ and $x - x_1 \in \mathfrak{m}^\alpha$ and $y - y_1 \in \mathfrak{m}^\alpha$. Then $x_1 \notin \mathfrak{m}^\alpha$ and $y_1 \notin \mathfrak{m}^\alpha$. Consequently A is not an integral domain.
(ii) Follows immediately from Proposition 5.1.11 and (i). $\square$

Theorem 5.1.13 *Let $(A, \mathfrak{m}, k)$ be a Noetherian local AP ring. Then the formal fibres of A are geometrically integral domains.*

Proof Clearly, from Proposition 5.1.11 it results that it is enough to show that if A is an integral domain with AP, the formal fibre of A in (0) is geometrically an integral domain. Let $Q(A)$ be the fraction field of A and K be a field, finite extension of $Q(A)$. Let $\{e_1, \ldots, e_n\}$ be a basis of K over $Q(A)$. Multiplying with a non-zero element of A, we can assume that the elements $e_1, \ldots, e_n$ are integral over A. Then $B = A[e_1, \ldots, e_n]$ is a finite A-algebra, and since A is a Henselian local ring, B is a finite product of local rings. But B is an integral domain, so B is local. From Proposition 5.1.11 we get B has AP, and from Proposition 5.1.12 it follows that $\widehat{B}$ is an integral domain. Then from the isomorphisms

$$K \otimes_A \widehat{A} \cong Q(B) \otimes_A \widehat{A} \cong Q(B) \otimes_B (B \otimes_A \widehat{A}) \cong Q(B) \otimes_B \widehat{B}$$

it follows that $K \otimes_A \widehat{A}$ is an integral domain. $\square$

Corollary 5.1.14 *Let $(A, \mathfrak{m}, k)$ be a Noetherian local AP ring. Then A is a universally catenary Nagata ring.*

Proof It follows from Theorem 5.1.13 and Theorem 3.1.35. $\square$

Proposition 5.1.15 *Let A be a Noetherian local AP ring. Then A has geometrically normal formal fibres.*

Proof By Corollary 5.1.14 it results that A is a Nagata ring; hence we may assume that A is a normal local domain, and we have to show that $\widehat{A}$ is normal. By Proposition 5.1.12 we know that $\widehat{A}$ is a domain. Let $a, b \in \widehat{A}, b \neq 0$, be such that if we put $\alpha = \frac{a}{b}$, we have a relation of integral dependence

$$\alpha^n + c_{n-1}\alpha^{n-1} + \ldots + c_0 = 0, \ c_0, \ldots, c_{n-1} \in \widehat{A}.$$

Then we obtain the relation

$$a^n + c_{n-1}a^{n-1}b + \ldots + c_0b^n = 0.$$

Consider the equation over A

$$X^n + Z_{n-1}X^{n-1}Y + \ldots + Z_0Y^n = 0.$$

Because A has the approximation property, for any $k \in \mathbb{N}$, there exist elements $a_k, b_k, c_{ik} \in A$ such that

$$a_k^n + c_{n-1,k}a_k^{n-1}b_n + \ldots + c_{0k}b_k^n = 0$$

and

$$a_k \equiv a(\bmod \mathfrak{m}^k), \ b_k \equiv b(\bmod \mathfrak{m}^k).$$

Since A is normal, we get that $a_k \in Ab_k, \forall k \in \mathbb{N}, ;$ hence $a \in \widehat{A}b + \mathfrak{m}^k, \forall k \in \mathbb{N}$. Thus $\alpha \in \widehat{A}$, that is, $\widehat{A}$ is normal.   □

We close this section with an interesting property of AP rings. It is well-known that the completion of a Noetherian local UFD must not be a UFD [336, Ch. V, Théorème]. The following result shows that such a ring cannot be an AP ring.

Proposition 5.1.16 *Let $(A, \mathfrak{m})$ be a Noetherian local AP ring. If A is a UFD, then $\widehat{A}$ is a UFD too.*

Proof Suppose that $\widehat{A}$ is not a UFD. Let us first observe that by Proposition 5.1.12, the ring $\widehat{A}$ is a domain. Therefore, there exist an irreducible element $x \in \widehat{A}$ that is not prime. Let $a_1, \ldots, a_n \in \mathfrak{m}$ be a system of generators of $\mathfrak{m}$. Because x is irreducible, the equation $\sum_{i,j=1}^{n} a_i a_j U_i T_j = x$ has no solution in $\widehat{A}$, and there exist $y, z \in \widehat{A} \setminus x\widehat{A}$ such that $yz \in x\widehat{A}$.

Therefore the equation $YZ = XV$ has a solution (y, z, x, v) in $\widehat{A}$, and the equations $y = xU$ and $z = xV$ have no solutions in $\widehat{A}$. Hence there exist $y_0, z_0, x_0, v_0 \in A$ such that $y_0 z_0 = x_0 v_0$ and the equations $\sum\limits_{i,j=1}^{n} a_i a_j U_i T_j = x_0$, $y_0 = x_0 U$, $z_0 = x_0 V$ have no solution in A. Thus x_0 is irreducible but not prime in A. Contradiction!   □

5.2   Néron Desingularization

The main aim of this section is to present the difficult result Theorem 5.2.56 of D. Popescu stating that any regular morphism between Noetherian rings is a filtered inductive limit of smooth morphisms of finite type. This result was first proved by Popescu in [289–291]. A gap in the initial proof was fixed by Ogoma [279], while simplifications of the complicated proof were given by André [13] and especially by Swan [361], this last paper being so far considered as the most readable presentation of the result. Another proof, but still complicated, was given by Spivakovski [358]. A very nice and clear exposition of Popescu's proof, following closely Swan's paper, is contained in the nice recent book by Majadas and Rodicio [231, ch. 5]. Some further simplifications can be found in The Stacks Project [359, Ch. 16]. We shall follow closely the presentation in [231], using some notations that are meant to make the things easier.

Notation 5.2.1 Let A be a ring and $B = A[X_1, \ldots, X_n]/\mathfrak{a}$ be a finitely presented A-algebra, that is, $\mathfrak{a}$ is a finitely generated ideal. Let $S \subset B$ be a multiplicatively closed system and $C = S^{-1}B$. Then we can write

$$C = S^{-1}(A[X_1, \ldots, X_n]/\mathfrak{a}) = T^{-1}A[X_1, \ldots, X_n]/T^{-1}\mathfrak{a},$$

where T is a multiplicatively closed system in $A[X_1, \ldots, X_n]$. Let now $\mathfrak{b}$ be a finitely generated ideal contained in $\mathfrak{a}$, and put $B_\mathfrak{b} := A[X_1, \ldots, X_n]/\mathfrak{b}$. Let s be the minimal number of generators of $\mathfrak{b}$ and $g_1, \ldots, g_s$ be a minimal system of generators of $\mathfrak{b}$. Then we set

$$\Delta_{B_\mathfrak{b}/A} = \Delta(g_1, \ldots, g_s),$$

which is the ideal of $B_\mathfrak{b}$ generated by the $s \times s$ minors of the Jacobian matrix $\left(\frac{\partial g_i}{\partial X_j}\right)$. Then we denote

$$H_{C/A}(B) := \sqrt{\sum_\mathfrak{b} \Delta_{B_\mathfrak{b}/A} \cdot (\mathfrak{b} : \mathfrak{a})C},$$

where the sum is taken over all finitely generated ideals $\mathfrak{b}$ contained in $\mathfrak{a}$.

Remark 5.2.2 With the above notations, if R is a multiplicatively closed system in C, since $\mathfrak{a}$ is finitely generated, it is easy to see that $R^{-1}H_{C/A}(B) = H_{R^{-1}C/A}(B)$.

Proposition 5.2.3 *Let* $C := S^{-1}(A[X_1, \ldots, X_n]/\mathfrak{a})$ *be an essentially finitely presented* A*-algebra,* $B = A[X_1, \ldots, X_n]/\mathfrak{a}$ *and* $\mathfrak{q} \in \mathrm{Spec}(C)$. *Then* $C_\mathfrak{q}$ *is smooth over* A *if and only if* $\mathfrak{q} \not\supseteq H_{C/A}(B)$.

Proof Clearly we may assume that C is a local ring and $\mathfrak{q} = \mathfrak{m}$ is the maximal ideal of C; hence we have to show that C is smooth over A if and only if $H_{C/A}(B) = C$. Set $A[X] = A[X_1, \ldots, X_n]$. Suppose that C is smooth over A. Then by [249, Th. 28.6], we have a split exact sequence

$$0 \to \mathfrak{a}C/\mathfrak{a}^2C = S^{-1}(\mathfrak{a}/\mathfrak{a}^2) \to \Omega_{A[X]/A} \otimes_{A[X]} C \to \Omega_{C/A} \to 0.$$

Since C is local, $S^{-1}(\mathfrak{a}/\mathfrak{a}^2)$ is a free direct summand of $\Omega_{A[X]/A} \otimes_{A[X]} C$. We have

$$\mathrm{rk}(S^{-1}(\mathfrak{a}/\mathfrak{a}^2)) = \mu((T^{-1}\mathfrak{a})/(T^{-1}\mathfrak{a})^2) = \mu(T^{-1}\mathfrak{a}) := r.$$

Let $f_1, \ldots, f_r$ be elements of $\mathfrak{a}$ mapping to a system of generators of $T^{-1}\mathfrak{a}$. Let $\mathfrak{b} = (f_1, \ldots, f_r)$; hence $T^{-1}\mathfrak{a} = T^{-1}\mathfrak{b}$, and it follows that $(\mathfrak{b} : \mathfrak{a})C = C$. The above exact sequence becomes

$$0 \to (T^{-1}\mathfrak{b})/(T^{-1}\mathfrak{b})^2 \to \Omega_{A[X]/A} \otimes_{A[X]} C \to \Omega_{C/A} \to 0.$$

Then the rank of the matrix $(\frac{\partial f_i}{\partial X_j})$ is r; hence $\Delta_{B_\mathfrak{b}/A}(\mathfrak{b} : \mathfrak{a})C = C$, so that $H_{C/A}(B) = C$.

Conversely, suppose that $H_{C/A}(B) = C$. Since C is a local ring, there exists an ideal $\mathfrak{b} \subset \mathfrak{a}$ such that $\Delta_{B_\mathfrak{b}/A}(\mathfrak{b} : \mathfrak{a})C = C$. Then $\Delta_{B_\mathfrak{b}/A}C = C$ and $(\mathfrak{b} : \mathfrak{a})C = C$. Consider $\mathfrak{n} \in \mathrm{Max}(T^{-1}A[X])$ such that $\mathfrak{n}/T^{-1}\mathfrak{a} = \mathfrak{m}$. As $(\mathfrak{b} : \mathfrak{a})C = C$, we get that

$$(T^{-1}\mathfrak{b} : T^{-1}\mathfrak{a}) = T^{-1}(\mathfrak{b} : \mathfrak{a}) \not\subseteq \mathfrak{n}$$

and furthermore

$$(T^{-1}\mathfrak{b})_\mathfrak{n} : (T^{-1}\mathfrak{a})_\mathfrak{n} = (T^{-1}\mathfrak{b} : T^{-1}\mathfrak{a})_\mathfrak{n} = (T^{-1}A[X])_\mathfrak{n}.$$

This means that $(T^{-1}\mathfrak{b})_\mathfrak{n} = (T^{-1}\mathfrak{a})_\mathfrak{n}$. Thus, replacing T by the inverse image of $T^{-1}A[X] \setminus \mathfrak{n}$ in $A[X]$, we may assume that $T^{-1}\mathfrak{a} = T^{-1}\mathfrak{b}$. Then

$$S_\mathfrak{b}^{-1}B_\mathfrak{b} = T^{-1}A[X]/T^{-1}\mathfrak{b} = C,$$

where $S_\mathfrak{b}$ is the image of T in $B_\mathfrak{b}$. Hence we have

$$S_\mathfrak{b}^{-1} \Delta_{B_\mathfrak{b}/A} = \Delta_{B_\mathfrak{b}/A} S_\mathfrak{b}^{-1} B_\mathfrak{b} = \Delta_{B_\mathfrak{b}/A} C = C$$

and we have a split exact sequence

$$0 \to (T^{-1}\mathfrak{b})/(T^{-1}\mathfrak{b})^2 = S_\mathfrak{b}(\mathfrak{b}/\mathfrak{b}^2) \to \Omega_{A[X]/A} \otimes_{A[X]} C \to \Omega_{C/A} \to 0.$$

Therefore C is smooth over A.   □

Corollary 5.2.4 *Let $C := S^{-1}(A[X_1, \ldots, X_n]/\mathfrak{a})$ be an essentially finitely presented A-algebra and $B = A[X_1, \ldots, X_n]/\mathfrak{a}$. Then C is smooth over A if and only if $H_{C/A}(B) = C$.*

Corollary 5.2.5 *Let $C := S^{-1}(A[X_1, \ldots, X_n]/\mathfrak{a})$ be an A-algebra essentially of finite presentation and $B = A[X_1, \ldots, X_n]/\mathfrak{a}$. Then $H_{C/A}(B)$ does not depend on the presentation B.*

Notation 5.2.6 Let $C := S^{-1}(A[X_1, \ldots, X_n]/\mathfrak{a})$ be an essentially finitely presented A-algebra. Then, according to 5.2.5, we put $H_{C/A} = H_{C/A}(B)$, where $B = A[X_1, \ldots, X_n]/\mathfrak{a}$ is a finite presentation of C.

Definition 5.2.7 Let $A \to C$ be an A-algebra essentially of finite presentation, and pick an element $a \in H_{C/A}$. We say that a is **standard with respect to a given presentation** $C = S^{-1}(A[X_1, \ldots, X_n]/\mathfrak{a})$, if there exists an ideal $\mathfrak{b} = (f_1, \ldots, f_r) \subseteq \mathfrak{a}$ such that $a \in \sqrt{\Delta_{B_\mathfrak{b}/A}(\mathfrak{b} : \mathfrak{a})C}$. We say that a is a **strictly standard element** if $a \in \Delta_{B_\mathfrak{b}/A}(\mathfrak{b} : \mathfrak{a})C$. Furthermore we say that a is **standard** if a is standard with respect to some finite presentation.

We recall the definition of stably free modules, a notion that is important in algebraic K-theory. For more informations the interested reader can look at [324, Ex. 4.92] or [359, Ch. 15, Section 15.3].

Definition 5.2.8 Let A be a ring. A finitely generated A-module M is called **stably free** if there exist finitely generated free A-modules F and G such that $M \oplus F \cong G$.

Lemma 5.2.9 (Elkik [102]) *Let $B = A[X_1, \ldots, X_n]/\mathfrak{a}$ be a finitely presented algebra and let $a \in H_{B/A}$. If $(\mathfrak{a}/\mathfrak{a}^2)_a$ is a free B_a-module, then a is standard.*

Proof Let $g_1, \ldots, g_s$ be a basis of $(\mathfrak{a}/\mathfrak{a}^2)_a$ and let $h \in A[X]$ representing a. Then we have $\mathfrak{a}_h = (g_1, \ldots, g_s)A[X]_h + \mathfrak{a}_h^2$; therefore there exists $\alpha \in \mathfrak{a}_h$ such that

$$(1 + \alpha)\mathfrak{a}_h \subseteq (g_1, \ldots, g_s)_h^2.$$

Suppose that $\alpha = \frac{g}{h^t}$, with $g \in \mathfrak{a}$, for some $t \in \mathbb{N}$. Then there exists a natural number $e \in \mathbb{N}$ such that

$$h^e(h^t + g)\mathfrak{a} \subseteq (g_1, \ldots, g_s),$$

that is

$$h^e(h^t + g) \in (g_1, \ldots, g_s) : \mathfrak{a},$$

so that

$$\mathfrak{a}^{t+e} \in (g_1, \ldots, g_s) : \mathfrak{a}B.$$

On the other hand, the map $d : \mathfrak{a}_h/\mathfrak{a}_h^2 \to \overset{n}{\underset{i=1}{\oplus}} B_a dX_i$ is a split injection, because $A \to B_a$ is smooth. The map d is actually a map $A_a^m \to A_a^n$, between free modules, given by the matrix $(\frac{\partial f_i}{\partial X_j})$, so that the $r \times r$ minors of this matrix generate the unit ideal in A_a. Thus a power of a lies in the corresponding Δ.   □

Proposition 5.2.10 *Let A be a ring, let C be an A-algebra essentially of finite presentation and a* $\in H_{C/A}$. *Then a is standard if and only if either a is nilpotent or C_a is smooth over A and $\Omega_{C_a/A}$ is stably free.*

Proof Assume that a is standard and $a \notin N(C)$. Let $\mathfrak{b} \subset \mathfrak{a}$ be a finitely generated ideal such that $a \in \sqrt{\Delta_{B_\mathfrak{b}/A}(\mathfrak{b} : \mathfrak{a})C}$. Let us denote $\mathfrak{c} := \Delta_{B_\mathfrak{b}/A}(\mathfrak{b} : \mathfrak{a})C$ and $\mathfrak{r} := \sqrt{\mathfrak{c}}$. Then $\mathfrak{r} \cdot C_a = C_a$, hence $\mathfrak{c} \cdot C_a = C_a$. Therefore $\Delta_{B_\mathfrak{b}/A} \cdot C_a = C_a$ and $(\mathfrak{b} : \mathfrak{a}) \cdot C_a = C_a$. As in the final part of the proof of Proposition 5.2.3, we obtain that $S_\mathfrak{b}^{-1}(\mathfrak{b}/\mathfrak{b}^2)_a$ is a free C_a-module and that the sequence

$$0 \to S_\mathfrak{b}^{-1}(\mathfrak{b}/\mathfrak{b}^2)_a \to \Omega_{A[X]/A} \otimes_{A[X]} C_a \to \Omega_{C_a/A} \to 0$$

is split exact, whence $\Omega_{C_a/A}$ is stably free.

For the converse, let us first observe that if a is nilpotent, then a is obviously standard. Assume that C_a is smooth over A and $\Omega_{C_a/A}$ is stably free. Let S be a multiplicatively closed system such that $C_a = S^{-1}(A[X]/\mathfrak{a})$. We have a split exact sequence

$$0 \to S^{-1}(\mathfrak{a}/\mathfrak{a}^2) \to \Omega_{A[X]/A} \otimes_{A[X]} C_a \to \Omega_{C_a/A} \to 0.$$

Since $\Omega_{C_a/A}$ is stably free, it follows by [324, Ex. 4.92, iv)] that also $S^{-1}(\mathfrak{a}/\mathfrak{a}^2)$ is stably free. Hence there exists $r \in \mathbb{N}$ such that $S^{-1}(\mathfrak{a}/\mathfrak{a}^2) \oplus C_a^r$ is a free C_a-module. Replacing the presentation $B = A[X]/\mathfrak{a}$ by $B = A[X, Y_1, \ldots, Y_r]/(\mathfrak{a}, Y_1, \ldots, Y_r)$, we get an exact sequence

$$0 \to S^{-1}(\mathfrak{a}/\mathfrak{a}^2) \oplus C_a^r \to (\Omega_{A[X]/A} \otimes_{A[X]} C_a) \oplus C_a^r \to \Omega_{C_a/A} \to 0.$$

Thus we may assume that $S_{\mathfrak{b}}^{-1}(\mathfrak{b}/\mathfrak{b}^2)$ is a free C_a-module, and we apply Lemma 5.2.9.
□

Lemma 5.2.11 *Let A be a ring, $B = A[X_1, \ldots, X_r]/\mathfrak{a}$, where $\mathfrak{a}$ is a finitely generated ideal and $C = S^{-1}B$ be a ring of fractions of B. Let A' be an A-algebra, $B' = B \otimes_A A'$, $\mathfrak{a}' = \mathfrak{a}A'[X_1, \ldots, X_r]$ and $C' = C \otimes_A A'$. Let $a \in C$ be a strictly standard element for the above presentation of C. Then the image of a in C' is a strictly standard element for the presentation $C' = S^{-1}(A'[X_1, \ldots, X_r]/\mathfrak{a}')$ of C' over A'.*

Proof Let $\mathfrak{b} \subseteq \mathfrak{a}$ be a finitely generated ideal such that $a \in \Delta_{B_{\mathfrak{b}}/A}(\mathfrak{b} : \mathfrak{a})C$ and let $\mathfrak{b}' = \mathfrak{b}A'[X_1, \ldots, X_r]$. Then, since $\mu(\mathfrak{b}') \leq \mu(\mathfrak{b})$, we have

$$\Delta_{B_{\mathfrak{b}}/A}B'_{\mathfrak{b}'} \subseteq \Delta_{B'_{\mathfrak{b}'}/A'}.$$

On the other hand, $(\mathfrak{b} : \mathfrak{a})A'[X_1, \ldots, X_r] \subseteq (\mathfrak{b}' : \mathfrak{a}')$; hence

$$\Delta_{B_{\mathfrak{b}}/A}(\mathfrak{b} : \mathfrak{a})C' \subseteq \Delta_{B'_{\mathfrak{b}'}/A'}(\mathfrak{b}' : \mathfrak{a}')C'.$$
□

Lemma 5.2.12 *Let A be a ring and $f : B \to C$ be an essentially of finite type morphism of A-algebras essentially of finite type. Then $(H_{B/A}C) \cap H_{C/B} \subseteq H_{C/A}$.*

Proof Let $\mathfrak{q} \in \operatorname{Spec}(C)$ be such that $\mathfrak{q} \not\supseteq (H_{B/A}C) \cap H_{C/B}$. Then $H_{B/A} \not\subseteq \mathfrak{p} = \mathfrak{q} \cap B$ and $H_{C/B} \not\subseteq \mathfrak{q}$. It follows that $B_{\mathfrak{p}}$ is smooth over A and $C_{\mathfrak{q}}$ is smooth over B, hence $C_{\mathfrak{q}}$ is smooth over A. Thus $H_{C/A} \not\subseteq \mathfrak{q}$.
□

Lemma 5.2.13 *Let A be a ring, B an A-algebra essentially of finite presentation, C an A-algebra and $D = B \otimes_A C$. Then $H_{B/A}D \subseteq H_{D/C}$.*

Proof If $a \in H_{B/A}$, then B_a is smooth over A, hence by [249, Th. 28.2] we get that D_a is smooth over C.
□

Lemma 5.2.14 *Let A be a ring and C be a smooth A-algebra essentially of finite presentation. Then there exists a smooth C-algebra of finite presentation D, such that $\Omega_{D/A}$ is a free D-module.*

Proof Let $C = R^{-1}(A[X]/\mathfrak{a})$ be a presentation of C. We have a split exact sequence

$$0 \to \mathfrak{a}/\mathfrak{a}^2 \to \Omega_{A[X]/A} \otimes_{A[X]} C \to \Omega_{C/A} \to 0.$$

Let D be the symmetric algebra of the C-module $\mathfrak{a}/\mathfrak{a}^2$. Then, by [131, Prop.19.3.2], D is smooth over C, hence also over A, so that we have a split exact sequence

$$0 \to \Omega_{C/A} \otimes_C D \to \Omega_{D/A} \to \Omega_{D/C} \to 0.$$

It follows that

$$\Omega_{D/A} = (\Omega_{C/A} \otimes_C D) \oplus \Omega_{D/C} = (\Omega_{C/A} \otimes_C D) \oplus (\mathfrak{a}/\mathfrak{a}^2 \otimes_C D) =$$

$$= \Omega_{A[X]/A} \otimes_{A[X]} D,$$

which is a free D-module.   □

Corollary 5.2.15 *Let A be a ring and C be an A-algebra of finite presentation. Then there exists a C-algebra of finite presentation D such that:*

(i) *$H_{C/A}D \subseteq H_{D/A}$.*
(ii) *$\Omega_{D_a/A}$ is a free D_a-module, for any $a \in H_{C/A}$.*
(iii) *D has a presentation such that the image in D of any element $a \in H_{C/A}$ is a standard element.*
(iv) *For every A-algebra morphism $u : C \to E$, there exists a factorization $C \to D \to E$.*

Proof Let D be the symmetric algebra of the C-module $\mathfrak{a}/\mathfrak{a}^2$. Since $\mathfrak{a}/\mathfrak{a}^2$ is a C-module of finite presentation, D is a C-algebra of finite presentation. The assertions (i) and (ii) follow from the fact that C_a is smooth over A and D_a is the symmetric algebra of $\mathfrak{a}_a/\mathfrak{a}_a^2$. The assertion (iii) follows from 5.2.10 and (iv) is clear.   □

We introduce some notations, in order to make things easier for the reader.

Definition 5.2.16 Let A be a ring. An element $a \in A$ is called **square-regular** if $(0 : a) = (0 : a^2)$.

Remark 5.2.17 An element a is square-regular if from $\alpha \cdot a^2 = 0$, with $\alpha \in A$, it follows that $\alpha \cdot a = 0$.

Definition 5.2.18 Let $f : A \to B$ be a ring morphism. An **NP-factorization** of f will be a triple (A, C, B), where C is an A-algebra of finite type and B is a C-algebra, such that f factorizes as $A \to C \to B$.

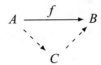

If (A, C, B) and (A, D, B) are two NP-factorizations of f, we say that the NP-factorization (A, D, B) is **bigger** than the NP-factorization (A, C, B) if there exists a morphism of A-algebras $u : C \to D$ making the following diagram commutative:

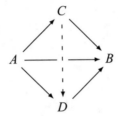

Notation 5.2.19 In order to shorten the notations, we will denote a NP-factorization (A, C, B) of the morphism $f : A \to B$ simply by C, whenever there is no possibility of confusion.

The proofs of the two following lemmas are quite difficult. In order to make things easier to be followed, we introduce some more notation.

Notation 5.2.20 In Lemmas 5.2.21, 5.2.22, 5.2.23, 5.2.24, etc., for any A-algebra X and any fixed element $a \in A$, we will denote $X_1 := X/aX$, $X_4 := X/a^4X$, $X_8 := X/a^8X$ or, more generally, $X_c := X/a^cX$, for $c \in \mathbb{N}$, when there will be no possibility of confusion.

Lemma 5.2.21 (Lifting Lemma [289, Lemma 8.1], see also [231, Lemma 5.2.1]) *Let $f : A \to B$ be a morphism of Noetherian rings and $a \in A$. Assume that both a and $f(a)$ are square-regular. Let $A_1 := A/aA$, $B_1 := B/aB$, $A_2 := A/a^2A$ and $B_2 := B/a^2B$, according to Notation 5.2.20. Let $f_1 : A_1 \to B_1$ and $f_2 : A_2 \to B_2$ be the induced morphisms, respectively. Let C be an NP-factorization of f_2. Then there exists a NP-factorization D of f and a morphism of A_2-algebras $C \to D_1 = D/aD$, such that:*

(i) The following diagram commutes:

$$
\begin{array}{ccc}
C & \longrightarrow & B_2 \\
\downarrow & & \downarrow \\
D_1 & \longrightarrow & B_1
\end{array}
$$

(ii) $\pi^{-1}(H_{C/A_2}B_2) \subset \sqrt{H_{D/A}B}$, where $\pi : B \to B_2$ is the canonical map.

Proof The situation is illustrated as much as possible in the diagram below:

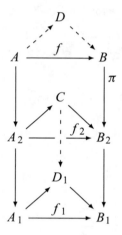

(i) Suppose that $C = A_2[X_1, \ldots, X_r]/\mathfrak{a}_2$, where $\mathfrak{a}_2$ is an ideal of $A_2[X_1, \ldots, X_r]$. To simplify the notations, set $X := (X_1, \ldots, X_r)$ and let $\mathfrak{a}$ be an ideal of $A[X]$ such that $a^2 \in \mathfrak{a}$ and $\mathfrak{a}_2 = \mathfrak{a}/(a^2 A[X])$. Then $C = A[X]/\mathfrak{a}$. By Definition 5.2.1 of H_{C/A_2}, we can choose finitely many elements of $\mathfrak{a}$, say $\{h_{ij}\}_{i \in \Gamma, \, j \in \Lambda_i}$, and finitely many polynomials with coefficients in A, say $\{w_i\}_{i \in \Gamma}$, with the following property: if we denote by $h_{ij,2}$ and $w_{i,2}$, the images in $A_2[X]$ of h_{ij} and w_i, respectively, setting for any $i \in \Gamma$ the ideals $\mathfrak{b}_i = (a^2, h_{ij}; j \in \Lambda_i)$ and $\mathfrak{b}_{i,2} = (h_{ij,2}; j \in \Lambda_i)$ the image of $\mathfrak{b}_i$ in $A_2[X]$, we have

$$
H_{C/A_2} = \sqrt{\sum_{i \in \Gamma} p_{i,2} C}.
$$

We may assume that for any $i \in \Gamma$ the elements $p_{i,2}$ are the images in $A_2[X]$ of some elements $p_i \in A[X]$ of the form $p_i = m_i \cdot w_i$, where, if c_i is the number of elements of Λ_i, each m_i is a $c_i \times c_i$-minor of the matrix $(\frac{\partial h_{ij}}{\partial X_t})_{j,t}$. We also have $w_{i,2}\mathfrak{a}_2 \subseteq \mathfrak{b}_{i,2}$; therefore $w_i \mathfrak{a} \subseteq (a^2 + \mathfrak{b}_i)$. Now, if necessary, we enlarge the set $\{h_{ij}\}_{i \in \Gamma, j \in \Lambda_i}$ to a system of generators of $\mathfrak{a}$ of the form $\{h_{ij}\}_{i \in \Gamma', j \in \Lambda_i}$, where $\Gamma' \supseteq \Gamma$. Consider an A-algebra morphism $v : A[X] = A[X_1, \ldots, X_r] \to B$ that makes commutative the following diagram:

$$
\begin{array}{ccc}
A[X_1,\ldots,X_r] & \xrightarrow{\ \text{v}\ } & B \\
\downarrow & & \downarrow \\
C & \longrightarrow & B_2
\end{array}
$$

We have $v(h_{ij}) \in a^2 B$, for any $i \in \Gamma'$ and $j \in \Lambda_i$. Let $\xi_{ij} \in aB$ be such that $v(h_{ij}) = a\xi_{ij}$, $\forall\, i \in \Gamma', j \in \Lambda_i$. Let $Z = \{Z_{ij},\ i \in \Gamma',\ j \in \Lambda_i\}$ be a family of indeterminates and let $g_{ij} = h_{ij} - aZ_{ij} \in A[X_1,\ldots,X_r,Z] = A[X,Z]$. For every $k \in \Gamma$, we have $p_k a \subseteq b_k = (a^2, h_{kj}; j \in \Lambda_k)$, hence for any $i \in \Gamma'$ and any $j \in \Lambda_i$, we have

$$
p_k h_{ij} = \sum_{j_0 \in \Lambda_k} H_{ijkj_0} h_{kj_0} + a^2 G_{ijk},\quad H_{ijkj_0},\ G_{ijk} \in A[X,Z],
$$

where, if $i = k$, we choose $H_{ijkj_0} = p_k \delta_{j,j_0}$ and $G_{ijk} = 0$. Set

$$
F_{ijk} = p_k Z_{ij} - \sum_{j_0 \in \Delta_k} H_{ijkj_0} Z_{kj_0} - a G_{ijk}.
$$

Then $F_{ijk} = 0$ for $i = k$ and

$$
a F_{ijk} = p_k(h_{ij} - g_{ij}) - \sum_{j_0 \in \Lambda_k} H_{ijkj_0}(h_{kj_0} - g_{kj_0}) - a^2 G_{ijk} =
$$

$$
= -p_k g_{ij} + \sum_{j_0 \in \Lambda_k} H_{ijkj_0} g_{kj_0} \in g := (g_{ij}; i \in \Gamma', j \in \Lambda_i).
$$

Thus $F_{ijk} \in (g : a) \cap ((Z) + (a))$ for any $i \in \Gamma', j \in \Lambda_i, k \in \Gamma$. Let

$$
f := (\{F_{ijk}\}, i \in \Gamma', j \in \Lambda_i, k \in \Gamma),
$$

$$
c := g + f,\quad D := A[X,Z]/c.
$$

The morphism $v : A[X] \to B$ extends to a morphism $v' : A[X,Z] \to B$ by mapping $Z_{ij} \mapsto \xi_{ij}$. Thus

$$
v'(g_{ij}) = v'(h_{ij} - aZ_{ij}) = v(h_{ij}) - a\xi_{ij} = 0.
$$

Let $\beta \in (g : a) \cap (Z + (a))$. Then $a\beta \in g$; hence $av(\beta) = v(a\beta) = 0$. Also $v(Z_{ij}) = a\xi_{ij} \in aB$. Thus $v(\beta) \in aB$; hence $v(\beta) \in aB \cap (0 : a)$. Therefore $v(\beta) = a \cdot \lambda$, with

$\lambda \in B$ and $0 = a \cdot v(\beta) = a^2 \cdot \lambda = 0$, hence $v(\beta) = 0$ since $f(a)$, the image of a in B, is square-regular by hypothesis. We have shown that

$$v((\mathfrak{g} : a) \cap ((Z) + (a))) = (0).$$

But then $v(\mathfrak{f}) = (0)$, hence $v(\mathfrak{c}) = (0)$; therefore v' induces a morphism $D \to B$. We have $\mathfrak{a} = (h_{ij}, i \in \Gamma', j \in \Lambda_i) \subseteq \mathfrak{g} + (a) \subset A[X, Z]$; therefore there exists a morphism $C \to A[X]/\mathfrak{a} \to D_1 = D/aD$. Obviously this morphism makes the diagram of the assertion (i) commute.

(ii) Since the relations $g_{ij} = h_{ij} - a Z_{ij}$ imply that $D_a \cong A_a[X, Z]/\mathfrak{c} = A_a[X]$, it follows that $a \in H_{D/A}$. Choose an ordering on the sets Γ' and Λ_i, for all $i \in \Gamma'$, and let h be the column vector $(h_{ij})_{i \in \Gamma', j \in \Lambda_i}$ with the ordering

$$(i, j) \le (i_1, j_1) \text{ if } i < i_1 \text{ or } i = i_1, j \le j_1.$$

Let $k \in \Gamma$. Then

$$p_k h_{ij} = \sum_{j_0 \in \Lambda_k} H_{ijkj_0} h_{kj_0} + a^2 G_{ijk} =$$

$$= \sum_{i_0 \in \Gamma', j_0 \in \Lambda_{i_0}} H_{ijki_0 j_0} h_{i_0 j_0} + a^2 G_{ijk},$$

where

$$H_{ijki_0 j_0} = \begin{cases} H_{ijkj_0}, & \text{if } i_0 = k \\ 0, & \text{if } i_0 \ne k. \end{cases}$$

Then, if we consider the square matrix $H_k = (H_{ijki_0 j_0})_{(i,j) \in \Gamma' \times \Lambda_i, (i_0, j_0) \in \Gamma' \times \Lambda_{i_0}}$, where ij is the row and $i_0 j_0$ is the column, we have

$$p_k h \equiv H_k h \pmod{a}.$$

Let us denote $S_{k_1, k_2} := p_{k_2} H_{k_1} - H_{k_2} H_{k_1}$. We have

$$S_{k_1, k_2} h \equiv p_{k_2} p_{k_1} h - H_{k_2} p_{k_1} h \equiv p_{k_2} p_{k_1} h - p_{k_2} p_{k_1} h \equiv 0 \pmod{a}.$$

Differentiating with respect to each variable, we obtain

$$S_{k_1, k_2} \frac{\partial h}{\partial X_t} \in (a) + \mathfrak{a} = (a) + \mathfrak{g}, \quad t = 1, \dots, r,$$

that is, all the coordinates of $S_{k_1,k_2}\frac{\partial h}{\partial X_t}$ are in the ideal $(a) + \mathfrak{g} \subset A[X, Z]$. Since $p_{k_1} = m_{k_1} w_{k_1}$, where m_{k_1} is a $c_1 \times c_1$-minor of the matrix $(\frac{\partial h_{k_1 j}}{\partial X_t})_{j,t}$, reordering Γ' such that k_1 is the first element and reordering $X_1, \ldots, X_r$ if necessary, we may assume that $m_{k_1} = \det(M)$, where M is the $c_1 \times c_1$-submatrix of the upper left corner of the matrix $(\frac{\partial h_{ij}}{\partial X_t})_{(i,j),t}$. Since $H_{ijk_1 i_0 j_0} = 0$ if $i_0 \neq k_1$, it follows that $(S_{k_1,k_2})_{(ij)(i_0 j_0)} = 0$ for $i_0 \neq k_1$; hence $(S_{k_1,k_2}) = (S|0)$ where S is the matrix formed by the first c_1 columns. It follows that we can write the relation

$$S_{k_1,k_2}(\frac{\partial h}{\partial X_t}) \in (a) + \mathfrak{g}, \text{ for all } t = 1, \ldots, r,$$

as

$$(S|0)(\frac{M}{*}) \in (a) + \mathfrak{g}.$$

Therefore $SM \in (a) + \mathfrak{g}$. Multiplying first by the adjoint matrix of M, then by w_{k_1} and using the relation $p_{k_1} = m_{k_1} w_{k_1}$, we get that

$$p_{k_1} S_{k_1,k_2} \in (a) + \mathfrak{g}.$$

We view Z as a column vector. Moreover let $\mathfrak{f}_{(k_1)}$ be the ideal generated by the elements $\{F_{ijk_1}; i \in \Gamma', j \in \Lambda_i\}$ and F_{k_2} be the column vector $(F_{ijk_2})_{ij}$. Then we have

$$H_{k_1} Z \equiv p_{k_1} Z \bmod (a, \mathfrak{f}_{(k_1)}),$$

so that

$$S_{k_1,k_2} Z = p_{k_2} H_{k_1} Z - H_{k_2} H_{k_1} Z \equiv$$

$$\equiv p_{k_2} p_{k_1} Z - p_{k_1} H_{k_2} Z \equiv p_{k_1}(p_{k_2} Z - H_{k_2} Z) \equiv p_{k_1} F_{K_2} \bmod (a, \mathfrak{f}_{(k_1)}).$$

Let $E_{k_1} := \left(A[X_1, \ldots, X_r, Z]/(\mathfrak{g} + \mathfrak{f}_{(k_1)})\right)_{p_{k_1}}$. We will show that E_{k_1} is smooth over A. We have

$$a F_{ijk_1} = -p_{k_1} g_{ij} + \sum_{j_0 \in \Lambda_{k_1}} H_{ijk_1 j_0} g_{k_1 j_0},$$

hence each g_{ij} is 0 in $\left(A[X_1, \ldots, X_r, Z]/((g_{k_1 j})_{j \in \Lambda_{k_1}} + \mathfrak{f}_{(k_1)})\right)_{p_{k_1}}$. Then

$$E_{k_1} = \left(A[X_1, \ldots, X_r, Z]/((g_{k_1 j})_{j \in \Lambda_{k_1}} + \mathfrak{f}_{(k_1)})\right)_{p_{k_1}}.$$

We have

$$F_{ijk_1} = p_{k_1} Z_{ij} - \sum_{j_0 \in \Lambda_{k_1}} H_{ijk_1 j_0} Z_{k_1 j_0} - a G_{ijk_1},$$

so that we can solve for Z_{ij} and we finally get that

$$E_{k_1} = \left(A[X_1, \ldots, X_r, \{Z_{k_1 j}\}_{j \in \Lambda_{k_1}}]/\left((g_{k_1 j})_{j \in \Lambda_{k_1}}\right)\right)_{p_{k_1}}.$$

Since $g_{ij} = h_{ij} - a Z_{ij}$, it follows that

$$\left(\frac{\partial g_{k_1 j}}{\partial (X_t, Z_{k_1 j_0})}\right)_{j \in \Lambda_{k_1}, 1 \le t \le r, j_0 \in \Lambda_{k_1}} = \left((\frac{\partial h_{k_1 j}}{\partial X_t}) - a \cdot \mathbb{I}_{c_{k_1}}\right),$$

where $\mathbb{I}_u$, for some $u \in \mathbb{N}$, is the unit $u \times u$-matrix. By the definition of p_{k_1}, some $c_{k_1} \times c_{k_1}$-minor of $(\frac{\partial h_{k_1 j}}{\partial X_t})$ divides p_{k_1}, whence E_{k_1} is smooth and consequently flat over A. We consider the exact sequences of A-modules

$$0 \to (0 : a) \to A \xrightarrow{a} A$$

and

$$0 \to (0 : a^2) \to A \xrightarrow{a^2} A$$

and tensorize them with the flat A-module E_{k_1}. It results that a is square-regular as an element of E_{k_1}. Since $p_{k_1} S_{k_1, k_2} \in (a) + \mathfrak{g}$, it follows that $(S_{k_1, k_2})_{ij}$ maps to 0 in $E_{k_1}/a E_{k_1}$. It results that $F_{ijk_2} \in a E_{k_1}, \forall i, j$. Since $a F_{ijk_2} \in \mathfrak{g}$, it results that $a F_{ijk_2} = 0$ in E_{k_1}. Since a is square-regular in E_{k_1}, it follows that $F_{ijk_2} = 0$ in E_{k_1}; hence $E_{k_1} = D_{p_{k_1}}$ and therefore $p_{k_1} \in H_{D/A}$. Since

$$\pi^{-1}(H_{C/A_2} B_2) \subseteq \sqrt{(a) + (p_k; k))}$$

we get that

$$\pi^{-1}(H_{C/A_2} B_2) \subseteq \sqrt{H_{D/A} B}.$$

$\square$

Lemma 5.2.22 (Desingularization   Lemma   [289,   Prop. 7.1],   see   also [231, Lemma 5.2.2])   *Let $f : A \to B$ be a morphism of Noetherian rings and*

$a \in A$ be such that $f(a)$ is square-regular. Let C be a NP-factorization of f and $\mathfrak{a} := \operatorname{Ann}_A\left(\frac{\operatorname{Ann}_A(a^2)}{\operatorname{Ann}_A(a)}\right)$. Assume that:

(i) The image of $\mathfrak{a}$ in C is strictly standard.
(ii) There exists a retract of $A_4 = A/a^4 A$-algebras, $g : C_4 = C/a^4 C \to A_4 = A/a^4 A$, such that the diagram

commutes.

Then there exists a NP-factorization D of f, bigger than C, such that $\mathfrak{a}D \subseteq H_{D/A}$.

Proof Let $C = A[X_1, \ldots, X_r]/\mathfrak{b}$ be a presentation of C such that the image a' of a in C is strictly standard. For any $i = 1, \ldots, r$, let $X_{i,4}$ be the image of X_i in C_4 and $y_{i,4} = g(X_{i,4}) \in A_4$. Consider also elements $y_i \in A$ representing $y_{i,4}$, for $i = 1, \ldots, r$. Let $\mathfrak{c} \subseteq \mathfrak{b}$ be an ideal of $A[X] := A[X_1, \ldots, X_r]$ such that $a' \in \Delta_{C_{\mathfrak{c}}/A}(\mathfrak{c} : \mathfrak{b})C$, where $C_{\mathfrak{c}} = A[X]/\mathfrak{c}$, and let $P(X) = P(X_1, \ldots, X_r) \in \Delta_{C_{\mathfrak{c}}/A}(\mathfrak{c} : \mathfrak{b})$ be an element representing a' in $A[X]$, that is, $P(X) - a' \in \mathfrak{b}$. Then $P(y_4) - a_4 = 0$ in A_4, that is, $P(y) - a \in a^4 A$. In particular, $P(y) = as$ where $s \in A$ is such that $s \equiv 1 \mod aA$. Let $f = (f_1, \ldots, f_u)$ be a minimal set of generators of $\mathfrak{c}$ and let $\{m_v\}_v$ be the $u \times u$-minors of the matrix $(\frac{\partial f}{\partial X})$. Then $P(X) = \sum_v l_v m_v$. Reordering, if necessary, the variables $X_1, \ldots, X_r$ and the polynomials $f_1, \ldots, f_u$, we may suppose that $\det(H_1) = m_1$, where

$$H_1 := \begin{pmatrix} \frac{\partial f_1}{\partial X_1} & \cdots & \frac{\partial f_1}{\partial X_r} \\ & \cdots & \\ \frac{\partial f_u}{\partial X_1} & \cdots & \frac{\partial f_u}{\partial X_r} \\ 0 & \cdots & \mathbb{I}_{r-u} \end{pmatrix}.$$

For each v we define in the same way H_v, keeping the first u rows and changing the $r - u$ last rows in such a way that $\det(H_v) = m_v$. Let G'_v be the adjoint matrix of H_v and $G_v(X) := l_v G'_v$. Then

$$G_v H_v = H_v G_v = l_v m_v \cdot \mathbb{I}_r,$$

hence

$$P(X) \cdot \mathbb{I}_r = \sum_v G_v H_v$$

and in particular

$$as \cdot \mathbb{I}_r = P(y) \cdot \mathbb{I}_r = \sum_v G_v(y) H_v(y).$$

Denote by ξ_i the image of X_i in B and by η_i the image of y_i in B, for $i = 1, \ldots, r$. Then

$$\xi_i \equiv \eta_i \bmod a^4 B,$$

hence

$$\xi_i = \eta_i + a^3 \varepsilon_i, \quad \varepsilon_i \in aB.$$

Let ε be the column vector ${}^t(\varepsilon_1, \ldots, \varepsilon_r)$, and consider similarly the column vectors ξ and η. Let $\tau^v = {}^t(\tau_1^v, \ldots, \tau_r^v)$ be the column vector $H_v(y)\varepsilon$. Note that $\tau_i^v \in aB$, for any $i = 1, \ldots, r$. We have

$$\sum_v G_v(y)\tau^v = P(y)\varepsilon = as\varepsilon,$$

hence

$$s(\xi - \eta) = sa^3 \varepsilon = a^2 \cdot \left(\sum_v G_v(y)\tau^v \right).$$

Since all the matrices H_v have the same first u rows, the same is true for τ^v, so that we have $\tau^v = (\tau_1, \ldots, \tau_u, \tau_{u+1}^v, \ldots, \tau_r^v)$. Let $W_1, \ldots, W_r$ be variables and $T^v = (T_1, \ldots, T_u, T_{u+1}^v, \ldots, T_r^v)$, where $\{T_j\}_{1 \le j \le u}$ and $\{T_j^v\}_{u+1 \le j \le r}$ are variables too. Let

$$h_j(X, W, T) := s \cdot (X_j - y_j) - a^3 W_j - a^2 \cdot \left(\sum_v G_v(y)T^v \right)_j \in A[X, W, T],$$

$$(5.1)$$

$$1 \le j \le r.$$

Let

$$f_i(X) - f_i(y) = \sum_j \frac{\partial f_i}{\partial X_j}(y)(X_j - y_j) + \text{ terms of higher degree in } (X_j - y_j).$$

Since

$$s \cdot (X_j - y_j) \equiv \left(a^3 W_j + a^2 \left(\sum_v G_v(y)T^v \right)_j \right) \bmod h_j,$$

if we set $m := \max\{\deg(f_i)\}$, we have

$$s^m \cdot \left(f_i(X) - f_i(y)\right) \equiv \left(\sum_j s^{m-1} \frac{\partial f_i}{\partial X_j}(y) \cdot \left(a^3 W_j + a^2 \left(\sum_v G_v(y) T^v\right)_j\right) + a^4 Q_i'\right) \bmod \mathfrak{h},$$

where $\mathfrak{h} = (\{h_j, 1 \le j \le r\})$ and $Q_i' \in (W, T)^2 A[W, T]$. It follows that

$$s^m \cdot f_i(X) - s^m \cdot f_i(y) \equiv \left(s^{m-1} a^2 \sum_j \frac{\partial f_i}{\partial X_j}(y) \left(\sum_v G_v(y) T^v\right)_j + a^3 Q_i\right) \bmod \mathfrak{h}$$

with $Q_i \in \sum_j A W_j + a \cdot (W, T)^2 A[W, T]$. Since $i \le u$ we have

$$\sum_j \frac{\partial f_i}{\partial X_j}(y) \cdot (G_v(y) T^v)_j = (H_v(y) G_v(y) T^v)_i = \left(l_v(y) m_v(y) T^v\right)_i =$$

$$= l_v(y) m_v(y) T_i$$

and so

$$\sum_j \frac{\partial f_i}{\partial X_j}(y) \left(\sum_v G_v(y) T^v\right)_j = \sum_v l_v(y) m_v(y) T_i = a \cdot s \cdot T_i.$$

It results that

$$s^m \cdot f_i(X) - s^m \cdot f_i(y) \equiv \left(s^m \cdot a^3 \cdot T^i + a^3 \cdot Q_i\right) \bmod \mathfrak{h}.$$

But $f_i(y) \in a^4 A$; hence we can write $f_i(y) = a^3 c_i$, for some $c_i \in aA$. Let

$$g_i = s^m c_i + s^m T_i + Q_i \in A[W, T], \ 1 \le i \le u. \tag{5.2}$$

Then

$$a^3 g_i = s^m a^3 c_i + s^m a^3 T_i + a^3 Q_i =$$

$$= s^m f_i(y) + s^m a^3 T_i + a^3 Q_i \equiv s^m f_i(X) \bmod \mathfrak{h}.$$

Let $D := A[X, W, T]/(\mathfrak{b} + \mathfrak{g} + \mathfrak{h}) = C[W, T]/(\mathfrak{g} + \mathfrak{h})$, where $\mathfrak{g} = (\{g_i\}, 1 \le i \le u)$. We have a morphism of A-algebras

$$\varphi : A[X, W, T] \to B,$$

$$\varphi(X_j) = \xi_j, \ \varphi(W_j) = 0, \ j = 1, \ldots, r$$

$$\varphi(T_j) = \tau_j, \ j = 1, \ldots, u, \ \varphi(T_j^v) = \tau_j^v, \ j = u + 1, \ldots, r.$$

Then $\varphi(\mathfrak{h}) = \varphi(\mathfrak{a}) = (0)$, and the above congruence modulo $\mathfrak{h}$ shows that

$$a^3 \varphi(g_i) = s^m \varphi(f_i(X)) = 0,$$

hence

$$\varphi(g_i) = (0 : f(a)^3) = (0 : f(a)).$$

Since $c_i \in aA$, $\tau_j^v \in aB$, $Q_i \in \sum_j AW_j + a(W, T)^2 A[W, T]$ and $\varphi(W_j) = 0$, by the definition (5.2) of the elements g_i, we have $\varphi(g_i) \in aB$, that is, $\varphi(g_i) = ab, b \in B$. But $a\varphi(g_i) = 0$, hence $b \in (0 : f(a)^2) = (0 : f(a))$, so that $\varphi(g_i) = 0$. Thus φ induces a morphism $D \to B$. As $a^3 g_i \equiv s^m f_i(X) \bmod \mathfrak{h}$, it follows that $g_i = 0$ in $C_a[W, T]/\mathfrak{h}$; hence $D_a = C_a[W, T]/\mathfrak{h}$. We can solve in this ring the equations $h_i = 0$ for the W_i and then $D_a = C_a[T]$. It follows that D_a is smooth over C_a and therefore D_a is smooth over A, that is, $a \in H_{D/A}$.

By the definition (5.1) of the elements h_j, we have

$$X_j \equiv y_j \bmod (a^2 + \mathfrak{h})A_s[X, T, W].$$

In particular, if $q(X) \in \mathfrak{b}$, then

$$q(X) \equiv q(y) \equiv 0 \bmod (a^2 + \mathfrak{h})A_s[X, T, W],$$

hence

$$\mathfrak{b}A_s[X, W, T] \subseteq (a^2 + \mathfrak{h})A_s[X, W, T].$$

Also

$$P(X) \equiv P(y) \equiv a \cdot s \bmod (a^2 + \mathfrak{h})A_s[X, W, T],$$

that is, there exists $\lambda \in \mathbb{N}$ such that

$$s^\lambda P(X) \equiv a \cdot s^{\lambda+1} \bmod (a^2 + \mathfrak{h})A[X, W, T].$$

Since $s \equiv 1 \bmod aA$, we have $a \cdot s \equiv a \bmod a^2 A$. Therefore

$$s^\lambda P(X) \equiv a \bmod (a^2 + \mathfrak{h})A[X, W, T]$$

and then there exists $t \in A[X, W, T]$, such that $1 - t \in aA[X, W, T]$ and

$$s^\lambda P(X) \equiv t \cdot a \bmod \mathfrak{h}.$$

Let $E = A[X, W, T]/(\mathfrak{g} + \mathfrak{h})$. Then $D = E/\mathfrak{c}E$. From the relation

$$a^3 g_i \equiv s^m f_i(X) \bmod \mathfrak{h}$$

we get that $f_i = 0$ in E_s. Since $P(X)\mathfrak{b} \subseteq \mathfrak{c} = (f_1, \ldots, f_u)$ we have $at\mathfrak{b}E_s = (0)$, whence $E_{ast} = D_{ast}$. Since D_a is smooth over A, it results that E_{ast} is smooth over A, that is, $ast \in H_{E/A}$. In $E_s = A_s[X, W, T]/((\mathfrak{g} + \mathfrak{h})A_s[X, W, T])$, as one can see from (5.1), we can solve the equations $h_j = 0$ for the variables X_j, so that $E_s = A_s[W, T]/\mathfrak{g}A_s[W, T]$. For each $i, j = 1, \ldots, u$, since

$$Q_i \in \sum_j AW_j + a \cdot (W, T)^2 A[W, T],$$

we have

$$\frac{\partial g_i}{\partial T_j} \equiv \delta_{ij} s^m \bmod a A[W, T].$$

By taking determinants we have

$$s^{mu} + ae \in H_{E_s/A_s}, \text{ for some } e \in E_s.$$

Then

$$s^{mu+1}t + stae \in H_{E/A}.$$

Since $ast \in H_{E/A}$, it results that $s^{mu+1}t \in H_{E/A}$; hence $st \in H_{E/A}$.

Because $\mathfrak{c}A_s[X, W, T] \subseteq (a^2 + \mathfrak{h})A_s[X, W, T]$, it follows that $\mathfrak{c} \cdot E_{st} \subseteq a \cdot E_{st}$; hence $\mathfrak{c} \cdot E_{st} = a \cdot \mathfrak{j}$, where $\mathfrak{j}$ is an ideal of E_{st}. Since $at\mathfrak{c}E_s = 0$, we have $a\mathfrak{c}E_{st} = 0$, so that

$a^2 j = 0$. It follows that, denoting by $\tilde{a}$ the image of a in E_{st} and using the flatness of E_{st} over A, we have

$$\mathfrak{j} \subseteq \mathrm{Ann}_{E_{st}}(\tilde{a}^2) \subseteq \mathrm{Ann}_{E_{st}}\left(\tilde{a} \cdot (\mathrm{Ann}(\frac{\mathrm{Ann}(\tilde{a}^2)}{\mathrm{Ann}(\tilde{a})}))\right) = \mathrm{Ann}_{E_{st}}(a\mathfrak{a}).$$

Thus we get $a\mathfrak{a}\mathfrak{j} = 0$, that is, $\mathfrak{a}\mathfrak{b}E_{st} = 0$. Then for any $\alpha \in \mathfrak{a}$, we have $\mathfrak{b}E_{st\alpha} = 0$. It follows that $D_{st\alpha} = E_{st\alpha}$, so that $st\alpha \in H_{D/A}$. Since $a \in H_{D/A}$ and since $s \equiv 1 \bmod a$ and $t \equiv 1 \bmod a$, it results that $1 - st = \gamma \cdot a \in H_{D/A}$, for some γ. Then $\alpha - st\alpha \in H_{D/A}$; therefore $\alpha \in H_{D/A}$.   □

Lemma 5.2.23 *Let $f : A \to B$ be a morphism of Noetherian rings and $a \in A$ be an element such that a and $f(a)$ are both square-regular. Let C and D be NP-factorizations of f. Assume that there exists a morphism of $A_4 = A/a^4 A$-algebras $\tilde{\varphi} : C_4 = C/a^4 C \to D_4 = D/a^4 D$, such that the diagram*

commutes. If the image of a in C is strictly standard, there exists a NP-factorization E of f, bigger than both C and D, such that the diagram

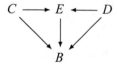

commutes and $H_{D/A} E \subseteq H_{E/A}$.

Proof Let $F = C \otimes_A D$, $F_4 = F/a^4 F$ and let $\rho : F_4 \to D_4$ be the morphism defined by $\rho(c_4 \otimes d_4) = \tilde{\varphi}(c_4) d_4$. Let also $\theta : D_4 \to F_4$ be the canonical morphism. Then $\rho \circ \theta = 1_{D_4}$ and the diagram

$$F_4 \xrightarrow{\rho} D_4$$
$$\downarrow \;\; \swarrow$$
$$B_4$$

commutes. By Lemma 5.2.11 the image of a in F is a strictly standard element. Let $\mathfrak{a} := \mathrm{Ann}_D\left(\frac{\mathrm{Ann}_D(a^2)}{\mathrm{Ann}_D(a)}\right)$. From the Desingularization Lemma 5.2.22 applied to the NP-factorization F of the morphism $D \to B$ and to the element a, we obtain that there exists a NP-factorization E of $D \to B$, bigger than F and such that $\mathfrak{a}E \subseteq H_{E/D}$. Let $x \in H_{D/A}$. Then $A \to D_x$ is smooth; therefore flat. It results that a is also square-regular in D_x. Then $\mathfrak{a}D_x = D_x$; hence $E_x \subseteq (H_{E/D})_x = H_{E_x/D_x}$. Thus $D_x \to E_x$ is smooth; hence $A \to E_x$ is smooth, so that $x \in H_{E/A}$.   □

Having in mind the notations considered in 5.2.20, we have:

Lemma 5.2.24 *Let $f : A \to B$ be a morphism of Noetherian rings and C a NP-factorization of f. Let $a \in A$ be an element such that a and $f(a)$ are square-regular and the image of a in C is strictly standard. Let $c \in \mathbb{N}$ be such that $c \geq 8$. Assume that we have a NP-factorization F of the induced morphism $A_c \to B_c = B$, bigger than C_c. Then there exists a NP-factorization E of f, bigger than C, such that $\pi^{-1}(H_{F/A_c}B_c) \subseteq \sqrt{H_{E/A}B}$, where $\pi : B \to B_c$ is the canonical map.*

Proof Let $c = 8$. By Lemma 5.2.21 applied for a^4, it follows that there exists a NP-factorization D of f and a morphism $F \to D_4 = D/a^4D$ such that the diagram

$$C_8 = C/a^8C \longrightarrow B_8 = B/a^8B$$
$$\downarrow \qquad\qquad\qquad \downarrow$$
$$D_4 = D/a^4D \longrightarrow B_4 = B/a^4B$$

commutes and $\pi^{-1}(H_{F/A_8}B_8) \subseteq \sqrt{H_{D/A}B}$. Now the morphism $C_8 \to F$ induces a morphism $C_4 \to F_4$, and the morphism $F_8 \to D_4$ induces a morphism $F_4 \to D_4$, so that we get a morphism $C_4 \to D_4$. Applying Lemma 5.2.23 to the NP-factorizations C and D of f, we obtain a NP-factorization E of f, bigger than C and such that $H_{D/A}E \subseteq H_{E/A}$.

If $c > 8$ we have $H_{F/A_c}F_8 \subseteq H_{F_8/A_8}$ and the assertion follows from the case $c = 8$.   □

Definition 5.2.25 Let $f : A \to B$ be a ring morphism, C a NP-factorization of f and $\mathfrak{q} \in \mathrm{Spec}(B)$. We say that $\mathfrak{q}$ is **resolvable** with respect to the NP-factorization C if there exists a NP-factorization D of f, bigger than C, and such that $H_{C/A}B \subseteq \sqrt{H_{D/A}B} \not\subseteq \mathfrak{q}$.

Proposition 5.2.26 *Let $f : A \to B$ be a morphism of Noetherian rings and C a NP-factorization of f. Let $a \in A$ be an element that is strictly standard on C and $\mathfrak{q} \in \mathrm{Spec}(B)$.*

(i) If $H_{C/A}B \not\subseteq \mathfrak{q}$, then $\mathfrak{q}$ is resolvable with respect to C.
(ii) If $H_{C/A}B \subseteq \mathfrak{q}$, let $c \geq 8$ and $\bar{\mathfrak{q}} = \mathfrak{q}/a^cB$. Assume that a and $f(a)$ are square-regular. If $\bar{\mathfrak{q}}$ is resolvable with respect to $C_c = C/a^cC$, then $\mathfrak{q}$ is resolvable with respect to C.

Proof (i) Obvious.

(ii) Let D be a NP-factorization of $f_c : A_c = A/a^cA \to B_c = B/a^cB$ such that

$$H_{C_c/A_c}B_c \subseteq H_{D/A_c}B_c \nsubseteq \bar{\mathfrak{q}}.$$

By Lemma 5.2.24 there exists a NP-factorization E of f bigger than C and such that $\pi^{-1}(H_{D/A_c}B_c) \subseteq \sqrt{H_{E/A}B}$, where $\pi : B \to B_c$ is the canonical morphism. But $H_{C/A}C_c \subseteq H_{C_c/A_c}$; hence $H_{C/A}B_c \subseteq H_{C_c/A_c}B_c$, so that

$$H_{C/A}B \subseteq \pi^{-1}(H_{C_c/A_c}B_c) \subseteq \pi^{-1}(H_{D/A_c}B_c) \subseteq \sqrt{H_{E/A}B}.$$

But $H_{D/A_c}B_c \nsubseteq \bar{\mathfrak{q}}$, so that $\pi^{-1}(H_{D/A_c}B_c) \nsubseteq \pi^{-1}(\bar{\mathfrak{q}}) = \mathfrak{q}$; hence $\sqrt{H_{E/A}B} \nsubseteq \mathfrak{q}$. $\qquad\square$

Proposition 5.2.27 *Let $f : A \to B$ be a morphism of Noetherian rings, C be a NP-factorization of f and $\mathfrak{q}$ be a prime ideal of B containing $H_{C/A}B$. Let $e \geq 1$ and $c = 8 \cdot e$. Let $a_1, \ldots, a_r \in A$ be such that the image of each a_i^e in C is strictly standard, and for each i the image of a_i and $f(a_i)$ in $A/(a_1^c, \ldots, a_{i-1}^c)$ and $B/(a_1^c, \ldots, a_{i-1}^c)B$ resp. is square-regular. For any A-algebra X, similarly as before, set $X_c := X/(a_1^c, \ldots, a_r^c)X$. If $\bar{\mathfrak{q}} = \mathfrak{q}/(a_1^c, \ldots, a_r^c)$ is resolvable with respect to C_c, then $\mathfrak{q}$ is resolvable with respect to C.*

Proof Replacing each a_i by a_i^e, we may assume that $e = 1$, and we proceed by induction on r. If $r = 1$ we apply Lemma 5.2.26. Suppose that the assertion is true for $r - 1$. Let $\tilde{A} = A/(a_1^c, \ldots, a_{r-1}^c)$, $\tilde{B} = B/(a_1^c, \ldots, a_{r-1}^c)B$ and $\tilde{C} = C/(a_1^c, \ldots, a_{r-1}^c)C$. By Lemma 5.2.11 the image of a_r in $\tilde{C}$ is strictly standard. Now we apply Lemma 5.2.26 to the NP-factorization $\tilde{C}$ of the induced morphism $\tilde{f} : \tilde{A} \to \tilde{B}$ and to $\tilde{\mathfrak{q}} = \mathfrak{q}/(a_1^c, \ldots, a_{r-1}^c)$. $\qquad\square$

Lemma 5.2.28 *Let $f : A \to B$ be a morphism of finite presentation and let $S \subset B$ be a multiplicatively closed system. If $A \to S^{-1}B$ is smooth, there exists $u \in S$ such that $A \to B_u$ is smooth.*

Proof We have $S^{-1}H_{B/A} = H_{S^{-1}B/A} = S^{-1}B$; hence there exist $x \in H_{B/A}$ and $s \in S$ such that $\frac{x}{s} = 1$ in $S^{-1}B$. Thus there exist $t \in S$ such that $(x - s)t = 1$ in B. Then, with $u = st$ we have $(H_{B/A})_u = B_u = H_{B_u/A}$. $\qquad\square$

Lemma 5.2.29 *Let $A \to C \to B$ be ring morphisms such that C is an A-algebra of finite presentation. If there exists a factorization $A \to C \to E \to B$ such that E is smooth and essentially of finite presentation over A, then there exists a factorization $A \to C \to D \to B$ such that D is smooth and of finite presentation over A.*

Proof Let $E = S^{-1}F$ where F is of finite presentation over A. By Lemma 5.2.28 there exists $s \in S$ such that $A \to F_s$ is smooth. Let $C = A[X_1, \ldots, X_n]/(f_1, \ldots, f_r)$ and let

X_i map to $\frac{c_i}{t}$ in E. Define $\phi : A[X_1, \ldots, X_n] \to F_{st}$, $\phi(X_n) = \frac{c_i}{t}$. Then $\phi(f_j) = 0$ in $E = S^{-1}F$; hence there is an element $u \in S$ such that $\phi(f_j) = 0$ in F_{stu}. We get the factorization $A \to C \to F_{stu} \to B$ and $D := F_{stu}$ is smooth and of finite presentation over A.   □

Remark 5.2.30 Let A be a ring and let $C = T^{-1}(A[X_1, \ldots, X_n])/\mathfrak{a}$ be a smooth A-algebra essentially of finite presentation. There is a split exact sequence

$$0 \to \mathfrak{a}/\mathfrak{a}^2 \to \Omega_{A[X_1,\ldots,X_n]/A} \otimes_{A[X_1,\ldots,X_n]} C \to \Omega_{C/A} \to 0.$$

Let $D = S_C(\mathfrak{a}/\mathfrak{a}^2)$ be the symmetric algebra of the projective C-module $\mathfrak{a}/\mathfrak{a}^2$. Then by [131, Prop. 19.3.2] D is smooth over C; hence smooth over A, and we have a split exact sequence

$$0 \to \Omega_{C/A} \otimes_C D \to \Omega_{D/A} \to \Omega_{D/C} \to 0.$$

Thus

$$\Omega_{D/A} = (\Omega_{C/A} \otimes_C D) \oplus \Omega_{D/C} = (\Omega_{C/A} \otimes_C D) \oplus (\mathfrak{a}/\mathfrak{a}^2 \otimes_C D) =$$

$$= \Omega_{A[X_1,\ldots,X_n]/A} \otimes_{A[X_1,\ldots,X_n]} D,$$

therefore $\Omega_{D/A}$ is a free D-module. It follows that if C is an A-algebra essentially of finite presentation, there exists a C-algebra of finite presentation D such that $H_{C/A}D \subset H_{D/A}$ and $\Omega_{D_a/A}$ is a free D_a-module for all $a \in H_{C/A}$.

Lemma 5.2.31 *Let* $f : A \to B$ *be a morphism of Noetherian rings and* C *be a NP-factorization of* f.

(i) *Let* $a \in A$ *be such that* $f(a) \in \sqrt{H_{C/A}B}$. *Then there exists a NP-factorization* D *of* f *bigger than* C, *such that* $H_{C/A}D \subseteq H_{D/A}$ *and the image of* a *in* D *is standard with respect to a finite presentation of* D.

(ii) *If* $H_{C/A}B = B$, *there exists a NP-factorization* D *of* f, *bigger than* C, *such that* D *is smooth over* A.

Proof

(i) Let $r \in \mathbb{N}$ be such that $a^r \in H_{C/A}B$. Assume that $H_{C/A} = (c_1, \ldots, c_n)$, so that $a^r = \sum_{i=1}^n b_i c_i$, with $b_1, \ldots, b_n \in B$. Let $E = C[X_1, \ldots, X_n]/(a^r - \sum_{i=1}^n c_i X_i)$ and let $\varphi : E \to B$ be the morphism of C-algebras such that $\varphi(X_i) = b_i$, $i = 1, \ldots, n$. Since we have $E_{c_i} = C_{c_i}[X_1, \ldots, X_{i-1}, X_{i+1}, \ldots, X_n]$ and $A \to E_{c_i}$ is smooth, it follows that $c_1, \ldots, c_n \in H_{E/A}$. Thus $H_{C/A}C \subseteq H_{E/A}$. By the definition of E, it follows that the image of a^r in E is actually in $H_{C/A}E$, so that $a \in H_{E/A} = \sqrt{H_{E/A}}$. By Remark 5.2.30 there exists an E-algebra of finite type D such that $H_{E/A}D \subseteq H_{D/A}$ and the image of a in D is standard for some presentation of D. Moreover the morphism $E \to B$ factors through D.

(ii) It is enough to take $a = 1$ in (i).

$\square$

Remark 5.2.32 In the situation of Lemma 5.2.31, (i), it follows that there exists $s \geq 1$ such that the image of a^s in D is strictly standard.

Proposition 5.2.33 *Let* $f : A \to B$ *be a morphism of Noetherian rings, C be a NP-factorization of f and $q \in \operatorname{Min}(H_{C/A}B)$. If qB_q is resolvable with respect to the NP-factorization C of the morphism $A \to B_q$, then there exists a NP-factorization D of f, bigger than C and such that $H_{D/A}B \not\subseteq q$.*

Proof By assumption there exists a NP-factorization E of $A \to B_q$, bigger than C, such that $H_{E/A}B_q \not\subseteq qB_q$, that is, $H_{E/A}B_q = B_q$. By Lemma 5.2.31,ii) we may assume that E is smooth over A. Let $E = C[X_1, \ldots, X_n]/(F_1, \ldots, F_m) = C[x_1, \ldots, x_n]$. There exists $t \in B \setminus q$ such that the morphism $E \to B_q$ factors through B_t. Replacing t by some power of t, we may assume that the image of each variable X_i in B_t is of the form $\frac{z_i}{t}$, for some $z_i \in B$. Consider the commutative diagram

$$
\begin{array}{ccc}
C[X_1, \ldots, X_n, Y] & \xrightarrow{\varphi} & E[Y] = C[x_1, \ldots, x_n, Y] \\
\downarrow{\scriptstyle\lambda} & & \downarrow{\scriptstyle\chi} \\
B & \xrightarrow{\psi} & B_t
\end{array}
$$

where the morphism φ is defined by $\varphi(X_i) = Yx_i$, $i = 1, \ldots, n$ and $\varphi(Y) = Y$. Moreover we have $\lambda(X_i) = z_i$ for $i = 1, \ldots, n$, $\lambda(Y) = s$ and $\chi(Y) = s$. Since $\psi(\lambda(\ker(\varphi))) = 0$, there exists a power $h \in \mathbb{N}$ such that $t^h \lambda(\ker(\varphi)) = 0$. Let $D = C[X_1, \ldots, X_n, Y]/(Y^h \cdot \ker(\varphi))$. It is obvious that λ induces a C-algebra morphism $D \to B$. If y is the class of Y in D, then $D_y = E[Y]_y = E[Y, \frac{1}{Y}]$ which is smooth over E, hence also over A. Then $t = \lambda(Y) \in H_{D/A}B$, whence $H_{D/A}B \not\subseteq q$.

$\square$

Proposition 5.2.34 *Let* $f : A \to B$ *be a morphism of Noetherian rings, C be a NP-factorization of f and $q \in \operatorname{Min}(H_{C/A}B) \cap \operatorname{Min}(B)$. If there exists a NP-factorization D of f bigger than C and such that $H_{D/A}B \not\subseteq q$, then there exists a NP-factorization E of f bigger than D and such that $H_{C/A}B \subseteq \sqrt{H_{E/A}B} \not\subseteq q$.*

Proof If $H_{C/A}B$ is a nilpotent ideal, we can take $E = D$. Assume that $H_{C/A}B$ is not a nilpotent ideal. Then $\mathfrak{q}$ is not nilpotent and (0) is not a primary ideal in B. Then (0) is not irreducible, and we can write $(0) = \mathfrak{a} \cap \mathfrak{b}$, with $\mathrm{Ass}_B(B/\mathfrak{a}) = \{\mathfrak{q}\}$ and $\mathfrak{q} \notin \mathrm{Ass}_B(B/\mathfrak{b})$. As $\sqrt{\mathfrak{a}} = \{\mathfrak{q}\}$ and $\mathfrak{q}$ is not nilpotent, we have $\mathfrak{b} \neq (0)$. Let $\mathfrak{a} = (z_1, \ldots, z_r)$ and $\mathfrak{b} = (y_1, \ldots, y_s)$. Suppose that

$$D = C[X_1, \ldots, X_n]/(F_1, \ldots, F_m)$$

and let

$$E = C[X_1, \ldots, X_n, Y_1, \ldots, Y_s, Z_1, \ldots, Z_r]/(\{Y_i F_j\}_{i,j}, \{Y_k Z_l\}_{k,l}).$$

Consider the morphism of $C[X]$-algebras $\rho : C[X, Y, Z] \to B$ given by $\rho(Y_i) = y_i$, $i = 1, \ldots, s$ and $\rho(Z_j) = z_j$, $j = 1, \ldots, r$. Then $\rho(Y_i F_j) = y_i F_j = 0$ and $\rho(Y_k Z_l) = y_k z_l \in \mathfrak{b}\mathfrak{a} \subseteq \mathfrak{a} \cap \mathfrak{b} = (0)$. Hence ρ factors through a morphism $\varphi : E \to B$. Since $\mathfrak{b} \not\subseteq \mathfrak{q}$, reordering $y_1, \ldots, y_s$ we can assume that $y_1 \notin \mathfrak{q}$. Let $\bar{Y}_1$ be the image of Y_1 in E. Then $\bar{Y}_1 \notin \mathfrak{p} := \varphi^{-1}(\mathfrak{q})$ and $E_{\bar{Y}_1} = (C[X, Y]/(F_1, \ldots, F_m))_{\bar{Y}_1}$ which is smooth over D; hence $H_{E/D} \not\subseteq \mathfrak{p}$. By assumption $H_{D/A}B \not\subseteq \mathfrak{q}$; hence from Lemma 5.2.12, it results $H_{E/A} \not\subseteq \mathfrak{p}$ and furthermore $H_{E/A}B \not\subseteq \mathfrak{q}$. Let $\mathfrak{n}$ be a prime ideal of B such that $\mathfrak{q} \not\subseteq \mathfrak{n}$. We have $\mathfrak{a} \not\subseteq \mathfrak{n}$ so, after reordering if necessary the elements $z_1, \ldots, z_r$, we can suppose that $z_1 \notin \mathfrak{n}$. Let $\bar{Z}_1$ be the image of Z_1 in E and let $\mathfrak{m} = \varphi^{-1}(\mathfrak{n})$. Then $\bar{Z}_1 \notin \mathfrak{m}$; hence $E_{\bar{Z}_1}$ is isomorphic to a localization of $C[X, Z]$, and therefore is smooth over C. This shows that if $\mathfrak{q} \not\subseteq \mathfrak{n}$, then $H_{E/C} \not\subseteq \mathfrak{m} = \varphi^{-1}(\mathfrak{n})$. Thus, if we take a prime ideal $\mathfrak{n}$ of B such that $\mathfrak{n} \not\supseteq H_{C/A}B$, we have $\mathfrak{q} \not\subseteq \mathfrak{n}$; hence $H_{E/C} \not\subseteq \mathfrak{m} = \varphi^{-1}(\mathfrak{n})$. On the other hand, $H_{C/A}B \not\subseteq \mathfrak{n}$ implies that $H_{C/A}E \not\subseteq \mathfrak{m}$. By Lemma 5.2.12 we have $H_{E/A} \not\subseteq \mathfrak{m}$ and so $H_{E/A}B \not\subseteq \mathfrak{n}$. Thus $H_{C/A}B \subseteq \sqrt{H_{E/A}B}$. □

Theorem 5.2.35 *Let $f : A \to B$ be a morphism of Noetherian rings and C a NP-factorization of f. Let $\mathfrak{q}$ be a prime ideal of B and $\mathfrak{p} = f^{-1}(\mathfrak{q})$. Assume that:*

(i) $\mathfrak{q} \in \mathrm{Min}(H_{C/A}B)$ and $\mathfrak{p} \in \mathrm{Min}(f^{-1}(H_{C/A}B))$.
(ii) The morphism $A_{\mathfrak{p}} \to B_{\mathfrak{q}}$ is formally smooth.
(iii) The residue field extension $k(\mathfrak{p}) \to k(\mathfrak{q})$ is separable.

Then there exists a NP-factorization D of f, bigger than C, such that $H_{C/A}B \subseteq \sqrt{H_{D/A}B} \not\subseteq \mathfrak{q}$.

Proof Case 1: $\mathrm{ht}(\mathfrak{p}) = 0$. We proceed by induction on $\mathrm{ht}(\mathfrak{q})$. If $\mathrm{ht}(\mathfrak{q}) = 0$, by Proposition 5.2.33 and Proposition 5.2.34, we may assume that A and B are Artinian local rings and f is a local morphism. Then $\mathfrak{p}$ is the maximal ideal of A, and the closed fibre $B/\mathfrak{p}B$ is a regular local ring, whence a field and therefore $\mathfrak{p}B \in \mathrm{Max}(B)$. Then B is what is usually called a Cohen A-algebra [131, Déf. 19.8.1], that is, B is a local complete flat A-algebra, such that the closed fibre is the residue field of B. Since the residue field extension is separable, by [57, Ch. IX, App.] it results that B is a filtered inductive limit $B = \varinjlim_i B_i$ where B_i are local A-subalgebras of B such that for any index i, $A \to B_i$ is a morphism essentially of finite type, formally smooth by the assumption (iii), hence smooth by Corollary 3.2.10. If $C = A[X_1, \ldots, X_n]/(h_1, \ldots, h_r)$, then there exists an index i such that the morphism $A[X_1, \ldots, X_n] \to B$ factors through B_i and we can take i large enough such that all the elements $h_1, \ldots, h_r$ map to 0. Finally, we can find a factorization E of f, bigger than C and such that E is smooth and essentially of finite type over A. By Lemma 5.2.29 there exists a factorization D of f, bigger than C, and such that D is smooth and of finite type over A, so that $H_{D/A} = D$.

Assume now that $ht(\mathfrak{q}) > 0$. Since $\mathfrak{q} \in Min(H_{C/A}B)$, we get $\sqrt{H_{C/A}B} \cdot B_\mathfrak{q} = \mathfrak{q}B_\mathfrak{q}$. Since by assumption (ii) $B_\mathfrak{q}/\mathfrak{p}B_\mathfrak{q}$ is a regular local ring, there exists $\xi \in \sqrt{H_{C/A}B}$ whose image in $B_\mathfrak{q}/\mathfrak{p}B_\mathfrak{q}$ is a part of a regular system of parameters. Let $g : G := A[X] \to B$ be the morphism of A-algebras such that $g(X) = \xi$. Let $\mathfrak{r} = g^{-1}(\mathfrak{q}) = (\mathfrak{p}, X)A[X]$. Then there is a local morphism $g_\mathfrak{q} : G_\mathfrak{r} \to B_\mathfrak{q}$, and this induces furthermore a morphism $G_\mathfrak{r}/\mathfrak{p}G_\mathfrak{r} = k(\mathfrak{p})[X]_{(X)} \to B_\mathfrak{q}/\mathfrak{p}B_\mathfrak{q}$. The situation is illustrated in the following commutative diagram.

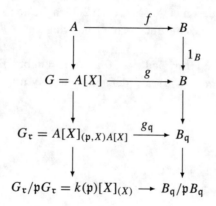

Since ξ is a part of a regular system of parameters, the fibre $B_\mathfrak{q}/\mathfrak{r}B_\mathfrak{q}$ is a regular local ring; hence by the separability assumption, it follows that the local morphism $G_\mathfrak{r} \to B_\mathfrak{q}$ is formally smooth. Consider the NP-factorization of g given by $G = A[X] \to C[X] \to B$. Lemma 5.2.13 implies that $\xi \in \sqrt{H_{C[X]/A[X]}B}$. From 5.2.31 we obtain a NP-factorization E of g, bigger than $C[X]$, such that $H_{C[X]/G}E \subseteq H_{E/G}$, and the image of X in E is a standard element. Let $s \in \mathbb{N}$ be large enough such that the element $a = X^s \in A[X]$ is square-regular in $A[X]$ and in B and the image of a in E is strictly standard. Denote by $U_8 := U/a^8U$ for any G-algebra U. Since a is not a zero-divisor in G and $G \to B_\mathfrak{q}$ is flat, a is not a zero-divisor on $B_\mathfrak{q}$ too. Thus $ht(\mathfrak{q}_8) = ht(\mathfrak{q}/a^8B) < ht(\mathfrak{q})$. If $H_{E_8/G_8}B \not\subseteq \mathfrak{q}$, it follows that $\mathfrak{q}_8$ is resolvable with respect to (G_8, E_8, B_8). Assume that $H_{E_8/G_8} \subseteq \mathfrak{q}$. By Lemma 5.2.13 we have $H_{E/G}E_8 \subseteq H_{E_8/G_8}$, and since $H_{C[X]/G}E \subseteq H_{E/G}$, it results that $\mathfrak{q}_8 \in Min(H_{E_8/G_8}B)$. Applying the induction assumption, we obtain that $\mathfrak{q}_8$ is resolvable with respect to (G_8, E_8, B_8). Then by Proposition 5.2.26, it follows that $\mathfrak{q}$ is resolvable with respect to E. Thus there exists a NP-factorization D of $G \to B$, bigger than E, such that $H_{E/G}B \subseteq \sqrt{H_{D/G}B} \not\subseteq \mathfrak{q}$. Since $A \to G$ is smooth, by Lemma 5.2.12 we have $H_{D/G} \subseteq H_{D/A}$. Then

$$\sqrt{H_{C/A}B} \subseteq \sqrt{H_{C[X]/G}B} \subseteq \sqrt{H_{E/G}B} \subseteq \sqrt{H_{D/G}B} \subseteq \sqrt{H_{D/A}B} \not\subseteq \mathfrak{q}.$$

It follows that $\mathfrak{q}$ is resolvable with respect to C.

Case 2: $\mathrm{ht}(\mathfrak{p}) > 0$. Since by assumption $\mathfrak{p}$ is minimal over $f^{-1}(H_{C/A}B)$, there exists an element $\alpha \in f^{-1}(H_{C/A}B)$ such that $\alpha \notin \mathfrak{n}$, for any prime ideal $\mathfrak{n} \subset \mathfrak{p}$ with $\mathrm{ht}(\mathfrak{n}) = 0$. Then $\mathrm{ht}(\alpha A) \geq 1$, therefore $\mathrm{ht}(\mathfrak{p}/(\alpha A)) < \mathrm{ht}(\mathfrak{p})$. By Lemma 5.2.31 there exists a NP-factorization D of f, bigger than C, such that $H_{C/A}D \subseteq H_{D/A}$ and the image of α in D is a standard element. Replacing α by some power α^t if necessary, we can assume that the image of α in D is strictly standard and that α is square-regular in A and in B. If $H_{D/A}B \not\subseteq \mathfrak{q}$ we are done. If $H_{D/A}B \subseteq \mathfrak{q}$, since $\mathfrak{q} \in \mathrm{Min}(H_{C/A}B)$ and $\mathfrak{p} \in \mathrm{Min}(f^{-1}(H_{D/A}B))$, we get that $\mathfrak{q} \in \mathrm{Min}(H_{D/A}B)$. Now the assertion follows by induction on $\mathrm{ht}(\mathfrak{p})$, by applying Proposition 5.2.26 to the NP-factorization D and to $\mathfrak{q}$. $\square$

We want to prove a result similar to Theorem 5.2.35 in the non-separable case. We need to remind some features of the construction of Cohen rings.

Definition 5.2.36 Let A be a ring and $p > 0$ a prime number. For an ideal $\mathfrak{a}$ of A, we define $\tilde{\mathfrak{a}} = \{a^p + pb \mid a, b \in \mathfrak{a}\}$. We define inductively $\mathfrak{a}(0) = \mathfrak{a}$ and for any $n \geq 1$, $\mathfrak{a}(n) = \tilde{\mathfrak{a}}_{n-1}$.

Remark 5.2.37

(i) $\tilde{A}$ is a subring of A and $\tilde{\mathfrak{a}}$ is an ideal of $\tilde{A}$.
(ii) In general $\tilde{A}/\tilde{\mathfrak{a}} \not\cong \widetilde{A/\mathfrak{a}}$.

Proof

(i) Indeed, $a^p + pb - (c^p + pd) = (a - c)^p - p\alpha + p(b - d)$, where $\alpha \in \mathfrak{a}$ and $(a^p + pb)(c^p + pd) = (ac)^p + p\beta$, with $\beta \in \mathfrak{a}$. Obviously $\tilde{\mathfrak{a}}$ is an ideal of $\tilde{A}$.
(ii) Take, for example, $A = \mathbb{F}_p[X]/(X^{p+1}) = \mathbb{F}_p[x]$ and $\mathfrak{a} = (x^p)$. Then $\widetilde{A/\mathfrak{a}} = \mathbb{F}_p$ and $\tilde{A}/\tilde{\mathfrak{a}} = \mathbb{F}_p[x^p]$.

                                    $\square$

Lemma 5.2.38 *Let A be a ring, $\mathfrak{a}$ an ideal of A and $p > 0$ a prime number. If $p \in \mathfrak{a}$, then $\mathfrak{a}(n) \subseteq \mathfrak{a}^{n+1}, \forall n \geq 0$.*

Proof We have $\mathfrak{a}(0) = \mathfrak{a} \subseteq \mathfrak{a}^1 = \mathfrak{a}$. Assume that $\mathfrak{a}(n-1) \subseteq \mathfrak{a}^n$. Then

$$\mathfrak{a}(n) = \widetilde{\mathfrak{a}(n-1)} \subseteq \widetilde{\mathfrak{a}^n} \subseteq (\mathfrak{a}^n)^p + p\mathfrak{a}(n) \subseteq \mathfrak{a}^{n+1}.$$

$\square$

Remark 5.2.39 In 5.2.40, 5.2.41 and 5.2.42, the ring A is not necessarily Noetherian.

Lemma 5.2.40 Let $(A, \mathfrak{m}, k)$ be a local ring such that $\text{char}(k) = p > 0$ and such that there exists $r \in \mathbb{N}$ with $\mathfrak{m}^{r+1} = (0)$. Let D be a subring of A such that $D + \mathfrak{m} = A$. Then for any $n \geq r$, we have $D(n) = A(n)$.

Proof Let $x \in A$. Then $x = d + u$, where $d \in D$ and $u \in \mathfrak{m}$; hence $x^p = d^p + (u^p + pv)$, with $v \in \mathfrak{m}$, so that $x^p \in \widetilde{D} + \widetilde{\mathfrak{m}}$. In the same way, it follows that $px = pd + pu \in \widetilde{D} + \widetilde{\mathfrak{m}}$. Similarly we have $D(n) + \mathfrak{m}(n) = A(n)$ for any $n \geq 0$. But by Lemma 5.2.38, we get $\mathfrak{m}(r) \subseteq \mathfrak{m}^{r+1} = (0)$; hence $D(r) = A(r)$. $\square$

Lemma 5.2.41 Let $(A, \mathfrak{m}, k)$ be a local ring such that $\text{char}(k) = p > 0$ and such that there exists $r \in \mathbb{N}$ with $\mathfrak{m}^{r+1} = (0)$. Let $\{\alpha_i\}_{i \in I}$ be a family of elements of A such that, denoting by $\bar{\alpha}_i$ the image of α_i in k, we have $k^p(\bar{\alpha}_i, i \in I) = k$. Then $A(n)[\alpha_i, i \in I] = A(n+1)[\alpha_i, i \in I]$, for any $n \geq r$.

Proof Let $\pi : A \to k$ be the canonical surjection. Then $\pi(A(1)) = k^p$; hence $\pi(A(n+1)) = k^{p^{n+1}}$. From the assumption we get that $k^p = k^{p^2}(\bar{\alpha}_i^p, i \in I)$. So $\pi(A(n+1)[\alpha_i, i \in I]) = k^{p^{n+1}}(\bar{\alpha}_i^p, i \in I)$. It follows that

$$k = k^p(\bar{\alpha}_i, i \in I) = k^{p^2}(\bar{\alpha}_i^p, i \in I)(\bar{\alpha}_i, i \in I) = k^{p^2}(\bar{\alpha}_i, i \in I).$$

Therefore, in general, $k = k^{p^n}(\bar{\alpha}_i, i \in I)$; hence $A = A(n+1)[\alpha_i, i \in I] + \mathfrak{m}$ and by Lemma 5.2.40 we have $A(n) = (A(n+1)[\alpha_i, i \in I])(n) \subseteq A(n+1)[\alpha_i, i \in I]$. Then $A(n)[\alpha_i, i \in I] \subseteq A(n+1)[\alpha_i, i \in I]$. This concludes the proof. $\square$

Proposition 5.2.42 Let $(A, \mathfrak{m}, K)$ be a local ring such that $\text{char}(K) = p > 0$ and such that there exists a power $r \in \mathbb{N}$ with $\mathfrak{m}^{r+1} = (0)$. Let $\{\alpha_i\}_{i \in I}$ be a family of elements of A such that, denoting by $\bar{\alpha}_i$ the image of α_i in K, the family $\{\bar{\alpha}_i \mid i \in I\}$ is a p-basis of K. Set $C = A(r)[\alpha_i, i \in I]$. Then:

(i) C is an Artinian local subring of A.
(ii) pC is the maximal ideal of C and $C/pC \cong K$.
(iii) If A is Artinian, then it is a finite C-module.

Proof (i) A is integral over $\widetilde{A}$, hence also over C; therefore C is a local ring with maximal ideal $\mathfrak{n} := \mathfrak{m} \cap C$. Let $\pi : A \to K$ be the canonical surjection, and consider the commutative diagram

$$
\begin{array}{ccc}
C & \hookrightarrow & A \\
\downarrow & & \downarrow{\scriptstyle \pi} \\
C/\mathfrak{n} & \longrightarrow & K
\end{array}
$$

We get that $\pi(C) = C/\mathfrak{n} = K^{p^r}[\bar{\alpha}_i, i \in I] = K$. Since $p \in \mathfrak{m} \cap C = \mathfrak{n}$, there exists a commutative diagram

$$
\begin{array}{ccc}
\mathbb{Z}_{(p)} & \longrightarrow & C \\
\downarrow & & \downarrow \\
\mathbb{F}_p & \longrightarrow & K
\end{array}
$$

It follows that we have the following Jacobi-Zariski exact sequence associated with the morphisms $\mathbb{Z}_{(p)} \to \mathbb{F}_p \to K$

$$0 = H_2(\mathbb{F}_p, K, K) \to H_1(\mathbb{Z}_{(p)}, \mathbb{F}_p, K) \to H_1(\mathbb{Z}_{(p)}, K, K) \to H_1(\mathbb{F}_p, K, K) = 0,$$

hence

$$H_1(\mathbb{Z}_{(p)}, K, K) \cong H_1(\mathbb{Z}_{(p)}, \mathbb{F}_p, K) \cong p\mathbb{Z}_{(p)}/p^2\mathbb{Z}_{(p)} \otimes_{\mathbb{F}_p} K.$$

We have a commutative diagram with exact upper row:

$$
\Omega_{A(r+1)/\mathbb{Z}_{(p)}} \otimes_{A(r+1)} K \xrightarrow{\ \beta\ } \Omega_{C/\mathbb{Z}_{(p)}} \otimes_C K \xrightarrow{\ \gamma\ } \Omega_{C/A(r+1)} \otimes_C K \longrightarrow 0
$$

$$
\Omega_{A(r)/\mathbb{Z}_{(p)}} \otimes_{A(r)} K
$$

with ϵ between them.

The elements $\{da^p, d(pb), a, b \in A(r)\}$ generate $\Omega_{A(r+1)/\mathbb{Z}_{(p)}}$, so that in $\Omega_{A(r)/\mathbb{Z}_{(p)}} \otimes_{A(r)} K$ we have

$$\epsilon(da^p \otimes 1) = da^p \otimes 1 = pa^{p-1}da \otimes 1 = 0;$$

$$\epsilon(d(pb) \otimes 1) = d(pb) \otimes 1 = pdb \otimes 1 = 0.$$

Then $\epsilon = 0$, hence $\beta = 0$; therefore γ is an isomorphism. Since $C = A(r+1)[\alpha_i, i \in I]$, it follows that $\Omega_{C/A(r+1)}$ is generated by $\{d\alpha_i, i \in I\}$. Hence $\Omega_{C/\mathbb{Z}_{(p)}} \otimes_C K$ is generated by $\{d\alpha_i \otimes 1, i \in I\}$.

Consider the morphism $\lambda : \Omega_{C/\mathbb{Z}_{(p)}} \otimes_C K \to \Omega_{K/\mathbb{Z}_{(p)}}$. We have $\lambda(d\alpha_i \otimes 1) = d\bar{\alpha}_i$, which is a basis of $\Omega_{K/\mathbb{Z}_{(p)}} = \Omega_{K/\mathbb{F}_p}$. It follows that λ is injective. We consider the Jacobi-Zariski exact sequence associated with $\mathbb{Z}_{(p)} \to C \to K$:

$$H_1(\mathbb{Z}_{(p)}, K, K) = p\mathbb{Z}_{(p)}/p^2\mathbb{Z}_{(p)} \otimes_{\mathbb{F}_p} K \xrightarrow{\eta} H_1(C, K, K) = \mathfrak{n}/\mathfrak{n}^2 \xrightarrow{\lambda} \Omega_{K/\mathbb{Z}_{(p)}} \to 0.$$

Since λ is injective, η is surjective; hence $\mathfrak{n}/\mathfrak{n}^2$ is generated by $\eta(p)$. It results that $C/\mathfrak{n}^2$ is a complete local ring with finitely generated maximal ideal; hence is Noetherian. Assume that $C/\mathfrak{n}^i$ is Noetherian for some $i \geq 2$, and we show that $C/\mathfrak{n}^{i+1}$ is Noetherian. Since

$$\dim_K H_1(\mathbb{Z}_{(p)}, K, K) = \dim_K p\mathbb{Z}_{(p)}/p^2\mathbb{Z}_{(p)} \otimes_{\mathbb{F}_p} K < \infty,$$

from the exact sequence

$$H_1(\mathbb{Z}_{(p)}, K, K) \to H_1(\mathbb{Z}_{(p)}/p^i\mathbb{Z}_{(p)}, K, K) \to H_0(\mathbb{Z}_{(p)}, \mathbb{Z}_{(p)}/p^i\mathbb{Z}_{(p)}, K) = 0$$

we get that $\dim_K H_1(\mathbb{Z}_{(p)}/p^i\mathbb{Z}_{(p)}, K, K) < \infty$. Since $C/\mathfrak{n}^i$ is Noetherian, we have also that $\dim_K H_2(C/\mathfrak{n}^i, K, K) < \infty$ so that from the exact sequence

$$H_2(C/\mathfrak{n}^i, K, K) \to H_1(\mathbb{Z}_{(p)}/p^i\mathbb{Z}_{(p)}, C/\mathfrak{n}^i, K) \to H_1(\mathbb{Z}_{(p)}/p^i\mathbb{Z}_{(p)}, K, K)$$

we obtain that $\dim_K H_1(\mathbb{Z}_{(p)}/p^i\mathbb{Z}_{(p)}, C/\mathfrak{n}^i, K) < \infty$. Using the exact sequence

$$H_1(\mathbb{Z}_{(p)}, \mathbb{Z}_{(p)}/p^i\mathbb{Z}_{(p)}, K) \to H_1(\mathbb{Z}_{(p)}, C/\mathfrak{n}^i, K) \to H_1(\mathbb{Z}_{(p)}/p^i\mathbb{Z}_{(p)}, C/\mathfrak{n}^i, K)$$

we get that $\dim_K H_1(\mathbb{Z}_{(p)}, C/\mathfrak{n}^i, K) < \infty$. Consider now the exact sequence

$$H_1(\mathbb{Z}_{(p)}, C/\mathfrak{n}^i, K) \xrightarrow{\eta_i} H_1(C, C/\mathfrak{n}^i, K) \to \Omega_{C/\mathbb{Z}_{(p)}} \otimes_C K \xrightarrow{\lambda_i}$$

$$\xrightarrow{\lambda_i} \Omega_{(C/\mathfrak{n}^i)/\mathbb{Z}_{(p)}} \otimes_{C/\mathfrak{n}^i} K \to 0.$$

We have proved above that the map

$$\lambda : \Omega_{C/\mathbb{Z}_{(p)}} \otimes_C K \xrightarrow{\lambda_i} \Omega_{(C/\mathfrak{n}^i)/\mathbb{Z}_{(p)}} \to \Omega_{K/\mathbb{Z}_{(p)}}$$

is injective, whence also λ_i is injective. Then η_i is surjective; hence we have $\dim_K H_1(C, C/\mathfrak{n}^i, K) < \infty$. It results that $\dim_K \mathfrak{n}/\mathfrak{n}^{i+1} < \infty$. Thus $\mathfrak{n}/\mathfrak{n}^{i+1}$ is a finitely generated $C/\mathfrak{n}^{i+1}$-module. Since $C/\mathfrak{n}^i$ is Noetherian, $\mathfrak{n}/\mathfrak{n}^i$ is a $C/\mathfrak{n}^i$-module of finite type; hence a $C/\mathfrak{n}^{i+1}$-module of finite type. From the exact sequence

$$0 \to \mathfrak{n}^i/\mathfrak{n}^{i+1} \to \mathfrak{n}/\mathfrak{n}^{i+1} \to \mathfrak{n}/\mathfrak{n}^i \to 0$$

it follows that the maximal ideal of $C/\mathfrak{n}^{i+1}$ is finitely generated; hence $C/\mathfrak{n}^{i+1}$ is Noetherian. By induction it follows that $C/\mathfrak{n}^s$ is Noetherian for any $s \in \mathbb{N}$; hence C is Noetherian, because $\mathfrak{n}$ is nilpotent.

(ii) Since $\mathfrak{n}/\mathfrak{n}^2$ is generated by the image of p, using Nakayama's Lemma and the fact that C is Noetherian, we obtain that $\mathfrak{n} = (p)$.

iii) Assume that A is Artinian. Then $\dim_K \mathfrak{m}^i/\mathfrak{m}^{i+1} < \infty$; hence $\mathfrak{m}^i/\mathfrak{m}^{i+1}$ is a finitely generated C-module, because K is the residue field of C. Since $\mathfrak{m}$ is nilpotent, it follows that $\mathfrak{m}$ is a finitely generated C-module. As also $A/\mathfrak{m} = K$ is a finitely generated C-module, it results that A is a finitely generated C-module. $\square$

Theorem 5.2.43 (Popescu [291, Lemma 8]) *Let $u : (A, \mathfrak{m}, K) \to (B, \mathfrak{n}, L)$ be a local injective morphism of Artinian local rings such that $\mathrm{char}(K) = \mathrm{char}(L) = p > 0$ and $\dim_L H_1(K, L, L) < \infty$. Then there exists a filtered family $\mathcal{D}$ of local subrings of B and local inclusions between the elements of $\mathcal{D}$, such that:*

(i) *Any $D \in \mathcal{D}$ contains A and is an A-algebra essentially of finite type.*
(ii) *$B = \varinjlim_{D \in \mathcal{D}} D$.*
(iii) *If $(D_1, \mathfrak{m}_1) \subset (D_2, \mathfrak{m}_2)$ are elements in $\mathcal{D}$, then $\mathfrak{m}_1 D_2 = \mathfrak{m}_2$ and $\mathfrak{m}_1 B = \mathfrak{n}$.*
(iv) *B is a flat D-algebra for any $D \in \mathcal{D}$.*
(v) *D_2 is a flat D_1-algebra for any $D_1, D_2 \in \mathcal{D}$ such that $D_1 \subseteq D_2$.*
(vi) *For any $D \in \mathcal{D}$ with residue field E, if $x \in B$ and $\bar{x}$ is the image of x in L, then for any large enough power q of p, there exists $D' \in \mathcal{D}, D \subset D'$, with residue field $E' = E(\bar{x}^q)$, such that $x^q \in D'$.*

Proof ([231, Th. 5.5.9]) We have the Jacobi-Zariski exact sequence

$$H_1(K, L, L) \to \Omega_{K/\mathbb{F}_p} \otimes_K L \to \Omega_{L/\mathbb{F}_p} \to \Omega_{L/K} \to 0.$$

Let $\{\alpha_i\}_{i \in I}$ be elements of A such that their images $\{\bar{\alpha}_i\}_{i \in I}$ form a p-basis of K. Then by [131, Cor. 21.2.5], we know that $\{d\bar{\alpha}_i\}_{i \in I}$ is a basis of $\Omega_{K/\mathbb{F}_p}$. Choose a finite basis of $H_1(K, L, L)$ and represent the images of these elements in $\Omega_{K/\mathbb{F}_p} \otimes_K L$ as linear combinations of $\{d\bar{\alpha}_i \otimes 1\}_{i \in I}$. We remove from the family $\{d\bar{\alpha}_i\}_{i \in I}$ the elements $d\bar{\alpha}_i$ occurring in these representations. We obtain a family $\{d\bar{\alpha}_j\}_{j \in J_0}$, where J_0 is a subset of I such that $I \setminus J_0$ is finite and such that the images of the elements $\{d\bar{\alpha}_j\}_{j \in J_0}$ in $\Omega_{L/\mathbb{F}_p}$ are linearly independent. We can choose a basis of $\Omega_{L/\mathbb{F}_p}$ of the form $\{d\bar{\alpha}_j\}_{j \in J_0} \cup \{d\bar{\beta}_j\}_{j \in J_1}$, where $\{\beta_j\}_{j \in J_1} \subseteq B$.

Using the notations from Definition 5.2.36, let $B_a = B(r)[\{\alpha_j\}_{j \in J_0} \cup \{\beta_j\}_{j \in J_1}]$ where r is such that $\mathfrak{n}^{r+1} = 0$. By Proposition 5.2.42 we see that B_a is an Artinian local subring of B with maximal ideal pB_a, residue field L and such that B is a finite B_a-module. We define similarly $A_a = A(r)[\{\alpha_i\}_{i \in I}]$. Let now E be a field such that $K \subseteq E \subseteq L$ and E is a finitely generated field extension of K. We have an exact sequence

$$\Omega_{K/\mathbb{F}_p} \otimes_K E \to \Omega_{E/\mathbb{F}_p} \to \Omega_{E/K} \to 0.$$

The set $\{d\bar{\alpha}_j \otimes 1\}_{j \in J_0}$ generates a subspace of finite codimension in the K-vector space $\Omega_{K/\mathbb{F}_p} \otimes_K E$. Since $\dim_E \Omega_{E/K} < \infty$, the images of $\{d\bar{\alpha}_j \otimes 1\}_{j \in J_0}$ in $\Omega_{E/\mathbb{F}_p}$ generate a subspace of finite codimension in $\Omega_{E/\mathbb{F}_p}$ and are E-linearly independent, because they are L-linearly independent in $\Omega_{L/\mathbb{F}_p}$ and $L \supseteq E$. Let $\{\gamma_h\}_{h \in H_E}$ be finitely many elements of B_a such that $\{d\bar{\alpha}_j\}_{j \in J_0} \cup \{d\bar{\gamma}_h\}_{h \in H_E}$ is a basis of $\Omega_{E/\mathbb{F}_p}$. We consider all possible choices of $\{\gamma_h\}_{h \in H_E}$. Let $\pi : B \to L$ be the canonical morphism. Then $\pi^{-1}(E)$ is a local subring of B containing A and with maximal ideal $\mathfrak{n}$. Let $B_{H_E} = \pi^{-1}(E)(r)[\{\alpha_j\}_{j \in J_0} \cup \{\gamma_h\}_{h \in H_E}]$. By Proposition 5.2.42 B_{H_E} is an Artinian local subring of $\pi^{-1}(E)$ with maximal ideal pB_{H_E} and residue field E. Since $\{\gamma_h\}_{h \in H_E} \subset B_a$, it follows that $B_{H_E} \subset B_a$. Since H_E is finite, it results that B_{H_E} is a finitely generated algebra over $\pi^{-1}(E)(r)[\{\alpha_j\}_{j \in J_0}]$.

For each E as above, namely, E is a finitely generated field extension of K contained in L, set $M_E = \{j \in J_1 \mid \bar{\beta}_j \in E\}$. Then the set $\{d\bar{\alpha}_j\}_{j \in J_0} \cup \{d\bar{\beta}_j\}_{j \in M_E}$ is E-linearly independent in $\Omega_{E/\mathbb{F}_p}$; hence M_E is a finite set. Now let us define

$$C_E := \pi^{-1}(E)(r)[\{\alpha_j\}_{j \in J_0} \cup \{\beta_j\}_{j \in M_E}].$$

Clearly $(C_E)_E$ is a filtered direct system and $\varinjlim_E C_E = B_a$. Then, since we have

$$\pi^{-1}(E)(r)[\{\alpha_j\}_{j \in J_0}] \subseteq C_E$$

and B_{H_E} is a finitely generated $\pi^{-1}(E)(r)[\{\alpha_j\}_{j \in J_0}]$-algebra, there exists a subfield $F \supset E$ such that $B_{H_E} \subset C_F$. Conversely, given F a finitely generated subextension of L containing K, some finite subset H_F satisfies $M_F = \{j \in J_1 \mid \bar{\beta}_j \in F\} \subseteq \{\gamma_h\}_{h \in H_F}$ and thus $C_F \subseteq B_{H_F}$. This shows that also B_{H_E} is a filtered direct system with direct limit B_a. Since $B_{H_E} \subseteq B_a$ and the maximal ideals of both rings are generated by p, from local criterion of flatness [249, Th. 22.3], it follows that the morphism $B_{H_E} \to B_a$ is flat, hence faithfully flat. By Proposition 5.2.42 B is a B_a-algebra of finite type of the form $B = B_a[Y_1, \ldots, Y_n]/(F_1, \ldots, F_s)$, where the image of each variable Y_i in B belongs to the maximal ideal $\mathfrak{n}$. Since $\mathfrak{n}^{r+1} = (0)$, we may assume that $Y_i^{r+1} \in \{F_1, \ldots, F_s\}$ for any $i = 1, \ldots, n$. Let $\mathcal{H} := \{H_E \mid B_{H_E}$ contains all coefficients of $F_1, \ldots, F_s\}$. Then $\{B_{H_E}\}_{H_E \in \mathcal{H}}$ is a filtered subsystem of $\{B_{H_E}\}$ with limit B_a. For each $H_E \in \mathcal{H}$, let $D_{H_E} = B_{H_E}[Y_1, \ldots, Y_n]/(F_1, \ldots, F_s)$ and let $\mathcal{G} = \{D_{H_E} \mid H_E \in \mathcal{H}\}$. Let y_i be the images of Y_i in D_{H_E}. Then $y_i^{r+1} = 0$. The rings D_{H_E} are local rings with maximal ideal

$(p, y_1, \ldots, y_n)$, and the morphisms $D_{H_E} \to D_{H_F}$ are local morphisms. Since $B_{H_E} \to B_a$ are faithfully flat, it results that $D_{H_E} \to B_a \otimes_{B_{H_E}} D_{H_E}$ is faithfully flat. In particular $D_{H_E} \subseteq B$, so that by [56, Ch. I, §3, no. 4, Prop. 7] the morphisms $D_{H_E} \to D_{H_F}$ are faithfully flat. So we already have the properties (ii), (iii), (iv) and (v) for the family $\mathcal{G}$. By Proposition 5.2.42 we know that A is a finite A_a-algebra; hence A is a $A(r)[\{\alpha_j\}_{j \in J_0}]$-algebra of finite type. Therefore, since for any H_E we have $A(r)[\{\alpha_j\}_{j \in J_0}] \subseteq D_{H_E}$ and $\{D_{H_E}\}$ is a filtered system whose limit contains A, there exists $H_E \in \mathcal{H}$ such that $A \subseteq D_{H_E}$, and this inclusion is a local morphism, because u is a local morphism. Set $\mathcal{D} = \{D_{H_E} \mid H_E \in \mathcal{H} \text{ and } A \subseteq D_{H_E}\}$. The morphism $A \to D_{H_E}$ is a local morphism whose residue field extension is the finitely generated field extension $K \subseteq E$. It follows that E is an A-algebra essentially of finite type. Since D_{H_E} is Artinian, we get that D_{H_E} is an A-algebra essentially of finite type. Obviously we continue to have the properties (ii), (iii), (iv) and (v) for the family $\mathcal{D}$, and moreover also condition (i) is fulfilled. It remains to show that the family $\mathcal{D}$ satisfies condition (vi). Let $D_{H_E} \in \mathcal{D}$, let $x \in B$ and let $\bar{x}$ be the image of x in L. Assume that $\bar{x}$ is algebraic over K. Replacing x by x^q where q is a large enough power of p, we may assume that $\bar{x}$ is separable over E and $x \in B(r) \subseteq B_a$. Let $F = E(\bar{x})$. Then $\Omega_{E/\mathbb{F}_p} \otimes_E F \cong \Omega_{F/\mathbb{F}_p}$ so that $\{d\bar{\alpha}_j\}_{j \in J_0} \cup \{d\bar{\gamma}_h\}_{h \in H_E}$ is a basis of $\Omega_{F/\mathbb{F}_p}$. We take $H_F = H_E$ and $B_{H_F} = \pi^{-1}(F)(r)[\{\alpha_j\}_{j \in J_0} \cup \{\gamma_h\}_{h \in H_E}]$. Since $\bar{x} \in F$, we have $x \in \pi^{-1}(F)$, so that $x^{p^r} \in \pi^{-1}(F)(r) \subseteq B_{H_F} \subseteq D_{H_F}$. Moreover, since $\bar{x}$ is separable over E, we have $F = E(\bar{x}) = E(\bar{x}^{p^r})$.

Assume now that $\bar{x}$ is transcendental over E. Replacing x by x^q, we may assume that $x \in B_a$. Let $F = E(\bar{x})$. Then $\{d\bar{\alpha}_j\}_{j \in J_0} \cup \{d\bar{\gamma}_h\}_{h \in H_E} \cup \{d\bar{x}\}$ is a basis of $\Omega_{F/\mathbb{F}_p}$. Setting $H_F = H_E \cup \{w\}$, where $\gamma_w = x$, we have

$$x \in B_{H_F} = \pi^{-1}(F)(r)[\{\alpha_j\}_{j \in J_0} \cup \{\gamma_h\}_{h \in H_E} \cup \{x\}] \subseteq D_{H_F}.$$

$\square$

Lemma 5.2.44 *Let $u : (A, \mathfrak{m}, K) \to (B, \mathfrak{n}, L)$ be a formally smooth morphism of Noetherian local rings. Then $\dim_L H_1(K, L, L) < \infty$.*

Proof Using Corollary 3.2.4, the Jacobi-Zariski exact sequence associated with the morphisms $A \to B \to L$ is

$$0 = H_1(A, B, L) \to H_1(A, L, L) \to H_1(B, L, L) = \mathfrak{n}/\mathfrak{n}^2,$$

hence $\dim_L H_1(A, L, L) < \infty$. Looking at the exact sequence

$$H_1(A, L, L) \to H_1(K, L, L) \to H_0(A, K, L) = \Omega_{K/A} \otimes_K L = 0$$

we conclude the proof.

$\square$

Lemma 5.2.45 ([231, Lemma 5.4.4]) *Let* $A \xrightarrow{u} B \xrightarrow{v} C$ *be morphisms of Noetherian local rings such that u and vu are flat and let $\mathfrak{a}$ be a proper ideal of A.*

(i) *If $C/\mathfrak{a}C$ is a flat $B/\mathfrak{a}B$-module, then v is flat.*
(ii) *If $C/\mathfrak{a}C$ is a formally smooth $B/\mathfrak{a}B$-algebra, then v is formally smooth.*

Proof

(i) Follows from [249, Th. 22.3].
(ii) Since v is flat by (i), we can apply [6, Prop. IV.54], and we obtain

$$H_1(B, C, L) \cong H_1(B/\mathfrak{a}B, C/\mathfrak{a}C, L)$$

where L is the residue field of C. Now the assertion follows from Corollary 3.2.4. □

Proposition 5.2.46 *Let $f : A \to B$ be a morphism of Noetherian rings, let $\mathfrak{q}$ be a prime ideal of B and $\mathfrak{p} = f^{-1}(\mathfrak{q})$. Let $K = k(\mathfrak{p})$ and $L = k(\mathfrak{q})$ be the residue fields in $\mathfrak{p}$ and $\mathfrak{q}$, respectively. Suppose that $\mathrm{char}(K) = p > 0$ and that the morphism $A_\mathfrak{p} \to B_\mathfrak{q}$ is formally smooth. Let $n > 0$ be an integer, $\widetilde{B} = B_\mathfrak{q}/\mathfrak{q}^n B_\mathfrak{q}$ and $\widetilde{A}$ the image of $A_\mathfrak{p}$ in $\widetilde{B}$. Let G be a finite subset of $\widetilde{B}$. Then there exists a morphism $f' : A' = A[Y_1, \ldots, Y_m] \to B$ extending f and a family $\widetilde{\mathcal{D}}$ of local subrings of $\widetilde{B}$ such that, if we denote $\mathfrak{p}' = (f')^{-1}(\mathfrak{q})$ and $\widetilde{A}' = A'_{\mathfrak{p}'}/(\mathfrak{p}')^n A'_{\mathfrak{p}'}$, we have:*

(i) $\mathfrak{p}' B_\mathfrak{q} = \mathfrak{q} B_\mathfrak{q}$.
(ii) *The induced morphism $A'_{\mathfrak{p}'} \to B_\mathfrak{q}$ is flat.*
(iii) *For any $\widetilde{D} \in \widetilde{\mathcal{D}}$ we have:*
    (iii_1) $\widetilde{A}' \subset \widetilde{D}$.
    (iii_2) $G \subset \widetilde{D}$.
    (iii_3) $\widetilde{A}' \to \widetilde{D}$ *is a smooth morphism essentially of finite type.*
    (iii_4) $\widetilde{D} \to \widetilde{B}$ *is a flat morphism.*
    (iii_5) $\mathfrak{p}'\widetilde{D}$ *is the maximal ideal of $\widetilde{D}$.*
(iv) *If $\widetilde{D} \in \widetilde{\mathcal{D}}$ and $\widetilde{x} \in \widetilde{B}$, there exists a power q of p and $\widetilde{D}' \in \widetilde{\mathcal{D}}$ such that $\widetilde{D} \subset \widetilde{D}'$ and $\widetilde{x}^q \in \widetilde{D}'$.*

Proof Suppose that $\mathfrak{q} = (b_1, \ldots, b_r)$. Enlarging G if necessary, we can suppose that G contains the images of $b_1, \ldots, b_r$ in $\widetilde{B}$. By Lemma 5.2.44 we know that $\dim_K H_1(K, L, L) < \infty$. Since the residue field of $\widetilde{A}$ is K, the morphism $\widetilde{A} \to \widetilde{B}$ satisfies the assumptions of Theorem 5.2.43. Therefore there exists a family $\mathcal{D}$ of local subrings of $\widetilde{B}$ satisfying the conditions (i)–(vi) of Theorem 5.2.43. Because G is a finite set, there exists $D \in \mathcal{D}$ such that $G \subset D$. Let E be the residue field of D and let $e_1, \ldots, e_t$ be elements of D such that $\{d\bar{e}_1, \ldots, d\bar{e}_t\}$ is a basis $\Omega_{E/K}$, where $\bar{e}_j$ is the image of e_j in E.

We want to show first that we may always assume that the images of $e_1, \ldots, e_t$ in $\tilde{B}$ are images of some elements $y_1, \ldots, y_t \in B$.

Let $y_1, \ldots, y_t \in B$ and $s \in B \setminus \mathfrak{q}$ be such that e_i is the image of $\frac{y_i}{s} \in B_\mathfrak{q}$, for any $i = 1, \ldots, t$. Let us denote by $\bar{s}$ the image of s in L. There are two cases to consider.

Case 1.   If $\bar{s}$ is algebraic over E, replacing s by s^q and y_i by $\frac{s^{q-1} y_i}{s^q}$, where q is some power of p, we can assume that $\bar{s}$ is separable over E. By Theorem 5.2.43,vi), replacing again s by some p-power s^{p^t}, it follows that there exists $D' \in \mathcal{D}$, with $D \subseteq D'$, such that $\tilde{s}$, the image of s in $\tilde{B}$, belongs to D' and the residue field of D' is $E' = E(\bar{s})$. Replacing once more s by s^p if necessary, we can assume that $d\bar{s} = 0 \in \Omega_{E'/K}$. Using [6, Lemma VII.3] we have the exact sequence

$$0 = H_1(E, E', E') \to \Omega_{E/K} \otimes_E E' \to \Omega_{E'/K} \to \Omega_{E'/E} = 0.$$

It results that $\{d\bar{e}_1, \ldots, d\bar{e}_t\}$ is also a basis of $\Omega_{E'/K}$. Since $\bar{y}_i = \bar{s}\bar{e}_i \in E'$ we have in $\Omega_{E'/K}$

$$d\bar{y}_i = \bar{s}d\bar{e}_i + \bar{e}_i d\bar{s} = \bar{s}d\bar{e}_i,$$

hence $\{d\bar{y}_1, \ldots, d\bar{y}_t\}$ is a basis of $\Omega_{E'/K}$. So, replacing $e_1, \ldots, e_t$ by $se_1, \ldots, se_t$ and D by D', we can assume that $e_1, \ldots, e_t$ are images of the elements $y_1, \ldots, y_t \in B$.

Case 2.   Suppose now that $\bar{s} \in L$ is transcendental over E. Replacing s by s^q, where q is, as above, some power of p, let $D' \in \mathcal{D}$, with $D \subset D'$ be such that $\tilde{s} \in D'$ and the residue field of D' is $E' = E(\bar{s})$ as in Theorem 5.2.43,v) and where $\bar{s}$ and $\tilde{s}$ are as in Case 1. Again using [6, Lemma VII.3], we have an exact sequence

$$0 = H_1(E, E', E') \to \Omega_{E/K} \otimes_E E' \to \Omega_{E'/K} \to \Omega_{E'/E} \to 0,$$

hence $\{d\bar{e}_1, \ldots, d\bar{e}_t, d\bar{s}\}$ is a basis of $\Omega_{E'/K}$. Also as in case 1, we have

$$d\bar{y}_i = \bar{s}d\bar{e}_i + \bar{e}_i d\bar{s},$$

whence $\{d\bar{y}_1, \ldots, d\bar{y}_t, d\bar{s}\}$ is a basis of $\Omega_{E'/K}$. Replacing D by D' and $\{e_1, \ldots, e_t\}$ by $\{se_1, \ldots, se_t, s\}$, we may assume that $e_1, \ldots, e_t$ are images of some elements $y_1, \ldots, y_t \in B$.

From the cases 1 and 2, we deduce that we can always assume that $e_1, \ldots, e_t$ are images of the elements $y_1, \ldots, y_t \in B$.

Let $\varphi : A[Z_1, \ldots, Z_t] \to B$ be the A-algebra morphism such that $\varphi(Z_i) = y_i$ for $i = 1, \ldots, t$ and let $\mathfrak{p}'' = \varphi^{-1}(\mathfrak{q})$. We want to show that $A[Z_1, \ldots, Z_t]_{\mathfrak{p}''} \to B_{\mathfrak{q}}$ is a formally smooth morphism. We show first that $H_1(K[Z_1, \ldots, Z_t], B \otimes_A K, L) = (0)$. Let $K[Z] := K[Z_1, \ldots, Z_t] \to E$ be the K-algebra morphism such that $Z_i \to e_i$, $i = 1, \ldots, t$. From the morphisms $K \to K[Z] \to E$, we obtain the Jacobi-Zariski exact sequence

$$0 = H_1(K, K[Z], L) \to H_1(K, E, L) \to H_1(K[Z], E, L) \to$$

$$\to \Omega_{K[Z]/K} \otimes_{K[Z]} L \to \Omega_{E/K} \otimes_E L.$$

Because $\{d\bar{e}_1, \ldots, d\bar{e}_t\}$ is a basis of $\Omega_{E/K}$, the last morphism is injective; hence $H_1(K, E, L) \cong H_1(K[Z], E, L)$.

We have a commutative diagram with exact rows and columns:

$$
\begin{array}{ccccc}
& & & & H_2(E, L, L) \\
& & & & \downarrow \\
H_1(K, B \otimes_A K, L) & \longrightarrow & H_1(K, E, L) & \longrightarrow & H_1(B \otimes_A K, E, L) \\
& & \downarrow & & \downarrow \\
0 \longrightarrow H_1(K[Z], B \otimes_A K, L) & \to & H_1(K[Z], L, L) & \to & H_1(B \otimes_A K, L, L) \\
& & \downarrow & & \downarrow \\
& & H_1(E, L, L) & = & H_1(E, L, L)
\end{array}
$$

Since

$$H_1(K, B \otimes_A K, L) = H_1(A_{\mathfrak{p}}, B_{\mathfrak{q}}, L) = 0,$$

$$H_2(B \otimes_A K, L, L) = 0,$$

$$H_1(K, E, L) = H_1(K[Z], E, L),$$

the above diagram becomes

$$
\begin{array}{ccccc}
& & & & 0 \\
& & & & \downarrow \\
0 & \longrightarrow & H_1(K[Z], E, L) & \to & H_1(B \otimes_A K, E, L) \\
& & \downarrow & & \downarrow \\
0 \longrightarrow H_1(K[Z], B \otimes_A K, L) & \to & H_1(K[Z], L, L) & \to & H_1(B \otimes_A K, L, L) \\
& & \downarrow & & \downarrow \\
& & H_1(E, L, L) & = & H_1(E, L, L)
\end{array}
$$

showing by a simple computation that $H_1(K[Z], B \otimes_A K, L) = (0)$. Then

$$H_1((A/\mathfrak{p}[Z])_{\mathfrak{p}''}, (B/\mathfrak{p}B)_\mathfrak{q}, L) = H_1(K[Z]_{\mathfrak{p}''}, (B/\mathfrak{p}B)_\mathfrak{q}, L) = (0),$$

hence the morphism $A[Z]_{\mathfrak{p}''}/\mathfrak{p}A[Z]_{\mathfrak{p}''} \to B_\mathfrak{q}/\mathfrak{p}B_\mathfrak{q}$ is formally smooth. Lemma 5.2.45 implies that $A[Z]_{\mathfrak{p}''} \to B_\mathfrak{q}$ is formally smooth. In particular $B_\mathfrak{q}/\mathfrak{p}''B_\mathfrak{q}$ is a regular local ring.

Therefore let $\{u_1, \ldots, u_l\} \subseteq \{b_1, \ldots, b_r\}$ be a set of elements whose image in $B_\mathfrak{q}/\mathfrak{p}''B_\mathfrak{q}$ is a regular system of parameters. Let $A' = A[Z_1, \ldots, Z_t, U_1, \ldots, U_l] := A[Z, U]$ and let $A' \to B$ be the morphism extending φ by mapping U_i to u_i, for $i = 1, \ldots, l$ and let $\mathfrak{p}' = \mathfrak{q} \cap A'$. We consider the morphisms $A_\mathfrak{p} \to A'_{\mathfrak{p}'} \to B_\mathfrak{q}$. The induced morphism of regular local rings $A'_{\mathfrak{p}'}/\mathfrak{p}A'_{\mathfrak{p}'} \to B_\mathfrak{q}/\mathfrak{p}B_\mathfrak{q}$ is flat by [249, Th. 23.1]. Applying Lemma 5.2.45 for the ideal $\mathfrak{p}A_\mathfrak{p}$, we deduce that the morphism $A'_{\mathfrak{p}'} \to B_\mathfrak{q}$ is flat. Since $\mathfrak{p}'B_\mathfrak{q} = \mathfrak{q}B_\mathfrak{q}$, by base change the morphism $\widetilde{A} = A'_{\mathfrak{p}'}/\mathfrak{p}''^n A'_{\mathfrak{p}'} \to \widetilde{B} = B_\mathfrak{q}/\mathfrak{q}^n B_\mathfrak{q}$ is flat; hence faithfully flat and consequently injective. Since the images of $y_1, \ldots, y_t$ and $u_1, \ldots, u_l$ in $\widetilde{B}$ are contained in D, we have an injective morphism $\widetilde{A}' \to D$. From $\mathfrak{p}'D \subseteq \mathfrak{m}_D$, where by $\mathfrak{m}_D$ we mean the maximal ideal of D, it follows that $\mathfrak{p}'\widetilde{B} \subseteq \mathfrak{m}_D\widetilde{B} = \mathfrak{q}B_\mathfrak{q}/\mathfrak{q}^n B_\mathfrak{q}$. But $\mathfrak{p}'\widetilde{B} = \mathfrak{q}B_\mathfrak{q}/\mathfrak{q}^n B_\mathfrak{q}$, whence $\mathfrak{p}'\widetilde{B} = \mathfrak{m}_D\widetilde{B}$. As $\widetilde{B}$ is faithfully flat over D, we obtain that $\mathfrak{p}'D = \mathfrak{m}_D$. Because $\widetilde{B}$ is faithfully flat over $\widetilde{A}'$ and over D, it results that D is faithfully flat over $\widetilde{A}'$. Let $\widetilde{\mathcal{D}}$ be the family of those $\widetilde{D} \in \mathcal{D}$ such that $D \subset \widetilde{D}$, and the residue field of $\widetilde{D}$ is separable over E. Then the properties (i), (ii) and (iii$_1$)–(iii$_4$) are clearly satisfied, and the property (iii$_5$) follows from Theorem 5.2.43,(iv). Let K' be the residue field of $\widetilde{A}'$. We have the exact sequence

$$\Omega_{K'/K} \otimes_{K'} E \xrightarrow{\theta} \Omega_{E/K} \to \Omega_{E/K'} \to 0.$$

Since $\Omega_{E/K}$ is generated by $\{d\bar{y}_1, \ldots, d\bar{y}_t\}$, we deduce that θ is surjective and then $\Omega_{E/K'} = 0$. Then by Cartier's equality [249, Th. 26.10], it results that E is separable over K', and hence H, the residue field of $\widetilde{D}$, is separable over K'. The local morphism of Noetherian local rings $\widetilde{A}' \to \widetilde{D}$ is flat, and its closed fibre $K' \subseteq H$ is a separable field extension, since $\mathfrak{p}'D = \mathfrak{m}_D$. Then $\widetilde{A}' \to \widetilde{D}$ is formally smooth and iii$_3$) is fulfilled too.

Let $\widetilde{D} \in \widetilde{\mathcal{D}}$ with residue field H and let $\tilde{x} \in \widetilde{B}$. By Theorem 5.2.43,v) there exists $\widetilde{D}'$ with $\widetilde{D} \subset \widetilde{D}'$ and residue field $M = H(\bar{x}^q)$, such that $\tilde{x}^q \in \widetilde{D}'$. Taking q large enough so that $\tilde{x}^q$ is separable over H, it follows that M is separable over H, so that $\widetilde{D}' \in \widetilde{\mathcal{D}}$. □

Lemma 5.2.47 *Let A be a Noetherian local ring and $0 \to M' \to M \to M'' \to 0$ be an exact sequence of finitely generated A-modules. Let $a_1, \ldots, a_s$ be elements of A that are a M'-regular and M''-regular sequence. Then $a_1, \ldots, a_s$ is also a M-regular sequence.*

Proof Consider the commutative diagram:

$$
\begin{array}{ccccccccc}
0 & \longrightarrow & M' & \longrightarrow & M & \longrightarrow & M'' & \longrightarrow & 0 \\
& & \downarrow a_1 & & \downarrow a_1 & & \downarrow a_1 & & \\
0 & \longrightarrow & M' & \longrightarrow & M & \longrightarrow & M'' & \longrightarrow & 0
\end{array}
$$

By the snake lemma, it follows that a_1 is M-regular, and we have an exact sequence $0 \to M'/a_1 M' \to M/a_1 M \to M''/a_1 M'' \to 0$. Now we repeat the argument. $\quad\square$

Lemma 5.2.48 *Let A be a Noetherian ring, M a finitely generated A-module and S a multiplicative system in A. Let a be an element of A which is square-regular on $S^{-1}M$. Then there exists an element $t \in S$ such that at^n is square-regular on M, for any $n > 0$.*

Proof We have

$$
S^{-1}\left(\frac{(0 : a^2)_M}{(0 : a)_M} \right) = \frac{(0 : a^2)_{S^{-1}M}}{(0 : a)_{S^{-1}M}} = (0).
$$

Then there exists $u \in S$ such that $u(0 : a^2)_M \subseteq (0 : a)_M$. Let n_0 be a large enough natural number such that $(0 : u^m)_M = (0 : u^{m+1})_M$ for any $m \geq n_0$ and let $t = u^{n_0}$ and $s = t^n$. Then $s(0 : a^2)_M \subseteq (0 : a)_M$ and $(0 : s)_M = (0 : s^2)_M$. Let $x \in (0 : (as)^2)_M$, whence $xa^2 s^2 = 0$. Then $xa^2 \in (0 : s^2)_M = (0 : s)_M$. It follows that $xs \in (0 : a^2)_M$; hence $xs^2 \in (0 : a)_M$ and then $ax \in (0 : s^2)_M = (0 : s)_M$. It results that $x \in (0 : as)_M$. $\quad\square$

Lemma 5.2.49 *Let A be a ring, $\mathfrak{p} \in \mathrm{Spec}(A)$ and $d_1, \ldots, d_r \in \mathfrak{p}$ a regular $A_{\mathfrak{p}}$-sequence. Let also $c > 0$ be a natural number. There exist $s_1, \ldots, s_r \in A \setminus \mathfrak{p}$ such that in the ring $A[T] = A[T_1, \ldots, T_r]$ we have for any i*

$$
\left(\left((d_1 s_1 T_1)^c, \ldots, (d_{i-1} s_{i-1} T_{i-1})^c \right) : d_i s_i T_i \right) =
$$

$$
= \left(\left((d_1 s_1 T_1)^c, \ldots, (d_{i-1} s_{i-1} T_{i-1})^c \right) : (d_i s_i T_i)^2 \right).
$$

Proof From [249, Th. 16.1], we know that for any i, the elements $d_1^c, \ldots, d_{i-1}^c, d_i$ form a regular sequence in $A_{\mathfrak{p}}$. Moreover $d_1 T_1$ is a regular element in $A_{\mathfrak{p}}[T]$. Now it follows easily that $d_1 T_1, \ldots, d_r T_r$ is a regular sequence in $A_{\mathfrak{p}}[T]$; hence again by [249, Th. 16.1] we obtain that $d_1^c T_1^c, \ldots, d_{i-1}^c T_{i-1}^c, d_i T_i$ is a regular sequence in $A_{\mathfrak{p}}[T]$. Let $S = A \setminus \mathfrak{p}$. If we have already obtained $s_1, \ldots, s_{i-1}$, then defining $M_i = A[T]/\left((d_1 s_1 T_1)^c, \ldots, (d_{i-1} s_{i-1} T_{i-1})^c \right)$, we have

$$
S^{-1}M_i = A_{\mathfrak{p}}[T]/\left((d_1 T_1)^c, \ldots, (d_{i-1} T_{i-1})^c \right)
$$

and then $d_i T_i$ is square-regular on $S^{-1} M_i$. The result follows by Lemma 5.2.48. $\square$

Now we can prove a result similar to Theorem 5.2.35 in the case of positive characteristic.

Theorem 5.2.50 *Let $f : A \to B$ be a morphism of Noetherian rings and C be a NP-factorization of f. Let $\mathfrak{q}$ be a prime ideal of B and $\mathfrak{p} = f^{-1}(\mathfrak{q})$. Assume that:*

(i) $\mathfrak{q} \in \mathrm{Min}(H_{C/A}B)$ and $\mathfrak{p} \in \mathrm{Min}(f^{-1}(H_{C/A}B))$.
(ii) The morphism $A_{\mathfrak{p}} \to B_{\mathfrak{q}}$ is formally smooth.
(iii) $\mathrm{char}(k(\mathfrak{p})) = p > 0$.

Then there exists a NP-factorization D of f, bigger than C, such that $H_{C/A}B \subseteq \sqrt{H_{D/A}B} \not\subseteq \mathfrak{q}$.

Proof Let $r = \mathrm{ht}(\mathfrak{q})$ and let $\mathfrak{n}_{B_{\mathfrak{q}}}$ be the nilradical of $B_{\mathfrak{q}}$. Since $\mathfrak{q} \in \mathrm{Min}(H_{C/A}B)$, there exists a power $N \in \mathbb{N}$, such that $\mathfrak{n}_{B_{\mathfrak{q}}}^N = 0$ and $\mathfrak{q}^N B_{\mathfrak{q}} \subseteq H_{C/A}B_{\mathfrak{q}}$. Let $c_1, \ldots, c_M$ be a set of generators of $H_{C/A}$, let

$$C[X, Z] := C\left[X_1, \ldots, X_r, Z_{11}, \ldots, Z_{1M}, \ldots, Z_{r1}, \ldots, Z_{rM}\right]$$

and let

$$C' = C[X, Z]/\left(X_i^{2N} - \sum_{j=1}^M c_j Z_{ij}, \ 1 \le i \le r\right).$$

Since for each $j = 1, \ldots, M$ we have

$$C'_{c_j} \cong C_{c_j}[X_1, \ldots, X_r, \{Z_{ih}, h \ne j\}],$$

it follows that $C_{c_j}[X_1, \ldots, X_r] \to C'_{c_j}$ is a smooth morphism. But $c_j \in H_{C/A}$; therefore $A[X] \to C'_{c_j}$ is smooth. It results that the image of each c_j in C' belongs to $H_{C'/A[X]}$; hence $H_{C/A}C' \subset H_{C'/A[X]}$. But $X_i^{2N} \in H_{C/A}C'$ for any $i = 1, \ldots, r$, so that it results that $X_i^{2N} \in H_{C'/A[X]}$. Let $\mathfrak{a}$ be the kernel of a finite free presentation of the $A[X]$-algebra C' and let $C'' = S_{C'}(\mathfrak{a}/\mathfrak{a}^2)$ be the symmetric algebra of the C'-module $\mathfrak{a}/\mathfrak{a}^2$. Let also $\rho : C'' \to C'$ be the augmentation morphism. As in Remark 5.2.30, we get that $H_{C'/A[X]}C'' \subset H_{C''/A[X]}$ and so $H_{C/A}C'' \subset H_{C''/A[X]}$. Moreover C'' has a presentation such that the image in C'' of any element of $H_{C'/A[X]}$ is standard. Let $e \in \mathbb{N}$ be such that the image of X_i^e in C'' is strictly standard over $A[X]$ for any $i = 1, \ldots, r$ and let $c = 8e$. Applying Proposition 5.2.46 for $n = N + rc$ and $G \subseteq \tilde{B} := B_{\mathfrak{q}}/\mathfrak{q}^n B_{\mathfrak{q}}$ containing the images of a set of generators of C over A, it follows that there exist a morphism

$f' : A' := A[Y_1, \ldots, Y_m] \to B$ and a family $\tilde{\mathcal{D}}$ of local subrings of $\tilde{B}$ satisfying the conditions (i)–(iv) from Proposition 5.2.46, that is, if we denote $\mathfrak{p}' = (f')^{-1}(\mathfrak{q})$ and $\tilde{A}' = A'_{\mathfrak{p}'}/(\mathfrak{p}')^n A'_{\mathfrak{p}'}$, we have:

(i) $\mathfrak{p}' B_\mathfrak{q} = \mathfrak{q} B_\mathfrak{q}$.

(ii) The induced morphism $A'_{\mathfrak{p}'} \to B_\mathfrak{q}$ is flat.

(iii) For any $\tilde{D} \in \tilde{\mathcal{D}}$ we have:

(iii_1) $\tilde{A}' \subset \tilde{D}$.

(iii_2) $G \subset \tilde{D}$.

(iii_3) $\tilde{A}' \to \tilde{D}$ is a smooth morphism essentially of finite type.

(iii_4) $\tilde{D} \to \tilde{B}$ is a flat morphism.

(iii_5) $\mathfrak{p}'\tilde{D}$ is the maximal ideal of $\tilde{D}$.

(iv) If $\tilde{D} \in \tilde{\mathcal{D}}$ and $\tilde{x} \in \tilde{B}$, there exist a power q of p and $\tilde{D}' \in \tilde{\mathcal{D}}$ such that $\tilde{D} \subset \tilde{D}'$ and $\tilde{x}^q \in \tilde{D}'$.

We note that by the choice of G the image of C in $\tilde{B}$ is contained in each subring $\tilde{D} \in \tilde{\mathcal{D}}$.

Claim 5.2.51 Let $d_1, \ldots, d_r \in \mathfrak{p}'$. Then there exist $\tilde{D} \in \tilde{\mathcal{D}}$ and a morphism $\theta : C' \to B$, such that $\theta(X_i) = \varepsilon_i d_i$, with $\varepsilon_i \in B \setminus \mathfrak{q}$ for $i = 1, \ldots, r$ and such that the image of C' in $\tilde{B}$ is contained in $\tilde{D}$.

Proof Let $\tilde{D} \in \tilde{\mathcal{D}}$. Since $\mathfrak{q}^N B_\mathfrak{q} \subseteq H_{C/A} B_\mathfrak{q}$, it follows that $d_i^N \in H_{C/A} B_\mathfrak{q}$ for all $i = 1, \ldots, r$. Thus, if $\tilde{d}_i$ is the image of d_i in $\tilde{B}$, we have $\tilde{d}_i^N \in H_{C/A}\tilde{B}$. As $\tilde{B}$ is faithfully flat over $\tilde{D}$, it results that $\tilde{d}_i^N \in H_{C/A}\tilde{B} \cap \tilde{D} = H_{C/A}\tilde{D}$; hence for any $i = 1, \ldots, r$ there exist $\tilde{z}'_{ij} \in \tilde{D}$, $j = 1, \ldots, M$, such that $\tilde{d}_i^N = \sum_{j=1}^{M} c_j \tilde{z}'_{ij}$. Let $z'_{ij} \in B_\mathfrak{q}$ be elements representing $\tilde{z}_{ij}'$. Then

$$d_i^N - \sum_{j=1}^{M} c_j z'_{ij} \in \mathfrak{q}^n B_\mathfrak{q} \subseteq \mathfrak{q}^{n-N} H_{C/A} B_\mathfrak{q}, \ i = 1, \ldots, r$$

hence

$$d_i^N = \sum_{j=1}^{M} c_j (z'_{ij} + z''_{ij}), \text{ with } z''_{ij} \in \mathfrak{q}^{n-N} B_\mathfrak{q}, \ i = 1, \ldots, r.$$

Since $\tilde{d}_i^N$, $\tilde{z}'_{ij} \in \tilde{D}$, the image of $d_i^N z'_{ij}$ in $\tilde{B}$ belongs to $\tilde{D}$ for each $i = 1, \ldots, r$ and $j = 1, \ldots, M$. Moreover,

$$d_i^N z''_{ij} \in \mathfrak{q}^N \mathfrak{q}^{n-N} B_{\mathfrak{q}} = \mathfrak{q}^n B_{\mathfrak{q}}, i = 1, \ldots, r \text{ and } j = 1, \ldots, M,$$

hence the image of $d_i^N z''_{ij}$ in $\tilde{B}$ is 0. Let $z_{ij} = d_i^N z'_{ij} + d_i^N z''_{ij}$. Then the image $\tilde{z}_{ij}$ of z_{ij} in $\tilde{B}$ belongs to $\tilde{D}$. Thus $d_i^{2N} = \sum_{j=1}^{M} c_j z_{ij}$, where the image $\tilde{z}_{ij}$ of z_{ij} in $\tilde{B}$ belongs to $\tilde{D}$.

Let $z_{ij} = \frac{w_{ij}}{s_i}$, where $w_{ij} \in B$ and $s_i \in B \setminus \mathfrak{q}$. Replacing, if necessary, s_i by a sufficiently large power and enlarging $\tilde{D}$, by Proposition 5.2.46,iv) we may assume that $\tilde{s}_i \in \tilde{D}$ and so $\tilde{w}_{ij} = \tilde{z}_{ij} \tilde{s}_i \in \tilde{D}$. Let $t_1, \ldots, t_r \in B \setminus \mathfrak{q}$ be such that

$$t_i\left((s_i d_i)^{2N} - \sum_{j=1}^{M} c_j s_i^{2N-1} w_{ij}\right) = 0, \ i = 1, \ldots, r.$$

Replacing, if necessary, t_i by a sufficiently large power and enlarging $\tilde{D}$, we can assume that $\tilde{t}_1, \ldots, \tilde{t}_r \in \tilde{D}$. Set $\varepsilon_i = s_i t_i$, for $i = 1, \ldots, r$. Then the morphism

$$\theta : C' \to B,$$

$$\theta(X_i) = d_i \varepsilon_i, \ i = 1, \ldots, r$$

$$\theta(Z_{ij}) = s_i^{2N-1} t_i^{2N} w_{ij} \ i = 1, \ldots, r, \ j = 1, \ldots M$$

is well-defined, and its image in $\tilde{B}$ is contained in $\tilde{D}$. Thus Claim 5.2.51 is proved.   □

We continue the proof of Theorem 5.2.50. Using the proof of Theorem 5.2.35, case 2), it is clear that we may assume that $\mathrm{ht}(\mathfrak{p}) = 0$. The morphism $A_{\mathfrak{p}} \to A'_{\mathfrak{p}'}$ is formally smooth; therefore $A'_{\mathfrak{p}'}/\mathfrak{p}A'_{\mathfrak{p}'}$ is a regular local ring. Since $\dim(A_{\mathfrak{p}}) = 0$, it results that $A'_{\mathfrak{p}'}$ is Cohen-Macaulay. Let $d_1, \ldots, d_s \in \mathfrak{p}'$ be such that their classes in $A'_{\mathfrak{p}'}/\mathfrak{p}A'_{\mathfrak{p}'}$ form a regular system of parameters. Since $\mathfrak{p}A'_{\mathfrak{p}'}$ is nilpotent, we get that $d_1, \ldots, d_s$ is a system of parameters of $A'_{\mathfrak{p}'}$; hence a regular sequence in $A'_{\mathfrak{p}'}$. Because $A'_{\mathfrak{p}'} \to B_{\mathfrak{q}}$ is flat, the images of the elements $d_1, \ldots, d_s$ in $B_{\mathfrak{q}}$ form a regular sequence. We have $\dim(A'_{\mathfrak{p}'}) = s \leq \dim(B_{\mathfrak{q}})$. Since $d_1, \ldots, d_s$ is a system of parameters in $A'_{\mathfrak{p}'}$, there exists a power $t \geq 1$ such that

$$(\mathfrak{p}')^t A'_{\mathfrak{p}'} \subseteq (d_1, \ldots, d_s) A'_{\mathfrak{p}'}.$$

By Proposition 5.2.46,i) we get that

$$\mathfrak{q}^t B_{\mathfrak{q}} \subseteq (d_1, \ldots, d_s) B_{\mathfrak{q}},$$

whence $r = \dim(B_q) = s$. Now we replace $d_1, \ldots, d_r$ by $d_1 s_1, \ldots, d_r s_r$ with $s_1, \ldots, s_r \in A' \setminus \mathfrak{p}$ obtained by applying Lemma 5.2.49. Keeping the notations $d_1, \ldots, d_r$ for these elements, we may assume that

$$\left((d_1 T_1)^c, \ldots, (d_{i-1} T_{i-1})^c\right) : d_i T_i\right) = \left((d_1 T_1)^c, \ldots, (d_{i-1} T_{i-1})^c\right) : (d_i T_i)^2\right).$$

Claim 5.2.52 We can enlarge $\tilde{D} \in \mathcal{D}$ in such a way that there exist elements $\delta_1, \ldots, \delta_r \in B \setminus \mathfrak{q}$ with:

(i) $\tilde{\delta}_1, \ldots, \tilde{\delta}_r \in \tilde{D}$.

(ii) There exists a morphism $\xi : C' \to B$, $\xi(X_i) = d_i \delta_i$, $i = 1, \ldots, r$, such that the image of C' in $\tilde{B}$ is contained in $\tilde{D}$.

(iii) For all $i = 1, \ldots, r$ we have in B

$$\left((d_1 \delta_1)^c, \ldots, (d_{i-1} \delta_{i-1})^c\right) : d_i \delta_i\right)_B = \left((d_1 \delta_1)^c, \ldots, (d_{i-1} \delta_{i-1})^c\right) : (d_i \delta_i)^2\right)_B.$$

Proof Let $\varepsilon_1, \ldots, \varepsilon_r \in B \setminus \mathfrak{q}$ be the elements obtained by applying Claim 5.2.51. Assume that we already found elements $\eta_1, \ldots, \eta_{i-1} \in B \setminus \mathfrak{q}$ such that $\delta_j = \varepsilon_j \eta_j$ for $j < i$ satisfying (iii) and that, applying Proposition 5.2.46,iv), we have enlarged $\tilde{D}$ in such a way that $\tilde{\eta}_1, \ldots, \tilde{\eta}_{i-1} \in \tilde{D}$. Since $d_1 \varepsilon_1, \ldots, d_r \varepsilon_r$ is a regular sequence in B_q, by Lemma 5.2.48 it follows that there exist elements $\eta_1, \ldots, \eta_r \in B \setminus \mathfrak{q}$ such that $d_i \varepsilon_i \eta_i^q$ satisfies (iii) for all $q > 0$. Applying again Proposition 5.2.46,iv), we can enlarge $\tilde{D}$ such that $\tilde{\eta}_i^q \in \tilde{D}$ for some q. Now we replace η_i by η_i^q. By the above proof of Claim 5.2.51, we obtain a map

$$\mu : C' \to B,$$

$$\mu(X_i) = d_i \varepsilon_i, \ \mu(Z_{ij}) = \zeta_{ij},$$

$$i = 1, \ldots, r, \ j = 1, \ldots, M$$

for some $\zeta_{ij} \in B$. Then we can define

$$\xi : C' \to B$$

such that

$$\xi(X_i) = d_i \varepsilon_i \eta_i, \ \xi(Z_{ij}) = \eta_i^{2N} \zeta_{ij}$$

$$i = 1, \ldots, r, \ j = 1, \ldots, M$$

and this concludes the proof of 5.2.52. $\square$

Claim 5.2.53 We have $q^n B_q \subseteq (d_1^c, \ldots, d_r^c) B_q$ and $(\mathfrak{p}')^n A'_{\mathfrak{p}'} \subseteq (d_1^c, \ldots, d_r^c) A'_{\mathfrak{p}'}$.

Proof We have chosen the elements $d_1, \ldots, d_r$ such that

$$(d_1, \ldots, d_r) + \mathfrak{p} A'_{\mathfrak{p}'} = \mathfrak{p}' A'_{\mathfrak{p}'}.$$

But $\mathfrak{p}' A'_{\mathfrak{p}'}$ is nilpotent; hence

$$q B_q = \mathfrak{p}' B_q = (d_1, \ldots, d_r) B_q + \eta_{B_q}.$$

Because $\eta_{B_q}^N = 0$, it results that

$$q^n B_q = \sum_{i=0}^{N-1} (d_1, \ldots, d_r)^{n-i} (\eta_{B_q})^i \subseteq (d_1, \ldots, d_r)^{n-(N-1)} B_q =$$

$$= (d_1, \ldots, d_r)^{rc+1} B_q \subseteq (d_1^c, \ldots, d_r^c) B_q.$$

Since $A'_{\mathfrak{p}'} \to B_q$ is faithfully flat, the second relation follows from the first one by tensorizing with B_q over $A'_{\mathfrak{p}'}$.   □

We continue the proof of Theorem 5.2.50. Let $A'' := A[T] = A[T_1, \ldots, T_r]$, and define the morphism

$$h : A[X] = A[X_1, \ldots, X_r] \to A'',$$

$$h(X_i) = d_i T_i, \ i = 1, \ldots, r.$$

Let $f'' : A'' \to B$ be the map extending $f' : A' \to B$ by $f''(T_i) = \delta_i, \ i = 1, \ldots, r$. Let $\Gamma := A'' \otimes_{A[X]} C''$. We define $\alpha : \Gamma \to B$ by $\alpha(a'' \otimes b) = f''(a'') \rho(\xi(b))$. We have

$$H_{C''/A[X]} \Gamma \subseteq H_{\Gamma/A''} \text{ and } H_{C/A} C'' \subseteq H_{C''/A[X]},$$

hence

$$H_{C/A} \Gamma \subseteq H_{\Gamma/A''}.$$

Assume that q can be resolved with respect to Γ, that is, there exists a NP-factorization E of $A'' \to B$, bigger than Γ, such that $H_{\Gamma/A''} B \subseteq \sqrt{H_{E/A''} B} \not\subseteq q$. Since A'' is smooth over A, it results that $H_{E/A''} \subseteq H_{E/A}$; hence

$$H_{C/A}B \subseteq H_{\Gamma/A''}B \subseteq \sqrt{H_{E/A''}B} \subseteq \sqrt{H_{E/A}B} \nsubseteq \mathfrak{q}.$$

This means that $\mathfrak{q}$ can be resolved, by the same E, also with respect to C. Thus it suffices to resolve $\mathfrak{q}$ with respect to Γ. We apply Proposition 5.2.27 to the elements $a_i := d_i T_i$, $i = 1, \dots, r$. The assumption on the colon ideals holds in A'' as shown in Lemma 5.2.49, and by Claim 5.2.52 it holds also in B. By the choice of e the image of X_i^e in C'' is strictly standard over $A[X]$. By Lemma 5.2.11 the element a_i^e, which is the image of X_i^e in Γ, is strictly standard over A''. Let $\mathfrak{a} = \big((d_1 T_1)^c, \dots, (d_r T_r)^c\big)$. Then by Proposition 5.2.27 it is enough to show that $\mathfrak{q}/\mathfrak{a}B$ is resolvable with respect to $\Gamma/\mathfrak{a}\Gamma$. Note that the images of T_i in $B_\mathfrak{q}$ are invertible, hence by Claim 5.2.53 we have $\mathrm{ht}(\mathfrak{q}/\mathfrak{a}B) = 0$. Then by Proposition 5.2.33 and Proposition 5.2.34, it is enough to show that $\mathfrak{q}B_\mathfrak{q}/\mathfrak{a}B_\mathfrak{q}$ is resolvable with respect to the NP-factorization $\Gamma/\mathfrak{a}\Gamma$ of $A''/\mathfrak{a} \to B_\mathfrak{q}/\mathfrak{a}B_\mathfrak{q}$. It is enough to find a NP-factorization E' of $A''/\mathfrak{a} \to B_\mathfrak{q}/\mathfrak{a}B_\mathfrak{q}$, that is, E' is a smooth finitely generated $A''/\mathfrak{a}$-algebra. Set $\mathfrak{p}'' = (f'')^{-1}(\mathfrak{q})$ and $R = A'' \setminus \mathfrak{p}''$. By Lemma 5.2.29 it is enough to show that there exists a NP-factorization E of $R^{-1}(A''/\mathfrak{a}) \to B_\mathfrak{q}/\mathfrak{a}B_\mathfrak{q}$, bigger than $R^{-1}(\Gamma/\mathfrak{a}\Gamma)$. Thus E must be smooth and finitely generated over $R^{-1}(A''/\mathfrak{a})$.

Let $\mathfrak{b} = (d_1^c, \dots, d_r^c) \subset A'$. Since $T_1, \dots, T_r \in R$, we have $\mathfrak{a}R^{-1}A'' = \mathfrak{b}R^{-1}A''$. Therefore we can rewrite

$$R^{-1}(A''/\mathfrak{a}) \to R^{-1}(\Gamma/\mathfrak{a}\Gamma) \to E \to B_\mathfrak{q}/\mathfrak{a}B_\mathfrak{q}$$

as

$$R^{-1}(A''/\mathfrak{b}A'') \to R^{-1}(\Gamma/\mathfrak{b}\Gamma) \to E \to B_\mathfrak{q}/\mathfrak{b}B_\mathfrak{q}.$$

Set $S = A' \setminus \mathfrak{p}' \subseteq R$. By flatness it is enough to show that there exists a NP-factorization F of the morphism $S^{-1}(A''/\mathfrak{b}A'') \to B_\mathfrak{q}/\mathfrak{b}B_\mathfrak{q}$, bigger than $S^{-1}(\Gamma/\mathfrak{b}\Gamma)$, that is,

$$S^{-1}(A''/\mathfrak{b}A'') \to S^{-1}(\Gamma/\mathfrak{b}\Gamma) \to F \to B_\mathfrak{q}/\mathfrak{b}B_\mathfrak{q}$$

such that F is a smooth finitely generated $S^{-1}(A''/\mathfrak{b}A'')$-algebra. Actually this means that we have to show that there exists a NP-factorization F of the morphism $A'_{\mathfrak{p}'}[T]/\mathfrak{b}A'_{\mathfrak{p}'}[T] \to B_\mathfrak{q}/\mathfrak{b}B_\mathfrak{q}$ bigger than $A'_{\mathfrak{p}'}[T]/\mathfrak{b}A'_{\mathfrak{p}'}[T] \otimes_{A'[T]} \Gamma$ that is,

$$A'_{\mathfrak{p}'}[T]/\mathfrak{b}A'_{\mathfrak{p}'}[T] \to A'_{\mathfrak{p}'}[T]/\mathfrak{b}A'_{\mathfrak{p}'}[T] \otimes_{A'[T]} \Gamma \to F \to B_\mathfrak{q}/\mathfrak{b}B_\mathfrak{q}$$

such that F is smooth and finitely generated over $A'_{\mathfrak{p}'}[T]/\mathfrak{b}A'_{\mathfrak{p}'}[T]$. But by Claim 5.2.53 we know that $(\mathfrak{p}')^n A'_{\mathfrak{p}'}[T] \subseteq \mathfrak{b}A'_{\mathfrak{p}'}[T]$; hence we may replace $\mathfrak{b}$ by $(\mathfrak{p}')^n$. Thus we have to show that there exists a NP-factorization F of the morphism $A'_{\mathfrak{p}'}[T]/(\mathfrak{p}')^n A'_{\mathfrak{p}'}[T] \to$

$B_q/(\mathfrak{p}')^n B_q$ bigger than $A'_{\mathfrak{p}'}[T]/(\mathfrak{p}')^n A'_{\mathfrak{p}'}[T] \otimes_{A'[T]} \Gamma$, that is,

$$A'_{\mathfrak{p}'}[T]/(\mathfrak{p}')^n A'_{\mathfrak{p}'}[T] \to A'_{\mathfrak{p}'}[T]/(\mathfrak{p}')^n A'_{\mathfrak{p}'}[T] \otimes_{A'[T]} \Gamma \to F \to B_q/(\mathfrak{p}')^n B_q$$

such that $A'_{\mathfrak{p}'}[T]/(\mathfrak{p}')^n A'_{\mathfrak{p}'}[T] \to F$ is smooth and finitely generated, that is,

$$\widetilde{A}'[T] \to \widetilde{A}' \otimes_{A'[T]} \Gamma \to F \to \widetilde{B}$$

or what is the same

$$\widetilde{A}'[T] \to \widetilde{A}' \otimes_{A'[T]} C'' \to F \to \widetilde{B},$$

such that F is smooth and finitely generated over $\widetilde{A}'[T]$.

The images of $\widetilde{A}'[T]$ and C'' in $\widetilde{B}$ are contained in $\widetilde{D}$ by Claim 5.2.52. By Proposition 5.2.46 it results that the morphism $\widetilde{A}'[T] \to \widetilde{D}[T]$ is smooth and essentially of finite type. So applying Lemma 5.2.29, it is enough to show that the morphism $\widetilde{A}'[T] \otimes_{A[X]} C'' \to \widetilde{D}$ factors through the morphism

$$\pi : \widetilde{D}[T] \to \widetilde{D}$$

given by

$$\pi(T_i) = \tilde{\delta}_i, \ i = 1, \ldots, r.$$

Let $w : C' = C[X_1, \ldots, X_r, \{Z_{ij}\}]/(\{X_i^{2N} - \sum_j c_j Z_{ij} \xi_i\}) \to B$ be the morphism given by Claim 5.2.52, $w(X_i) = d_i \delta_i, \ i = 1, \ldots, r$, and let ζ_{ij} be the image of Z_{ij} in B. The images $\tilde{\zeta}_{ij}$ of these elements in $\widetilde{B}$ are contained in $\widetilde{D}$ by Claim 5.2.52. The morphism $A[X] \to A'' = A'[T]$ maps X_i to $d_i T_i, \ i = 1, \ldots, r$ so that we have an $A[X]$-algebra morphism $C[X] \to \widetilde{D}[T]$ mapping X_i to $\tilde{d}_i T_i$, for $i = 1, \ldots, r$. The elements $\tilde{d}_1, \ldots, \tilde{d}_r$ are invertible in $\widetilde{B}$, and the maximal ideal of $\widetilde{D}$ generates the maximal ideal of $\widetilde{B}$, whence $\tilde{d}_1, \ldots, \tilde{d}_r$ are invertible in $\widetilde{D}$. Therefore we can define a morphism of $A[X]$-algebras

$$u : C' \to \widetilde{D}[T], \ u(Z_{ij}) = (\frac{T_i}{\tilde{d}_i})^{2N} \tilde{\zeta}_{ij}.$$

This morphism makes the following diagram commute:

$$
\begin{array}{ccc}
C' & \longrightarrow & \widetilde{D} \\
\downarrow{\scriptstyle u} & \nearrow & \\
\widetilde{D}[T] &
\end{array}
$$

Composing with the augmentation morphism $C'' \to C'$, we obtain a morphism $C'' \to \widetilde{D}[T]$, which induces a morphism

$$\widetilde{A}'[T] \otimes_{A[X]} C'' \to \widetilde{D}[T].$$

This concludes the proof of Theorem 5.2.50.

Theorem 5.2.54 *Let* $f : A \to B$ *be a morphism of Noetherian rings and* C *a NP-factorization of* f. *Let* $\mathfrak{q} \in \mathrm{Min}(H_{C/A}B)$ *and* $\mathfrak{p} = f^{-1}(\mathfrak{q})$. *Assume that* $\mathfrak{p} \in \mathrm{Min}(f^{-1}(H_{C/A}B))$ *and that the morphism* $A_{\mathfrak{p}} \to B_{\mathfrak{q}}$ *is formally smooth. Then* $\mathfrak{q}$ *is resolvable with respect to* C.

Proof Follows from Theorem 5.2.35 and Theorem 5.2.50. □

Theorem 5.2.55 (Popescu [289–291]) *Let* $f : A \to B$ *be a regular morphism of Noetherian rings. Assume that we have a NP-factorization* C *of* f *and let* $g : C \to B$ *be the structural morphism. Then there exists a NP-factorization* D *of* g, *bigger than* C, *such that* $A \to D$ *is smooth.*

Proof Since B is Noetherian, we can choose a NP-factorization E of f bigger than C, such that $\sqrt{H_{E/A}B}$ is maximal. Applying Lemma 5.2.31,ii) it is enough to show that $H_{E/A}B = B$. Assume that this is not true and let $\mathfrak{q} \in \mathrm{Min}(H_{E/A}B)$. It is enough to obtain a NP-factorization D of f, bigger than E, such that

$$H_{E/A}B \subseteq \sqrt{H_{D/A}B} \not\subseteq \mathfrak{q}.$$

Let $\mathfrak{a} = f^{-1}(H_{E/A}B)$ and $\mathfrak{p} \in \mathrm{Min}(\mathfrak{a})$. Then we can choose $\mathfrak{q} \in \mathrm{Min}(H_{E/A}B)$ such that $\mathfrak{p} = f^{-1}(\mathfrak{q})$. Indeed, if

$$\sqrt{H_{E/A}B} = \mathfrak{q}_1 \cap \ldots \cap \mathfrak{q}_n,$$

then

$$\sqrt{\mathfrak{a}} = f^{-1}(\mathfrak{q}_1) \cap \ldots \cap f^{-1}(\mathfrak{q}_n) \subseteq \mathfrak{p}$$

and therefore for some i we have $f^{-1}(\mathfrak{q}_i) \subseteq \mathfrak{p}$. Now we apply Theorem 5.2.54. □

We are now able to state and even to prove the main result of the section and one of the most important results of the book:

Theorem 5.2.56 (Popescu [290, Th. 1.8]) *Let* $f : A \to B$ *be a regular morphism of Noetherian rings. Then* f *is a filtered inductive limit of smooth morphisms of finite type.*

Proof Any A-algebra is a direct limit of its A-subalgebras of finite type; hence we have anyway $f = \varinjlim_{i \in I} \phi_i$, where $\phi_i : A \to C_i$ is a morphism of finite type for any $i \in I$, so that

for any $i \in I$ there is a factorization $A \xrightarrow{\phi_i} C_i \xrightarrow{\psi_i} B$. Then by Theorem 5.2.55 there exists a NP-factorization D_i of ψ_i, where D_i is smooth over A.

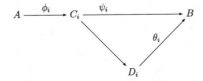

Therefore we obtain a surjective morphism $\theta : \varinjlim_{i \in I} D_i \to B$. Since D_i is by assumption an A-algebra of finite type, there exists some $j \geq i$ such that the morphism $D_i \to B$ factors as $D_i \to C_j \to B$. Let $x \in \varinjlim_{i \in I} D_i$ be an element such that $\theta(x) = 0$. Let $i \in I$ be such that $x \in D_i$ and let $k \geq j$ be such that x maps to 0 in C_k. It follows that x maps to 0 in D_k; hence $x = 0$. Thus θ is an isomorphism. $\square$

5.3   Excellent Henselian Rings

This section is dedicated to the characterization of local excellent Henselian rings as exactly the rings with Artin approximation. One direction is an easy consequence of the structure theorem 5.2.56 of regular morphisms.

Theorem 5.3.1 (Popescu [290, Th. 1.3]) *Let A be an excellent Henselian local ring. Then A has the approximation property.*

Proof Let $F = (F_1, \ldots, F_r)$ be a system of polynomial equations in the variables $Z = (Z_1, \ldots, Z_s)$ over A and let $\hat{z} \in \hat{A}^n$ be a solution of F. Consider the morphism $v : B := A[Z]/(F) \to \hat{A}$ given by $v(Z) = \hat{z}$. Then B is a NP-factorization of the regular morphism $u : A \to \hat{A}$. Now Theorem 5.2.55 tells us that there exists a bigger smooth standard NP-factorization of u, say $C = (A[Y]/(H))_g$, with $g \in \Delta_F$. Thus we have a commutative diagram

$$A \longrightarrow B \xrightarrow{\ v\ } \hat{A}$$

Then $\hat{y} = w(Y)$ is a solution of F such that $g(\hat{y}) = w(g) \notin m\hat{A}$. Therefore $\Delta_H(\hat{y}) \not\subseteq m\hat{A}$. By Lemma 5.1.7 it results that there exists a solution $y \in A^n$ of H such that

$y \equiv \widehat{y}(\mathrm{mod}\,\mathfrak{m}\widehat{A})$; hence $g(y) \equiv g(\widehat{y})(\mathrm{mod}\,\mathfrak{m}\widehat{A})$. Thus we obtain an A-algebra morphism $\rho : C \to A$ such that $\rho(Y) = y$. It follows that $z = \rho(q(Z))$ is a solution of F in A. $\square$

In order to prove the converse of Theorem 5.3.1, we need several preparations. We start with an easy lemma.

Lemma 5.3.2 *Let $u : (A, \mathfrak{m}) \to (B, \mathfrak{n})$ be a flat local morphism of Noetherian local rings. Assume that B is a regular local ring and that $\mathfrak{m}B = \mathfrak{n}$. Let $n = \dim(A)$ and let $\mathfrak{b}$ be an ideal of B of height $n - k$. Then there exist elements $a_1, \ldots, a_k \in \mathfrak{m}$ such that $\mathfrak{b} + (a_1, \ldots, a_k)B$ is an $\mathfrak{n}$-primary ideal.*

Proof By assumption $\dim(B) = n$. Since B is a regular local ring, it follows that $\dim(B/\mathfrak{b}) = k$. Now the assertion follows by induction on k, the case $k = 0$ being obvious.

$\square$

The next notion is a special case of a very well-known one [100, Ch. I, §5, pag. 147].

Definition 5.3.3 Let $(B, \mathfrak{n})$ be a Noetherian local ring. If $a \neq 0$ is an element of B, we put $v(a) := \min\{k \in \mathbb{N} \mid a \in \mathfrak{n}^k \text{ and } a \notin \mathfrak{n}^{k+1}\}$. We denote by a^* the image of a in $\mathfrak{n}^{v(a)}/\mathfrak{n}^{v(a)+1}$ and call it the **initial form** of a. In this way we obtain a map

$$B \setminus \{0\} \to \mathrm{gr}_{\mathfrak{n}}(B) = \overset{\infty}{\underset{i=0}{\oplus}} \mathfrak{n}^i/\mathfrak{n}^{i+1},$$

$$a \mapsto a^*.$$

If $\mathfrak{b}$ is an ideal of B, we set $\mathfrak{b}^* = (a^* \mid a \in \mathfrak{b})$ and call it the **initial ideal** of $\mathfrak{b}$.

In the following we consider a local flat morphism of Noetherian regular local rings $u : (A, \mathfrak{m}, K) \to (B, \mathfrak{n}, L)$ such that $\mathfrak{m}B = \mathfrak{n}$. In particular this means that $\dim(A) = \dim(B)$.

Let $\{x_1, \ldots, x_n\}$ be a regular system of parameters of A. It follows that $\{x_1, \ldots, x_n\}$ is also a regular system of parameters in B. We denote by Γ_k the set of monomials of degree k in $x_1, \ldots, x_n$. It is well-known that Γ_k has $r_k := \binom{n+k-1}{k}$ elements; therefore, in order to simplify notations, we may write

$$\Gamma_k := \{x_1^{e_1} \ldots x_n^{e_n} \mid \sum_{i=1}^{n} e_i = k\} = \{m_j^k \mid 1 \le j \le r_k\},$$

where we understand that m_j^k is a monomial of degree k. We do not consider any ordering on Γ_k. It is well-known that Γ_k is a basis of the K-vector space $\mathfrak{m}^k/\mathfrak{m}^{k+1}$, and in our special situation that $\{x_1, \ldots, x_n\}$ is also a regular system of parameters in B, it is also a basis of the L-vector space $\mathfrak{n}^k/\mathfrak{n}^{k+1}$.

Consider now $\mathfrak{q} \subset \mathfrak{p}$ prime ideals of B such that $(B/\mathfrak{q})_\mathfrak{p} = B_\mathfrak{p}/\mathfrak{q}B_\mathfrak{p}$ is not a regular local ring. Let $S = \{q_1, \ldots, q_m\} \subseteq \mathfrak{q}$ be a finite set that is a system of generators of $\mathfrak{q}$ and such that $\{q_1^*, \ldots, q_m^*\}$ is a system of generators of the ideal $\mathfrak{q}^*$ of the ring $\mathrm{gr}_\mathfrak{n}(B)$. One should note that we cannot pick any system of generators of $\mathfrak{q}$, but there are systems of generators with this property. Let $t := \max\{v(q_i) \mid 1 \le i \le m\}$ and let $k \le t$. We can arrange the elements of S as follows:

$$S = \{q_1, \ldots, q_s, q_{s+1}, \ldots, q_u, q_{u+1}, \ldots, q_m\}$$

where

$$v(q_1) = \ldots = v(q_s) = k,$$

$$v(q_i) < k, \ i = s+1, \ldots, u,$$

$$v(q_j) > k, \ j = u+1, \ldots, m.$$

We consider the elements

$$q_1^*, \ldots, q_s^*, \{q_{s+1}^* m_j^{k-v(q_{s+1})}\}_{1 \le j \le r_{k-v(q_{s+1})}}, \ldots, \{q_u^* m_j^{k-v(q_u)}\}_{1 \le j \le r_{k-v(q_u)}}.$$

These elements generate a subspace of the L-vector space $\mathfrak{n}^k/\mathfrak{n}^{k+1}$ that will be denoted by $W(k)$.

Remark 5.3.4

(i) Summarizing the above considerations, it follows that for every $k \le t$ there exists a finite set of elements of B

$$\Pi_k := \left\{ q_1, \ldots, q_s, \{q_{s+1} m_j^{k-v(q_{s+1})}\}_{1 \le j \le r_{k-v(q_{s+1})}}, \ldots, \{q_u m_j^{k-v(q_u)}\}_{1 \le j \le r_{k-v(q_u)}} \right\},$$

such that $v(q_1) = \ldots = v(q_s) = k, v(q_{s+1}) < k, \ldots, v(q_u) < k$ and such that the set of their initial forms, that is,

$$\Pi_k^* := \left\{ q_1^*, \ldots, q_s^*, \{q_{s+1}^* m_j^{k-v(q_{s+1})}\}_{1 \le j \le r_{k-v(q_{s+1})}}, \ldots, \{q_u^* m_j^{k-v(q_u)}\}_{1 \le j \le r_{k-v(q_u)}} \right\},$$

is a basis of the L-vector space $W(k)$.

(ii) For every $k \le t$, we pick a finite subset $\Theta_k \subseteq \Gamma_k$ such that $\Pi_k^* \cup \Theta_k$ is a basis of $\mathfrak{n}^k/\mathfrak{n}^{k+1}$. Then $|\Pi_k| + |\Theta_k| = \dim_L(\mathfrak{n}^k/\mathfrak{n}^{k+1})$, where by $|X|$ we mean the cardinality of the set X. We may assume, for simplicity of the computations, that $\Theta_k = \{m^k_{j_1}, \ldots, m^k_{j_h}\}$.

(iii) For any $i = 1, \ldots, m$, we have relations describing the degrees of the generators of the ideal q

$$q_i = \sum_{j=1}^{r_{v(q_i)}} f_{ij} \cdot m_j^{v(q_i)}, \text{ with } f_{ij} \in B. \tag{5.3}$$

(iv) Since Π_k is a basis of $W(k)$, for any $k \le t$ and for any monomial $m^k_j \in \Gamma_k \setminus \Theta_k$, that is, for any $j \in \{1, \ldots, r_k\} \setminus \{j_1, \ldots, j_h\}$, we have a relation

$$m^k_j = \sum_{i=1}^{s} b_{kji} q_i + \sum_{i=s+1}^{u} \sum_{l=1}^{r_{k-v(q_l)}} c_{kil} q_i \cdot m_l^{k-v(q_l)} +$$

$$+ \sum_{l=1}^{h} d_{kl} \cdot m^k_{j_l} + \sum_{j=1}^{r_{k+1}} e_{kj} \cdot m^{k+1}_j, \tag{5.4}$$

where the coefficients b_{kji}, c_{kil}, d_{kl} and e_{kj} are elements of B.

We consider the sets of variables corresponding to the relations (5.3) and (5.4), that is:

$$\{Q_i, \ 1 \le i \le m\},$$

$$\{F_{ij}, \ 1 \le i \le m, 1 \le j \le r_{v(q_i)}\},$$

$$\{B_{kji}, \ 1 \le k \le t, 1 \le j \le r_k, j \ne j_1, \ldots, j \ne j_h, 1 \le i \le s\},$$

$$\{C_{kil}, \ s+1 \le i \le n, 1 \le k \le t, 1 \le l \le r_{k-v(q_i)}\},$$

$$\{D_{kl}, \ 1 \le k \le t, 1 \le l \le h\},$$

$$\{E_{kj}, \ 1 \le k \le t, 1 \le j \le r_{k+1}\}.$$

We define

$$A_1 = A[Q_i, F_{ij}, B_{kji}, C_{kil}, D_{kl}, E_{kj}]$$

where for simplicity we did not mention the sets where the indices of the variables are running, but we understand that these sets are exactly those in the above description of the variables. We also define

$$B_1 = A_1/\mathfrak{a} = A[Q_i, F_{ij}, B_{kji}, C_{kil}, D_{kl}, E_{kj}]/\mathfrak{a},$$

where $\mathfrak{a}$ is the ideal of A_1 described below

$$\mathfrak{a} = \Big(Q_i - \sum_{j=1}^{r_{\nu(q_i)}} F_{ij} \cdot m_j^{\nu(q_i)}, \ m_j^k - \sum_{i=1}^{s} B_{kji} Q_i + \sum_{i=s+1}^{u} \sum_{l=1}^{r_{k-\nu(q_l)}} C_{kil} Q_i \cdot m_l^{k-\nu(q_l)} + $$

$$+ \sum_{l=1}^{k} D_{kl} \cdot m_{jl}^k + \sum_{j=1}^{r_{k+1}} E_{kj} \cdot m_j^{k+1} \Big).$$

Proposition 5.3.5 *Let* $\varphi : B_1 \to B$ *be an* A-*algebra morphism such that* $\varphi(Q_i) \in \mathfrak{q}$ *for all* $i = 1, \ldots, m$. *Then* $\{\varphi(Q_1), \ldots, \varphi(Q_m)\}$ *is a system of generators of* $\mathfrak{q}$.

Proof Since the elements $\varphi(Q_1), \ldots, \varphi(Q_m)$ satisfy the relations (5.3) for some elements $\varphi(F_{ij}) \in B$, it results that for any $i = 1, \ldots, m$ we have the inequality $\deg(\varphi(Q_i)) \geq \deg(q_i)$. From the relations (5.4), it follows that the initial forms of the elements $\varphi(Q_i)$ generate $W(k)$ for $k \leq t$, as a L-vector space. Let $\tilde{\mathfrak{q}}$ be the ideal generated by $\{\varphi(Q_1), \ldots, \varphi(Q_m)\}$. Then $\tilde{\mathfrak{q}} \subseteq \mathfrak{q}$ and $\tilde{\mathfrak{q}}^* = \mathfrak{q}^*$, hence $\tilde{\mathfrak{q}} = \mathfrak{q}$.   □

Remark 5.3.6

(i) Since $\mathfrak{q} \subseteq \mathfrak{p}$ we can enlarge $\{q_1, \ldots, q_m\}$ to a finite system of generators of $\mathfrak{p}$, say $\Omega = \{q_1, \ldots, q_m, q_{m+1}, \ldots, q_t\}$. By enlarging Ω if necessary, we may assume that there are elements $q_{i_1}, \ldots, q_{i_l} \in \Omega$ such that:

$$\text{there exists } s \in B \setminus \mathfrak{p} \text{ such that } s \cdot \mathfrak{p} \subseteq (q_{i_1}, \ldots, q_{i_l})B; \tag{5.5}$$

$$\mathrm{ht}(q_{i_1}, \ldots, q_{i_l})B = l. \tag{5.6}$$

(ii) Because B is a regular local ring of dimension n, the relation (5.6) is equivalent to the fact that there exist elements $L_{1,1}, \ldots, L_{1,n-l} \in B$ such that the ideal $(q_{i_1}, \ldots, q_{i_l}, L_{1,1}, \ldots, L_{1,n-l})$ is $\mathfrak{n}$-primary. By Lemma 5.3.2 we may assume that $L_{1,1}, \ldots, L_{1,n-l} \in \mathfrak{m}$. There exists $k \in \mathbb{N}$ such that

$$\mathfrak{n}^k = (x_1, \ldots, x_n)^k \subseteq (q_{i_1}, \ldots, q_{i_l}, L_{1,1}, \ldots, L_{1,n-l}).$$

We consider the following relations in B :

$$v_j = q_{i_j}, \ j = 1, \ldots, l. \tag{5.7}$$

$$s \cdot q_j = \sum_{i=1}^{l} g_{ji} v_i, \ 1 \le j \le t; \tag{5.8}$$

$$m_j^k = \sum_{i=1}^{l} q_{j_i} v_i + \sum_{s=1}^{n-l} h_{js} L_{1,s}, \ 1 \le j \le r_k. \tag{5.9}$$

As above we consider the sets of variables corresponding to the relations (5.8) and (5.9), that is:

$$\{Q_i, 1 \le i \le t\}, S,$$

$$\{V_j, 1 \le j \le l\},$$

$$\{G_{ji}, 1 \le i \le l, 1 \le j \le t\},$$

$$\{H_{js}, 1 \le j \le r_k, 1 \le s \le n-l\}.$$

Keeping the same convention about the indices as in the definition of A_1, we define

$$A_2 := A_1[V_j, Q_i, S, G_{ji}, H_{js}]$$

and

$$B_2 = A_2/\mathfrak{b},$$

where $\mathfrak{b}$ is the ideal of A_2 generated by

$$\mathfrak{b} = (\mathfrak{a}A_2, V_j - Q_{i_j}, SQ_j - \sum_{i=1}^{l} G_{ji} Q_{ji}, m_j^k - \sum_{i=1}^{l} Q_{ji} V_i + \sum_{j=1}^{n-l} H_{js} L_{1,j}).$$

Proposition 5.3.7 *Let* $\varphi : B_2 \to B$ *be an A-algebra morphism such that* $\varphi(S) \notin (\varphi(Q_{i_1}), \ldots, \varphi(Q_{i_l}))$. *Then there exists* $\mathfrak{p}_\varphi \in \mathrm{Min}(\varphi(Q_1), \ldots, \varphi(Q_t))$ *such that* $\mathrm{ht}(\mathfrak{p}_\varphi) = \mathrm{ht}(\mathfrak{p}) = l$.

Proof From the relations (5.9), it follows that

$$\mathfrak{n}^k \subseteq (\varphi(V_1), \ldots, \varphi(V_l), L_{1,1}, \ldots, L_{1,n-l})B),$$

hence $\varphi(V_1), \ldots, \varphi(V_l)$ is a part of a system of parameters of B and

$$\mathrm{ht}(\varphi(V_1), \ldots, \varphi(V_l))B = l.$$

In particular, since B is a regular local ring, $(\varphi(V_1), \ldots, \varphi(V_l))$ is an unmixed ideal of B, that is, all the associated primes of $(\varphi(V_1), \ldots, \varphi(V_l))$ have the same height. Let $\mathfrak{c}$ be the ideal generated by $\{\varphi(Q_1), \ldots, \varphi(Q_t)\}$. Then $\varphi(S)$ does not belong to the ideal $(\varphi(V_1), \ldots, \varphi(V_l))$; hence there exists a primary component $\mathfrak{r}$ of $(\varphi(V_1), \ldots, \varphi(V_l))$ such that $\varphi(S) \notin \mathfrak{r}$. By the relations (5.8) we have $\varphi(S)\mathfrak{c} \subseteq \mathfrak{r}$; hence there is $h \in \mathbb{N}$ such that $\mathfrak{c}^h \subseteq \mathfrak{r}$. Let $\mathfrak{p}_\varphi = \sqrt{\mathfrak{r}}$. Then $\mathfrak{c} \subseteq \mathfrak{p}_\varphi$. Since $(\varphi(V_1), \ldots, \varphi(V_l))$ is unmixed, it results that $\mathrm{ht}(\mathfrak{p}_\varphi) = \mathrm{ht}(\varphi(V_1), \ldots, \varphi(V_l))$. In particular, since $\varphi(V_1), \ldots, \varphi(V_l) \in \mathfrak{c}$ and $\mathfrak{p}_\varphi \in \mathrm{Min}(\varphi(V_1), \ldots, \varphi(V_l))$, we obtain that $\mathfrak{p}_\varphi \in \mathrm{Min}(\mathfrak{c})$. $\square$

Notation 5.3.8 In the following we denote by S_m the symmetric group on m elements.

Remark 5.3.9

(i) Since $B_\mathfrak{p}/\mathfrak{q}B_\mathfrak{p}$ is not a regular local ring, the ideal $\mathfrak{q}B_\mathfrak{p}$ is not generated by a part of a regular system of parameters of $B_\mathfrak{p}$. This means that the elements $\frac{q_1}{1}, \ldots, \frac{q_m}{1}$ are not linearly independent in $\mathfrak{p}B_\mathfrak{p}/\mathfrak{p}^2B_\mathfrak{p}$. Thus there exists an index s such that $\frac{q_{s+1}}{1} \in (\frac{q_1}{1}, \ldots, \frac{q_s}{1})B_\mathfrak{p}$, whence there exists an element $\theta \in B \setminus \mathfrak{p}$, with $\theta \cdot q_{s+1} \in \mathfrak{p}^2 + (q_1, \ldots, q_s)$.

(ii) Let us now pick $\pi \in S_m$. We use the same argument as in (i), which is actually an example for the case $\pi = 1$. We can consider that the system of generators of the ideal $\mathfrak{q}$ is $\mathfrak{q} = (q_{\pi(1)}, \ldots, q_{\pi(m)})$, and we repeat the reasoning above. Since $B_\mathfrak{p}/\mathfrak{q}B_\mathfrak{p}$ is not a regular ring, the ideal $\mathfrak{q}B_\mathfrak{p}$ is not generated by a part of a regular system of parameters of $B_\mathfrak{p}$. This means that the elements $\frac{q_{\pi(1)}}{1}, \ldots, \frac{q_{\pi(m)}}{1}$ are not linearly independent in $\mathfrak{p}B_\mathfrak{p}/\mathfrak{p}^2B_\mathfrak{p}$. Thus there exists $k_\pi \in \mathbb{N} \cup \{0\}$, such that $k_\pi < r = \mathrm{ht}(\mathfrak{q})$, and there exists $\theta_\pi \in B \setminus \mathfrak{p}$ such that $\theta_\pi \cdot q_{\pi(k_\pi+1)} \in \mathfrak{p}^2 + (q_{\pi(1)}, \ldots, q_{\pi(k_\pi)})$. Now let us define $\theta := \prod_{\pi \in S_m} \theta_\pi \in B \setminus \mathfrak{p}$. Then for all $\pi \in S_m$, we have the following relations in B :

$$\theta \cdot q_{\pi(k_\pi+1)} = \sum_{i,j=1}^{t} g_{\pi ij} q_i q_j = \sum_{h=1}^{k_\pi} \lambda_{\pi h} \cdot q_{\pi(h)}. \tag{5.10}$$

As ht($\mathfrak{p}$) $= l$ and $\theta \notin \mathfrak{p}$, we have ht($\mathfrak{p} + \theta B$) $= l + 1$. Therefore there are $n - l - 1$ elements $L_{2,1}, \ldots, L_{2,n-l-1} \in \mathfrak{m}$ such that there exists $s \in \mathbb{N}$ with

$$\mathfrak{n}^s \subseteq \mathfrak{p} + \theta B + (L_{2,1}, \ldots, L_{2,n-l-1})B.$$

This gives for every $m_j^s \in \Gamma_s$ the following relations in B :

$$m_j^s = \sum_{h=1}^{t} o_{jh} \cdot q_h + r_j \cdot \theta + \sum_{i=1}^{n-l-1} u_{ji} L_{2,i}. \qquad (5.11)$$

As above we consider the sets of variables corresponding to the relations (5.10) and (5.11), that is:

$$\{Q_i, \ 1 \le i \le t\}, T,$$

$$\{V_i, \ 1 \le i \le l\},$$

$$\{G_{\pi ij}, \ 1 \le i, j \le t, \pi \in S_m\},$$

$$\{\Lambda_{\pi h}, \ \pi \in S_m, \ 1 \le h \le k_\pi\},$$

$$\{O_{jh}, \ 1 \le j \le r_k, 1 \le h \le t\},$$

$$\{R_j, \ 1 \le j \le r_k\},$$

$$\{U_{ji}, \ 1 \le i \le n - l - 1, 1 \le j \le r_k\}.$$

We define

$$A_3 = A_2[Q_i, T, V_i, G_{\pi ij}, \Lambda_{\pi h}, O_{jh}, R_j, U_{ji}]$$

and

$$B_3 = A_3/\mathfrak{c},$$

where $\mathfrak{c}$ is the ideal of A_3 generated by

$$\mathfrak{c} = \left(\mathfrak{b}A_3,\, T\,Q_{\pi(k_\pi+1)} - \sum_{i,j=1}^{t} G_{\pi ij}\, Q_i\, Q_j,\, T\,Q_{\pi(k_\pi+1)} - \sum_{h=1}^{k_\pi} \Lambda_{\pi h}\, Q_{\pi(h)},\, m_j^s - \right.$$

$$\left. - \sum_{h=1}^{t} O_{jh}\, Q_h - R_j\, T - \sum_{i=1}^{n-l-1} U_{ji}\, L_{2,i}\right).$$

Proposition 5.3.10 *Let* $\varphi : B_3 \to B$ *be an A-algebra morphism such that:*

(i) $\varphi(Q_i) \in \mathfrak{q},\ 1 \le i \le m.$
(ii) $\varphi(S) \notin (\varphi(V_1), \ldots, \varphi(V_l))B.$

Then there exists $\mathfrak{p}_\varphi \in \mathrm{Min}\big((\varphi(Q_1), \ldots, \varphi(Q_t))\big) \cap V(\mathfrak{q})$ *such that* $\mathrm{ht}(\mathfrak{p}_\varphi) = l$ *and* $(B/\mathfrak{q})_{\mathfrak{p}_\varphi}$ *is not a regular local ring.*

Proof Let $\mathfrak{d} = \big(\varphi(Q_1), \ldots, \varphi(Q_t)\big)$. From Proposition 5.3.5 and from the assumption that $\varphi(Q_1), \ldots, \varphi(Q_m) \in \mathfrak{q}$, we get that $\mathfrak{q} = \big(\varphi(Q_1), \ldots, \varphi(Q_m)\big)$; hence $\mathfrak{q} \subseteq \mathfrak{d}$. Since $\varphi(S) \notin \big(\varphi(V_1), \ldots, \varphi(V_l)\big)B$, by Proposition 5.3.7 it results that there exists $\mathfrak{p}_\varphi \in \mathrm{Min}(\mathfrak{d})$ such that $\mathrm{ht}(\mathfrak{p}_\varphi) = l$. It remains to show that $(B/\mathfrak{q})_{\mathfrak{p}_\varphi} = B_{\mathfrak{p}_\varphi}/\mathfrak{q}B_{\mathfrak{p}_\varphi}$ is not a regular local ring. By the relations (5.11) we have

$$\mathfrak{n}^s \subseteq (\mathfrak{p}_\varphi, \varphi(T)) + (L_{2,1}, \ldots, L_{2,n-l-1})B,$$

hence $\mathrm{ht}(\mathfrak{p}_\varphi, \varphi(T)) \ge l + 1$. Since $\mathrm{ht}(\mathfrak{p}_\varphi) = l$, it results that $\varphi(T) \notin \mathfrak{p}_\varphi$. By the relations (5.10), for every $\pi \in S_m$ there exists $k(\pi) < r$ such that

$$\varphi(T)\varphi(Q_{\pi(k(\pi)+1)}) \in \mathfrak{p}_\varphi^2 + \big(\varphi(Q_{\pi(1)}), \ldots, \varphi(Q_{\pi(k(\pi))})\big)B.$$

This means that for every $\pi \in S_m$, $\{\varphi(Q_{\pi(1)}), \ldots, \varphi(Q_{\pi(r)})\}$ is not a part of a regular system of parameters of $B_{\mathfrak{p}_\varphi}$. Since $\mathrm{ht}(\mathfrak{q}) = r$, this means that the elements $\{\varphi(Q_1), \ldots, \varphi(Q_m)\}$, which are a system of generators of $\mathfrak{q}$, do not contain a part of a regular system of parameters of $B_{\mathfrak{p}_\varphi}$; hence $(B/\mathfrak{q})_{\mathfrak{p}_\varphi}$ is not a regular local ring. □

Remark 5.3.11 Assume that $\mathrm{Reg}(B/\mathfrak{q})$ is open in $\mathrm{Spec}(B/\mathfrak{q})$. Then $\mathrm{Sing}(B/\mathfrak{q})$ has finitely many minimal elements. Suppose that $\mathfrak{p} \in \mathrm{Min}\big(\mathrm{Sing}(B/\mathfrak{q})\big)$, and assume also that moreover $\mathfrak{p}$ has not minimal height among all primes in $\mathrm{Min}\big(\mathrm{Sing}(B/\mathfrak{q})\big)$. Let $\mathfrak{r}_1, \ldots, \mathfrak{r}_r \in \mathrm{Spec}(B)$ be the traces in B of the minimal prime ideals of $\mathrm{Sing}(B/\mathfrak{q})$ such that $\mathrm{ht}(\mathfrak{r}_i) < l = \mathrm{ht}(\mathfrak{p})$, and take an element $c_0 \in \bigcap_{i=1}^{r} \mathfrak{r}_i \setminus \mathfrak{p}$. Then $\mathrm{ht}(\mathfrak{p} + c_0 B) = l + 1$; hence there exist

$L_{3,1}, \ldots, L_{3,n-l-1} \in \mathfrak{m}$ such that

$$\mathfrak{n}^p \subseteq \mathfrak{p} + c_0 B + (L_{3,1}, \ldots, L_{3,n-l-1}) B.$$

Then for any $m_r^p \in \Gamma_p$, we get a relation

$$m_r^p = \sum_{j=1}^{t} v_{rj} q_j + w_r c_0 + \sum_{i=1}^{n-l-1} z_{ri} L_{3,i}, \ 1 \le r \le r_p. \tag{5.12}$$

As we did before, we consider the sets of variables corresponding to the relations (5.12), that is:

$$\{V_{rj}, \ 1 \le j \le r_p\}, \ C,$$

$$\{W_r, \ 1 \le r \le r_p\},$$

$$\{Z_{ri}, \ 1 \le r \le r_p, \ 1 \le i \le n - l - 1\}.$$

Let

$$A_4 = A_3[V_{rj}, W_r, C, Z_{ri}]/c_4,$$

where

$$c_4 = c_3 A_4 + \left(m_r^p - \sum_{j=1}^{l} V_{rj} Q_j + W_r C + \sum_{i=1}^{n-l-1} Z_{ri} L_{3,i} \right).$$

Proposition 5.3.12 *Let $\varphi : B_4 \to B$ be an A-algebra morphism satisfying:*

(a) $\varphi(Q_i) \in \mathfrak{q}$, for any $i \in \{1, \ldots, m\}$.
(b) $\varphi(S) \notin \big(\varphi(V_1), \ldots, \varphi(V_l) \big) B$.
(c) $\varphi(C) \equiv c_0 (\bmod \mathfrak{q})$.

Then there exists a prime ideal $\mathfrak{p}_\varphi \in \mathrm{Min}\big((\varphi(Q_1), \ldots, \varphi(Q_l)) \big)$ such that:

(i) $\mathfrak{q} \subset \mathfrak{p}_\varphi$.
(ii) $\mathrm{ht}(\mathfrak{p}_\varphi) = l$.
(iii) $(B/\mathfrak{q})_{\mathfrak{p}_\varphi}$ is not a regular local ring.
(iv) $\mathfrak{r}_i \not\subset \mathfrak{p}_\varphi$, for any $i = 1, \ldots, r$.

Proof *(i)*, *(ii)* and *(iii)* follow clearly from *(a)* and *(b)* as it was shown in Proposition 5.3.10. In order to show *(iv)*, let $\mathfrak{s}$ be the ideal generated by $\{\varphi(Q_1), \ldots, \varphi(Q_t)\}$, and suppose that there exists some $i \in \{1, \ldots, r\}$ such that $\mathfrak{r}_i \subseteq \mathfrak{p}_\varphi$. Since $c_0 \notin \mathfrak{r}_i$ and $\mathfrak{q} \subset \mathfrak{r}_i$, by *c)* we have that $\varphi(C) \in \mathfrak{p}_\varphi$. On the other hand, from the relations (5.11), we obtain that $\mathrm{ht}(\mathfrak{s}, \varphi(C)) \geq l + 1$. But this contradicts the fact that $\mathrm{ht}(\mathfrak{p}_\varphi) = l$ and $(\mathfrak{s}, \varphi(C)) \subseteq \mathfrak{p}_\varphi$.   □

Using the previous preparations, we are now able to prove the main result:

Theorem 5.3.13 (Rotthaus [331, Th. 1]) *Let A be a Noetherian local ring with Artin approximation property. Then A is excellent.*

Proof Because A is universally catenary by Corollary 5.1.14, we have only to show that A has geometrically regular formal fibres, and therefore by Theorem 1.6.23, it is enough to prove that for any finite A-algebra B which is a domain and for any $\mathfrak{p} \in \mathrm{Spec}(\widehat{B})$ such that $\mathfrak{p} \cap B = (0)$, the ring $\widehat{B}_\mathfrak{p}$ is a regular local ring. Since by Proposition 5.1.3 and 5.1.11 we know that B is a Henselian ring, it is a local domain with Artin approximation property. Replacing A by B, we may assume that $(A, \mathfrak{m}, K)$ is a local domain with Artin approximation property, and we have to show that $\mathfrak{p} \cap A \neq (0)$ for any prime ideal $\mathfrak{p} \in \mathrm{Sing}(\widehat{A})$. Suppose that there is a prime ideal $\mathfrak{p}$ in $\widehat{A}$ such that $\widehat{A}_\mathfrak{p}$ is not a regular ring and $\mathfrak{p} \cap A = (0)$. Then there exists such a prime ideal $\mathfrak{p} \in \mathrm{Min}(\mathrm{Sing}(\widehat{A}))$, with $\mathrm{ht}(\mathfrak{p}) \leq \mathrm{ht}(\mathfrak{q})$ for any prime ideal $\mathfrak{q} \in \mathrm{Sing}(\widehat{A})$, such that $\mathfrak{q} \cap A = (0)$. Set

$$P := \begin{cases} \text{the prime field of } A, & \text{if } A \text{ contains a field} \\ \\ \mathbb{Z}_{(p)}, \ p \text{ prime}, & \text{if } A \text{ doesn't contain a field and } \mathbb{Z}_{(p)} \subseteq A. \end{cases}$$

Then $\widehat{A} = W[[X_1, \ldots, X_m]]/\mathfrak{q}$, where W is a field or a p-Cohen ring [131, Th. 19.8.8]. Let $x_1, \ldots, x_m$ be a system of generators of $\mathfrak{m}$. If $u : A \to \widehat{A}$ is the canonical completion morphism, we may assume that u maps the elements $x_1, \ldots, x_m$ to the classes modulo $\mathfrak{q}$ of the elements $X_1, \ldots, X_m$, respectively. Thus we obtain a commutative diagram

$$\begin{array}{ccc} C := P[X_1, \ldots, X_m] & \xrightarrow{\psi} & D := W[[X_1, \ldots, X_m]] \\ \sigma \downarrow & & \downarrow \rho \\ A & \xrightarrow{u} & \widehat{A} = W[[X_1, \ldots, X_m]]/\mathfrak{q} \end{array}$$

where $\sigma(X_i) = x_i$ and $\rho(X_i) = X_i \pmod{\mathfrak{q}}$. Because the morphism $P \to W$ is a regular one, it follows that the morphism ψ is regular; hence by Theorem 5.2.56 it results that ψ is a filtered inductive limit of smooth morphisms of finite type. Remember that $\mathfrak{p}$ is minimal in $\mathrm{Sing}(\widehat{A})$ and that $\mathrm{ht}(\mathfrak{p}) \leq \mathrm{ht}(\mathfrak{q})$ for any prime ideal $\mathfrak{q} \in \mathrm{Sing}(\widehat{A})$ with $\mathfrak{q} \cap A = (0)$. We need to consider two cases:

Case 1: $\mathfrak{p}$ *is an ideal of minimal height in* $\mathrm{Sing}(\widehat{A})$, that is, $\mathrm{ht}(\mathfrak{p}) \leq \mathrm{ht}(\mathfrak{q})$, for any prime ideal $\mathfrak{q} \in \mathrm{Sing}(\widehat{A})$. Then we denote by E the finitely generated C-algebra B_3 constructed in Proposition 5.3.10.

Case 2: $\mathfrak{p}$ *is not of minimal height in* $\mathrm{Sing}(\widehat{A})$. Let $\mathfrak{r}_1, \ldots, \mathfrak{r}_r$ be the prime ideals in $\mathrm{Min}(\mathrm{Sing}(\widehat{A}))$ such that $\mathrm{ht}(\mathfrak{r}_i) < \mathrm{ht}(\mathfrak{p})$, for $i = 1, \ldots, r$. Then we know that $\mathfrak{r}_i \cap A \neq (0)$ for $i = 1, \ldots, r$. Because A is a domain, we can pick $c \in \bigcap_{i=1}^{r} (\mathfrak{r}_i \cap A)$, $c \neq 0$ and let c_0 be a preimage of c in C. Since $\mathfrak{p} \cap A = (0)$, it results that $c_0 \notin \mathfrak{p}$. Let us now denote by E the finitely generated C-algebra B_4 from Proposition 5.3.12.

We continue in both cases with E defined above. Then we have a canonical C-algebra morphism $\tau : E \to D$. Because ψ is a filtered inductive limit of smooth morphisms of finite type, the morphism τ factors as $E \to G \xrightarrow{\varphi} D$, where G is a smooth C-algebra of finite type.

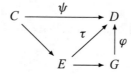

Let $F = G \otimes_C A$. Then φ induces an A-algebra morphism $v_0 : F \to \widehat{A}$ satisfying the conditions

$$v_0(Q_i) = 0, 1 \leq i \leq m \text{ and } v_0(C) = c, \ c \in A. \tag{5.13}$$

Hence, if we denote $M := F/(Q_1, \ldots, Q_m, C - c)$, the morphism v_0 factors through M as follows:

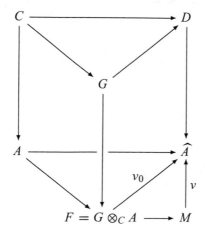

For every $n \in \mathbb{N}$, we pick elements $s_n, q_{jn} \in A$, for $m + 1 \leq j \leq t$ such that

$$v(S) - s_n \in \mathfrak{m}_{\widehat{A}}^n \tag{5.14}$$

and

$$v(Q_j) - q_{jn} \in \mathfrak{m}_{\widehat{A}}^n, \text{ for all } j = m + 1, \ldots, t. \tag{5.15}$$

Since A has AP, by Lemma 5.1.4 it follows that for every $n \in \mathbb{N}$ there exists an A-algebra morphism $v_n : M \to A$ such that

$$v_n(S) - s_n \in \mathfrak{m}_A^n \tag{5.16}$$

and

$$v_n(Q_j) - q_{jn} \in \mathfrak{m}_A^n, \text{ for all } j = m + 1, \ldots, t. \tag{5.17}$$

Moreover, by the construction of M, we get:

$$\cdot \; v_n(Q_j) = 0, \text{ for any } j = 1, \ldots, m \tag{5.18}$$

and

$$v_n(C) = c. \tag{5.19}$$

Clearly every v_n induces a C-algebra morphism $\psi_n : G \to A \subseteq \widehat{A} = D/\mathfrak{q}$. Since G is smooth over C and D is complete, by [131, Prop. 19.3.10] every ψ_n lifts to a C-algebra morphism $\phi_n : G \to D$, and we have a commutative diagram

$$
\begin{array}{ccc}
G & \xrightarrow{\psi_n} & A \subseteq \widehat{A} = D/\mathfrak{q} \\
{\scriptstyle \phi_n} \downarrow & \nearrow & \\
D & &
\end{array}
$$

with the properties

$$\phi_n(Q_j) \in \mathfrak{q}, \text{ for all } j = 1, \ldots, m \tag{5.20}$$

and

$$\phi_n(C) \equiv c_0 \pmod{\mathfrak{q}}. \tag{5.21}$$

Suppose that $\phi_n(S) \in \big(\phi_n(V_1), \ldots, \phi_n(V_l)\big)$ for infinitely many $n \in \mathbb{N}$. Then $\psi_n(S) \in \big(\psi_n(V_1), \ldots, \psi_n(V_l)\big)$, so that for infinitely many $n \in \mathbb{N}$ we have

$$\nu_n(S) \in \big(\nu_n(V_1), \ldots, \nu_n(V_l)\big) \subseteq \big(\nu_n(Q_1), \ldots, \nu(Q_t)\big) \subseteq \widehat{A}.$$

By the relations (5.17), we have $\nu_n(S) \in \mathfrak{p} + \mathfrak{m}_{\widehat{A}}^n \subseteq \widehat{A}$, and hence by the relations (5.16), we get that $\nu(S) \in \mathfrak{p} + \mathfrak{m}_{\widehat{A}}^n \subseteq \widehat{A}$ for infinitely many $n \in \mathbb{N}$; therefore $\nu(S) \in \mathfrak{p} \subseteq \widehat{A}$. But by the construction of ν resp. ϕ, it holds that $\phi(S) \notin \mathfrak{p} \subseteq D$. Contradiction! Therefore for sufficiently large $n \in \mathbb{N}$, we have $\phi_n(S) \notin \big(\phi_n(V_1), \ldots, \phi_n(V_l)\big)$.

By Proposition 5.3.10 in Case 1 and by Proposition 5.3.12 in Case 2, we get that for $n \geq n_0$ there exists $\mathfrak{p}_n \in \mathrm{Min}\big(\phi_n(Q_1), \ldots, \phi_n(Q_t)\big)$ such that $\mathfrak{q} \subseteq \mathfrak{p}_n$, $\mathrm{ht}(\mathfrak{p}_n) = l$ and $D_{\mathfrak{p}_n}/\mathfrak{q}D_{\mathfrak{p}_n}$ is not a regular local ring. Moreover, in Case 2, that is, if $\mathfrak{p}$ is not of minimal height in $\mathrm{Sing}(\widehat{A})$, we also have that $\mathfrak{p}_n \not\supseteq \mathfrak{r}_i$, for $i = 1, \ldots, r$, where we denoted also by $\mathfrak{r}_i$ the trace of $\mathfrak{r}_i$ in D. Let $\tilde{\mathfrak{p}}_n = \mathfrak{p}_n/\mathfrak{q}$ be the image of $\mathfrak{p}_n$ in $\widehat{A}$. Obviously $\tilde{\mathfrak{p}}_n \in \mathrm{Sing}(\widehat{A})$ and $\tilde{\mathfrak{p}}_n \cap A \neq (0)$, since $\tilde{\mathfrak{p}}_n \in \mathrm{Min}\big((\nu_n(Q_1), \ldots \nu(Q_l))\big)$. Because $\mathrm{ht}(\tilde{\mathfrak{p}}_n) = \mathrm{ht}(\mathfrak{p})$ and because $\tilde{\mathfrak{p}}_n \not\supseteq \mathfrak{r}_i$ for $i = 1, \ldots, r$, it follows that $\tilde{\mathfrak{p}}_n \in \mathrm{Min}(\mathrm{Sing}(\widehat{A}))$. But $\mathrm{Min}(\mathrm{Sing}(\widehat{A}))$ is a finite set, whence $\tilde{\mathfrak{p}}_n = \tilde{\mathfrak{p}}_{n+k}$ for infinitely many $k \in \mathbb{N}$. On the other hand, from the relations (5.17), we have $\mathfrak{p} \subseteq \tilde{\mathfrak{p}}_n + \mathfrak{m}^n$; hence $\mathfrak{p} \subseteq \tilde{\mathfrak{p}}_n$. But $\mathrm{ht}(\mathfrak{p}) = \mathrm{ht}(\tilde{\mathfrak{p}}_n)$, hence $\mathfrak{p} = \tilde{\mathfrak{p}}_n$. This contradicts the fact that $\mathfrak{p} \cap A = (0)$.   □

Corollary 5.3.14 *Let A be a Noetherian local ring. Then A has Artin approximation property if and only if A is a Henselian excellent ring.*

Proof Apply Theorem 5.3.1 and Th. 5.3.13.   □

Corollary 5.3.15 *Let $(A, \mathfrak{m})$ be a Noetherian local ring with Artin approximation property, $\mathfrak{a}$ an ideal of A and A^* the completion of A in the $\mathfrak{a}$-adic topology. Then A^* has the Artin approximation property.*

Proof By Theorem 5.3.13 A is an excellent Henselian local ring. By Theorem 4.3.6 and Corollary 3.1.62, also A^* is an excellent Henselian ring. Then Theorem 5.3.1 shows that A has Artin approximation property.   □

Further Results on Classes of Good Rings

This chapter is dedicated to miscellaneous results connected with the topics of the book. Mainly this refers to morphisms whose fibres are complete intersections and to the behaviour of integral closure to base change. The Evolution problem of Mazur and Eisenbud-Mazur conjecture are described.

6.1   Structure of Local Homomorphisms: Cohen Factorizations

A fundamental and very useful result in Commutative Algebra is the so-called Cohen's theorem on the structure of complete local rings. This was proved in a paper published in 1946 by the famous American mathematician I. S. Cohen (1917–1955) and says, roughly speaking, that a complete local ring may be presented as an homorphic image of a complete regular local ring, or more precisely of a formal power series over a complete discrete valuation ring [57, ch. IX, §2, n. 5] or [131, Th. 19.8.8]. A recent generalization of Cohen's theorem was given by Gabber [171, Exp. IV]. The proof is quite intricate, but for an elementary proof in the equicharacteristic $p > 0$ case, one can see [204]. We shall not deal with this generalization. This section contains a relativization of Cohen's structure theorem, proved by Avramov, Foxby and B. Herzog in a paper published in 1994 [31].

Definition 6.1.1 Let $u : (A, \mathfrak{m}, k) \to (B, \mathfrak{n}, K)$ be a morphism of local rings. We say that u is **factorizable**, if there are morphisms of local rings $f : A \to C$ and $g : C \to B$ such that $u = g \circ f$, g is surjective, f is flat and $C/\mathfrak{m}C$ is a regular local ring.

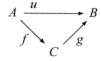

C. Ionescu, *Classes of Good Noetherian Rings*, Frontiers in Mathematics,
https://doi.org/10.1007/978-3-031-22292-4_6

Such a decomposition will be called a **regular factorization** of u and will be shortly denoted by $A \xrightarrow{f} C \xrightarrow{g} B$. A regular factorization of u is called a **Cohen factorization**, if C is a complete local ring.

Remark 6.1.2 If $u : A \to B$ has a Cohen factorization, then B is a complete local ring.

Example 6.1.3

(i) Any local morphism essentially of finite type is factorizable.
(ii) Let $(A, \mathfrak{m}, k)$ be a Noetherian local ring and let $\mathrm{char}(k) := p \geq 0$. Then there is a local morphism $\eta_A : \mathbb{Z}_{(p)} \to A$ obtained by localization from the canonical morphism $\mathbb{Z} \to A$. Cohen's structure theorem [131, Th. 19.8.8] says that there exists a complete discrete valuation ring (D, pD, k) and a local ring morphism $\theta : D \to A$ inducing the identity on the residue fields. Moreover, if $y_1, \ldots, y_m$ is a set of generators of $\mathfrak{m}$, then θ can be uniquely extended to a surjective morphism

$$\theta' : D[[Y]] = D[[Y_1, \ldots, Y_m]] \to A,$$

$$\theta'(Y_i) = y_i, \; i = 1, \ldots, m.$$

Thus Cohen's structure theorem can be expressed as follows in terms of Cohen factorizations: Let A be a complete local ring. Then the local structure morphism $\eta_A : \mathbb{Z}_{(p)} \to A$ has a Cohen factorization.

Definition 6.1.4 If A is a complete local ring, a **coefficient morphism** for A is a local morphism $\theta : D \to A$, where (D, pD, k) is a complete discrete valuation ring, inducing the identity on the residue fields.

The following result extends Cohen's structure theorem to any local morphism to a complete local ring. We must notice that if $u : A \to B$ is a local morphism of Noetherian local rings, then there is a morphism $u' : A \to \widehat{B}$ induced by u, namely, the composition of u with the canonical completion morphism $\tilde{u} : B \to \widehat{B}$, where $\widehat{B}$ is the completion of B in the topology given by the maximal ideal. The morphism u' will be called the **semi-completion** of u.

Remark 6.1.5 If $u : A \to B$ is a local morphisms of Noetherian local rings, by a **Cohen factorization** of u we will mean a Cohen factorization of the semi-completion $u' : A \to \widehat{B}$.

With these preparations we can state the result.

Theorem 6.1.6 (Avramov, Foxby and B. Herzog [31, Th. 1.1]) *Let* $u : (A, \mathfrak{m}, k) \to$ *(B, $\mathfrak{n}$, K) be a local morphisms of Noetherian local rings. Then the semi-completion morphism* $u' : A \to \widehat{B}$ *has a Cohen factorization.*

Proof Let $\widehat{u} : \widehat{A} \to \widehat{B}$ be the completion of u and assume that $\widehat{u}$ has a Cohen factorization $\widehat{u} = g \circ f$.

$$
\begin{array}{ccc}
A & \xrightarrow{\ u'\ } & \widehat{B} \\
\phi \downarrow & \widehat{u} \ \ \nearrow & \uparrow g \\
\widehat{A} & \xrightarrow{\ f\ } & \widehat{C}
\end{array}
$$

Then obviously $u = g \circ (f \circ \phi)$ is a Cohen factorization of u'. In other words, we may assume that A and B are complete local rings. Let $y_1, \ldots, y_n$ be a system of generators of $\mathfrak{n}$. Consider the commutative diagram of local rings and morphisms:

$$
\begin{array}{ccc}
\mathbb{Z}_{(p)} & \xrightarrow{\ \eta_{D[[Y]]}\ } & D[[Y]] \\
\eta_C \downarrow & \xi' \ \nearrow & \downarrow \sigma' \\
C & \xrightarrow{v'} A \xrightarrow{u} & B,
\end{array}
$$

where C is a Cohen ring for A, $v' : C \to A$ is a coefficient morphism for A and

$$
\sigma' : D[[Y]] = D[[Y_1, \ldots, Y_n]] \to B,
$$

$$
\sigma'(Y_i) = y_i, \ i = 1, \ldots, n.
$$

is a surjective morphism given by Cohen's structure theorem. From [131, Th. 19.8.6, i)] we know that there exists a morphism $\xi' : C \to D[[Y]]$ preserving the commutativity of the diagram. Let $x_1, \ldots, x_m$ be a system of generators of $\mathfrak{m}$ and let

$$
v : C[[X]] = C[[X_1, \ldots, X_m]] \to A,
$$

defined by

$$
v|_C = v', \ v(X_i) = x_i, \ i = 1, \ldots, m.
$$

Furthermore we can extend σ' to a morphism

$$\sigma : D[[X, Y]] \to B,$$

$$\sigma|_{D[[Y]]} = \sigma', \ \sigma(X_i) = u(x_i), \ i = 1, \ldots, m$$

and we can extend ξ' to a morphism

$$\xi : C[[X]] \to D[[X, Y]],$$

$$\xi|_C = \xi', \ \xi(X_i) = X_i, \ i = 1, \ldots, m.$$

Thus we have the commutative diagram

$$
\begin{array}{ccc}
C[[X]] & \xrightarrow{\xi} & D[[X, Y]] \\
v \downarrow & & \downarrow \sigma \\
A & \xrightarrow{u} & B.
\end{array}
$$

We observe that v and σ are surjective. Clearly ξ' is a flat morphism, since C is a discrete valuation domain and $D[[Y]]$ is a torsion-free C-module. Consequently also ξ is flat.

Let $T := A \otimes_{C[[X]]} D[[X, Y]]$ and let

$$f : A \to T, \ f(a) = a \otimes 1,$$

$$g : T \to B, \ g(a \otimes h) = u(a)\sigma(h).$$

Clearly $u = g \circ f$. We will prove that this a Cohen factorization of u. Since σ is surjective, it follows that g is surjective. Also, the flatness of ξ together with the fact that $f = \xi \otimes_{C[[X]]} 1_A$, proves the flatness of f. We have

$$T/\mathfrak{m}T \cong (A/\mathfrak{m}) \otimes_A (A \otimes_{C[[X]]} D[[X, Y]]) \cong$$

$$\cong (C[[X]]/(p, X)C[[X]]) \otimes_{C[[X]]} D[[X, Y]] \cong$$

$$\cong D[[X, Y]]/(p, X)D[[X, Y]] \cong (D/pD)[[Y]],$$

which is a regular local ring. We also have

$$T \cong D[[X, Y]]/(\ker(v)D[[X, Y]]),$$

hence T is a complete local ring. The proof is complete.   □

It is clear from the preceding proof that there is no chance for the uniqueness of Cohen factorizations. However, we shall at least try to relate Cohen factorizations.

Definition 6.1.7 Let $u : A \to B$ be a morphism of Noetherian local rings and $A \xrightarrow{f_1} C_1 \xrightarrow{g_1} \widehat{B}$ and $A \xrightarrow{f_2} C_2 \xrightarrow{g_2} \widehat{B}$ be two Cohen factorizations of u. A local morphism $v : C_1 \to C_2$ is called a **comparison** of the two Cohen factorizations, if $v \circ f_1 = f_2$ and $g_2 \circ v = g_1$.

Remark 6.1.8 In other words, v is a comparison if and only if the following diagram commutes:

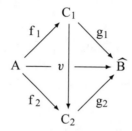

Definition 6.1.9 Let $u : A \to B$ be a morphism of Noetherian local rings and $A \xrightarrow{f_1} C_1 \xrightarrow{g_1} \widehat{B}$ and $A \xrightarrow{f_2} C_2 \xrightarrow{g_2} \widehat{B}$ be two Cohen factorizations of u. If a comparison $v : C_1 \to C_2$ is surjective, we say that $g_2 \circ f_2$ is a **reduction** of $g_1 \circ f_1$ or that $g_1 \circ f_1$ is a **deformation** of $g_2 \circ f_2$.

Theorem 6.1.10 (Avramov, Foxby and B. Herzog [31, Th. 1.2]) *Let $u : (A, \mathfrak{m}) \to B$ be a morphism of Noetherian local rings and $A \xrightarrow{f_1} C_1 \xrightarrow{g_1} \widehat{B}$ and $A \xrightarrow{f_2} C_2 \xrightarrow{g_2} \widehat{B}$ be two Cohen factorizations of u. Then:*

(i) *there exists a common deformation $A \xrightarrow{f} P \xrightarrow{g} \widehat{B}$ of the two given Cohen factorizations;*

(ii) *Let $v : C_1 \to C_2$ be a deformation of the Cohen factorization $A \xrightarrow{f_2} C_2 \xrightarrow{g_2} \widehat{B}$ to the Cohen factorization $A \xrightarrow{f_1} C_1 \xrightarrow{g_1} \widehat{B}$, where $\mathfrak{m}'$ is the maximal ideal of C_1. Then $\ker(v) = (a_1, \ldots, a_n)$, where $a_1, \ldots, a_n \in \mathfrak{m}'/(\mathfrak{m}'^2 + \mathfrak{m}C_1)$ is a C_1-regular sequence and $a_1, \ldots, a_n$ are linearly independent over $C_1/\mathfrak{m}'$.*

Proof

(i) Let us consider the fibre product of C_1 and C_2 over A [131, 18.1.2]

$$Q := \{(c_1, c_2) \in C_1 \times C_2 \mid f_1(c_1) = f_2(c_2)\}.$$

Then from [134, Lemme 19.3.2.1] it follows that Q is a Noetherian complete local ring. From the canonical projections $C_1 \leftarrow C_1 \times C_2 \rightarrow C_2$, we get surjective morphisms

$$C_1 \xleftarrow{\pi_1} Q \xrightarrow{\pi_2} C_2$$

such that $f_1 \circ \pi_1 = f_1 \circ \pi_2$. Let $A \xrightarrow{f} P \xrightarrow{v} Q$ be a Cohen factorization of the morphism

$$A \rightarrow Q, \; a \mapsto (f_1(a), f_2(a))$$

and put

$$v_1 = \pi_1 \circ v : P \rightarrow C_1,$$

$$v_2 = \pi_2 \circ v : P \rightarrow C_2,$$

$$g = g_1 \circ v_1 = g_2 \circ v_2 : P \rightarrow \widehat{B}.$$

It is easy to see that $A \xrightarrow{f} P \xrightarrow{g} \widehat{B}$ is a Cohen factorization of u and that v_1 and v_2 are reductions of this factorization to $g_1 \circ f_1$ and $g_2 \circ f_2$, respectively.

(ii) Let $a_1, \ldots, a_r$ be a minimal system of generators of $\ker(v)$. We have an exact sequence

$$0 \rightarrow \ker(v) \rightarrow C_1 \rightarrow C_2 \rightarrow 0.$$

From the flatness of C_2 over A we get an exact sequence

$$0 \rightarrow \ker(v)/\mathfrak{m} \cdot \ker(v) \rightarrow C_1/\mathfrak{m}C_1 \xrightarrow{\bar{v}} C_2/\mathfrak{m}C_2 \rightarrow 0.$$

Therefore $\bar{a}_1 := a_1 + \mathfrak{m}C_1, \ldots, \bar{a}_r := a_r + \mathfrak{m}C_1$ is a minimal system of generators of $\ker(\bar{v})$. From [249, Th. 14.2] we obtain that $\bar{a}_1, \ldots, \bar{a}_r$ is a $C_1/\mathfrak{m}C_1$-regular sequence and the elements $\bar{a}_1, \ldots, \bar{a}_r$ are linearly independent over $C_1/\mathfrak{m}C_1$ modulo $(\mathfrak{m}'/\mathfrak{m}C_1)^2$. This means that $a_1, \ldots, a_r$ are linearly independent modulo $(\mathfrak{m}'^2 + \mathfrak{m}C_1)$. The flatness of f implies by [249, Cor. to Th. 22.5] that the sequence $a_1, \ldots, a_r$ is C_1-regular.   □

Lemma 6.1.11 *Let* $u : (A, \mathfrak{m}, k) \to (B, \mathfrak{n}, K)$ *be a local morphism of Noetherian local rings and* $A \xrightarrow{f} C \xrightarrow{g} \widehat{B}$ *be a Cohen factorization of* u. *Then*

$$\mathrm{edim}(B/\mathfrak{m}B) \leq \dim(C) - \dim(A).$$

Proof From the surjective morphism g we get a surjection $\overline{g} : C/\mathfrak{m}C \to \widehat{B}/\mathfrak{m}\widehat{B}$ and this implies that

$$\mathrm{edim}(B/\mathfrak{m}B) = \mathrm{edim}(\widehat{B}/\mathfrak{m}\widehat{B}) \leq \mathrm{edim}(C/\mathfrak{m}C).$$

On the other hand, $C/\mathfrak{m}C$ is regular and f is flat, hence

$$\mathrm{edim}(C/\mathfrak{m}C) = \dim(C/\mathfrak{m}C) = \dim(C) - \dim(A).$$

$\square$

Definition 6.1.12 Let $u : (A, \mathfrak{m}) \to B$ be a morphism of Noetherian local rings. A Cohen factorization $A \xrightarrow{f} C \xrightarrow{g} \widehat{B}$ is called a **minimal Cohen factorization** if $\mathrm{edim}(B/\mathfrak{m}B) = \dim(C) - \dim(A)$.

Remark 6.1.13 From the definition and from the fact that $\ker(\overline{g}) = \ker(g)(C/\mathfrak{m}C)$, it follows that a Cohen factorization is minimal if and only if $\ker(g) \not\subseteq \mathfrak{m}'^2 + \mathfrak{m}C$, where $\mathfrak{m}'$ is the maximal ideal of C.

Example 6.1.14 (Avramov, Foxby and B. Herzog [31, Ex. 1.8])   Not any two Cohen factorizations can be compared in the sense of Definition 6.1.7. To see this, let p be a prime number and $\mathbb{F}_p$ be the prime field with p elements. Let $k = \mathbb{F}_p(t)$ be the field of rational fractions in the variable t and $K := k\left(\sqrt[p^3]{t}\right)$. Consider indeterminates X, Y, Z and let $A := k,\ A_1 := K[[X]]$ and $A_2 = K[[Y]]$. Consider the K-algebra morphisms

$$g_1 : A_1 \to B := K[[Z]]/(Z^p),\ g_1(X) = Z + (Z^p)$$

and

$$g_2 : A_2 \to B := K[[Z]]/(Z^p),\ g_2(Y) = Z + (Z^p).$$

Consider also the morphism $f_1 : A = k \to A_1$ given by the composition $k \subset K \to K[[X]]$. It is easy to see that $A \xrightarrow{f_1} A_1 \xrightarrow{g_1} B$ is a minimal Cohen factorization of the local morphism $u = g_1 \circ f_1 : A \to B$. Now consider the morphism of $\mathbb{F}_p$-algebras $\theta : \mathbb{F}_p[t] \to K[[Y]]$, sending t to $t + Y^{p^2}$. The image of a non-zero polynomial $f \in \mathbb{F}_p[t]$

is a polynomial in Y with constant term $f(t) \neq 0$, hence is invertible in $K[[Y]]$. This shows that θ gives a morphism of $\mathbb{F}_p$-algebras $f_2 : A = \mathbb{F}_p(t) \to A_2 = K[[Y]]$. We have $g_2 \circ f_2(t) = g_2(t + Y^{p^2}) = t$; hence we have a Cohen factorization $A \xrightarrow{f_2} A_2 \xrightarrow{g_2} B$ that is easily seen to be minimal. Suppose that there exists a comparison $v : A_1 \to A_2$ and let $a = v(\sqrt[p^3]{t})$. Then

$$g_2(a) = g_2\big(v(\sqrt[p^3]{t})\big) = g_1(\sqrt[p^3]{t}) = g_2(\sqrt[p^3]{t}).$$

Then Y divides $a - \sqrt[p^3]{t} \in K[[Y]]$; hence Y^{p^3} divides $(a - \sqrt[p^3]{t})^{p^3} = a^{p^3} - t$. On the other hand

$$a^{p^3} = v\big((\sqrt[p^3]{t})^{p^3}\big) = v(t) = v(f_1(t)) = f_2(t) = t + Y^{p^2},$$

hence $a^{p^3} - t = Y^{p^2}$. Contradiction!

Theorem 6.1.15 (Avramov, Foxby and B. Herzog [31, Prop. 1.5]) *Let $u : (A, \mathfrak{m}) \to B$ be a morphism of Noetherian local rings. Then:*

(i) any Cohen factorization of u has a reduction to a minimal one.

(ii) any reduction of a minimal Cohen factorization of u is an isomorphism.

Proof

(i) Let $A \xrightarrow{f} C \xrightarrow{g} \widehat{B}$ be a Cohen factorization, which is supposed not minimal and let $\mathfrak{m}'$ be the maximal ideal of C. Then there exists $x \in \ker(g) \setminus (\mathfrak{m}'^2 + \mathfrak{m}C)$. Let $\bar{x}$ be the image of x in $C/\mathfrak{m}C$. Clearly $\bar{x} \neq 0$. Let $v : C \to C/(x)$ be the canonical morphism and $\phi = v \circ f : A \to C/(x)$. Then from [249, Th. 22.5] we get that x is a non-zero divisor on C and that ϕ is flat. Since $\bar{x}$ is not in the square of the maximal ideal of $C/\mathfrak{m}C$, it follows that $(C/\mathfrak{m}C)/x(C/\mathfrak{m}C)$ is a regular local ring. Thus the composition $A \xrightarrow{\phi} C/(x) \xrightarrow{\psi} \widehat{B}$ is a Cohen factorization of u. It is clear that this Cohen factorization is obtained as a reduction of the original one. Moreover,

$$\dim(C/(x)) - \dim(A) - \mathrm{edim}(B/\mathfrak{m}B) < \dim(C) - \dim(A) - \mathrm{edim}(B/\mathfrak{m}B).$$

After a finite number of iterations of this procedure, we will obtain a minimal Cohen factorization.

(ii) Let now $A \xrightarrow{f} C \xrightarrow{g} \widehat{B}$ be a minimal Cohen factorization and $v : C \to C_1$ be a reduction. From Theorem 6.1.10 we have that $\ker(v)$ is generated by elements which are linearly independent in $\mathfrak{m}'/(\mathfrak{m}'^2 + \mathfrak{m}C_1)$. On the other hand, from Remark 6.1.13 we have $\ker(v) \subseteq (\mathfrak{m}'^2 + \mathfrak{m}C_1)$. This implies that $\ker(v) = 0$. $\qquad\square$

Theorem 6.1.16 (Avramov, Foxby and B. Herzog [31, Th. 1.6]) *Let* $\pi : (P, \mathfrak{m}) \to$
$(A, \mathfrak{n})$ *and* $f : (A, \mathfrak{n}) \to B$ *be local morphisms of Noetherian local rings. Assume that* π
is surjective and that B *is a complete local ring such that* $B/\mathfrak{n}B$ *is a regular local ring.*
Then there exists a commutative diagram of local morphisms

$$
\begin{array}{ccc}
P & \xrightarrow{\psi} & C \\
{\scriptstyle\pi}\downarrow & & \downarrow{\scriptstyle\rho} \\
A & \xrightarrow{f} & B
\end{array}
$$

with the properties:

(i) *C is a complete local ring such that $C/\mathfrak{m}C$ is a regular local ring;*
(ii) *ρ is surjective;*
(iii) *the induced map $v : A \otimes_P C \to B$ is an isomorphism.*

Proof Let $P \xrightarrow{\psi} C \xrightarrow{\rho} B$ be a minimal Cohen factorization of $f \circ \pi$. Clearly the morphism
$v : A \otimes_P C \to B$ induced by the universal property of the tensor product is surjective.
Moreover the morphism $1_A \otimes \psi : A \to A \otimes_P C$ is flat and the closed fibre is isomorphic
to the regular local ring $C/\mathfrak{m}C$. Hence $v \circ (1_A \otimes \psi)$ is a Cohen factorization of f. We have

$$
\dim(A \otimes_P C) - \dim(A) = \dim(C) - \dim(P) = \dim(C/\mathfrak{m}C) = \mathrm{edim}(C/\mathfrak{m}C).
$$

But from the minimality of $\rho \circ \psi$, we obtain $\mathrm{edim}(C/\mathfrak{m}C) = \mathrm{edim}(B/\mathfrak{m}B)$. Thus $v \circ$
$(1_A \otimes \psi)$ is a minimal Cohen factorization of f. From the commutative diagram

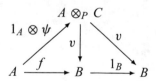

we get that v is a reduction of $v \circ 1_A \otimes \psi$ to $1_B \circ f$. Now we apply Theorem 6.1.15. □

We will illustrate the power of Cohen factorization by using this device to give a proof
of the Localization Theorem 4.2.10 for some properties. Let us first remind the following
notion.

Definition 6.1.17 (see [77, Def. 1.4.15]) If A is a Noetherian ring, a finitely generated
A-module M is called a **perfect A-module** if $\mathrm{pd}_A(M) = \mathrm{grade}_A(M)$. An ideal $\mathfrak{a}$ of A is
called a **perfect ideal** if $A/\mathfrak{a}$ is a perfect A-module.

Definition 6.1.18 ([27]) Let B be a Noetherian local ring, $\mathfrak{a}$ be an ideal of B such that $\mathrm{pd}_B(A) < \infty$, where $A := B/\mathfrak{a}$. Then we define the **imperfection** of the B-module A to be the positive integer $\mathrm{imp}_B(A) := \mathrm{pd}_B(A) - \mathrm{grade}_B(A)$.

Remark 6.1.19 Let B be a Noetherian local ring, $\mathfrak{a}$ be an ideal of B such that $\mathrm{pd}_B(A) < \infty$, where $A := B/\mathfrak{a}$. Let $\mathfrak{q} \in \mathrm{Spec}(A)$ and $\mathfrak{p} = \mathfrak{q} \cap B$. Then

$$\mathrm{imp}_{\widehat{B}}(\widehat{A}) = \mathrm{imp}_B(A)$$

and

$$\mathrm{imp}_{B_{\mathfrak{p}}}(A_{\mathfrak{q}}) \leq \mathrm{imp}_B(A).$$

Proof While the first relation is clear, the second one follows from the known inequalities

$$\mathrm{pd}_{B_{\mathfrak{p}}}(A_{\mathfrak{q}}) \leq \mathrm{pd}_B(A) \text{ and } \mathrm{grade}_{B_{\mathfrak{p}}}(A_{\mathfrak{q}}) \geq \mathrm{grade}_B(A).$$

$\square$

Proposition 6.1.20 *Let $u : (A, \mathfrak{m}, k) \to (B, \mathfrak{n})$ and $v : B \to C$ be surjective morphisms of local rings such that $\ker(u)$ is generated by an A-regular sequence and $\mathrm{pd}_B(C) < \infty$. Then $\mathrm{imp}_A(C) = \mathrm{imp}_B(C)$.*

Proof Using an induction argument, we may assume that $B = A/(x)$, where $x \in \mathfrak{m}$ is a non-zero divisor. We have a change of rings spectral sequence [359, Ch. 15, Ex. 15.61.2]

$$E^2_{pq} = \mathrm{Tor}^B_p(k, \mathrm{Tor}^A_q(B, C)) \Rightarrow \mathrm{Tor}^A_{p+q}(k, C).$$

Since by assumption $\mathrm{Tor}^A_q(B, C) = (0)$ for $q \geq 2$ and $\mathrm{Tor}^A_1(B, C) = \mathrm{Tor}^A_0(B, C) \cong C$, it follows that $\mathrm{pd}_A(C) = \mathrm{pd}_B(C) + 1$. Since we also obviously have $\mathrm{grade}_A(C) = \mathrm{grade}_B(C) + 1$, we are done. $\square$

Proposition 6.1.21 *Let $u : (A, \mathfrak{m}, k) \to (B, \mathfrak{n}, K)$ be a local flat morphism and let $A \xrightarrow{f} C \xrightarrow{g} \widehat{B}$ be a Cohen factorization. Then:*

(i) $\mathrm{pd}_C(\widehat{B}) < \infty$;
(ii) $\mathrm{cmd}(B/\mathfrak{m}B) = \mathrm{imp}_C(\widehat{B})$;
(iii) $\mathrm{cid}(B/\mathfrak{m}B) = \mu(\ker(g)) - \mathrm{grade}_C(\widehat{B})$.

Proof Of course we may assume that B is complete.

(i) Let $\bar{g} := 1_k \otimes g : C/\mathfrak{m}C \to B/\mathfrak{m}B$ and let $\mathbb{F}$ be a free resolution of B over C. The flatness of f and u implies that $H_i(k \otimes_A \mathbb{F}) = \mathrm{Tor}_i^A(k, B) = (0)$ for any $i > 0$. Therefore $k \otimes_A \mathbb{F}$ is a free resolution of $B/\mathfrak{m}B$ over $C/\mathfrak{m}C$. Then for any integer $n \in \mathbb{Z}$ we have

$$\mathrm{Tor}_n^C(K, B) = H_n(K \otimes_C \mathbb{F}) \cong H_n(K \otimes_{C/\mathfrak{m}C} (k \otimes_A \mathbb{F})) =$$

$$= \mathrm{Tor}_n^{C/\mathfrak{m}C}(K, B/\mathfrak{m}B).$$

This implies that $\mathrm{pd}_C(B) = \mathrm{pd}_{C/\mathfrak{m}C}(B/\mathfrak{m}B)$. But the ring $C/\mathfrak{m}C$ is a regular local ring, hence $\mathrm{pd}_C(B) < \infty$.

(ii) Now we choose a $C/\mathfrak{m}C$-regular sequence in $\ker(\bar{g})$ of length $l := \mathrm{grade}_{C/\mathfrak{m}C}$ $(B/\mathfrak{m}B)$. Since $\ker(\bar{g}) = \ker(g)C/\mathfrak{m}C$, this sequence can be lifted to a sequence of length l in $\ker(g)$. The flatness of C over A implies, using [249, 22.5 Corollary], that this sequence is C-regular. Therefore $\mathrm{grade}_C(B) \geq l = \mathrm{grade}_{C/\mathfrak{m}C}(B/\mathfrak{m}B)$. On the other hand, because C and B are flat over A and $C/\mathfrak{m}C$ is a Cohen-Macaulay ring, we have the relations

$$\mathrm{grade}_C(B) \leq \dim(C) - \dim(B) =$$

$$= \big(\dim(A) + \dim(C/\mathfrak{m}C)\big) - \big(\dim(A) + \dim(B/\mathfrak{m}B)\big) =$$

$$= \dim(C/\mathfrak{m}C) - \dim(B/\mathfrak{m}B) = \mathrm{grade}_{C/\mathfrak{m}C}(B/\mathfrak{m}B).$$

Finally we obtain

$$\mathrm{grade}_C(B) = \mathrm{grade}_{C/\mathfrak{m}C}(B/\mathfrak{m}B)$$

and consequently

$$\mathrm{imp}_C(B) = \mathrm{imp}_{C/\mathfrak{m}C}(B/\mathfrak{m}B) = \mathrm{cmd}(B/\mathfrak{m}B).$$

(iii) Follows from [249, Th. 21.1, iii)].   □

Corollary 6.1.22 *Let $u : (A, \mathfrak{m}, k) \to (B, \mathfrak{n}, K)$ be a local flat morphism of complete local rings such that $B/\mathfrak{m}B$ is a complete intersection and let $A \xrightarrow{f} C \xrightarrow{g} B$ be a Cohen factorization of u. Then $\ker(g)$ is generated by a regular sequence.*

Lemma 6.1.23 *Let $u : (A, \mathfrak{m}, k) \to (B, \mathfrak{n}, K)$ be a local flat morphism of complete local rings such that $B/\mathfrak{m}B$ is a regular local ring, let $\mathfrak{q} \in \mathrm{Spec}(B)$ and $\mathfrak{p} = \mathfrak{q} \cap A$. Then $B_\mathfrak{q}/\mathfrak{p}B_\mathfrak{q}$ is a complete intersection.*

Proof If A and B are both regular local rings, then $\mathfrak{p}A_\mathfrak{p}$ is generated by a $A_\mathfrak{p}$-regular sequence, hence flatness implies that $\mathfrak{p}B_\mathfrak{q}$ is generated by a $B_\mathfrak{q}$-regular sequence. Therefore, since $B_\mathfrak{q}$ is a regular local ring, it follows that $B_\mathfrak{q}/\mathfrak{p}B_\mathfrak{q}$ is a complete intersection. In the general case, by Cohen's structure theorem, there exists a regular local ring P and a surjective morphism $\pi : P \to A$. From Theorem 6.1.16, there exists a commutative diagram of local morphisms

$$
\begin{array}{ccc}
P & \xrightarrow{\;v\;} & C \\
{\scriptstyle \pi}\downarrow & & \downarrow{\scriptstyle \chi} \\
A & \xrightarrow{\;u\;} & B
\end{array}
$$

where C is a complete local ring, v is flat with regular closed fibre, χ is surjective and $B = A \otimes_P C$. If we denote $\mathfrak{r} := \mathfrak{q} \cap P$, we have

$$A_\mathfrak{p}/\mathfrak{p}A_\mathfrak{p} \otimes_A B \cong A_\mathfrak{p}/\mathfrak{p}A_\mathfrak{p} \otimes_A A \otimes_P C \cong P_\mathfrak{r}/\mathfrak{r}P_\mathfrak{r} \otimes_P C,$$

hence

$$C_{\mathfrak{q} \cap C}/\mathfrak{r}C_{\mathfrak{q} \cap C} \cong B_\mathfrak{q}/\mathfrak{p}B_\mathfrak{q},$$

that is a complete intersection by the first case. $\square$

Theorem 6.1.24 (Avramov and Foxby [27, Main Th.]) *Let $u : (A, \mathfrak{m}, k) \to (B, \mathfrak{n}, K)$ be a local flat morphism of local rings, let $\mathfrak{q} \in \mathrm{Spec}(B)$ and $\mathfrak{p} = \mathfrak{q} \cap A$. Then*

$$\mathrm{cmd}(B_\mathfrak{q}/\mathfrak{p}B_\mathfrak{q}) + \mathrm{cmd}(k(\mathfrak{q}) \otimes_B \widehat{B}) \le \mathrm{cmd}(B/\mathfrak{m}B) + \mathrm{cmd}(k(\mathfrak{p}) \otimes_A \widehat{A}).$$

Proof Assume first that A and B are complete local rings. Let $A \xrightarrow{f} C \xrightarrow{g} B$ be a Cohen factorization of u, let $\mathfrak{r} := \mathfrak{q} \cap C$ and let

$$A_\mathfrak{p} \xrightarrow{f_\mathfrak{r}} C_\mathfrak{r} \xrightarrow{g_\mathfrak{q}} B_\mathfrak{q}$$

be the induced factorization of u_q that gives by completion in the respective maximal ideals the local flat morphisms

$$\widehat{A_p} \xrightarrow{\psi} \widehat{C_r} \xrightarrow{\nu} \widehat{B_q}.$$

Let us remark that the closed fibre of $\nu\psi = \widehat{u_q}$ is actually $\widehat{B_q}/\mathfrak{p}\widehat{B_q} = \widehat{B_q/\mathfrak{p}B_q}$ that is the completion of the closed fibre of u_q, hence

$$\mathrm{cmd}(B_q/\mathfrak{p}B_q) = \mathrm{cmd}(\widehat{B_q}/\mathfrak{p}\widehat{B_q}).$$

Now consider a Cohen factorization of $\psi : \widehat{A_p} \to \widehat{C_r}$, say

$$\psi : \widehat{A_p} \xrightarrow{\theta} D \xrightarrow{\eta} \widehat{C_r}$$

and notice that $\widehat{A_p} \xrightarrow{\theta} D \xrightarrow{\nu\eta} \widehat{B_q}$ is a Cohen factorization of $\widehat{u_q}$. Therefore Proposition 6.1.21 shows that

$$\mathrm{cmd}(\widehat{B_q}/\mathfrak{p}\widehat{B_q}) = \mathrm{imp}_D(\widehat{B_q}).$$

The closed fibre of ψ is $\widehat{C_r}/\mathfrak{p}\widehat{C_r}$ that is the completion of the closed fibre $C_r/\mathfrak{p}C_r$ of f_r. By Lemma 6.1.23, it follows that they are both complete intersections and Corollary 6.1.22 implies that $\ker(\eta)$ is generated by a D-regular sequence. By Proposition 6.1.20 and Remark 6.1.19 we obtain

$$\mathrm{imp}_D(\widehat{B_q}) = \mathrm{imp}_{\widehat{C_r}}(\widehat{B_q}) = \mathrm{imp}_{C_r}(B_q) \le \mathrm{imp}_C(B) = \mathrm{cmd}(B/\mathfrak{m}B).$$

Consequently we get

$$\mathrm{cmd}(B_q/\mathfrak{p}B_q) \le \mathrm{cmd}(B/\mathfrak{m}B).$$

Consider now the general case, that is, A and B are not necessarily complete. Take a prime ideal q^* in $\widehat{B}$ lying over q and let $\mathfrak{p}^* = q^* \cap \widehat{A}$. Then

$$\mathrm{cmd}(B_q/\mathfrak{p}B_q) + \mathrm{cmd}(\widehat{B}_{q^*}/q\widehat{B}_{q^*}) =$$

$$= \left(\mathrm{cmd}(B_q) - \mathrm{cmd}(A_p)\right) + \left(\mathrm{cmd}(\widehat{B}_{q^*}) - \mathrm{cmd}(B_q)\right) = \mathrm{cmd}(\widehat{B}_{q^*}) - \mathrm{cmd}(A_p) =$$

$$= \left(\mathrm{cmd}(\widehat{A}_{p^*}) - \mathrm{cmd}(A_p)\right) + \left(\mathrm{cmd}(\widehat{B}_{q^*}) - \mathrm{cmd}(\widehat{A}_{p^*})\right) =$$

$$= \mathrm{cmd}(\widehat{A}_{p^*}/\mathfrak{p}\widehat{A}_{p^*}) + \mathrm{cmd}(\widehat{B}_{q^*}/\mathfrak{p}\widehat{B}_{q^*}).$$

But let us note that $\widehat{A}_{\mathfrak{p}^*}/\mathfrak{p}\widehat{A}_{\mathfrak{p}^*}$ is a localization of $k(\mathfrak{p}) \otimes_A \widehat{A} = \widehat{A}_{\mathfrak{p}}/\mathfrak{p}\widehat{A}_{\mathfrak{p}}$, hence by Lemma 1.1.22 we obtain that

$$\left(\mathrm{cmd}(\widehat{A}_{\mathfrak{p}^*}/\mathfrak{p}\widehat{A}_{\mathfrak{p}^*})\right) \leq \mathrm{cmd}\big(k(\mathfrak{p}) \otimes_A \widehat{A}\big).$$

But the case of complete local rings shows that

$$\mathrm{cmd}(\widehat{B}_{\mathfrak{q}^*}/\mathfrak{p}\widehat{B}_{\mathfrak{q}^*}) \leq \mathrm{cmd}(\widehat{B}/\mathfrak{m}\widehat{B}) = \mathrm{cmd}(B/\mathfrak{m}B).$$

Therefore we have

$$\mathrm{cmd}(B_{\mathfrak{q}}/\mathfrak{p}B_{\mathfrak{q}}) + \mathrm{cmd}(\widehat{B}_{\mathfrak{q}^*}/\mathfrak{q}\widehat{B}_{\mathfrak{q}^*}) \leq \mathrm{cmd}(B/\mathfrak{m}B) + \mathrm{cmd}(k(\mathfrak{p}) \otimes_A \widehat{A}).$$

Taking the supremum over the prime ideals $\mathfrak{q}^*$ of B lying over $\mathfrak{q}$, we get the desired assertion.   □

Corollary 6.1.25 (Localization Theorem for Cohen-Macaulay [27, Th. 4.1]) *Let $u :$ $(A, \mathfrak{m}, k) \to (B, \mathfrak{n}, K)$ be a flat local morphism of local rings. Assume that the formal fibres of A are Cohen-Macaulay and that the closed fibre of u is Cohen-Macaulay. Then all the fibres of u are Cohen-Macaulay.*

Proof The assumptions mean that

$$\mathrm{cmd}(B/\mathfrak{m}B) = \mathrm{cmd}\big(k(\mathfrak{p}) \otimes_A \widehat{A}\big) = 0,$$

hence Theorem 6.1.24 gives

$$\mathrm{cmd}(B_{\mathfrak{q}}/\mathfrak{p}B_{\mathfrak{q}}) = \mathrm{cmd}\big(k(\mathfrak{q}) \otimes_B \widehat{B}\big) = 0.$$

□

Remark 6.1.26

(i) Comparing with the proof of Theorem 4.2.10, one can observe that the above proof relies only on algebraic arguments.

(ii) A proof quite similar to the above one, involving essentially the same computations about the complete intersection defect and the Cohen-Macaulay type instead of the Cohen-Macaulay defect, leads to a proof of the Localization Theorem for **P**=complete intersection and **P**=Gorenstein. One can look at the comments in Remark 4.2.11.

(iii) Using the device of Cohen factorizations, Avramov [24] and Avramov and Foxby [26, 29] succeeded to generalize the notions of Cohen-Macaulay, Gorenstein and complete intersection morphisms from the case of flat morphisms to the case of morphisms with finite flat dimension.

6.2 Morphisms with Complete Intersection Fibres and Around

In this section, we deal with several aspects of the complete intersection morphisms, especially connections with other types of morphisms, for example, reduced ones.

Proposition 6.2.1 (Avramov [24, Lemma 1.7]) *Let* $u : (A, \mathfrak{m}, k) \to (B, \mathfrak{n}, K)$ *be a morphism of Noetherian local rings and let* $A \xrightarrow{f} C \xrightarrow{g} \widehat{B}$ *be a Cohen factorization of* u. *Then there is a canonical isomorphism* $H_n(A, B, K) \cong H_n(C, \widehat{B}, K)$, *for any* $n \geq 2$.

Proof From [6, Prop. IV.54] it follows that for any natural number n we have

$$H_n(B, \widehat{B}, K) \cong H_n(K, \widehat{B} \otimes_B K, K) = H_n(K, K, K) = (0).$$

The Jacobi-Zariski sequence associated to $A \to B \to \widehat{B}$ gives

$$(0) = H_{n+1}(B, \widehat{B}, K) \to H_n(A, B, K) \xrightarrow{\cong} H_n(A, \widehat{B}, K) \to H_n(B, \widehat{B}, K) = (0).$$

Let $D := C/\mathfrak{m}C$. The Jacobi-Zariski exact sequence of $k \to D \to K$ gives the exact sequence

$$H_{n+1}(D, K, K) \to H_n(k, D, K) \to H_n(k, K, K).$$

Since $H_n(k, K, K) = (0)$ for $n \geq 2$ and $H_{n+1}(D, K, K) = (0)$ for $n \geq 1$ because D is regular, we get that

$$H_n(k, D, K) = (0), \ n \geq 2.$$

Applying [6, Prop. IV.54] we obtain $H_n(A, C, K) \cong H_n(k, D, K)$ for any natural number n, so that from the Jacobi-Zariski sequence associated to $A \to C \to \widehat{B}$, we have the exact sequence

$$H_n(A, C, K) \to H_n(A, \widehat{B}, K) \to H_n(C, \widehat{B}, K) \xrightarrow{\theta_n} H_{n-1}(A, C, K).$$

Thus θ_n is an isomorphism for $n \geq 3$ and injective for $n = 2$. If we consider also the Jacobi-Zariski exact sequence associated to $k \to D \to K$, we obtain a commutative diagram with exact rows

$$H_2(A, \widehat{B}, K) \longrightarrow H_2(C, \widehat{B}, K) \overset{\lambda}{\longrightarrow} H_1(A, C, K) \longrightarrow H_1(A, \widehat{B}, K)$$

$$H_2(D, K, K) \longrightarrow H_1(k, D, K) \longrightarrow H_1(k, K, K)$$

This implies that $\lambda = 0$; hence the map $H_2(A, \widehat{B}, K) \to H_2(C, \widehat{B}, K)$ is surjective.   □

Corollary 6.2.2 *Let $u : (A, \mathfrak{m}, k) \to (B, \mathfrak{n}, K)$ be a morphism of Noetherian local rings, let $A \overset{f}{\to} C \overset{g}{\to} \widehat{B}$ be a Cohen factorization of u and let $\mathfrak{a} = \ker(g)$. Then $\mathfrak{a}$ is generated by a regular sequence if and only if $H_2(A, B, K) = (0)$. When this is the case, $H_n(A, B, K) = (0)$ for any $n \geq 2$.*

Proof Follows from Proposition 6.2.1 and [6, Th. VI.25].   □

Corollary 6.2.3 *Let $u : (A, \mathfrak{m}, k) \to (B, \mathfrak{n}, K)$ be a morphism of Noetherian local rings, let $g : C \to \widehat{B}$ and $h : D \to \widehat{B}$ be two Cohen factorizations of u and let $\mathfrak{a} = \ker(g)$ and $\mathfrak{b} = \ker(h)$. Then $\mathfrak{a}$ is generated by a regular sequence if and only if $\mathfrak{b}$ is generated by a regular sequence.*

Theorem 6.2.4 (Avramov [24, Th. 1.2]) *Let $u : A \to B$ be a morphism of Noetherian rings. The following are equivalent:*

(i) $H_2(A, B, -) = (0)$.

(ii) *for any prime ideal $\mathfrak{q}$ of B and for any Cohen factorization $A_{\mathfrak{q} \cap A} \to C \overset{h}{\to} \widehat{B_{\mathfrak{q}}}$ of the localization of u with respect to $\mathfrak{q}$, the kernel of h is generated by a regular sequence.*

Proof $(i) \Rightarrow (ii)$: We know that for any B-module E we have

$$H_n(A, B, E)_{\mathfrak{q}} \cong H_n(A_{\mathfrak{q} \cap A}, B_{\mathfrak{q}}, E_{\mathfrak{q}}).$$

Now we apply Corollary 6.2.2.

$(ii) \Rightarrow (i)$: From the assumption and Corollary 6.2.2 we have $H_n(A, B, k(\mathfrak{q})) = (0)$, for any prime ideal $\mathfrak{q}$ of B and for any natural number $n \geq 2$. But then [6, Prop. S.29] implies that $H_2(A, B, -) = (0)$.   □

Corollary 6.2.5 *Let $u : A \to B$ be a morphism of Noetherian local rings such that $H_2(A, B, -) = (0)$. Then $\mathrm{fd}_A(B) < \infty$.*

Proof If $A \to C \to \widehat{B}$ is a Cohen factorization of the semi-completion of u, then

$$\mathrm{fd}_A(B) = \mathrm{fd}_{\widehat{A}}(\widehat{B}) \le \mathrm{fd}_C(\widehat{B}) = \mathrm{pd}_C(\widehat{B}) < \infty$$

because the kernel of the surjection $C \to \widehat{B}$ is generated by a regular sequence.   □

In the case of flat morphisms, the vanishing of the second André-Quillen homology group characterizes the fibres of the morphism. We begin with a lemma.

Lemma 6.2.6 (Marot [244, Lemma 4.1]) *Let k be a field and $(A, \mathfrak{m}, K)$ a Noetherian local k-algebra. The following are equivalent:*

(i) *A is a complete intersection ring;*
(ii) *$H_n(k, A, E) = (0)$ for any K-module E and any $n \ge 2$;*
(iii) *$H_2(k, A, K) = (0)$.*

Proof $(i) \Rightarrow (ii)$: From the morphisms $k \to A \to K$ we obtain the long exact sequence

$$H_{n+1}(k, K, K) \to H_{n+1}(A, K, K) \to H_n(k, A, K) \to H_n(k, K, K).$$

But for $n \ge 2$ we know that $H_n(k, K, K) = (0)$ by [6, Prop. VII.4] and that $H_{n+1}(A, K, K) = (0)$ by [6, Cor. X.20].
$(iii) \Rightarrow (i)$: From the above exact sequence we get $H_3(A, K, K) = (0)$. Now apply [6, Cor. X.20].   □

Proposition 6.2.7 (Marot [244, Lemma 4.2]) *Let $u : A \to B$ be a flat morphism of Noetherian rings. The following are equivalent:*

(i) *u is a complete intersection morphism (that is the fibres of u are complete intersections);*
(ii) *$H_2(A, B, -) = (0)$;*
(iii) *$H_3(A, B, -) = (0)$;*
(iv) *$H_n(A, B, -) = (0)$ for any natural number $n \ge 2$;*
(v) *$H_2(A, B, B/\mathfrak{q}) = (0)$ for any prime ideal $\mathfrak{q}$ of B;*
(vi) *$H_2(A, B, k(\mathfrak{q})) = (0)$ for any prime ideal $\mathfrak{q}$ of B;*
(vii) *there exists a natural number n such that $H_m(A, B, -) = (0)$ for any $m \ge n$.*

Proof $(i) \Leftrightarrow (vi)$: The morphism u is a complete intersection if and only if the local ring $B_\mathfrak{q}/\mathfrak{p}B_\mathfrak{q}$ is a complete intersection for every prime ideal $\mathfrak{q}$ of B, where $\mathfrak{p} = \mathfrak{q} \cap A$. From Lemma 6.2.6, this is equivalent to $H_2\big(k(\mathfrak{p}), B_\mathfrak{q} \otimes_{A_\mathfrak{p}} k(\mathfrak{p}), k(\mathfrak{q})\big) = (0)$. But by [6, Prop.

IV.54 and Prop. V.27] we have

$$H_2\big(k(\mathfrak{p}), B_\mathfrak{q} \otimes_{A_\mathfrak{p}} k(\mathfrak{p}), k(\mathfrak{q})\big) = H_2\big(A_\mathfrak{p}, B_\mathfrak{q}, k(\mathfrak{q})\big) = H_2\big(A, B, k(\mathfrak{q})\big).$$

$(vi) \Rightarrow (iv)$: By [6, Prop. S.29] it is enough to prove that $H_n\big(A, B, k(\mathfrak{q})\big) = (0)$, for any $n \geq 2$ and any prime ideal $\mathfrak{q}$ of B. But as we saw above

$$H_n\big(A, B, k(\mathfrak{q})\big) \cong H_n\big(k(\mathfrak{p}), B_\mathfrak{q} \otimes_{A_\mathfrak{p}} k(\mathfrak{p}), k(\mathfrak{q})\big).$$

Now apply Lemma 6.2.6.
$(iv) \Rightarrow (v)$ and $(iv) \Rightarrow (vii)$: Obvious.
$(v) \Rightarrow (vi)$: Follows from [6, Prop. IV.59].
$(vii) \Rightarrow (i)$: Let $\mathfrak{q}$ be a prime ideal of B and $\mathfrak{p} = \mathfrak{q} \cap A$. As above we have

$$H_m\big(k(\mathfrak{p}), B_\mathfrak{q} \otimes_{A_\mathfrak{p}} k(\mathfrak{p}), k(\mathfrak{q})\big) = H_m\big(A_\mathfrak{p}, B_\mathfrak{q}, k(\mathfrak{q})\big) = H_m\big(A, B, k(\mathfrak{q})\big) = (0).$$

Now we apply 6.2.6 and Remark 1.1.9.
$(iv) \Rightarrow (ii)$: Obvious.
$(ii) \Rightarrow (i)$: Let $\mathfrak{q}$ be a prime ideal of B and $\mathfrak{p} = \mathfrak{q} \cap A$. Then by [6, Prop. IV.54, Prop. V.27] and the assumption we have

$$H_2\big(k(\mathfrak{p}), B_\mathfrak{q} \otimes_{A_\mathfrak{p}} k(\mathfrak{p}), k(\mathfrak{q})\big) = H_2\big(A_\mathfrak{p}, B_\mathfrak{q}, k(\mathfrak{q})\big) = H_2\big(A, B, k(\mathfrak{q})\big) = 0.$$

By Lemma 6.2.6 it follows that $B_\mathfrak{q}/\mathfrak{p}B_\mathfrak{q}$ is a complete intersection ring.
$(iv) \Rightarrow (iii)$: Obvious.
$(iii) \Rightarrow (i)$: By [6, Prop. S.29] it is enough to prove that $H_n(A, B, k(\mathfrak{q})) = (0)$, for any $n \geq 2$ and any prime ideal $\mathfrak{q}$ of B. But

$$H_n(A, B, k(\mathfrak{q})) \cong H_n(k(\mathfrak{p}), B_\mathfrak{q} \otimes_{A_\mathfrak{p}} k(\mathfrak{p}), k(\mathfrak{q}))$$

and we apply Lemma 6.2.6 and Remark 1.1.9. $\square$

In the last years, flat morphisms with complete intersection fibres that are also finitely presented are called *syntomic morphisms* [359, Def. 10.136.1]. The word was invented and introduced by Barry Mazur, explained by the author on htpps://mathoverflow.net as follows: *I'm thinking of "local complete intersection" as being a way of cutting out a (sub-)space from an ambient surrounding space; the fact that it is flat over the parameter space means that each such "cutting" as you move along the parameter space, is—more or less—cut out similarly. I'm also thinking of the word "syntomic" as built from the verb temnein (i.e., to cut) and the prefix "syn" which I take in the sense of "same" or "together". So I think it fits.*

In the case of morphisms of finite presentation, one can characterize the morphisms having complete intersection fibres in a nice way:

Theorem 6.2.8 (see [221, Cor. 3.2.2]) *Let* $u : A \to B$ *be a morphism of finite type of Noetherian rings. The following are equivalent:*

(i) *for any presentation* $A \to C = A[X_1, \ldots, X_n] \to C/\mathfrak{a}$ *of* B, *the ideal* $\mathfrak{a}$ *is locally generated by a regular sequence;*

(ii) *there exists a presentation* $A \to C = A[X_1, \ldots, X_n] \to C/\mathfrak{a}$ *of* B, *such that the ideal* $\mathfrak{a}$ *is locally generated by a regular sequence;*

(iii) $H^2(A, B, -) = (0)$;

(iv) $H_2(A, B, B) = (0)$ *and for any presentation* $A \to C = A[X_1, \ldots, X_n] \to C/\mathfrak{a}$ *of* B, *the ideal* $\mathfrak{a}/\mathfrak{a}^2$ *is locally free;*

(v) $H_2(A, B, -) = (0)$.

Proof Let C be a polynomial presentation of B, that is $B = C[X_1, \ldots, X_n]/\mathfrak{a}$. Then from the Jacobi-Zariski sequence associated to $A \to C \to B$, it follows that $H_2(A, B, E) \cong H_2(C, B, E)$ and $H^2(A, B, E) \cong H^2(C, B, E)$ for any B-module E. Hence the equivalence $i \Leftrightarrow (ii) \Leftrightarrow (iii) \Leftrightarrow (v)$ follows from [6, Th. VI.25].

$(iii) \Rightarrow (iv)$: Since (iii) and (v) are equivalent, $H_2(A, B, B) = (0)$. Since $H^2(A, B, -) = (0)$, it follows that the functor $H^1(A, B, -) \cong \mathrm{Hom}_B(\mathfrak{a}/\mathfrak{a}^2, -)$ is right-exact; hence $\mathfrak{a}/\mathfrak{a}^2$ is projective.

$(iv) \Rightarrow (v)$: Because $H_2(A, B, B) = (0)$, from [4, Prop. 16.1] it follows that

$$H_2(A, B, E) \cong \mathrm{Tor}_1^B(H_1(A, B, B), E) \cong \mathrm{Tor}_1^B(\mathfrak{a}/\mathfrak{a}^2, E) = (0),$$

because $\mathfrak{a}/\mathfrak{a}^2$ is flat.   □

Definition 6.2.9 Let $u : A \to B$ be a morphism of finite type of Noetherian rings. We say that B *is a complete intersection over* A if there exists a presentation $A \to C = A[X_1, \ldots, X_n] \to C/\mathfrak{a}$ of B, such that the ideal $\mathfrak{a}$ is locally generated by a regular sequence.

Corollary 6.2.10 (Rodicio [318, Th. 2.1]) *Let* A *be a Noetherian ring and* B *be an* A-algebra of finite type. The following are equivalent:*

(i) B *is a complete intersection over* A;

(ii) $H_n(A, B, -) = (0)$ *for any* $n \geq 2$;

(iii) $\mathrm{fd}_A(B) < \infty$ *and* $H_3(A, B, -) = (0)$;

(iv) $\mathrm{fd}_A(B) < \infty$ *and the functor* $H_3(A, B, -)$ *is right-exact.*

Proof $(i) \Leftrightarrow (ii)$: Let $C = A[X_1, \ldots, X_n]$ and let $\mathfrak{a}$ be an ideal of C such that $B = C/\mathfrak{a}$. Then $\mathfrak{a}$ is locally generated by a regular sequence. Let $\mathfrak{q} \in \mathrm{Spec}(B)$ and $\mathfrak{p} = \mathfrak{q} \cap C$. By [6, Cor. IV.59 and Cor. V.27] it follows that for any $n \geq 1$ we have

$$H_n(A, B, E)_{\mathfrak{q}} = H_n(C, B, E)_{\mathfrak{q}} = H_n(C_{\mathfrak{p}}, B_{\mathfrak{q}}, E_{\mathfrak{q}}) = (0).$$

The assertion follows from [6, Th. VI.25].

$(i) \Rightarrow (iii)$: Theorem 6.2.4 implies that $\mathrm{fd}_A(B) < \infty$.

$(iii) \Rightarrow (iv)$: Obvious.

$(iv) \Rightarrow (i)$: Let $C, \mathfrak{q}$ and $\mathfrak{p}$ be as above. Then $\mathrm{fd}_{C_{\mathfrak{p}}}(B_{\mathfrak{q}}) < \infty$ and the functor $H_3(C_{\mathfrak{p}}, B_{\mathfrak{q}}, -)$ is right-exact. The assertion follows from [6, Th. XVII.11]. $\square$

Corollary 6.2.11 (Rodicio [318, Th. 2.6]) *Let A be a Noetherian complete intersection ring and B an A-algebra of finite type. The following are equivalent:*

(i) B is complete intersection over A;

(ii) $\mathrm{fd}_A(B) < \infty$ and $H_n(A, B, -) = (0)$ for sufficiently large n.

Proof $(i) \Rightarrow (ii)$: Follows from 6.2.10.

$(ii) \Rightarrow (i)$: Let $C = A[X_1, \ldots, X_n]$ and $\mathfrak{a}$ be an ideal of C such that $B = C/\mathfrak{a}$. Let $\mathfrak{q} \in \mathrm{Spec}(B)$ and $\mathfrak{p} = \mathfrak{q} \cap C$. Then $C_{\mathfrak{p}}$ is a local complete intersection ring. Therefore for $n \geq 3$ the Jacobi-Zariski sequence associated to $C_{\mathfrak{p}} \to B_{\mathfrak{q}} \to B_{\mathfrak{q}}/\mathfrak{q}B_{\mathfrak{q}} = k(\mathfrak{q})$ is

$$(0) = H_{n+1}\big(C_{\mathfrak{p}}, k(\mathfrak{q}), k(\mathfrak{q})\big) \to H_{n+1}\big(B_{\mathfrak{q}}, k(\mathfrak{q}), k(\mathfrak{q})\big) \to$$

$$\to H_n\big(C_{\mathfrak{p}}, B_{\mathfrak{q}}, k(\mathfrak{q})\big) \to H_n\big(C_{\mathfrak{p}}, k(\mathfrak{q}), k(\mathfrak{q})\big) = (0).$$

Hence $B_{\mathfrak{q}}$ is a complete intersection and we apply Corollary 6.2.10. $\square$

We will now study a class of morphisms that are close to smooth and regular morphisms, namely, the morphisms having the fibres reduced complete intersection. For this we need to remember the almost smooth morphisms. Recall that a morphism $u : A \to B$ is called almost smooth if $H_1(A, B, B) = 0$, see Definition 3.2.6.

Theorem 6.2.12 *Let $u : A \to B$ be a morphism of Noetherian rings. The following conditions are equivalent:*

(i) u is almost smooth;

(ii) $H^1(A, B, E) \cong \mathrm{Ext}_B^1(\Omega_{B/A}, E)$, for any B-module E;

(iii) $H_1(A, B, E) \cong \mathrm{Tor}_1^B(\Omega_{B/A}, E)$, for any B-module E;

(iv) $H_1(A, B, E) = (0)$, for any flat B-module E;

(v) $H^1(A, B, E) = (0)$, for any injective B-module E.

Proof The equivalence of (i), (ii) and (iii) is just for memory, being contained in Lemma 3.2.5. The equivalence with (iv) and (v) follows at once.   □

Example 6.2.13

(i) From Theorem 3.2.8 we see that any regular morphism is almost smooth. Consequently any smooth morphism is almost smooth.
(ii) If A is a ring and $n \geq 2$ is any natural number, the morphism $A \to B := A[X]/(X^n)$ is never almost smooth. Indeed, we have the exact sequence

$$0 \to H_1(A, B, B) \to \frac{(X^n)}{(X^{2n})} \xrightarrow{\delta} \Omega_{A[X]/A} \otimes_A B \to \Omega_{B/A} \to 0.$$

Since $\delta(x^{n+1}) = dx^{n+1} \otimes 1 = (n+1)X^n dX \otimes 1 = (n+1)dX \otimes X^n = 0$, where d is the universal derivation, it follows that δ is not injective.
(iii) Later on, in Example 6.2.35, we will see an example of an almost smooth morphism that is not smooth.

Proposition 6.2.14 (André [10, Prop. 1]) *Let k be a field and A a Noetherian k-algebra.*

(i) if A is geometrically reduced over k and is a complete intersection, then A is almost smooth over k;
(ii) If A is almost smooth over k and $\mathrm{fd}_A \Omega_{A/k} \leq 1$, then A is geometrically reduced over k.
(iii) If A is a complete intersection and is almost smooth over k, then $\mathrm{fd}_A \Omega_{A/k} \leq 1$.

Proof

(i) Let $\mathfrak{q}_1, \ldots, \mathfrak{q}_t$ be the minimal prime ideals of A. Since A is geometrically reduced over k, for any $i = 1, \ldots, t$ the ring $L_i := A_{\mathfrak{q}_i}$ is a field, separable extension of k and moreover $A \subseteq \bigoplus_{i=1}^{t} L_i$. By Lemma 6.2.6 we know that $H_2(k, A, -) = (0)$, hence the functor $H_1(k, A, -)$ is left-exact. Therefore we have an injection

$$0 \to H_1(k, A, A) \to H_1(k, A, \bigoplus_{i=1}^{t} L_i) \cong \bigoplus_{i=1}^{t} H_1(k, L_i, L_i) = 0.$$

(ii) Since $H_1(k, A, E) \cong \mathrm{Tor}_1^A(\Omega_{A/k}, E)$ for any A-module E and $\mathrm{fd}_A \Omega_{A/k} \leq 1$, it follows that $H_1(k, A, -)$ is left-exact. Therefore, for any prime ideal $\mathfrak{q} \in \mathrm{Ass}(A)$ we have $H_1(k, A, A/\mathfrak{q}) = 0$. Let $\mathfrak{n}$ be the nilradical of A. Assume that $\mathfrak{n} \neq (0)$ and let

$q \in \mathrm{Ass}_A(\mathfrak{n}) \subseteq \mathrm{Ass}_A(A)$. Then

$$H_1(k, A_\mathfrak{q}, A_\mathfrak{q}/\mathfrak{q}A_\mathfrak{q}) = H_1(k, A, A/\mathfrak{q})_\mathfrak{q} = (0),$$

hence $k \to A_\mathfrak{q}$ is formally smooth. This implies that $A_\mathfrak{q}$ is a regular local ring, that is, $\mathfrak{n}A_\mathfrak{q} = (0)$. Contradiction, whence $\mathfrak{n} = 0$, that is, A is reduced. If $k \subseteq K$ is a finite field extension, we have

$$H_1(k, A, A) \otimes_k K = H_1(k, A, A \otimes_k K) = H_1(K, A \otimes_k K, A \otimes_k K)$$

and we use the same argument as above.

(iii) We have $H_1(k, A, A) = H_2(k, A, A) = (0)$; hence by [6, Lemme III.19] and because of the complete intersection property, we obtain

$$H_2(k, A, -) = \mathrm{Tor}_2^A(\Omega_{A/k}, -) = (0).$$

$\square$

Theorem 6.2.15 (André [10, Th. 2]) *Let $u : A \to B$ be a local flat morphism of Noetherian local rings. If u is a reduced and complete intersection morphism, then u is almost smooth and $\Omega_{B/A}$ is a flat A-module.*

Proof Let $\mathfrak{q} \in \mathrm{Spec}(A)$ and $\mathfrak{p} = \mathfrak{q} \cap A$. By assumption $B_\mathfrak{q}/\mathfrak{p}B_\mathfrak{q}$ is a complete intersection that is geometrically reduced over $k(\mathfrak{p})$. By Proposition 6.2.14, (i) we obtain that

$$H_1(k(\mathfrak{p}), B_\mathfrak{q}/\mathfrak{p}B_\mathfrak{q}, B_\mathfrak{q}/\mathfrak{p}B_\mathfrak{q}) = (0),$$

hence

$$H_1(A, B, B_\mathfrak{q}/\mathfrak{p}B_\mathfrak{q}) = H_1(A, B, B/\mathfrak{p}B)_\mathfrak{q} = (0),$$

for any prime ideal $\mathfrak{q}$ of B. Now fixing $\mathfrak{p}$ we get that

$$H_1(A, B, B \otimes k(\mathfrak{p})) = (0),$$

for any prime ideal $\mathfrak{p} \in \mathrm{Spec}(A)$. Since u is a complete intersection morphism, the functor $H_1(A, B, -)$ is left-exact, whence $H_1(A, B, B \otimes A/\mathfrak{p}) = 0$, for any prime ideal $\mathfrak{p} \in \mathrm{Spec}(A)$. Thus $H_1(A, B, B \otimes_A -) = (0)$, hence $H_1(A, B, B) = (0)$ and

$$\mathrm{Tor}_1^A(\Omega_{B/A}, A/\mathfrak{p}) = H_1(A, B, B/\mathfrak{p}B) = (0)$$

for any $\mathfrak{p} \in \mathrm{Spec}(A)$, that is $\Omega_{B/A}$ is flat over A.

$\square$

Theorem 6.2.16 (André [10, Th. 2]) *Let* $u : A \to B$ *be a local flat morphism of Noetherian local rings. Assume that* u *is almost smooth,* $\Omega_{B/A}$ *is a flat* A-*module and* $\mathrm{fd}_B(\Omega_{B/A}) \leq 1$. *Then* u *is a reduced morphism.*

Proof Let $\mathfrak{q} \in \mathrm{Spec}(A)$ and $\mathfrak{p} = \mathfrak{q} \cap A$. By assumption $H_1(A, B, B) = (0)$. From [6, Prop. IV.54, Cor. IV.59 and Cor. V.27] it follows that

$$H_1\big(k(\mathfrak{p}), B_\mathfrak{q} \otimes k(\mathfrak{p}), B_\mathfrak{q} \otimes k(\mathfrak{p})\big) = (0).$$

Also we see that $\mathrm{fd}_{B_\mathfrak{q} \otimes k(\mathfrak{p})}(\Omega_{B_\mathfrak{q} \otimes k(\mathfrak{p})/k(\mathfrak{p})}) \leq 1$. By Proposition 6.2.14 it follows that $B_\mathfrak{q} \otimes k(\mathfrak{p})$ is geometrically reduced over $k(\mathfrak{p})$ for any prime ideals $\mathfrak{q} \in \mathrm{Spec}(B)$ and $\mathfrak{p} = \mathfrak{q} \cap A$. But this means exactly that u is reduced. $\square$

Theorem 6.2.17 (André [10, Th. 2]) *Let* $u : A \to B$ *be a local flat morphism of Noetherian local rings. Assume that* u *is almost smooth and a complete intersection morphism. Then* $\mathrm{fd}_B(\Omega_{B/A}) \leq 1$.

Proof By Proposition 6.2.7 we know that $H_2(A, B, -) = (0)$. Since by assumption we have $H_1(A, B, B) = (0)$, applying [6, Lemme III.19] we obtain

$$\mathrm{Tor}_2^B(\Omega_{B/A}, E) = H_2(A, B, E) = (0),$$

for any B-module E, hence $\mathrm{fd}_B(\Omega_{B/A}) \leq 1$. $\square$

The class of reduced complete intersection morphisms was considered by Michel André who called them *morphismes pséudo-réguliers*.

Definition 6.2.18 (André [10, Déf. 3]) A flat morphism $u : A \to B$ is called a **pseudo-regular morphism** if the fibres of u are reduced and complete intersection. A Noetherian local ring A is called **pseudo-excellent** if the formal fibres of A are reduced and complete intersection.

Remark 6.2.19 A Noetherian local ring A is pseudo-excellent if and only if the completion morphism $A \to \widehat{A}$ is pseudo-regular.

The next theorem shows the subtle narrow difference between quasi-excellent and pseudo-excellent rings.

Theorem 6.2.20 (André [10, Prop. 8]) *Let A be a Noetherian local ring. Then:*

(i) *A is pseudo-excellent if and only if A is a Nagata ring, $\mathrm{fd}_A(\Omega_{\widehat{A}/A}) = 0$ and $\mathrm{fd}_{\widehat{A}}(\Omega_{\widehat{A}/A}) \leq 1$;*

(ii) *A is quasi-excellent if and only if A is a Nagata ring and $\mathrm{fd}_{\widehat{A}}(\Omega_{\widehat{A}/A}) = 0$.*

Proof

(i) If A is pseudo-excellent, it is a Nagata ring by definition, while from Theorem 6.2.15 it follows that $\Omega_{\widehat{A}/A}$ is a flat A-module and that the canonical completion morphism $u : A \to \widehat{A}$ is almost smooth. Now Theorem 6.2.17 tells us that $\mathrm{fd}_{\widehat{A}}(\Omega_{\widehat{A}/A}) \leq 1$. Conversely, the completion morphism $u : A \to \widehat{A}$ is reduced by Theorem 2.3.6; hence we have only to show that the fibres of u are complete intersections. Let $\mathfrak{q}$ be a prime ideal of $\widehat{A}$ and $\mathfrak{p} = \mathfrak{q} \cap A$. First let us note that by assumption there is an exact sequence

$$0 \to F \to G \to \Omega_{\widehat{A}/A} \to 0$$

where F and G are flat $\widehat{A}$-modules. Tensorizing with $A/\mathfrak{p}$ over A, we obtain an exact sequence

$$0 = \mathrm{Tor}_1^A(\Omega_{\widehat{A}/A}, A/\mathfrak{p}) \to F/\mathfrak{p}F \to G/\mathfrak{p}G \to \Omega_{\widehat{A}/A} \otimes_A A/\mathfrak{p} \to 0,$$

because $\Omega_{\widehat{A}/A}$ is flat over A and furthermore an exact sequence

$$0 \to F_\mathfrak{q}/\mathfrak{p}F_\mathfrak{q} \to G_\mathfrak{q}/\mathfrak{p}G_\mathfrak{q} \to (\Omega_{\widehat{A}/A} \otimes_A A/\mathfrak{p})_\mathfrak{q} \to 0,$$

where $F_\mathfrak{q}/\mathfrak{p}F_\mathfrak{q}$ and $G_\mathfrak{q}/\mathfrak{p}G_\mathfrak{q}$ are flat $\widehat{A}_\mathfrak{q}/\mathfrak{p}\widehat{A}_\mathfrak{q}$-modules. This means that

$$\mathrm{fd}_{\widehat{A}_\mathfrak{q}/\mathfrak{p}\widehat{A}_\mathfrak{q}}(\Omega_{(\widehat{A}_\mathfrak{q}/\mathfrak{p}\widehat{A}_\mathfrak{q})/k(\mathfrak{p})}) \leq 1.$$

Thus we see that all the fibres of u have the properties from (i). By the classical Cohen structure theorem [131, Th. 19.8.8] (or by Cohen's factorization 6.1.6), there exists a regular local ring R, a surjective morphism $R \to \widehat{A} = R/\mathfrak{a}$ and a regular morphism $\mathbb{P} \to R$, where $\mathbb{P} = \mathbb{Q}$ or $\mathbb{P} = \mathbb{Z}_{(p)}$, for a certain prime number p. Let $\mathfrak{r}$ be the prime ideal of R such that $\mathfrak{q} = \mathfrak{r}/\mathfrak{a}$. There are two possibilities:

1. $\mathfrak{r} \cap \mathbb{P} = (0)$. Then $\mathbb{P} = \mathbb{Q}$ and the $\mathbb{Q}$-algebra $R_\mathfrak{r}$ is formally smooth, hence a regular local ring. Therefore $\widehat{A}_\mathfrak{q}$ is the quotient of a regular local ring containing a field, namely, $R_\mathfrak{r}$; hence also $\widehat{A}_\mathfrak{q}/\mathfrak{p}\widehat{A}_\mathfrak{q}$ is the quotient of a regular local ring containing a field.

2. $\mathfrak{r} \cap \mathbb{P} \neq (0)$. Then $\mathbb{P} = \mathbb{Z}_{(p)}$ for some prime number p and $\mathfrak{r} \cap \mathbb{P} = p\mathbb{Z}_{(p)}$. In this case the integral domains $\widehat{A}/\mathfrak{q}$ and $R/\mathfrak{r}$ both contain a field of characteristic $p > 0$. Moreover the morphism $\mathbb{F}_p \to R_{\mathfrak{r}}/pR_{\mathfrak{r}}$ is formally smooth; hence $\widehat{A}_{\mathfrak{q}}/\mathfrak{p}\widehat{A}_{\mathfrak{q}}$ is the quotient of the regular local ring containing a field, namely, $R_{\mathfrak{r}}/pR_{\mathfrak{r}}$.

Anyway, in all cases we may assume that the fibre $C := \widehat{A}_{\mathfrak{q}}/\mathfrak{p}\widehat{A}_{\mathfrak{q}}$ is the quotient of an equicharacteristic regular local ring. Let P be the prime field of $k(\mathfrak{p}) := A_{\mathfrak{p}}/\mathfrak{p}A_{\mathfrak{p}}$. We have a commutative diagram

$$\begin{array}{ccccccccc}
H_1(k(\mathfrak{p}), C, C) & \xrightarrow{\alpha} & \Omega_{k(\mathfrak{p})/P} \otimes_{k(\mathfrak{p})} C & \longrightarrow & \Omega_{C/P} & \longrightarrow & \Omega_{C/k(\mathfrak{p})} & \longrightarrow & 0 \\
\downarrow & & \downarrow{\scriptstyle \beta} & & & & & & \\
\displaystyle\sum_{i=1}^{s} H_1(k(\mathfrak{p}), C, L_i) & \longrightarrow & \displaystyle\sum_{i=1}^{s} \Omega_{k(\mathfrak{p})/P} \otimes L_i & & & & & &
\end{array}$$

where $L_1, \ldots, L_s$ are the residue fields of the reduced ring C in the minimal prime ideals of C. Because for all indices i we have $H_1(k(\mathfrak{p}), C, L_i) = (0)$, it follows that $\beta \circ \alpha = 0$. But β is an injection, hence $\alpha = 0$. Consequently we have an exact sequence

$$0 \to \Omega_{k(\mathfrak{p})/P} \otimes_{k(\mathfrak{p})} C \to \Omega_{C/P} \to \Omega_{C/k(\mathfrak{p})} \to 0,$$

where the left-side module is free. But then $\mathrm{fd}_C(\Omega_{C/P}) \leq 1$, by [393, Ex. 4.1.2]. This shows that we may assume that $k(\mathfrak{p}) := k$ is the prime field. The ring C is a quotient of a regular equicharacteristic local ring U, say $C = \widehat{A}_{\mathfrak{q}}/\mathfrak{p}\widehat{A}_{\mathfrak{q}} = U/\mathfrak{b}$. Consider a minimal prime ideal $\mathfrak{r}$ of C and let $\mathfrak{s}$ be the corresponding prime ideal of U. We have an exact sequence

$$0 \to T \to \Omega_{U/k} \otimes_U C \to \Omega_{C/k} \to 0.$$

Since U is formally smooth over k, the module $\Omega_{U/k} \otimes_U C$ is a flat C-module. But $\mathrm{fd}_C(\Omega_{C/k}) \leq 1$; hence T is a flat C-module. But T is a quotient of the finitely generated C-module $\mathfrak{b}/\mathfrak{b}^2 = H_1(U, C, C)$, therefore it is a finitely generated free C-module. Let t denote its rank. There is an exact sequence

$$0 \to H_1(U, C, C_{\mathfrak{r}}) \to \Omega_{U/k} \otimes_U C_{\mathfrak{r}} \to \Omega_{C/k} \otimes_C C_{\mathfrak{r}} \to 0,$$

because $H_1(k, C, C_{\mathfrak{r}}) = (0)$. This means that

$$T_{\mathfrak{r}} \cong H_1(B, C, C_{\mathfrak{r}}) \cong H_1(B_{\mathfrak{s}}, C_{\mathfrak{r}}, C_{\mathfrak{r}}).$$

On the other hand, $U_\mathfrak{s}$ is a regular local ring and $C_\mathfrak{r}$ is its residue field. We have

$$t = \dim(U_\mathfrak{s}) = \dim(U) - \dim(U/\mathfrak{s}) =$$

$$= \dim(U) - \dim(C/\mathfrak{r}) \geq \dim(U) - \dim(C).$$

Now we apply [6, Lemme S.24] and we obtain: either $\mathfrak{b}$ is generated by a regular sequence of t elements and consequently C is a complete intersection or $\mathfrak{b}$ contains a regular sequence of $t + 1$ elements. In the second case, we have

$$\dim(C) \leq \dim(U) - t - 1 \leq \dim(C) - 1.$$

Thus C is a complete intersection.

(ii) If A is quasi-excellent, then by Proposition 3.3.7 it is a Nagata ring and by Theorem 3.2.8 $\Omega_{\widehat{A}/A}$ is a flat A-module. Conversely, we already know that A is pseudo-excellent; hence by Theorem 6.2.15 the completion morphism $A \to \widehat{A}$ is almost smooth, that is, $H_1(A, \widehat{A}, \widehat{A}) = (0)$. Then

$$H_1(A, \widehat{A}, -) \cong \mathrm{Tor}_1^{\widehat{A}}(\Omega_{\widehat{A}/A}, -) = (0),$$

because $\Omega_{\widehat{A}/A}$ is a flat $\widehat{A}$-module. Therefore A is quasi-excellent.   □

Example 6.2.21 There exist Noetherian local rings that are pseudo-excellent and not quasi-excellent, as one can deduce from Example 3.3.10.

The final part of this section is devoted to an interesting problem in Commutative Algebra. We need some preparations.

Notation 6.2.22 Let $u : A \to B$ be a morphism of Noetherian rings and $\mathfrak{a}$ an ideal of B. We denote by $D(\mathfrak{a}) := \{f \in \mathfrak{a} \mid d(f) \in \mathfrak{a}, \forall d \in \mathrm{Der}_A(B)\}$.

Lemma 6.2.23 *Let* $u : A \to B$ *be a morphism of Noetherian rings and* $\mathfrak{a}$ *be an ideal of* B. *Then* $D(\mathfrak{a})$ *is an ideal of* B *such that* $\mathfrak{a}^2 \subseteq D(\mathfrak{a}) \subseteq \mathfrak{a}$.

Proof It is easy to prove that $D(\mathfrak{a})$ is an ideal and by definition $D(\mathfrak{a}) \subseteq \mathfrak{a}$. If $x, y \in \mathfrak{a}$, then $d(xy) = xd(y) + yd(x) \in \mathfrak{a}$ for any $d \in \mathrm{Der}_A(B)$. This implies that $\mathfrak{a}^2 \subseteq D(\mathfrak{a})$.   □

Definition 6.2.24 Let $u : A \to B$ be a morphism of Noetherian rings and $\mathfrak{a}$ be an ideal of B. The ideal $D(\mathfrak{a})$ is called the **primitive ideal** of $\mathfrak{a}$.

Lemma 6.2.25 $u : A \to B$ *be a morphism of Noetherian rings and* $\mathfrak{q}$ *be a primary ideal of* B. *Then also* $D(\mathfrak{q})$ *is a primary ideal.*

Proof Let $x, y \in B$ be such that $xy \in D(\mathfrak{q})$, that is $xy \in \mathfrak{q}$ and $d(xy) \in \mathfrak{q}$ for any $d \in \mathrm{Der}_A(B)$ and assume that $x \notin D(\mathfrak{q})$. Then either $x \notin \mathfrak{q}$ or $x \in \mathfrak{q}$ and $x \notin D(\mathfrak{q})$. If $x \notin \mathfrak{q}$, there is some $n \geq 1$ such that $y^n \in \mathfrak{q}$ and then for any derivation $d \in \mathrm{Der}_A(B)$ we have $d(y^{n+1}) = (n+1)y^n d(y) \in \mathfrak{q}$, that is $y^{n+1} \in D(\mathfrak{q})$. If $x \in \mathfrak{q}$ and there is some $d_0 \in \mathrm{Der}_A(B)$ such that $d_0(x) \notin \mathfrak{q}$, we have $yd_0(x) = d_0(xy) - xd_0(y) \in \mathfrak{q}$, so that there exists some $n \geq 1$ such that $y^n \in \mathfrak{q}$. But we have already seen that this implies $y^{n+1} \in D(\mathfrak{q})$. $\qquad\square$

Before going further, we need to recall a well-known notion.

Definition 6.2.26 Let A be a ring, n a natural number and $\mathfrak{a}$ an ideal of A. The n-th **symbolic power** of $\mathfrak{a}$ is

$$\mathfrak{a}^{(n)} := \bigcap_{\mathfrak{q} \in \mathrm{Ass}(A/\mathfrak{a})} \left(\mathfrak{a}^n A_\mathfrak{q} \cap A \right) = \{x \in A \mid x \in \mathfrak{a}^n A_\mathfrak{q}, \forall \, \mathfrak{q} \in \mathrm{Ass}(A/\mathfrak{a})\}.$$

Remark 6.2.27

(i) If $\mathfrak{p}$ is a prime ideal of the ring A, then $\mathfrak{p}^{(n)} = \mathfrak{p}^n A_\mathfrak{p} \cap A$. Thus $\mathfrak{p}^n = \mathfrak{p}^{(n)}$ iff $\mathfrak{p}^n$ is a $\mathfrak{p}$-primary ideal.
(ii) If $\mathfrak{a} = \mathfrak{p}_1 \cap \ldots \cap \mathfrak{p}_t$ is a radical ideal and $\mathfrak{p}_1, \ldots, \mathfrak{p}_t$ are its minimal primes, then $\mathfrak{a}^{(n)} = \mathfrak{p}_1^{(n)} \cap \ldots \cap \mathfrak{p}_t^{(n)}$.
(iii) It is easy to see that $\mathfrak{a}^n \subseteq \mathfrak{a}^{(n)}$. On the other hand, a basic example in any lecture on primary decomposition is an example of a prime ideal having a power that is not primary, hence strictly contained in the corresponding symbolic power [18, Ex. 3), p. 51].

Lemma 6.2.28 *Let* $u : A \to B$ *be a morphism of Noetherian rings and* $\mathfrak{a}$ *an ideal of* B. *Then* $\mathfrak{a}^{(2)} \subseteq D(\mathfrak{a})$.

Proof Let $z \in \mathfrak{a}^{(2)} = \bigcap_{\mathfrak{q} \in \mathrm{Ass}(B/\mathfrak{a})} (\mathfrak{a}^2 B_\mathfrak{q} \cap B) = \{x \in B \mid x \in \mathfrak{a}^2 B_\mathfrak{q}, \forall \, \mathfrak{q} \in \mathrm{Ass}(B/\mathfrak{a})\}$.

Then there exists an element $s \in B$ which is a non-zero divisor on $B/\mathfrak{a}$ such that $s \cdot z \in \mathfrak{a}^2$, so that $z \in \mathfrak{a}$. Now for any derivation $d \in \mathrm{Der}_A(B)$ we have

$$s \cdot d(z) + z \cdot d(s) \in d(\mathfrak{a}^2) \subseteq \mathfrak{a},$$

hence $s \cdot d(z) \in \mathfrak{a}$ and since s is a non-zero divisor on $B/\mathfrak{a}$ we get $d(z) \in \mathfrak{a}$. $\qquad\square$

Remark 6.2.29 Let A be a ring and $u : A \to C$ a ring morphism. If

$$A \to B = A[X_i, i \in I] \to C = B/\mathfrak{a}$$

is a polynomial presentation of C, there exists an exact sequence [6, Th. V.1 and Cor. III.36]

$$0 \to H_1(A, C, C) \to \mathfrak{a}/\mathfrak{a}^2 \xrightarrow{\delta_{C/B/A}} \Omega_{B/A} \otimes_B C \xrightarrow{v} \Omega_{C/A} \to 0,$$

where $\delta_{C/B/A}$ is induced by the universal A-derivation $d : B \to \Omega_{B/A}$. More precisely $\delta_{C/B/A}(x + \mathfrak{a}^2) = d(x) \otimes 1$. Thus there exists a canonical morphism $\gamma_{C/B/A} : \mathfrak{a}/\mathfrak{a}^2 \to \ker(v)$. We shall denote for simplicity, when there is no possible confusion, $\delta := \delta_{C/B/A}$ and $\gamma := \gamma_{C/B/A}$. The same argument is valid if B is not necessarily a polynomial ring over A but is smooth or at least $\mathfrak{a}$-smooth over A.

Proposition 6.2.30 *Let $u : A \to B$ be a morphism of Noetherian rings. Let $\mathfrak{a}$ be an ideal of B and let $C := B/\mathfrak{a}$. Suppose that u is $\mathfrak{a}$-smooth and that $\Omega_{B/A}$ is a free B-module. Then $\ker(\delta_{C/B/A}) = D(\mathfrak{a})/\mathfrak{a}^2$.*

Proof Since u is $\mathfrak{a}$-smooth, by Theorem 3.2.3 we have $H_1(A, B, C) = 0$; hence the sequence of C-modules

$$0 \to H_1(A, C, C) \to \mathfrak{a}/\mathfrak{a}^2 \xrightarrow{\delta} \Omega_{B/A} \otimes_B C \xrightarrow{v} \Omega_{C/A} \to 0$$

is exact. Suppose that $\Omega_{B/A}$ has a basis $\{\omega_i\}_{i \in I}$ and let $\{d_i\}_{i \in I}$ be the dual basis in $\mathrm{Hom}_B(\Omega_{B/A}, B) \cong \mathrm{Der}_A(B)$. Thus

$$d_i(\omega_j) = \delta_{ij} = \begin{cases} 0, & \text{if } i \neq j \\ 1, & \text{if } i = j. \end{cases}$$

Let $\partial \in \mathrm{Der}_A(B)$. Then ∂ has a unique factorization $\partial = v \circ d$, where $d : \Omega_{B/A} \to B$ is the universal A-derivation of B and $v \in \mathrm{Hom}_B(\Omega_{B/A}, B)$. Therefore we have $\partial(f) = \sum_{i \in I} v(\omega_i) d_i(f)$ and we get

$$D(\mathfrak{a}) = \{f \in \mathfrak{a} \mid d_i(f) \in \mathfrak{a} \text{ for all } i \in I\}.$$

But then one can see immediately that since

$$\delta(f \bmod \mathfrak{a}^2) = d(f) \otimes 1 = \Big(\sum_{i \in I}(d_i(f)\omega_i)\Big) \otimes 1 = \sum_{i \in I} \omega_i \otimes \big(\phi(d_i(f))\big),$$

where $\phi : B \to C$ is the canonical surjection, we have

$$H_1(A, C, C) = \ker(\delta) = \{f + \mathfrak{a}^2 \mid \partial(f) \in \mathfrak{a}, \ \forall \partial \in \mathrm{Der}_A(B)\} = D(\mathfrak{a})/\mathfrak{a}^2.$$

$\square$

Corollary 6.2.31 *Let $u : A \to B$ be a morphism of Noetherian rings. Let $\mathfrak{a}$ be an ideal of B and let $C := B/\mathfrak{a}$. Suppose that u is $\mathfrak{a}$-smooth and that $\Omega_{B/A}$ is a free B-module. The following are equivalent:*

(i) *C is almost smooth over A;*

(ii) *the sequence of C-modules $0 \to \mathfrak{a}/\mathfrak{a}^2 \xrightarrow{\delta} \Omega_{B/A} \otimes_B C \to \Omega_{C/A} \to 0$ is exact;*

(iii) *$D(\mathfrak{a}) = \mathfrak{a}^2$.*

Proof As in the proof of Proposition 6.2.30 the sequence of C-modules

$$0 \to H_1(A, C, C) \to \mathfrak{a}/\mathfrak{a}^2 \xrightarrow{\delta} \Omega_{B/A} \otimes_B C \to \Omega_{C/A} \to 0$$

is exact.

(i) $\Leftrightarrow$ (ii): Follows from the above exact sequence.
(i) $\Leftrightarrow$ (iii): Follows from Proposition 6.2.30. $\square$

Corollary 6.2.32 *Let $u : A \to B$ be a morphism of Noetherian rings. The following are equivalent:*

(i) *u is almost smooth;*

(ii) *for any polynomial presentation $B = A[X_i, i \in I]/\mathfrak{a}$, we have $D(\mathfrak{a}) = \mathfrak{a}^2$;*

(iii) *there exists a polynomial presentation $B = A[X_i, i \in I]/\mathfrak{a}$, such that $D(\mathfrak{a}) = \mathfrak{a}^2$;*

Proof Everything follows from Corollary 6.2.31 because $A \to A[X_i, i \in I]$ is smooth, hence $\mathfrak{a}$-smooth for any ideal $\mathfrak{a}$ and $\Omega_{A[X_i, i \in I]/A}$ is free. $\square$

Corollary 6.2.33 *Let $u : A \to C$ be a morphism of Noetherian rings, let $B = A[X_i, i \in I]$ and let $\mathfrak{p}$ be a prime ideal in B such that $C := B/\mathfrak{p}$. If C is almost smooth over A, then $\mathfrak{p}^{(2)} = \mathfrak{p}^2$.*

Proof Since $\mathfrak{p}$ is prime, by Lemma 6.2.25 the ideal $D(\mathfrak{p})$ is primary and Corollary 6.2.32 implies that $D(\mathfrak{p}) = \mathfrak{p}^2$. Hence $\mathfrak{p}^2$ is $\mathfrak{p}$-primary. $\square$

Proposition 6.2.34 (Eisenbud-Mazur [101, Th. 3]) *Let A be a Noetherian regular ring, B a localization of a polynomial ring in finitely many variables over A and $C = B/\mathfrak{a}$, where $\mathfrak{a}$ is a radical ideal of B. If C is generically separable over A, then $\ker(\delta) = \mathfrak{a}^{(2)}/\mathfrak{a}^2$.*

Proof Let $\mathfrak{q}$ be a minimal prime of $\mathfrak{a}$ and $\mathfrak{p} = \mathfrak{q} \cap A$. We have a commutative diagram

$$
\begin{array}{ccccccccc}
0 & \longrightarrow & H_1(A,C,C) = \ker(\delta) & \longrightarrow & \mathfrak{a}/\mathfrak{a}^2 & \overset{\delta}{\longrightarrow} & \Omega_{B/A} \otimes_B C & \overset{v}{\longrightarrow} & \Omega_{C/A} \longrightarrow 0 \\
& & \downarrow & & \downarrow & & \downarrow & & \downarrow \\
(0) = & & H_1(A_\mathfrak{p}, C_\mathfrak{q}, C_\mathfrak{q}) = \ker(\delta_\mathfrak{q}) & \longrightarrow & \mathfrak{a}C_\mathfrak{q}/\mathfrak{a}^2 C_\mathfrak{q} & \overset{\delta_\mathfrak{q}}{\longrightarrow} & \Omega_{B_\mathfrak{q}/A_\mathfrak{p}} \otimes_B C_\mathfrak{q} & \overset{v_\mathfrak{q}}{\longrightarrow} & \Omega_{C_\mathfrak{q}/A} \longrightarrow 0
\end{array}
$$

Since $\mathfrak{a}$ is a radical ideal, C is reduced, hence $C_\mathfrak{q}$ is a field, separable over the domain $A_\mathfrak{p}$ by generic separability. Therefore

$$\ker(\delta_\mathfrak{q}) = H_1(A_\mathfrak{p}, C_\mathfrak{q}, -) = (0).$$

Because $\Omega_{B/A} \otimes_B C$ is free over C, the map

$$\Omega_{B/A} \otimes_B C \to \bigoplus_{\mathfrak{q} \in \mathrm{Min}(\mathfrak{a})} \Omega_{B_\mathfrak{q}/A_\mathfrak{p}} \otimes_{B_\mathfrak{q}} C_\mathfrak{q}.$$

is an injection. This means that in the commutative diagram below, the maps ψ, $\delta_\mathfrak{q}$ and $\beta \circ \psi$ are injective.

$$
\begin{array}{ccc}
\mathfrak{a}/\mathfrak{a}^2 & \overset{\delta}{\longrightarrow} & \Omega_{B/A} \otimes_B C \\
\phi \downarrow & & \downarrow \psi \\
\mathfrak{a}C_\mathfrak{q}/\mathfrak{a}^2 C_\mathfrak{q} & \overset{\delta_\mathfrak{q}}{\longrightarrow} & \Omega_{B_\mathfrak{q}/A_\mathfrak{p}} \otimes_B C_\mathfrak{q} \\
\alpha \downarrow & & \downarrow \beta \\
\bigoplus_{\mathfrak{q} \in \mathrm{Min}(\mathfrak{a})} \mathfrak{a}C_\mathfrak{q}/\mathfrak{a}^2 C_\mathfrak{q} & \overset{\oplus \delta_\mathfrak{q}}{\longrightarrow} & \bigoplus_{\mathfrak{q} \in \mathrm{Min}(\mathfrak{a})} \Omega_{B_\mathfrak{q}/A_\mathfrak{p}} \otimes_{B_\mathfrak{q}} C_\mathfrak{q}
\end{array}
$$

Therefore, $\ker(\delta) = \ker(\alpha \circ \psi) = \mathfrak{a}^{(2)}/\mathfrak{a}$, as desired. $\square$

Example 6.2.35 Let k be a perfect field and let $C = k[X, Y, Z, T]$. We consider the ideal $\mathfrak{p} = (X^2 - YZ, Y^2 - XT, XY - ZT) \subset C$. Then $\mathfrak{p}$ is a prime ideal of C and $A = C/\mathfrak{p}$ is not smooth over k, since it is not a regular ring. One can also see that $\mathfrak{p}^2$ is $\mathfrak{p}$-primary. Then Proposition 6.2.34 implies that $H_1(k, A, A) = (0)$, that is, A is almost smooth over k.

The evolutions have been introduced by B. Mazur [255] in connection with the work of Wiles on Galois deformations and his proof of Fermat's last theorem. They have been previously studied by Scheja and Storch [339] and Böger [50].

Definition 6.2.36 (Mazur [101,255]) Let A be a Noetherian ring and C a local A-algebra essentially of finite type. A local A-algebra B, essentially of finite type, together with a surjective morphism $B \to C$ is called an **evolution of C over A** if the canonical morphism $\Omega_{B/A} \otimes_B C \to \Omega_{C/A}$ is an isomorphism. The evolution is called **trivial** if $B \to C$ is an isomorphism. The A-algebra C is called **evolutionarily stable** if it has no non-trivial evolution.

It is again interesting to see how B. Mazur motivates the term **evolution**: *The motivation for the term is just that, thinking of* $\mathrm{Spec}(C)$ *as a closed subscheme of* $\mathrm{Spec}(B)$, *the larger scheme* $\mathrm{Spec}(B)$ *possesses no new infinitesimal directions that can be seen from the vantage-point of* $\mathrm{Spec}(C)$, *i.e. whatever "growth" occurs, going from the smaller scheme to the larger, follows in already set-down directions: it is an "evolution"* [255, p. 97].

Remark 6.2.37 Let us consider the Jacobi-Zariski exact sequence associated to $A \to B \to C = B/\mathfrak{a}$:

$$\mathfrak{a}/\mathfrak{a}^2 \xrightarrow{\delta} \Omega_{B/A} \otimes_B C \to \Omega_{C/A} \to 0.$$

Then B is an evolution of C if and only if $\mathrm{Im}(\delta) = (0)$.

Example 6.2.38 Let k be a Noetherian domain and $A = k[X_1, \ldots, X_n]_{(X_1,\ldots,X_n)}$. Let $f \in A$ and let $\mathfrak{a} = (\partial f/\partial X_1, \ldots, \partial f/\partial X_n)A$, be the ideal generated by the partial derivatives of f. Let $C := A/(\mathfrak{a}, f)$ and $B := A/\mathfrak{a}$. Then B is an evolution of C, that is non-trivial if $f \notin \mathfrak{a}$.

Proof We have the exact sequence

$$\mathfrak{c}/\mathfrak{c}^2 \xrightarrow{\delta} \Omega_{B/A} \otimes_B C \to \Omega_{C/A} \to 0$$

where $C = B/\mathfrak{c}$, hence $\mathfrak{c} = (\mathfrak{a}, f)/\mathfrak{a}$. Let $x = a + bf \in (\mathfrak{a}, f)$, with $a \in \mathfrak{a}$ and $b \in A$. Then $a = 0$ in $\mathfrak{c}$ and

$$\delta(af) = 1 \otimes ad(f) + 1 \otimes fd(a) = a \otimes d(f) + f \otimes d(a) = 0,$$

since $d(f) = \sum_{i=1}^{n} \frac{\partial f}{\partial X_i} dX_i \in \mathfrak{a}\Omega_{C/A}$ and $f \in (\mathfrak{a}, f)$. $\square$

On the other hand, the next result shows that homogeneous polynomials, at least in the characteristic zero case, do not produce evolutions in this way.

Proposition 6.2.39 *Let $f \in \mathbb{C}[X_1, \ldots, X_n]$ be an homogeneous polynomial. Then*

$$f \in \left(\frac{\partial f}{\partial X_1}, \ldots, \frac{\partial f}{\partial X_n}\right)\mathbb{C}[X_1, \ldots, X_n].$$

Proof Suppose that the degree of f is k. If $f = X_1^{e_1} \ldots X_n^{e_n}$ is a monomial of total degree k, that is, $e_1 + \ldots + e_n = k$, then

$$X_1\frac{\partial f}{\partial X_1} + \ldots + X_n\frac{\partial f}{\partial X_n} = (e_1 + \ldots + e_n)f = kf,$$

hence $f \in \left(\frac{\partial f}{\partial X_1}, \ldots, \frac{\partial f}{\partial X_n}\right)\mathbb{C}[X_1, \ldots, X_n]$. For general homogeneous f, the proof follows.   □

In [255] Mazur asked the following question:

Question 6.2.40 (Evolution problem [255]) Let k be a field or a DVR of mixed characteristic. Let A be a reduced, local, essentially of finite type and flat k-algebra. Is it true that A is evolutionarily stable?

First we will see a criterion of H. Lenstra characterizing evolutionarily stable algebras. To make things simpler, we need the following definition.

Definition 6.2.41 Let A be a ring. An epimorphism of A-modules $u : M \to N$ is called a **minimal epimorphism** if there is no proper submodule $E \subsetneq M$ such that $u(E) = N$.

Lemma 6.2.42 *Let A be a ring and $u : A \to C$ a ring morphism. If $A \to B \to C = B/\mathfrak{a}$ is a polynomial presentation of C and $\mathfrak{b} \subseteq \mathfrak{a}$ is an ideal of B, then $B/\mathfrak{b}$ is an evolution of $C = B/\mathfrak{a}$ if and only if $\gamma(\mathfrak{a}/\mathfrak{a}^2) = \gamma\big((\mathfrak{b} + \mathfrak{a}^2)/\mathfrak{a}^2\big)$, where $\gamma = \gamma_{C/B/A}$.*

Proof Assume that $B/\mathfrak{b}$ is an evolution of $C = B/\mathfrak{a}$. It is clear that

$$\gamma(\mathfrak{a}/\mathfrak{a}^2) \supseteq \gamma\big((\mathfrak{b} + \mathfrak{a}^2)/\mathfrak{a}^2\big).$$

To prove the other inclusion, let $x \in \mathfrak{a}$. We need to find $z \in \mathfrak{b} + \mathfrak{a}^2$ such that $\gamma(z + \mathfrak{a}^2) = \gamma(x + \mathfrak{a}^2)$. Let $\pi : B/\mathfrak{b} \to B/\mathfrak{a} = C$. Then $\ker(\pi) = \mathfrak{a}/\mathfrak{b}$, so that

$$(\mathfrak{a}/\mathfrak{b})/(\mathfrak{a}/\mathfrak{b})^2 = \mathfrak{a}/(\mathfrak{a}^2 + \mathfrak{b})$$

and since $B/\mathfrak{b}$ is an evolution of C, we have

$$\bar{\delta}(\mathfrak{a}/(\mathfrak{a}^2 + \mathfrak{b})) = (0),$$

where

$$\bar{\delta} = \delta_{C/(B/\mathfrak{b})/A} : \mathfrak{a}/(\mathfrak{a}^2 + \mathfrak{b}) \to \Omega_{(B/\mathfrak{b})/A} \otimes_{B/\mathfrak{b}} C.$$

Therefore we know that $dx \otimes 1 = 0$ in $\Omega_{(B/\mathfrak{b})/A} \otimes_{B/\mathfrak{b}} C$. If $dx \otimes 1 = 0$ in $\Omega_{B/A} \otimes_B C$, we take $z = 0$. If $dx \otimes 1 \neq 0$ in $\Omega_{B/A} \otimes_B C$, then $x \in \mathfrak{b}$.
Conversely, assume that $\gamma(\mathfrak{a}/\mathfrak{a}^2) = \gamma((\mathfrak{b} + \mathfrak{a}^2)/\mathfrak{a}^2)$. Let $x \in \mathfrak{a}$. Then $\bar{\delta}(x + (\mathfrak{a}^2 + \mathfrak{b})) = d(x) \otimes 1$ in $\Omega_{(B/\mathfrak{b})/A} \otimes_{B/\mathfrak{b}} C$. On the other hand, in $\Omega_{B/A} \otimes_B C$, we have $d(x) \otimes 1 = d(\mathfrak{b}) \otimes 1$, for some $\mathfrak{b} \in \mathfrak{b}$, that is zero in $\Omega_{(B/\mathfrak{b})/A} \otimes_{B/\mathfrak{b}} C$.   □

Proposition 6.2.43 (H. Lenstra cf. [255, Lemma, §4, p.98]) *Let A be a Noetherian ring and C be a local A-algebra essentially of finite type. The following are equivalent:*

(i) *C is evolutionary stable over A;*
(ii) *for any presentation $C = B/\mathfrak{a}$, where B is a localization of a polynomial ring over A, the canonical map $\gamma : \mathfrak{a}/\mathfrak{a}^2 \to \ker\left(\Omega_{B/A} \otimes_B C \to \Omega_{C/A}\right)$ is minimal;*
(iii) *there exists a presentation $C = B/\mathfrak{a}$, where B is a localization of a polynomial ring over A, such that the canonical map $\gamma : \mathfrak{a}/\mathfrak{a}^2 \to \ker\left(\Omega_{B/A} \otimes_B C \to \Omega_{C/A}\right)$ is minimal.*

Proof $(i) \Leftrightarrow (ii)$: For any presentation $A \to B \to C = B/\mathfrak{a}$ and any ideal $\mathfrak{b} \subseteq \mathfrak{a}$, using Nakayama's Lemma in the local ring B we have

$$\mathfrak{b} = \mathfrak{a} \Leftrightarrow \mathfrak{b} + \mathfrak{a}^2 = \mathfrak{a} \Leftrightarrow (\mathfrak{b} + \mathfrak{a}^2)/\mathfrak{a}^2 = \mathfrak{a}/\mathfrak{a}^2.$$

Applying the Lemma 6.2.42 we get the assertion.
$(iii) \Rightarrow (ii)$: It is enough to show that the minimality of the epimorphism γ does not depend on the polynomial presentation. The family of presentations is filtered; hence it is enough to show that if $C = B/\mathfrak{a}$ is a presentation, D is a localization of a polynomial ring in one variable X over B, and $C = D/\mathfrak{c}$ is a presentation extending $C = B/\mathfrak{a}$, then the two maps γ and γ' corresponding to the two presentations are simultaneously minimal. Let $g \in B$ be an element having the same image in C as X, that is, $X - g \in \mathfrak{c}$. Replacing X by $X - g$, we may assume that $g = 0$. Then $\mathfrak{c}/\mathfrak{c}^2 = \mathfrak{a}/\mathfrak{a}^2 \oplus CX$. It is easy to see that $\mathrm{Im}(\gamma') = \mathrm{Im}(\gamma) \oplus CdX$. If γ is minimal, let E be a submodule of $\mathfrak{c}/\mathfrak{c}^2$ such that $\gamma'(E) = \mathrm{Im}(\gamma) \oplus CdX$. We can write $E = E_1 \oplus E_2$, where $E_1 \subseteq \mathfrak{a}/\mathfrak{a}^2$ and $E_2 \subseteq CX$. Then $\gamma(E_1) = \mathrm{Im}(\gamma)$ and since γ is minimal, $E_1 = \mathfrak{a}/\mathfrak{a}^2$. But $E_2 = aCX$, hence by Nakayama's Lemma $aCdX = CdX$ iff $a = 1$. Thus $E_2 = CX$. Conversely, if γ' is minimal and $E \subseteq \mathfrak{a}/\mathfrak{a}^2$ is such that $\gamma(E) = \mathrm{Im}(\gamma)$, then $\gamma'(E \oplus CX) = CdX \oplus \mathrm{Im}(\delta) = \mathrm{Im}(\gamma')$, hence $E = \mathfrak{a}/\mathfrak{a}^2$.   □

Theorem 6.2.44 (Hübl [167, Th. 1.1]) *Let A be a Noetherian ring and $(C, \mathfrak{m})$ be a local A-algebra essentially of finite type. The following are equivalent:*

  (i) C is evolutionary stable over A;

  (ii) for any local smooth A-algebra essentially of finite type B and any ideal $\mathfrak{a}$ of B such that $C = B/\mathfrak{a}$, we have $D(\mathfrak{a}) \subseteq \mathfrak{m} \cdot \mathfrak{a}$;

  (iii) there exists a local smooth A-algebra essentially of finite type B and an ideal $\mathfrak{a}$ of B such that $C = B/\mathfrak{a}$ and such that $D(\mathfrak{a}) \subseteq \mathfrak{m} \cdot \mathfrak{a}$;

Proof $(i) \Rightarrow (ii)$: Suppose that there is an element $f \in D(\mathfrak{a})$, such that $f \notin \mathfrak{m} \cdot \mathfrak{a}$. There exists an ideal $\mathfrak{b} \subsetneq \mathfrak{a}$ such that $\mathfrak{b} + fB = \mathfrak{a}$. Let $D := B/\mathfrak{b}$ and $\mathfrak{c} = \mathfrak{a}/\mathfrak{b}$. Since $f \in D(\mathfrak{a})$ it follows easily that $\delta(f) \in \mathfrak{a} \cdot \Omega_{B/A}$. We have the exact sequence

$$\mathfrak{c}/\mathfrak{c}^2 \cong (f + \mathfrak{b})/\mathfrak{b} \overset{\alpha}{\to} \Omega_{B/A}/\mathfrak{c} \cdot \Omega_{B/A} = \Omega_{D/A} \otimes_D C \overset{\beta}{\to} \Omega_{C/A} \to 0,$$

where $\alpha(f + \mathfrak{c}^2) = d(f) + \mathfrak{c} \cdot \Omega_{D/A}$. The B-module $\Omega_{B/A}$ is free, because B is smooth and essentially of finite type over A. Since $f \in D(\mathfrak{a})$, we see that $\partial(f) \in \mathfrak{a}$ for any $\partial \in \mathrm{Der}_A(C) \cong \mathrm{Hom}_C(\Omega_{C/A}, C)$, hence $d(f) \in \mathfrak{a}\Omega_{C/A}$. This means that $\alpha(f + \mathfrak{c}^2) = 0$, hence β is an isomorphism.

$(ii) \Rightarrow (i)$: Suppose that there exists a non-trivial evolution $\epsilon : D \to C$ and assume that $D = B/\mathfrak{c}$ and $C = B/\mathfrak{a}$, where $(B, \mathfrak{m})$ is a smooth A-algebra. This means in particular that $\mathfrak{c} \subsetneq \mathfrak{a}$. Without loss of generality, we may assume that it is clear that for any ideal $\mathfrak{c}'$ such that $\mathfrak{c} \subseteq \mathfrak{c}' \subsetneq \mathfrak{a}$, the ring $B/\mathfrak{c}'$ is a non-trivial evolution; therefore we may assume that $\mathfrak{a} = \mathfrak{c} + (f)$, for some element $f \notin \mathfrak{a}$, such that $f\mathfrak{m} \subseteq \mathfrak{c}$. We have $f \notin \mathfrak{m}\mathfrak{a}$, otherwise $\mathfrak{a} = \mathfrak{c} + (f) \subseteq \mathfrak{c} + \mathfrak{m}\mathfrak{a} \subseteq \mathfrak{c}$. Let us remark that

$$\Omega_{C/A} \cong \Omega_{B/A}/(\mathfrak{a} \cdot \Omega_{B/A} + B \cdot d\mathfrak{a}) \cong \Omega_{B/A}/(\mathfrak{c} \cdot \Omega_{B/A} + f\Omega_{B/A} + B \cdot d\mathfrak{a}),$$

and on the other hand, since $D \to C$ is an evolution,

$$\Omega_{C/A} \cong \Omega_{D/A}/f \cdot \Omega_{D/A} \cong \Omega_{B/A}/(\mathfrak{c} \cdot \Omega_{B/A} + B \cdot d\mathfrak{c} + f \cdot \Omega_{B/A}).$$

Therefore

$$d(f) \in \mathfrak{c} \cdot \Omega_{B/A} + B \cdot d(\mathfrak{c}) + f \cdot \Omega_{B/A} = \mathfrak{a} \cdot \Omega_{B/A} + B \cdot d(\mathfrak{c}).$$

This means that there exist elements $g_1, \ldots, g_l \in \mathfrak{c}$ and $r_1, \ldots, r_l \in B$ such that

$$df - \sum_{i=1}^{l} r_i dg_i = \eta \in \mathfrak{a} \cdot \Omega_{B/A}.$$

If we denote $h := f - \sum_{i=1}^{l} r_i g_i$, we have $\mathfrak{a}/\mathfrak{c} = h \cdot \mathfrak{m}$, hence $h \notin \mathfrak{m} \cdot \mathfrak{a}$ and

$$dh = df - \sum_{i=1}^{l} r_i dg_i - \sum_{i=1}^{l} g_i dr_i = \eta - \sum_{i=1}^{l} g_i dr_i \in \mathfrak{a} \cdot \Omega_{B/A}.$$

Thus $h \in D(\mathfrak{a})$.

$(iii) \Rightarrow (ii)$: The proof follows the same lines as the proof of the implication $(iii) \Rightarrow (ii)$ of 6.2.43. We let the details to the reader. $\qquad \square$

Proposition 6.2.45 *Let A be a Noetherian regular ring, B a localization of a polynomial ring in finitely many variables over A and $C = B/\mathfrak{a}$, where $\mathfrak{a}$ is a radical ideal of B. If C is generically separable over A, then every evolution of C is trivial iff $\mathfrak{a}^{(2)} \subseteq \mathfrak{m} \cdot \mathfrak{a}$, where $\mathfrak{m}$ is the maximal ideal of C.*

Proof It follows from Theorem 6.2.44 and Proposition 6.2.34. $\qquad \square$

Because of the above results, Eisenbud and Mazur conjectured the following:

Conjecture 6.2.46 (Eisenbud-Mazur Conjecture [101]) Let $(A, \mathfrak{m})$ be a Noetherian regular local ring containing a field of characteristic zero and $\mathfrak{a}$ an unmixed (that is all the associated primes of $\mathfrak{a}$ are minimal) radical ideal of A. Then $\mathfrak{a}^{(2)} \subseteq \mathfrak{m} \cdot \mathfrak{a}$.

Example 6.2.47 (i) If the ring A is not regular, Conjecture 6.2.46 is not true.

Proof Indeed, let $A := k[[X, Y, Z]]/(X^2 - YZ)$, where k is a field of any characteristic different from 2 and let $\mathfrak{p} = (x, y)A$. Since $y = \frac{x^2}{z} \in \mathfrak{p}^{(2)}$, it follows that $\mathfrak{p}^{(2)} = (y)$. But obviously $y \notin \mathfrak{m} \cdot \mathfrak{p}$. $\qquad \square$

(ii) Also in the positive characteristic case Conjecture 6.2.46 is not true. (Kunz [101])

Proof Let k be a field of characteristic $p > 0$ and consider the map

$$\phi : k[X_1, X_2, X_3, X_4] \to k[T],$$

given by

$$\phi(X_1) = T^{p^2}, \quad \phi(X_2) = T^{p(p+1)}, \quad \phi(X_3) = T^{p^2+p+1}, \quad \phi(X_4) = T^{(p+1)^2}.$$

Let $p = \ker(\phi)$. Then $k[X_1, X_2, X_3, X_4]/\mathfrak{p}$ is isomorphic to a subdomain of $k[T]$, hence $\mathfrak{p}$ is a prime ideal. Let

$$f := X_1^{p+1} X_2 - X_2^{p+1} - X_1 X_3^p + X_4^p.$$

Consider also the polynomials

$$g_1 = X_1^{p+1} - X_2^p, \quad g_2 = X_1 X_4 - X_2 X_3, \quad g_3 = X_1^p X_2 - X_3^p.$$

It is a simple computation to see that $f, g_1, g_2, g_3 \in \mathfrak{p}$. It is also easy to see that

$$x_1^p f = g_1 g_3 + g_2^p$$

and since $x_1 \notin \mathfrak{p}$, this shows that $f \in \mathfrak{p}^{(2)}$. We will see that $f \notin (X_1, X_2, X_3, X_4) \cdot \mathfrak{p}$. Since X_4^p is a term of f, it suffices to show that no element of $\mathfrak{p}$ has a term of the form X_4^a, $0 < a < p$. Since $\mathfrak{p}$ is generated by binomials, it is enough to show that in $\mathfrak{p}$ there is no binomial of the form $X_4^a - X_3^b X_2^c X_1^d$. This means that the equation

$$a(p + 1)^2 = b(p^2 + p + 1) + cp(p + 1) + dp^2 \tag{6.1}$$

cannot be satisfied by non-negative integers a, b, c, d such that $0 < a < p$. Anyway, the above equation implies that $a \equiv b \bmod p$, that is, $b = a + np$, for some $n \geq 0$. If $n \geq 1$, the above equation becomes

$$0 = (np(p^2 + p + 1) - ap) + cp(p + 1) + dp^2,$$

and because $c, d \geq 0$ this implies

$$ap \geq np(p^2 + p + 1).$$

Since $0 < a < p$ we obtain

$$(p - 1)p \geq ap \geq np(p^2 + p + 1)$$

that is impossible. Therefore $a = b$. From the equation (6.1) we get

$$p(p - 1) \geq ap = cp(p + 1) + dp^2$$

and the right-hand side is either 0 or greater than p^2. Contradiction. $\square$

(iii) In mixed characteristic Conjecture 6.2.46 is not true (Kurano - Roberts [203, Rem. 3.3,(b)]).

Proof Let $(E, \mathfrak{n})$ be a regular local ring of dimension 3 not containing a field and let $\{x, y, z\}$ be a regular system of parameters of E. Assume that char$(E/\mathfrak{n}) = 2$ and that 2 divides x. Consider the following morphisms between rings of formal power series over E :

$$\theta : A = E[[S, T, U, V]] \to E[[W]],$$

defined by

$$\theta(S) = x^3 W, \quad \theta(T) = y^3 W,$$

$$\theta(U) = z^3 W, \quad \theta(V) = (xyz)^2 W.$$

Let $\mathfrak{p} = \ker(\theta)$. Then one can see, for example by using SINGULAR, that $\mathfrak{p}$ is generated by the following elements:

$$g_1 = y^3 S - x^3 T, \; g_2 = z^3 T - y^3 U, \; g_3 = x^3 U - z^3 S, \; g_4 = xV - y^2 z^2 S,$$

$$g_5 = yV - z^2 x^2 T, \; g_6 = zV - x^2 y^2 U, \; g_7 = V^2 - xyz^4 ST.$$

Thus, if we denote by $\mathfrak{m}$ the maximal ideal of A, that is, $\mathfrak{m} = (x, y, z, S, T, U, V)A$, we have $\mathfrak{p} \subseteq \mathfrak{m}^2$. One can also easily see that

$$\frac{1}{x^2}(g_4^2 + yzg_1 \cdot g_3) = V^2 + \frac{2}{x}y^2 z^2 SV + xy^4 zSU + xyz^4 ST + x^4 yzTU.$$

Note that $\mathfrak{p} \cap E = (0)$, hence $x^2 \notin \mathfrak{p}$. Therefore the left-hand side belongs to $\mathfrak{p}^2 A_\mathfrak{p}$ and the right-hand side is in A and does not belong to $\mathfrak{m}^3$. Therefore $\mathfrak{p}^{(2)} \not\subseteq \mathfrak{m}\mathfrak{p}$. $\quad\square$

We will now show a connection between the almost smooth morphisms and Mazur's evolution problem 6.2.40. We begin with the following consequence of Proposition 6.2.43.

Corollary 6.2.48 ([182, Prop. 5.4]) *Let* $u : A \to C$ *be an almost smooth morphism of Noetherian rings. Then* C *is evolutionarily stable over* A.

Proof Take a polynomial presentation $A \to B \to C = B/\mathfrak{a}$. Then we have the Jacobi-Zariski exact sequence

$$0 = H_1(A, C, C) \to \mathfrak{a}/\mathfrak{a}^2 \xrightarrow{\delta} \Omega_{B/A} \otimes_B C \xrightarrow{v} \Omega_{C/A} \to 0.$$

This means that $\gamma : \mathfrak{a}/\mathfrak{a}^2 \to \ker(v)$ is the identity map. Now we apply Proposition 6.2.43. $\quad\square$

Proposition 6.2.49 ([182, Prop. 5.7]) *Let* $u : A \to C$ *be a morphism of Noetherian rings, let* $A \to B \to C = B/\mathfrak{a}$ *be a polynomial presentation of* C *and let* $v : \Omega_{B/A} \otimes_B C \to \Omega_{C/A}$ *be the canonical morphism. Assume that:*

(i) u is not almost smooth;
(ii) $H_1(A, C, C)$ is an injective C-module or $\ker(v)$ *is a projective C-module.*

Then C is not evolutionarily stable over A.

Proof Considering a polynomial presentation $A \to B \to C = B/\mathfrak{a}$, we have the Jacobi-Zariski exact sequence

$$0 \to H_1(A, C, C) \to \mathfrak{a}/\mathfrak{a}^2 \xrightarrow{\delta} \Omega_{B/A} \otimes_B C \xrightarrow{v} \Omega_{C/A} \to 0$$

that induces an exact sequence

$$0 \to H_1(A, C, C) \to \mathfrak{a}/\mathfrak{a}^2 \xrightarrow{\gamma} \ker(v) = \mathrm{Im}(\delta) \to 0.$$

Therefore

$$\mathfrak{a}/\mathfrak{a}^2 \cong H_1(A, C, C) \oplus \ker(v) = H_1(A, C, C) \oplus \mathrm{Im}(\delta).$$

Now we apply Proposition 6.2.43 to get the conclusion. $\qquad\square$

Corollary 6.2.50 *Let A be a Noetherian ring and C be a local A-algebra essentially of finite type. Suppose that:*

(i) C is not almost smooth over A;
(ii) $\Omega_{C/A}$ is a projective C-module.

Then C is not evolutionarily stable over A.

Proof Considering again a polynomial presentation $A \to B \to C = B/\mathfrak{a}$ and keeping the above notations, since $\Omega_{C/A}$ is projective, it follows that $\ker(v)$ is a direct summand of the free module $\Omega_{B/A} \otimes_B C$; hence it is projective. Now apply Proposition 6.2.49. $\qquad\square$

Remark 6.2.51 It is well-known that if A contains the field of rationals $\mathbb{Q}$ and C is a local A-algebra essentially of finite type such that $\Omega_{C/A}$ is a projective C-module, then C is smooth over A. Thus Corollary 6.2.50 cannot be applied in this case. On the other hand, if A is a DVR of mixed characteristic, the situation seems not clear; therefore the following question makes sense.

Question 6.2.52 ([182]) Does there exist a DVR of mixed characteristic A and a local reduced flat A-algebra essentially of finite type C such that $\Omega_{C/A}$ is a projective C-module and C is not a smooth A-algebra?

Remark 6.2.53 A positive answer to the above problem would provide a counter- example to the evolution problem in the mixed characteristic case. The question 6.2.52 can be considered also in a more general setting, for example, without asking C to be local, reduced, flat or essentially of finite type over A.

6.3 Hochschild Homology and Regular Morphisms

Hochschild (co-)homology is a (co-)homology theory for associative algebras invented by G. Hochschild (1915–2010) in 1945 for algebras over a field and generalized to algebras over more general rings in 1956 by H. Cartan (1904–2008) and S. Eilenberg (1913–1998). Hochschild (co-)homology provided many applications in Commutative Algebra in the last decades. We use several properties of Hochschild homology and cohomology, so that we will shortly remember the basic things.

Let us first shortly remind the main things about differential graded algebras. The interested reader can look at [23, 137] for more informations.

Definition 6.3.1 Let k be a commutative ring. A **differential graded A-algebra**, shortly a **DG-algebra over A**, is a chain complex $A_\bullet$ of k-modules together with k-bilinear maps

$$A_n \times A_m \to A_{n+m}, \ (a,b) \mapsto ab$$

such that $\underset{i \geq 0}{\oplus} A_i$ becomes an associative and unital k-algebra and

$$d_{n+m}(ab) = d_n(a)b + (-1)^n a d_m(b).$$

A DG-algebra is **commutative** if

$$ab = (-1)^{\deg(a)\,\deg(b)} ba, \ \text{for any } a, b \in A$$

and is called **strictly commutative** if $a^2 = 0$, for any $a \in A$ of odd degree.

Definition 6.3.2 A **morphism of DG-algebras** $\phi : (A, d) \to (B, d)$ is a k-algebra morphism $\phi : A \to B$, compatible with the differentials and the gradings.

Definition 6.3.3 Let k be a ring and $(A, d), (B, d)$ be two DG-algebras. The **tensor product of DG-algebras** is the tensor product $A \otimes_k B$ endowed with the multiplication

$$(a \otimes b)(c \otimes d) = (-1)^{\deg(c)\,\deg(b)} ac \otimes bd$$

and the differential

$$d(a \otimes b) = d(a) \otimes b + (-1)^{\deg(a)} a \otimes d(b).$$

Now we shall remind the main definitions and results about Hochschild homology in the case of commutative algebras. The reader can find the detailed definitions and properties in [79, 222, 393] and [394].

Let A be a ring, B be an A-algebra (remember that everything is commutative) and M a B-module. Let

$$C_n(B, M) := M \otimes_A B^{\otimes n} = M \otimes_A \underbrace{B \otimes_A \ldots \otimes_A B}_{n-\text{times}}.$$

Let also

$$d_n : C_n(B, M) \to C_{n-1}(B, M),$$

$$d_n(m \otimes b_1 \otimes \ldots \otimes b_n) = (mb_1) \otimes b_2 \otimes \ldots \otimes b_n +$$

$$+ \sum_{i=1}^{n-1} (-1)^i m \otimes b_1 \otimes \ldots \otimes b_i b_{i+1} \otimes \ldots \otimes b_n + (-1)^n (b_n m) \otimes b_1 \otimes \ldots \otimes b_{n-1}.$$

Remark 6.3.4 In the general case, B is not necessarily commutative and in that case M must be a $B - B$-bimodule. This explains the multiplication of $m \in M$ by b_1 and b_n in the above formula. Even if in the commutative case this is not important, we kept the formula in order to illustrate the basic ideas.

Remark 6.3.5 $C_\bullet(B, M) := (C_n(B, M), d_n)_{n \in \mathbb{N}}$ is a complex of B-modules:

$$C_\bullet(B, M) := \ldots \to M \otimes_A B^{\otimes n} \xrightarrow{d_n} M \otimes_A B^{\otimes(n-1)} \xrightarrow{d_{n-1}} \ldots \xrightarrow{d_1} M \otimes_A B \xrightarrow{d_0} M.$$

This complex is called the **Hochschild complex** of the A-algebra B with coefficients in M [222, Lemma 1.1.2].

Definition 6.3.6 The homology of the Hochschild complex is called the **Hochschild homology** of the A-algebra B with coefficients in M, that is

$$HH_n(B, M) := H_n(C_\bullet(B, M)) = H_n(C_n(B, M), d_n).$$

We shall deal mainly with the Hochschild homology with coefficients in B, that is $HH_n(B) := HH_n(B, B)$.

Remark 6.3.7

(i) If B is a flat A-module, then $HH_n(B, M) \cong \operatorname{Tor}_n^{B \otimes_A B}(M, B)$ [393, Cor. 9.1.5].
(ii) $HH_\bullet(B) := \bigoplus_{i=0}^{\infty} HH_i(B)$ has a structure of a differential graded algebra over A given
    by the shuffle product [222, Cor. 4.2.7];
(iii) There is a natural isomorphism of B-modules $\Omega_{B/A} \cong HH_1(B)$ given by $db \to$
    $\operatorname{cls}(1 \otimes b)$ [222, Prop. 1.1.10]. This isomorphism extends to a homomorphism of
    graded B-algebras [222, Prop. 3.4.4]

$$\gamma : \Omega_{B/A}^\bullet = \bigwedge \Omega_{B/A} \to HH_\bullet(B)$$

We will consider now the case of topological algebras in order to extend to this case the classical definition of Hochschild (co)-homology (see [165]). For this, consider a commutative ring A with a linear topology τ_0 and a commutative topological A-algebra B. This means that B is a commutative A-algebra with a linear topology τ, such that the structural morphism $u : A \to B$ is continuous. We will always use the following definition in the case when the topology τ on B is the $\mathfrak{a}$-adic topology, where $\mathfrak{a}$ is an ideal of B.

Definition 6.3.8 ([165, Def. 1.9]) Let (A, τ_0) be a topological ring and (B, τ) a topological A-algebra. Then we denote by

$$T_n(B|A, \tau) := \underbrace{B \widehat{\otimes} B \widehat{\otimes} \dots \widehat{\otimes} B}_{n+1-\text{times}},$$

the n-fold complete tensor product of the topological algebra B [135, Sec. 7.7.1]. The **Hochschild complex of the topological algebra B** $(T_\bullet(B|A), d_\bullet, \tau)_{n \in \mathbb{N}}$ is the component-wise projective limit of the projective system of Hochschild complexes $(C_\bullet(B/\mathfrak{b}), d_\mathfrak{b})$ where $\mathfrak{b}$ runs over the open ideals of B.

Definition 6.3.9 Let (A, τ_0) be a topological ring and (B, τ) a topological A-algebra. Then

$$HH_n(B, \tau) := H_n(T_\bullet(B|A), d_\bullet, \tau)$$

is called the **Hochschild homology of the topological A-algebra B**.

Notation 6.3.10 Let (A, τ_0) be a topological ring and (B, τ) a topological A-algebra, where τ is the $\mathfrak{a}$-adic topology, for some ideal $\mathfrak{a}$ of B. Let $\mathfrak{b}$ be the kernel of the canonical surjection $B \widehat{\otimes}_A B \rightarrow \widehat{B}$ induced by the multiplication $\mu : B \otimes_A B \rightarrow B$. We put $\Omega_{(B/A,\tau)} := \mathfrak{b}/\mathfrak{b}^2$.

Remark 6.3.11

(i) Let (A, τ_0) be a topological ring and (B, τ) a topological A-algebra, where τ is the $\mathfrak{a}$-adic topology, for some ideal $\mathfrak{a}$ of B. If $C_n(B)$ is Noetherian, then $T_n(B|A, \tau)$ is Noetherian [165, Rem. 1.8, iii)].

(ii) $HH_\bullet(B, \tau)$ has a structure of a graded $\widehat{B}$-module and of an anti-commutative differential graded $\widehat{B}$-algebra structure.

(iii) The Hochschild homology $HH_n(B, \tau)$ does not depend on the topology τ_0 on A [165, Rem. 1.5, i)].

(iv) If τ is the discrete topology, then $HH_n(B, \tau) \cong HH_n(B)$.

(v) If u is flat and τ is the discrete topology on B (that is $\mathfrak{a} = (0)$) or if u is flat and $T_n(B|A)$ is Noetherian for any $n \in \mathbb{N}$, then by [165, Prop.1.13]

$$HH_n(B, \tau) \cong \mathrm{Tor}_n^{B \widehat{\otimes}_A B}(\widehat{B}, \widehat{B}).$$

(vi) As in the discrete case, there is a natural isomorphism of B-modules

$$\Omega_{(B/A,\tau)} \cong HH_1(B, \tau).$$

This isomorphism extends to a homomorphism of graded B-algebras

$$\gamma_\tau : \Omega_{(B/A,\tau)}^\bullet = \bigwedge \Omega_{(B/A,\tau)} \rightarrow HH_\bullet(B, \tau).$$

We start by proving a basic useful result.

Theorem 6.3.12 (Quillen [294, Prop. 8.13]) *Let A be a Noetherian ring and $\mathfrak{a}$ be an ideal of A. The following are equivalent:*

(i) $\mathfrak{a}/\mathfrak{a}^2$ *is a projective $A/\mathfrak{a}$-module and* $\overset{q}{\bigwedge} \mathfrak{a}/\mathfrak{a}^2 \cong \mathrm{Tor}_A^q(A/\mathfrak{a}, A/\mathfrak{a})$, *for any $q \geq 0$;*

(ii) $\mathfrak{a}/\mathfrak{a}^2$ *is a flat $A/\mathfrak{a}$-module and* $\overset{q}{\bigwedge} \mathfrak{a}/\mathfrak{a}^2 \cong \mathrm{Tor}_A^q(A/\mathfrak{a}, A/\mathfrak{a})$, *for any $q \geq 0$;*

(iii) $\mathfrak{a}/\mathfrak{a}^2$ *is a projective $A/\mathfrak{a}$-module and the canonical map* $\overset{2}{\bigwedge} \mathfrak{a}/\mathfrak{a}^2 \rightarrow \mathrm{Tor}_A^2(A/\mathfrak{a}, A/\mathfrak{a})$ *is surjective;*

(iv) $\mathfrak{a}/\mathfrak{a}^2$ *is a projective $A/\mathfrak{a}$-module and* $\mathrm{Sym}(\mathfrak{a}/\mathfrak{a}^2) \cong \mathrm{gr}_\mathfrak{a}(A)$;

(v) $\mathfrak{a}A_\mathfrak{p}$ *is generated by a regular sequence, for any $\mathfrak{p} \in \mathrm{Max}(A) \cap V(\mathfrak{a})$.*

Proof $(i) \Leftrightarrow (ii)$: Since A is Noetherian, $\mathfrak{a}/\mathfrak{a}^2$ is flat iff it is projective. We may assume henceforth that A is a local ring and $\mathfrak{a}$ is contained in the maximal ideal, because all the conditions are local.

$(i) \Rightarrow (iii)$: Obvious.

$(iii) \Rightarrow (v)$: Let $f := \{f_1, \ldots, f_r\}$ be a minimal system of generators of $\mathfrak{a}$. Then the Koszul complex $K_\bullet(f, A)$ is a free differential graded algebra over A with an augmentation to $B = A/\mathfrak{a}$; hence there exists a canonical morphism of graded B-algebras $\theta_* : H(f; B) \to \mathrm{Tor}^A(B, B)$, where $H_i(f, B)$ are the groups of Koszul homology. There is a spectral sequence [324, Th. 10.48]

$$E^2_{p,q} = \mathrm{Tor}^A_p(H_q(f; A), B) \Rightarrow H_{p+q}(f; B)$$

whose associated five-term sequence is

$$H_2(f; B) \xrightarrow{\theta_2} \mathrm{Tor}^A_2(B, B) \to H_1(f; A) \otimes_A B \to H_1(f; B) \xrightarrow{\theta_1} \mathrm{Tor}^A_1(B, B) \to 0.$$

But θ_1 is an isomorphism, because $\mathfrak{a}/\mathfrak{a}^2$ is free since A is local. Since we have $H_2(f; B) \cong \bigwedge^2 H_1(f; B)$ and θ_* is an algebra morphism, it follows that actually θ_2 is the map $\bigwedge^2 \mathfrak{a}/\mathfrak{a}^2 \to \mathrm{Tor}^2_A(A/\mathfrak{a}, A/\mathfrak{a})$ which is surjective by assumption. Hence $H_1(f; A) \otimes_A B = 0$. By Nakayama's Lemma we get that $H_1(f; A) = 0$, that is $\{f_1, \ldots, f_r\}$ is a regular sequence.

$(v) \Rightarrow (i)$: Assume that $\mathfrak{a} = (f_1, \ldots, f_r)$, where $f = \{f_1, \ldots, f_r\}$ is a regular sequence. Then the Koszul complex $K_\bullet(f, A)$ is a free resolution of $B = A/\mathfrak{a}$, hence

$$\mathrm{Tor}^A_i(B, B) = H_i(f; B) \cong \bigwedge^i H_1(f; B) \text{ for any natural number } i \geq 0$$

and in particular

$$\mathfrak{a}/\mathfrak{a}^2 \cong \mathrm{Tor}^A_1(B, B) \cong H_1(f; B) \cong B^r.$$

Thus $\mathfrak{a}/\mathfrak{a}^2$ is projective.

$(iv) \Leftrightarrow (v)$: Follows from [131, Cor. 15.1.11]. □

Theorem 6.3.13 (Hochschild-Kostant-Rosenberg Theorem [7, 162]) *Let* $u : A \to B$ *be a morphism of Noetherian rings. The following are equivalent:*

(i) u is regular;

(ii) $\Omega_{B/A}$ is a flat B-module and $\bigwedge \Omega_{B/A} \cong \mathrm{Tor}^{B \otimes_A B}(B, B)$.

Proof $(i) \Rightarrow (ii)$: We have a commutative diagram

$$
\begin{array}{ccc}
A & \xrightarrow{\ u\ } & B \\
\downarrow & & \downarrow \\
B & \longrightarrow & B \otimes_A B \xrightarrow{\ \mu\ } B
\end{array}
$$

where μ is the canonical multiplication and $\Omega_{B/A} = \mathfrak{b}/\mathfrak{b}^2$, where $\mathfrak{b} = \ker(\mu)$. Since u is regular, by Theorem 3.2.8 it follows that $\Omega_{B/A}$ is flat and $H_1(A, B, -) = 0$. Using the Jacobi-Zariski exact sequence associated to $B \to B \otimes_A B \to B$, it is easy to see that $H_2(B \otimes_A B, B, -) = 0$. This implies that $\mathfrak{b}$ is locally generated by a regular sequence [6, Th. VI.25]. Now apply Theorem 6.3.12.

$(ii) \Rightarrow (i)$: Follows in the same way using Theorem 6.3.12 and Theorem 3.2.8. $\qquad\square$

Proposition 6.3.14 *Let* $u : A \to B$ *be a flat morphism of Noetherian rings such that* $B \otimes_A B$ *is Noetherian and let* $\mathfrak{a}$ *be an ideal of* B. *On* B *we consider the* $\mathfrak{a}$*-adic topology. Then for any* $n \in \mathbb{N}$ *and for any* $\widehat{B}$*-module* W, *we have an isomorphism*

$$
H_n(A, B, W) \cong H_{n+1}(B\widehat{\otimes}_A B, \widehat{B}, W).
$$

Proof From the Jacobi-Zariski exact sequence associated to $B \to B \otimes_A B \to B$, it follows that $H_n(A, B, W) \cong H_{n+1}(B \otimes_A B, B, W)$. Let $\mu : B \otimes_A B \to B$ be the multiplication map and $\mathfrak{b} := \mathrm{Im}(\mathfrak{a} \otimes_A B) + \mathrm{Im}(B \otimes_A \mathfrak{a})$. Then $\mu(\mathfrak{b}) = \mathfrak{a}$ and [4, Prop. 21.1] yields an isomorphism $H_{n+1}(B \otimes_A B, B, W) \cong H_{n+1}(B\widehat{\otimes}_A B, \widehat{B}, W)$. $\qquad\square$

Theorem 6.3.15 ([178]) *Let* $u : A \to B$ *be a morphism of Noetherian rings,* $\mathfrak{a}$ *an ideal of* B *and* τ *the* $\mathfrak{a}$*-adic topology on* B. *Assume that* $B \otimes_A B$ *is Noetherian. If* u *is regular, then there is an isomorphism of DG-algebras*

$$
\bigwedge \Omega_{(B/A, \tau)} \cong \mathrm{Tor}^{B\widehat{\otimes}_A B}(\widehat{B}, \widehat{B}).
$$

Proof First let us note that $B\widehat{\otimes}_A B$ is Noetherian and remember that $\Omega_{(B/A,\tau)} \cong \mathfrak{b}/\mathfrak{b}^2$, where $\mathfrak{b}$ is the kernel of the morphism $B\widehat{\otimes}_A B \to \widehat{B}$ induced by the multiplication $\mu : B \otimes_A B \to B$. Let $\mathfrak{m} \in \mathrm{Max}(B\widehat{\otimes}_A B) \cap V(\mathfrak{b})$ and denote by $\mathfrak{n} = \mathfrak{m}/\mathfrak{b}$ the corresponding maximal ideal of $\widehat{B}$. By Proposition 6.3.14, [6, Cor. V.27] and Theorem 3.2.8 we have

$$
H_2((B\widehat{\otimes}_A B)_\mathfrak{n}, \widehat{B}_\mathfrak{n}, k(\mathfrak{n})) \cong H_2(B\widehat{\otimes}_A B, \widehat{B}, k(\mathfrak{n})) \cong H_1(A, B, k(\mathfrak{n})) = (0).
$$

This means that $\mathfrak{b}$ is locally generated by a regular sequence [6, Th. VI.25]; therefore the conclusion follows by Theorem 6.3.12. $\qquad\square$

For the discrete topology, we can strengthen this result using Theorem 5.2.56.

Corollary 6.3.16 *Let $u : A \to B$ be a regular morphism of Noetherian rings. Then there is an isomorphism of DG-algebras*

$$\bigwedge \Omega_{B/A} \cong \operatorname{Tor}^{B \otimes_A B}(B, B).$$

Proof By Theorem 5.2.56 we know that u is a filtered inductive limit of finite type smooth morphisms. Since Hochschild homology commutes with inductive limits [165, Prop. 1.17], we may assume that u is of finite type and smooth. Now we can apply Theorem 6.3.15.
□

Theorem 6.3.17 (Generalized Hochschild-Kostant-Rosenberg Theorem [178]) *Let $u : A \to B$ be a flat morphism of Noetherian rings, $\mathfrak{a}$ an ideal of B and τ the $\mathfrak{a}$-adic topology on B. Assume that $B \widehat{\otimes}_A B$ is a Noetherian ring and that B is $\mathfrak{a}$-smooth over A. Then there is an isomorphism of DG-algebras,*

$$\bigwedge \Omega_{(B/A, \tau)} \cong \operatorname{Tor}^{B \widehat{\otimes}_A B}(\widehat{B}, \widehat{B}).$$

Proof Let $\mathfrak{b}$ be the kernel of the morphism $B \widehat{\otimes}_A B \to \widehat{B}$, so that $\Omega_{(B/A, \tau)} \cong \mathfrak{b}/\mathfrak{b}^2$. Let $\mathfrak{m} \in \operatorname{Max}(B \widehat{\otimes}_A B) \cap V(\mathfrak{b})$ and denote by $\mathfrak{n} = \mathfrak{m}/\mathfrak{b}$ the corresponding maximal ideal of $\widehat{B}$. Since u is $\mathfrak{a}$-smooth over A, by [131, Prop. 19.3.5] it results that $B \otimes_A B$ is $\mathfrak{c}$-smooth over B, where $\mathfrak{c} := \operatorname{Im}(\mathfrak{a} \otimes B) + \operatorname{Im}(B \otimes \mathfrak{a})$, hence by [131, Prop. 19.3.6] we get that $B \widehat{\otimes}_A B$ is $\mathfrak{c} \cdot B \widehat{\otimes}_A B$-smooth over $\widehat{B}$. Let $\mathfrak{p} := \mathfrak{m} \cap \widehat{B}$. Then by [131, Prop. 19.3.5 and Prop. 19.3.8], it follows that $(B \widehat{\otimes}_A B)_\mathfrak{m}$ is $\mathfrak{m}(B \widehat{\otimes}_A B)_\mathfrak{m}$- smooth over $\widehat{B}_\mathfrak{p}$. Thus Corollary 3.2.4 yields $H_1(\widehat{B}_\mathfrak{p}, (B \widehat{\otimes}_A B)_\mathfrak{m}, k(\mathfrak{m})) = (0)$. This implies

$$H_2((B \widehat{\otimes}_A B)_\mathfrak{m}, \widehat{B}_\mathfrak{n}, k(\mathfrak{n})) = H_2(B \widehat{\otimes}_A B, \widehat{B}, k(\mathfrak{m})) = H_1(\widehat{B}, B \widehat{\otimes}_A B, k(\mathfrak{m})) =$$

$$= H_1(\widehat{B}_\mathfrak{p}, (B \widehat{\otimes}_A B)_\mathfrak{m}, k(\mathfrak{m})) = (0).$$

This means that $\mathfrak{b}_\mathfrak{m}$ is generated by a regular sequence by [6, Th. VI.25]; hence the conclusion follows by Theorem 6.3.12.
□

In the following, we will show several connections between smoothness, regularity and Hochschild homology. We start with a result of Rodicio. One should note that the ring $B \otimes_A B$ in Theorem 6.3.19 is not necessarily Noetherian.

Lemma 6.3.18 *Let $u : A \to B$ be a flat morphism of commutative rings, let K be an A-algebra and let $C := K \otimes_A B$. If $\operatorname{fd}_{B \otimes_A B}(B) < \infty$, then $\operatorname{fd}_{C \otimes_K C}(C) < \infty$.*

Proof It follows at once from [393, Th. 9.1.7].   □

Theorem 6.3.19 (Rodicio [321, Th. 1]) *Let $u : A \to B$ be a flat morphism of Noetherian rings such that $\mathrm{fd}_{B \otimes_A B}(B) < \infty$. Then u is regular.*

Proof Let $\mathfrak{q}$ be a prime ideal of B and $\mathfrak{p} = \mathfrak{q} \cap A$. We need to prove that the induced morphism $A_{\mathfrak{p}} \to B_{\mathfrak{q}}$ is formally smooth. Since by [324, Prop. 7.17] for any $B \otimes_A B$-module M, we have

$$\mathrm{Tor}_n^{B_{\mathfrak{q}} \otimes_{A_{\mathfrak{p}}} B_{\mathfrak{q}}}(B_{\mathfrak{q}}, M_{\mathfrak{q}}) = \left(\mathrm{Tor}_n^{B \otimes_A B}(B, M) \right)_{\mathfrak{q}},$$

we may assume that A and B are local rings and that u is a local morphism, and we must prove that u is formally smooth. If K is the residue field of A, we must prove that $C := B \otimes_A K$ is geometrically regular over K. From Lemma 6.3.18 we know that $\mathrm{fd}_{C \otimes_K C}(C) < \infty$. But C is a Noetherian local ring and by [79, Ch. IX, Cor. 4.4] we have

$$\mathrm{Tor}_n^{C \otimes C}(C, M \otimes_K N) \cong \mathrm{Tor}_n^C(M, N),$$

hence $\mathrm{gldim}(C) < \infty$, that is, C is a regular local ring. If L is a field, finite extension of K, the same proof shows that $B \otimes_A L$ is a regular ring.   □

Corollary 6.3.20 *Let $K \subseteq L$ be a field extension such that $\mathrm{fd}_{L \otimes_K L}(L) < \infty$. Then L is separable over K.*

Remark 6.3.21 Let $u : A \to B$ be a flat morphism of Noetherian rings such that $\mathrm{fd}_{B \otimes_A B}(B) < \infty$. It follows by Theorem 6.3.13 and Theorem 6.3.19 that there exists a natural number $n \in \mathbb{N}$, such that $\bigwedge^{n+i} \Omega_{B/A} = 0$, $\forall i \geq 0$.

Example 6.3.22 The converse of 6.3.20 is not true.

Proof Indeed, let $K \subseteq L$ be a separable field extension such that $\Omega_{L/K}$ is not finitely generated. For example, one can take for K any field of characteristic zero and $L = K((X))$. Then $K \subseteq L$ is a regular morphism, but $\bigwedge^n \Omega_{L/K} \neq 0$ for any natural number n, hence $\mathrm{fd}_{L \otimes_K L}(L) = \infty$.   □

Proposition 6.3.23 (Rodicio [318, Prop. 4.2]) *Let $u : A \to B$ be a flat ring morphism such that $B \otimes_A B$ is a Noetherian ring. The following are equivalent:*

(i) $H_1(A, B, -) = (0)$;

(ii) $\mathrm{fd}_{B \otimes_A B}(B) < \infty$ *and* $H_2(A, B, -) = (0)$.

Proof $(i) \Rightarrow (ii)$: From $H_1(A, B, -) = (0)$ we obtain that $H_2(B \otimes_A B, B, -) = (0)$, hence Theorem 6.2.10 implies that $\mathrm{fd}_{B \otimes_A B}(B) < \infty$ and from Theorem 3.2.8 it results that $H_2(A, B, -) = (0)$.

$(ii) \Rightarrow (i)$: Follows from Theorem 3.2.8 and Theorem 6.3.19.   □

Corollary 6.3.24 *Let* $u : A \to B$ *be a morphism of Noetherian rings such that* $B \otimes_A B$ *is Noetherian. The following are equivalent:*

 (i) *u is regular;*
(ii) *u is flat and* $\mathrm{fd}_{B \otimes_A B}(B) < \infty$.

Proof $(i) \Rightarrow (ii)$: u is flat by assumption. Now apply Proposition 6.3.23.

$(ii) \Rightarrow (i)$: Follows from Theorem 6.3.19.   □

Lemma 6.3.25 *Let* A *be a Noetherian ring and* $\mathfrak{a}$ *an ideal of* A *locally generated by a regular sequence. Then* $\mathrm{fd}_A(A/\mathfrak{a}) = \sup\{\mathrm{rk}(\mathfrak{a}A_{\mathfrak{p}}/\mathfrak{a}^2 A_{\mathfrak{p}}) \mid \mathfrak{p} \in V(\mathfrak{a})\}$.

Proof Since $\mathrm{fd}_A(A/\mathfrak{a}) = \sup\{\mathrm{fd}_{A_{\mathfrak{q}}}(A_{\mathfrak{q}}/\mathfrak{a}A_{\mathfrak{q}}) \mid \mathfrak{q} \in \mathrm{Spec}(A)\}$, we may assume that A is local and $\mathfrak{a}$ is generated by a regular sequence. Then

$$\mathrm{fd}_A(A/\mathfrak{a}) = \mu(\mathfrak{a}) = \mathrm{rk}(\mathfrak{a}/\mathfrak{a}^2).$$

                □

In the proof of Proposition 6.3.31, we need a nice result of Ferrand, result important in its own. We need some preparations about ring epimorphisms.

Lemma 6.3.26 *Let* $f : A \to B$ *and* $g : B \to C$ *be ring morphisms.*

 (i) *If* $g \circ f$ *is an isomorphism and* f *is an epimorphism, then* g *is an isomorphism.*
(ii) *If* $g \circ f$ *is an epimorphism, then* g *is an epimorphism.*

Proof

(i) The morphism g is the composition of the morphisms

$$v : B \to C \otimes_A B, \quad v(b) = 1 \otimes g(b)$$

and

$$u : B \otimes_A C \to C, \quad u(b \otimes c) = bc.$$

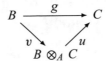

We have a commutative cocartesian diagram of ring morphisms

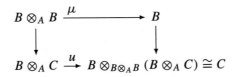

Since μ is an isomorphism, it follows that u is an isomorphism. Moreover v is obtained from the morphism $g \circ f$ by the base change $A \to B$, hence v is an isomorphism. Therefore u and v are isomorphisms; hence g is an isomorphism.

(ii) Let $v : C \otimes_B C \to C$ be the canonical multiplication and $w : C \otimes_A C \to C \otimes_B C$ be the canonical morphism. We have the commutative diagram

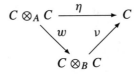

Clearly w is surjective, hence an epimorphism. This means that the multiplication morphism $\xi : (C \otimes_B C)_{C \otimes_A C}(C \otimes_B C) \to (C \otimes_B C)$ is an isomorphism. By assumption η is an isomorphism. Since $\eta = v \circ w$, it follows that v is an isomorphism, so that g is an epimorphism. □

Lemma 6.3.27 *Let $f : (A, \mathfrak{m}, k) \to (B, \mathfrak{n})$ be a local epimorphism of local rings. If A is a Noetherian complete local ring, then f is surjective.*

Proof Since f is an epimorphism, it follows that $k \to B/\mathfrak{m}B$ is an epimorphism, hence an isomorphism by Corollary 3.7.14. Thus $\mathfrak{m}B = \mathfrak{n}$. Let $\mathfrak{b} := \bigcap_{n \geq 0} \mathfrak{m}^n B$ and $\mathfrak{a} = \mathfrak{b} \cap A = f^{-1}(\mathfrak{b})$. Then $B/\mathfrak{b} = B/\bigcap_{n \geq 0} \mathfrak{m}^n B$, hence $B/\mathfrak{b}$ is $\mathfrak{n}$-adically separated. Then the composition $A/\mathfrak{a} \to B/\mathfrak{a}B \to B/\mathfrak{b}$ is an isomorphism. Since $A/\mathfrak{a} \to B/\mathfrak{a}B$ is an epimorphism, by Lemma 6.3.26 we get that $B/\mathfrak{a}B \to B/\mathfrak{b}$ is an isomorphism, hence $A/\mathfrak{a} \to B/\mathfrak{a}B$ is an isomorphism. But this implies that $A/\mathfrak{a}^2 \to B/\mathfrak{a}^2 B$ is surjective, hence $B/\mathfrak{a}^2 B$ is Noetherian and consequently $\mathfrak{m}$-adically separated. It follows that $\mathfrak{b} = \mathfrak{a}B = \mathfrak{a}^2 B$. But $\mathfrak{a}B$ is finitely generated, because A is Noetherian. By Nakayama Lemma, $\mathfrak{a}B = (0)$, hence B is $\mathfrak{m}$-adically separated. Now we apply [56, ch. III, §2, n. 8] □

Theorem 6.3.28 (Ferrand [108, Th. 3.5]) *Let* $(A, \mathfrak{m}, k)$ *be a Noetherian local ring,* $(B, \mathfrak{n})$ *a local ring and* $f : A \to B$ *a local epimorphism. Then* f *is essentially finite over* A.

Proof Let $g : \widehat{A} \to E := \widehat{A} \otimes_A B$ be the epimorphism induced by f. Then $\widehat{A}/\mathfrak{m}\widehat{A} \to E/\mathfrak{m}E$ is a faithfully flat epimorphism, hence an isomorphism. Thus $\mathfrak{n} = \mathfrak{m}B$ and $\mathfrak{r} := \mathfrak{m}E$ is the unique prime ideal of E lying over $\mathfrak{m}\widehat{A}$. The morphism $\theta : \widehat{A} \to E_{\mathfrak{r}}$ is a local epimorphism, hence surjective by Lemma 6.3.27. Let $\mathfrak{a}$ be the kernel of θ, that is, $E_{\mathfrak{r}} \cong \widehat{A}/\mathfrak{a}$. Since $\widehat{A}/\mathfrak{a} \to E/\mathfrak{a}E$ is an epimorphism and the composed map $\widehat{A}/\mathfrak{a} \to E/\mathfrak{a}E \to E_{\mathfrak{r}}$ is an isomorphism, by Lemma 6.3.26,i) it follows that $E/\mathfrak{a}E \cong E_{\mathfrak{r}}$. But $\mathfrak{a}$ is a finitely generated ideal and $\mathfrak{a}E_{\mathfrak{r}} = 0$, hence there exists $t \in E \setminus \mathfrak{r}$ such that $\mathfrak{a}E_t = 0$. It follows that the morphism $E_t \to E_{\mathfrak{r}}$ is injective. But the composed map $E \to E_t \to E_{\mathfrak{r}}$ is surjective, hence $E_t \cong E_{\mathfrak{r}}$. Since $t \in E = B \otimes_A \widehat{A}$, we can write $t = \sum_{i=1}^{s} \alpha_i \otimes t_i$, with $\alpha_i \in \widehat{A}$ and $t_i \in B$. Let $C = A[t_1, \ldots, t_s] \subseteq B$. Since $\widehat{A}$ is flat over A, the map $D = \widehat{A} \otimes_A C \to E$ is injective. We have the commutative diagram:

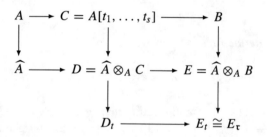

As the map $\widehat{A} \to E_t$ is surjective, it follows that the map $D_t \to E_t$ is surjective and consequently bijective. Since $C \to D_t$ is flat and $B \to E_{\mathfrak{r}}$ is faithfully flat, by [56, ch. I, §3, no. 4, Prop. 7] it follows that $C \to B$ is flat. By Lemma 6.3.26,ii) it follows that $C \to B$ is a flat epimorphism. Let $\mathfrak{p} := \mathfrak{n} \cap C$. By Proposition 3.7.16 we obtain $C_{\mathfrak{p}} \cong B$, so that B is essentially of finite type over A. Since $A \to C_{\mathfrak{p}}$ is an epimorphism, the map induced on the spectra is injective; hence the ideal $\mathfrak{p}$ is a minimal prime ideal of $\mathfrak{m}C$. On the other hand, the composition $A/\mathfrak{m} \to C/\mathfrak{p} \to B/\mathfrak{n}$ is an isomorphism, so that the injection $C/\mathfrak{p} \hookrightarrow B/\mathfrak{n}$ is actually an isomorphism. This means that $\mathfrak{p}$ is a maximal ideal of C. Therefore Zariski's main theorem [306, Ch. IV, Th. 1] implies that there exists a finite A-subalgebra $F \subseteq C$ such that, with $\mathfrak{q} = \mathfrak{p} \cap F$, the morphism $F_{\mathfrak{q}} \to C_{\mathfrak{p}}$ is an isomorphism. Consequently $F_{\mathfrak{q}} \cong B$ and the proof is complete. □

The next theorem was proved by D. Ferrand and shows that the Noetherianity of $B \otimes_A B$ cannot happen too often.

Theorem 6.3.29 (Ferrand [108, Cor. 3.6]) *Let* $u : A \to B$ *be a local morphism of local rings, such that* A *and* $B \otimes_A B$ *are Noetherian. Then* B *is an* A-*algebra essentially of finite type.*

Proof Let $\mathfrak{a}$ be the kernel of the multiplication map $\mu : B \otimes_A B \to B$. Then $\mathfrak{a}$ is a finitely generated ideal of $B \otimes_A B$. This implies that there exists an A-algebra of finite type C and an epimorphism of A-algebras $\psi : C \to B$. From Theorem 6.3.28 it follows that B is essentially of finite type over A. $\qquad\square$

Example 6.3.30 As a nice application of the above theorem, we can give another proof of the well-known fact that if k is an infinite field, the tensor product $k[[X]] \otimes_k k[[X]]$ is not a Noetherian ring. Indeed, by Theorem 6.3.29 this would imply that $k[[X]]$ is a finitely generated k-algebra. On the other hand, $\dim_k (k[[X]]) = \dim_k (k^{\mathbb{N}})$ is uncountable.

Proposition 6.3.31 (Majadas and Rodicio [230, Prop. 2.5]) *Let* $u : A \to B$ *be a morphism of Noetherian rings. Assume that:*

(i) u is regular;
(ii) $B \otimes_A B$ is a Noetherian ring.

Then $\mathrm{fd}_{B \otimes_A B}(B) = \sup\{\mathrm{tr.deg}_{k(\mathfrak{q} \cap A)}(B_{\mathfrak{q}} \otimes k(\mathfrak{q} \cap A)) \mid \mathfrak{q} \in \mathrm{Spec}(B)\}$.

Proof Let $\mathfrak{a}$ be the kernel of the multiplication map $p : B \otimes_A B \to B$. By (i) we have $H_2(B \otimes_A B, B, -) = H_1(A, B, -) = (0)$, so that $\mathfrak{a}$ is locally generated by a regular sequence [6, Th. VI.25]. From Lemma 6.3.25 we obtain that

$$\mathrm{fd}_{B \otimes_A B}(B) = \sup\{\mathrm{rk}(\Omega_{B_{\mathfrak{q}}/A_{\mathfrak{q} \cap A}}) \mid \mathfrak{q} \in \mathrm{Spec}(B)\}.$$

Thus we may assume that A and B are local rings and we must prove that $\mathrm{rk}(\Omega_{B/A}) = \mathrm{tr.deg}_k (B \otimes_A k)$, where k is the residue field of A. Since

$$\Omega_{B \otimes_A k/k} \cong \Omega_{B/A} \otimes_B (B \otimes_A k),$$

we may assume that $A = k$ is a field. Because u is regular, B is a regular local ring, hence an integral domain. Therefore let $L = Q(B)$ be the field of fractions of B. By Theorem 6.3.29 we get that B is a geometrically regular k-algebra essentially of finite type; hence $k \subseteq L$ is a finitely generated separable field extension. Then

$$\mathrm{rk}(\Omega_{B/k}) = \dim_L (\Omega_{B/k} \otimes_B L) = \dim_L (\Omega_{L/k}) = \mathrm{tr.deg}_k L.$$

$\qquad\square$

Corollary 6.3.32 (Majadas and Rodicio [230, Cor. 2.6]) *Let k be a field and A be a geometrically regular k-algebra essentially of finite type which is an integral domain. Then $\mathrm{fd}_{A \otimes_k A}(A) = \mathrm{tr.deg}_k A$.*

Lemma 6.3.33 *Let $u : A \to B$ be a flat morphism of Noetherian rings, where $(B, \mathfrak{n}, L)$ is a local ring. If $\mathrm{fd}_{B \otimes_A B}(B) < \infty$, then $\dim_L \Omega_{L/A} < \infty$.*

Proof By Remark 6.3.21 there exists $n \in \mathbb{N}$ such that $\bigwedge^n \Omega_{B/A} = 0$. Then by [54, ch. III, §7, n. 5, Prop. 8] we know that $\bigwedge^n (\Omega_{B/A} \otimes_A L) = 0$, hence $\dim_L (\Omega_{B/A} \otimes_A L) < \infty$. Since $\Omega_{L/A}$ is a quotient of $\Omega_{B/A} \otimes_A L$, we get the conclusion. $\square$

To prove the nice Theorem 6.3.45, we need several module-theoretic preparations. Let us first remind the following definition.

Definition 6.3.34 (see [390]) Let A be a ring and M an A-module. The **trace ideal** of M, denoted by $\tau_A(M)$, is the image of the map

$$\varphi : M \otimes_A M^* \to A, \quad \varphi(m, f) = f(m),$$

that is $\tau_A(M) := \mathrm{Im}(\varphi) = \sum_{f \in M^*} f(M)$.

Remark 6.3.35

(i) If M is a projective module and $M \oplus N = F$ is a free module, then $\tau_A(M)$ is the ideal generated by the coordinates of all elements of M in a chosen basis of F. Indeed, let $\mathcal{E}$ be a basis of F and let $\{\varphi_e\}_{e \in \mathcal{E}}$ be the dual basis. Clearly for any $e \in \mathcal{E}$ and any $m \in M$, we have $\varphi_e(m) \in \tau_A(M)$. Conversely, let us first notice that M^* is a quotient of $F^* = A^{\mathcal{E}}$. Let $f = (f_e)_{e \in \mathcal{E}} \in A^{\mathcal{E}}$ and let $m = a_1 e_1 + \ldots + a_t e_t$, with $a_1, \ldots, a_t \in A$ and $e_1, \ldots, e_t \in \mathcal{E}$. Then

$$f(m) = f(a_1 e_1 + \ldots + a_t e_t) = a_1 f(e_1) + \ldots + a_t f(e_t) =$$

$$= a_1 f_{e_1}(m) + \ldots + a_t f_{e_t}(m).$$

(ii) If M is projective, $M \oplus N = F$ is a free module and $u : A \to B$ is a ring morphism, then

$$(M \otimes_A B) \oplus (N \otimes_A B) = F \otimes_A B,$$

hence it follows from (i) that

$$\tau_B(M \otimes_A B) = \tau_A(M)B.$$

In particular, if M is projective and $\mathfrak{p}$ is a prime ideal of A, then

$$\tau_A(M)A_{\mathfrak{p}} = \tau_{A_{\mathfrak{p}}}(M_{\mathfrak{p}}).$$

(iii) If M is projective, using the notations from (i), let $m = a_1 e_1 + \ldots + a_t e_t \in M$. For any $f \in M^*$ we have

$$f(m) = f\left(\sum_{i=1}^{t} a_i e_i\right) = f\left(\sum_{i=1}^{t} a_i f_{e_i}(m)\right) = \sum_{i=1}^{t} f(a_i) f_{e_i}(m),$$

hence $\tau_A(M) = \tau_A(M)^2$.

Proposition 6.3.36 (Vasconcelos [390, Prop. 1.3]) *Let A be a Noetherian ring and M a projective module of constant finite rank. Then M is finitely generated.*

Proof (Katz [193]) Let $n := \mathrm{rk}_{\mathfrak{p}}(M)$ be the constant finite rank of M. We prove the assertion by induction on n. If $n = 0$, then M is locally (0), hence $M = (0)$. Assume that $n > 0$ and that the assertion is true for all projective modules over any commutative ring, of constant rank less than n. Since M is projective, there exist A-modules K and F, such that $M \oplus K = F$ and F is free. Consider the trace ideal $\tau_A(M)$. Suppose that $\tau_A(M) \subsetneq A$; hence there exists a maximal ideal $\mathfrak{m}$ containing $\tau_A(M)$. Let $x \in M$. Then $x = a_1 e_1 + \ldots + a_t e_t$, where $a_1, \ldots, a_t \in \tau_A(M) \subseteq \mathfrak{m}$ and $e_1, \ldots, e_t$ are elements of a basis of F. Then $x \in \mathfrak{m}F$, hence $M \subseteq \mathfrak{m}F = \mathfrak{m}M \oplus \mathfrak{m}K$. This means that $M \subseteq \mathfrak{m}M$, hence $M = \mathfrak{m}M$. Therefore $M_{\mathfrak{m}} = \mathfrak{m}M_{\mathfrak{m}}$. Since $M_{\mathfrak{m}}$ is finitely generated, Nakayama's Lemma implies that $M_{\mathfrak{m}} = (0)$, contradicting the fact that $n > 0$. Therefore $\tau_A(M) = A$. This means that there exist elements $f_1, \ldots, f_r \in M^*$ and $m_1, \ldots, m_r \in M$ such that

$$f_1(m_1) + \ldots + f_r(m_r) = 1. \tag{6.2}$$

If there exists an index i such that $f_i(m_i)$ is nilpotent, then $1 - f_i(m_i)$ is invertible and we get a similar relation with $r - 1$ terms. Thus we may assume that no $f_i(m_i)$ is nilpotent. Let S_i denote the multiplicative system $\{1, f_i(m_i), f_i(m_i)^2, \ldots\}$. For any index $i = 1, \ldots, r$ we have an induced morphism

$$\theta_i : S_i^{-1}M \to S_i^{-1}A, \quad \theta_i\left(\frac{m}{f_i(m_i)^k}\right) = \frac{f_i(m)}{f_i(m_i)^k}.$$

These morphisms are surjective; therefore for each i, there exists a projective $S_i^{-1}A$-module K_i such that

$$S_i^{-1}M = S_i^{-1}A \oplus K_i.$$

Since M has constant rank n, it follows that K_i has rank $n - 1$ as a $S_i^{-1}A$-module. By the induction assumption K_i is finitely generated; hence $S_i^{-1}M$ is a finitely generated S_iA^{-1}-module. Now the relation (6.2) shows that M is a finitely generated A-module. □

Theorem 6.3.37 (Vasconcelos [392, Th. 1.3]) *Let A be a commutative Noetherian ring and M be an A-module. Assume that there exists $n \in \mathbb{N}$ such that $\bigwedge^n M$ is a projective module of rank 1. Then M is a finitely generated projective A-module of rank n.*

Proof From Proposition 6.3.36 it follows that $P := \bigwedge^n M$ is finitely generated, so that we can consider a system of generators of P, say $\{m_1^{(i)} \wedge \ldots \wedge m_n^{(i)} \mid 1 \le i \le p\}$. We shall prove that M is locally generated by one of the sets $\{m_1^{(i)}, \ldots, m_n^{(i)}\}$. This implies that M is finitely generated. Since exterior powers commute with localization, we can assume that A is a local ring. Then we may choose one of the elements above, say $m_1 \wedge \ldots \wedge m_n$, to be a generator of P. Suppose that we have a relation $\sum_{i=1}^{n} a_i m_i = 0$, where $a_1, \ldots, a_n \in A$. Multiplying by $m_1 \wedge \ldots \wedge m_{k-1} \wedge m_{k+1} \wedge \ldots \wedge m_n$ we obtain $a_k(m_1 \wedge \ldots \wedge m_n) = 0$, hence $a_k = 0$. This shows that $\{m_1, \ldots, m_n\}$ generate a free submodule F of M. Let

$$S := \{m \in M \mid m \wedge m_1 \wedge \ldots \wedge m_{k-1} \wedge m_{k+1} \wedge \ldots \wedge m_n = 0, \ k = 1, \ldots, n\}.$$

Thus S is a submodule of M and $F \cap S = (0)$. Let $m \in M$. Then for any $k = 1, \ldots, n$ we can write

$$m_1 \wedge \ldots \wedge m_{k-1} \wedge m \wedge m_{k+1} \wedge \ldots \wedge m_n = \alpha_k(m_1 \wedge \ldots \wedge m_n), \ \alpha_k \in A.$$

Let $r := \sum_{i=1}^{n} \alpha_i m_i$. Then $m - r \in S$, hence $M = F \oplus S$. Taking exterior powers we obtain

$$A \cong \bigwedge^n M = \bigwedge^n F \oplus ((\bigwedge^{n-1} F) \otimes S) \oplus \ldots$$

It follows that $S = (0)$, that is, M is a finitely generated flat module, hence projective [249, Cor. to Th. 7.11]. □

Proposition 6.3.38 *Let A be a Noetherian ring and M be a flat A-module which is a submodule of a projective module F. Then:*

(i) $\mathfrak{a} = \mathrm{Ann}_A(M)$ *is generated by an idempotent element e;*
(ii) $\mathrm{Supp}_A(M) = V(e)$ *is a closed and open subset of* $\mathrm{Spec}(A)$.

Proof

(i) Let $a \in \mathfrak{a}$ and $m \in M$. Then $am = 0$ and since M is flat, by [56, ch. I, §2, n. 11, Cor. 1] we can write $m = \sum_{i=1}^{n} r_i m_i$ where $r_1, \ldots, r_n \in A$, $m_1, \ldots, m_n \in M$ and $ar_1 = \ldots = ar_n = 0$. Thus $m \in \mathrm{Ann}(a) \cdot M$. Let $\mathfrak{a} = (a_1, \ldots, a_s)$. Then the flatness of M implies that

$$M = \big(\mathrm{Ann}(a_1) \cdot M\big) \cap \ldots \cap \big(\mathrm{Ann}(a_s) \cdot M\big) = \big(\mathrm{Ann}(a_1) \cap \ldots \cap \mathrm{Ann}(a_s)\big) \cdot M.$$

If $\mathfrak{b} = \mathrm{Ann}(\mathfrak{a})$, we therefore have $M = \mathfrak{b}M = \mathfrak{b}^t M$, $t \geq 1$. Since M is a submodule of the projective module F, we have

$$M \subseteq \bigcup_{t=1}^{\infty}(\mathfrak{b}^t \cdot F) = \Big(\bigcup_{t=1}^{\infty} \mathfrak{b}^t\Big) \cdot F.$$

By the Krull intersection theorem [249, Th. 8.9], there exists $\alpha \in \bigcap_{t=1}^{\infty} \mathfrak{b}^t$ such that $(1+\alpha)\big(\bigcap_{t=1}^{\infty} \mathfrak{b}^t\big) = 0$. This means that $1 + \alpha \in \mathfrak{a}$, so that

$$\mathfrak{a} + \mathfrak{b} = A = \mathfrak{a} \oplus \mathfrak{b}.$$

This implies that $\mathfrak{a}$ is generated by an idempotent $e \in A$.

(ii) Let $\mathfrak{p} \in \mathrm{Spec}(A)$. If $M = \mathfrak{p}M$, as in the proof of (i) there exists an element $p \in \mathfrak{p}$ such that $(1 + p)M = 0$, hence $M_\mathfrak{p} = 0$. Thus $M_\mathfrak{p} = 0$ iff $\mathfrak{p} \supseteq \mathfrak{b} = \mathrm{Ann}(\mathfrak{a})$, so that $\mathrm{Supp}_A(M) = V(e)$.

$\square$

Lemma 6.3.39 *Let A be a ring and M a flat A-module. Then the exterior algebra $\bigwedge M$ and the tensor algebra TM are flat A-modules.*

Proof By Lazard's theorem [324, Th. 5.40], we can write $M = \varinjlim_i F_i$ where F_i are finitely generated free A-modules and applying [54, Ch. III, §5, n.4, Prop. 6 and §7, n. 6, Prop. 9] we may assume that M is free. But then $\bigwedge M$ and TM are free over A.

$\square$

Lemma 6.3.40 *Let A be a ring and M a flat A-module. Then the canonical map θ_n : $\bigwedge^n M \to T^n M$ is injective.*

Proof As above we may assume that M is free and finitely generated. Consider $\{e_1, \ldots, e_p\}$ a basis of M. Then $\{e_{i_1} \wedge \ldots \wedge e_{i_n} \mid 1 \le i_1 < \ldots < i_n \le p\}$ and $\{e_{i_1} \otimes \ldots \otimes e_{i_n} \mid 1 \le i_1, \ldots, i_n \le p\}$ are bases of $\bigwedge^n M$ and $T^n M$, respectively. The rows of the matrix of θ_n with respect to these bases have at most one non-zero element, which is equal to 1. Therefore θ_n is injective. $\qquad\square$

Proposition 6.3.41 (Lazard [215, Ch. III, Prop. 2.6]) *Let A be a ring, M and N be flat A-modules and $u : M \to N$ be an injective A-linear map. Then $\bigwedge u : \bigwedge M \to \bigwedge N$ is injective.*

Proof $\bigwedge^n M$ and $T^n M$ are flat A-modules, as follows from Lemma 6.3.39. Then the maps $u_i := T^{i-1} M \otimes u \otimes T^{n-i} N : T^i M \otimes T^{n-i} M \to T^{i-1} M \otimes T^{n-i+1} M$ are injective; hence $T^n u = u_1 \circ u_2 \circ \ldots \circ u_n$ is injective. Look at the commutative diagram

$$
\begin{array}{ccc}
\bigwedge^n M & \xrightarrow{\wedge^n u} & \bigwedge^n N \\
\downarrow{\theta_M^n} & & \downarrow{\theta_N^n} \\
T^n M & \xrightarrow{T^n u} & T^n N
\end{array}
$$

By Proposition 6.3.40 we know that θ_M^n is injective. It follows that $\bigwedge^n u$ is injective. $\qquad\square$

Corollary 6.3.42 *Let A be a Noetherian ring and M a flat submodule of a free module F. Then the set $\{\mathfrak{p} \in \operatorname{Spec}(A) \mid \operatorname{rk}_{\mathfrak{p}}(M) \ge n\}$ is an open-closed subset of $\operatorname{Spec}(A)$, for any $n \in \mathbb{N}$.*

Proof From Proposition 6.3.41 we have $\bigwedge M \subseteq \bigwedge F$. Now we apply Proposition 6.3.38. $\qquad\square$

Theorem 6.3.43 (Vasconcelos [392, Th. 3.1]) *Let $(A, \mathfrak{m}, k)$ be a Noetherian local ring and M a flat A-module of constant rank. Then M is finitely generated.*

Proof Let $n = \operatorname{rk}(M)$. We will prove that $\bigwedge^n M \cong A$ and then apply Theorem 6.3.37. Changing M with $\bigwedge^n M$ we may assume that $\operatorname{rk}(M) = 1$, so that $M/\mathfrak{m}M \cong k$. Let $m \in M \setminus \mathfrak{m}M$. We will prove by induction on $\dim(A)$, that $M = Am$. Replacing A by $A/N(A)$ we may also assume that A is reduced. Finally we assume that m generates $M/\mathfrak{q}M$ as an

$A/\mathfrak{q}$-module for any $\mathfrak{q} \in \mathrm{Spec}(A) \setminus \mathrm{Min}(A)$ and set $N = Am$. We have an exact sequence

$$0 \to N \to M \to C := M/N \to 0.$$

Let $d \in A \setminus \bigcup_{\mathfrak{q} \in \mathrm{Min}(A)} \mathfrak{q}$, so that d is not a zero divisor and $\dim(A/dA) \leq \dim(A) - 1$. Then we obtain an exact sequence

$$0 \to S := 0 :_C d \to N/dN \to M/dM \to C/dC \to 0.$$

By the induction hypothesis, $M/dM \cong A/dA$ and is generated by the image of m, hence $M/dM \cong N/dN$, so that $C = dC$ and $S = (0)$. This means that C has a natural structure of a $Q(A)$-module. But $Q(A)$ is semisimple, being a finite product of fields; hence C is a flat $Q(A)$-module [393, Th. 4.2.2]. It follows by [249, p. 46] that C is flat over A. Because M is flat over A, by [249, Th. 7.9] we get that N is a flat A-module, hence free. Then we obtain the exact sequence

$$0 \to N \otimes_A Q(A) \to M \otimes_A Q(A) \to C \otimes_A Q(A) \to 0$$

and so $C = (0)$. $\square$

We need the following generalization of Theorem 6.3.43 to the case that A is not necessarily a local ring.

Theorem 6.3.44 (Vasconcelos [392, Th. 3.2]) *Let A be a Noetherian ring and M a flat A-module, locally of finite rank at each prime, that is, a submodule of a free module F. Then M is finitely generated.*

Proof Since open-closed subsets of $\mathrm{Spec}(A)$ correspond to idempotents of A, by Corollary 6.3.42 we can decompose $A = \bigoplus_i Ae_i$ and $M = \bigoplus_i Me_i$ where e_i are idempotents and $\mathrm{rk}_{Ae_i}(Me_i)$ is constant. Thus, replacing A by Ae_i we may assume that M is of constant rank. Let $K = Q(A)$ be the total ring of fractions of A. Since K is a semilocal ring, Theorem 6.3.43 implies that $M \otimes_A K$ is a finitely generated K-module. Let $(f_i)_{i \in I}$ be a basis of F and $\{\frac{m_1}{s_1}, \ldots, \frac{m_r}{s_r}\}$ be a system of generators of $M \otimes_A K$. Then $m_1, \ldots, m_r$ can be written as linear combinations of a finite subset of elements of the basis, that is,

$$m_i = \sum_{j=1}^{k} \alpha_i^j f_i, \text{ with } \alpha_i^j \in A.$$ Note that we put together all the elements of the basis occurring in the elements $m_1, \ldots, m_r$, that is, some of the coefficients can be 0. Let

$m \in M$. Then

$$\frac{m}{1} = \frac{a_1}{t_1} \cdot \frac{m_1}{s_1} + \ldots + \frac{a_r}{t_r} \cdot \frac{m_r}{s_r} = \frac{a_1}{t_1} \cdot \frac{\sum_{i=1}^{k} \alpha_i^1 f_i}{s_1} + \ldots + \frac{a_r}{t_r} \cdot \frac{\sum_{i=1}^{k} \alpha_i^r f_i}{s_r},$$

that gives after computations a relation

$$l \cdot m = \sum_{i=1}^{k} b_i f_i, \quad \text{where } l, b_1, \ldots, b_k \in A.$$

Since l is not a zero divisor, this shows that any element of M can be written as a linear combination of $f_1, \ldots, f_k$. Therefore M is finitely generated as an A-module. □

Theorem 6.3.45 (Majadas and Rodicio [230, Th. 6.2]) *Let $u : A \to B$ be a flat morphism of Noetherian rings such that* $\mathrm{fd}_{B \otimes_A B}(B) < \infty$. *The following are equivalent:*

  (i) u is smooth;
  (ii) $\Omega_{B/A}$ is a projective B-module;
  (iii) $\Omega_{B/A}$ is a finitely generated B-module.

Proof $(i) \Rightarrow (ii)$: This is always true, see Theorem 3.2.7.
$(ii) \Rightarrow (i)$: From Theorem 6.3.19 it follows that u is regular, hence $H_1(A, B, B) = (0)$. Now apply Theorem 3.2.7.
$(ii) \Rightarrow (iii)$: Let $\mathfrak{q} \in \mathrm{Spec}(B)$ and $\mathfrak{p} = \mathfrak{q} \cap A$. The morphism $A_{\mathfrak{p}} \to B_{\mathfrak{q}}$ is flat and $\mathrm{fd}_{B_{\mathfrak{q}} \otimes_{A_{\mathfrak{p}}} B_{\mathfrak{q}}}(B_{\mathfrak{q}}) < \infty$; hence by Theorem 6.3.19 the morphism $A_{\mathfrak{p}} \to B_{\mathfrak{q}}$ is regular. From Remark 6.3.21 it results that $\bigwedge^n \Omega_{B_{\mathfrak{q}}/A_{\mathfrak{p}}} = (0)$, for sufficiently large n. It follows that the free $B_{\mathfrak{q}}$-module $\Omega_{B_{\mathfrak{q}}/A_{\mathfrak{p}}}$ has finite rank. Thus $\Omega_{B/A}$ is a projective B-module, locally of finite rank. By Theorem 6.3.44 we get that $\Omega_{B/A}$ is finitely generated.
$(iii) \Rightarrow (ii)$: $\Omega_{B/A}$ is flat by Theorems 6.3.19 and 3.2.7 and finitely generated by assumption, hence is projective. □

Example 6.3.46 (Majadas and Rodicio [230, Rem. p. 80]) There exists a smooth morphism $u : K \to L$ such that $\Omega_{L/K}$ is finitely generated and $\mathrm{fd}_{L \otimes_K L}(L) = \infty$.

Proof Indeed, let p be a prime number, $K = \mathbb{F}_p$ be the prime field of characteristic p (or more generally any perfect field of characteristic p), $E = K(X_1, \ldots, X_n, \ldots)$ and $L = E^{p^{-\infty}}$. Then L is smooth over K and $\Omega_{L/K} = 0$. But $\mathrm{fd}_{L \otimes_K L}(L) = \mathrm{tr.deg}_K L = \infty$. □

Lemma 6.3.47 *Let* $(A, \mathfrak{m})$ *be a local ring and* $u : P \to M$ *an* A*-linear map. Assume that* P *is projective,* M *is flat and* $u \otimes 1_{A/\mathfrak{m}}$ *is injective. Then* P *is a pure submodule of* M.

Proof The projective module P is a direct summand in a free module $P \oplus Q$. Replacing P by $P \oplus Q$, we may assume that P is free. Replacing if necessary P by its free finitely generated factors, we may assume that P is free and finitely generated. Using Lazard's theorem [324, Th. 5.40], we may furthermore assume that P and M are free and finitely generated. Now we apply [131, Cor. 19.1.11]. $\qquad \square$

Lemma 6.3.48 *Let* $(A, \mathfrak{m})$ *be a local ring,* M *a flat* A*-module and* d *a cardinal number such that* $0 \leq d \leq \mathrm{rk}_{\mathfrak{m}}(M)$. *Then there exists an exact sequence*

$$0 \to F \to M \to L \to 0$$

where F *is free of rank* d *and* L *is flat.*

Proof Let $\{\bar{e}_i\}_{i \in I}$ be a linearly independent set of $M/\mathfrak{m}M$ and consider the free A-module $F = A^{(I)} = \bigoplus_{i \in I} Ae_i$. Let $u : F \to M$ be the morphism defined by $u(e_i) = \bar{e}_i$, for any $i \in I$. Thus we have a commutative diagram

$$
\begin{array}{ccc}
F & \xrightarrow{\ u\ } & M \\
\downarrow & & \downarrow \\
F/\mathfrak{m}F & \xrightarrow{\ \bar{u}\ } & M/\mathfrak{m}M
\end{array}
$$

The morphism $\bar{u}$ is injective and Lemma 6.3.47 implies that $u(F)$ is a pure submodule of M. Then $\mathrm{Coker}(u) = L = M/u(F)$ is flat [209, Cor. 4.86]. $\qquad \square$

Proposition 6.3.49 (Cox - Rush [86, Th. 2.3]) *Let* A *be a ring,* M *a flat* A*-module and* $\mathfrak{p} \subseteq \mathfrak{q}$ *prime ideals of* A. *Then* $\mathrm{rk}_{\mathfrak{q}}(M) \leq \mathrm{rk}_{\mathfrak{p}}(M)$.

Proof Localizing in $\mathfrak{q}$ we may assume that A is local and $\mathfrak{q}$ is the maximal ideal. Let $d = \mathrm{rk}_{\mathfrak{q}}(M)$. From Lemma 6.3.48 we get an exact sequence

$$0 \to F \to M \to L \to 0$$

where F is free of rank d and L is flat. Then the map $F \otimes_A k(\mathfrak{p}) \to M \otimes_A k(\mathfrak{p})$ is injective, so that $\mathrm{rk}_{\mathfrak{p}}(M) \geq \mathrm{rk}_{\mathfrak{p}}(F) = \mathrm{rk}_{\mathfrak{q}}(F) = \mathrm{rk}_{\mathfrak{q}}(M)$. $\qquad \square$

Theorem 6.3.45 has a more general version in the case of local algebras over a field.

Proposition 6.3.50 (Majadas and Rodicio [230, Prop. 6.3]) *Let K be a field and $(A, \mathfrak{m}, L)$ be a Noetherian local K-algebra such that $\mathrm{fd}_{A \otimes_K A}(A) < \infty$. Let $Q = Q(A)$ be the total quotient ring of A. The following are equivalent:*

(i) A is smooth over K;
(ii) $\Omega_{A/K}$ is a free A-module;
(iii) $\Omega_{A/K}$ is a finitely generated A-module;
(iv) if $s := \max\{t \mid \bigwedge^t \Omega_{A/K} \neq (0)\}$, then $\bigwedge^t \Omega_{A/K}$ is a faithfully flat A-module;
(v) $\Omega_{A/K}$ is a submodule of a free A-module;
(vi) $\dim_Q(\Omega_{Q/K}) = \dim(A) + \dim_L(\Omega_{L/K}) - \dim_L H_1(K, L, L)$.

__Proof__ $(i) \Leftrightarrow (ii) \Leftrightarrow (iii)$: This is proved in Theorem 6.3.45.
$(v) \Rightarrow (iii)$: Follows from Theorem 6.3.44.
$(ii) \Rightarrow (v)$: Obvious.

$(iv) \Rightarrow (iii)$: Let $s := \max\{t \mid \bigwedge^t \Omega_{A/K} \neq (0)\}$. Then from Remark 6.3.21 we obtain that $s < \infty$. Since

$$\left(\bigwedge^j \Omega_{A/K}\right) \otimes Q \cong \bigwedge^j (\Omega_{A/K} \otimes Q) \cong \bigwedge^j \Omega_{Q/K}$$

and $\bigwedge^j \Omega_{A/K}$ is flat, we get that $\dim_Q(\Omega_{Q/K}) = s$. On the other hand,

$$\left(\bigwedge^j \Omega_{A/K}\right) \otimes_A L \cong \bigwedge^j (\Omega_{A/K} \otimes_A L) = (0) \text{ for } j > s,$$

hence $\dim_L(\Omega_{A/K} \otimes_A L) \leq s$. Because $\bigwedge^s \Omega_{A/K}$ is faithfully flat, we have

$$\bigwedge^s (\Omega_{A/K} \otimes_A L) \cong \left(\bigwedge^s \Omega_{A/K}\right) \otimes_A L \neq (0).$$

By Proposition 6.3.49 we see that $\Omega_{A/K}$ has constant rank at each prime $\mathfrak{p} \in \mathrm{Spec}(A)$, hence by Theorem 6.3.43 we obtain that $\Omega_{A/K}$ is finitely generated.
$(ii) \Rightarrow (iv)$: Obvious.
$(vi) \Leftrightarrow (ii)$: Since A is regular, by [6, Prop. VI.1] we have $\dim(A) = \dim_L H_1(A, L, L)$. Applying Lemma 6.3.33 we have an exact sequence of finite dimensional L-vector spaces

$$0 \to H_1(K, L, L) \to H_1(A, L, L) \to \Omega_{A/K} \otimes_A L \to \Omega_{L/K} \to 0.$$

Therefore

$$\dim_L(\Omega_{A/K} \otimes_A L) = \dim(A) + \dim_L(\Omega_{L/K}) - \dim_L H_1(K, L, L).$$

Thus (vi) holds iff $\dim_Q(\Omega_{Q/K}) = \dim_L(\Omega_{A/K} \otimes_A L)$, that is equivalent to condition (ii).   □

Our next goal is to prove some results connected with Propositions 3.2.24 and 3.3.15.

Proposition 6.3.51 (Majadas and Rodicio [230, Prop. 6.5]) *Let k be a field of characteristic $p > 0$ such that $[k : k^p] < \infty$ and let $(A, \mathfrak{m}, L)$ be a Noetherian local k-algebra with $\mathrm{fd}_{A \otimes_k A}(A) < \infty$. Then A is an excellent ring if and only if A is smooth over k.*

Proof If A is smooth over k, then A is excellent by Theorem 3.3.15. Conversely assume that A is an excellent ring. By Theorem 6.3.50 it suffices to show that $\Omega_{A/k}$ is a finitely generated A-module. Now $\dim_k \Omega_k = [k : k^p] < \infty$ and by Theorem 6.3.33 we have $\dim_L \Omega_{L/k} < \infty$. Thus the exact sequence

$$\Omega_k \otimes_k L \to \Omega_L \to \Omega_{L/k} \to 0$$

yields $\dim_L \Omega_L < \infty$, hence $[L : L^p] < \infty$. As A is excellent, by Theorem 3.5.34 and Theorem 3.5.41 we get that Ω_A is finitely generated. But $\Omega_{A/k}$ is a homomorphic image of Ω_A, therefore is finitely generated.   □

Question 6.3.52 (Majadas and Rodicio [230]) Let $u : A \to B$ be a smooth morphism of Noetherian rings. If A is quasi-excellent, does it follow that B is quasi-excellent?

Let us see some partial answer to this question.

Proposition 6.3.53 (Majadas and Rodicio [230, Prop. 6.9]) *Let $u : A \to B$ be a smooth morphism of Noetherian rings with perfect residue fields and let $\mathbf{P}$ be a property of Noetherian local rings satisfying $(\mathbf{A_1})$, $(\mathbf{A_2})$, $(\mathbf{A_3})$ and $(\mathbf{A_5})$. If A is $\mathbf{P}$-2 then B is $\mathbf{P}$-2.*

Proof It is enough to prove that $\mathbf{P}(B/\mathfrak{a})$ is open for any ideal $\mathfrak{a} \subset B$ and by Lemma 1.2.7 it suffices to see that for any prime ideal $\mathfrak{q}$ of B, the $\mathbf{P}$-locus $\mathbf{P}(B/\mathfrak{q})$ contains a non-empty open set. Let $\mathfrak{q} \in \mathrm{Spec}(B)$ and $\mathfrak{p} = \mathfrak{q} \cap A$. The morphism $A/\mathfrak{p} \to B/\mathfrak{p}B$ is injective and smooth and the morphism $A/\mathfrak{p} \to B/\mathfrak{q}$ is injective. Therefore we may assume that A is a domain, u is injective, $\mathfrak{q} \cap A = (0)$ and that the induced morphism $A \to B/\mathfrak{q}$ is injective. There exists $f \in A$, $f \neq 0$ such that A_f has the property $\mathbf{P}$. Replacing A by A_f and B by B_f, we may assume that A has $\mathbf{P}$. Let $C = B/\mathfrak{q}$. Since u is smooth, by [248, Th. 64] the set $S(C) := \{\mathfrak{p} \in \mathrm{Spec}(C) \mid A \to C_\mathfrak{p} \text{ is smooth}\}$ is open in $\mathrm{Spec}(C)$

and $S(C) \subseteq \mathrm{Reg}(C) \subseteq \mathbf{P}(C)$. Since the residue fields are perfect, $(0) \in S(C)$, so that $S(C) \neq \emptyset$. $\qquad\square$

Corollary 6.3.54 *Let $u : A \to B$ be a smooth morphism of Noetherian rings with perfect residue fields. If A is quasi-excellent then B is quasi-excellent.*

Proof Follows from Proposition 3.2.24 and Proposition 6.3.53. $\qquad\square$

The last part of this section is dedicated to the proof of Theorem 6.3.70. This originated in the following conjecture proposed in [321].

Conjecture 6.3.55 (Rodicio's conjecture [321]) Let k be a field of characteristic zero and let A be a k-algebra of finite type. If $HH_n(A) = (0)$ for sufficiently large $n \in \mathbb{N}$, then A is a smooth k-algebra.

After several partial results, the conjecture was proved to be true, in a little bit more general frame, by L. Avramov and M. Vigué-Poirrier [41], who showed the following:

Theorem 6.3.56 (Avramov and Vigué-Poirrier [41, Th.1]) *Let k be any field and let A be a k-algebra of finite type. If there exist an odd positive $i \in \mathbb{N}$ and an even positive $j \in \mathbb{N}$ such that $HH_i(A) = HH_j(A) = (0)$, then A is a smooth k-algebra.*

We shall prove a more general result, given by Rodicio [322], in the frame of supplemented algebras. We follow the proof in [36]. We need some preparations that the interested reader can find with more details in the nice survey of Avramov [23] or in the classical book of Gulliksen and Levin [137]. A detailed presentation of the divided power algebras can be found in [359, Ch. 23] or [137, Ch. I, §7]. Anyway, we remind some of the main features about divided powers.

Notation 6.3.57 Let k be a ring and $A = \bigoplus_{d \geq 0} A_d$ be a graded k-algebra whose components in negative degrees are 0. We will denote

$$A_+ = \bigoplus_{d > 0} A_d, \quad A_{even} = \bigoplus_{d \geq 0} A_{2d} \text{ and } A_{odd} = \bigoplus_{d \geq 0} A_{2d+1}.$$

Definition 6.3.58 Let k be a ring and $A = \bigoplus_{d \geq 0} A_d$ be a graded k-algebra that is strictly commutative. A system of maps $\gamma_n : A_{even,+} \to A_{even,+}$, for any $n > 0$ is called a **divided power structure** on A if:

(i) if $x \in A_{2d}$, then $\gamma_n(x) \in A_{2nd}$;
(ii) $\gamma_1(x) = x$, for all $x \in A_{even,+}$;

(iii) $\gamma_n(x)\gamma_m(x) = \frac{(n+m)!}{n!m!}\gamma_{n+m}(x)$;

(iv) $\gamma_n(xy) = x^n\gamma_n(y)$, for all $x \in A_{even}$ and $y \in A_{even,+}$;

(v) if x and y are homogeneous elements of odd degree and $n > 1$, then $\gamma_n(xy) = 0$;

(vi) if $x, y \in A_{even,+}$, then $\gamma_n(x+y) = \sum_{i=0}^{n} \gamma_i(x)\gamma_{n-i}(y)$;

(vii) $\gamma_n(\gamma_m(x)) = \frac{(nm)!}{n!(m!)^n}$ for any $x \in A_{even,+}$.

Remark 6.3.59 If A is an algebra over $\mathbb{Q}$, then it is well-known and easy to see that one must have $\gamma_n(x) = \frac{x^n}{n!}$.

Definition 6.3.60 Let k be a ring and $A = \bigoplus_{d=0}^{\infty}$ be a strictly commutative DG k-algebra. A divided power structure on A is **compatible with the differential graded structure** if for any $x \in A_{even,+}$ we have $d(\gamma_n(x)) = d(x)\gamma_{n-1}(x)$.

Remark 6.3.61 If A is a DG-algebra with a divided power structure compatible with the DG-structure and if x is a boundary in A, it doesn't follow that $\gamma_n(x)$ is a boundary. This means that in general there is not an induced divided power structure on the homology algebra $H(A)$.

Lemma 6.3.62 *Let k be a ring and let $f : A \to B$ be a surjective morphism of strictly commutative DG-algebras endowed with divided power structure compatible with the differential graded structures. If $H_k(A) = (0)$ for any $k > 0$, there is an induced divided power structure on the homology algebra $H(B)$.*

Proof Let $x, x' \in B_{2d}$ define the same homology class in $H(B)$, say $x - x' = d(w)$. Let $y \in A_{2d}$ and $z \in A_{2d+1}$ be such that $f(y) = x$ and $f(z) = w$. Then $f(y + d(z)) = x'$. Therefore, setting $y' = y + d(z)$, we have

$$\gamma_n(y') = \sum_{i=0}^{n} \gamma_i(y)\gamma_{n-i}(d(z)) = \gamma_n(y) + \sum_{i=0}^{n-1} \gamma_i(y)\gamma_{n-i}(d(z)).$$

But $H_k(A) = 0$ for $k > 0$ and $d(\gamma_l(d(z))) = 0$ for any l, hence we can write, for some elements $z_0, \ldots, z_{n-1} \in A$,

$$\gamma_n(y') = \gamma_n(y) + \sum_{i=0}^{n-1} \gamma_i(y)d(z_i).$$

For any index $i = 1, \ldots, n-1$ we have

$$d(\gamma_i(y)z_i) = d(\gamma_i(y))z_i + \gamma_i(y)d(z_i) = d(y)\gamma_{i-1}(y)z_i + \gamma_i(y)d(z_i).$$

Because $d(y)$ maps to 0 in B, we get that $\gamma_n(x)$ and $\gamma_n(x')$ have the same image in $H(B)$, thus obtaining a well-defined map $\gamma_n : H_{2d}(B) \to H_{2nd}(B)$, for any $d > 0$ and $n > 0$. The reader can easily prove that in this way we get a divided power structure on $H(B)$.

$\square$

Remark 6.3.63 (see [36]) Let $u : A \to B$ be a surjective ring morphism. Then there exists a factorization $A \xrightarrow{\phi} A \langle U \rangle \xrightarrow{\psi} B$, where $A \langle U \rangle$ is a DG-algebra obtained from A by adjoining exterior variables in odd degrees and divided powers variables in even degrees and ψ is a quasi-isomorphism. This is the so-called **Tate resolution** of A (see [137, Th. 1.2.3] or [23, Prop. 6.1.4]). Let U_{odd} and U_{even} be the sets of odd and even variables, respectively, and let $\gamma_s(v)$ be the s-th divided power of $v \in U_{even}$. Then, as a graded A-module, $A \langle U \rangle$ is free and generated by

$$\{u_{i_1} \ldots u_{i_m} \cdot \gamma_{s_1}(v_{j_1}) \ldots \gamma_{s_n}(v_{j_n}) \mid u_{i_k} \in U_{odd}, \ v_{j_l} \in U_{even}, \ m, n \geq 0, \ s_l \geq 0\}.$$

In order to get a basis, we have to order the variables in U and take the ordered monomials. Let us denote by $\mathfrak{a}^{(2)} A \langle U \rangle$ the A-module generated by the monomials $u_{i_1} \ldots u_{i_m} \cdot \gamma_{s_1}(v_{j_1}) \ldots \gamma_{s_n}(v_{j_n})$, with $m + \sum_{l=1}^{n} s_l \geq 2$.

Remark 6.3.64 It is proved in [137, Prop. 1.7.6] that the divided powers of the variables in U extend to a system of divided powers of $A \langle U \rangle$. Therefore for every element of even positive degree $a \in A \langle U \rangle$ and for any $s \geq 0$, an element $\gamma_s(a) \in A \langle U \rangle$ of degree $s \cdot \deg(a)$ is defined. These elements satisfy

$$\gamma_0(a) = 1; \ (s!)\gamma_s(a) = a^s, \ \forall s \geq 1; \ \gamma_s(a+b) = \sum_{i=0}^{s} \gamma_{s-i}(a)\gamma_i(b). \tag{6.3}$$

If ∂ is the differential of $A \langle U \rangle$, then

$$\partial(ab) = \partial(a)b + (-1)^{\deg a} a \partial(b), \text{ for any } a \text{ and } b;$$
$$\partial(\gamma_s(a)) = \partial(a)\gamma_{s-1}(a), \text{ for any } a \text{ of even positive degree and for any } s \geq 1. \tag{6.4}$$

Proposition 6.3.65 *Let k be a commutative ring and let A and B be commutative k-algebras. Then there is a canonical structure of a strictly graded commutative DG-algebra with divided powers on $\mathrm{Tor}^k_*(A, B)$.*

Proof Let $k \to T \to A$ be a Tate resolution of A over k. Because $T \to A$ is a quasi-isomorphism and T_d is a free k-module for each $d \in \mathbb{N}$, it follows that the DG-algebra $D := T \otimes_k B$ computes the groups $\mathrm{Tor}^k_i(A, B)$. Consider a surjective morphism $k[Y_j, j \in J] \to B$. Then D is a quotient of the DG-algebra $T[Y_j, j \in J]$, whose homology is concentrated in degree 0. It is easy to see that D and $T[Y_j, j \in J]$ have divided power structures compatible with the differentials. Now Lemma 6.3.62 shows that there is a divided power structure on $H(D)$. The divided power structure does not depend on the choice of the resolution, as can be shown by an easy computation left to the reader.

□

Following the idea of Avramov and Iyengar [36], we use the terminology of supplemented algebras from the classical book of Cartan and Eilenberg [79].

Definition 6.3.66 ([79, Ch. X]) Let A be a ring and $u : A \to B$ be an A-algebra. We say that B is a **supplemented** A-algebra if there is a ring homomorphism $v : B \to A$ such that $v \circ u = 1_A$.

Remark 6.3.67 Usually the morphism v from the Definition 6.3.66 is called the **augmentation** of the supplemented A-algebra B. If $\mathfrak{a} = \ker(v)$, then it is clear that, as A-modules, $B = A \oplus \mathfrak{a}$.

Remark 6.3.68 Let $(A, \mathfrak{m}, k)$ be a local ring and $u : A \to B$ be a surjective ring morphism. Constructing a Tate resolution $A \langle U \rangle$ we can proceed in such a way that $\partial(U_1)$ is a minimal system of generators of $\ker(u)$ and that $\partial(U_n)$ is a minimal system of generators of $H_{n-1}(A \langle U \rangle)$ for any $n \geq 2$. What we get is what is called an **acyclic closure** of B over A [23, Construction 6.3.1]. A Tate resolution is an acyclic closure iff

$$\partial(u) \in \mathfrak{m}U + \mathfrak{a}^{(2)} A \langle U \rangle , \ \forall u \in U \text{ (see [23])},$$

where $\mathfrak{a}^{(2)} A \langle U \rangle$ is defined in Remark 6.3.63. If $A \langle U \rangle$ is an acyclic closure of k over A, then by a theorem of Gulliksen and Schoeller [23, Th. 6.3.5], the DG-algebra $A \langle U \rangle$ is a minimal resolution of k over A, that is

$$\partial(A \langle U \rangle) \subseteq \mathfrak{m}A \langle U \rangle .$$

We can generalize the above Remark as follows:

Lemma 6.3.69 ([36, Lemma 3.6]) *Let $(A, \mathfrak{m}, k)$ be a local ring that is a supplemented B-algebra and let $A \langle U \rangle$ be an acyclic closure of B over A. Then*

$$\partial(A \langle U \rangle) \subseteq \mathfrak{m}A \langle U \rangle .$$

Proof Let us first note that B is a local ring with residue field k. Let $B \langle Y \rangle$ be an acyclic closure of k over B. Then $B \langle Y \rangle$ is a bounded below complex of free B-modules and

$A \langle U \rangle \to B$ is a quasi-isomorphism, hence also

$$A \langle U, Y \rangle = A \langle U \rangle \otimes_B B \langle Y \rangle \to B \otimes_B B \langle Y \rangle = B \langle Y \rangle$$

is a quasi-isomorphism. Also $B \langle Y \rangle \to k$ is a quasi-isomorphism; therefore $A \langle U, Y \rangle$ is a Tate resolution of k over A. If $A \langle X \rangle$ is an acyclic closure of k over A, then

$$H_n(k \langle U, Y \rangle) = H_n(A \langle U, Y \rangle \otimes_A k) = \mathrm{Tor}_n^A(k, k) =$$

$$= H_n(k \langle X \rangle) = H_n(A \langle X \rangle \otimes_A k).$$

By Proposition 6.3.65 there is an isomorphism of divided powers algebras

$$H_\bullet(k \langle U, Y \rangle) \cong H_\bullet(k \langle X \rangle) = k \langle X \rangle .$$

The second equality holds by the theorem of Gulliksen and Schoeller (see Remark 6.3.68). The same result implies that $\partial(Y) = 0$ in $k \langle U, Y \rangle$. Observing that $\partial(U_1) = 0$, assume by induction that $\partial(U_{\le n}) = 0$ for some $n \ge 1$. Then the differential of the divided power algebra $k \langle U_{\le n}, Y_{\le n} \rangle$ is trivial. By the first isomorphism above, the inclusion $k \langle U_{\le n}, Y_{\le n} \rangle \hookrightarrow k \langle U, Y \rangle$ induces an injective map in homology. Therefore $\partial(U_{n+1}) = 0$.

□

We can now prove the main result.

Theorem 6.3.70 (Rodicio [322, Th. 1]) *Let A be a Noetherian ring, B a supplemented A-algebra and $\mathfrak{a}$ be the kernel of the augmentation. The following are equivalent:*

(i) $\mathfrak{a}$ is locally generated by a regular sequence;
(ii) there exists an integer $m > 0$ such that $\mathrm{Tor}_n^A(B, B) = (0)$ for any $n \ge m$;
(iii) there exist an even integer $i > 0$ and an odd integer $j > 0$ such that

$$\mathrm{Tor}_i^A(B, B) = \mathrm{Tor}_j^A(B, B) = (0).$$

Proof ([36, Proof of Th. 3.1]) $(i) \Rightarrow (ii)$: See Theorem 6.3.12.
$(ii) \Rightarrow (iii)$: Obvious.
$(iii) \Rightarrow (i)$: We may assume that $(A, \mathfrak{m}, k)$ is a local ring. Choose a minimal set of generators of $\mathfrak{a}$, say $\{y_1, \ldots, y_r\}$ and let $K_\bullet = K_\bullet(y_1, \ldots, y_r)$ be the Koszul complex on $y_1, \ldots, y_r$. Let $A \langle U \rangle$ be an acyclic closure of B over A. This means that $U_1 = \{u_1, \ldots, u_r\}$, with $\partial(u_i) = y_i$, $i = 1, \ldots, r$ and $U_2 = \{v_1, \ldots, v_t\}$ where the classes of the elements $\partial(v_1), \ldots, \partial(v_t)$ are a minimal system of generators of $H_1(A \langle U_1 \rangle)$. Let us remark that in $B \langle U \rangle = B \otimes_A A \langle U \rangle$ we have $\partial(U_1) = 0$, because $\partial(u_i) = y_i \in \mathfrak{a}$ and

by the relations (6.4) we have $\partial(U_2) \subseteq BU_1$, hence

$$Z = B \langle u_1, \ldots, u_r \rangle \oplus B \langle v_1, \ldots, v_t \rangle u_1 \ldots u_r \subseteq B \langle U \rangle$$

is a submodule of cycles. By Lemma 6.3.69 we have $\partial(A \langle U \rangle) \subseteq \mathfrak{m}A \langle U \rangle$, hence the composition

$$Z \otimes_B k \to H_\bullet(B \langle U \rangle) \otimes_B k \to H_\bullet(B \langle U \rangle \otimes_B k) = k \langle U \rangle$$

is injective. Thus either $t = 0$ or $\operatorname{Tor}_n^A(B, B) = H_n(B \langle U \rangle) \neq (0)$, for any $n \in \mathbb{N}$ with $n \equiv r \bmod 2$. By assumption the second situation cannot occur, hence $A \langle U_1 \rangle \cong K_\bullet$, that is $y_1, \ldots, y_r$ is a regular sequence. $\qquad \square$

Corollary 6.3.71 (Rodicio [322, Cor. 2]) *Let $u : A \to B$ be a flat morphism such that $B \otimes_A B$ is Noetherian. If there exist an even integer $i > 0$ and an odd integer $j > 0$ such that $HH_i(B) = HH_j(B) = 0$, then u is smooth.*

Proof Follows immediately from Theorem 6.3.70. $\qquad \square$

Corollary 6.3.72 *Let $u : A \to B$ be a flat morphism such that $B \otimes_A B$ is Noetherian. The following are equivalent:*

   (i) *u is smooth;*
  (ii) *$\bigwedge \Omega_{B/A} \cong HH_\bullet(B)$;*
 (iii) *the B-algebra $HH_\bullet(B)$ is generated by its elements of degree 1;*
  (iv) *there exists $n \in \mathbb{N}$ such that $HH_i(B) = 0$, for any $i \geq n$.*

Proof Follows from Theorem 6.3.13 and Corollary 6.3.71. $\qquad \square$

6.4   About Normality and Formal Fibres

This last section is dedicated to some miscellaneous results concerning normal morphisms and the behaviour of integral closure with respect to base change.

Lemma 6.4.1 *Let A be a ring and $\mathfrak{a}$ be an ideal of A.*

   (i) *the integral closure of $A + \mathfrak{a}XA[X]$ in $A[X]$ is $A + \sqrt{\mathfrak{a}}XA[X]$;*
  (ii) *A is reduced if and only if A is integrally closed in $A[X]$.*

Proof

(i) Clearly any polynomial of the form aX^r with $a \in \sqrt{\mathfrak{a}}$ is integral over A. Conversely, let $f = c_0 X^r + \ldots + c_{r-1} X + c_r \in A[X]$ where $c_0, \ldots, c_r \in A$. If f is integral over $A + \mathfrak{a} X A[X]$, then we have an equation of the form

$$f^N + \left(a_{N-1} + X P_{N-1}(X)\right) f^{N-1} + \ldots + \left(a_1 + X P_1(X)\right) f + \left(a_0 + X P_0(X)\right) = 0,$$

where $a_0 \ldots, a_{N-1} \in A$ and the polynomials $P_0, \ldots, P_{N-1}$ have coefficients in $\mathfrak{a}$. The coefficient of X^{rN} is zero, but this coefficient is $c_0 + \alpha$, for some $\alpha \in \mathfrak{a}$. Hence $c_0 \in \sqrt{\mathfrak{a}}$. Now we continue by induction on r.

(ii) Follows from (i) applied for $\mathfrak{a} = (0)$. $\qquad\square$

The reader should note that in the following theorem, because A^* is not necessarily Noetherian, the morphism f is not what we called a *reduced* morphism, but a morphism *with geometrically reduced fibres* (see Definition 1.6.6). It should be reminded also that the property of being *absolutely reduced* was explained in Definition 1.6.7.

Theorem 6.4.2 (Lazarus [216, Théorème]) *Let A be a Noetherian ring and let $f : A \to A^*$ be a ring morphism, where A^* is not necessarily a Noetherian ring. The following are equivalent:*

(i) *f is flat and with geometrically reduced fibres;*

(ii) *f is flat and for any reduced A-algebra C, the ring $C \otimes_A A^*$ is reduced, that is f is absolutely reduced;*

(iii) *for any morphism of A-algebras $B \to C$, if D is the integral closure of B in C, then $D \otimes_A A^*$ is the integral closure of $B \otimes_A A^*$ in $C \otimes_A A^*$.*

Proof $(i) \Rightarrow (ii)$: Write $C = \varinjlim_{i \in I} C_i$, where C_i are A-algebras of finite type. Then $C_i \otimes_A A^*$ are reduced rings and $C \otimes_A A^* = \varinjlim_{i \in I} C_i \otimes_A A^*$, hence $C \otimes_A A^*$ is reduced.

$(ii) \Rightarrow (i)$: Obvious.

$(ii) \Rightarrow (iii)$: We can write $B = \varinjlim_{i \in I} B_i$, where B_i are the finitely generated A-subalgebras of B. Since the morphisms $B_i \to B_i \otimes_A A^*$ inherit the properties of f, it is clear that we may assume that $A = B$ and is a subring of C. Let $\mathfrak{n} = N(C) = \{x \in C \mid x^n = 0 \text{ for some } n \in \mathbb{N}\}$ be the nilradical of C. Then $\mathfrak{n} \subseteq D$. Let $T := C/\mathfrak{n} = C_{red}$ and $E = B'_T$ be the integral closure of B in T. Pick an element $e \in E \subseteq T$, hence e satisfies a relation

$$e^m + b_1 e^{m-1} + \ldots + b_m \in \mathfrak{n}$$

with some elements $b_1, \ldots, b_m \in B$. This means that there exists an exponent $k \in \mathbb{N}$ such that

$$(e^m + b_1 e^{m-1} + \ldots + b_m)^k = 0.$$

Therefore we may assume that $E \subseteq D/\mathfrak{n}$. Suppose that we have proved the assertion for T instead of C and let $x \in B \otimes_A A^*$ integral over A^*. Under the natural map $C \otimes_A A^* \to D \otimes_A A^*$, the image of x is integral over A^*, hence it belongs to $E \otimes_A A^*$. Since $E \subseteq D/\mathfrak{n}$, there exists an element $d \in D \otimes_A A^*$ such that $x - d \in \mathfrak{n} \otimes_A A^* \subseteq D \otimes_A A^*$, whence $d \in D \otimes_A A^*$. This shows that it is enough to prove the assertion for C reduced. By a direct limit argument as above, we may furthermore assume that C is reduced and finitely generated over A. In particular C is Noetherian and has only finitely many minimal prime ideals, say $\mathrm{Min}(C) = \{\mathfrak{q}_1, \ldots, \mathfrak{q}_r\}$. Let $\mathfrak{r}_i = \mathfrak{q}_i \cap D$, $i = 1, \ldots, r$. We denote

$$C_1 = \prod_{i=1}^r C/\mathfrak{q}_i, \quad L = \prod_{i=1}^r k(\mathfrak{q}_i), \quad D_1 = \prod_{i=1}^r D/\mathfrak{r}_i$$

and let D_2 be the integral closure of D in L. Since C is reduced, we know that C is a subring of C_1 and L is the total quotient ring of C. The rings C, D, C_1, D_1 and D_2 are all of them subrings of L. We have also $C \cap D_2 = D$ and $D_2 = \prod_{i=1}^r E_i$, where E_i is the integral closure of $D/\mathfrak{r}_i$ in $k(\mathfrak{q}_i)$. Actually we have the following commutative diagram:

$$
\begin{array}{ccc}
B = A \hookrightarrow & D \hookrightarrow & C \\
\downarrow & & \downarrow \\
D_1 = \prod_{i=1}^r D/\mathfrak{r}_i \hookrightarrow & C_1 = \prod_{i=1}^r C/\mathfrak{q}_i \\
\cap & & \cap \\
D_2 = \prod_{i=1}^r E_i \hookrightarrow & L = \prod_{i=1}^r k(\mathfrak{q}_i)
\end{array}
$$

By flatness $D \otimes_A A^* = (D_2 \otimes_A A^*) \cap (C \otimes_A A^*)$, so that to prove that $D \otimes_A A^*$ is the integral closure of $B \otimes_A A^*$ in $C \otimes_A A^*$, it is enough to show that $D_2 \otimes_A A^*$ is integrally closed in $L \otimes_A A^*$. If D_i is the integral closure of B in $C/\mathfrak{q}_i$ and if we prove that $D_i \otimes_A A^*$ is the integral closure of A^* in $C/\mathfrak{q}_i \otimes_A A^*$, then the assertion follows easily. This means that we may furthermore assume that C is a domain, finitely generated over A. Let K be the field of fractions of A and L be the field of fractions of C, which is therefore a finitely generated field extension of K. Let F be the algebraic closure of K in L. It is well-known [362, Prop. 3.3.2] that F is a finite algebraic extension of K. Thus, changing if necessary B with the integral closure of B in L or with F, we may reduce to prove the assertion (iii) on one part when B is a Noetherian domain and C is its field of fractions, on the other hand when B is a field.

Case (i): $A = B$ is a Noetherian domain and $C = Q(A)$.

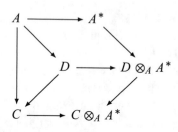

From Mori's theorem [266, Th. 33.10], it follows that D is a Krull ring. Thus there exists a family of discrete valuation rings $(V_i)_{i \in I}$ such that $D = \bigcap_{i \in I} V_i$ and such that any element of C belongs to all but a finite number of the rings V_i. Then

$$D \otimes_A A^* = \bigcap_{i \in I} (V_i \otimes_A A^*),$$

hence it is enough to consider the case when A is a DVR and consequently $D = A$. Let t be a parameter of A and let $x \in C \otimes_A A^*$ be an element integral over A^*. Since $C = A_t$, we can write $x = \frac{\alpha}{t^s}$, with $\alpha \in C \otimes_A A^*$. Assume first that $s = 1$. There is a relation

$$x^n + \beta_{n-1} x_{n-1} + \ldots + \beta_1 x + \beta_0 = 0, \text{ with } \beta_0, \ldots, \beta_{n-1} \in A^*.$$

It follows that $\alpha^n \in tA^*$. By assumption A^*/tA^* is a reduced ring, hence $\alpha = t \cdot \gamma$ for some element $\gamma \in A^*$, whence $t \cdot x = t \cdot \gamma$. By flatness, t is not a zero divisor on A^*, hence $x = \gamma \in A^*$. Thus the case $s = 1$ is proved and we finish the argument by induction on s.

Case (ii): $B = A := K$ is a field. Remember that also $C = L$ is a field, extension of finite type of K. If H is a subalgebra of $K \otimes_K A^* = A^*$, then also H verifies (ii), therefore we may assume that A^* is a K-algebra of finite type. Let $K_1, \ldots, K_m$ be the residue fields of $L \otimes_K A^*$ in its minimal primes. The assumption (ii) implies that each K_i is a separable finitely generated field extension of L. If the extension $L \subseteq L \otimes_K A^*$ is separable and finitely generated, we can decompose it as $L \subseteq E \subseteq L \otimes_K A^*$, where E is a pure transcendental extension of L and $L \otimes_K A^*$ is a finite separable extension of E. Thus we may assume that the extension $L \subseteq L \otimes_K A^*$ is finite and separable. Finally we have to show that if $D \to D \otimes_A A^*$ is free and étale and D is integrally closed in L, then $D \otimes_K A^*$ is integrally closed in $L \otimes_K A^*$. But this follows from Corollary 1.4.10. Note that we have the morphisms

$$L \to L \otimes_A A^* \to \prod_{i=1}^{m} K_i = Q(L \otimes_A A^*)$$

and what we actually have proved is that the conclusion from (iii) is valid for the flat composed morphism $A = B = K \to \prod_{i=1}^{m} K_i$. Since $(\prod_{i=1}^{m} K_i)/B \otimes_A A^*$ is flat over B and $\prod_{i=1}^{m} K_i$ is flat over $B \otimes_A A^*$, by applying again [56, ch. V, §1, n. 7, Cor. of Prop. 19] it follows that the conclusion is valid for the morphism $B \to B \otimes_A A^*$.

$(iii) \Rightarrow (ii)$: If f verifies (iii) and B is a reduced A-algebra, then by Lemma 6.4.1,(ii) we get that B is integrally closed in $B[X]$, hence $B \otimes_A A^*$ is integrally closed in $B \otimes_A A^*[X]$. Thus $B \otimes_A A^*$ is reduced. Let now $\mathfrak{p} \in \mathrm{Spec}(A)$. Then $B := A + \mathfrak{p}A[X]$ is integrally closed in $A[X]$ by Lemma 6.4.1. Hence by assumption $B \otimes_A A^*$ is a subring of $A^*[X]$. But the morphism $B \to A[X]$ identifies to the injective A-linear map $A \oplus \mathfrak{p}^{(\mathbb{N})} \to A \oplus A^{(\mathbb{N})}$. Then the map $\mathfrak{p} \otimes_A A^* \to A^*$ is injective, that is $\mathrm{Tor}_1^A(A/\mathfrak{p}, A^*) = (0)$. Since A is Noetherian it follows that $\mathrm{Tor}_1^A(M, A^*) = (0)$, for any finitely generated A-module M. Hence f is flat.   □

As a first application, we have the classical result from EGA [132]. We mention first an easy lemma.

Lemma 6.4.3 *Let $f : A \to A^*$ be a flat morphism of Noetherian rings and let $B \subseteq C$ be A-algebras that are integral domains. Then $B^* := B \otimes_A A^* \subseteq C^* := C \otimes_A A^*$ and for any minimal prime ideal $\mathfrak{q}$ of C^*, the ideal $\mathfrak{p} = \mathfrak{q} \cap B^*$ is a minimal prime ideal of B^*.*

Proof The inclusion $B^* \subseteq C^*$ follows from the flatness of f. Let us notice first that $C^* \cong C \otimes_B (B \otimes_A A^*) = C \otimes_B B^*$ and that the induced morphism $B \to B^*$ is flat. Every non-zero element of B is a non-zero divisor on B, B^* and C^*. If $\mathfrak{q}^*$ is a minimal prime ideal of C^*, then $\mathfrak{q}^* \cap C = (0)$, because every non-zero element of C is a non-zero divisor on C^*. Therefore we may assume that B and C are fields. But then $B^* \subseteq C^*$ is flat, hence satisfies the Going-Down property. Now the assertion follows easily.   □

Corollary 6.4.4 ([132, Prop. 6.14.1]) *Let $f : A \to A^*$ be a normal morphism of locally Noetherian rings and let B be a normal A-algebra. Then $B^* := B \otimes_A A^*$ is a normal ring.*

Proof We have to prove that $(B^*)_{\mathfrak{q}'}$ is normal, for each prime ideal $\mathfrak{q}'$ of B^*. Let $\mathfrak{p}' := \mathfrak{q}' \cap A^*$, $\mathfrak{q} = \mathfrak{q}' \cap B$ and $\mathfrak{p} := \mathfrak{q} \cap A$. The local ring $(B^*)_{\mathfrak{q}'}$ is a localization of $B_{\mathfrak{q}} \otimes_{A_{\mathfrak{p}}} (A^*)_{\mathfrak{p}'}$, so that it is enough to prove that $B_{\mathfrak{q}} \otimes_{A_{\mathfrak{p}}} (A^*)_{\mathfrak{p}'}$ is normal. This shows that we may assume that A and A^* are Noetherian local rings and that B is a local ring. Then B is a local normal ring, hence an integrally closed domain with field of fractions K. We can write $B = \varinjlim_{i \in I} B_i$, where B_i are the A-subalgebras of finite type of B. Thus B_i are integral domains and let $K_i = Q(B_i)$ be the field of fractions of B_i and C_i be the integral closure

of B_i. Then $K = \varinjlim_{i \in I} K_i$ and by assumption $B = \varinjlim_{i \in I} C_i$. Let also $K_i^* := K_i \otimes_A A^*$,
for any index $i \in I$. Then K_i^* is a localization of B_i^*, which is finitely generated over A^*. Thus K_i^* is a Noetherian normal ring; hence it is a direct product of finitely many integral domains. By Theorem 6.4.2 it follows that $C_i^* = C_i \otimes_A A^*$ is integrally closed in K_i^*; hence it is also a finite direct product of integrally closed domains. In particular $C_i^* = C_i \otimes_A A^*$ is normal and then, applying Proposition 1.3.8 and Lemma 6.4.3, also $B^* = \varinjlim_{i \in I} C_i^* = \varinjlim_{i \in I} C_i \otimes_A A^*$ is normal. $\qquad\square$

As another application, we will strengthen a result about inductive limits of Nagata rings. We need some preparations.

Lemma 6.4.5 (Greco [121, Lemma 1]) *Let $f : A \to B$ be a flat morphism of Noetherian rings, K and L be the total quotient rings of A and B, respectively. Then the canonical morphism $K \otimes_A B \to L$ is injective.*

Proof If S and T are the sets of non-zero divisors of A and B, respectively, by flatness we have $f(S) \subseteq T$, hence $f(S)^{-1}B \subseteq T^{-1}B$. But $f(S)^{-1}B = K \otimes_A B$ and $T^{-1}B = L$. $\qquad\square$

We can prove now a case when the inductive limit of Nagata rings is a Nagata ring and the rings are not necessarily local (compare with Theorem 3.7.6 and Example 3.7.7).

Theorem 6.4.6 (Doretti [91, Th. 2.2]) *Let $(A_i, \varphi_{ij})_{i \in I}$ be an inductive system of Noetherian rings. Assume that:*

(i) $A = \varinjlim_i A_i$ is Noetherian;

(ii) $\varphi_i : A_i \to A$ is a normal morphism, for any $i \in I$;

(iii) A_i is a Nagata ring, for any $i \in I$.

Then A is a Nagata ring.

Proof We have to prove that for any prime ideal $\mathfrak{p}$ of A, the ring $A/\mathfrak{p}$ is a Japanese ring. Let $B = A/\mathfrak{p}$, $\mathfrak{p}_i := \mathfrak{p} \cap A_i$ and $B_i = A_i/\mathfrak{p}_i$. By Proposition 1.3.2 we know that $B = \varinjlim_i B_i$.

Since A is a Noetherian ring, there exists an index $i_0 \in I$ such that for any $i \geq i_0$, we have $B \cong B_i \otimes_{A_i} A$. Let $K := Q(B)$ be the field of fractions of B. Then from Proposition 1.3.5 we know that $K = \varinjlim_i K_i$, where K_i is the field of fractions of B_i for any $i \in I$. Let $L = K(a_1, \ldots, a_r)$ be a finite extension of K. We may assume that $a_1, \ldots, a_r$ are integral over B. Then there exists an index $j_0 \in I$, such that $a_1, \ldots, a_r$ are integral over B_j for any $j \geq j_0$. For any $i \geq j_0$, set $L_i := K_i[a_1, \ldots, a_r]$. Then L_i is a finite extension of K_i and:

(a) $L \cong \varinjlim_{i \geq j_0} L_i$;

(b) $L_i = L_{j_0} \otimes_{K_{j_0}} K_i$ for any $i \geq j_0$;

(c) $L_j = L_i \otimes_{K_i} K_j$ for any $j \geq i \geq j_0$.

Let $i \in I$ be such that $i \geq i_0$ and $i \geq j_0$. The canonical morphism $B_i \to B_i \otimes_{A_i} A \cong B$ is normal. Moreover, we have

$$L_i \otimes_{K_i} K \cong L_i \otimes_{K_i} (\varinjlim_{j \geq i} K_j) \cong \varinjlim_{j \geq i}(L_i \otimes_{K_i} K_j) = \varinjlim_{j \geq i} L_j = L.$$

Let C_i be the integral closure of B_i in K_i. By Theorem 6.4.2 we know that

$$C_i \otimes_{B_i} B \cong C_i \otimes_{B_i} (B_i \otimes_{A_i} A)$$

is the integral closure of B in L. But B_i is a Japanese ring, being a finite A_i-algebra; hence C_i is finite over B_i. Thus the integral closure of B in L is finite.   □

For Example 6.4.11, we need the following nice result of the French mathematician Pierre Samuel (1921–2009). Let us remind first the notion of an isobaric polynomial.

Definition 6.4.7 Let A be a ring (not necessarily Noetherian). A polynomial $f \in A[X_1, \ldots, X_n]$ is called **isobaric of weight s**, if assigning positive weights $s_1, \ldots, s_n$ to the variables f becomes homogeneous of degree s. We say that s is the weight of f with respect to the given weights $s_1, \ldots, s_n$.

Remark 6.4.8 It is easy to see that if A is a domain and $g, h \in A[X_1, \ldots, X_n]$ are such that $g \cdot h$ is isobaric, then g and h are isobaric.

Lemma 6.4.9 (Samuel [337, Ch. 1, Sec. 8, Th. 8.1]) *Let A be a UFD and $f \in A[X_1, \ldots, X_n]$ be a prime polynomial that becomes isobaric of weight s when we assign weights $s_1, \ldots, s_n$, respectively, to the variables $X_1, \ldots, X_n$. If $p \equiv 1 \pmod{s}$ is a natural number prime to s, then the ring $B := A[X_1, \ldots, X_n, T]/(T^p - f)$ is a UFD.*

Proof Let $x_1, \ldots, x_n$ and t be the images of the variables in B. By Eisenstein' criterion, the polynomial $T^p - f$ is irreducible, so that the ring B is an integral domain. Moreover, $B/tB \cong A[X_1, \ldots, X_n]/(f)$, hence t is a prime element in B. If we show that $B[t^{-1}]$ is a UFD, then by Nagata's theorem [336, Th. 5] it follows that B is a UFD. Because

$t^P = t \cdot t^{ms}$, using the isobaricity of f we get

$$t = t^{-ms} f(x_1, \ldots, x_n) = f(t^{-ms_1} x_1, \ldots, t^{-ms_n} x_n).$$

If we put $y_i = x_i t^{-ms_i}$ for $i = 1, \ldots, n$, from $t = f(y_1, \ldots, y_n)$ we obtain

$$B[t^{-1}] = A[x_1, \ldots, x_n, t, t^{-1}] = A[y_1, \ldots, y_n, f(y_1, \ldots, y_n)^{-1}].$$

Suppose that there exists a polynomial $P \in A[U_1, \ldots, U_n]$ such that $P(y_1, \ldots, y_n) = 0$. Then there exists a non-zero $h \in \mathbb{N}$, such that $P(y_1, \ldots, y_n) = t^{-mh} H$, where H is a polynomial in $x_1, \ldots, x_n$ and t^m such that, assigning the weights $s_1, \ldots, s_n, 1$ to $x_1, \ldots, x_n, t^m$, respectively, it results that H is isobaric. But since

$$H = (T^P - f(X_1, \ldots, X_n)) \cdot R(X_1, \ldots, X_n, T)$$

it follows that $T^P - f(X_1, \ldots, X_n)$ must be isobaric too. Therefore $\frac{P}{m} = s$. Contradiction! This means that $y_1, \ldots, y_n$ are algebraically independent over A and this implies that $B[t^{-1}]$ is a UFD. $\qquad\square$

Corollary 6.4.10 *Let A be a UFD and let $a, b, c \geq 2$ be natural numbers such that a and b are relatively prime and ab divides $c - 1$. Then $A[X, Y, Z]/(X^a + Y^b - Z^c)$ is a UFD.*

Proof We apply Lemma 6.4.9 with $n = 2$ and $f(X, Y) = X^a + Y^b$. Assigning to X the weight b and to Y the weight a, the polynomial f becomes isobaric of weight $a \cdot b$. We need only to show that f is irreducible in $A[X, Y]$. If $f = u \cdot v$, where $u, v \in A[X, Y]$, it follows that u and v are isobaric of weights m and respectively, n and $m + n = a \cdot b$. We write $u = \sum \alpha_{ij} X^i Y^j$, with $\alpha_{ij} \neq 0$ and we get that $X^i Y^j$ is isobaric of weight m, that is, $a \cdot i + b \cdot j = m$, for any indices $i \leq a$ and $j \leq b$ occurring in the above writing of u. Moreover u must contain a term of the form $\alpha_{p0} X^p$, with $p \leq a$. Hence $a \cdot p = m$, that is $a \cdot i + b \cdot j = a \cdot p$. It follows that $b \mid j$ and $a \mid p - i$. Then $j = 0$ or $j = b$ and $p - i = 0$ or $p - i = a$. If $j = 0$, then $i = p$. If $j = b$, then $p - i = a$, that is $p = a$ and $i = 0$. Thus either $u = \alpha X^p$ and then $u \cdot v \neq f$, or u is associated with f. Therefore f is irreducible. $\qquad\square$

With these preparations we can display the desired example.

Example 6.4.11 (Marot [240, Prop. 5.2]) Let k be a field. There exists an inductive system $(A_i, \phi_{ij})_{i \in I}$ of semilocal Noetherian k-algebras that are integral domains such that:

(i) the ring $A := \varinjlim A_i$ is a Noetherian domain of dimension 2;
(ii) the morphisms ϕ_{ij} are faithfully flat for any indices $i \leq j$;

(iii) for any index $i \in I$, the ring A_i is excellent;

(iv) the ring A is a Nagata non-excellent ring;

(v) for any maximal ideal $\mathfrak{m}$ of A, the localization $A_\mathfrak{m}$ is an excellent ring.

Proof For any natural number n, let $B_n := k[X_n, Y_n, Z_n]/(X_n^a + Y_n^b - Z_n^c)$, where $a, b, c \geq 2$ are natural numbers such that a and b are relatively prime and ab divides $c - 1$ and let $\mathfrak{m}_n$ be the maximal ideal of B_n generated by the images of X_n, Y_n, Z_n. We define inductively $R_1 := B_1, R_2 := R_1 \otimes_k B_2, \ldots, R_n = R_{n-1} \otimes_k B_n = k[X_1, Y_1, Z_1, \ldots, X_n, Y_n, Z_n]/(X_1^a + Y_1^b - Z_1^c, \ldots, X_n^a + Y_n^b - Z_n^c)$ and let $\phi_{m,n} : R_m \to R_n$, $m \leq n$ be the canonical morphisms. Then an easy induction shows that R_n is an excellent integral domain. It is also easy to see that the morphisms $\phi_{m,n}$ are faithfully flat. If we apply Proposition 1.2.1, we get an integral domain $R = \underset{i \in \mathbb{N}}{\otimes} R_i$. For any integers $i \leq n$ the ideal $\mathfrak{m}_i R_n$, where $\mathfrak{m}_i = (X_i, Y_i, Z_i)$ is the ideal generated by the variables, is a prime ideal of R_n. Let $S_n := R_n \setminus \bigcup_{i=1}^{n} \mathfrak{m}_i R_n$. Then S_n is a multiplicatively closed system in R_n and $\phi_{mn}^{-1}(S_n) = S_m$. Let $S := \varinjlim S_n$ and $A_n = S_n^{-1} R_n$. Thus we have an inductive system $(A_n, \phi_{mn})_n$ and we denote $A := \varinjlim A_i$. The assertions (i), (ii), (iii) and (v) are clear. Since $A_{\mathfrak{m}_i A}$ is not a regular local ring for any index $i \in \mathbb{N}$, it follows that $\mathrm{Reg}(A)$ is not an open subset of $\mathrm{Spec}(A)$; hence A is not an excellent ring. It remains to show that A is a Nagata ring. Let $\mathfrak{p}$ be a prime ideal of A. Suppose first that $\mathfrak{p} \neq (0)$. Then $A/\mathfrak{p}$ is a semilocal ring such that $(A/\mathfrak{p})_\mathfrak{m}$ is excellent for any maximal ideal $\mathfrak{m}$. Therefore $A/\mathfrak{p}$ is an excellent ring. If $\mathfrak{p} = (0)$ and $\mathrm{char}(k) = 0$, then A is Japanese, because it is integrally closed. Assume now that $\mathrm{char}(k) > 0$. Let K be the field of fractions of A and of R and L a field, finite extension of K. By [132, Lemme 1.8.4.2] there exists an index m and a finite field extension L_m of K_m such that $L = L_m \otimes_K L$. Then for any $n \geq m$ we have that $L_n = L_m \otimes_{K_m} K_n$ is a finite extension of K_n and moreover we have $L = \varinjlim L_n$. Let C_n be the integral closure of R_n in L_n and C be the integral closure of R in L, so that $C = \varinjlim C_n$. We see that

$$C_m \otimes_{R_m} R_{m+1} = C_m \otimes_{R_m} (R_m \otimes_k B_{m+1}) = C_m \otimes_k B_{m+1}.$$

Since R_m is excellent, C_m is a finite R_m-module, hence is a Noetherian ring. The rings $C_m \otimes_k B_{m+1}$ and $L_m \otimes_k B_{m+1}$ are integral domains and have the same field of fractions. Since $L_{m+1} = L_m \otimes_{K_m} K_{m+1}$ is a ring of fractions of $C_m \otimes_{R_m} R_{m+1}$, the field of fractions of $C_m \otimes_k B_{m+1}$ and $L_m \otimes_k B_{m+1}$ is exactly L_{m+1}. On the other hand, Corollary 6.4.10 shows that B_m is a geometrically normal k-algebra; hence by Corollary 6.4.4 the ring $C_m \otimes_k B_{m+1}$ is normal. Because $C_m \otimes_k B_{m+1}$ contains R_{m+1} and its field of fractions is L_{m+1}, it follows that $C_m \otimes_k B_{m+1} = C_{m+1}$. This means that the canonical morphism $C_m \otimes_{R_n} R_m \to C_n$, $m \leq n$ is an isomorphism. Taking the inductive limit we get

$$C = \varinjlim C_n = C_m \otimes_{R_m} \varinjlim R_n = C_m \otimes_{R_m} C.$$

Since C_m is finite over R_m, it follows that C is finite over R. It follows easily that A is Japanese. □

Lemma 6.4.12 *Let A be a ring, t an element of A and consider $\mathfrak{p}$ and $\mathfrak{q}$ two prime ideals of A. Assume that:*

(i) $t \notin \mathfrak{p}$;
(ii) $\mathfrak{q} \subseteq t\mathfrak{q}$;
(iii) $\mathfrak{p}A_t + \mathfrak{q}A_t = A_t$.

Then $\mathfrak{p} + \mathfrak{q} = A$.

Proof By assumption there exists $\alpha \in \mathbb{N}$ such that $t^\alpha \in \mathfrak{p} + \mathfrak{q}$. We prove the assertion by induction on α. If $\alpha = 0$ we are done. Assume $\alpha \geq 1$. Then we have a relation $t^\alpha = a + b$ with $a \in \mathfrak{p}$ and $b \in \mathfrak{q}$. Since $\mathfrak{q} \subseteq t\mathfrak{q}$ it follows that $b = c \cdot t$ for some $c \in \mathfrak{q}$, hence $(t^{\alpha-1} - c) \cdot t = a \in \mathfrak{p}$. Since $t \notin \mathfrak{p}$, it follows that $t^{\alpha-1} - c \in \mathfrak{p}$, hence $t^{\alpha-1} \in \mathfrak{p} + \mathfrak{q}$. By induction on α we get the conclusion. □

Theorem 6.4.13 (Lazarus [217, Th. 1.2]) *Let $f : (A, \mathfrak{m}, k) \to (C, \mathfrak{n}, L)$ be a local morphism of local rings. Assume that:*

(i) A is Noetherian;
(ii) f is flat and the fibres of f are geometrically locally integral domains.

Then $\mathrm{Min}(C)$ is a finite set.

Proof We prove the assertion by induction on $n = \dim(A)$.

If $n = 0$, so that A is an Artinian ring, then $\mathfrak{m}$ is a nilpotent ideal and $C/\mathfrak{m}C$ is a local ring, hence a domain, because the fibres of f are locally integral domains. This means that $\mathfrak{m}C$ is a prime ideal of C. It follows easily that $\mathrm{Min}(C)$ has only one element.

Assume that $n = 1$. Let $\mathfrak{a} = N(A)$ be the nilradical of A, so that $A/\mathfrak{a}$ is a reduced ring. The fibres of the flat morphism $A/\mathfrak{a} \to C/\mathfrak{a}C$ are geometrically locally integral domains and $\dim(A/\mathfrak{a}) = \dim(A) = 1$. If we have proved that $\mathrm{Min}(C/\mathfrak{a}C)$ is finite, let $\mathfrak{p} \in \mathrm{Min}(C)$. Then $\mathfrak{p} \cap A \subseteq \mathfrak{a}$, hence $\mathfrak{p} \in \mathrm{Min}(\mathfrak{a}C)$ and it follows that $\mathrm{Min}(C)$ is finite. This shows that we may assume that A is reduced.

Assume now that A is reduced and let $\mathfrak{p}_1, \ldots, \mathfrak{p}_r$ be the minimal prime ideals of A. The canonical morphism $A \to A/\mathfrak{p}_1 \times \ldots \times A/\mathfrak{p}_r$ is finite and injective; hence also the morphism $C \to (\prod_{i=1}^r A/\mathfrak{p}_i) \otimes_A C = \prod_{i=1}^r C/\mathfrak{p}_i C$ is finite and injective. Therefore if $C/\mathfrak{p}_i$ has only finitely many minimal prime ideals for any index i, the same is true for C. Since $A/\mathfrak{p}_i$ an integral domain of dimension at most 1, it follows that we may assume that A is an integral domain.

If A is a domain, let A' be the integral closure of A. From [131, Cor. 23.2.5] we know that there exists a finite extension B of A, contained in A', such that the morphism $v : B \to A'$ is radicial. Moreover, since A is a Noetherian local domain of dimension 1, it follows that A' is a semilocal Dedekind domain. Because f is flat, if we denote by $u : A \to B$ the inclusion, the morphism $u \otimes_A 1_C$ is injective and of course finite, hence any element from $\mathrm{Min}(C)$ is the trace of an element from $\mathrm{Min}(B \otimes_A C)$. The morphism v is radicial, hence also the morphism $v \otimes_A 1_C$ is radicial, so that there is a bijection $\mathrm{Min}(A' \otimes_A C) \cong \mathrm{Min}(B \otimes_A C)$. Thus, if we prove that $\mathrm{Min}(A' \otimes_A C)$ is finite, it will follow that $\mathrm{Min}(C)$ is finite. Let $\mathfrak{q}$ be a maximal ideal of $A' \otimes_A C$ and $\mathfrak{p} = \mathfrak{q} \cap A'$. We have the commutative diagram

$$
\begin{array}{ccccccc}
A & \xrightarrow{\ u\ } & B & \xrightarrow{\ v\ } & A' & \longrightarrow & A'_{\mathfrak{p}} \\
\Big\downarrow{\scriptstyle f} & & \Big\downarrow & & \Big\downarrow & & \Big\downarrow \\
C & \hookrightarrow & B \otimes_A C & \hookrightarrow & A' \otimes_A C & \longrightarrow (A' \otimes_A C)_{\mathfrak{p}} \longrightarrow & (A' \otimes_A C)_{\mathfrak{q}}
\end{array}
$$

The morphism $A'_{\mathfrak{p}} \to (A' \otimes_A C)_{\mathfrak{q}}$ satisfies the same conditions as f and $A'_{\mathfrak{p}}$ is a DVR. If we can show that $(A' \otimes_A C)_{\mathfrak{q}}$ has only finitely many minimal prime ideals, then any maximal ideal of $A' \otimes_A C$ contains only finitely many minimal primes. But C is local, the morphism $C \to B \otimes_A C$ is finite and the morphism $B \otimes_A C \to A' \otimes_A C$ is radicial, whence $A' \otimes_A C$ is a semilocal ring. This shows that $A' \otimes_A C$ has only finitely many minimal prime ideals. Therefore we may assume that A is a DVR.

Let π be a parameter of A and $K := A_\pi$ be its field of fractions. By assumption the ring $C/\pi C$ is a domain, being local, hence πC is a prime ideal of C. Let $\mathfrak{p}_0$ be a minimal prime ideal of C contained in πC. If $\pi \in \mathfrak{p}_0$, then $\mathfrak{p}_0 = \pi C$ is a minimal prime ideal of C. Because f is flat, it follows that $\pi C \cap A = \mathfrak{m} = \pi A \in \mathrm{Min}(A)$, hence $\dim(A) = 0$. Contradiction! This shows that $\pi \notin \mathfrak{p}_0$. Let $x \in \mathfrak{p}_0$. Then $x = \pi \cdot \alpha$ and since $\pi \notin \mathfrak{p}_0$, it follows that $\alpha \in \mathfrak{p}_0$ and therefore $\mathfrak{p}_0 \subseteq \pi \mathfrak{p}_0$. Suppose that C has a minimal prime $\mathfrak{p} \neq \mathfrak{p}_0$. Obviously $\pi \notin \mathfrak{p}$. Then $\mathfrak{p}_0 C_\pi$ and $\mathfrak{p} C_\pi$ are distinct minimal prime ideals of C_π. By assumption the local rings of C_π are integral domains, hence $C_\pi = \mathfrak{p}_0 C_\pi + \mathfrak{p} C_\pi$. From Lemma 6.4.12 it follows that $\mathfrak{p}_0 + \mathfrak{p} = C$. This contradicts the fact that C is a local ring. Thus $\mathfrak{p}_0$ is the only minimal prime ideal of C.

Assume now that $\dim(A) = n > 1$ and that the assertion is proved for rings of dimension strictly smaller than n. Replacing A by its completion $\widehat{A}$, it is easy to see that we may assume that A is a Noetherian complete local ring. If $\mathfrak{p}_1, \ldots, \mathfrak{p}_r$ are the minimal prime ideals of A, by flatness each minimal prime ideal of C lies over a certain $\mathfrak{p}_i$. Then we can replace A by $A/\mathfrak{p}_i$ and we may moreover assume that A is a domain. But then A' is a complete local ring, finite over A and with the same residue field as A. Then it suffices to show that the semilocal ring $A' \otimes_A C$ has only finitely many minimal primes. Thus we may finally assume that A is a Noetherian complete normal local domain. By Proposition 4.5.16 there exists a non-zero element $x \in \mathfrak{m}$ such that A/xA is reduced and by the induction hypothesis, since $\dim(A/xA) < n$, the set $\mathrm{Min}(C/xC)$ is finite. Suppose that $\mathrm{Min}(C/xC) = \{q'_1, \ldots, q'_r\}$, so that $q'_1 \cap \ldots \cap q'_r = \sqrt{xC} = xC$. Let $q_i = q'_i \cap A$ for $i = 1, \ldots, r$. By flatness it results that $q_1, \ldots, q_r$ are minimal prime ideals of xA. Hence $\mathrm{ht}(q_i) = 1$ and A_{q_i} is a DVR, for each $i = 1, \ldots, r$. Let K be the field of fractions of A which is also the field of fractions of A_{q_i}, for $i = 1, \ldots, r$. Consider the commutative diagram

$$
\begin{array}{ccc}
A_{q_i} & \longrightarrow & C_{q'_i} \\
\downarrow & & \downarrow \\
K = Q(A_{q_i}) & \longrightarrow & C_{q'_i} \otimes K
\end{array}
$$

We have $\mathrm{Min}(C_{q'_i} \otimes K) = \mathrm{Min}(C_{q'_i})$ which is finite by the induction hypothesis. By the assumption on f, the ring $C_{q'_i} \otimes K$ is locally an integral domain, hence [190, Th. 168] implies that $C_{q'_i} \otimes K$ is a finite product of integral domains. Since the morphism f is reduced, by Theorem 6.4.2 the ring $C_{q'_i}$ is integrally closed in $C_{q'_i} \otimes K$. Then all the idempotents of $C_{q'_i} \otimes K$ belong to $C_{q'_i}$, which is a local ring; hence $C_{q'_i} \otimes K$ is an integral domain and a fortiori $C_{q'_i}$ is a domain. This shows that each prime ideal q'_i contains exactly one prime ideal $p'_i \in \mathrm{Min}(C)$, for $i = 1, \ldots, r$. We will show that $\mathrm{Min}(C) = \{p'_1, \ldots, p'_r\}$ and this will finish the proof. By flatness we have $p'_i \cap A = (0)$; hence for any index $i = 1, \ldots, r$ we obtain that $x \notin p'_i$. But

$$
p'_1 \cap \ldots \cap p'_r \subseteq q'_1 \cap \ldots \cap q'_r = xC,
$$

so that

$$
p'_1 \cap \ldots \cap p'_r \subseteq x \cdot (p'_1 \cap \ldots \cap p'_r).
$$

Suppose that there exists a minimal prime ideal p' of C different from $p'_1, \ldots, p'_r$. Then $x \notin p'$ and $p'C_x, p'_1C_x, \ldots, p'_rC_x \in \mathrm{Min}(C_x)$. Now $\dim(A_x) < n$ and is integrally closed, so using the same argument used to show that $C_{q'_i}$ is a domain, we see that C_x is locally an integral domain. Therefore any two minimal prime ideals of C_x are coprime. This means that $p'C_x + p_iC_x = C_x$, for any $i = 1, \ldots, r$ and consequently

$$
p'C_x + (p'_1 \cap \ldots \cap p'_r)C_x = C_x.
$$

From Lemma 6.4.12 one gets that

$$
p' + (p'_1 \cap \ldots \cap p'_r) = C.
$$

But this contradicts the fact that C is a local ring. $\square$

Corollary 6.4.14 (Lazarus [217, Cor. 2.1.1]) *Let* $f : A \to C$ *be a ring morphism, let* B *be an A-algebra and* $D = C \otimes_A B$. *Assume that:*

(i) *A is a Noetherian ring;*
(ii) *f is flat and its fibres are geometrically integral domains;*
(iii) *$(B, \mathfrak{r})$ is a geometrically unibranched local domain.*

Let $\mathfrak{n}' \in \mathrm{Spec}(D)$ *lying over the maximal ideal of B. Then $D_{\mathfrak{n}'}$ is a domain.*

Proof Assume first that B is normal. Then, by an inductive limit argument, we may assume that B is the integral closure of a finitely generated A-algebra, which therefore is a Noetherian ring. By localizing we may assume that A and C are local rings, f is a local morphism and B is the integral closure of A. Let $K = Q(A) = Q(B)$. By [131, Cor. 23.2.5] there exists a finite A-subalgebra E of B such that the morphism $E \to B$ is radicial. Since $E \otimes_A C$ is a semilocal ring, from Theorem 6.4.13 it results that $\mathrm{Min}(E \otimes_A C)$ is a finite set. Because the morphism $E \otimes_A C \to D = B \otimes_A C$ is radicial, we obtain that $\mathrm{Min}(D)$ is finite. Let F be a localization of D in a prime ideal, and let $F_K = F \otimes_B K$. Then $\mathrm{Min}(F_K)$ is finite. By assumption F_K is locally an integral domain, hence $F_K = E_1 \times \ldots \times E_n$, where $E_1, \ldots, E_n$ are integral domains. Since B is integrally closed in K and f is reduced, by Theorem 6.4.2 we see that F is integrally closed in F_K; hence F contains the idempotents of F_K. But F is a local ring and then $n = 1$. It follows that F_K is a domain and by flatness also F is a domain.

Let us now consider the case B is geometrically unibranched and let B' be the integral closure of B. Let k, k' and L be the residue fields of $B, D_{\mathfrak{n}'}$ and B', respectively. Because the extension $k \subseteq L$ is purely inseparable, the ring $k' \otimes_k L$ is local. Let $G := D_{\mathfrak{n}'} \otimes_B B'$ and let $\mathfrak{q} \in \mathrm{Max}(G)$. Since G is integral over $D_{\mathfrak{n}'}$, we have $\mathfrak{q} \cap D_{\mathfrak{n}'} = \mathfrak{n}'D_{\mathfrak{n}'}$. Because $\mathfrak{n}'D_{\mathfrak{n}'} \cap B = \mathfrak{r}$, the maximal ideal of B, it results that $\mathfrak{q} \cap B' = \overline{\mathfrak{m}}$, where $\overline{\mathfrak{m}}$ is the maximal ideal of B'. It follows that $\mathfrak{q} \in \mathrm{Spec}(k' \otimes_k L)$, hence G is a local ring. On the other hand, the local rings of $B \otimes_A C$ are integral domains by assumption. Then G is an integral domain and consequently $D_{\mathfrak{n}'}$ is an integral domain.   □

Corollary 6.4.15 (Lazarus [217, Cor. 2.1.2]) *Let* $f : A \to C$ *be a ring morphism, B an A-algebra and $D = C \otimes_A B$. Assume that:*

(i) *A is a Noetherian ring;*
(ii) *f is flat and its fibres are geometrically normal;*
(iii) *$(B, \mathfrak{n})$ is a local domain.*

Let $\mathfrak{n}' \in \operatorname{Spec}(D)$ be such that $\mathfrak{n}' \cap B = \mathfrak{n}$.

(1) If B is geometrically unibranched, then $D_{\mathfrak{n}'}$ is a geometrically unibranched local domain.

(2) If B is normal, then $D_{\mathfrak{n}'}$ is normal.

Proof

(1) We keep the notations used throughout the proof of 6.4.14. Let $M = Q(B)$. In the proof of 6.4.14, we saw that $G = D_{\mathfrak{n}'} \otimes B'$ is an integral domain. Since B' is integrally closed and f is reduced we get that $B' \otimes_B D$ is integrally closed in $D_M := D \otimes_B M$. Then, by localization, G is integrally closed in $G_M := G \otimes_B M$. We have the following commutative diagram:

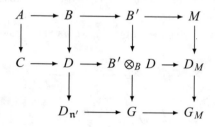

Since f is normal, D_M is a normal ring and then also G_M is normal; hence G is integrally closed. Then G is the integral closure of $D_{\mathfrak{n}'}$. It remains to show that the field extension $k' \subseteq N$ is purely inseparable, where N is the residue field of G. But the extension $k \subseteq L$ is purely inseparable and N is also the residue field of $k' \otimes_k L$. Then $k' \otimes_k L$ is also a purely inseparable extension of k'.

(2) Since B is normal, that is, $B = B'$, it follows that $D_{\mathfrak{n}'} = G$, which is a normal ring.

$\square$

Corollary 6.4.16 ([132, Th. 6.14.4], [217, Théorème]) *Let $f : A \to B$ be a ring morphism and C be an A-algebra. Assume that:*

(i) A is a Noetherian local ring;
(ii) f is flat and has geometrically normal fibres;
(iii) B is normal.

Then $B \otimes_A C$ is a normal ring.

Proof It is contained in 6.4.15, 2).

$\square$

For the last results of this section, we need some preparations.

Definition 6.4.17 A ring morphism $f : A \to B$ is called **almost absolutely integral** if for any A-algebra C which is a domain, the ring $C \otimes_A B$ is either the zero ring, or an integral domain.

The reader should compare the above definition with Definition 1.6.7.

Remark 6.4.18

(i) A flat epimorphism is easily seen to be almost absolutely integral.
(ii) f is absolutely integral iff f is almost absolutely integral and has the lying-over property.

Lemma 6.4.19 *Let* $f : A \to B$ *and* $g : B \to C$ *be ring morphisms. If* $g \circ f$ *is (almost) absolutely integral and* g *is faithfully flat, then* f *is (almost) absolutely integral.*

Proof For any A-algebra D which is a domain, we have the commutative diagram:

$$
\begin{array}{ccccc}
A & \xrightarrow{\; f \;} & B & \xrightarrow{\; g \;} & C \\
\downarrow & & \downarrow & & \downarrow \\
D & \xrightarrow{1_D \otimes f} & D \otimes_A B & \xrightarrow{1_D \otimes g} & D \otimes_A C.
\end{array}
$$

By assumption $D \otimes_A C = (0)$ or is an integral domain. If $D \otimes_A C = (0)$, the faithful flatness of g implies that $D \otimes_A B = (0)$. If $D \otimes_A C$ is an integral domain, since g is faithfully flat, we have that $1_D \otimes g$ is injective; hence also $D \otimes_A B$ is an integral domain. $\qquad\square$

Proposition 6.4.20 (Picavet [285, Prop. 4.9]) *Let A be a Noetherian local ring such that the canonical morphism $A \to A^h$ is absolutely integral. Then A is a Henselian ring.*

Proof Let $\mathfrak{p} \in \mathrm{Spec}(A)$. Then $A_{\mathfrak{p}}^h / \mathfrak{p} A_{\mathfrak{p}}^h \cong \prod_{i=1}^{s} L_i$, where L_i is a separable field extension of $k(\mathfrak{p})$, for any index $i = 1, \ldots, s$. Since the canonical henselization morphism $A \to A^h$ is absolutely integral, there is only one component in this direct product; in other words, for any prime ideal $\mathfrak{p}$ of A, $k(\mathfrak{p}) \otimes_A A^h$ is a field, isomorphic to $k(\mathfrak{p})$. Hence $A_{\mathfrak{p}}^h / \mathfrak{p} A_{\mathfrak{p}}^h \cong L_1 = k(\mathfrak{p})$. It follows that $A \to A^h$ is a radicial morphism. But $A^h \cong \varinjlim_j A_j$, where A_j are

essentially local étale A-algebras and $(A_j)^h \cong A^h$, for any index j [134, 18.6.4–18.6.7]. Let $\mathfrak{p} \in \mathrm{Spec}(A)$ be such that $\mathfrak{p} A_j \neq A_j$. Then, since there exists an unique prime ideal $\mathfrak{q} \in \mathrm{Spec}(A^h)$ such that $\mathfrak{q} \cap A = \mathfrak{p}$ and $\mathrm{Spec}(A^h) \to \mathrm{Spec}(A_j)$ is surjective, there exists an unique prime ideal $\mathfrak{q}' \in \mathrm{Spec}(A_j)$ such that $\mathfrak{q}' \cap A = \mathfrak{p}$. Moreover, any residue field extension of $A \to A_j$ is contained in a residue field extension of $A \to A^h$. But the residue field extensions of the henselization morphism are isomorphisms, as seen above. Hence the residue field extensions of $A \to A_j$ are isomorphisms. Let $\mu_j : A_j \otimes_A A_j \to A_j$ be the diagonal multiplication morphism of any of the components of the inductive limit above. If $\mathfrak{a}_j := \ker(\mu_j)$, then $\mathfrak{a}_j$ is a finitely generated ideal and $\Omega_{A_j/A} = (0)$. By Proposition 3.7.15 we get that $A \to A_j$ is an epimorphism. Since A^h is faithfully flat over A and over A_j, it follows that $A \to A_j$ is faithfully flat. Thus, by Lemma 3.7.13 we obtain that $A \cong A_j$, hence $A \cong A^h$. □

Proposition 6.4.21 *Let* $f : A \to B$ *be a ring morphism. Assume that:*

(i) for any A-algebra C which is an integral domain, the morphism $C \to C \otimes_A B$ *is flat;*
(ii) the fibres of f are (almost) geometrically integral domains.

Then f is (almost) absolutely integral.

Proof Let $\mathfrak{p} \in \operatorname{Spec}(A)$. From the assumptions, it follows that $B/\mathfrak{p}B$ is flat and consequently torsion-free over $A/\mathfrak{p}$. Hence the morphism $B/\mathfrak{p}B \to B \otimes_A k(\mathfrak{p}) = B_\mathfrak{p}/\mathfrak{p}B_\mathfrak{p}$ is injective. It follows that $\mathfrak{p}B_\mathfrak{p} \cap B = \mathfrak{p}B$. Let C be a finite A-algebra. Then the morphism $C \to C \otimes_A B$ has the same properties as f; hence we may assume that $C = A$. Let then $\mathfrak{p} \in \operatorname{Spec}(A)$. Then $B \otimes_A k(\mathfrak{p}) = B_\mathfrak{p}/\mathfrak{p}B_\mathfrak{p}$ is either (0) or an integral domain, so that $\mathfrak{p}B_\mathfrak{p}$ is either $B_\mathfrak{p}$ or a prime ideal of $B_\mathfrak{p}$. It follows that $\mathfrak{p}B$ is either B or a prime ideal of B. So, if C is a finite A-algebra, $\mathfrak{p}(B \otimes_A C)$ is either $B \otimes_A C$ or a prime ideal of $B \otimes_A C$. Suppose that $C = A[X_i; i \in I]/\mathfrak{q}$, with $\mathfrak{q} \in \operatorname{Spec}(A[X_i; i \in I])$. Then $B \otimes_A C = B[X_i; i \in I]/\mathfrak{q}B[X_i; i \in I]$. Because the morphism $A[X_i; i \in I] \to B[X_i; i \in I]$ has the same properties as f, we get that $\mathfrak{q}B[X_i; i \in I]$ is either $B[X_i; i \in I]$ or a prime ideal of $B[X_i; i \in I]$. This easily implies that f is (almost) absolutely integral. □

One implication of the last result is well-known from Grothendieck's EGA [134, Th. 18.9.1].

Theorem 6.4.22 (Grothendieck [134, Th. 18.9.1] and Picavet [285, Prop. 4.10]) *Let A be a Noetherian local ring with geometrically normal formal fibres. The following are equivalent:*

(i) A is Henselian;
(ii) the canonical morphism $A \to \widehat{A}$ *is absolutely integral.*

Proof $(i) \Rightarrow (ii)$: Let $\mathfrak{p} \in \operatorname{Spec}(A)$. Clearly we may assume that $\mathfrak{p} = (0)$. Because A is a Nagata ring, on the one hand the integral closure A' is finite over A; hence A' is a Henselian local domain. On the other hand, it follows that $\widehat{A}$ is reduced. Let $\operatorname{Min}(\widehat{A}) = \{\mathfrak{p}_1, \ldots, \mathfrak{p}_s\}$ and let C be the integral closure of $Q(A) \otimes_A \widehat{A}$ in $Q(\widehat{A})$. Then $C \cong \prod_{i=1}^{s} C_i$, where C_i is the integral closure of $\widehat{A}/\mathfrak{p}_i$. But $C \cong \widehat{A} \otimes_A A'$ so that C is a local ring. It follows that C is a domain and consequently $\widehat{A}$ is a domain. Any finite extension of A has the same properties as A; hence the fibres of the morphism $A \to \widehat{A}$ are geometrically integral domains. By Proposition 6.4.21 the morphism $A \to \widehat{A}$ is absolutely integral.

$(ii) \Rightarrow (i)$: Consider the morphisms $A \xrightarrow{u} A^h \xrightarrow{v} \widehat{A}$. By Lemma 6.4.19 the morphism u is absolutely integral. From Proposition 6.4.20 we obtain that A is a Henselian ring.   □

Corollary 6.4.23 *Any Henselian local ring with geometrically normal formal fibres is universally catenary.*

References

1. K. Ahmad-Amoli, M. Tousi, On the singular sets of modules. Commun. Algebra **24**, 3839–3844 (1996)
2. Y. Akizuki, Einige Bemerkungen über primäre Integritätsbereiche mit Teilerkettensatz. Proc. Phys. Math. Soc. Jpn. **17**, 327–336 (1935)
3. S. Alvite, N. Barral, J. Majadas, *On a theorem of Gulliksen on the homology of local rings.* Preprint. arXiv:2207.02153
4. M. André, *Méthode simpliciale en algèbre homologique et en algèbre commutative.* Lecture Notes in Mathematics, vol. 32 (Springer, Berlin, Heidelberg, New York, 1967)
5. M. André, Cohomologie des algèbres commutatives topologiques. Comment. Math. Helv. **43**, 235–255 (1968)
6. M. André, *Homologie des algèbres commutatives* (Springer, Berlin, Heidelberg, New York, 1974)
7. M. André, Algèbres Graduées Associées et Algèbres Symétriques Plates. Comment. Math. Helv. **49**, 277–301 (1974)
8. M. André, Localisation de la lissité formelle. Manuscripta Math. **13**, 297–307 (1974)
9. M. André, Quasi-excellence and formal series. Manuscripta Math. **45**, 289–292 (1984)
10. M. André, Morphismes pseudo-réguliers. Commun. Algebra **15**, 2129–2142 (1987)
11. M. André, Module des différentielles en caractéristique p. Manuscripta Math. **62**, 477–502 (1988)
12. M. André, Homologie de Frobenius. Math. Ann. **290**, 129–181 (1991)
13. M. André, Cinq exposés sur la désingularisation. Manuscript (1991)
14. M. André, Homomorphismes réguliers en caractéristique p. C. R. Acad. Sci. Paris **316**, 643–646 (1993)
15. M. André, Autre démonstration du théorème liant régularité et platitude en caractéristique p. Manuscripta Math. **82**, 363–379 (1994)
16. M. Artin, On the solutions of analytic equations. Invent. Math. **5**, 277–291 (1968)
17. M. Artin, Algebraic approximation of structures over complete local rings. Publ. Math. Inst. Hautes Études Sci. **36**, 23–58 (1969)
18. M.F. Atiyah, I.G. MacDonald, *Introduction to Commutative Algebra* (Addison-Wesley, Reading, MA, 1969)
19. L.L. Avramov, Flat morphisms of complete intersections. Dokl. Akad. Nauk. SSSR **225**, 11–14 (1975)
20. L.L. Avramov, Homology of local flat extensions and complete intersection defects. Math. Ann. **228**, 27–37 (1977)

© The Author(s), under exclusive license to Springer Nature Switzerland AG 2023
C. Ionescu, *Classes of Good Noetherian Rings*, Frontiers in Mathematics,
https://doi.org/10.1007/978-3-031-22292-4

21. L.L. Avramov, Descente des déviations par homomorphismes locaux et génération des idéaux de dimension projective finie. C. R. Acad. Sci. Paris **295**, 665–668 (1982)

22. L.L. Avramov, Local rings of finite simplicial dimension. Bull. Am. Math. Soc. **10**, 289–291 (1984)

23. L.L. Avramov, *Infinite Free Resolutions*. Six Lectures on Commutative Algebra (Bellaterra, 1996), Progr. Math., vol. 166 (Birkhäuser, Basel, 1998), pp. 1–118

24. L.L. Avramov, Locally complete intersection homomorphisms and a conjecture of Quillen on the vanishing of cotangent homology. Ann. Math. **150**, 455–487 (1999)

25. L.L. Avramov, H.-B. Foxby, Gorenstein local homomorphisms. Bull. Am. Math. Soc. **23**, 145–150 (1990)

26. L.L. Avramov, H.-B. Foxby, Locally Gorenstein homomorphisms. Am. J. Math. **114**, 1007–1047 (1992)

27. L.L. Avramov, H.-B. Foxby, *Grothendieck's Localization Problem*. Commutative Algebra: Syzygies, Multiplicities and Birational Algebra, Contemp. Math., vol. 159 (Amer. Math. Soc., Providence, RI, 1994), pp. 1–13

28. L.L. Avramov, H.-B. Foxby, Ring homomorphisms and finite Gorenstein dimension. Proc. Lond. Math. Soc. **75**, 241–270 (1997)

29. L.L. Avramov, H.-B. Foxby, Cohen-Macaulay properties of ring homomorphisms. Adv. Math. **133**, 54–95 (1998)

30. L.L. Avramov, H.-B. Foxby, S. Halperin, Descent and ascent of local properties along homomorphisms of finite flat dimension. J. Pure Appl. Algebra **38**, 167–185 (1985)

31. L.L. Avramov, H.-B. Foxby, B. Herzog, Structure of local homomorphisms. J. Algebra **164**, 124–145 (1994)

32. L.L. Avramov, H.-B. Foxby, J. Lescot, Bass series of local ring homomorphisms of finite flat dimension. Trans. Am. Math. Soc. **335**, 497–523 (1993)

33. L.L. Avramov, V. Gasharov, I. Peeva, Complete intersection dimension. Publ. Math. Inst. Hautes Études Sci. **86**, 67–114 (1997)

34. L.L. Avramov, S. Halperin, On the non-vanishing of cotangent cohomology. Comment. Math. Helv. **62**, 169–184 (1987)

35. L.L. Avramov, S. Iyengar, Finite generation of Hochschild homology algebras. Invent. Math. **140**, 143–170 (2000)

36. L.L. Avramov, S. Iyengar, *Homological Criteria for Regular Homomorphisms and for Locally Complete Intersection Homomorphism*. Algebra, Arithmetic, and Geometry (Mumbai 2000) T.I.F.R. Studies in Math., 16 vol. I (Narosa, New Delhi, 2002), pp. 97–122

37. L.L. Avramov, S. Iyengar, André-Quillen homology of algebra retracts. Ann. Sci. École Norm. Sup. **36**, 431–462 (2003)

38. L.L. Avramov, S. Iyengar, Gaps in Hochschild cohomology imply smoothness for commutative algebras. Math. Res. Lett. **12**, 789–804 (2005)

39. L.L. Avramov, S. Iyengar, Gorenstein algebras and Hochschild cohomology. Michigan Math. J. **57**, 17–35 (2008)

40. L.L. Avramov, S. Iyengar, C. Miller, Homology over local homomorphisms. Am. J. Math. **128**, 23–90 (2006)

41. L.L. Avramov, M. Vigué-Poirrier, Hochschild homology criteria for smoothness. Int. Math. Res. Not. IMRN **65**, 17–25 (1992)

42. Ş. Basarab, V. Nica, D. Popescu, Approximation properties and existential completeness for ring morphisms. Manuscripta Math. **33**, 227–282 (1981)

43. Ş. Bărcănescu, A. Brezuleanu, Formal power series algebraically independent over polynomials. Rev. Roumaine Math. Pures Appl. **XXV**, 147–155 (1989)

44. B. Bellaccini, Morfismi propri, fibre formali e desingolarizzazioni. Matematiche (Catania) **35**, 173–183 (1980)
45. B. Bellaccini, Proper morphisms and excellent schemes. Nagoya Math. J. **89**, 109–118 (1983)
46. R. Berger, R. Kiehl, E. Kunz, H.-J. Nastold, *Differentialrechnung in der analytischen Geometrie*. Lecture Notes in Mathematics, vol. 38 (Springer, Berlin, Heidelberg, New York, 1967)
47. M.H. Bijan-Zadeh, On the singular sets of a module. Commun. Algebra **21**, 4629–4639 (1993)
48. A. Blanco, J. Majadas, Sur l'annulation de la cohomologie de Hochschild des anneaux de Gorenstein. Commun. Algebra **24**, 1177–1183 (1996)
49. A. Blanco, J. Majadas, On the morphisms of complete intersection in characteristic p. J. Algebra **208**, 35–42 (1998)
50. E. Böger, Differentielle und ganz-algebraische Abhängigkeit bei Idealen analytischer Algebren. Math. Z. **121**, 188–189 (1971)
51. S. Bosch, *Algebraic Geometry and Commutative Algebra* (Springer, London, 2013)
52. S. Bosch, *Lectures on Formal and Rigid Geometry* (Springer, Cham, Heidelberg, New York, Dordrecht, London, 2014)
53. S. Bouchiba, S. Kabbaj, Tensor products of Cohen-Macaulay rings. Solution to a problem of Grothendieck. J. Algebra **252**, 65–73 (2002)
54. N. Bourbaki, *Algèbre*, ch. I–III (Hermann, Paris, 1970)
55. N. Bourbaki, *Algèbre*, ch. X, Algèbre homologique (Hermann, Paris, 1980)
56. N. Bourbaki, *Algèbre Commutative*, ch. I-VII (Hermann, Paris, 1961–1967)
57. N. Bourbaki, *Algèbre Commutative*, ch. VIII–IX (Springer, 2006)
58. N. Bourbaki, *Algèbre Commutative*, ch. X (Springer, 2007)
59. A. Brezuleanu, Anwendungen des kotangentialen Komplexes eines Morphismus. Rev. Roumaine Math. Pures Appl. **XVI**, 1031–1045 (1971)
60. A. Brezuleanu, Smoothness and regularity. Compos. Math. **24**, 1–10 (1972)
61. A. Brezuleanu, T. Dumitrescu, N. Radu, *Local Algebras* (Ed. Univ. Bucureşti, 1993)
62. A. Brezuleanu, C. Ionescu, On the localization theorems and the completion of P-rings. Rev. Roumaine Math. Pures Appl. **XXIX**, 371–380 (1984)
63. A. Brezuleanu, G. Pfister, Excellent Henselian rings not containing their residue field. Rev. Roumaine Math. Pures Appl. **XXVI**, 1295–1298 (1981)
64. A. Brezuleanu, N. Radu, Excellent rings and good separation of the module of differentials. Rev. Roumaine Math. Pures Appl. **XXIII**, 1455–1470 (1978)
65. A. Brezuleanu, N. Radu, *Lecţii de algebră* III, (Ed. Univ. Bucureşti, Mimeographed, 1982)
66. A. Brezuleanu, N. Radu, On the Jacobian criterion of Nagata-Grothendieck. Rev. Roumaine Math. Pures Appl. **XXXI**, 513–517 (1986)
67. A. Brezuleanu, N. Radu, Stability of geometric regularity as consequence of Nagata's jacobian criterion. Stud. Cerc. Mat. **40**, 457–462 (1988)
68. A. Brezuleanu, N. Radu, *Algebre locale* (Ed. Academiei Române, Bucureşti, 1998)
69. A. Brezuleanu, C. Rotthaus, Eine bemerkung über Ringe mit geometrisch normalen formalen fasern. Arch. Math. **39**, 19–27 (1982)
70. A. Brezuleanu, C. Rotthaus, Completion to local domains with bad sets of formal prime divisors. Rev. Roumaine Math. Pures Appl. **XXX**, 605–612 (1985)
71. M. Brodmann, Eine Bemerkung zur Theorie der P-Ringe. Math. Z. **160**, 235–240 (1978)
72. M. Brodmann, A peculiar class of regular domains. J. Algebra **54**, 366–373 (1978)
73. M. Brodmann, C. Rotthaus, Uber den regulären Ort in ausgezeichnete Ringe. Math. Z. **175**, 81–85 (1980)
74. M. Brodmann, C. Rotthaus, Local domains with bad sets of formal prime divisors. J. Algebra **75**, 386–394 (1982)

75. M. Brodmann, C. Rotthaus, A peculiar unmixed domain. Proc. Am. Math. Soc. **87**, 596–600 (1983)

76. M.L. Brown, Artin approximation and formal fibres of local rings. Math. Z. **183**, 221–228 (1983)

77. W. Bruns, J. Herzog, *Cohen-Macaulay Rings* (Cambridge University Press, Cambridge, 1998)

78. M.R. Cangemi, M. Imbesi, Some problems on **P**-morphisms in codimension and codepth k. Stud. Cerc. Mat. **45**, 19–30 (1993)

79. H. Cartan, S. Eilenberg, *Homological Algebra* (Princeton University Press, Princeton, 2007)

80. K. Česnavičius, Macaulayfication of Noetherian schemes. Duke Math. J. **170**, 1419–1455 (2021)

81. L. Chiantini, p-basi e basi differenziali di un anello. Rend. Semin. Mat. Univ. Politec. Torino **37**, 103–121 (1979)

82. G. Chiriacescu, On a theorem of Tate. Atti Accad. Naz. Lincei Rend. Cl. Sci. Fis. Mat. Nat. **LXIX**, 117–119 (1980)

83. G. Chiriacescu, On the theory of Japanese rings. Rev. Roumaine Math. Pures Appl. **XXVII**, 945–948 (1982)

84. I.S. Cohen, Length of prime ideal chains. Am. J. Math. **76**, 654–668 (1954)

85. B. Conrad, Irreducible components of rigid spaces. Ann. Inst. Fourier (Grenoble) **49**, 473–541 (1999)

86. S.H. Cox Jr., D. Rush, Finiteness in flat modules and algebras. J. Algebra **32**, 44–50 (1974)

87. M. Crupi, Condizioni Jacobiane in anelli non contenenti corpi. Atti Accad. Peloritana Pericolanti Cl. Sci. Fis. Mat. Natur. **LXVI**, 367–380 (1988)

88. R. Datta, K. Smith, Excellence in prime characteristic. Local and Global Methods in Algebraic Geometry. Contemp. Math. **712**, 105–116 (2018)

89. J. Dieudonné, *Topics in Local Algebra* (Univ. Notre-Dame Press, Notre Dame, Indiana, 1967)

90. L. Doretti, **P**-morfismi e prodotti tensoriali completati. Rend. Semin. Mat. Univ. Politec. Torino **41**, 129–147 (1983)

91. L. Doretti, Una nota sul limite induttivo piatto di **P**-anelli. Boll. U.M.I. **II**, 29–39 (1983)

92. A. Ducros, Les espaces de Bercovich sont excellents. Ann. Inst. Fourier (Grenoble) **59**, 1443–1552 (2009)

93. T. Dumitrescu, On a theorem of N. Radu and M. André. Stud. Cerc. Mat. **46**, 445–447 (1994)

94. T. Dumitrescu, Reducedness, formal smoothness and approximation in characteristic p. Commun. Algebra **23**, 1787–1795 (1995)

95. T. Dumitrescu, Regularity and finite flat dimension in characteristic $p > 0$. Commun. Algebra **24**, 3387–3401 (1996)

96. T. Dumitrescu, On some examples of atomic domains and of G-rings. Commun. Algebra **28**, 1115–1123 (2000)

97. T. Dumitrescu, On a theorem of N. Radu and M. André (II). Math. Rep. (Bucur.) **3**, 159–161 (2001)

98. T. Dumitrescu, C. Ionescu, Regularity and finite injective dimension in characteristic $p > 0$. Studia Sci. Math. Hungar. **47**, 108–112 (2010)

99. T. Dumitrescu, C. Ionescu, A locally F-finite Noetherian domain that is not F-finite. J. Commut. Algebra **14**, 177–180 (2022)

100. D. Eisenbud, *Commutative Algebra with a View Toward Algebraic Geometry* (Springer, New York, 1995)

101. D. Eisenbud, B. Mazur, Evolutions, symbolic squares and fitting ideals. J. Reine Angew. Math. **488**, 189–201 (1997)

102. R. Elkik, Solutions d'équations à coefficients dans un anneau hensélien. Ann. Sci. Ec. Norm. Supér. **6**, 553–604 (1973)

103. I. Emmanouil, The cotangent complex of complete intersections. C. R. Acad. Sci. Paris **321**, 21–25 (1995)
104. I. Emmanouil, The cyclic homology of affine algebras. Invent. Math. **121**, 1–19 (1995)
105. G. Faltings, Über Macaulayfizierung. Math. Ann. **238**, 175–192 (1978)
106. G. Faltings, Ein einfacher beweis dass geometrische Regularität formale Glattheit impliziert. Arch. Math. **30**, 284–285 (1978)
107. D. Ferrand, Autour de la platitude. Bull. Soc. Math. France **97**, 81–128 (1969)
108. D. Ferrand, Monomorphismes et morphismes absolument plats. Bull. Soc. Math. France **100**, 97–128 (1972)
109. D. Ferrand, M. Raynaud, Fibres formelles d'un anneau local noethérien. Ann. Sci. Ec. Norm. Supér. **3**, 427–428 (1970)
110. M. Flexor-Mangeney, Étude de l'Assassin du complété d'un anneau local noethérien. Bull. Soc. Math. France **98**, 117–125 (1970)
111. H. Flenner, Die Sätze von Bertini für lokale Ringe. Math. Ann. **229**, 97–111 (1977)
112. H.-B. Foxby, n-Gorenstein rings. Proc. Am. Math. Soc. **42**, 67–72 (1974)
113. A. Frankild, Quasi Cohen-Macaulay properties of local homomorphisms. J. Algebra **235**, 214–242 (2001)
114. M. Furuya, On Noetherian rings with integrable derivations and p-basis. TRU Math. **16**, 13–21 (1980)
115. O. Gabber, A property of non excellent rings. Manuscripta Math. **91**, 543–546 (1996)
116. O. Gabber, *Notes on Some t-Structures*, Geometric Aspects of Dwork Theory (Walter de Gruyter, Berlin, 2004), pp. 711–734
117. O. Gabber, Lettre à Yves Laszlo du 23 mai 2007
118. K.R. Goodearl, T.H. Lenagan, Constructing bad Noetherian local domains using derivations. J. Algebra **123**, 478–495 (1989)
119. A.R. Grandjean, M.J. Vale, Almost smooth algebras. Cah. Topol. Géom. Différ. Catég. **XXXII**, 131–138 (1991)
120. S. Greco, Sugli omomorfismi quasi étale e gli anelli eccellenti. Ann. Mat. Pura Appl. **XC**, 281–296 (1971)
121. S. Greco, Una generalizzazione del Lemma di Hensel. Symposia Math. **VIII**, 379–386 (1972)
122. S. Greco, Two theorems on excellent rings. Nagoya Math. J. **60**, 139–149 (1976)
123. S. Greco, *Normal Varieties* (Institutiones Mathematicae, IV, INdAM, Academic Press, London and New York, 1978)
124. S. Greco, A note on universally catenary rings. Nagoya Math. J. **87**, 95–100 (1982)
125. S. Greco, M.G. Marinari, Nagata's Criterion and Openness of loci for Gorenstein and Complete Intersection. Math. Z. **160**, 207–216 (1978)
126. S. Greco, P. Salmon, *Topics in $\mathfrak{m}$-adic Topologies* (Springer, Berlin, Heidelberg, New York, 1971)
127. A. Grothendieck, J. Dieudonné, *Eléments de géometrie algébrique I*, vol. 4 (Publ. Math. Inst. Hautes Études Sci., 1960)
128. A. Grothendieck, J. Dieudonné, *Eléments de géometrie algébrique II*, vol. 8 (Publ. Math. Inst. Hautes Études Sci., 1961)
129. A. Grothendieck, J. Dieudonné, *Eléments de géometrie algébrique III*, vol. 11 (Publ. Math. Inst. Hautes Études Sci., 1961)
130. A. Grothendieck, J. Dieudonné, *Eléments de géometrie algébrique III*, vol. 17 (Publ. Math. Inst. Hautes Études Sci., 1963)
131. A. Grothendieck, J. Dieudonné, *Eléments de géometrie algébrique IV*, vol. 20 (Publ. Math. Inst. Hautes Études Sci., 1964)

132. A. Grothendieck, J. Dieudonné, *Eléments de géometrie algébrique IV*, vol. 24 (Publ. Math. Inst. Hautes Études Sci., 1965)

133. A. Grothendieck, J. Dieudonné, *Eléments de géometrie algébrique IV*, vol. 28 (Publ. Math. Inst. Hautes Études Sci., 1966)

134. A. Grothendieck, J. Dieudonné, *Eléments de géometrie algébrique IV*, vol. 32 (Publ. Math. Inst. Hautes Études Sci., 1967)

135. A. Grothendieck, J. Dieudonné, *Eléments de géometrie algébrique* (Springer, Berlin, Heidelberg, New York, 1971)

136. A. Grothendieck, M. Raynaud, *Séminaire de Géométrie Algébrique du Bois Marie - 1962 - Cohomologie locale des faisceaux cohérents et théorèmes de Lefschetz locaux et globaux*, (North-Holland Publishing Company, Amsterdam, 1968)

137. T. Gulliksen, G. Levin, *Homology of Local Rings*. Queen's Papers in Pure and Applied Math., vol. 20 (Kingston, Ontario, 1969)

138. J.A. Guccione, J.J. Guccione, Hochschild homology of complete intersections and smoothness. J. Pure Appl. Algebra **95**, 131–150 (1994)

139. J.A. Guccione, J.J. Guccione, Cyclic homology of smooth algebras. Algebra Colloq. **2**, 193–202 (1995)

140. J. Hall, R. Sharp, Dualizing complexes and flat homomorphisms of commutative Noetherian rings. Math. Proc. Camb. Phil. Soc. **84**, 37–45 (1978)

141. S. Halperin, The non-vanishing of the deviations of a local ring. Comment. Math. Helv. **62**, 1646–653 (1987)

142. R. Hartshorne, *Algebraic Geometry* (Springer, New York, Heidelberg, 1977)

143. M. Hashimoto, *Cohen-Macaulay F-Injective Homomorphisms*, Geometric and Combinatorial Aspects of Commutative Algebra (Messina 1999) Lecture Notes in Pure and Appl. Math., 217, (Dekker, New York, 2001), pp. 231–244

144. M. Hashimoto, F-pure homomorphisms, strong F-regularity and F-injectivity. Commun. Algebra **38**, 4569–4596 (2010)

145. M. Hashimoto, F-finiteness of homomorphisms and its descent. Osaka J. Math. **52**, 205–215 (2015)

146. H. Hauser, The classical Artin approximation theorems. Bull. Am. Math. Soc. **54**, 595–633 (2017)

147. K. Heinrich, Some remarks on biequidimensionality of topological spaces and Noetherian schemes. J. Commut. Algebra **9**, 49–63 (2017)

148. W. Heinzer, C. Rotthaus, S. Wiegand, Noetherian domains inside a homomorphic image of a completion. J. Algebra **215**, 666–681 (1999)

149. W. Heinzer, C. Rotthaus, S. Wiegand, Intermediate rings between a local domain and its completion. Ill. J. Math. **43**, 19–46 (1999)

150. W. Heinzer, C. Rotthaus, S. Wiegand, Corrigendum to our paper Intermediate rings between a local domain and its completion. Ill. J. Math. **44**, 927–928 (2000)

151. W. Heinzer, C. Rotthaus, S. Wiegand, Intermediate rings between a local domain and its completion, II. Ill. J. Math. **45**, 965–979 (2001)

152. W. Heinzer, C. Rotthaus, S. Wiegand, *Building Noetherian and Non-Noetherian Integral Domains Using Power Series*. Ideal Theoretic Methods in Commutative Algebra. Lect. Notes Pure Appl. Math., vol. 220 (University of Missouri, Marcel Dekker, 2001), pp. 251–264

153. W. Heinzer, C. Rotthaus, S. Wiegand, *Catenary Local Rings with Geometrically Normal Formal Fibers*. Algebra, Arithmetic and Geometry with Applications (Purdue University, Springer, Berlin, 2003), pp. 497–510

154. W. Heinzer, C. Rotthaus, S. Wiegand, *Integral Domains Inside Noetherian Power Series Rings: Constructions and Examples*. Math. Surveys Monogr., vol. 259 (Amer. Math. Soc., Providence, RI, 2021)

155. R. Heitmann, A non-catenary normal local domain. Rocky Mountain J. Math. **12**, 145–148 (1982)

156. R. Heitmann, Characterization of completions of unique factorization domains. Trans. Am. Math. Soc. **337**, 379–387 (1993)

157. R. Heitmann, A locally Nagata PID that is not Nagata. J. Pure Appl. Algebra **226**, 106985 (2022)

158. J. Herzog, E. Kunz, *Der kanonische Modul eines Cohen-Macaulay-Rings*. Lecture Notes in Mathematics, vol. 238 (Springer, Berlin, Heidelberg, New York, 1971)

159. V. Hinich, Rings with approximation property admit a dualizing complex. Math. Nachr. **163**, 289–296 (1993)

160. H. Hironaka, Resolution of singularities of an algebraic variety over a field of characteristic zero I,II. Ann. Math. **79**, 109–326 (1964)

161. Y. Hiyoshi, J. Nishimura, Chain conditions on prime ideals in ideal-adically complete Nagata rings. J. Math. Kyoto Univ. **34**, 273–283 (1994)

162. G. Hochschild, B. Kostant, A. Rosenberg, Differential forms on regular affine algebras. Trans. Am. Math. Soc. **102**, 383–408 (1962)

163. M. Hochster, Non-openness of loci in Noetherian rings. Duke Math. J. **40**, 215–219 (1973)

164. H. Hutchins, *Examples of Commutative Rings* (Polygonal Publishing House, Passaic, NJ, 1981)

165. R. Hübl, *Traces of Differential Forms and Hochschild Homology*. Lecture Notes in Mathematics, vol. 1368 (Springer, Berlin, Heidelberg, New York, 1988)

166. R. Hübl, A note on the Hochschild homology and cyclic homology of a topological algebra. Manuscripta Math. **77**, 63–70 (1992)

167. R. Hübl, Evolutions and valuations associated to an ideal. J. Reine Angew. Math. **517**, 81–101 (1999)

168. R. Hübl, E. Kunz, On algebraic varieties over fields of prime characteristic. Arch. Math. **62**, 88–96 (1994)

169. L. Illusie, *Complexe cotangent et déformations, I*. Lecture Notes in Mathematics, vol. 239 (Springer, Berlin, Heidelberg, New York, 1971)

170. L. Illusie, *Complexe cotangent et déformations, II*. Lecture Notes in Mathematics, vol. 283 (Springer, Berlin, Heidelberg, New York, 1972)

171. L. Illusie, Y. Laszlo, F. Orgogozo, Travaux de Gabber sur l'uniformisation locale et la cohomologie étale des schémas quasi-excellents. Astérisque, **363–364** (2014)

172. M. Imbesi, On a lifting problem in codimension and codepth k. Stud. Cerc. Mat. **47**, 51–59 (1995)

173. C. Ionescu, Three-dimensional local rings whose formal fibers are geometrically normal but not geometrically regular. Rev. Roumaine Math. Pures Appl. **XXX**, 265–267 (1985)

174. C. Ionescu, Sur les anneaux aux fibres formelles géométriquement régulières en codimension n. Rev. Roumaine Math. Pures Appl. **XXXI**, 599–603 (1986)

175. C. Ionescu, Finiteness of the integral closure of a Noetherian ring. Rend. Semin. Mat. Univ. Politec. Torino **47**, 153–162 (1989)

176. C. Ionescu, Some remarks on Japanese rings. An. Univ. Bucureşti Math. **XL**, 55–58 (1991)

177. C. Ionescu, Reduced morphisms and Nagata rings. Arch. Math. **60**, 334–338 (1993)

178. C. Ionescu, Hochschild homology of a topological algebra and formal smoothness. Commun. Algebra **22**, 4801–4805 (1994)

179. C. Ionescu, Hochschild (co)-homology in commutative algebra. A survey. An. Ştiinţ. Univ. Ovidius Constanţa Ser. Mat. **9**, 87–96 (2001)

180. C. Ionescu, Cohen-Macaulay fibres of a morphism. Atti Accad. Peloritana Pericolanti Cl. Sci. Fis. Mat. Natur. **LXXXVI**, 1–9 (2008)

181. C. Ionescu, Smoothness and differentials in positive characteristic. Rend. Circ. Mat. Palermo **58**, 45–50 (2009)

182. C. Ionescu, On the vanishing of the first André-Quillen homology. Algebra Colloq. **16**, 593–602 (2009)

183. C. Ionescu, A note on smoothness and differential bases in positive characteristic. Bull. Math. Soc. Sci. Math. Roumanie (N.S.) **52**, 421–426 (2009)

184. C. Ionescu, Finite generation of André-Quillen (co-)homology of F-finite algebras. Commun. Algebra **49**, 1548–1552 (2021)

185. C. Ionescu, G. Restuccia, Remarks on normal morphisms. Rend. Sem. Mat. Messina **4**, 76–81 (1996–1997)

186. C. Ionescu, G. Restuccia, A note on **P**-morphisms. Math. Rep. (Bucur.) **3**, 169–175 (2001)

187. B. Iversen, *Generic Local Structure of the Morphisms in Commutative Algebra*. Lecture Notes in Mathematics, vol. 310 (Springer, Berlin, Heidelberg, New York, 1973)

188. S. Iyengar, *André-Quillen Homology of Commutative Algebras*. Interactions Between Homotopy Theory and Algebra (Chicago 2004), Contemp. Math., vol. 436 (Amer. Math. Soc., Providence, RI, 2007)

189. A.J. de Jong, Smoothness, semi-stability and alterations. Publ. Math. Inst. Hautes Études Sci. **83**, 51–93 (1996)

190. I. Kaplansky, *Commutative Rings* (Univ. of Chicago Press, Chicago, London, 1974)

191. G. Karpilovsky, *Topics in Field Theory*. Notas de Matematica, vol. 155 (North Holland, 1989)

192. T. Kawasaki, On Macaulayfication of Noetherian schemes. Trans. Am. Math. Soc. **352**, 2517–2552 (2000)

193. D. Katz, Note on projective modules. Math. Pannon. **1**, 3–5 (1989)

194. T. Kikuchi, On the finiteness of the derived normal ring of an affine ring. J. Math. Soc. Jpn. **15**, 360–365 (1963)

195. R. Kiehl, Ausgezeichnete Ringe in der nichtarchimedischen analytischen Geometrie. J. Reine Angew. Math. **234**, 89–98 (1969)

196. T. Kimura, H. Niitsuma, A note on the Noetherian integral domain with a p-basis. TRU Mathematics **15**, 1–4 (1979)

197. T. Kimura, H. Niitsuma, Regular local ring of characteristic p and p-basis. J. Math. Soc. Jpn. **32**, 363–371 (1980)

198. J. Koh, K. Lee, Some restrictions on the maps in minimal resolutions. J. Algebra **202**, 671–689 (1998)

199. E. Kunz, Characterization of regular local rings of characteristic p. Am. J. Math. **91**, 772–784 (1969)

200. E. Kunz, On Noetherian rings of characteristic p. Am. J. Math. **98**, 999–1013 (1976)

201. E. Kunz, *Introduction to Commutative Algebra and Algebraic Geometry* (Birkhäuser, Boston, Basel, Stuttgart, 1985)

202. E. Kunz, *Kähler Differentials* (Friedr. Vieweg & Son, Braunschweig, Wiesbaden, 1986)

203. K. Kurano, P. Roberts, The positivity of intersection multiplicities and symbolic powers of prime ideals. Compos. Math. **122**, 165–182 (2000)

204. K. Kurano, K. Shimomoto, An elementary proof of Cohen-Gabber theorem in the equal characteristic $p > 0$ case. Tohoku Math. J. **70**, 377–389 (2018)

205. K. Kurano, K. Shimomoto, Ideal-adic completion of quasi-excellent rings (after Gabber). Kyoto J. Math. **61**, 707–722 (2021)

206. H. Kurke, G. Pfister, M. Roczen, *Henselsche Ringe und algebraische Geometrie* (VEB Deutscher Verlag der Wissenschaften, Berlin, 1975)

207. H. Kurke, T. Mostowski, G. Pfister, D. Popescu, M. Roczen, *Die Approximationseigenschaft lokaler Ringe*. Lecture Notes in Mathematics, vol. 634 (Springer, Berlin, Heidelberg, New York, 1978)

208. A. Lago, A.G. Rodicio, Some recent results on Hochschild homology of commutative algebras. Publ. Math. **39**, 161–172 (1995)

209. T.Y. Lam, *Lectures on Modules and Rings* (Springer, New York, Berlin, Heidelberg, 1999)

210. K. Langmann, Japanische und geometrische Globale Ringe. Math. Ann. **222**, 229–241 (1976)

211. K. Langmann, Japanische und ausgezeichnete lokale Ringe. Math. Ann. **232**, 99–108 (1978)

212. K. Langmann, Offenheit des normalen Ortes. Arch. Math. **55**, 139–142 (1990)

213. K. Langmann, C. Rotthaus, Über den regulären Ort bei Japanischen lokalen Ringen. Manuscripta Math. **27**, 19–30 (1979)

214. T. Larfeldt, C. Lech, Analytic ramifications and flat couples of local rings. Acta Math. **146**, 201–208 (1981)

215. D. Lazard, Autour de la platitude. Bull. Soc. Math. France **97**, 81–128 (1969)

216. M. Lazarus, Fermeture intégrale et changement de base. C. R. Acad. Sci. Paris **289**, 51–53 (1979)

217. M. Lazarus, Fermeture intégrale et changement de base. Ann. Fac. Sci. Toulouse **VI**, 103–120 (1984)

218. C. Lech, *A Method for Constructing Bad Noetherian Local Rings*. Algebra, Algebraic Topology and their Interactions. Lecture Notes in Mathematics, vol. 1183 (Springer, 1986), pp. 241–247

219. C. Lech, *Yet Another Proof of a Result by Ogoma*. Algebra, Algebraic Topology and Their Interactions. Lecture Notes in Mathematics, vol. 1183 (Springer, 1986), pp. 248–249

220. Y. Lequain, Catenarian property in a domain of formal power series. J. Algebra **65**, 110–117 (1980)

221. S. Lichtenbaum, M. Schlessinger, The cotangent complex of a morphism. Trans. Am. Math. Soc. **128**, 41–70 (1967)

222. J.-L. Loday, *Cyclic Homology*. Grundlehren der mathematischen Wissenschaften, vol. 301, 2nd edn. (Springer, Berlin, Heidelberg, 1998)

223. S. Loepp, Characterization of completions of excellent domains of characteristic zero. J. Algebra **265**, 221–228 (2003)

224. S. Loepp, C. Rotthaus, On the completeness of factor rings. Proc. Am. Math. Soc. **130**, 2189–2195 (2002)

225. S. Loepp, C. Rotthaus, S. Sword, A class of local Noetherian domains. J. Commut. Algebra **1**, 647–677 (2009)

226. L. Ma, T. Polstra, *F-Singularities: A Commutative Algebra Approach (Preliminary Version)*. https://www.math.purdue.edu/~ma326/F-singularitiesBook.pdf

227. J. Majadas, On the vanishing of the homology of the exterior powers of the cotangent complex. Arch. Math. **64**, 484–489 (1995)

228. J. Majadas, On tensor product of complete intersections. Bull. Lond. Math. Soc. **45**, 1281–1284 (2018)

229. J. Majadas, A descent theorem for formal smoothness. Nagoya Math. J. **229**, 113–140 (2018)

230. J. Majadas, A. Rodicio, Commutative algebras of finite Hochschild homological dimension. J. Algebra **149**, 68–84 (1992)

231. J. Majadas, A. Rodicio, *Smoothness, Regularity and Complete Intersection*. London Mathematical Society Lecture Notes Series, vol. 373 (Cambridge Univ. Press, 2010)

232. M.G. Marinari, Successioni di Gorenstein e proprietà G_n. Rend. Semin. Mat. Univ. Padova **48**, 67–93 (1972)

233. M.G. Marinari, Sul luogo di Gorenstein di un anello. Ann. Univ. Ferrara **XXIV**, 175–180 (1978)

234. M.G. Marinari, G. Niesi, On the locus where the Cohen-Macaulay type has a fixed bound. J. Lond. Math. Soc. **21**, 413–418 (1980)

235. J. Marot, Sur les anneaux universellement japonais. C. R. Acad. Sci. Paris **277**, 1029–1031 (1973)

236. J. Marot, Sur les anneaux universellement japonais. C. R. Acad. Sci. Paris **278**, 1169–1172 (1974)

237. J. Marot, Sur les anneaux universellement japonais. Bull. Soc. Math. Fr. **103**, 103–111 (1975)

238. J. Marot, Sur les anneaux universellement japonais. Séminaire Dubreil. Algèbre **28**(Exp. 24), 1–6 (1974-1975)

239. J. Marot, Sur les anneaux de séries formelles restreintes. C. R. Acad. Sci. Paris **287**, 105–107 (1978)

240. J. Marot, Limite inductive plate de **P**-anneaux. J. Algebra **57**, 484–496 (1979)

241. J. Marot, About Nagata rings. Manuscripta Math. **33**, 27–35 (1980)

242. J. Marot, Sur la complétion adique des **P**-anneaux. C. R. Acad. Sci. Paris **292**, 135–138 (1981)

243. J. Marot, Sur les homorphismes d'intérsection complète. C. R. Acad. Sci. Paris **294**, 381–384 (1982)

244. J. Marot, **P**-rings and **P**-homomorphisms. J. Algebra **87**, 136–149 (1984)

245. C. Massaza, Su alcuni aspetti assiomatici del criterio di Nagata. Rend. Circ. Mat. Palermo **XXX**, 365–377 (1981)

246. C. Massaza, P. Valabrega, Sull'apertura dei luoghi in uno schema localmente Noetheriano. Boll. U.M.I. **14-A**, 564–574 (1977)

247. H. Matsumura, *Criteri Jacobiani* (Quaderni CNR, Torino, 1975)

248. H. Matsumura, *Commutative Algebra*, 2nd edn. (Benjamin/Cummings, Reading, MA, 1980)

249. H. Matsumura, *Commutative Ring Theory* (Cambridge University Press, Cambridge, 1986)

250. H. Matsumura, *Formal Power Series Rings over Polynomial Rings, I*. Number Theory, Algebraic Geometry and Commutative Algebra, in honor of Y. Akizuki, Kinokunya, Tokyo, 1973, pp. 511–520

251. H. Matsumura, *Noetherian Rings with Many Derivations*. Contributions to Algebra, a collection of papers dedicated to Ellis Kolchin (Academic Press, New York, 1977), pp. 279–294

252. H. Matsumura, Quasi-coefficient rings of a local ring. Nagoya Math. J. **68**, 123–130 (1977)

253. H. Matsumura, Integrable derivations. Nagoya Math. J. **87**, 227–245 (1982)

254. H. Matsumura, T. Ogoma, Remarks on associated prime ideals. Rend. Semin. Mat. Univ. Politec. Torino **38**, 53–61 (1980)

255. B. Mazur, *Galois Deformations and Hecke Curves* (Course Notes, Harvard, 1993). https://people.math.harvard.edu/~mazur/papers/scanGalois.pdf

256. S. McAdam, Saturated chains in Noetherian rings. Indiana Univ. Math. J. **23**, 719–728 (1974)

257. S. McAdam, *Asymptotic Prime Divisors*. Lecture Notes in Mathematics, vol. 1023 (Springer, Berlin, Heidelberg, New York, 1983)

258. S. McAdam, *Primes Associated to an Ideal*. Contemp. Math., vol. 102 (Amer. Math. Soc., Providence, RI, 1989)

259. C. Miller, The Frobenius endomorphism and homological dimensions. Contemp. Math. **331**, 207–234 (2003)

260. T. Murayama, A uniform treatment of Grothendieck's localization problem. Compos. Math. **158**, 57–88 (2022)

261. M. Nagata, An example of normal local ring which is analytically ramified. Nagoya Math. J. **9**, 111–113 (1955)

262. M. Nagata, On the chain problem for prime ideals. Nagoya Math. J. **10**, 51–64 (1956)

263. M. Nagata, A Jacobian criterion of simple points. Ill. J. Math. **1**, 427–432 (1957)

264. M. Nagata, An example of a normal local ring which is analytically reducible. Mem. Coll. Sci. Univ. Kyoto **31**, 83–85 (1958)

265. M. Nagata, On the closedness of singular loci. Publ. Math. Inst. Hautes Études Sci. **2**, 29–36 (1959)

266. M. Nagata, *Local Rings* (Wiley, New York, 1962)

267. V. Nica, D. Popescu, *Inele cu proprietatea de aproximare* (Univ. Bucureşti, 1979)
268. J. Nishimura, Note on Krull domains. J. Math. Kyoto Univ. **15**, 397–400 (1975)
269. J. Nishimura, On the ideal-adic completion of Noetherian rings. J. Math. Kyoto Univ. **21**, 153–169 (1981)
270. J. Nishimura, *Symbolic Powers, Rees Algebras and Applications*, Commutative Ring Theory. Proceedings of the Fés International Conference (M. Dekker, New York, Basel, Hong Kong, 1994), pp. 205–213
271. J. Nishimura, A few examples of local rings, I. Kyoto J. Math. **52**, 51–87 (2012)
272. J. Nishimura, A few examples of local rings, II. Kyoto J. Math. **56**, 49–96 (2016)
273. J. Nishimura, T. Nishimura, *Ideal-Adic Completion of Noetherian Rings II*. Algebraic Geometry and Commutative Algebra; in honor of Masayoshi Nagata (Kinokunya, Tokyo, 1988), pp. 453–467
274. M. Nomura, *Formal Power Series Rings over Polynomial Rings, II*. Number Theory, Algebraic Geometry and Commutative Algebra, in honor of Y. Akizuki (Kinokunya, Tokyo, 1973), pp. 521–528
275. T. Ogoma, Some examples of rings with curious formal fibers. Mem. Fac. Sci. Kochi Univ. **1**, 17–22 (1980)
276. T. Ogoma, Non-catenary pseudo-geometric normal rings. Jpn. J. Math. **6**, 147–163 (1980)
277. T. Ogoma, Descent of **P**-property by proper surjective morphisms. Nagoya Math. J. **92**, 175–177 (1983)
278. T. Ogoma, Noetherian property of inductive limits of Noetherian local rings. Proc. Jpn. Acad. **67**, 68–69 (1991)
279. T. Ogoma, General Néron desingularization based on an idea of Popescu. J. Algebra **167**, 57–84 (1994)
280. J.P. Olivier, *Fermeture intégrale et changements de base absolument plats*. Colloque d'Algèbre commut. Rennes, vol. 9 (Publ. Sem. Math. Univ. Rennes, 1972), pp. 1–13
281. A. Ooishi, Openess of loci, **P**-excellent rings and modules. Hiroshima Math. J. **10**, 419–436 (1980)
282. M. Paugam, Condition G_q d'Ischebeck et condition S_q de Serre. Séminaire Dubreil. Algèbre **26**(Exp. 14), 1–12 (1972–1973)
283. M. Paugam, La condition (G_q) de Ischebeck. C. R. Acad. Sci. Paris **276**, 109–112 (1973)
284. G. Pfister, D. Popescu, Die strenge Approximationseigenschaft lokaler Ringe. Invent. Math. **30**, 145–174 (1975)
285. G. Picavet, Absolutely integral homomorphisms. J. Algebra **311**, 584–605 (2007)
286. D. Popescu, A strong approximation theorem over discrete valuation rings. Rev. Roumaine Math. Pures Appl. **XX**, 659–692 (1975)
287. D. Popescu, Algebraically pure morphisms. Rev. Roumaine Math. Pures Appl. **XXIV**, 947–977 (1979)
288. D. Popescu, A remark on two-dimensional local rings with the property of approximation. Math. Z. **173**, 235–240 (1980)
289. D. Popescu, General Neron desingularization. Nagoya Math. J. **100**, 97–126 (1985)
290. D. Popescu, General Neron desingularization and approximation. Nagoya Math. J. **104**, 85–115 (1986)
291. D. Popescu, General Neron desingularization. Letter to the editor. Nagoya Math. J. **118**, 45–53 (1990)
292. D. Popescu, *Artin Approximation*. Handbook of Algebra, vol. 2, ed. by M. Hazewinkel (Elsevier Science, 2000), pp. 321–356
293. D. Quillen, *Homotopical Algebra*. Lecture Notes in Math., vol. 43 (Springer, Berlin, Heidelberg, New York, 1967)

294. D. Quillen, *Homology of Commutative Rings*. Mimeographed Notes (MIT, 1968)
295. D. Quillen, On the (co-)homology of commutative rings. Appl. categorical Algebra. Proc. Sympos. Pure Math. **17**, 65–87 (1970)
296. N. Radu, *Inele locale*, vol. I-II, Ed. Academiei R.S.R. (Bucureşti, 1968)
297. N. Radu, Sur les algèbres dont le module des différentielles est plat. Rev. Roumaine Math. Pures Appl. **21**, 933–939 (1976)
298. N. Radu, Sur un critère de lissité formelle. Bull. Math. Soc. Sci. Math. Roumanie **21**, 133–135 (1977)
299. N. Radu, Une classe d'anneaux noethériens. Math. Rep. (Bucur.) **37**, 79–82 (1992)
300. A. Ragusa, On the openness of H_n locus and semicontinuity of n-th deviation. Proc. Am. Math. Soc. **80**, 201–209 (1980)
301. L.J. Ratliff, On quasi-unmixed semi-local rings and the altitude formula. Am. J. Math. **87**, 278–284 (1965)
302. L.J. Ratliff, On quasi-unmixed local domains, the altitude formula and the chain condition for prime ideals,(I). Am. J. Math. **91**, 508–528 (1969)
303. L.J. Ratliff, On quasi-unmixed local domains, the altitude formula and the chain condition for prime ideals,(II). Am. J. Math. **92**, 99–144 (1970)
304. L.J. Ratliff, Characterizations of catenary rings. Am. J. Math. **93**, 1070–1108 (1971)
305. L.J. Ratliff, Catenary rings and the altitude formula. Am. J. Math. **94**, 458–466 (1972)
306. M. Raynaud, *Anneaux Locaux Henséliens*. Lecture Notes in Mathematics, vol. 169 (Springer, Berlin, Heidelberg, New York, 1970)
307. I. Reiten, R. Fossum, Commutative n-Gorenstein rings. Math. Scand. **31**, 33–48 (1972)
308. D. Rees, A note on analytically unramified local rings. J. Lond. Math. Soc. **36**, 24–28 (1961)
309. G. Restuccia, Criteri jacobiani di regolarità ed eccellenza per anelli di serie ristrette. Ann. Univ. Ferrara **XXIII**, 1–10 (1977)
310. G. Restuccia, Anelli con molte derivazioni in caratteristica diseguale. Matematiche (Catania) **XXXII**, 323–342 (1977)
311. G. Restuccia, Anelli di tipo analitico su un campo di caratteristica positiva. Atti Accad. Peloritana Pericolanti Cl. Sci. Fis. Mat. Natur. **LX**, 65–83 (1982)
312. G. Restuccia, On the completion of excellent rings in characteristic $p > 0$. Rend. Circ. Mat. Palermo **XXXII**, 289–306 (1983)
313. G. Restuccia, Sur le lieu $U_R(A)$ d'un anneau noethérien. Bull. Sci. Math. **108**, 129–141 (1984)
314. G. Restuccia, Sur la completion adique d'un **P**-anneau non semilocal. Afrika Mat. **5**, 23–32 (1995)
315. G. Restuccia, H. Matsumura, Integrable derivations in rings of unequal characteristic. Nagoya Math. J. **93**, 173–178 (1984)
316. P. Roberts, Rings of type 1 are Gorenstein. Bull. Lond. Math. Soc. **15**, 48–50 (1983)
317. A. Rodicio, A characterization of smooth and regular algebras in characteristic zero. Extr. Math. **2**, 90–92 (1987)
318. A. Rodicio, Some characterizations of smooth, regular and complete intersection algebras. Manuscripta Math. **59**, 491–498 (1987)
319. A. Rodicio, On a result of Avramov. Manuscripta Math. **62**, 181–185 (1988)
320. A. Rodicio, On the Jacobian criterion of formal smoothness. Publ. Math. **33**, 339–343 (1989)
321. A. Rodicio, Smooth algebras and vanishing of Hochschild homology. Comment. Math. Helv. **65**, 474–477 (1990)
322. A. Rodicio, Commutative augmented algebras with two vanishing homology modules. Adv. Math. **111**, 162–165 (1995)
323. G. Rond, Artin approximation. J. Singul. **17**, 108–192 (2018)
324. J. Rotman, *An Introduction to Homological Algebra*, 2nd edn. (Springer, 2009)

325. C. Rotthaus, Nichausgezeichnete, universell japanische Ringe. Math. Z. **152**, 107–125 (1977)
326. C. Rotthaus, Universell japanische Ringe mit nicht offenem regulären Ort. Nagoya Math. J. **74**, 123–135 (1979)
327. C. Rotthaus, Komplettierung semilokaler quasiausgezeichneter Ringe. Nagoya Math. J. **76**, 173–180 (1979)
328. C. Rotthaus, Zur komplettierung ausgezeichnete Ringe. Math. Ann. **253**, 213–226 (1980)
329. C. Rotthaus, Potenzreihenerweiterung und formale Fasern in lokalen Ringen mit Approximationseigenschaft. Manuscripta Math. **42**, 53–65 (1983)
330. C. Rotthaus, On the approximation property of excellent rings. Invent. Math. **87**, 39–63 (1987)
331. C. Rotthaus, Rings with approximation property. Math. Ann. **287**, 455–466 (1990)
332. C. Rotthaus, Divisorial ascent in rings with the approximation property. J. Algebra **178**, 541–560 (1995)
333. C. Rotthaus, Homomorphic images of regular local rings. Commun. Algebra **24**, 445–476 (1996)
334. C. Rotthaus, Descent of the canonical module in rings with approximation property. Proc. Am. Math. Soc. **124**, 1713–1717 (1996)
335. C. Rotthaus, L. Şega, Open loci of graded modules. Trans. Am. Math. Soc. **283**, 4959–4980 (2005)
336. P. Samuel, *Anneaux factoriels* (Sociedade de Matematica de Sao Paulo, 1963)
337. P. Samuel, *Lectures On Unique Factorization Domains*. Notes by M. Pavman Murthy (Tata Institute of Fundamental Research, Bombay, 1964)
338. G. Scheja, *Differentialmoduln lokaler analytischer Algebren* (Schrift. Math. Inst. Univ. Fribourg, 1972)
339. G. Scheja, U. Storch, Über differentielle Abhängigkeit bei Idealen analytischer Algebren. Math. Z. **114**, 101–112 (1970)
340. G. Scheja, U. Storch, Differentielle Eigenschaften der Lokalisierungen analytischer Algebren. Math. Ann. **197**, 137–170 (1972)
341. P. Schenzel, A.-M. Simon, *Completion, Čech and Local Homology and Cohomology - Interactions Between Them*. Springer Monographs in Mathematics (Springer, 2018)
342. P. Seibt, Infinitesimal extensions of commutative algebras. J. Pure Appl. Algebra **16**, 197–206 (1980)
343. P. Seibt, *On the Vanishing of Certain Simplicial Homology Groups in Commutative Algebra*, Preprint, vol. 166 (Univ. Bielefeld., 1986)
344. J.P. Serre, *Algèbre Locale. Multiplicités*. Lecture Notes in Mathematics, vol. 11 (Springer, Berlin, Heidelberg, New York, 1965)
345. H. Seydi, Anneaux henséliens et conditions de chaînes. Bull. Soc. Math. France **98**, 9–31 (1970)
346. H. Seydi, Anneaux henséliens et conditions de chaînes, III. Bull. Soc. Math. France **98**, 329–336 (1970)
347. H. Seydi, La réciproque d'un théorème de Kikuchi. J. Math. Kyoto Univ. **11**, 415–424 (1971)
348. H. Seydi, Sur la théorie des anneaux excellents en caractéristique *p*. Bull. Sc. Math. **96**, 193–198 (1972)
349. H. Seydi, Sur le critère jacobien de Nagata. J. Math. Kyoto Univ. **13**, 213–216 (1972)
350. H. Seydi, Sur une note d'Ernst Kunz. C. R. Acad. Sci. Paris **274**, 714–716 (1972)
351. H. Seydi, Exemple d'un anneau local noethérien japonais normal qui n'est pas formellement réduit. C. R. Acad. Sci. Paris **274**, 1334–1337 (1972)
352. H. Seydi, Sur une note de Matsumura et Nomura. C. R. Acad. Sci. Paris **281**, 729–730 (1975)
353. H. Seydi, Sur la théorie des anneaux excellents en charactéristique *p* II. J. Math. Kyoto Univ. **20**, 155–167 (1980)

354. R. Sharp, Acceptable rings and homomorphic images of Gorenstein rings. J. Algebra **44**, 246–261 (1977)

355. R. Sharp, On the fibre rings of a formal power series extension. Quart. J. Math. **28**, 487–494 (1977)

356. R. Sharp, A commutative Noetherian ring which possesses a dualizing complex is acceptable. Math. Proc. Camb. Phil. Soc. **82**, 197–213 (1977)

357. J.M. Soto, Some remarks on relative Tor and regular homomorphisms. Rocky Mt. J. Math. **33**, 1095–1099 (2003)

358. M. Spivakovsky, A new proof of D. Popescu's theorem on smoothing of a ring homomorphism. J. Am. Math. Soc. **294**, 381–444 (1999)

359. The Stacks Project Authors, *The stacks project*. https://stacks.math.columbia.edu

360. R. Swan, On the number of generators of a module. Math. Z. **102**, 318–322 (1967)

361. R. Swan, *Néron-Popescu Desingularization*. Algebra and Geometry (National Taiwan University, 1996), pp. 135–192

362. I. Swanson, C. Huneke, *Integral Closure of Ideals, Rings, and Modules*. London Mathematical Society Lecture Note Series, vol. 363 (Cambridge Univ. Press, 2006)

363. M. Tabaâ, Sur les homomorphismes d'intérsection complète. C. R. Acad. Sci. Paris **298**, 437–439 (1984)

364. M. Tabaâ, Sur le produit tensoriel d'algèbres. Math. Scand. **119**, 5–13 (2016)

365. R. Takahashi, Nagata criterion for Serre's (R_n) and (S_n)-conditions. Math. J. Okayama Univ. **41**, 37–43 (1999)

366. H. Tanaka, Infinite dimensional excellent rings. Commun. Algebra **47**, 482–489 (2019)

367. H. Tanimoto, Some characterizations of smoothness. J. Math. Kyoto Univ. **23**, 695–706 (1983)

368. H. Tanimoto, Smoothness of Noetherian rings. Nagoya Math. J. **95**, 163–179 (1984)

369. H. Tanimoto, On the base field change for **P**-rings and **P**-2 rings. Nagoya Math. J. **90**, 77–83 (1983)

370. L.A. Tarrio, *Esquemas nórdicos y la semicontinuidad de los defectos de intersección completa*. Alxebra, vol. 52 (Universidad de Santiago de Compostela, 1989)

371. L.A. Tarrio, A.G. Rodicio, On the upper semicontinuity of the complete intersection defect. Math. Scand. **65**, 161–164 (1989)

372. J. Tate, Homology of Noetherian rings and local rings. Ill. J. Math. **1**, 14–27 (1957)

373. G. Tedeschi, Sull'apertura di luoghi negli spettri affine e projettivo di un dominio graduato Noetheriano. Ann. Univ. Ferrara **XXIV**, 167–173 (1978)

374. M. Temkin, Desingularization of quasi-excellent schemes in characteristic zero. Adv. Math. **219**, 488–522 (2008)

375. M. Tousi, S. Yassemi, Tensor products of some special rings. J. Algebra **268**, 672–676 (2003)

376. M. Tousi, S. Yassemi, Catenary, locally equidimensional and tensor products of algebras. Commun. Algebra **33**, 1023–1029 (2005)

377. V. Trivedi, A local Bertini theorem in mixed characteristic. Commun. Algebra **22**, 823–827 (1994)

378. V. Trivedi, Erratum: A local Bertini theorem in mixed characteristic. Commun. Algebra **25**, 1685–1686 (1997)

379. A. Tyç, Differential basis, p-basis and smoothness in characteristic $p > 0$. Proc. Am. Math. Soc. **103**, 389–394 (1988)

380. P. Valabrega, On two-dimensional regular local rings and a lifting problem. Ann. Sc. Norm. Super. Pisa Cl. Sci. **27**, 787–807 (1973)

381. P. Valabrega, Regular local rings and excellent rings. J. Algebra **26**, 440–445 (1973)

382. P. Valabrega, Regular local rings and excellent rings II. J. Algebra **26**, 446–450 (1973)

383. P. Valabrega, On the excellent property for rings of restricted power series. Boll. U.M.I. **9**, 486–494 (1974)

384. P. Valabrega, On the excellent property for power series rings over polynomial rings. J. Math. Kyoto Univ. **15**, 387–395 (1975)

385. P. Valabrega, A few theorems on completion of excellent rings. Nagoya Math. J. **61**, 127–133 (1976)

386. P. Valabrega, Scioglimenti di singolarità, criteri jacobiani, anelli eccellenti. Boll. U.M.I. **14**, 221–239 (1977)

387. P. Valabrega, Formal fibers and openness of loci. J. Math. Kyoto Univ. **18**, 199–208 (1978)

388. P. Valabrega, **P**-morfismi e prolungamento di fasci. Rend. Semin. Mat. Univ. Politec. Torino **36**, 313–330 (1978)

389. M.J. Vale, An application of the cotangent complex to regularity and complete intersection. Commun. Algebra **27**, 6167–6173 (1999)

390. W. Vasconcelos, On projective modules of finite rank. Proc. Am. Math. Soc. **22**, 430–433 (1969)

391. W. Vasconcelos, Quasi-normal rings. Ill. J. Math. **14**, 268–273 (1970)

392. W. Vasconcelos, Flat modules over commutative Noetherian rings. Trans. Am. Math. Soc. **152**, 137–143 (1970)

393. C. Weibel, *An Introduction to Homological Algebra* (Cambridge Univ. Press, Cambridge, 1995)

394. S. Witherspoon, *Hochschild Cohomology for Algebras*, GSM 204 (AMS, Providence, Rhode-Island, 2019)

395. N. Yamauchi, On algebras with universal finite module of differentials. Nagoya Math. J. **83**, 107–121 (1981)

396. C. Yu, Japanese Dedekind domains are excellent. Taiwanese J. Math. **26**, 483–500 (2022)

397. O. Zariski, The concept of a simple point of an abstract algebraic variety. Trans. Am. Math. Soc. **62**, 1–52 (1947)

398. O. Zariski, Sur la normalité analytique des variétés normales. Ann. Inst. Fourier **2**, 161–164 (1950)

399. O. Zariski, P. Samuel, *Commutative Algebra, I-II* (Springer, Berlin, Heidelberg, New York, 1958)

Index of Terminology

© The Author(s), under exclusive license to Springer Nature Switzerland AG 2023
C. Ionescu, *Classes of Good Noetherian Rings*, Frontiers in Mathematics,
https://doi.org/10.1007/978-3-031-22292-4

Authors Index

A

Akizuki, Y., vii, 50, 79
André, M., ix, 143, 145, 146, 167, 172, 252,
    257, 267–269, 314, 393–396
Artin, M., 307
Auslander, viii
Avramov, L.L., 139, 257, 373, 375, 377,
    379–381, 384, 386–388, 433, 436
Azumaya, G., 75

B

Böger, E., 402
Brezuleanu, A., viii, 50, 146, 190, 278, 296
Brodmann, M., 50, 60, 117, 276
Buchsbaum, viii

C

Cangemi, M.R., 258
Cartan, H., 411, 436
Česnavičius, K., 250, 257
Chevalley, vii
Chiriacescu, G., 88, 91, 95, 277
Cohen, I.S., 109, 114, 373
Cox, S.H., 430

D

Datta, R., 189, 196
de Jong, A.J., 249
Doretti, L., 233, 241, 242, 261, 443
Dumitrescu, T., 167, 168, 172, 175, 177, 272,
    289

E

Eilenberg, S., 411, 436
Eisenbud, D., 373, 402, 407
Elkik, R., 316

F

Faltings, G., 46, 257
Ferrand, D., 50, 421, 422
Flenner, H., ix, 245, 278, 279, 281, 282, 285
Foxby, H.-B., 257, 373, 375, 377, 379–381, 384,
    386
Furuya, M., 164

G

Gabber, O., ix, 180, 186, 245, 249, 254, 276,
    299, 300, 373
Granjean, 135
Greco, S., 15, 19–21, 23, 74, 110, 111, 131,
    132, 151, 443
Grothendieck, A., vii, ix, 29, 48, 62, 68, 72, 77,
    100, 101, 109, 149, 245, 248, 252,
    259, 278, 453
Gulliksen, T., 433, 436

H

Hall, J., 257
Heinzer, W., 50
Heitmann, R., 50, 51
Hensel, K., 75
Herzog, B., 373, 375, 377, 379–381
Hironaka, H., 248, 250, 254, 276, 299

List of Examples

© The Author(s), under exclusive license to Springer Nature Switzerland AG 2023
C. Ionescu, *Classes of Good Noetherian Rings*, Frontiers in Mathematics,
https://doi.org/10.1007/978-3-031-22292-4

Printed in the United States
by Baker & Taylor Publisher Services